ANNUAL REPORTS IN MEDICINAL CHEMISTRY

VOLUME 55

Editor-in-Chief

STEPHEN NEIDLE

School of Pharmacy, University College London,
London, United Kingdom

VOLUME FIFTY FIVE

ANNUAL REPORTS IN MEDICINAL CHEMISTRY

Medicinal Natural Products: A Disease-Focused Approach

Volume Editors

SATYAJIT D. SARKER

Centre for Natural Products Discovery (CNPD), School of Pharmacy and Biomolecular Sciences, Liverpool John Moores University, Liverpool, United Kingdom

LUTFUN NAHAR

Centre for Natural Products Discovery (CNPD), School of Pharmacy and Biomolecular Sciences, Liverpool John Moores University, Liverpool, United Kingdom; Laboratory of Growth Regulators, Institute of Experimental Botany ASCR & Palacký University, Olomouc, Czech Republic

ELSEVIER

ACADEMIC PRESS

An imprint of Elsevier

Academic Press is an imprint of Elsevier
50 Hampshire Street, 5th Floor, Cambridge, MA 02139, United States
525 B Street, Suite 1650, San Diego, CA 92101, United States
The Boulevard, Langford Lane, Kidlington, Oxford OX5 1GB, United Kingdom
125 London Wall, London, EC2Y 5AS, United Kingdom

First edition 2020

ISBN: 978-0-12-821019-2
ISSN: 0065-7743

For information on all Academic Press publications
visit our website at https://www.elsevier.com/books-and-journals

Publisher: Zoe Kruze
Acquisitions Editor: Jason Mitchell
Editorial Project Manager: Hilal Johnson
Production Project Manager: Denny Mansingh
Cover Designer: Mark Rogers

Typeset by SPi Global, India

Contents

Contributors

Preface

Natural products, usually defined as the products of secondary metabolism in living organisms, have played an important role in the discovery and development of several modern drugs since the early 19th century, when the well-known narcotic analgesic agent, morphine, was isolated from *Papaver somniferum* in its pure form and developed as a modern drug. However, the actual use of natural materials (not purified secondary metabolites), e.g., plants and plant products, mostly in their crude forms or extracts, for the treatment of various human ailments can be traced back to thousands of years, as far back as the time of the Neanderthals.

To date, several best-selling drugs of modern time, e.g., the anticancer drugs camptothecin, doxorubicin, paclitaxel, vinblastine, and vincristine, have been discovered from natural sources. One of the major contributions of natural products in modern drug discovery has been in the area of anti-infective drugs including several antibiotics, starting from the discovery of penicillin, and antimalarial drugs quinine and artemisinin, the discovery of the latter awarded Youyou Tu with the Nobel Prize in Physiology/Medicine in 2015. Owing to rich chemical diversity, natural products will continue to be one of the major sources of new drug leads for drug discovery and development covering several therapeutic areas in the years to come.

The book, *Medicinal Natural Products: A Disease-Focused Approach*, captures some of the recent developments and contributions of natural products in drug discovery and development, focusing on specific disease areas like cancer, diabetes, obesity, and so on. Although there are several books covering various aspects of natural products, this is the first disease-focused compilation of various medicinal natural products that are applicable to the treatment or potential treatment of human diseases as evident from recent clinical trials involving human subjects or preclinical trials utilizing various in vivo animal models.

All 16 chapters cover the history, current status, and future opportunities of natural products in the treatment and prevention of any particular ailment or medical conditions. While the primary focus is on the medicinal chemistry aspects of these natural products with therapeutic applications, structural modifications, and in silico approaches are also addressed wherever relevant. All chapters review recent literature mainly of last 10 years or so, and present an overview on the developments in respective areas.

This book starts with Chapter 1 providing an overview of the full content of this volume. Chapter 2 talks about anticancer natural products, while, Chapters 2–8 cover emerging antimicrobial, antimalarial and antiparasitic, antiinflammatory, neuroprotective, hepatoprotective, and nephroprotective drug candidates obtained from various natural sources. Chapter 9 presents a comprehensive coverage of natural products with cancer chemopreventive potential. Chapters 10–16 discuss the antipsoriatic, antiosteoporotic, antidiabetic, antiobesity, analgesic and antipyretic, fertility regulating and antipsychotic agents that may potentially be developed as modern drugs. All chapters have been contributed by internationally known active researchers in relevant areas.

This book aims to cater primarily for the needs of the natural products researcher community, especially those who have been driving their research efforts in search for new drug molecules from various natural sources. However, in general, this book will be a valuable reference to anybody who is interested in drugs from natural origins.

Satyajit D. Sarker
Centre for Natural Products Discovery (CNPD)
School of Pharmacy and Biomolecular Sciences
Liverpool John Moores University, Liverpool, United Kingdom

Lutfun Nahar
Centre for Natural Products Discovery (CNPD)
School of Pharmacy and Biomolecular Sciences
Liverpool John Moores University, Liverpool, United Kingdom
Laboratory of Growth Regulators
Institute of Experimental Botany ASCR & Palacký University,
Olomouc, Czech Republic

CHAPTER ONE

Medicinal natural products—An introduction

Lutfun Nahar[a,b,*], Satyajit D. Sarker[a]

[a]Centre for Natural Products Discovery (CNPD), School of Pharmacy and Biomolecular Sciences, Liverpool John Moores University, Liverpool, United Kingdom
[b]Laboratory of Growth Regulators, Institute of Experimental Botany ASCR & Palacký University, Olomouc, Czech Republic
*Corresponding author: e-mail addresses: drnahar@live.co.uk; l.nahar@ljmu.ac.uk

Contents

1. Introduction

Natural products generally refer to secondary metabolites, often small molecules (mol wt. <2000 amu), that are produced by living organisms, and they are well known for unique chemical diversity and medicinal

Annual Reports in Medicinal Chemistry, Volume 55
ISSN 0065-7743
https://doi.org/10.1016/bs.armc.2020.02.008

properties.[1] They have continued to offer templates for the development of novel scaffolds of drugs to treat various human ailments including cancers and tumors, inflammation, pain, fever, infections, metabolic disorders and complications in human organs, *e.g.*, the brain, the heart, the kidneys and the liver. Natural products are also effective in maintaining good health and well-being, as well as preventing various diseases, for example, preventing cancers (cancer chemoprevention). Natural products can be obtained from terrestrial or marine sources: plants (*e.g.*, paclitaxel from *Taxus brevifolia*), animals (*e.g.*, vitamins A and D from cod liver oil) or micro-organisms (*e.g.*, cephalosporins from *Acremonium* spp. *alt. Cephalosporium* spp.) (Fig. 1).[1]

A large number of natural products possess various biological and pharmacological properties, and some of them have medicinal values. Thus, medicinal natural products can be defined as the natural products with clinically (or through pre-clinical studies) proven medicinal properties. In fact, historically natural products have played a significant role in the development of modern medicine by providing unique chemical molecules as lead compounds for drug development, and have remained as one of the major sources of new and effective modern drugs.[2] About a third of Food and Drug Administration (FDA)-approved drugs over the past two decades are based on natural products or their derivatives.[3] Among the natural products, higher plants have remained to be one of the major sources of modern drugs over several decades. In fact, over 25% of all FDA and/or the European Medical Agency (EMA) approved drugs are of plant origins. The vast majority of all successful anticancer drugs and antibiotics (and anti-infective agents) is from natural origins, the latter being mainly from microbial sources, *e.g.*, penicillins from *Penicillium* spp., tetracyclines from *Streptomyces aureofaciens*, doxorubicin from *Streptomyces peucetius* and cyclosporine from *Tolypocladium inflatum* (Fig. 2).

This chapter aims to introduce the book, 'Medicinal Natural Products—A Disease Focused Approach,' and to present an overview of various topics that have been covered in various chapters in this book. Although there are several books covering various aspects of natural products, this is the first disease-focused compilation of various medicinal natural products that are applicable to the treatment or potential treatment of human diseases as evident from recent clinical trials involving human subjects or pre-clinical trials utilizing various *in vivo* animal models.

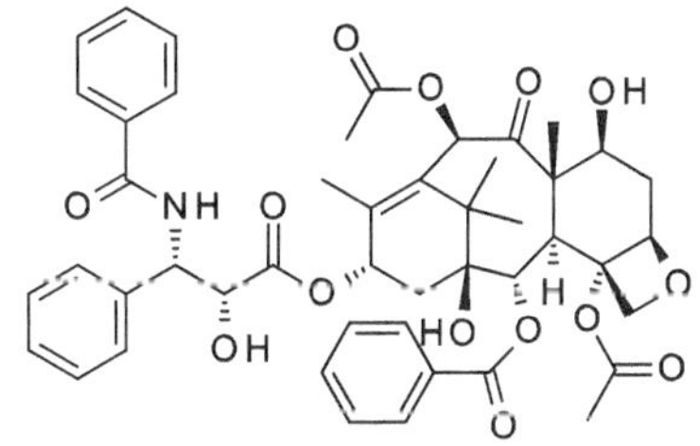

Paclitaxel – an anticancer drug from *Taxus brevifolia*

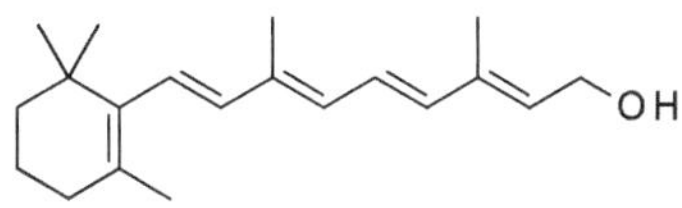

Retinol (vitamin A1)

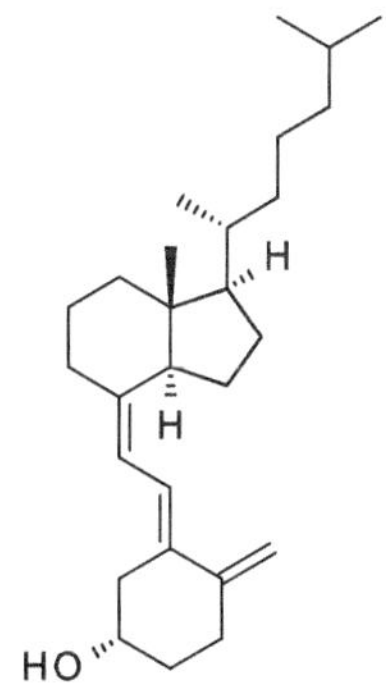

Vitamin D_3 (Colecalciferol)

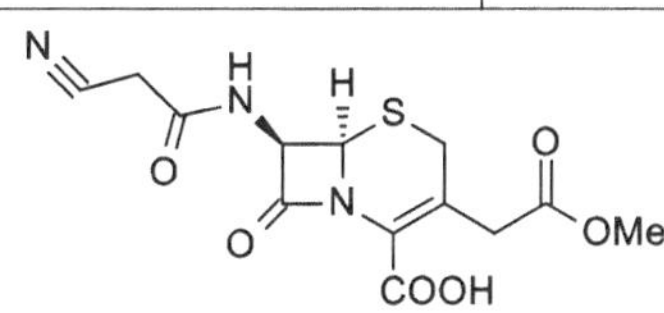

Cephalosporins: Cefacetril – an antimicrobial drug

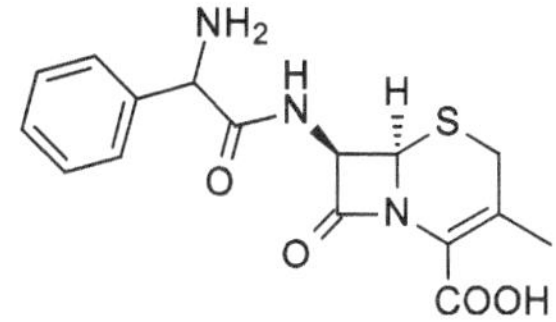

Cephalosporins: Cefadin – an antimicrobial drug

Fig. 1 Examples of some well-known medicinal natural products from various sources.

2. Analgesic and antipyretic natural products

The word 'analgesia' means the inability to feel pain, and analgesics, commonly known as painkillers, are drugs that are used to achieve analgesia or relief of pain, and work in several ways on the peripheral and the central

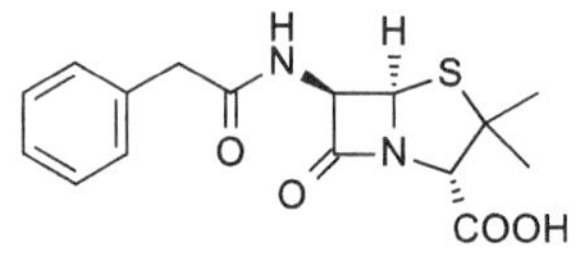

Penicillin G from *Penicillium notatum* (Trichocomaceae)

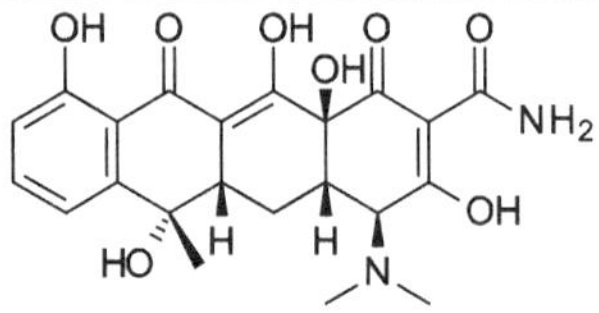

Tetracycline from *Streptomyces aureofaciens* (Streptomycetaceae)

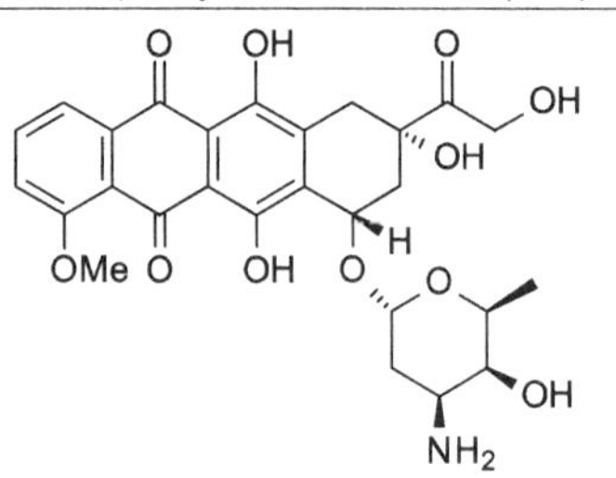

Doxorubicin from *Streptomyces peucetius* (Streptomycetaceae)

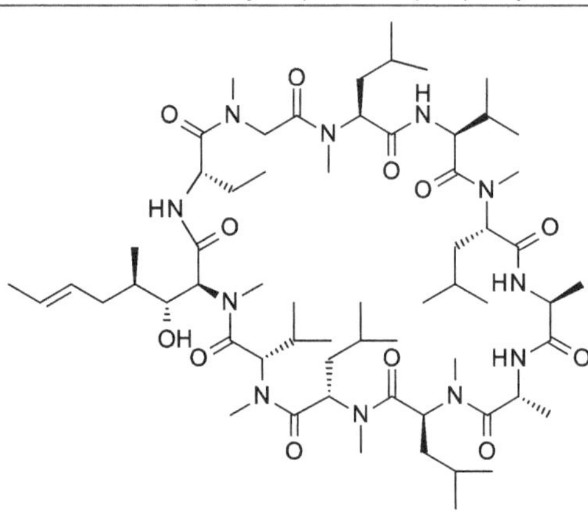

Cyclosporine from *Tolypocladium inflatum* (Ophiocordycipitaceae)

Fig. 2 Examples of some antimicrobial drugs from microbial sources.

nervous system (CNS). Many of the currently available analgesic agents are of natural origins or inspired by natural products templates, and can broadly be classified into two classes: narcotic analgesics (opioid analgesics) like codeine and morphine, and non-narcotic analgesics like aspirin, which derived from naturally occurring salicylic acid or salicin from *Salix alba* (Salicaceae), known as white willow (Fig. 3). In fact, morphine, produced

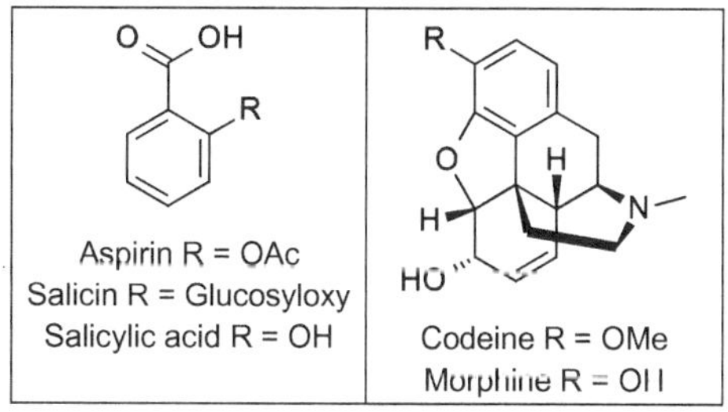

Fig. 3 Examples of medicinal natural products with analgesic property.

by *Papaver somniferum* (Papaveraceae), commonly known as opium poppy, is the first ever purified natural molecule developed as a modern drug in the early 19th century.[1] Generally, narcotic analgesics are strong analgesics, but habit forming, whereas aspirin like non-narcotic analgesics are not habit forming.

The mechanism of actions of the above analgesic drugs are well studied and well described in the literature. Aspirin inhibits the arachidonic acid pathway, thus prevents the synthesis of eicosanoids (particularly prostaglandins), which mediates pain signal through binding to cyclooxygenase (COX) group of enzymes. Aspirin also inhibits the cyclooxygenase (COX) enzymes and thus stops their binding to prostaglandins.[2] Opioid analgesics, like codeine and morphine, bind to receptors for endogenous opiates in the CNS. Thus, the painkilling activity of these compounds are mediated through their binding to opiate receptors. There have been a number of naturally occurring compounds shown to possess analgesic property, which are described in Chapter "Analgesic and antipyretic natural products" by Shilpi and Uddin.

Pyretic means febrile or fever, and antipyretic refers to a drug that can prevent or reduce fever, usually by causing the hypothalamus to override a prostaglandin-induced increase in temperature, and prompting the body to work to lower the temperature, which results in a reduction in fever. Fever means abnormal rise in body temperature above the normal range in response to a variety of infectious, immunological and neoplastic stimuli. Thus, antipyretic drugs do not necessarily treat the infections or immunological responses causing the fever, but reduce body temperature and allows the body to heal itself, if no other drugs are administered to treat the cause of the fever. Among the natural products derived antipyretic drugs, salicylates group, particularly aspirin (Fig. 3) is probably the best-known and most extensively used globally. There have been several medicinal plants and their crude products claimed to possess antipyretic property,[4] but not many

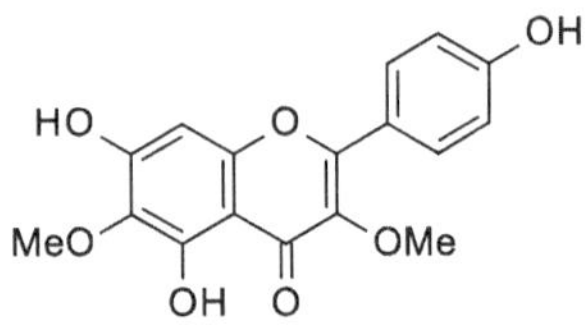

Fig. 4 Viscosine, a naturally occurring flavonoid with antipyretic property.

purified molecules from nature have been developed as antipyretic drugs to contribute to modern medicine. However, there have been a few molecules undergoing at least pre-clinical trials in animal models for the assessment of their potential antipyretic activity. For example, viscosine (Fig. 4), a flavonoid isolated from *Dodonaea viscosa* (Sapindaceae), has recently been evaluated by yeast-induced pyrexia rat model and utilizing molecular docking studies.[5] *Chapter 14* also looks into the current status of natural products derived antipyretic drugs that are under preclinical or clinical trials.

3. Anticancer natural products

One of the major therapeutic areas, where natural products have made significant contributions over the past few decades in the development of successful drugs, is cancer (a disease caused by uncontrolled division of abnormal cells in a part of the body, and can eventually invade nearby tissues) and tumor. In fact, higher plants have provided best-selling anticancer drugs like paclitaxel (Fig. 1), vinblastine, vincristine (Fig. 5) and several others (*see* Chapter "Anticancer natural products" by Sarker et al.). In fact, the earliest natural products-derived anticancer drugs approved by the FDA were the *Vinca* alkaloids: vincristine in 1963 and vinblastine in 1965 (Fig. 5), which were isolated from Madagascar periwinkle plants (*Catharanthus* spp.). Among the 98 anticancer drugs approved by the FDA during 1981–2010, 11 were purely natural compounds, 32 derived from natural products, 11 utilized the natural products templates for the synthesis, 16 were natural products mimics, and eight were linked to natural products, with only 20 of them were purely synthetic compounds.[6] During 1981–2014, a total of 136 new anticancer drugs were introduced, with 83% of these drugs being natural products or based on natural products, or analogues of natural products.[7] It is estimated that around 60% of all anticancer drugs that are currently in the market are natural products or inspired by natural products, confirming the nature still being the most attractive source for new anticancer drug discovery and development.

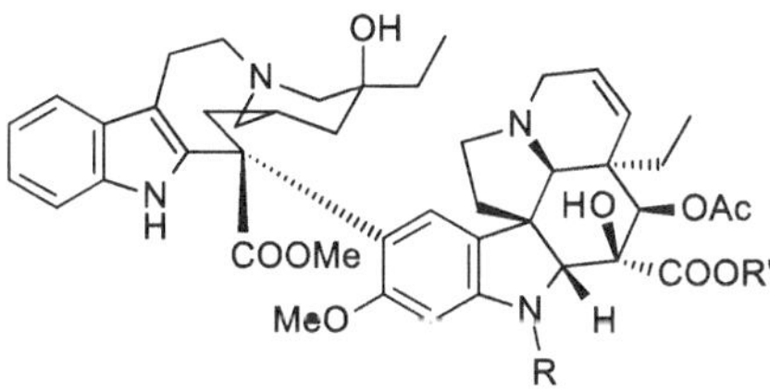

Vinblastine (R = R' = Me) and vincristine (R = CHO; R' = Me)
Source: *Catharanthus roseus* (Apocynaceae)

Fig. 5 Anticancer *Vinca* alkaloids – vinblastine and vincristine.

Natural products have remained as one of the major sources of anticancer lead molecules for future anticancer drug development. *Chapter 2* presents an overview of anticancer natural products that have been developed as commercially available drugs, and/or are at various developmental stages within the drug development pipeline for potential treatment of various forms of cancer. At the same time, *Chapter 2* provides a summary of information on possible mechanisms of actions of anticancer drugs from natural origins.

4. Antidiabetic natural products

Diabetes refers to a life-long medical condition that causes high levels of glucose in your blood, results predominantly from inadequate production of the hormone insulin in the pancreas. Diabetes leads to several life-threatening health problems including cardiovascular problems and nephropathy. There are two main types of diabetes—type 1 and type 2, the latter being the most prevalent one. Among the diabetes sufferers in the United Kingdom, there are only 8% type 1 and 92% type 2 diabetes. While type 2 diabetes is often linked to life-style, food intake, lack of exercise, obesity and family history, type 1 does not have anything to do with diet or lifestyle. While the precise cause of type 1 diabetes is not adequately known, it is often linked to malfunctioning of the immune system attacking and destroying the insulin-producing beta cells of the pancreas. It is also believed that type 1 diabetes is caused by genes and environmental factors, such as viruses, that might trigger the disease. Type 1 diabetes most commonly affects children and young adults, while type 2 diabetes is more common in older people, usually over 40 years of age. Type 2 diabetes is the third largest cause of human mortality and morbidity, just after cancer and cardiovascular diseases. There are 4.7 million people living with diabetes in the

United Kingdom, and this figure has trebled since 1996, when there were 1.4 million. It is estimated that by 2025, there will be about 5 million people with diabetes in the United Kingdom.[8]

Natural medicines in their crude forms have long been used in the treatment of diabetes, and some of those natural sources have helped discovering drugs like biguanides from the α-glucosidase inhibitors acarbose and galegine (Fig. 6) for treating patients with diabetes mellitus.[9] Biguanides, for example metformin (Fig. 6), are a group of oral type 2 diabetes drugs that work by preventing glucose production in the liver, enhancing insulin sensitivity and reducing absorption of sugar by the intestines. Acarbose is a natural product produced by the bacterial species, *Actinoplanes* spp. (Micromonosporaceae), and it can inhibit glycoside hydrolases, which are essential for carbohydrate digestion, and thus prevents and/or reduces the absorption of free glucose into the blood.

Galegine is an alkaloid with antidiabetic property, isolated from the plant *Galega officinalis* (Fabaceae), commonly known as goat's rue, French lilac, Italian fitch or professor-weed. This has been known since the Middle Ages for relieving the symptoms of diabetes mellitus, and used in the 1920s as an antidiabetic drug. However, because of its toxicity, the use was stopped, and replaced by other alternatives. Galegine helped the development of one of the most popular antidiabetic drugs of modern time, metformin (Fig. 6), for the treatment of type 2 diabetes. Galegine exerts its antidiabetic action by inducing weight loss, reduction in weight gain, reduction of blood glucose,

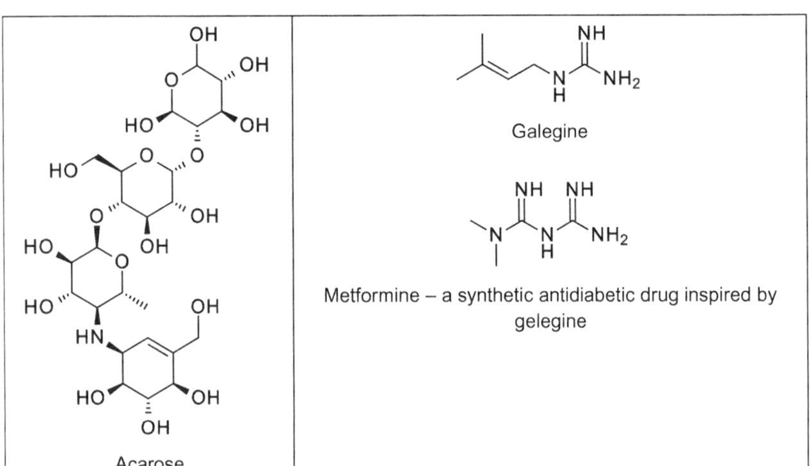

Fig. 6 Acarbose and galegine – natural antidiabetic drugs.

and a decrease in food intake. Galegine has been demonstrated to stimulate glucose uptake by both 3T3-L1 adipocytes and L6 myotubes, resulting in its *in vivo* hypoglycaemic action.

In recent years, there have been several natural products described in the literature to possess antidiabetic properties as demonstrated in *in vivo* animal trials, and those natural products mainly belong to the chemical classes of alkaloids, anthraquinones, coumarins, flavonoids, polysaccharides, saponins and terpenes.[10] Chapter "Antidiabetic natural products" by Mukherjee et al. is devoted to medicinal natural products for the treatment of diabetes.

5. Anti-inflammatory agents from natural origins

Simply, inflammation (Latin: inflammatio) is a process by which the white blood cells and the substances they produce protect us from microbial infections and invasion of any harmful foreign materials like allergens. However, in some diseases, for example arthritis, the body's immune system triggers an inflammatory response, when there are no foreign invaders to fight off. More precisely, inflammation can be defined as a part of the complex biological response of body tissues to harmful stimuli, such as pathogens, damaged cells, or irritants, and is a protective response involving immune cells, blood vessels, and molecular mediators.[11] There are two types of inflammations, acute and chronic inflammation. Acute inflammation is manifested by pain, redness and swelling, whereas the common symptoms of chronic inflammation may include fatigue, fever, mouth sores, rashes, abdominal pain and chest pain. Chronic inflammation is believed to be linked to diseases such as asthma, cancer, cardiovascular diseases, inflammatory bowel diseases, metabolic syndrome, neurodegenerative diseases and rheumatic conditions.

Drugs that can mitigate inflammation are called anti-inflammatory drugs, and can be classified into two major classes: steroidal and non-steroidal anti-inflammatory drugs (NSAIDs). Most currently available anti-inflammatory drugs, diclofenac, hydrocortisone, ibuprofen, naproxen, and several others have limitations associated with side effects and cost. Among the anti-inflammatory drugs available today, the natural salicylate derivative, acetylsalicylic acid (aspirin) (Fig. 3) is the least costly one, and has minimum side effects. However, the search for new more effective, less toxic and affordable anti-inflammatory drugs, particularly of natural origins, has continued over the past few decades. Chapter "Anti-inflammatory natural products" by Olajide and Sarker in this book presents the pharmacological actions of some naturally occurring, alkaloids, flavonoids and terpenes, which have shown promising benefits in both *in vitro* and *in vivo* experimental models reflecting the pathologies of inflammatory diseases. Because of unique chemical diversity that exists in natural products, and also due to known mechanisms of anti-inflammatory action, further clinical investigations on natural products, which have been explored recently, may result in the development of novel anti-inflammatory agents, or to unveil new mechanism of action. Examples of some promising anti-inflammatory agents of natural origins,[12,13] proven from various *in vitro* and *in vivo*, as well as *in silico* studies, are shown in Fig. 7.

There were at least eight different molecules from marine organisms or their analogues in the preclinical and clinical trials during 2016. Only one of them, 3-(2,4-dimethoxybenzylidene)-anabaseine (also known as GTS-21 or DMXBA) (Fig. 8), was successful, and studies on the other seven were discontinued because of lack of success in the earlier trials.[12] Anabaseine (Fig. 8) is a nemertine alkaloid toxin isolated from several marine worms in the Phylum Nemertea, and its synthetic analogue is 3-(2,4-dimethoxybenzylidene)-anabaseine.

6. Antimalarial and antiparasitic natural products

Malaria is a life-threatening disease, yet preventable and treatable, caused by *Plasmodium* parasites that are transmitted to people through the bites of infected female *Anopheles* mosquitoes.[14] It is one of the major global health problems affecting around 219 million people in 87 countries and causing 435,000 deaths in 2017. However, African countries are the worst affected areas with 92% of all malaria cases and 93% of malaria-related deaths of the world in 2017. Significant concerted global efforts have been made over the years to eliminate malaria, and a total funding for malaria control and elimination reached an estimated US$ 3.1 billion in 2017, with contributions of USD $900 million, representing 28% of total funding, from governments of malaria endemic countries.

There are five different species of *Plasmodium* that can cause malaria in humans, but *P. falciparum* and *P. vivax* are the most prevalent two. In 2017, while *P. falciparum* was responsible for 99.7% of estimated malaria cases in the WHO African Region, as well as in the majority of cases in the WHO regions of South-East Asia (62.8%), the Eastern Mediterranean (69%) and the Western Pacific (71.9%), *P. vivax* was the major parasite in the WHO Region of the Americas, representing 74.1% of malaria cases.[14]

From terrestrial plants	
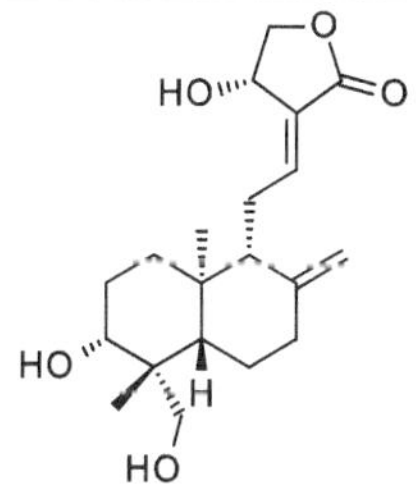 Andrographolide from *Andrographis paniculata* (Acanthaceae)	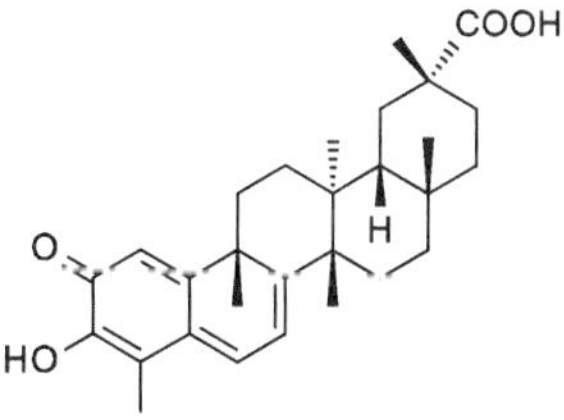 Celastrol from *Tripterygium wilfordii* (Celastraceae)
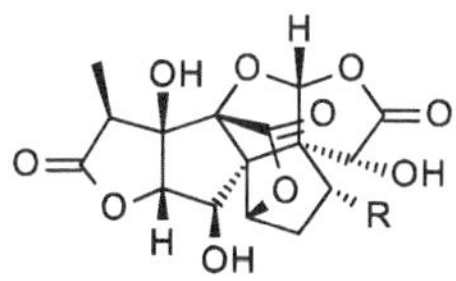 Ginkgolide B from *Ginkgo biloba* (Ginkgoaceae)	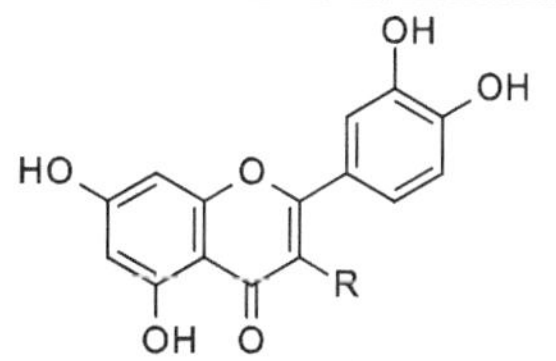 Luteolin R = H and quercetin R = OH from various fruits and vegetables
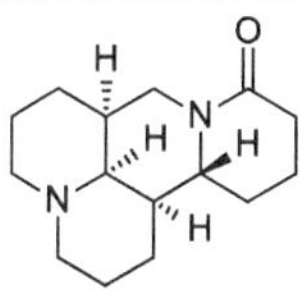 Matrine from *Sophora flavescens* (Fabaceae)	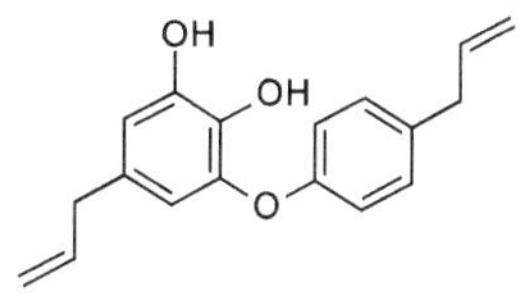 Obovatol from *Magnolia liliflora* (Magnoliaceae)
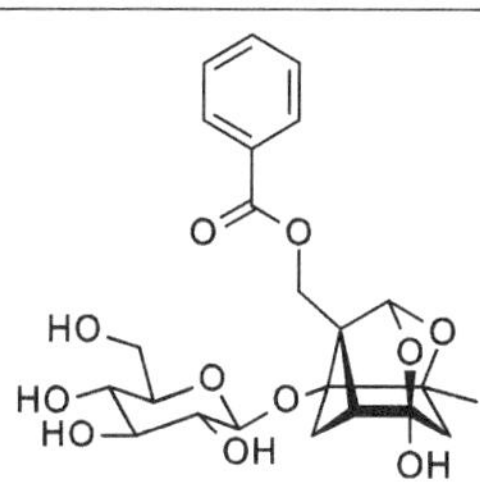 Paeoniflorin from *Paeonia lactiflora* Pall. (Paeoniaceae)	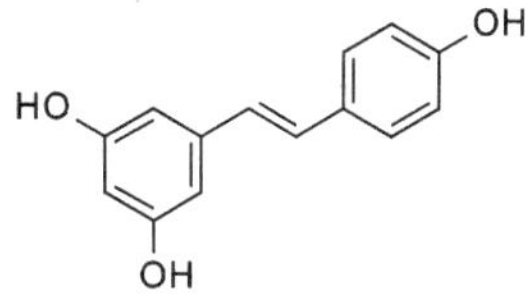 Resveratrol from *Vitex* species (Verbenaceae)
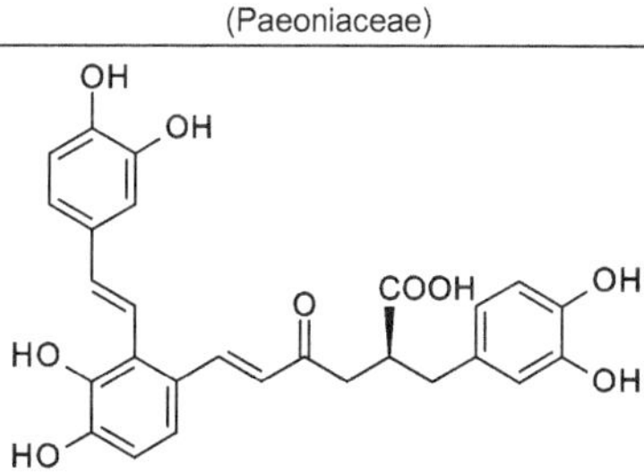 Salvianolic acid A from *Salvia miltiorrhiza* (Lamiaceae)	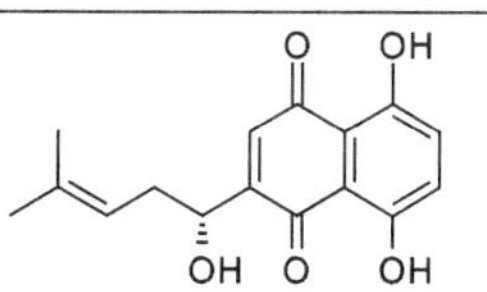 Shikonin from *Lithospermum erythrorhizon* (Boraginaceae)

Fig. 7 Examples of some promising anti-inflammatory agents of natural origins.

From marine sources: coral, sponges and microorganisms	
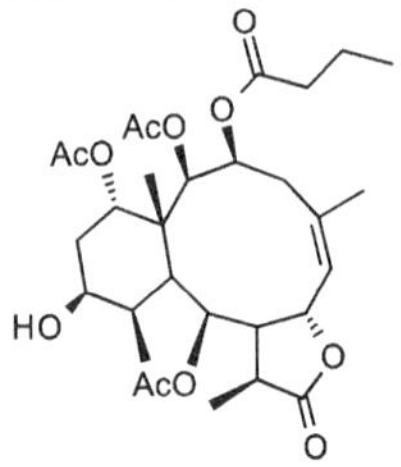 Briarane-type diterpenes from *Briareum excavatum* (Briareidae)	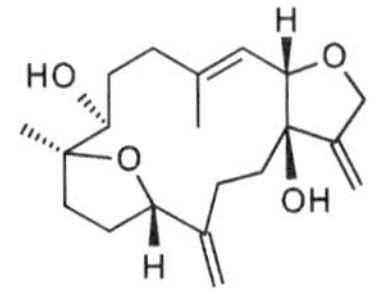 Crassumol E from *Lobophytum crissum* (Alcyoniidae)
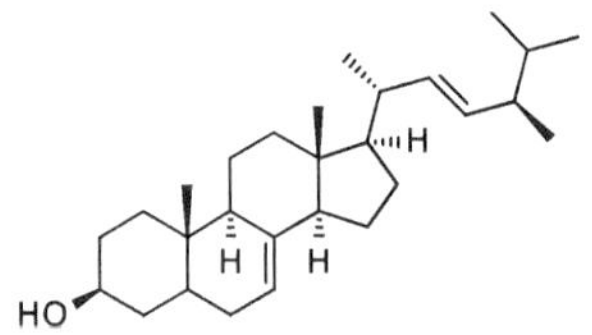 Ergosta-7,22-dien-3-ol from *Marthasterias glacialis* (Asteriidae)	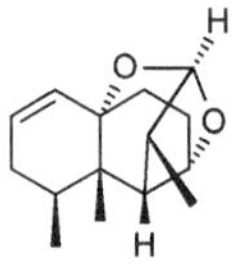 Flavalin A from *Lemnalia flava* (Nephtheidae)
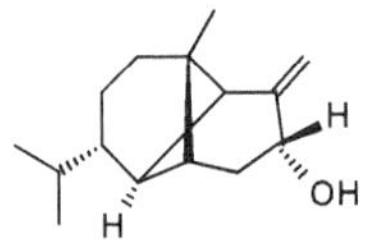 Lemnalol from *Lemnalia cervicorni* (Nephtheidae)	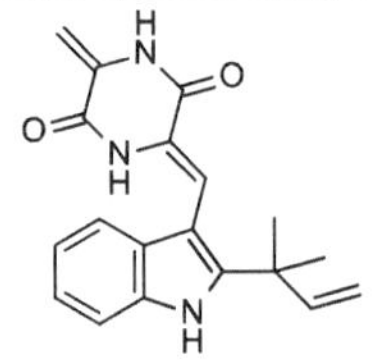 Neoechinulin B from *Eurotium* sp. SF-5989 (Trichocomaceae)
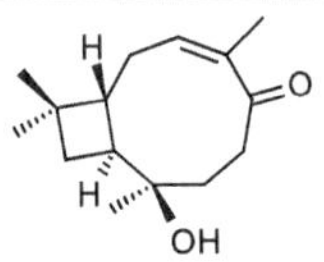 Rumphellol A from *Rumphella antipathies* (Gorgoniidae)	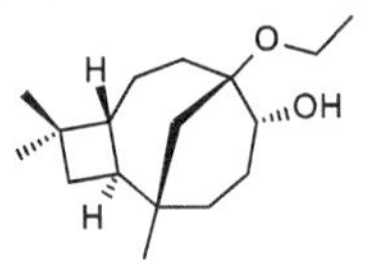 Rumphellol B from *Rumphella antipathies* (Gorgoniidae)
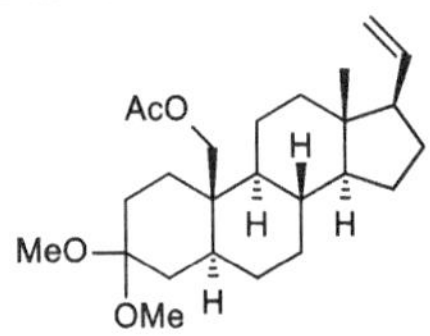 Sclerosteroid J from *Scleronephthya gracillimum* (Nephtheidae)	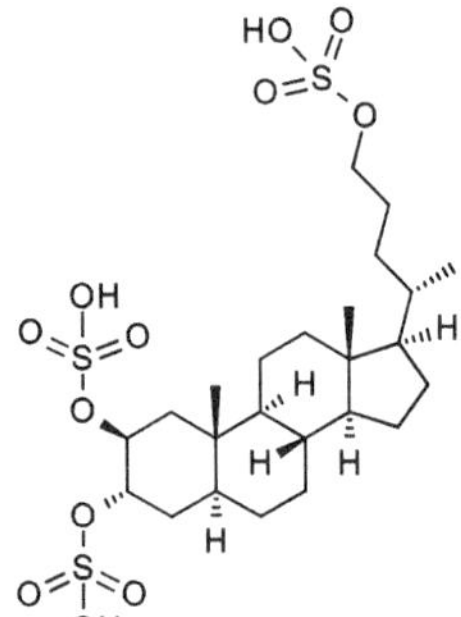 Solomonsterol A from *Theonella swinhoei* (Theonellidae)

Fig. 7—Cont'd

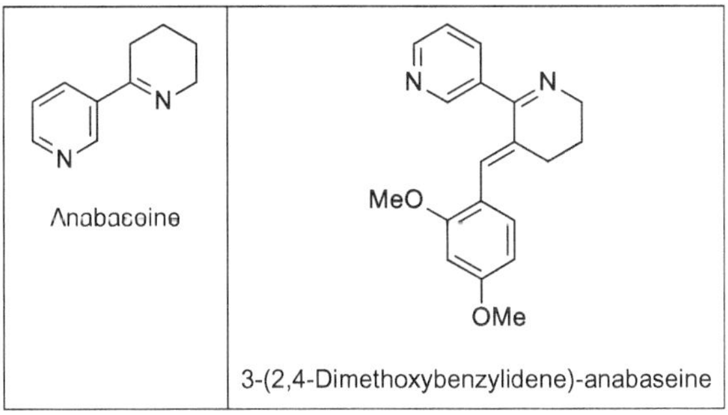

Fig. 8 Structures of anabaseine and 3-(2,4-dimethoxybenzylidene)-anabaseine.

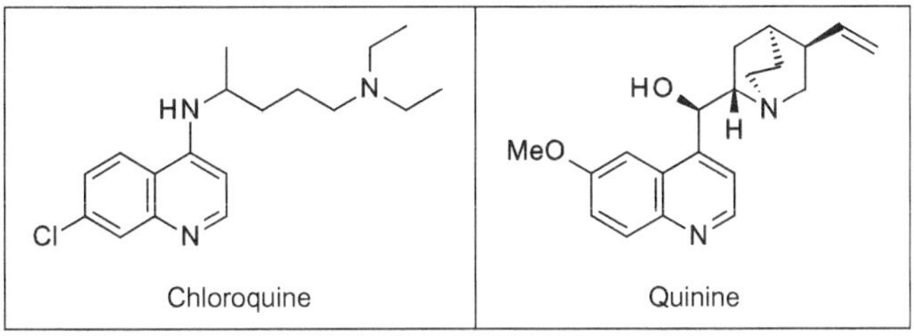

Fig. 9 Quinine, the first antimalarial natural product drug, and its analogue chloroquine.

Natural products have traditionally provided antimalarial 'lead' compounds to fight against malaria. Quinine (Fig. 9), an alkaloid isolated from the bark of *Cinchona succirubra* Pavon. (Rubiaceae) in 1920, was the first natural compound to enter the antimalarial drugs market, and soon became one of the essential medicines listed by the WHO. However, as of 2006, quinine is no longer recommended by the WHO because of ineffectiveness against resistant strains of malarial parasites, and also for its toxicities. However, quinine paved the way to the synthesis of various quinine analogues, *e.g.*, chloroquine (Fig. 9) synthesized in 1934, as antimalarial drugs, and has been used in the treatment or prevention of malaria from *P. vivax*, *P. ovale*, and *P. malariae*, recently excluding the malaria parasite *Plasmodium falciparum*, which started to develop widespread resistance to it. Quinine itself kills malarial parasite *P. falciparum* by interfering with parasite's ability to dissolve and metabolize hemoglobin, whereas its analogues are believed to exert their antimalarial actions by inhibiting haemozoin biocrystallization in haeme detoxification pathway, which facilitates the aggregation of cytotoxic haeme, which accumulates in the parasites, causing their deaths.[15] It is also hypothesized that quinine might be able to target purine nucleoside phosphorylase enzyme.

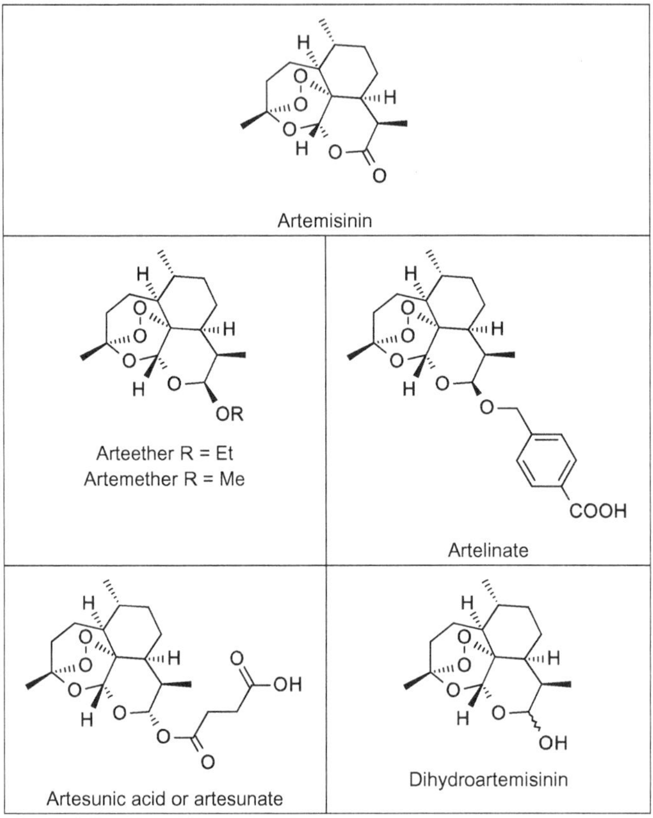

Fig. 10 Artemisinin and its semisynthetic analogues.

Artemisinin (Fig. 10), a sesquiterpene lactone that contains a unique peroxide bridge, also known as endoperoxy (1,2,4-trioxane ring) functionality, isolated from the Chinese medicinal plant, *Artemisia annua* (Asteraceae), common name, 'qinghao,' in 1972 by Tu Youyou, the winner of the 2015 Noble Prize in Medicine or Physiology, is the latest contribution from mother nature in the fight against malaria.[16] Nowadays artemisinin and its semisynthetic analogues[17] (Fig. 10) are in use against malarial infections caused by *P. falciparum*. It is believed that the endoperoxide 1,2,4-trioxane ring contributes to the drug's mechanism of action. However, the exact mechanism of action of artemisinin and its analogues was not known until 2018. It has now been demonstrated that artemisinins do not directly attack malarial parasites or cells, but they are required to undergo chemical transformations in the blood through activation of the functional group endoperoxide ring by cleavage of the endoperoxide ring. When artemisinin or its analogues comes in contact with the haeme, the iron(II) oxide breaks the endoperoxide ring, resulting in the production of free radicals that in turn damage susceptible proteins, leading to the death of the parasite.[18–20] Unlike other antimalarials, which are active only on a particular stage of the life cycle of malarial parasite, artemisinin is able to kill all the life cycle stages.

Despite the success stories in the fight against malaria due to the introduction of quinine and artemisinin based antimalarial drugs, resistance to antimalarial medicines is a recurring problem. Because of wide-spread resistance *P. falciparum* against quine, it is no longer recommended by the WHO as a treatment option for malaria. Resistance of *P. falciparum* malaria parasites to other generations of medicines, such as chloroquine (Fig. 9) and sulfadoxine-pyrimethamine became known in the 1950s and 1960s. Initially it was thought artemisinin and its analogues were not prone to resistance, but there have been several reports on the resistance against artemisinins published in the recent literature.[21] The mechanisms of resistance developed by the parasites against artemisinin compounds affect only one stage of the malaria parasite cycle in humans: the ring stage, causing 'partial resistance,' which is often the result of poor treatment practices, inadequate patient adherence to prescribed antimalarial regimens, and the widespread availability of oral artemisinin-based monotherapies and substandard forms of the drug. However, it is not known whether partial artemisinin resistance could further lead to complete resistance affecting other stages of the life cycle of the parasites. Artemisinin partial resistance appeared prior to 2001, and prior to the widespread deployment of artemisinin-based combination therapies (ACTs) in the Greater Mekong Subregion (GMS). To date, five countries of the GMS: Cambodia, the Lao People's Democratic Republic, Myanmar, Thailand and Viet Nam, have experienced artemisinin partial resistance. It is now known that partial artemisinin resistance has emerged independently in several locations in the GMS and spread within the subregion. However, artemisinin partial resistance has not been confirmed in Africa. Because of the incidence of resistance, artemisinin, being the last line of defense against malaria, is explicitly discouraged by the WHO as a monotherapy, and recommend therapies combining artemisinin or its analogues with some other antimalarial drugs for the treatment for malaria.

Because of the threat of resistance against available antimalarial therapies as outlined above, the search for new antimalarial drugs, especially from natural sources, still remains a valid task for natural products researchers. While

Plant-derived antimalarial compounds

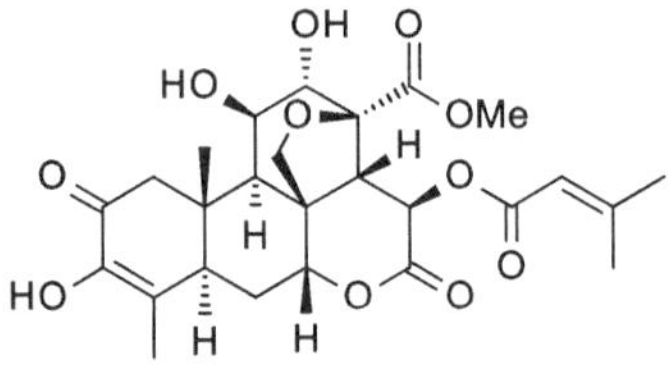

Brusatol from *Brucea javanica* (Simaroubaceae)

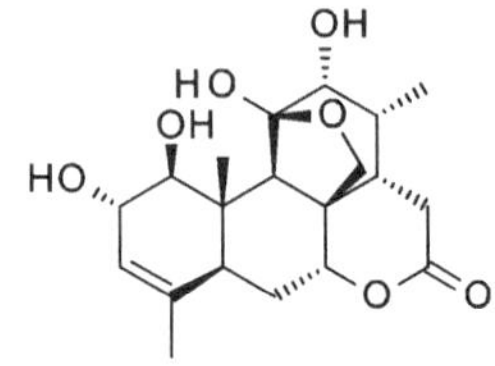

Chaparrin from *Castela nicholsonii* (Simaroubaceae)

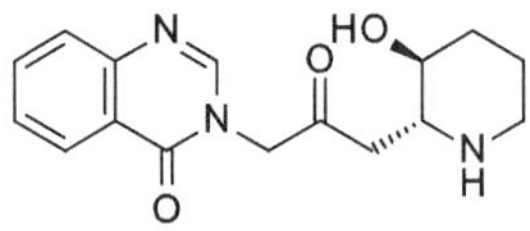

Febrifugine, isolated from *Dichroa febrifuga* (Hydrangeaceae)

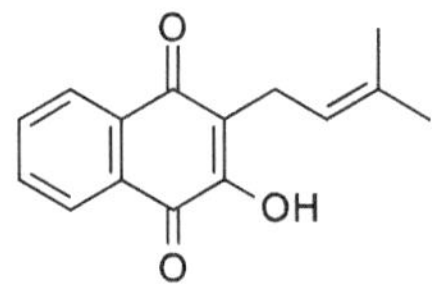

Lapachol, isolated from *Handroanthus impetiginosus* (Bigoniaceae), is currently used clinically as an antimalarial drug

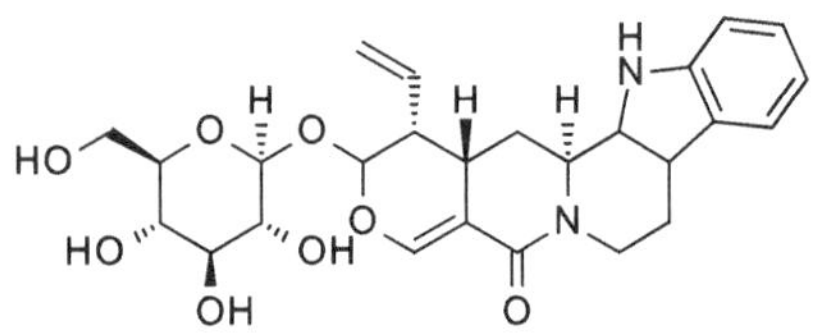

Strictosamide, isolated from *Nauclea pobeguinii* (Rubiaceae), is a prodrug that needs to be bioactivated in the gastrointestinal tract to produce antimalarial action

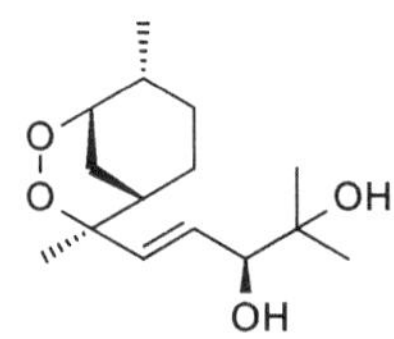

Yingzhaosu, isolated from *Artabotrys uncinatus*, also known as *Artabotrys hexapetalus,* (Annonaceae)

Antimalarial compounds from microbial origin

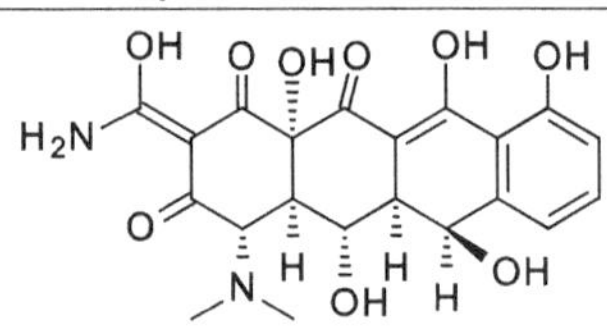

6-Demethyltetracycline, isolated from *Streptomyces* species (Actinomycetaceae)

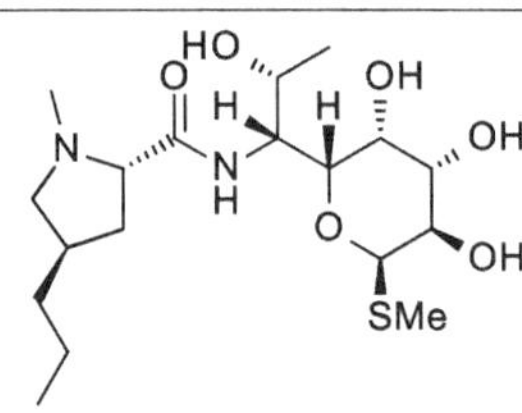

Lincomycin, isolated from *Streptomyces lincolnensis* (Actinomycetaceae)

Fig. 11 Recently reported antimalarial natural products.

Chapter "Antimalarial and antiparasitic natural products" by Ismail et al. deals with medicinal natural products for the treatment of malaria and other parasitic diseases, Fig. 11 displays a few examples of some of the recently reported natural molecules with antimalarial prospects.[22–24]

In addition to malaria, which is caused by the parasites from the genus *Falciparum*, there are several other parasitic diseases, some which can be quite difficult to treat. Some examples of parasitic diseases may include African trypanosomiasis (also known as African sleeping sickness) caused by *Trypanosoma brucci*, amoebiasis caused by *Entamoeba histolytica*, Chagas disease (also known as American trypanosomiasis) caused by *Trypanosoma cruzi*. Ascariasis (also known as intestinal wound worm infection) caused by *Ascaris lumbricoides*, Leishmaniasis (also known as Kala-azar or *Leishmania* Infection) caused by *Leishmania* parasites, and many more. Some of these diseases have long been neglected, and the number of drugs available for treating these neglected diseases is somewhat limited to a few. Efforts have been made in recent years to develop drugs from natural products to treat these diseases. Although *Chapter 4* presents details on the natural products derived remedies for these diseases, Fig. 12 exhibits just a few examples of antiparasitic natural molecules, for which, the efficacy has been demonstrated by various *in vitro* and *in vivo* studies.[25–28]

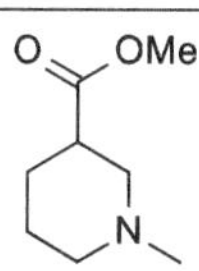

Arecoline from *Areca catechu* (Arecaceae) has anthelmintic property

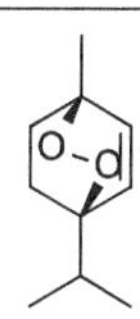

Ascaridole, an anthelmintic drug isolated from *Chenopodium ambrosioides* (Amaranthaceae)

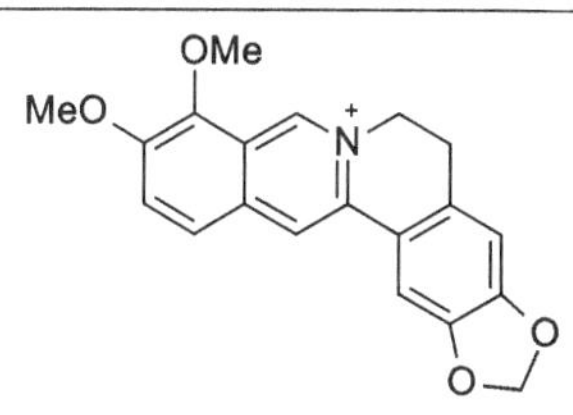

Berberine, an anti-leishmanial alkaloid from *Hydrastis canadensis* (Ranunculaceae)

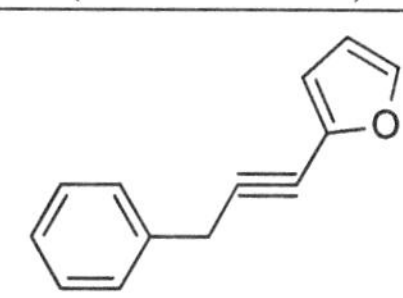

Carlina oxide, a antitrypanosomal (African trypanosomiasis) compound from *Carlina acaulis* (Asteraceae)

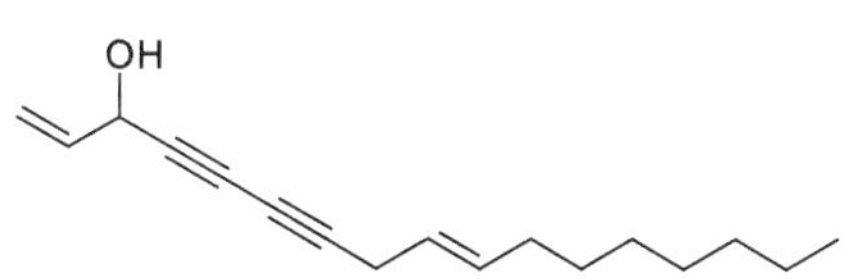

Polyacetylenes (*e.g.*, panaxynol) from ginseng, *Panax ginseng* (Araliaceae), are effective against African trypanosomiasis

Fig. 12 Examples of a few recently reported antiparasitic natural products.

7. Antimicrobial natural products

Traditionally, crude natural products, *e.g.*, cinnamon, clove, coriander, garlic, ginger, honey, oregano and thyme, from various sources have played a significant role in the fight against microbial infections, as well as food preservation. Until the discovery of penicillin G (Fig. 13) from the fungus *Penicillium chrysogenum or P. notatum* (Trichocomaceae) in 1928 by Alexander Fleming, which opened up a new perspective for the treatment of microbial infections, and led the generation of a series of semisynthetic penicillins, the most widely used group of antibiotics in the world today, and the dawn of the 'antibiotics era,' various plant-derived molecules, especially components of essential oils, *e.g.*, carvacrol, eugenol, thymol and vanillin, have been used routinely to treat microbial infections.[1] Several phytochemicals of the chemical classes, *e.g.*, alkaloids, alkenyl phenols, flavonoids, phorbol esters, saponins, tannins and terpenes, were also shown to considerable antimicrobial properties. Since

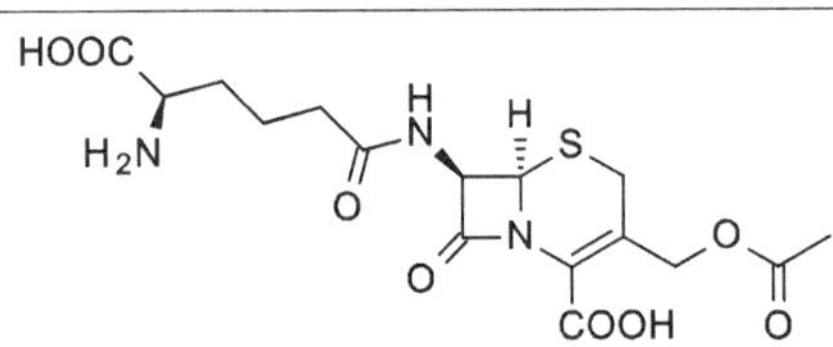

Cephalosporin C from the fungus *Acremonium alt. Cephalosporium* species (Hypocreaceae)

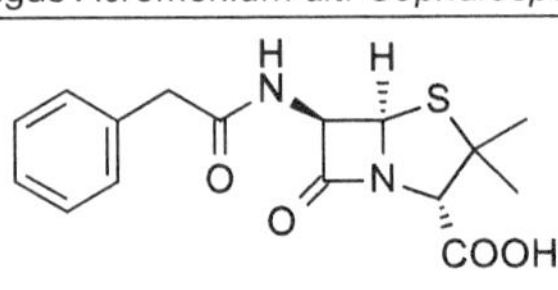

Penicillin G from *Penicillium chrysogenum* (Trichocomaceae)

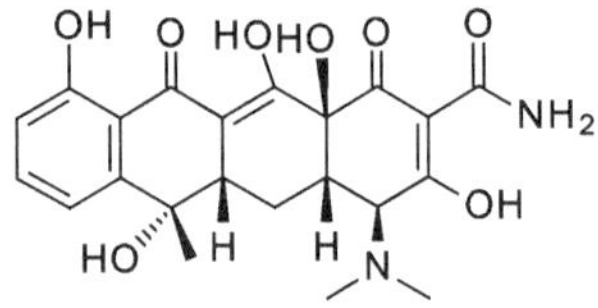

Tetracycline from Streptomyces species (Actinomycetaceae)

Fig. 13 Structures of popular antibiotics of natural origins.

the discovery of penicillins, much attention was given to discovery of new antibiotics from microbial sources, leading to the discovery of several successful antibiotics like cephalosporins (the natural one is cephalosporin C) and tetracyclines (the natural one is tetracycline) (Fig. 13). In fact, in modern drug discovery attempts, the major contributions from natural products still remain to be in the areas of anticancer and antimicrobial drug development. It is estimated that more than 70% of antimicrobial drugs that are in use today, are from natural products or natural product analogues of synthetic or semi-synthetic origins.[7]

Among the 326 antimicrobial drugs that were introduced during 1981–2014, about 40% drugs were unaltered natural products, natural product derivatives/analogues or natural product mimics.[7] There were 141 antibacterial drugs emerged during this period, and 83 of them were natural products or can be traced back to natural products. Similarly, of 32 antifungal drugs, six drugs were natural product analogues and/or natural product mimics. A total of 26 antiviral drugs of 138 introduced in the market, were natural product derivatives and natural product mimics. There were nine antibacterial drugs approved by the FDA during 2011–2014. While fidaxomicin was the unaltered natural product, ceftaroline and cetolozane were cephalosporin derivatives (Fig. 14).

Wu et al.[29] reviewed antimicrobial natural products reported from marine Actinomycetes during 2012–2017, and reported that arenimycin, mollemycin A, glycosylated macrolactins and xinghaiamine A (Fig. 15) could display significant antimicrobial activity against pathogenic bacterial strains, revealing their potential as new antimicrobial drug candidates. Similarly, lajollamycins, bahamaolides and glycosylated benzoxazine appeared to have antifungal potential. In fact, Actinomycetes has been one of the most productive sources of new antibiotics, and a number of the antibiotics currently in use are natural products or their analogues from Actinomycetes. Chaudhary et al.[30] highlighted several marine natural products and their synthetic analogues (Fig. 16), that have attracted attention as effective antimicrobial agents over during the same period. However, none of these newly discovered marine natural products has entered any clinical trials. A recent review article published by Khameneh et al.[31] covers antimicrobial phytochemicals and their mechanisms of actions.

Because of the severe threats posed by the drug resistant microorganisms, *e.g.,* methicillin-resistant *Staphylococcus aureus*, making almost all

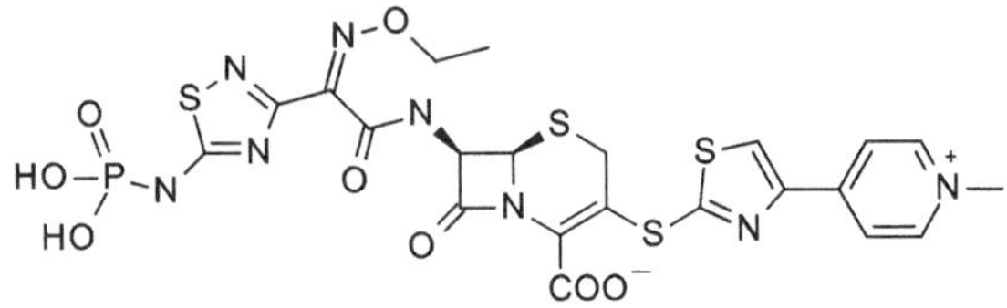

Ceftaroline, a semisynthetic analogue of cephalosporin

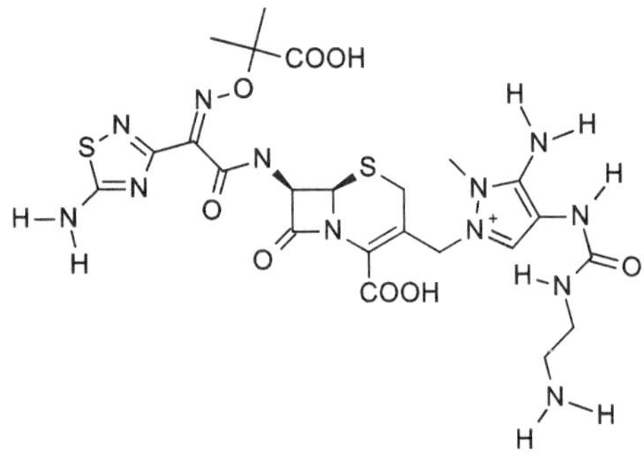

Cetolozane, a semisynthetic analogue of cephalosporin

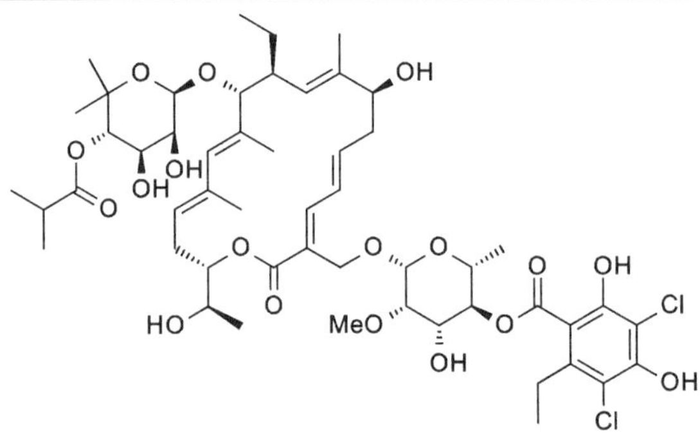

Fidaxomycin obtained from the actinomycete *Dactylosporangium aurantiacum* subspecies *hamdenesis* (Micromonosporaceae)

Fig. 14 Example of antimicrobial natural product drugs and their analogues reported during 2011–2014.

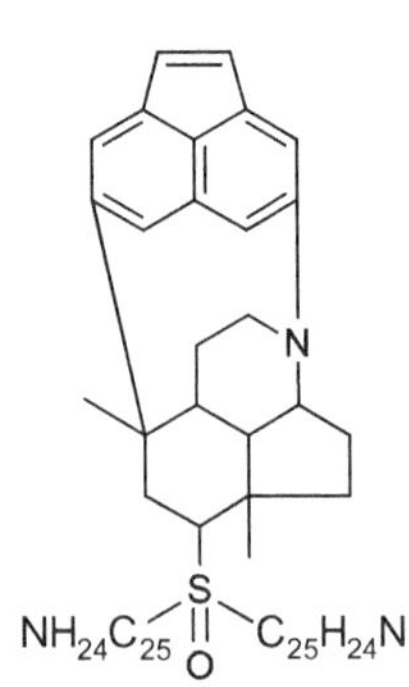

Fig. 15 Xinghaiamine, an example of Actinomycetes-derived antimicrobial agent.

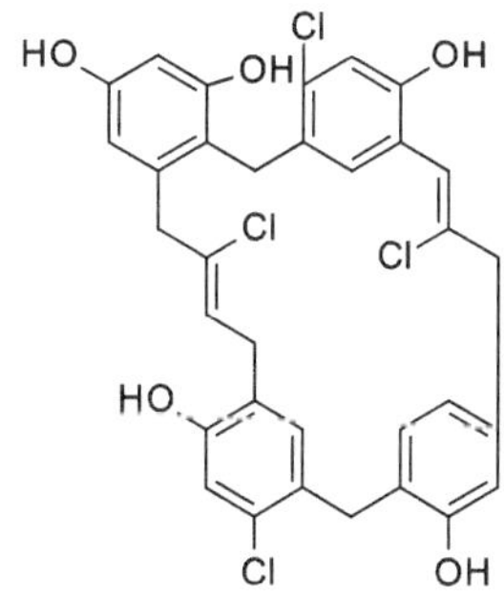

Chrysophaentin A, isolated from the chrysophyte alga *Chrysophaeum taylori* (Phaeothamniaceae)

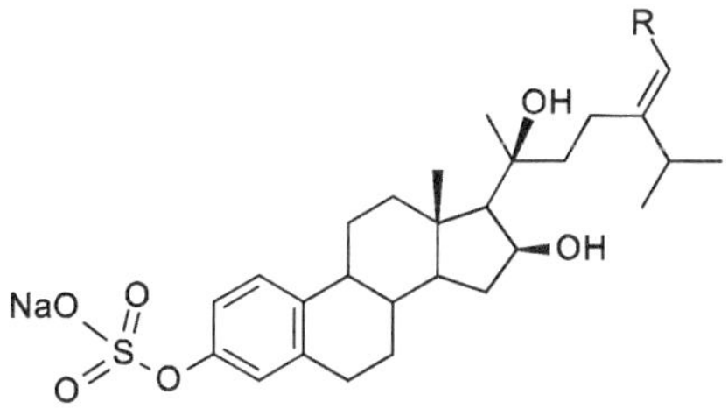

Geodisterol 3-*O*-sulfite (R = Me) and 29-demethylgeodisterol 3-*O*-sulfite (R = H) from the marine sponge *Topsentia* sp. (Halichondriidae)

Fig. 16 Examples of some marine natural products as antimicrobial agents.

available antibiotics either ineffective or less effective, it has been a constant battle for medicinal and natural products chemists to discover new chemical entities or scaffolds for new antimicrobial drug development. Antimicrobial resistance has been the cause of around 50,000 deaths each year only in Europe and the United States, and the number is expected to rise to 10 million worldwide.[32,33] Thus, the search for new effective antibiotics or antimicrobial agents will continue, and as natural products, especially from microbial origins, have historically provided new antibiotics classes, it is reasonable to assume that the antimicrobial drug discovery efforts will mainly be driven involving untapped natural resources.

The major mechanisms of actions of currently available antibiotics, synthetic or natural, involve inhibition of or interference with bacterial protein biosynthesis, cell-wall biosynthesis, nucleic acid synthesis, destruction of bacterial membrane, and inhibition of the efflux pump. Chapter

"Antimicrobial natural products" by Rahman covers recent advances in the discovery of new antimicrobial agents and their possible mechanisms of actions.

8. Antipsoriatic natural products

Chapter 10 presents a comprehensive coverage of antipsoriatic natural products, and discusses their mechanism of actions. Simply, psoriasis is an immune-mediated disease that causes raised, red, scaly patches to appear on the skin, and it affects about 3% population globally.[34] It is characterized by inflamed, red plaques with silvery scales and thickened epidermis (acanthosis) usually appearing on the elbows, knees, scalp, trunk and nails. Psoriasis is caused, at least in part, by the immune system mistakenly attacking healthy skin cells. Other causes may include various genetic (family history) and environmental factors such as drugs/medicines, smoking, alcohol intake, diet, insect or snake bites, mental stress and depression.[35] It is worth mentioning that psoriasis is not contagious and it cannot spread from person to person.

There are several modern drugs available for the treatment of psoriasis, some of them are of natural origins, and also there have been a number of natural products identified as potential 'leads' for antipsoriatic drug development. 8-Methoxypsoralen (8-MOP, also known as xanthotoxin) (Fig. 17), a well-known natural furanocoumarin mainly from plants of the families, Apiaceae and Rutaceae, has revolutionized psoriasis treatment; 8-MOP (oral or topical) combined with UVA exposure (photochemotherapy, PUVA) is one of the oldest and most effective therapy for psoriasis.[36,37]

Topical treatment (symptom relief) of psoriasis mainly include topical corticosteroids, retinoids, anthralin, vitamin D analogues, salicylates and moisturizer; of them vitamin D and salicylates are clearly natural products. Systemic treatments could include methotrexate, cyclosporine and

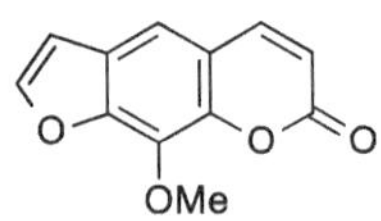

Fig. 17 Xanthotoxin (8-MOP), a plant-derived furanocoumarin, with antipsoriatic property.

retinoids; cyclosporine is a microbial natural product. Phototherapy is also effective in the treatment of psoriasis, especially in combination with antipsoriatic drugs like 8-MOP.

Fig. 18 presents a few examples of antipsoriatic natural products from both terrestrial and marine organisms. However, it is noteworthy that most of the antipsoriatic natural products that are currently under preclinical or clinical trials are from terrestrial plants, and include chemical classes like alkaloids, cyclopeptides, flavonoids, furanocoumarins, organic acid derivatives, polyphenols, sphingolipids and terpenoids.

The pathogenesis of psoriasis is complex and not fully understood. However, several hallmark features may include epidermal hyperplasia, abnormal keratinocyte differentiation, inflammatory cell infiltration and increase in dermal vascularization. Generally, mechanisms of actions of currently available antipsoriatic drugs or antipsoriatic natural products concern one or more of these hallmarks. T-cell dysregulation, activation of nuclear factor-kB (NF-κB) pathway, overexpression of vascular endothelial growth factor and vascular inflammatory markers are known to play a crucial role in inflammation and angiogenesis. Thus, antipsoriatic drugs aim to reverse these markers, *e.g.*, inhibition of activation of NF-κB.

9. Antipsychotic natural products

Psychosis can be defined as a mental health problem that causes people to perceive or interpret things differently from those around them, and often involve hallucinations or delusions.[38] There are several antipsychotic medications available to date, and they generally act by reducing the amounts of dopamine in the brain, or restore the balance between dopamine and other chemicals in the brain. They are classified into two groups: typical or 'first generation,' which have been used since 1950s, for example, benperidol, chlorpromazine, flupentixol, haloperidol and pericyazine, and atypical or 'second generation' antipsychotic drugs which have been in use since 1990s, for example, amisulpride, clozapine, olanzapine and paliperidone. Most of these antipsychotic drugs show mild to severe side effects, especially when they are used for a long term. Despite the availability of various antipsychotic drugs, most patients have a tendency to use natural antipsychotic medicines based on crude natural product extracts. There are some natural compounds, which have now been demonstrated to possess defined

From terrestrial plants and microbes	
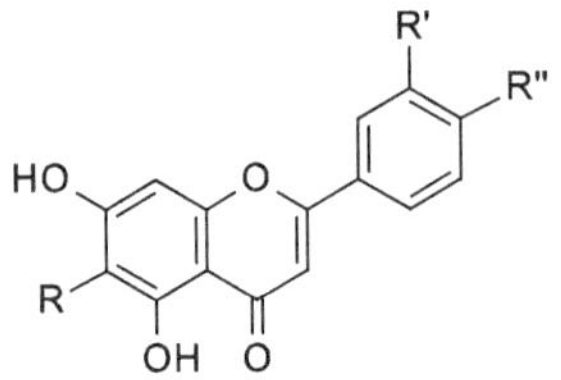 Baicalein (R = OH; R' = R" = H) from *Scutellaria baicalensis* (Lamiaceae) Luteolin (R = H; R' = R" = OH), widely distributed in the plant kingdom	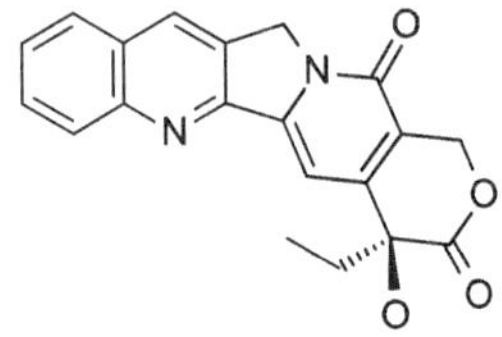 Camptothecin from *Camptotheca acuminata* (Nyssaceae)
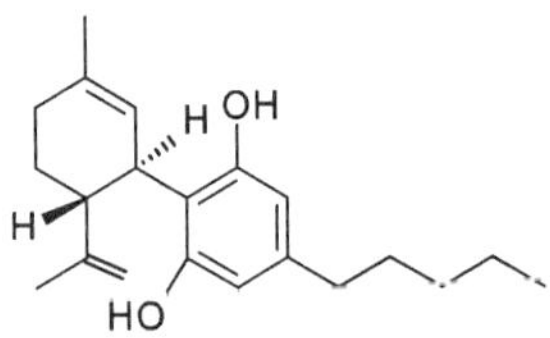 Cannabidiol (CBD), one of the major compounds of *Cannabis sativa* (Cannabaceae)	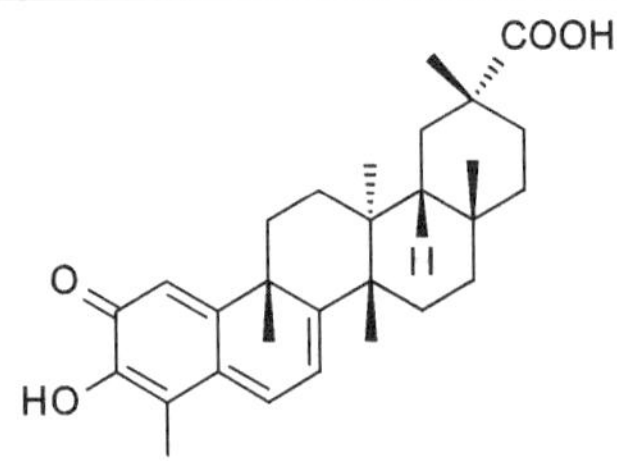 Celastrol, isolated from *Celastrus orbiculatus* (Celastraceae)
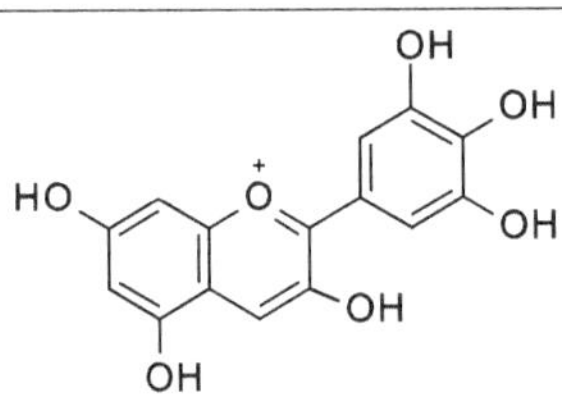 Delphinidin from the species of the genera, *Viola* (Violaceae) and *Delphinium* (Ranunculaceae)	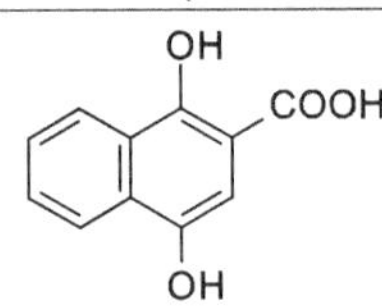 1,4-Dihydroxy-2-naphthoic acid isolated from *Rubia cordifolia* (Rubiaceae)
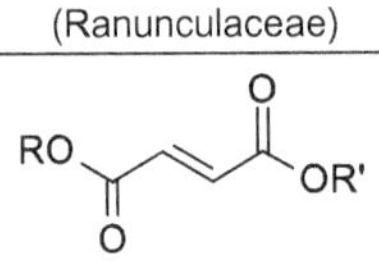 Fumaric acid esters Fumaric acid R = R' = H Dimethylfumarate R = R' = Me Monoethylfumarate R = H; R' = Et Fumaric acid was initially detected in a fungus, *Boletus pseudoignarius* (Boletaceae) and later isolated from *Fumaria officinalis* (Papaveraceae)	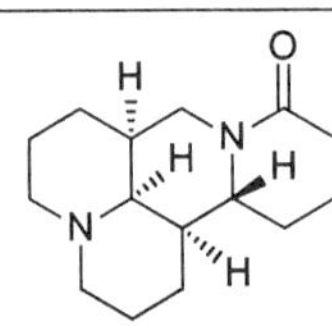 Matrine from *Sophora flavescens* (Fabaceae)

Fig. 18 A few examples of antipsoriatic natural products.

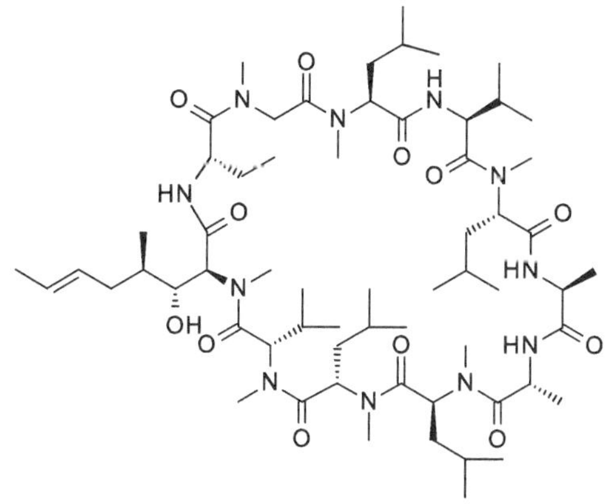

Cyclosporine, a microbial antipsoriatic drug, isolated from the fungus *Tolypocladium inflatum* (Ophiocordycipitaceae)

From marine organisms

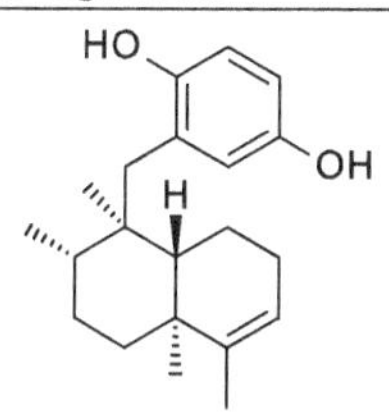

Avarol from the sponge *Dysidea avara* (Dysideidae)

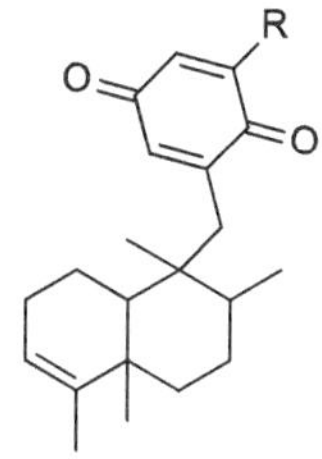

Avarol derivatives
Derivative 1, R = Benzylamine
Derivative 2
R = *p*-methoxybenzylamine

Fig. 18—Cont'd

antipsychotic properties as evident from preclinical and clinical trials.[39] More than a 100 randomized clinical trials provided evidence in support of significant antipsychotic property of some Chinese and Ayurvedic herbs, *e.g., Ginkgo biloba*, as well as purified natural products, *e.g., N*-acetylcysteine, estradiol, glycine, reserpine, sarcosine and vitamin B6 (Fig. 19).

In a recent study, cannabidiol (CBD) (Fig. 18), one of the nonpsychoactive secondary metabolites produced by *Cannabis sativa* (Cannabaceae), has been shown to improve cognition and social interaction deficits in a maternal immune activation model relevant to the etiology of

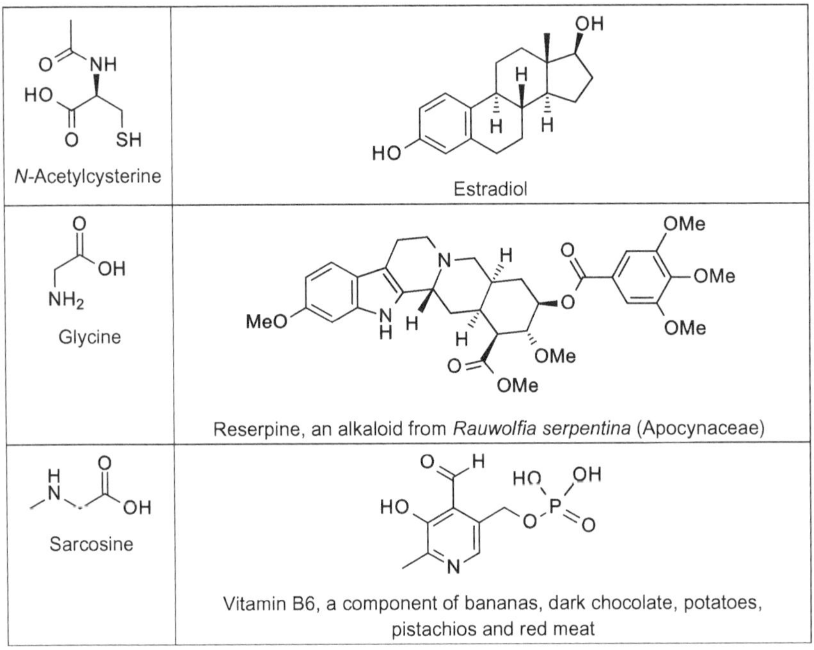

Fig. 19 Examples of some antipsychotic natural products based on clinical trials.

schizophrenia.[40] Deng et al.[41] demonstrated that deoxyvasicine (Fig. 20), one of the major compounds isolated from the aerial parts of *Peganum harmala* (Nitrariaceae), could ameliorate scopolamine-induced cognitive dysfunction by restoring cholinergic function in mice; it could alleviate learning and memory deficits in those mice. Chrysin, a well-known flavone distributed in various plants, could protect against changes in striatal dopamine, 3,4-di-hydroxyphenyl acetic acid, and homovanillic acid, resulting in significant improvement in behavioral, cognitive and neurochemical parameters in mice.[42] There are several other natural products, mainly of plant origin, currently in preclinical trials to assess their potential as antipsychotic agents for development of new antipsychotic drugs (Fig. 20).[43–49]

10. Fertility regulating natural products

Fertility regulation can be described as intentional methods of avoidance of pregnancy used by either male or female. There have been methods of fertility regulation at least as far back as the beginnings of recorded history, as families have resorted to control the number and timing of births.[50]

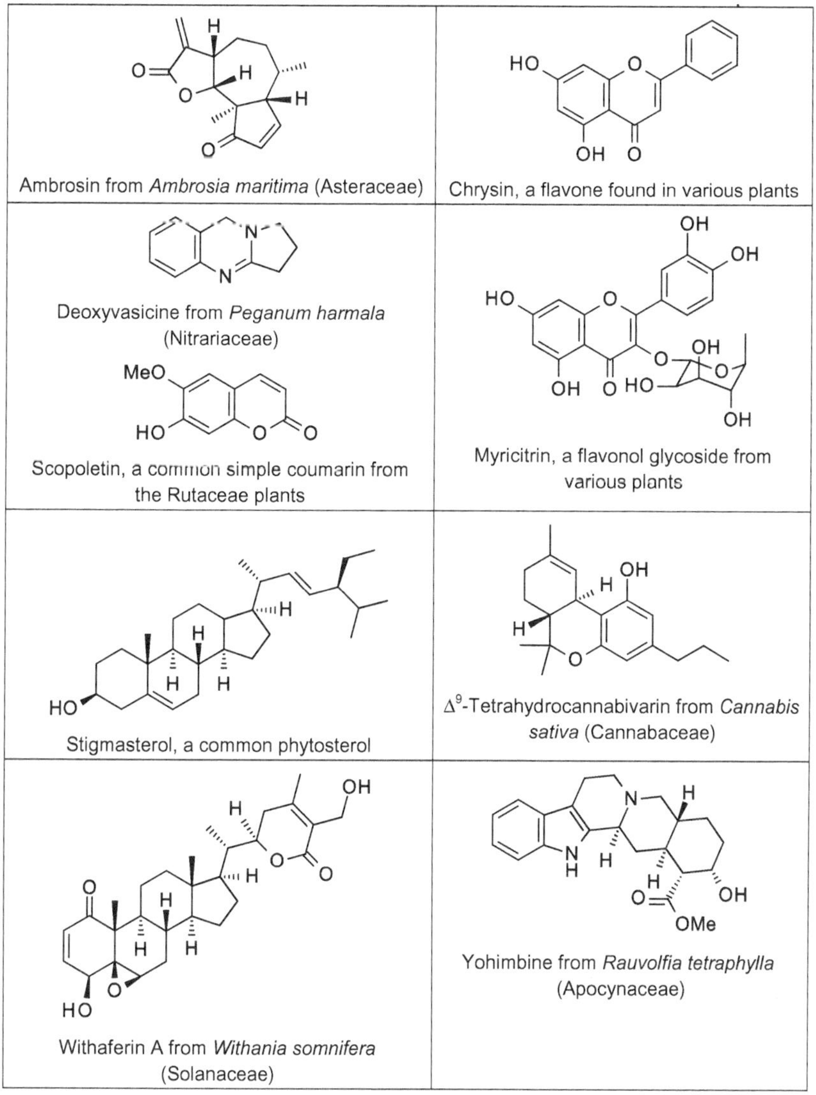

Ambrosin from *Ambrosia maritima* (Asteraceae)

Chrysin, a flavone found in various plants

Deoxyvasicine from *Peganum harmala* (Nitrariaceae)

Scopoletin, a common simple coumarin from the Rutaceae plants

Myricitrin, a flavonol glycoside from various plants

Stigmasterol, a common phytosterol

Δ^9-Tetrahydrocannabivarin from *Cannabis sativa* (Cannabaceae)

Withaferin A from *Withania somnifera* (Solanaceae)

Yohimbine from *Rauvolfia tetraphylla* (Apocynaceae)

Fig. 20 Examples of some recently reported antipsychotic natural products.

Although ending unwanted pregnancy or the process of abortion is still one of the most contentious issues of reproduction, the contribution of natural products in this area has been well documented. There are several plants and plant products still in use as abortifacients in many traditional medicine systems. Similarly, there are also several plants that possess various other fertility regulatory properties, *e.g.*, spermicidal property. The Egyptian Papyrus

(1550 BC) and the Kahun Papyrus (1850 BC) have earliest documentation of fertility regulations by using honey, acacia leaves and lint as spermicide.[51–53] The Egyptian Papyrus (1550 BC) outlined the use of ground dates, acacia tree bark, and honey together as a pessary. In medieval Europe, women used lily roots and rue to prevent pregnancy. The oldest condom made up of animal gut dated back to 1640 has been recovered in the ruins of Dudley Castle in England. Some religious communities still use lemon juice as locally applied spermicide. Hippocrates has described the use of Queen Anne's lace also known as wild carrot's seed as contraceptive.[51,52] Anti-implantation and antifertility properties of *Abroma angusta* (Sterculiaceae), *Achyranthus aspera* (Amaranthaceae), *Adhatoda vasica* (Acanthaceae), *Carica papaya* (Caricaceae), *Daucus carota* (Apiaceae), *Gossipium herbacium* (Malvaceae) and *Grewia asiatica* (Tiliaceae) are well-known in Indian traditional medicines including the Ayurveda.[54] Mishra et al. [53] showed the male fertility regulatory properties of *Allamanda cathartica, Azadirachta indica, Bacopa monnieri* and *Curcuma longa* in mice model. Plant-based fertility regulatory medications for females, covering the use of over 200 plant species, have recently been reviewed by Ansari et al..[55] Further descriptions on natural products-based fertility regulations are presented in *Chapter 15*, which also documents various purified natural products as fertility regulators. The secondary metabolites present in plants with fertility regulatory property are generally from the chemical classes—alkaloids, coumarins, flavonoids, glycosides, isoflavonoids, phenolics and terpenoids. Fig. 21 provides a few examples of natural products with fertility regulatory properties, as proven by various *in vitro* and *in vivo* experiments.[56–66]

Fertility regulatory agents, either synthetic or of natural origin, can be broadly classified as abortifacient (causes early abortion), anti-estrogenic (estrogen antagonists), antifertility (prevents fertilization), anti-implantation (blocks implantation), anti-ovulatory (inhibit ovulation), antispermatogenic (reduces sperm production), antisteroidogenic (reduces steroidogenesis) and spermicidal (destroys sperm). Some phytochemicals have specific effects on the ovary, the uterus, on hormonal levels, implantation pattern, sperm production and sperm morphology, and belong to one or more of the classes outlined above. Some of these compounds have been shown to create a protective layer around the egg to prevent fertilization.[58] Moreover, some of these phytochemicals can also modulate the fertility cycle, ovarian cells, steroidogenesis and sperm functions without blocking or preventing the process of fertilization. Furthermore, a few compounds help in assisted reproductive technologies.

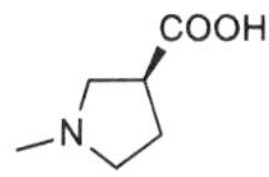 Achyranthine from *Achyranthes aspera* (Amranthaceae)	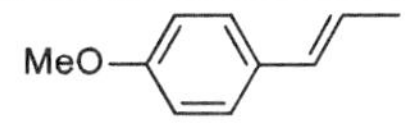 Anethole from *Croton zehntneri* (Euphorbiaceae)
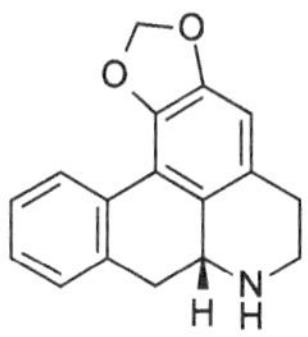 Anonaine from *Annona squamosa* (Annonaceae)	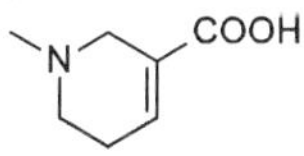 Arecaidine from *Areca catechu* (Arecaceae)
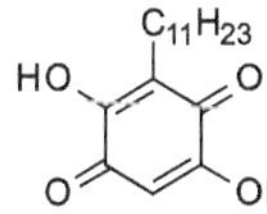 Embelin from *Embelia ribes* (Primulaceae)	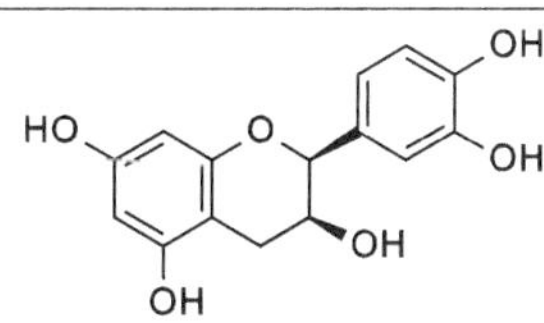 Epicatechin from *Amburana cearensis* (Fabaceae)
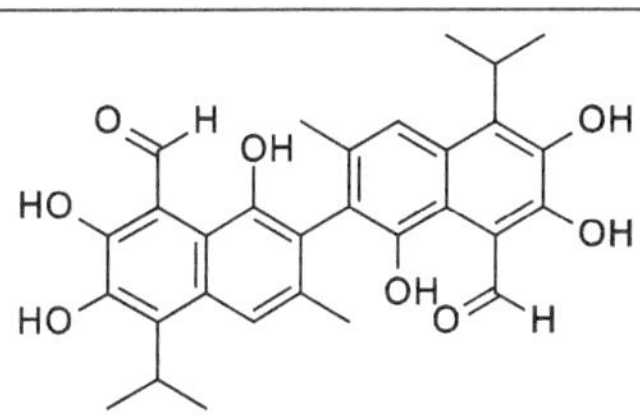 Gossypol from *Gossypium* species (Malvaceae)	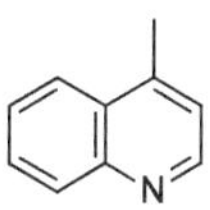 Lepidine from *Lepidium sativum* (Brassicaceae)
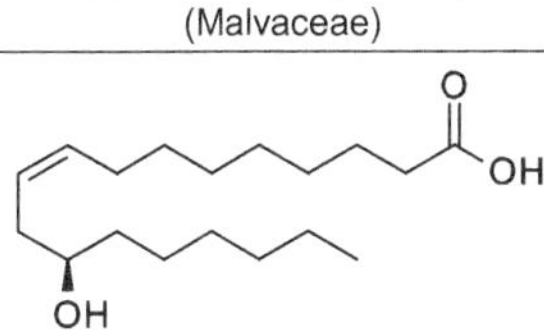 Ricinoleic acid from *Ricinus communis* (Euphorbiaceae),	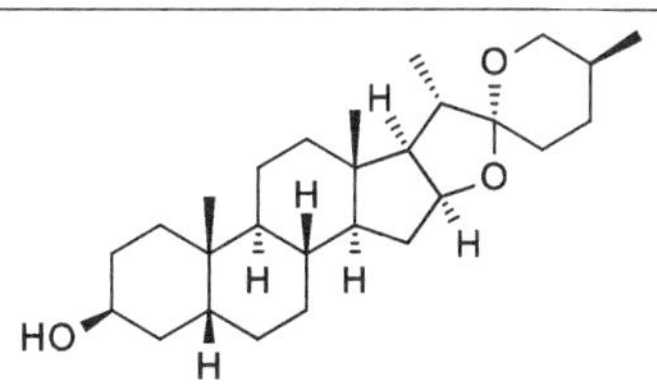 Sarsapogenin from *Yucca schidigera* (Asparagaceae)
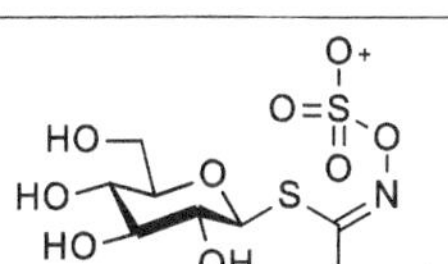 Sinigrin from *Carica papaya* (Caricaceae)	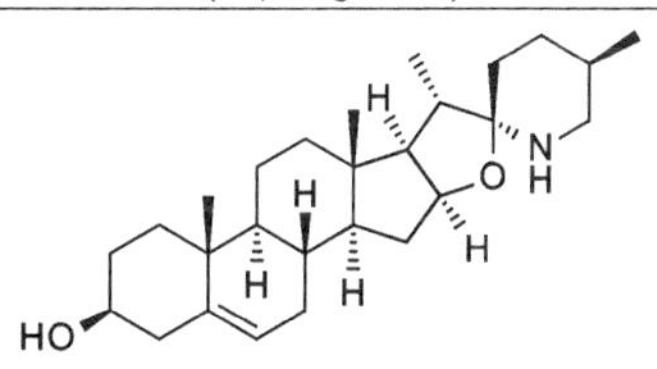 Solasodine from *Solanum xanthocarpum* (Solanaceae)

Fig. 21 Examples of some fertility regulatory natural products.

Although several natural products have been isolated from a wide range of plants and evaluated for their fertility regulatory potential, extensive randomized clinical trials on these compounds involving human subjects are rather extremely limited, almost none. Despite the fact that the literature available suggests on traditional uses of many plants to regulate fertility, not much systematic studies aiming at isolation of active phytochemicals from these plants, establishing their mechanism of action, optimizing their physicochemical properties for drug development have been performed to date.

11. Hepatoprotective, nephroprotective and neuroprotective natural products

Traditionally, several plants and other natural products have long been known to protect the liver, the kidneys and the nervous system from various injuries or assaults, caused by various internal and/or external stress factors, as well as offer protection against various infections and other diseases. In most cases, the protection offered by natural products may be linked to their inherent antioxidant, anti-inflammatory and immunoprotective properties. Following the traditional uses, several plants and other natural sources possessing hepatoprotective, nephroprotective and neuroprotective properties have been exploited to isolate and identify compounds, and to test those compounds in various *in vitro* and *in vivo* studies to establish their efficacy in protecting these organs/systems. There are also some reports, albeit limited, on clinical trials using various natural products in human models. Chapters "Neuroprotective natural products" by Talukdar et al.; "Hepatoprotective natural products" by Bachar et al.; "Nephroprotective natural products" by Jivishov et al. deal with, respectively, neuroprotective, hepatoprotective and nephroprotective properties of natural products.

11.1 Hepatoprotective natural products

The liver is one of the most vital organs, and regulates diverse processes, *e.g.*, metabolism, secretion, storage, and detoxification of endogenous and exogenous substances. Because of these functions, the liver is often exposed to internal and external stress causing damage and/or limiting its performance. Hepatic diseases continue to be among the main threats to public health globally. There are many modern drugs that may cause serious injuries to the liver, but there are hardly any modern drugs that are completely effective in stimulating hepatic function, offering complete protection of the organ, and/or helping regeneration of the hepatic cells. Thus, the search for effective hepatoprotective agents, particularly from

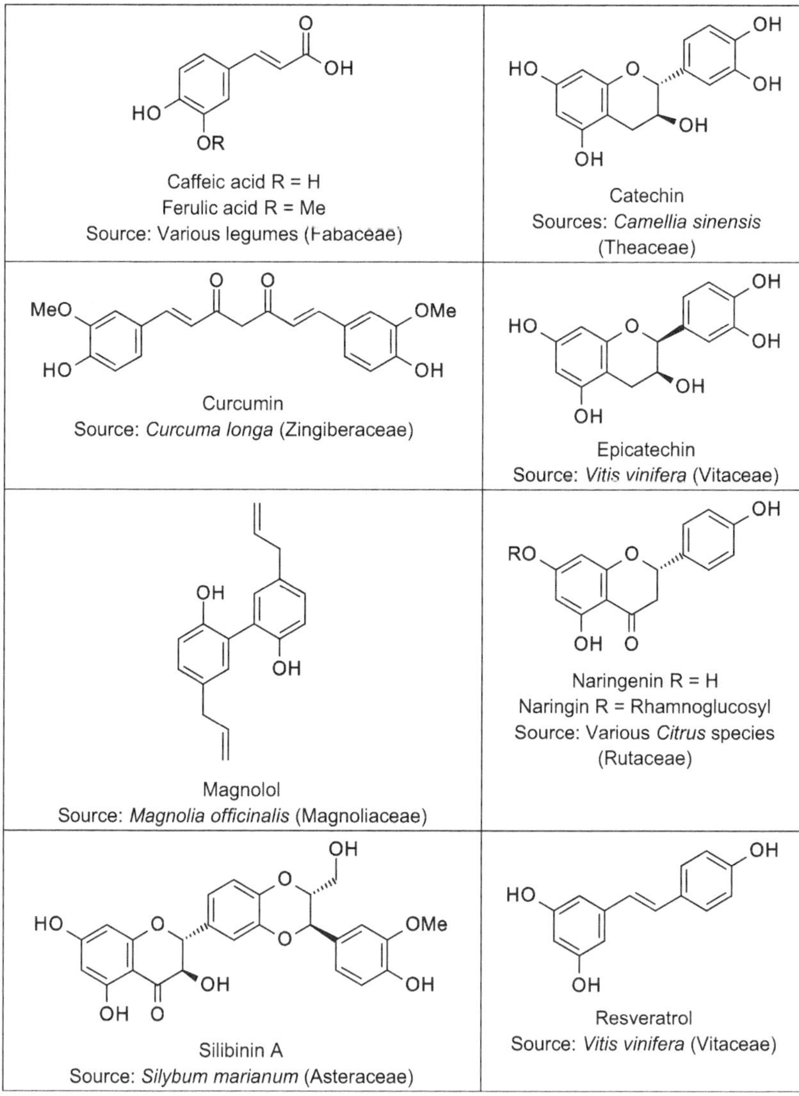

Fig. 22 Examples of some hepatoprotective natural products.

natural sources like fruits and vegetables, is in the main stay of natural products research. For example, some fruits, grapefruit (*Citrus paradisi*), cranberries (*Vaccinium* species) and grapes (*Vitis vinifera*),[67] which are consumed frequently by humans have hepatoprotective property. Fig. 22 shows a few examples of purified natural products that showed hepatoprotective activity in various *in vivo* experimental animal models and/or in human clinical trials.

Some of these hepatoprotective natural compounds, for example, naringin and naringenin (Fig. 22), were shown to exert their hepatoprotective action through enhancing ethanol and lipid metabolism as well as inducing the hepatic antioxidant defense system.[67] A recent review article, published be Saha et al.,[68] has outlined the hepatoprotective potential of various phenolic compounds, *e.g.*, baicalein (Fig. 18), caffeic acid, curcumin, ferulic acid, magnolol (Fig. 22) and so on, with possible mechanism of actions.

11.2 Nephroprotective natural products

The kidneys are important organs that remove waste and toxic materials from the blood, and maintain electrolyte balance, regulate homeostasis of the blood plasma, blood volume, blood pressure, and red blood cell genesis. The kidneys can be damaged or may become totally inactive by several factors, mainly by oxidative stress, inflammation, ischemia/reperfusion injury, diabetes, and nephrotoxic agents, most often various modern drugs that are in use today. Crude natural product extracts, particularly plant extracts, have played a significant role in traditional medicine in preventing kidney damages by virtue of their antioxidant and anti-inflammatory properties. In recent years, several naturally occurring compounds with significant nephroprotective properties have been discovered, and their possible mechanisms of actions have been established through appropriate *vitro*, *in vivo* (animal models) and *in silico* studies. *Chapter 8* describes potential nephroprotective natural products that have undergone preclinical and/or clinical trials, and provides an overview on the mechanisms of actions of nephroprotecting natural products.

Curcumin (Fig. 22) is probably one of the most studied natural products in relation to nephroprotective as well as hepatoprotective activity, mainly because of its well established antioxidative property. Generally, plant-derived phenolic (*e.g.*, ferulic acid) and polyphenolic compounds (*e.g.*, catechin and epicatechin) (Fig. 22) possess antioxidant property, and thus offer the liver and the kidneys against various injuries, usually caused by oxidative stress.

Not only the phenolic and polyphenolic compounds, but there are also a few non-phenolic natural products that have significant nephroprotective property; for example, dioscin from *Dioscorea* species (Dioscoreaceae) and steveoside from *Stevia rebaudiana* (Asteraceae) (Fig. 23).

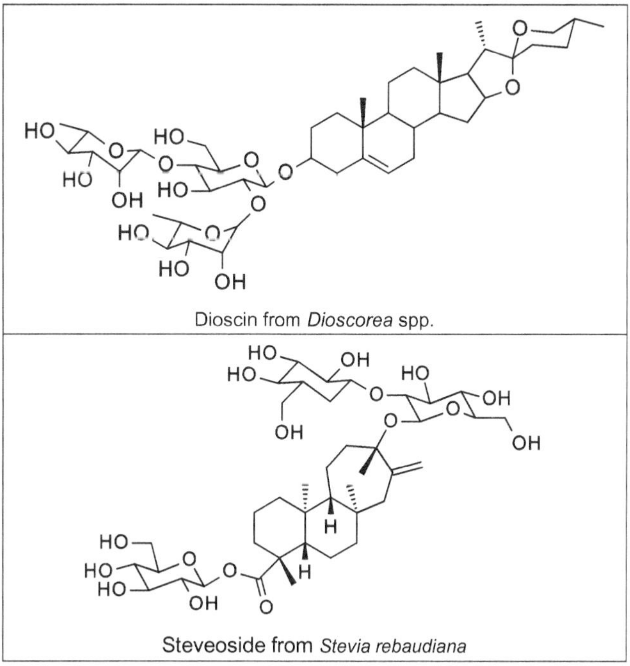

Fig. 23 Examples of non-phenolic natural products with nephroprotective property.

11.3 Neuroprotective natural products

Chapter 6 discusses various aspects of neuroprotective natural products, predominantly from plant origins, their plausible mechanisms of actions, and *in silico* approaches to identify natural products leads with potential neuroprotective and therapeutic applications in CNS disorders. Neuroprotection means safeguarding the central nervous system (CNS) from neuronal damages caused by chronic or acute neurodegenerative diseases. Not only the terrestrial plants that offer neuroprotection and neuroprotective compounds, several marine organisms, especially macroalgae, are also known to produce neuroprotective compounds. The underlying mechanisms of action of these natural neuroprotective agents are predominantly based on their antioxidative and anti-inflammatory properties.

Curcumin (Fig. 22) from *C. longa* again leads the way in terms of research into neuroprotective property of natural products. Several

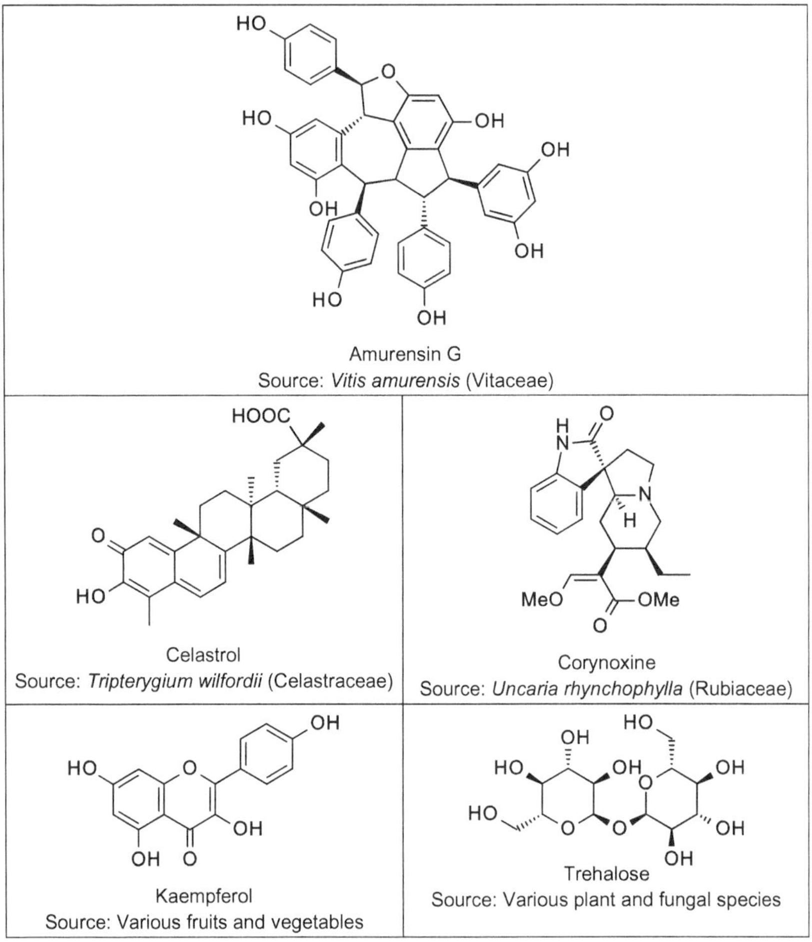

Fig. 24 Examples of a few autophagy-lysosome pathway regulating natural products.

phenolic and polyphenolic compounds, because of their inherent antioxidant property, have been shown to offer neuroprotection against various neurodegenerating disease like Parkinson's Disease.[69,70] Many of these neuroprotective natural products, *e.g.*, curcumin, naringenin and resveratrol (Fig. 22), have been shown to regulate autophagy-lysosome pathway, impairment of which is implicated to pathogenesis of neurodegerative diseases, *e.g.*, Parkinson's Disease, and thus offer neuroprotective effects *in vivo*. Fig. 24 presents a few examples of autophagy-lysosome pathway regulating natural products.

12. Natural products in cancer chemoprevention

Cancer chemoprevention is the process of using a chemical to prevent cancer, which is a complex interplay of a multitude of biological processes.[71] Chemicals, either natural or synthetic, with the ability to modulate any one of the cancer promoting processes are classed as cancer chemopreventive agents. Natural products, especially phytochemicals from various edible plants, for example, resveratrol (Fig. 22) from red grapes and curcumin (Fig. 22) from turmeric, dominate this area of research. *Chapter 9* covers the fundamentals of chemoprevention and proposes a new definition of 'complete chemopreventive' to allow the identification of the most promising natural products with cancer preventing properties.

In addition to curcumin and resveratrol, several flavonoids, apigenin (Fig. 25), kaempferol (Fig. 24), luteolin (Fig. 7) and quercetin, isoflavonoid genistein (Fig. 25), alkaloid capsaicin, carotenoids, β-carotene and lycopene (Fig. 25) are among the most often cited phytochemicals with cancer chemopreventive property, proven predominantly by various *in vitro* and *in vivo* animal studies. All these compounds were shown to suppress NF-κB activation, which is one of the establishes mechanisms of cancer chemoprevention through reduction of inflammation.

Several other phytochemicals, such as fisetin, piperlonguminine and pristimerin (Fig. 25), exert cancer chemoprevention by inducing apoptosis. It can be noted that the cancer cells lack the ability to undergo apoptosis due to mutations in the apoptotic transcription factor p53.

Oxidative stress can be generated by various internal cellular and external factors, resulting in the formation of reactive oxygen species (ROS) such as the superoxide anion which ultimately leads to the production of the hydroxyl ion. The hydroxyl ion is commonly accepted to be the most damaging ROS and can readily attack DNA causing mutations and increased cancer risk. Natural products have been shown to reduce such oxidative stress by inducing enzyme based cellular defense mechanisms. Much of this research focused on the master regulator molecule of antioxidant gene induction, Nrf2 (nuclear erythroid 2-related factor). Several phytochemicals, *e.g.*, capsaicin, curcumin and resveratrol, are known to induce activation of Nrf2 acellular antioxidant defense mechanism. Most of the cancer Chemopreventive natural products are strong antioxidants and anti-inflammatory agents.

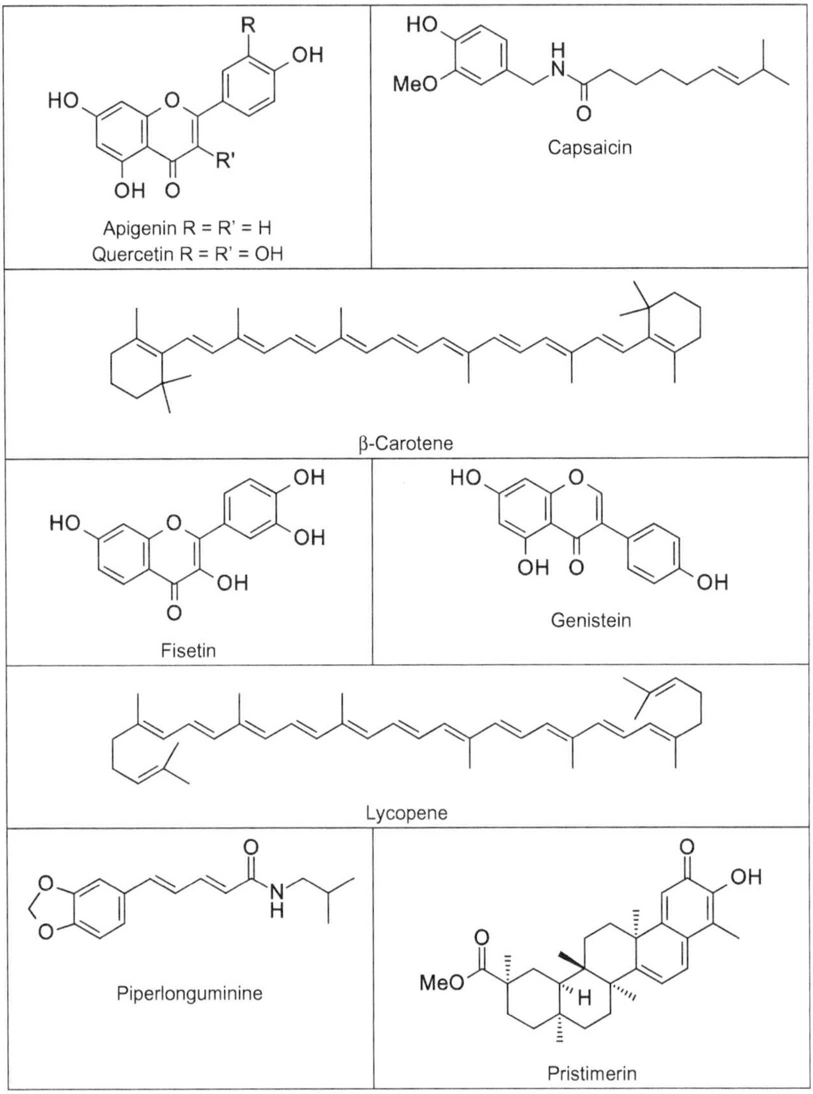

Fig. 25 Examples of a few cancer chemopreventive phytochemicals.

In 2016, the following natural product-derived agents were in clinical trials: metformin (seven trials), polyphenon E (nine trials), retinoids (20 trials), soy isoflavones (16 trials), vitamin C (10 trials), vitamin D (20 trials), and vitamin E (16 trials) (Fig. 26).[72] Despite several attempts, curcumin was not successful in any previously conducted clinical trials, mainly because of its limitations associated with absorption and metabolism However, its chemical derivatives (Fig. 27) were subjected to 12 trials in 2016 with some success.

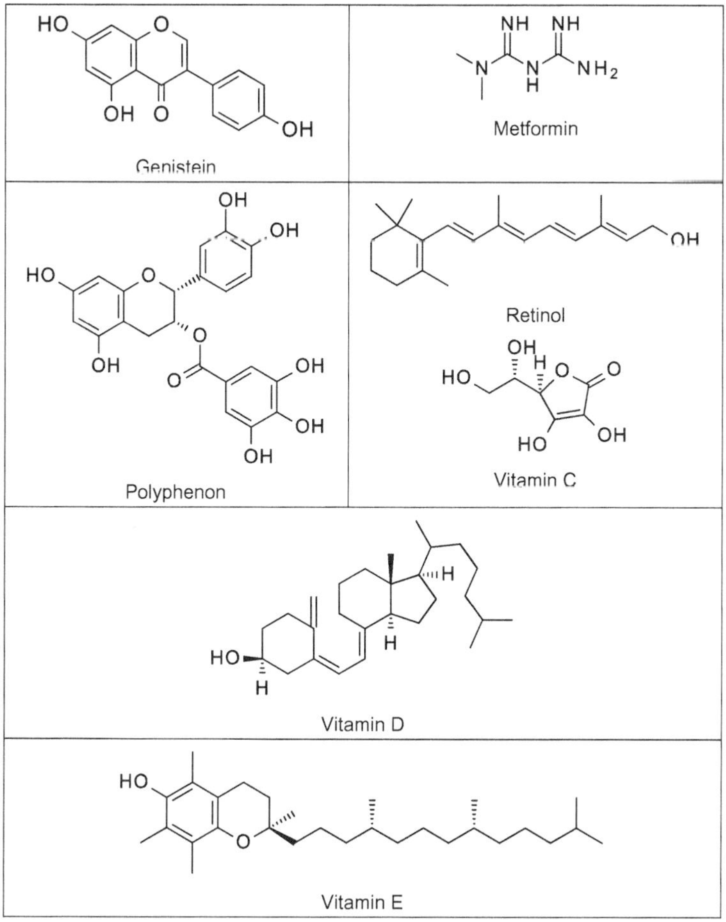

Fig. 26 Natural-product-derived cancer chemopreventive agents in clinical trials.

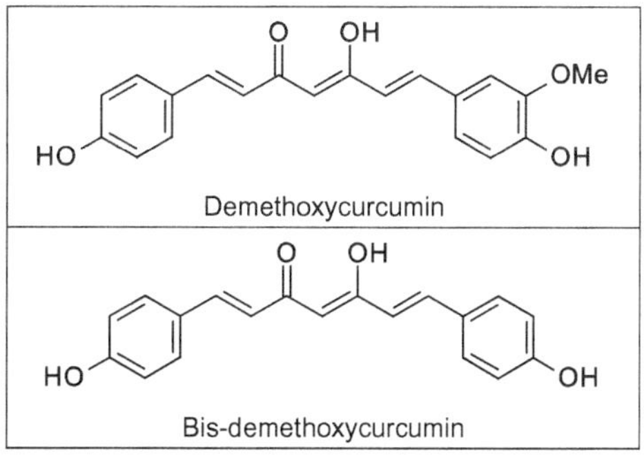

Fig. 27 Curcumin derivatives in clinical trials.

13. Natural products for the treatment of obesity

Obesity is an excessive body weight in the form of fat caused by an imbalance between energy intake and expenditure; it is a common, complex and growing global health problem, and one of the major contributors to other serious human ailments like diabetes, heart disease, osteoarthritis and cancer. It is estimated that there are more than 300 million obese people around the globe, and the number is increasing at an alarming rate throughout the world. Many approved anti-obesity medicines have recently been withdrawn from the market because of serious adverse effects, as well as their potential for drug abuse and dependency; this has opened the door of opportunities for new drug development to tackle obesity. Traditionally, plants, especially edible plants, have long been used to treat or prevent obesity, and several bioactive phytochemicals have led the way for new anti-obesity drug discovery and development processes. Many of these food or medicinal plants and their bioactive compounds have been subjected to clinical trials and have been proven to be effective in the treatment of obesity. Chapter "Anti-obesity natural products" by Arya et al. presents an overview on anti-obesity natural products, their mechanisms of actions, and evaluates their potential in novel anti-obesity drug development paradigm.

Anti-obesity products from green tea (*Camellia sinensis*) leaves are popular in functional food market. The major bioactive constituents in green tea, accounting for up to 35% of the dry weight, are polyphenols including flavonols, flavones, and flavan-3-ols (catechins). A number of clinical trials have revealed the beneficial effects of catechins (270–1200 mg/day), *e.g.*, reduced body weight, lowered serum leptin levels, and reduced absorption of fatty acid. Another bioactive constituent in tea leaves is caffeine, which influences somatic nervous system activity and acts synergistically with catechins to increase energy expenditure and fat oxidation. Most commonly cited phytochemicals as anti-obesity agents include capsaicin (Fig. 25), celastrol (Fig. 24) and curcumin (Fig. 22).

Among the natural products from microbial sources, lipstatin (Fig. 28), a pancreatic lipase inhibitor produced by *Streptomyces toxytricini*, is commonly used to combat obesity and diabetes by interfering with the gastrointestinal absorption of fat.[73]

Several natural products from marine organisms have also contributed in the search of anti-obesity agents; callyspongynic acid, dysidine and palinurin are just to name a few (Fig. 29).[74–81]

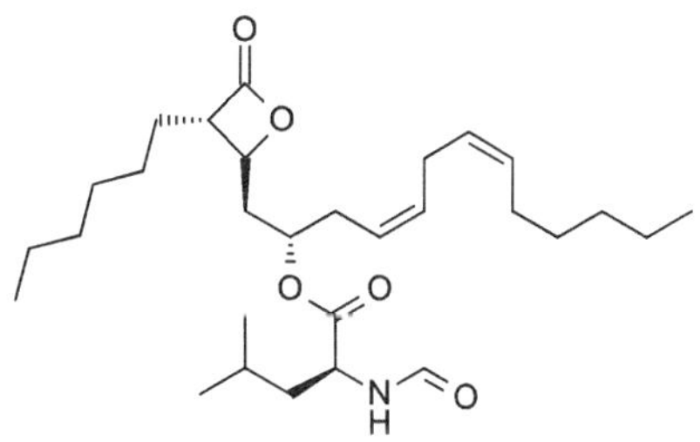

Fig. 28 Lipstatin, an anti-obesity agent from *Streptomyces toxytricini.*

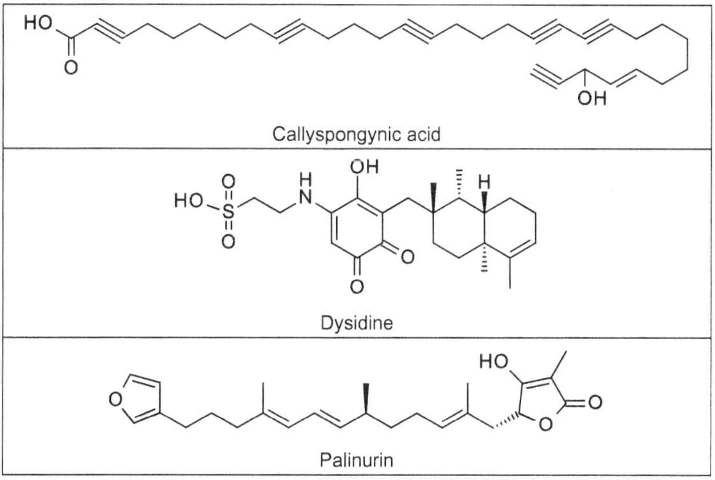

Fig. 29 A few examples of anti-obesity agents from marine sources.

14. Natural products for the treatment and prevention of osteoporosis

Osteoporosis, which develops over several years, is a health condition that weakens bones, making them fragile and more likely to break. It is often only diagnosed when a fall or sudden impact causes a bone to break, most commonly broken wrist, hip or spinal bones. Osteoporosis is normally associated with aging, and bone loss. Osteopenia is the just the condition before osteoporosis, when a bone density scan can reveal lower bone density than the average for any particular age, but not low enough to be classed as osteoporosis. However, osteopenia does not always lead to osteoporosis. Usually, osteoporosis is treated with bone strengthening medicines. There are over three million osteoporosis sufferers in the United Kingdom, and over half a million people need hospital treatment every year for bone fragility as a consequence of osteoporosis. In addition to aging, there are other factors that may cause osteoporosis; the factors include longer use of steroidal drugs, medical conditions like inflammation, malabsorption and hormone deficiency, genetic factor (family history), long term use of anti-estrogenic medications, anorexia or bulimia, low body mass index (BMI), lack of exercise and heavy drinking and smoking. Generally, women are more vulnerable to osteoporosis than men.

In addition to conventional medical treatment and medications for osteoporosis, alternative treatment options, especially by using herbal or phytotherapy, are also available, and they are proven to be effective based on several randomized control clinical trials using human volunteers.[82] Icariin, isopsoralen, psoralen and salidroside (Fig. 30) are some of the major bioactive compounds reported from herbs used in the treatment of osteoporosis.

The active ingredients derived from traditional Chinese medicines that promote osteoblasts proliferation and differentiation mainly belong to the classes of anthraquinones, diarylheptanoids), flavonoids, glycosides, coumarins, terpenoids, oligomeric stilbenoids, phenolic acids and phenols.[83] The bone formation effect induced by these natural products was demonstrated

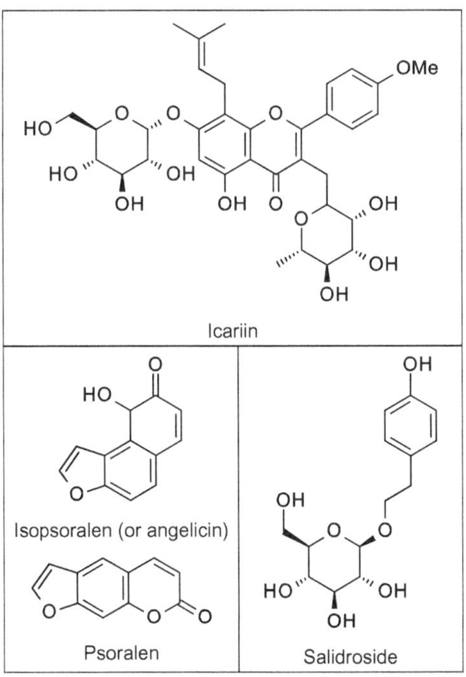

Fig. 30 Major bioactive constituents from herbs used in the treatment of osteoporosis.

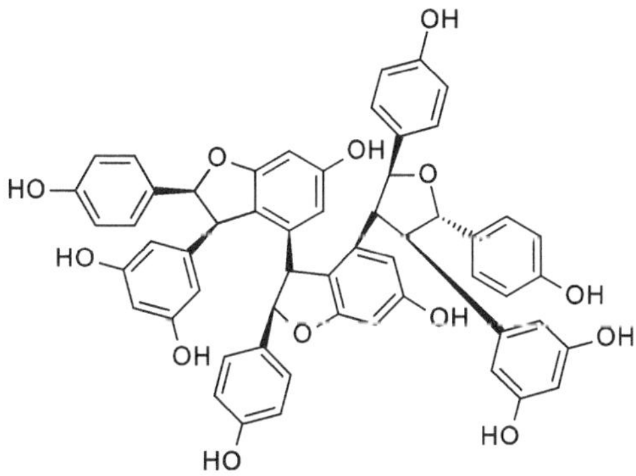

Fig. 31 Kobophenol A, the major bioactive compound of *Caragan sinica.*

to be regulated by the expressions of bone specific matrix proteins, transcription factors (Runx2, Cbfa1 and Osx), signal pathways (MAPK and BMP), local factors (ROS and NO), OPG/RANKL system of osteoblasts and estrogen-like biological activities. Kobophenol A (Fig. 31), a tetrameric stilbene isolated from *Caragan sinica* Rhed. (Fabaceae) was found to be effective in the treatment and management of osteoporosis in humans.

Chapter 11 deals with mechanistically well-studied plant-based extracts and compounds that possess clinical potential in the treatment of osteoporosis, and presents several specific examples.

15. Conclusions

Natural products have long been the main source of modern drugs for the treatment and prevention of various human ailments including cancer, diabetes, infections and so on, and will continue to remain so in the years to come. The natural products molecules that are currently in clinical trials and/or in the drug development chain, are expected to become commercially available modern drugs, and thus help in our constant fight against human diseases. Any effective drug discovery and development effort must include multidisciplinary and collaborative approach, incorporating natural product lead discovery linking with optimization through virtual screening, combinatorial and medicinal chemistry, total synthesis, and combinatorial biochemistry. Considering the unique chemical diversity that prevails in natural source materials, it can be hoped that new modern drug leads that are capable of interacting with most or possibly all therapeutic targets, will be discovered from natural products, or semisynthetic or synthetic products inspired by natural products scaffolds.

Acknowledgments

L. Nahar gratefully acknowledges the financial support of the European Regional Development Fund—Project ENOCH (No. CZ.02.1.01/0.0/0.0/16_019/0000868).

References

1. Sarker, S. D.; Nahar, L. *Natural Products Isolation*, 3rd ed.; Humana Press—Springer-Verlag: New Jersey, 2012.
2. Nahar, L.; Sarker, S. D. *Chemistry for Pharmacy Students*, 2nd ed.; Wiley and Sons: Chichester, 2019.
3. Thomford, N. E.; Senthebane, D. A.; Rowe, A.; Munro, D.; Seele, P.; Maroyi, A.; Dzobo, K. *Int. J. Mol. Sci.* **2018**, *19*, article number: 1578. https://doi.org/10.3390/ijms19061578.
4. Sultana, S.; Asif, H. M.; Akhtar, N.; Ahmad, K. *Asian Pac. J. Trop. Dis.* **2015**, *5*(1), S202–S208.
5. Muhammad, A.; Khan, B.; Iqbal, Z.; Khan, A. Z.; Khan, I.; Khan, K.; Alamzeb, M.; Ahmad, N.; Khan, K.; Badshah, S. L.; Ullah, A.; Muhammad, S.; Jan, M. T.; Nadeem, S.; Kabır, N. *ACS Omega* **2019**, *4*, 14188–14192.
6. Paller, C. J.; Denmeade, S. R.; Carucci, M. A. *Clin. Adv. Hematol. Oncol.* **2016**, *14*(6), 446–455.
7. Newman, D. J.; Cragg, G. M. *J. Nat. Prod.* **2016**, *79*(3), 629–661.
8. NHSinform https://www.nhsinform.scot/illnesses-and-conditions/diabetes/diabetes; 2019 (accessed on 22 December 2019).
9. Harvey, A. L. *Curr. Org. Chem.* **2010**, *14*(16), 1670–1677.
10. Qi, L. W.; Liu, E. H.; Chu, C.; Peng, Y. B.; Cai, H. X.; Li, P. *Curr. Top. Med. Chem.* **2010**, *10*(4), 434–457.
11. Ferrero-Miliani, L.; Nielsen, O. H.; Andersen, P. S.; Girardin, S. E. *Clin. Exp. Immunol.* **2006**, *147*, 227–235.
12. Cheung, R. C.; Ng, T. B.; Wong, J. H.; Chen, Y.; Chan, W. Y. *Appl. Microbiol. Biotechnol.* **2016**, *100*(4), 1645–1666.
13. Wang, Y.-H.; Zeng, K.-W. *Tradit. Med. Res.* **2019**, *4*(5), 257–268.
14. The World Health Organization (WHO) *Malaria*. https://www.who.int/news-room/fact-sheets/detail/malaria; 2019 (accessed on 22 December 2019).
15. Foley, M.; Tilley, L. *Int. J. Parasitol.* **1997**, *27*(2), 231–240.
16. Nahar, L.; Guo, M.; Sarker, S. D. *Phytochem. Anal.* **2020**, *31*(1), 5–14.
17. Woodrow, C. J.; Haynes, R. K.; Krishna, S. *Postgrad. Med. J.* **2005**, *81*, 71–78.
18. Winzeler, E. A.; Manary, M. J. *Genome Biol.* **2014**, *15*(11), 544. https://doi.org/10.1186/s13059-014-0544-6.
19. Cravo, P.; Napolitano, H.; Culleton, R. *Acta Trop.* **2015**, *148*, 1–7.
20. Tilley, L.; Straimer, J.; Gnädig, N. F.; Ralph, S. A.; Fidock, D. A. *Trends Parasitol.* **2016**, *32*(9), 682–696.
21. Suresh, N.; Haldar, K. *Curr. Opin. Pharmacol.* **2018**, *42*, 46–54.
22. Fernández-Álvaro, E.; Hong, W. D.; Nixon, G. L.; O'Neill, P. M.; Calderón, F. *J. Med. Chem.* **2016**, *59*, 5587–5603.
23. Mojab, F. *Avicenna J. Phytomed.* **2012**, *2*(2), 52–62.
24. Wells, T. N. C. *Malaria J.* **2011**, *10*(1), article number: 53. https://doi.org/10.1186/1475-2875-10-S1-S3.
25. Kayser, O.; Kiderlen, A. F.; Croft, S. L. In: *Studies in Natural Products Chemistry*; Atta-ur-Rahman, Ed.; Vol. 26; Elsevier Science, 2002; pp 779–848.
26. Paloque, L.; Triastuti, A.; Bourdy, G.; Haddad, M. In: *Natural Antimicrobial Agents*; Merillon, J.-M., Riviere, C., Eds.; Springer, 2018; pp 215–245.
27. Monzote, L. *Curr. Clin. Pharmacol.* **2014**, *9*, 181–186.

28. Herrmann, F.; Hamoud, R.; Sporer, F.; Tahrani, A.; Wink, M. *Planta Med.* **2011**, 77(17), 1905–1911.
29. Wu, Q.; Chen, J.; Li, X.; Chen, S.; Wang, H. *Mini Rev. Org. Chem.* **2017**, *14*, 501–515.
30. Chaudhary, A.; Naughton, L. M.; Montanchez, I.; Dobson, A. D. W.; Rai, D. K. *Mar. Drugs* **2017**, *15*, article number: 272. https://doi.org/10.3390/md15090272.
31. Khameneh, B.; Iranshahy, M.; Soheli, V.; Bazzaz, B. S. F. *Antimicrob. Resist. Infect. Control* **2019**, *8*, article number: 118. https://doi.org/10.1186/s13756-019-0559-6.
32. O'Neill, J. *Rev. Antimicrob. Resist.* **2014**, *20*, 1–16.
33. Singer, A. C.; Shaw, H.; Rhodes, V.; Hart, A. *Front. Microbiol.* **2016**, 7, 1728.
34. National Psoriasis Foundation. *Psoriasis.* https://www.psoriasis.org/about-psoriasis; 2019 (accessed on 26 December 2019).
35. Zeng, Z.; Luo, S.; Huang, Y.; Lu, Q. *J. Dermatol.* **2017**, *44*, 863–872.
36. Huang, T.-H.; Lin, C.-F.; Alalaiwe, A.; Yang, S.-C.; Fang, J.-Y. *Int. J. Mol. Sci.* **2019**, *20*, 2558.
37. Christensen, L. P. Polyphenols and polyphenol-derived compounds from plants and contact dermatitis. In: *Polyphenols—Prevention and Treatment of Human Disease*, 2nd ed.; Watson, R. R., Preedy, V. R., Zibadi, S., Eds.; San Diego: Elsevier Inc., 2018.
38. NHS *Overview: Psychosis.* https://www.nhs.uk/conditions/psychosis/; 2019 (accessed on 28 December 2019).
39. Hoenders, H. J. R.; Bartels-Velthuis, A. A.; Volbehr, N. K.; Btuggeman, R.; Knegtering, H.; de Jong, J. T. V. M. *J. Nerv. Ment. Dis.* **2018**, *206*, 81–101.
40. Osborne, A. L.; Solowij, N.; Babic, I.; Lum, J. S.; Newell, K. A.; Huang, X. F.; Weston-Green, K. *Progress Neuro-Psychopaharmacol. Biol. Psychiatry* **2019**, *95*, article number: UNSP 109666. https://doi.org/10.1016/j.pnpbp.2019.109666.
41. Deng, G.; Wu, C.; Rong, X. J.; Li, S. P.; Ju, Z. C.; Wang, Y. X.; Ma, C.; Ding, W. Z.; Guan, H. D.; Cheng, X. M.; Liu, W.; Wang, C. H. *Phytomedicine* **2019**, *63*, article number: UNSP 153007. https://doi.org/10.1016/j.phymed.2019.153007.
42. Del Fabbro, L.; Goes, A. R.; Jesse, C. R.; de Gomes, M. G.; Souza, L. C.; Ladd, F. V. L.; Ladd, A. A. B. L.; Arantes, A. V. N.; Simionato, A. R.; Oliveira, M. S.; Furian, A. F.; Boeira, S. P. *Neurosci. Lett.* **2019**, *706*, 158–163.
43. Khalil, M. N. A.; Choucry, M. A.; El Senousy, A. S.; Hassan, A.; El-Marasy, S. A.; El Awdan, S. A.; Omar, F. A. *PLoS One* **2019**, *14*(7), article number: e0219378. https://doi.org/10.1371/journal.pone.0219378.
44. Chengappa, K. N. R.; Brar, J. S.; Gannon, J. M.; Schlicht, P. J. *J. Clin. Psychiatry* **2018**, *79*(5), article number: 17m11826. https://doi.org/10.4088/JCP.17m11826.
45. Yadav, M.; Parle, M.; Jindal, D. K.; Dhingra, S. *Pharmacol. Rep.* **2018**, *70*(3), 591–599.
46. Pandy, V.; Vijeepallam, K. *Exp. Anim.* **2017**, *66*(4), 417–423.
47. Cascio, M. G.; Zamberletti, E.; Marini, P.; Parolaro, D.; Pertwee, R. G. *Brit. J. Pharmacol.* **2015**, *172*(5), 1305–1318.
48. Gupta, S.; Khanna, V. K.; Maurya, A.; Bawankule, D. U.; Shukla, R. K.; Pal, A.; Srivastava, S. K. *Fitoterapia* **2012**, *83*(6), 1092–1099.
49. Pereira, M.; Siba, I. P.; Chioca, L. R.; Correia, D.; Vital, M. A. B. F.; Pizzolatti, M.; Santos, A. R. S.; Andreatini, R. *Prog. Neuro-psychopharmacol. Biol. Psychiatry* **2011**, *357*, 1636–1644.
50. Newman, L.; In, F. *International Encyclopedia of the Social and Behavioral Sciences*, 2nd ed.; 2015; p 697–699.
51. Dias, D. A.; Urban, S.; Roessner, U. *Metabolites* **2012**, *2*(2), 303–336.
52. Castellanos, J. R. G.; Prieto, J. M.; Heinrich, M. *J. Ethnopharmacol.* **2009**, *121*(1), 1–13.
53. Mishra, R. K.; Singh, S.; Singh, S. K. *Ind. J. Med. Res.* **2018**, *148*(1), S107–S114.
54. Tamboli, S. A.; Konadawar, M. S. *Int. J. Pharm. Pharm. Sci.* **2013**, *5*(4), 128–129.
55. Ansari, A. S.; Sevliya, K.; Mohammad, I.; Badar, A.; Lohiya, N. K. *J. Pharmacol. Toxicol.* **2017**, *12*(2), 57–75.

56. Varuna, K. M.; Khan, M. U.; Sharma, P. K. *J. Pharm. Res.* **2010**, *3*(4), 714–717.
57. Sá, N.; Araújo, V.; Correia, H.; Ferreira, A.; Guerreiro, D.; Sampaio, A.; Escobar, E.; Santos, F.; Moura, A.; Lôbo, C. *Theriogenology* **2017**, *89*, 226–234.
58. Bala, K.; Arya, M.; Katare, D. P. *World J. Pharm. Pharm. Sci.* **2014**, *3*, 305–1326.
59. Shrestha, J.; Shanbhag, T.; Shenoy, S.; Amuthan, A.; Prabhu, K.; Sharma, S.; Banerjee, S.; Kafle, S. *Ind. J. Pharmacol.* **2010**, *42*(5), 306–311.
60. Kamal, R.; Gupta, R.; Lohiya, N. *Phytother. Res.* **2003**, *17*(6), 579–590.
61. Barberino, R.; Barros, V.; Menezes, V.; Santos, L.; Araújo, V.; Queiroz, M.; Almeida, J.; Palheta, R.; Matos, M. *Zygote* **2016**, *24*(2), 277–285.
62. Coutinho, E. M. *Contraception* **2002**, *65*(4), 259–263.
63. Pande, D.; Malik, S.; Bora, M.; Srivastava, P. *In Vitro Cell Dev. Biol. Plant* **2002**, *38*(5), 451–455.
64. Štochmaľová, A.; Kadasi, A.; Alexa, R.; Grossman, R.; Sirotkin, A. *Potravinarstvo Slovak J. Food Sci.* **2014**, *8*(1), 87–91.
65. Han, F.; Zhang, X.-X.; Pi, N.-N.; Wang, J.; Ding, X.; Su, D.-H.; Gao, P. *Sichuan J. Zool.* **2008**, *27*(3), 390–395.
66. Lakshman, J.; Changamma, C. *Int. J. Integr. Sci, Innov. Technol.* **2013**, *4*(2), 6–10.
67. Madrigal-Santillan, E.; Madrigal-Bujaidar, E.; Alvarez-Gonzales, I.; Sumaya-Martinez, M. T.; Gutierrez-Salinas, J.; Baitista, M.; Morales-Gonzalez, A.; Garcia-Luna, M.; Gonzalez-Rubio, Y.; Aguilar-Faisal, J. L.; Morales-Gonzalez, J. *World J. Gastroenterol.* **2014**, *20*(40), 14787–14804.
68. Saha, P.; Talukdar, A. D.; Nath, R.; Sarker, S. D.; Nahar, L.; Sahu, J.; Choudhury, M. D. *Front. Pharmacol.* **2019**, *10*, article number: 509. https://doi.org/10.3389/fphar.2019.00509.
69. Wang, Z.-Y.; Liu, J.-Y.; Yang, C.-B.; Malampati, S.; Huang, Y.-Y.; Li, M.-X.; Li, M.; Song, J.-X. *Phytother. Res.* **2017**, *31*, 1119–1127.
70. Rehman, M. U.; Wali, A. F.; Ahmad, A.; Shakeel, S.; Rasool, S.; Ali, R.; Rashid, S. M.; Madkhali, H.; Ganaie, M. A.; Khan, R. *Curr. Neuropharmacol.* **2019**, *17*(3), 247–267.
71. Mehta, R. G.; Murillo, G.; Naithani, R.; Peng, X. *Pharm. Res.* **2010**, *27*(6), 950–961.
72. Cragg, G. M.; Pezzuto, J. M. *Med. Princ. Pract.* **2016**, *25*(2), 41–59.
73. Pham, J. V.; Yilma, M. A.; Feliz, A.; Majid, M. T.; Ma, N.; Walker, J. R.; Kim, E.; Cho, H. J.; Reynolds, J. M.; Song, M. C.; Park, S. R.; Yoon, Y. J. *Front. Microbiol.* **2019**, *10*, article number: 1404. https://doi.org/10.3389/fmicb.2019.01404.
74. Bidon-Chanal, A.; Fuertes, A.; Alonso, D. *Eur. J. Med. Chem.* **2013**, *60*, 479–489.
75. Khanfar, M. A.; Asal, B. A.; Mudit, M.; Kaddoumi, A.; El Sayed, K. A. *Bioorg. Med. Chem.* **2009**, *17*(16), 6032–6039.
76. Nakao, Y.; Uehara, T.; Matunaga, S.; Fusetani, N.; van Soest, R. W. *J. Nat. Prod.* **2002**, *65*(6), 922–924.
77. Barde, S. R.; Sakhare, R. S.; Kanthale, S. B.; Chandak, P. G.; Jamkhande, P. G. *Asian Pac. J. Trop. Dis.* **2015**, *5*, S209–S213.
78. Pascual, I.; Lopéz, A.; Gómez, H.; Chappe, M.; Saroyan, A.; Gonzalez, Y.; Cisneros, M.; Charli, J. L.; Chavez, M. A. *Enzym. Microb. Technol.* **2007**, *40*(3), 414–419.
79. Giannini, C.; Debitus, C.; Posadas, I.; Paya, M.; D'Aura, M. V. *J. Nat. Prod.* **2001**, *64*(5), 612–615.
80. Li, Y.; Zhang, Y.; Shen, X.; Guo, Y.-W. *Bioorg. Med. Chem. Lett.* **2009**, *19*(2), 390–392.
81. Zhang, Y.; Li, Y.; Guo, Y.-W.; Jiang, H.-I.; Shen, X. *Acta Pharmacol. Sin.* **2009**, *30*(3), 333–345.
82. Leung, P.-C.; Siu, W.-S. *J. Tradit. Complement. Med.* **2013**, *3*(2), 82–87.
83. An, J.; Yang, H.; Zhang, Q.; Liu, C.; Zhao, J.; Zhang, L.; Chen, B. *Life Sci.* **2016**, *147*, 46–58.

CHAPTER TWO

Anticancer natural products

Satyajit D. Sarker[a,*], Lutfun Nahar[a], Anca Miron[b], Mingquan Guo[c]

[a]Centre for Natural Products Discovery (CNPD), School of Pharmacy and Biomolecular Sciences, Liverpool John Moores University, Liverpool, United Kingdom
[b]Department of Pharmacognosy, Faculty of Pharmacy, "Grigore T. Popa" University of Medicine and Pharmacy Iasi, Romania
[c]Wuhan Botanical Garden, Chinese Academy of Sciences, Wuhan, China
*Corresponding author: e-mail address: s.sarker@ljmu.ac.uk

Contents

1. Introduction

Cancer is a disease caused by uncontrolled division of abnormal cells in a part of the body, and can eventually invade nearby tissues.[1] There are more than 200 different types of cancer, of which breast, colon, lungs and prostate cancers are the most common ones. Cancers can be grouped into five major classes: brain tumor, carcinoma, leukemia, lymphoma and sarcoma. It has been envisaged that one in two people in the United Kingdom will have cancer in their lifetime, and currently cancer is the second largest cause of deaths globally.[2] It is estimated that there will be approximately 20 million new cancer cases globally by the end of the year 2025.[3,4] Thus, finding effective new treatment options for cancers still remains a major challenge for drug discovery researchers and pharmaceutical companies. However, current cancer treatment modalities include surgery, chemotherapy,

Annual Reports in Medicinal Chemistry, Volume 55
ISSN 0065-7743
https://doi.org/10.1016/bs.armc.2020.02.001

radiotherapy, hormone therapy, stem cell and bone marrow transplants and various anticancer drugs. Among the successful chemotherapeutic agents and anticancer drugs, historically, many could find their origins in natural products. In fact, many successful anticancer drugs, *e.g.*, camptothecin, etoposide, paclitaxel, vinblastine and vincristine (Fig. 1) that are in use today are natural products or their semi-synthetic analogues.[5] Natural products have played a pivotal role in cancer chemotherapy and chemoprevention (also see Chapter "Cancer chemopreventive natural products" by Ritchie and Sarker) for over half a century, and provided established anticancer drugs, new lead compounds for synthetic modifications, and for understanding cellular and molecular mechanisms of actions for their anticancer potentials.[6] The earliest natural products-derived anticancer drugs approved by the Food and Drug Administration (FDA) in the USA were the *Vinca* alkaloids: vincristine in 1963 and vinblastine in 1965, which were isolated from Madagascar periwinkle plants (*Catharanthus* spp.) growing in Jamaica and the Philippines.[7] Subsequently, their semisynthetic analogues, vinorelbine and vindesine, were approved as anticancer drugs.

Natural products have continued to remain as one of the important sources of anticancer lead molecules for future anticancer drug development. This chapter primarily discusses anticancer natural products that have been developed as commercially available drugs, and/or are at various developmental stages within the drug development pipeline for potential treatment of various forms of cancer. Moreover, this chapter provides a summary of information on possible mechanisms of actions of anticancer drugs from natural origins.

2. Anticancer natural products

During 1981–2010, there were 98 small molecule anticancer drugs approved by the FDA, and of them, 11 were purely natural compounds, 32 derived from natural products, 11 utilized the natural products templates for the synthesis, 16 were natural products mimics, and 8 were linked to natural products, with only 20 of them were purely synthetic compounds.[7] During 1981–2014, a total of 136 new anticancer drugs were introduced, with 83% of these drugs being natural products or based on natural products, or analogues of natural products.[8] From 2010 to 2014, at least 28 anticancer drugs directly of natural origins or inspired by natural products were introduced in the market (Table 1).[9] It is estimated that around 60% of all anticancer drugs that are currently in the market are natural products,

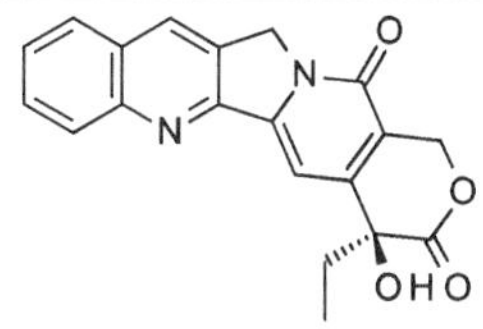

Camptothecin
Source: *Camptotheca acuminata* (Nyssaceae)

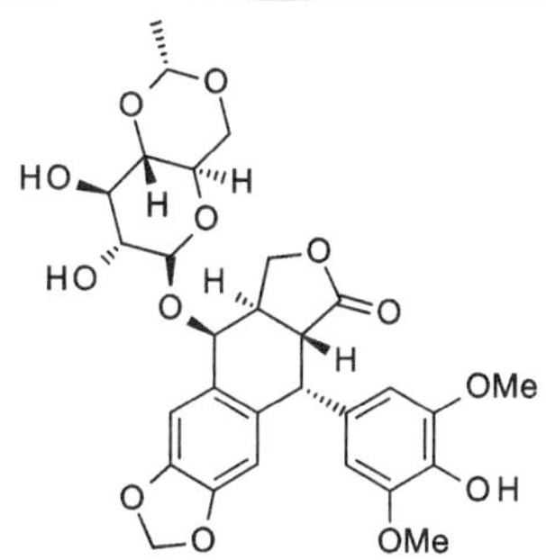

Etoposide
Source: A synthetic derivative of podophyllotoxin, isolated from *Podophyllum pellatum* (Berberidaceae)

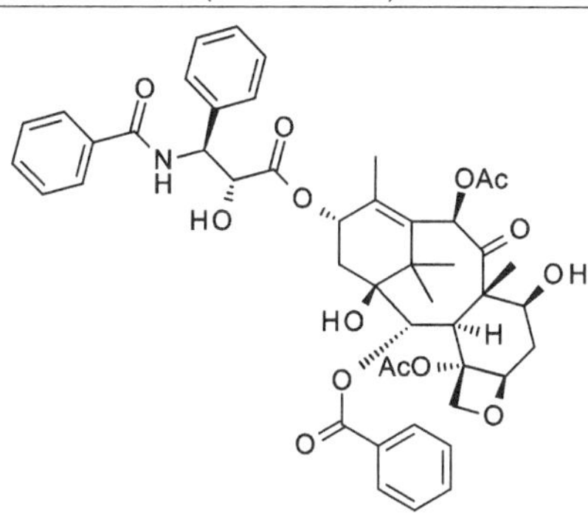

Paclitaxel
Source: *Taxus brevifolia* (Taxaceae)

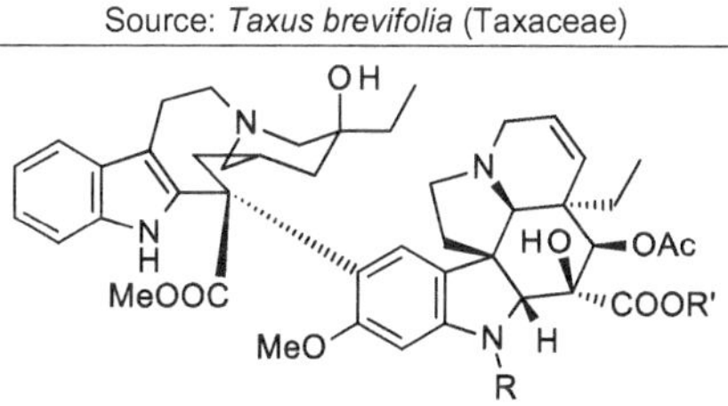

Vinblastine (R = R' = Me) and vincristine (R = CHO; R' = Me)
Source: *Catharanthus roseus* (Apocynaceae)

Fig. 1 Some examples of natural products and their analogues as anticancer drugs.

Table 1 Anticancer drugs directly of natural origins or inspired by natural products introduced during 2010–14.[9]

Year	Anticancer drugs
2010	Cabazitaxel, eribulin, mifamurtide, romidepsin and vinflunine
2011	Abiraterone, brentuximab, dabrafenib, ruxolitinib, vanetabnib and vemurafenib
2012	Axitinib, Bf-200 wing, bosutinib, cabozantinib, carfilzomib, pixantrone dimaleate, radotinib and regorafenib
2013	Afatinib, ibrutinib, ponatinib, trametinib DMSO, trastuzumab and emtansine
2014	Belinostat, idelalisib and cloridrato de tipiracil

confirming the nature still being the most attractive source for new anticancer drug discovery and development.

While several natural products have been developed as anticancer drugs in their original forms, several other natural compounds have paved the way for the synthesis of several synthetic anticancer and antitumor drugs, meaning that, those syntheses were inspired by natural products. Classic examples could include the development of etoposide from podophyllotoxin, a lignan isolated from *Podophyllum* species, and cytarabine synthesis, inspired by arabinonucleosides isolated from the marine sponge *Tectitethya crypta*[10] (Fig. 2). There have been numerous review articles describing anticancer natural products drugs from diverse sources published over the last decade or so, and all those articles demonstrate the current relevance of natural products in modern anticancer drug discovery and developmental scene.[6,11–16]

2.1 Plant-derived anticancer natural products

Plants have evolved to biosynthesize a variety of compounds, including quite simple molecules as well as compounds with significant steric complexities (*e.g.*, taxol), to survive and protect themselves against environmental challenges.[17] Many phytochemicals have potent activities that can be used to treat human ailments, including cancers and tumors. The history of using plants to treat cancer can be traced back to the Ebers papyrus in 1500 BC. However, the actual search for anticancer lead compounds from higher plants for modern drug discovery and development tentatively started in the early 20th century.

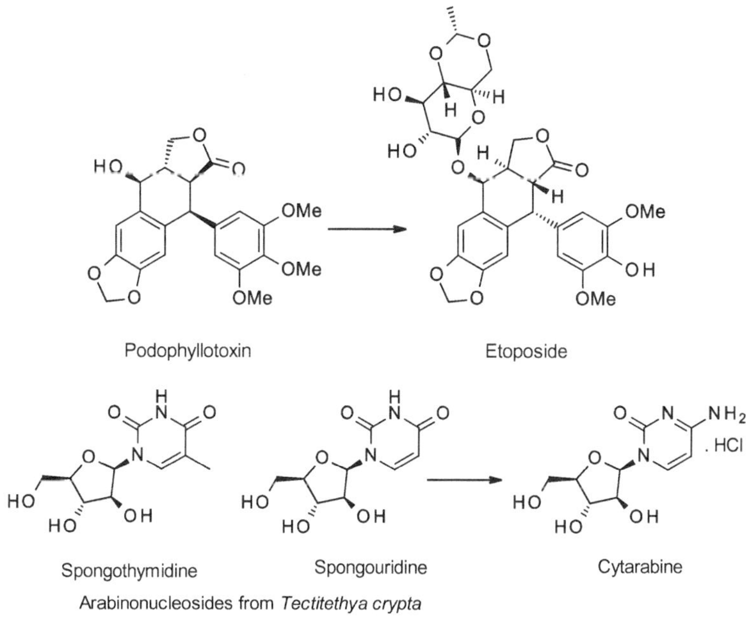

Fig. 2 Examples of synthetic anticancer drugs inspired by natural products.

Because of the relative ease of access to the plants, most of the earlier discoveries of anticancer drugs were from plants. In fact, plant-derived natural products have been a major source of several clinically relevant anticancer drugs, and will continue to offer new chemical entities in anticancer drug discovery as evidenced by the number of promising new agents in clinical development, based on selective activity against cancer-related molecular targets.[18] Among those discoveries, the drug obtained from the bark of *Taxus brevifolia*, paclitaxel, presents one of the most successful stories in plant-based anticancer drug discovery. Plants still remain as one of the main sources of new anticancer leads for cancer therapeutics. There are four main classes of higher plant-derived anticancer drugs used clinically in the USA and in Europe, and they include the *Vinca* (*Catharanthus*) bisindole alkaloids: vinblastine, vincristine (*see* Fig. 1), vinorelbine and vinflunine; the semi-synthetic *epi*-podophyllotoxins: etoposide (*see* Fig. 1), etoposide phosphate and teniposide; the taxanes: paclitaxel (*see* Fig. 1) and paclitaxel albumin-stabilized nanoparticle formulation, docetaxel and cabazitaxel; and the camptothecin derivatives: irinotecan and topotecan (Fig. 3).[6]

Since 2010, numerous scientific papers have been published covering anticancer plant materials, plants with anticancer potential, anticancer

Vinca alkaloids	
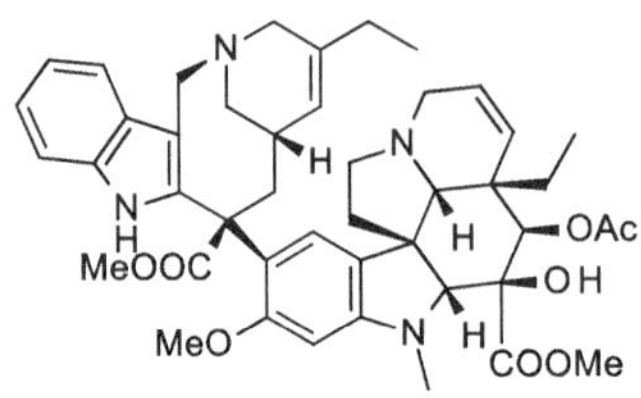 Vinorelbine (brand name: Navelbine®) Approved in the USA in 1994 to use against metastatic breast cancer and non-small cell lung cancer	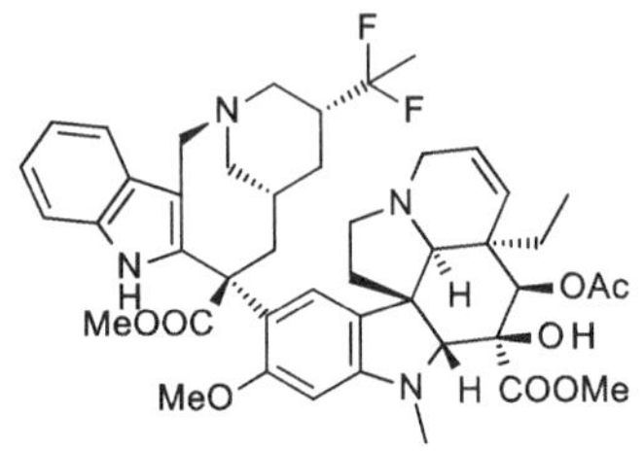 Vinflunine (brand name: Javlor®) Approved in Australia in 2012 to use against advanced cancer of the bladder and the urethra
Semi-synthetic *epi*-podophyllotoxins	
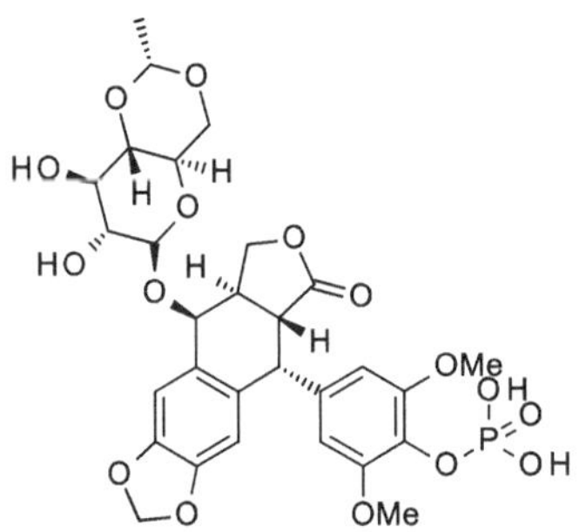 Etoposide phosphate (brand name: Etopophos®) To treat testicular cancer, lung cancer, lymphoma, leukaemia, neuroblastoma, and ovarian cancer	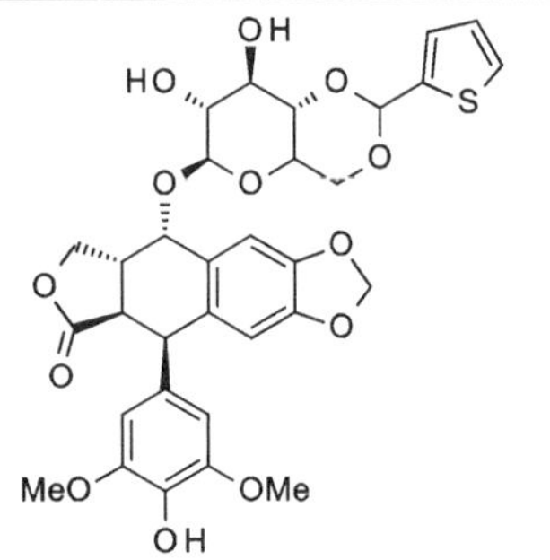 Teniposide (brand name: Vumon®) For the treatment of childhood acute lymphocytic leukaemia (ALL), Hodgkin's lymphoma, certain brain tumours, and other types of cancer
Taxanes	
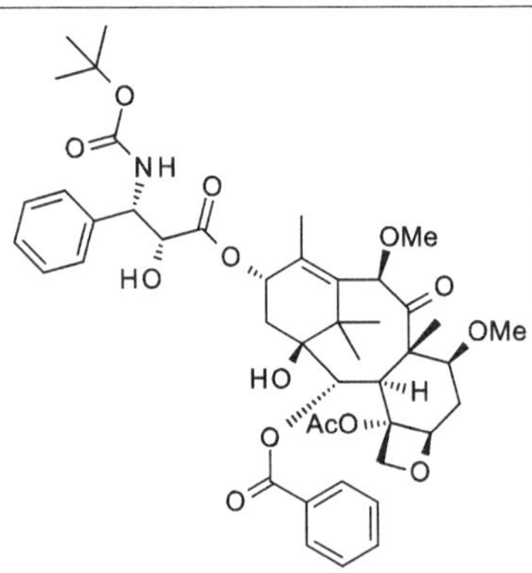 Cabazitaxel (brand name: Jevtana®) Approved in the USA in 2010 for hormone refractory prostate cancer	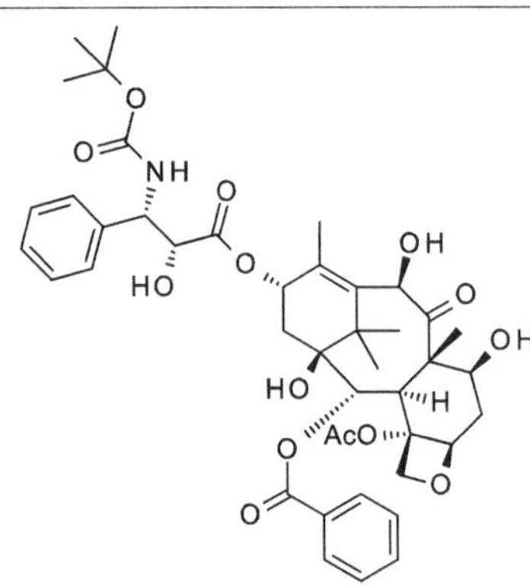 Docetaxel (brand name: Taxotere®) Approved in the USA (1995) for breast, head and neck cancer, stomach, prostate and non-small-cell lung cancers

Fig. 3 Examples of some plant-derived and inspired anticancer drugs.

Camptothecin derivatives

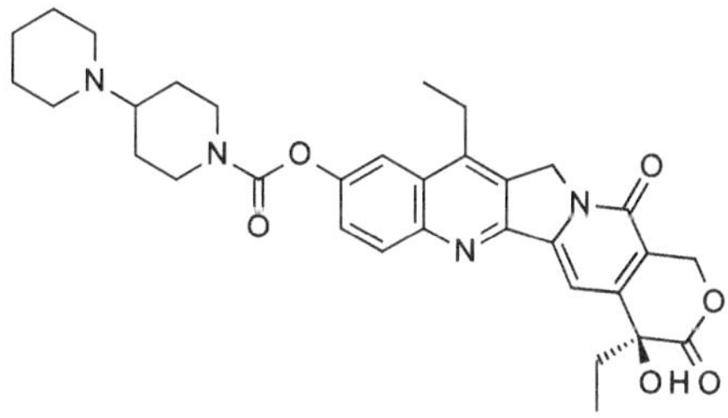

Irinotecan (brand name: Campto® or Camptosar®)
Approved in the USA (1996 for bowel, colorectal and small cell lung cancers

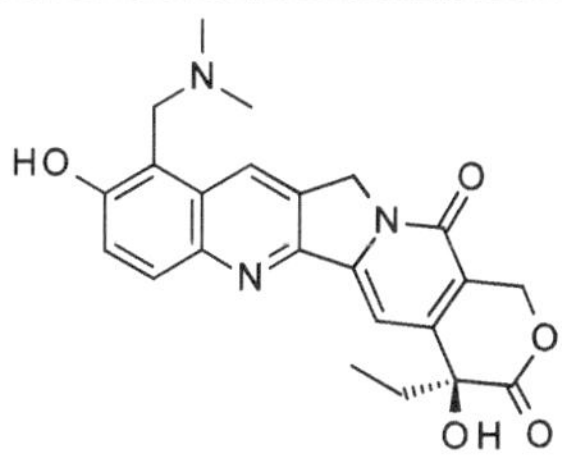

Topotecan (brand name: Hycamtin®)
Approved in the USA in 2007 for the treatment of ovarian and lung cancers

Fig.3—Cont'd

phytochemicals, plant activity against cancer cells and plant-derived anticancer drugs.[13–16] Most of these studies were at the preliminary screening level using various indirect *in vitro* assays or cell line assays using human cancer cells, whereas there were only a handful of studies carried out involving animal-based *in vivo* cancer models or human clinical trials. In fact, the number of publications on plant-derived anticancer drugs, either at the clinical studies or already in the market are only about 6% of all publications.

2.2 Anticancer natural products from microbial sources

There are many anticancer agents of microbial origins that have been evaluated through clinical trials, and are currently available as recognized anticancer and antitumor medications.[19] Since the discovery of actinomycin, an alkylating agent, composed of two cyclic peptides attached to a phenoxazine ring system, from *Streptomyces antibioticus* (family: Streptomycetaceae) in 1940[20] (which is also found in *Streptomyces parvulus* and other *Streptomyces* species), many microbial products have been approved as anticancer and antitumor drugs, including actinomycin D (also known as dactinomycin), anthracyclines, bleomycin, mitomycin C, enediynes and epothilones (Fig. 4).

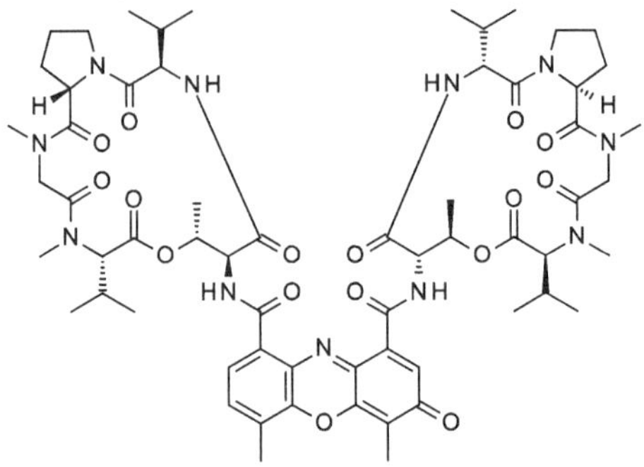

Actinomycin D (also known as dactinomycin)
Source: *Streptomyces parvulus* (Streptomycetaceae)

Anthracyclines

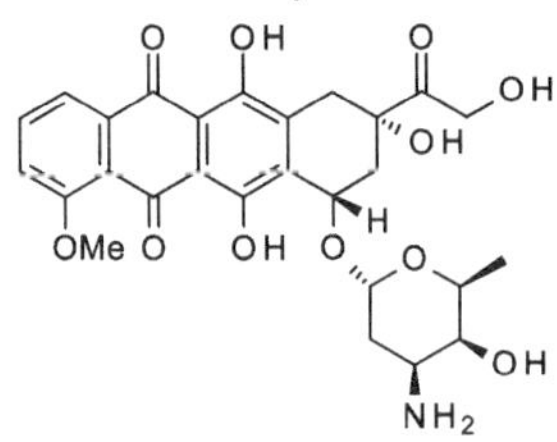

Doxorubicin
Source: *Streptomyces peucetius* (Streptomycetaceae)

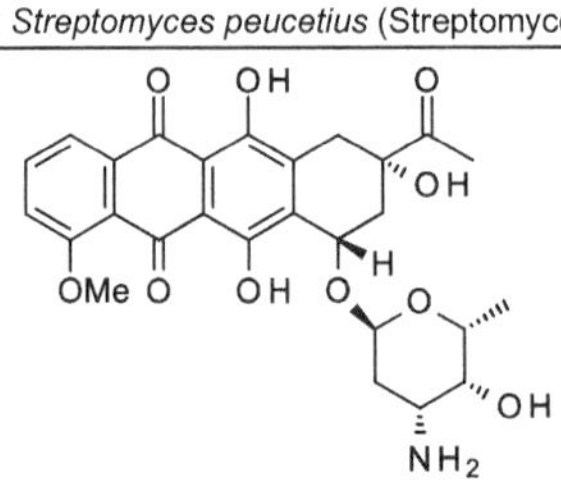

Daunorubicin
Source: *Streptomyces* sp. (Streptomycetaceae)

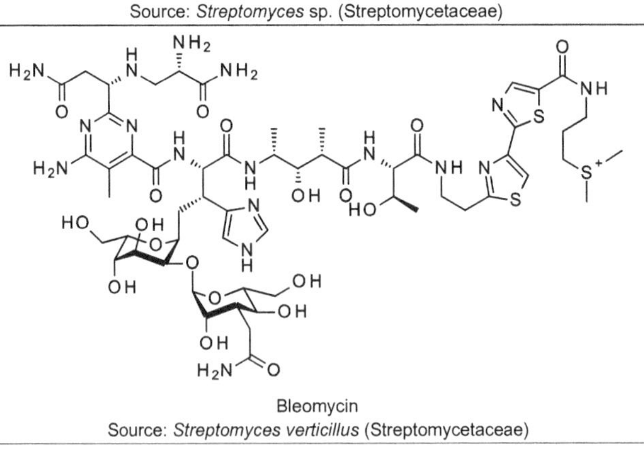

Bleomycin
Source: *Streptomyces verticillus* (Streptomycetaceae)

Fig. 4 Examples of some anticancer drugs from microbial sources (anticancer antibiotics).

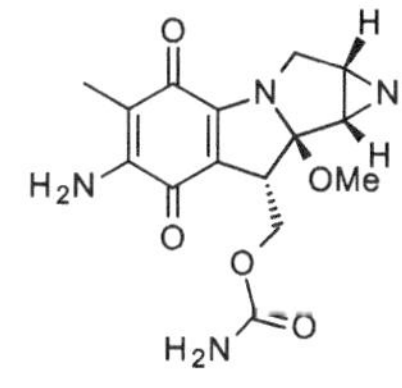

Mitomycin C
Source: *Streptomyces caespitosus* (Streptomycetaceae)

Enediyne

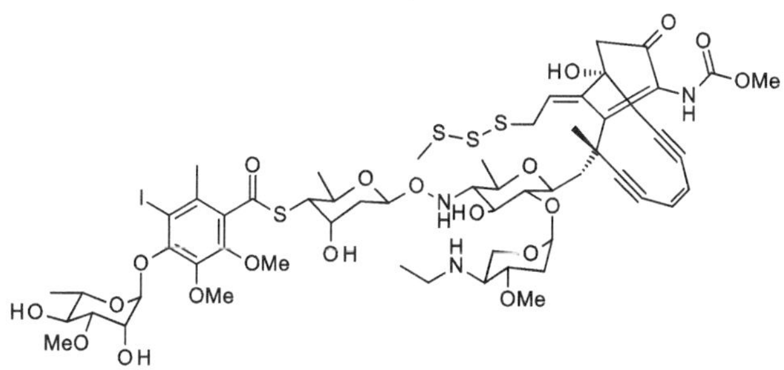

Calicheamicin
Source: *Micromonospora echinospora*

Epothilones

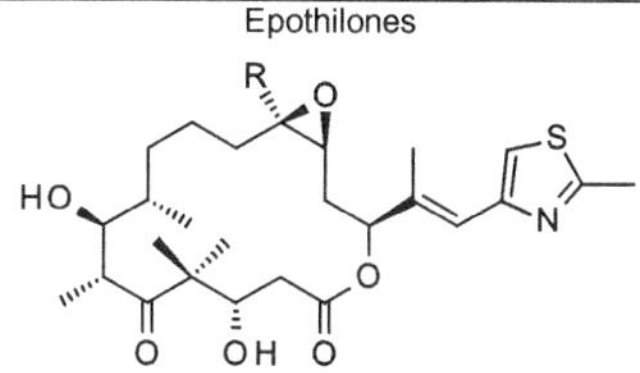

Epothilone A (R = H)
Epothilone B (R = Me)
Source: *Sorangium cellulosum* (Polyangiaceae)

Fig.4—Cont'd

Although actinomycin is the first anticancer antibiotics, because of its severe toxicity, it was not used long in cancer chemotherapy; related molecules and analogues were subsequently discovered, and are now widely used for cancer chemotherapy. Among these anticancer antibiotics, the most relevant are anthracyclines, which were isolated from *Streptomyces peucetius* in the early 1960s and commonly known as doxorubicin and daunorubicin (Fig. 4). Currently, doxorubicin is one of the most potent antineoplastic drugs prescribed alone or in combination with other anticancer drugs.[9] Enediynes, especially neocarzinostatin and calicheamicin (Fig. 4), were introduced as anticancer and antitumor drugs in 1980s. Later, several other anticancer agents of microbial origins have appeared which include esperamicin, dinemicina, kedarcidina, namenamycin, C-1027 chromophore, maturepeptina, N1999A2 and shishijimicin A, and most recently uncialamycin.[9]

Actinomycin D, used since 1954, is generally indicated in the chemotherapy of Wilm's tumor and Ewing's sarcoma. Wilms' tumor, named after the German surgeon Max Wilms who first described it, is also known as nephroblastoma, and is a cancer of the kidneys that generally occurs in children, but rarely in adults. On the other hand, Ewing's sarcoma is cancer of the bone or the soft tissue; the most common areas being the legs, pelvis, and chest wall. Simply, actinomycin D works by damaging the DNA of the cancer cells, and thus stopping cell divisions and cell growth. Bleomycin, discovered from *Streptomyces verticillus* in 1966 and commercially launched in Japan in 1969 and in the United States in 1973,[21] is a cytotoxic antibiotic that is used for the treatment of various types of cancer, mainly including Hodgkin's lymphoma, non-Hodgkin's lymphoma, testicular cancer, ovarian cancer, and cervical cancer. It also works in the same way as actinomycin D. Moreover, it generates free radicals inside the body, and thus damages DNA.

Mitomycin C was discovered in the 1950s from *Streptomyces caespitosus*.[22] This is an antitumor antibiotic that interferes and damages DNA by stopping its synthesis, and has been indicated for the treatment of anal, breast, bladder, cervical, colorectal, head and neck, liver, lung, pancreas, stomach and upper gastrointestinal (gullet) cancers as well as solid tumors.[23]

Doxorubicin, an anthracycline anticancer/antitumor antibiotic, isolated from *Streptomyces peucetius* and first commercially introduced in the USA in 1974, is one of the most successful anticancer drugs of microbial origin, and has been used for the treatment of various types of cancers, mainly breast, bladder cancers, Kaposi's sarcoma, lymphoma, and acute lymphocytic leukemia. It slows and/or stops the growth of cancer cells by blocking an enzyme topoisomerase 2, which is essential for division and growth of cancer cells.[24] On the other hand, another anthracycline anticancer antibiotic, daunorubicin (also known as daunomycin), isolated from a *Streptomyces* species, launched in 1979 in the USA, is used for the treatment of acute myeloid leukemia, acute lymphocytic leukemia, chronic myelogenous leukemia, and Kaposi's sarcoma. It can be mentioned here that Kaposi's sarcoma is a type of cancer that forms masses in the skin, lymph nodes, or other organs.

Enediynes are organic compounds that possess two triple bonds and one double bond, and known for their limited use as anticancer/antitumor antibiotics (Fig. 4).[25,26] They are efficient at inducing nonspecific apoptosis in cells, both normal and cancer cells. Calicheamicin, which was isolated from

Micromonospora echinospora in 1980s is an example of antitumor agent of this class.[27,28] This antibiotic has been in use for the treatment of non-solid tumor cancer acute myeloid leukemia. In 2017, a calicheamicin-linked monoclonal antibody, marketed as Besponsa, was approved by the FDA for use in the treatment of adults with relapsed or refractory B-cell precursor acute lymphoblastic leukemia.[29]

The epothilones are anticancer drugs, similar to taxanes, isolated from a myxobacterium (*Sorangium cellulosum*) at the beginning of the 1990s, and they are microtubule stabilizing antimitotic agents that prevent cancer cells from dividing by interfering with tubulin.[30] There are epothilones A to F discovered to date (Fig. 4).

2.3 Anticancer natural products from marine organisms

Despite the fact that terrestrial plant and microbial natural products have dominated the successful anticancer drug discovery and development scene over the years, marine organisms, especially marine sponges, have also provided a few anticancer drugs. The other two major marine sources for cytotoxic compounds are tunicates and sea hares. However, anticancer drug development from marine sources has been handicapped by the significant difficulties in obtaining adequate amounts of active compounds, and taxonomical complications related to source materials. The anticancer agent ecteinascidin-743 (known as trabectedin) (Fig. 5) was the first marine natural product with anticancer activity, isolated from *Ecteinascidia turbinata* (family: Perophoridae) in 1984. This marine organism is found in shallow waters in the Caribbean Sea, the east coast of Florida, Bermuda and the Gulf of Mexico. Trabectedin has now been developed as a chemotherapy drug and is also known by its brand name, Yondelis, for the treatment of advanced soft tissue sarcoma (approved in 2007), and sometimes, ovarian cancer

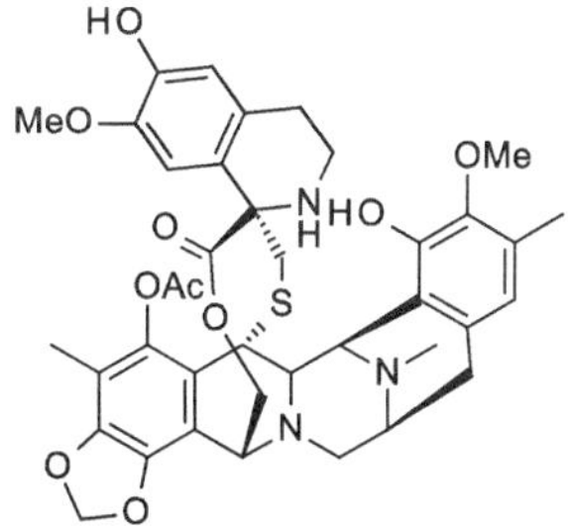

Fig. 5 Trabectedin—the first marine organism-derived anticancer drug.

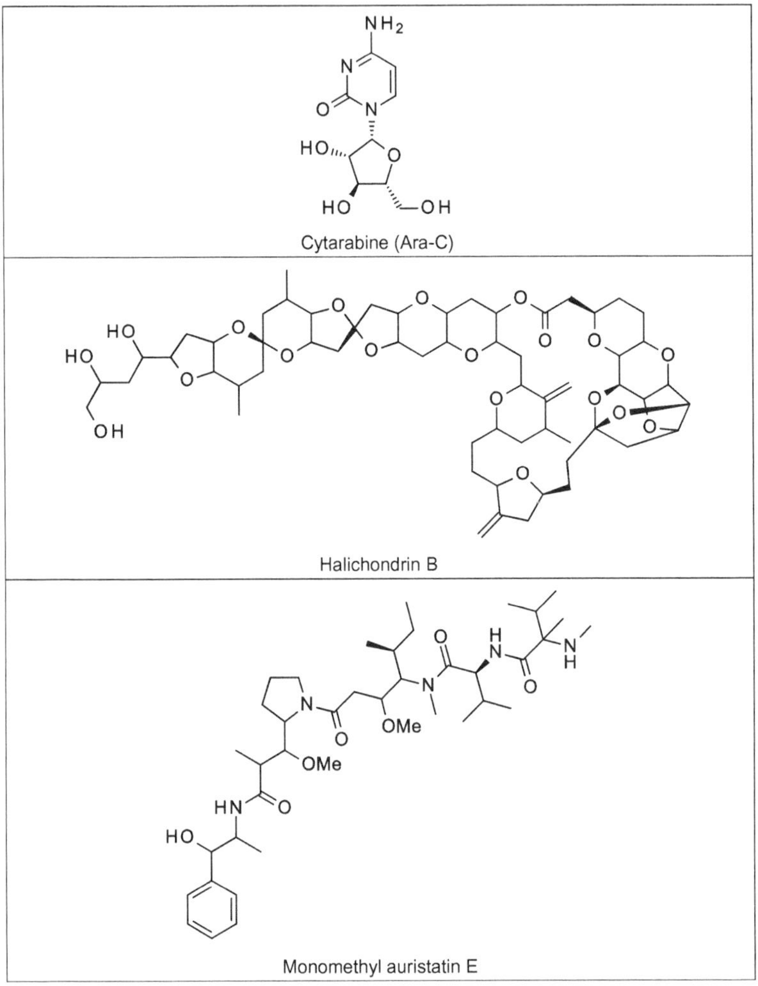

Fig. 6 Examples of some anticancer drugs from marine organisms.

(approved in 2009).[31,32] Simply, trabectedin exerts its action by sticking to the DNA in cells and damaging it.

Since the introduction of trabectedin (Fig. 5), at least three more antitumor drugs, cytarabine (Ara-C), halichondrin B and monomethyl auristatin E (Fig. 6), have recently appeared in the market, which were obtained from marine organisms. However, Ara-C is not exactly a natural product, but was synthesized using the nucleoside from the Caribbean sponge *Tectitethya crypta* as a model for the synthesis. Ara-C was approved as an antitumor and anticancer drug in 1972, mainly indicated for the treatment of leukemia and lymphoma.[33] The macrocyclic polyether halichondrin B (Fig. 6) was obtained from several species of marine sponges, *Axinella* spp. (fam: Axinellidae), *Halichondria okadai* (fam: Halichondriidae) and *Phakellia* spp. (fam: Axinellidae), and used as a model for the synthesis and development of the anticancer drug, eribulin mesylate (E7389), approved in 2010 and commercialized as Halaven® for metastatic breast cancer chemotherapy. Monomethyl auristatin E is a synthetic analogue of an antimitotic polypeptide belonging to the class of dolastatins, which were isolated from the sea hare *Dolabella auricularia* (Phylum Mollusca), and are also produced by Cyanobacteria. This polypeptide was conjugated with an antibody and found effective in the treatment of carcinoma, and subsequently the chimerical antibody conjugated with monomethyl auristatin E was given the generic name "brentuximab vedotin" and launched commercially in 2011 by Seattle Genetics Inc. as ADCETRIS®, indicated for the treatment of lymphomas.[33]

3. Mechanisms of action of anticancer natural products

A schematic summary of mechanisms of actions of anticancer drugs known to date is presented in Fig. 7. Most of the anticancer drugs of natural origins, *e.g.*, camptothecins, daunoribicine and doxorubicin, etoposide and teniposide, exert their actions through inhibiting topoisomerases, which are DNA processing enzymes that relieve supercoiling in DNA, and are essential for normal cell division. It has been observed that type 1B topoisomerase (Top1) is overexpressed in certain tumors, and this observation prompted numerous studies involving this enzyme as a target for cancer chemotherapy.[34] Several anticancer drugs act as 'poisons' of this enzyme's relegation step, producing Top1-DNA lesions, DNA breakage and cell death at the end.

Two other common mechanisms of action by natural anticancer drugs are interference with DNA structure by formation of DNA adduct, for example, mitomycin C, and inhibition of microtubules, for example, by epothilones, taxanes (paclitaxel) and *Vinca* alkaloids (vinblastine and vincristine).

The main mechanism of action of the taxane class of anticancer drugs is the disruption of microtubule function, which is essential for cell division.[35–37] Taxanes stabilize guanosine diphosphate (GDP)-bound tubulin in the microtubule, and thus inhibit the process of cell division by preventing depolymerization. On the other hand, the *Vinca* alkaloids

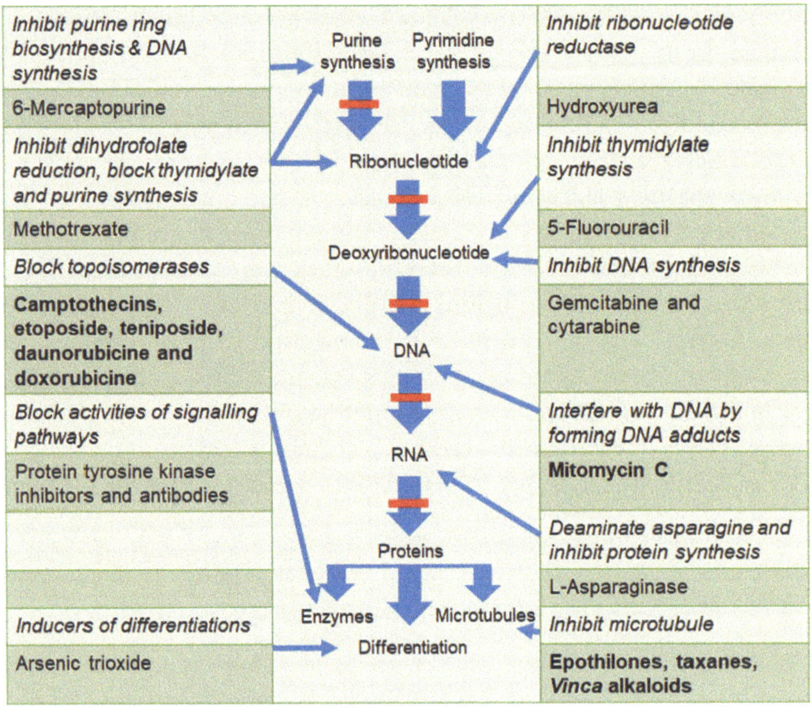

Fig. 7 Summary of mechanisms of action of anticancer drugs.

prevent mitotic spindle formation through inhibition of tubulin polymerization. Taxanes enhance the microtubules stabilization and inhibit the cellular mitosis and division. Moreover, taxanes prevent androgen receptor signaling by binding cellular microtubules and the microtubule-associated motor protein dynein, thus avoiding androgen receptor nuclear translocation. Like taxanes and *Vinca* alkaloids, the main mechanism of action of the epothilones is inhibition of microtubule function, and thus stopping proper cell divisions.

Many studies have attributed the antitumor activity of doxorubicin specifically to its ability to interleave in the DNA helix and/or to covalently bind the proteins involved in DNA replication and transcription, resulting in DNA breakdown, inhibition of RNA replication and transcription and consequently protein synthesis, leading to cell death. Doxorubicin and a few other natural anticancer drugs inhibit topoisomerases.[38,39] Doxorubicin and daunorubicin (Fig. 4) interact with DNA by intercalation and inhibition of macromolecular biosynthesis. They inhibit the progression of topoisomerase 2, an enzyme which is known to relax supercoils in DNA for transcription.

Doxorubicin stabilizes the topoisomerase 2 complex after it has broken the DNA chain for replication, preventing the DNA double helix from being resealed and thereby stopping the process of replication. It has also been shown that these anticancer drugs might also increase quinone-type free-radical production, and thus contribute to cytotoxicity. Through intercalation, doxorubicin can also induce histone eviction from transcriptionally active chromatin resulting in DNA damage response; epigenome and transcriptome are deregulated in doxorubicin-exposed cells.

Camptothecin and its analogues, like doxorubicin, are inhibitors of topoisomerases. Irinotecan, a semisynthetic analogue of camptothecin, is activated by hydrolysis to SN-38, an inhibitor of topoisomerase 1. This is then inactivated by glucuronidation by uridine diphosphate glucuronosyltransferase 1A1 (UGT1A1). The inhibition of topoisomerase 1 by the active metabolite SN-38 eventually leads to inhibition of both DNA replication and transcription.[40,41] The molecular mechanism of action of irinotecan takes place through trapping a subset of topoisomerase-1-DNA cleavage complexes, those with a guanine +1 in the DNA sequence.[38] One irinotecan molecule stacks against the base pairs flanking the topoisomerase-induced cleavage site and inactivates or inhibits the topoisomerase 1 enzyme.

Topotecan is another semisynthetic derivative of camptothecin, and it exerts its anticancer action in a similar way as other camptothecins do.[42] In physiological environments, topotecan is in equilibrium with its inactive carboxylate form, and its active lactone form intercalates between DNA bases in the topoisomerase 1 cleavage complex. It was shown that the binding of topotecan in the cleavage complex could prevent topoisomerase 1 from religating the nicked DNA strand after relieving the strain. This intercalation traps the topoisomerase 1 in the cleavage complex bound to the DNA, and when the replication-fork collides with the trapped topoisomerase 1, it damages DNA. The accumulation of trapped topoisomerase 1 complexes is a known response to apoptotic stimuli, and this disruption prevents DNA replication leading to cell death.

Antitumor antibiotics of the anthrax cycline class (anthracyclines), *e.g.*, idarubicin and valrubicin, also interfere with the action of the topoisomerases, the enzymes which are involved in DNA replication. These molecules are capable of exerting their action independently of the cell cycle phase, although they are more efficient in mitotic process cells.[43]

Mitomycin C is a chemotherapeutic drug that requires an enzymatic bioreduction to exert its anticancer effects. Upon reduction, this drug is converted to a highly reactive bis-electrophilic intermediate that alkylates cellular nucleophiles. The anticancer action of mitomycin C is mediated through formation of DNA adduct and interfering with DNA structure.[23] Alkylation of DNA is the most favored mechanism of action for mitomycin C, but other modes of action, such as redox cycling and inhibition of rRNA, could also contribute to the biological action of this drug.

The two main mechanisms of anticancer and antitumor actions of antimicrobial anticancer drugs, actinomycin D and bleomycin, are intercalation to DNA and the stabilization of cleavable complexes of topoisomerases 1 and 2 with DNA.[44] Additionally, the slow dissociation of actinomycin D from DNA complexes, its photodynamic activity and free radical formation might be other contributing factors for its anticancer property.

Enedynes anticancer agents produce their anticancer actions through the cyclization of the enediyne functional group, which creates a transient reactive 1,4-benzenoid diradical that acts as a nucleophile and attacks electrophiles in order to achieve a more stable form. In biological systems, once the diradical is positioned in the minor groove of double-stranded DNA, it abstracts two hydrogen atoms from the sugars opposite strands at either the C1, C4, or C5 positions. The DNA radicals can then cause interstrand crosslinks or react with O_2, leading to double- or single-stranded DNA cleavage, and ultimately cancer cell death.[45,46] Calicheamicins (Fig. 4) target DNA and cause strand scission. These drugs bind with DNA in the minor groove, undergo a reaction analogous to the Bergman cyclization to generate a diradical species (1,4-didehydrobenzene), which abstracts hydrogen atoms from the deoxyribose (sugar) backbone of DNA, leading to strand scission.[47] The specificity of binding of calicheamicins to the minor groove of DNA was shown to be due to the aryltetrasaccharide group of the molecule.

4. Virtual screening of anticancer natural products

Virtual screening of compounds, including various natural products, alkaloids, coumarins, flavonoids, lignans, quinones, tannins and terpenes, has become an indispensable tool for modern drug discovery programs, especially aiming at anticancer drug discovery.[48,49] Virtual screening has become popular as using this tool, one can develop alternative strategies that could help in choosing an appropriate set of compounds, avoiding unnecessary expenses with resources and generating unwanted compounds. In a recent review article, an overview on the application of different methods of virtual screening to select natural products with potential anticancer activity has been presented.[50]

Virtual screening has several advantages over traditional ways of screening (*e.g.*, HTS), some of which are highlighted below.[51]

i. Virtual screening offers higher hit rate than traditional screening methods.

ii. In virtual screening it is not necessary to collect and assay several natural compounds, which saves drug discovery time and cost.

iii. Virtual screening can play a unique role in achieving the task of examining the interaction of all existing natural products with all possible targets.

iv. Prediction of ADME/Tox and other "drug-like" properties can be incorporated in the pipeline of virtual screening, which increases the quality of hit compounds and reduces the failure rates in drug development step.

However, it can be mentioned that virtual screening results have to be backed up by appropriate laboratory experiments, both *in vitro* and *in vivo* screening, and clinical studies. A recent example of virtual screening of natural products for anticancer activity can be observed in the paper published by Alam and Khan,[52] where they carried out virtual screening, docking, ADMET and system pharmacology studies on caged xanthone derivatives produced by the genus *Garcinia*. Caged *Garcinia* xanthones are a family of natural products that are produced by tropical/subtropical trees of the genus *Garcinia*, and have a unique chemical architecture, defined by the presence of a caged scaffold at the C ring of a xanthone moiety (Fig. 8), and exhibit a broad range of biological activities, including anticancer and antitumor properties. Earlier, network pharmacology-based virtual screening for identification of anticancer therapeutics from the metabolites of *Clerodendrum* species, having cytotoxic properties, was reported.[53] In that study, a library comprising 53 compounds from these plants was created from the literature and three dimensional space analyses were performed to establish the drug-likeness of the library, and a natural products-cancer network was built based on docking studies, which eventually helped predict five compounds, apigenin 7-*O*-β-D-glucoside, hispidulin, scutellarein 7-*O*-β-D-glucuronate, acteoside and verbascoside, as potential binding therapeutics for cancer target proteins. Apigenin 7-glucoside and hispidulin displayed maximum binding interactions with 17 cancer drug targets in terms of docking weighted network pharmacological analysis.

Ryan et al.[14] used a set of 617 approved anticancer drugs, constituting the active domain, and a set of 2892 natural products, constituting the inactive domain, to build predictive models and to index natural products for their anticancer bioactivity, and utilizing the iterative stochastic elimination

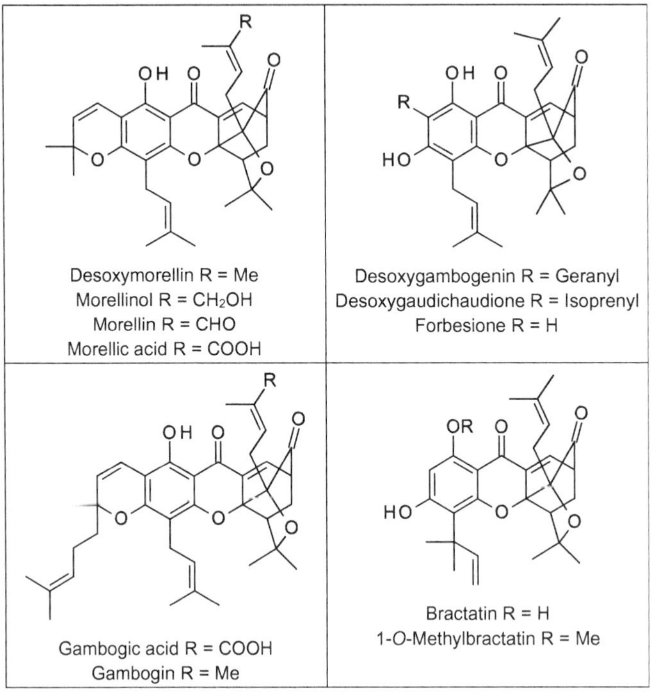

Fig. 8 Examples of some caged *Garcinia* xanthones.

optimization technique, a highly discriminative and robust model was created. Among the natural products screened in this study, only 12 scored highly as potential anticancer drugs candidates, and a few of those compounds, *e.g.*, colchicine, neoechinulin, and piperolactam A (Fig. 9), have already been found to possess anticancer potential in various *in vitro* and *in vivo* screening.

Because of tumor-suppressor function, p53 is considered as an attractive drug target, aiming at stabilization of the functional p53 or reactivation of mutated p53. A total of 181 natural products have recently been screened virtually, and one of those, torilin (Fig. 10), which was isolated from *Torilis japonica* as an inhibitor of testosterone 5α-reductase,[54] was found not only to improve p53 activity, but also to increase p21 protein expression level, which lies downstream of p53, thus suppress HCT116 cancer cell growth.[55] It was shown that L1/S3 of WT-p53 is a druggable pocket, and torilin has potential cytotoxicity through activating the p53 pathway. In another study, sigma 2 receptor was used in the virtual screening of several marine natural products for antitumor activity.[56] It can be noted that sigma 2

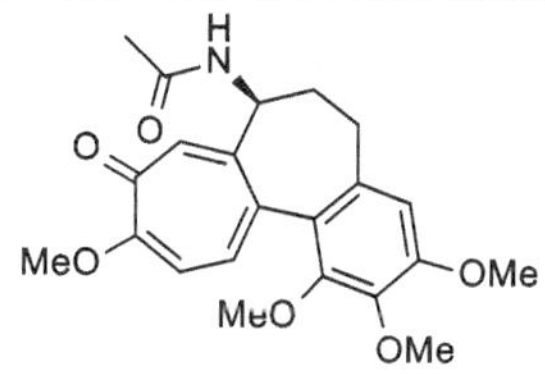

Colchicine
Source: *Colchicum autumnale* (Colchicaceae)

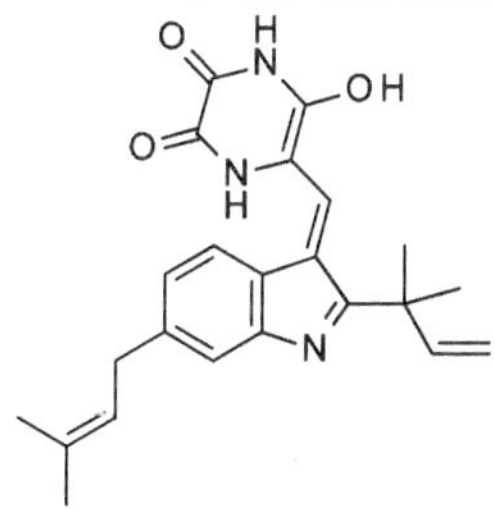

Neoechinulin
Source: *Eurotium rubrum* (Aspergillaceae)

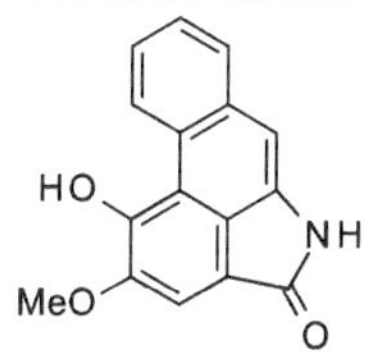

Piperolactam A
Source: *Piper betle* (Piperaceae)

Fig. 9 Hits from a virtual screening for anticancer activity.

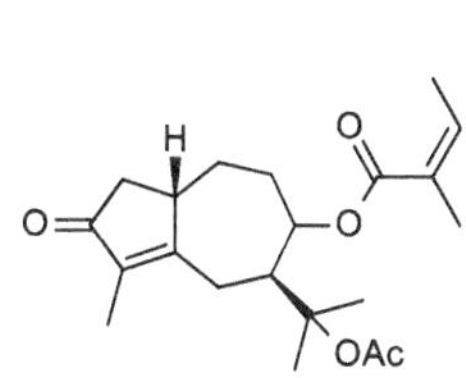

Fig. 10 Torilin—a hit from a virtual screening of natural products for anticancer property.

receptor ligands determine tumor cell death through apoptotic and non-apoptotic pathways, and the overexpression of sigma 2 receptors in several tumor cell lines.

Traditional Chinese Medicine (TCM) has been around for more than 5000 years, and has continued to be a major and integral part of the Chinese healthcare system. Various Chinese medicinal plants used in

TCM are considered as an alternative resource for novel anticancer drug discovery, and have led to the discovery of several successful modern drugs, for example, artemisinin. Dai et al.[57] attempted to identify potential anticancer compounds and plants from the TCM database by using cheminformatics, aiming at offering pertinent information for anticancer drug discovery. In that study, a total of 5278 anticancer compounds from TCM database were subjected to virtual screening. Just over 300 compounds were found to be highly potent against 60 cell lines, and about 75% of the 5278 compounds were highly similar to the approved anticancer drugs.

5. Anticancer natural products in clinical trials (in the drug development chain)

In 2013, about 100 unaltered natural products as well as their analogues were in clinical trials for various ailments, and a majority of them were for the treatment of various forms of cancers and tumors.[4] Some examples of natural products of plant origins that are currently at various stages of clinical trials on their route to emerging as new anticancer agents for cancer chemotherapy, or very recently introduced in the market, may include anhydrovinblastine, combretastatins, curcumin, indirubin, ingenol mebutate, paclitaxel poliglumex, salvicine and triptolide (Fig. 11).[2] In 2015, clinicaltrials.gov listed more than 1200 active clinical trials evaluating one or more of *Vinca* alkaloids, in combination with other drugs, as interventions in clinical trials to treat cancer.

Anhydrovinblastine, a *Vinca* alkaloid isolated from *Catharanthus roseus* (also known as *Vinca rosea*), which produces the well-known anticancer drugs, vincristine and vinblastine, is an investigational anticancer drug, currently in clinical trials studying the treatment of unspecified adult solid tumors, and belongs to the family of drugs called mitotic inhibitors.[58]

Combretastatins are diarylethanoids or dihydrostilbenoids, predominantly produced by *Combretum caffrum,* which is commonly known as South African Bush Willow. Among the natural combretastatins, combretastatins A-1 and A-4 are the most potent in terms of tubulin binding ability and cytotoxicity, and are currently in clinical trials.[59] These compounds bind to the β-subunit of tubulin, inhibit tubulin polymerization and thereby prevent cancer cells from producing microtubules, which are fundamental to cytoskeleton production, intercellular movement, cell movement, and formation of the mitotic spindle used in chromosome segregation and cellular division[59,60]; they are also vascular targeting agents.

Curcumin, a diarylheptanoid, isolated from *Curcuma longa*, commonly known as turmeric, has long been in various clinical trials without any fruitful outcome, mainly because of its poor bioavailability.[61] In fact, the US government spent several million dollars in research into curcumin through the National Center for Complementary and Integrative Health, and could not find any support for curcumin as a medical treatment. However, curcumin and its modified analogues are still in clinical trials for the potential treatment of various cancers.[62,63] A recent randomized Phase II clinical trial with curcumin compared to *placebo* in treating patients with prostate cancer that was removed by surgery suggested that curcumin might be able to stop the growth of tumor cells by blocking some of the enzymes needed for cell

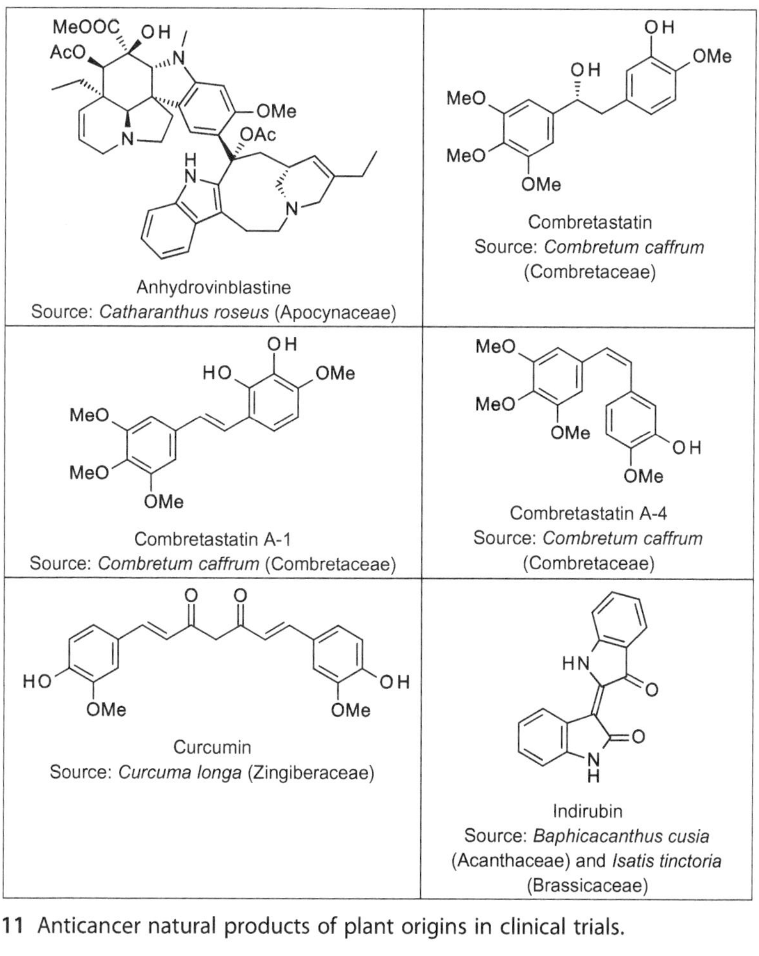

Fig. 11 Anticancer natural products of plant origins in clinical trials.

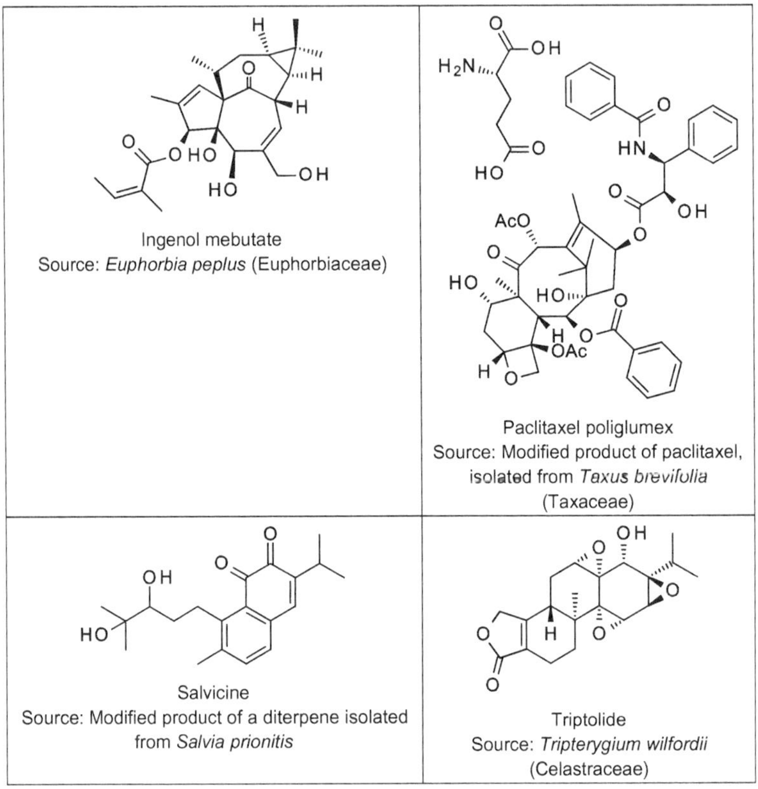

Ingenol mebutate
Source: *Euphorbia peplus* (Euphorbiaceae)

Paclitaxel poliglumex
Source: Modified product of paclitaxel, isolated from *Taxus brevifolia* (Taxaceae)

Salvicine
Source: Modified product of a diterpene isolated from *Salvia prionitis*

Triptolide
Source: *Tripterygium wilfordii* (Celastraceae)

Fig.11—Cont'd

growth and help to decrease or prevent prostate cancer from returning after surgery (also *see* Chapter 9).[64] Another randomized Phase IIb trial indicated that this compound, being a potent antioxidant, might be able to stop the growth of tumor cells by blocking some of the enzymes needed for cell growth. Curcumin has also been evaluated in a clinical trial for its application in combination with chemotherapy in advanced breast cancer, especially combining with paclitaxel.[63]

Ingenol mebutate, found in the sap of *Euphorbia peplus*, commonly known as milkweed and used traditionally for the treatment of dermatological lesions, is a cytotoxic drug that has been introduced for the treatment of actinic keratosis on the face of scalp, usually caused by excessive exposure to the sun or UV radiations.[65] This anticancer drug was developed and approved in Australia in 2013.

Indirubin, an investigational drug, is a constituent of a Traditional Chinese Medicine (TCM), indigo naturalis (also known as Qing Dai). Qing Dai is the leaf of *Baphicacanthus cusia* and *Isatis tinctoria.* Indirubin and related compounds are also obtained from *Indigofera tinctoria* (also known as *I. sumatrana),* but most often produced as a by-product of bacterial metabolism. Indirubin has been identified as the major active ingredient of the TCM formulation known as Danggui Longhui Wan, which has long been used for the treatment of chronic myelogenous leukemia in China. Currently, indirubin is under investigation in clinical trial for its antineoplastic applications, particularly against chronic myelocytic and chronic granulocytic leukemia.[66]

Paclitaxel poliglumex, a macromolecular taxane, is yet another investigational anticancer drug, which has undergone several clinical trials assessing its therapeutic potential against several forms of cancer including mainly advanced colorectal cancer, breast cancer, ovarian cancer, squamous cell carcinoma, lung cancer, and esophageal cancer.[67–69] Paclitaxel poliglumex originates from linking paclitaxel, isolated from *Taxus brevifolia,* to a biodegradable, water-soluble polyglutamate polymer. The polyglutamate residue enhances the water solubility of paclitaxel and thus allows delivery of higher doses than those achievable with paclitaxel alone, and at the same time it increases the therapeutic index of paclitaxel.

Salvicine, diterpenoid quinone obtained by structural modification of a natural product lead isolated from the Chinese herb, *Salvia prionitis,*[70] is a topoisomerase II inhibitor, and has been in clinical trials for possible anticancer applications.[71]

Triptolide, isolated from the thunder god vine (*Tripterygium wilfordii*) is an investigational drug for pancreatic cancer. It has been shown that triptolide can suppress the proliferation and promote apoptotic cell death in various cancers, especially prostate and pancreatic cancers.[46] However, because of its severe toxicity, a synthetic water-soluble prodrug (minnelide), based on triptolide, has been in various clinical trials.[72] Its Phase II clinical trials have just been concluded in February 2019.[73]

Flavopiridol, a synthetic compound, inspired by the phytochemical rohitukine, isolated from *Dysoxylum bunectariferum* (Meliaceae), showed growth inhibitory activity against breast and lung cancer cell lines both *in vitro* and *in vivo* (animal) studies. As a result, this compound being a pan-cyclin-dependent kinase (CDK) inhibitor that induces cell cycle arrest and apoptosis in many cancer cells, entered in the clinical trials at Phase I and Phase II, either alone or in combination with other chemotherapeutic agents for the potential treatment of leukemias, lymphomas and solid tumors, but with limited success.[74]

Since the approval of cytarabine about half a century ago, the marine-derived anticancer pharmaceutical pipeline incorporates four approved drugs, and at least 18 in clinical trials with six at a late developmental stage.[75] In fact, several cytotoxic compounds from marine sources are also currently in various stages of pre-clinical and clinical trials for their therapeutic applications as anticancer and antitumor drugs. For example, several dolastatins, isolated from *Dollabella auricularia*, are in the clinical trial, and one of them, monomethyl auristatin E, is already in the market as an antitumor drug.[33] Similarly, the depsipeptide dehydrodidemnin B, also known as aplidine or plitidepsin, isolated from the Mediterranean tunicate *Aplidium albicans* has been under clinical trials, initiated by PharmaMar for the development of Aplidin® for the treatment of multiple myeloma (Phase III), and for solid and hematological malignant neoplasias, like T-cell lymphoma (Phase II) (Fig. 12). Halimide (also known as phenylahistin) (Fig. 12), isolated from a marine *Aspergillus* spp. paved the way for the development of its synthetic analogue, plinabulin, which has been in the Phase I, II and III clinical trials against solid tumors and lymphomas, initiated by Nereus Pharmaceuticals, and continued by BeyondSpring Pharmaceuticals. Other marine compounds or their synthetic analogues that have been under clinical trials, especially Phase I, include marizomib, bryostatin 1, analogue of hemiasterlin and lurbinectedin (Fig. 12). In fact, bryostatin 1 has recently entered Phase II clinical trial against melanoma, non-Hodgkin's lymphoma, renal cancer and colorectal cancer.[16] ET-743, a tetrahydroisoquinilone alkaloid isolated from tunicate *Ecteinascidia turbinate,* entered Phase I clinical trials; the antiproliferative effect of this compound is produced through selective alkylation of guanine residues in the DNA minor groove, whereas dolastatin 10, a member of a peptide family isolated from the mollusk *Dolabella auricularia*, reached Phase II clinical trials. Dolastatin 10 inhibits microtubule assembly, which eventually leads to metaphase arrest in the cell cycle.[16]

Nearly five decades after the approval of cytarabine, over 1000 clinical trials have been listed in European and US databases, with more than 150 studies in Phase III.[75] Marine-derived natural products that are currently in clinical trials are shown in Table 2, grouped under various phases, Phase I to Phase III.

Among the antibiotic anticancer and antitumor agents, bleomycin, albeit has long been used in the treatment of various cancers, is currently in clinical trials for its potential applications in other forms of cancers. Similarly, some

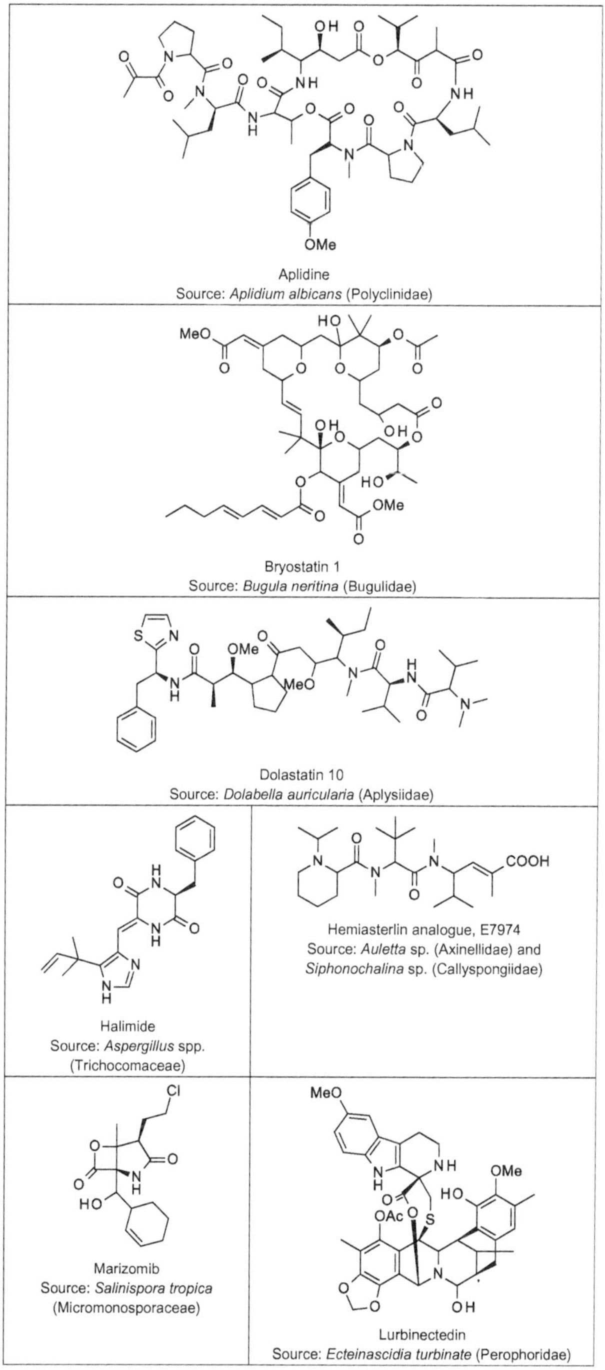

Fig. 12 Anticancer marine natural products or their analogues in clinical trials.

Table 2 Marine-derived natural products in Phase I, II and III clinical trials.

Compound name	Lead compound	Molecular target	Cancer conditions
Phase I			
ABBV-085	Dolastatin 10	LRRC15	Advanced solid tumors, sarcoma, squamous cell carcinoma of head and neck and breast cancer
AGS-67E		CD37 and microtubules	Refractory or relapsed lymphoid malignancy
AGS-15ME		SLITRK6 and microtubules	Advanced urothelial cancer
PF-06647020		PTK7 and microtubules	Advanced solid tumors and breast cancer
Phase II			
AGS16C3F	Dolastatin 10	ENPP3 and microtubules	Advanced renal cancer
ASG-22ME		Nectin-4 and microtubules	Urinary bladder, urologic, renal pelvis, ureteral and urethral neoplasms, and urothelial cancer
CDX-011		Microtubules	Breast cancer, osteosarcoma and lung cancer
GSK2857916		BCMA	Multiple myeloma
Ladiratuzumab vedotin		LIV-1 and microtubules	Breast cancer
Plocabulin		Microtubules	Advanced colorectal cancer
Telisotuzumab vedotin		Microtubules	Bladder, cervical, prostate, and esophagus cancers
Tisotumab vedotin		c-Met	Lung cancer

Table 2 Marine-derived natural products in Phase I, II and III clinical trials.—cont'd

Compound name	Lead compound	Molecular target	Cancer conditions
Phase III			
Depatuxizumab vedotin	Dolastatin 10	EGFR and microtubules	Early glioblastoma
Lurbinectedin	Trabectedin	DNA minor groove	Ovarian cancer
Plinabulin	Halimide	Microtubules	Non-small-cell lung cancer and neutropenia
Plitidepsin	Plitidepsin	JNK activation	Relapsed or refractory multiple myeloma
Polatuzumab vedotin	Dolastatin 10	Microtubules	Large B-cell lymphoma
Salinosporamide A	Salinosporamide A	20S proteasome	Newly diagnosed glioblastoma

Adopted from Perreira, R.B.; Evdokimov, N.M.; Lefranc, F.; Valentao, P.; Kornienko, A.; Perreira,-D.M.; Andrade, P.B.; Gomes, N.G.M. *Mar. Drugs Marine Drugs* **2019,** *17*(6), 329, DOI:10.3390/md17060329.

of the epothilones are also currently in clinical trials. Ixabepilone, an epothilone B analogue, developed by Bristol-Myers Squibb as an anticancer drug, was FDA approved in 2007 for the treatment of unresponsive aggressive metastatic or locally advanced breast cancer, and has recently been in Phase II trial in the treatment of recurrent/persistent carcinosarcoma of the uterus.[76]

Rapamycin, wortmannin, and geldanamycin (Fig. 13), because of their significant antiproliferative activities, are in clinical trials for the development of novel chemotherapeutic agents.[77] While rapamycin, obtained from *Streptomyces rapamycinicus,* possesses antitumor activity mediated by hindering proliferation, triggering apoptosis, and inhibiting angiogenesis, wortmannin, a fungal furanosteroid produced by *Penicillium funiculosum* also offer similar anticancer effects by acting as an effective selective inhibitor of phosphoinositide 3-kinases (PI3Ks) and PI3K- related enzymes. Geldanamycin, a benzoquinone ansamycin antitumor compound from *Streptomyces hygroscopicus* var. *geldanus*, prevents ATPase activity by binding to the heat shock protein and affecting the stability and function of oncogenic protein kinases involved in signal amplification cascade that controls

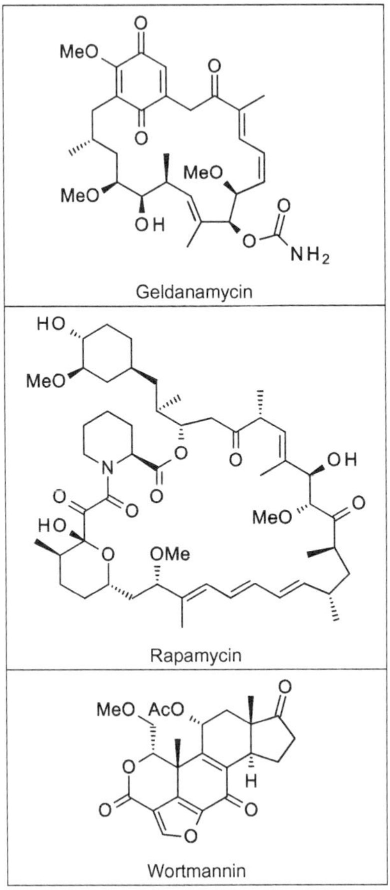

Fig. 13 Examples of anticancer microbial natural products in clinical trials.

proliferation and apoptosis. Geldanamycin and its analogues have been in clinical trials for multiple myeloma, breast, and prostate cancers.

6. Conclusions

Natural products have long been the main source of anticancer drugs, and will continue to remain so in the years to come. The natural products molecules that are currently in clinical trials and/or in the drug development chain, are expected to find their place in modern anticancer therapy, and thus help in our constant fight against cancer. Any effective drug discovery and development effort must include multidisciplinary and collaborative approach, incorporating natural product lead discovery linking with optimization through virtual screening, combinatorial and medicinal chemistry, total synthesis, and combinatorial biochemistry. Considering the unique chemical diversity that exists in natural source materials, it can be expected that new anticancer drug leads that are capable of interacting with most or possibly all therapeutic targets, will be discovered from natural products, or semisynthetic or synthetic products inspired by natural products scaffolds.

References

1. Cancer Research UK. https://www.cancerresearchuk.org/about-cancer/what-is-cancer; 2019, accessed on 11.11.2019.
2. Gezici, S.; Şekeroğlu, N. *Anticancer Agents Med. Chem.* **2019**, *19*(1), 101–111.
3. Bray, F. *World Cancer Report 2014*. International Agency for Research on Cancer: Lyon, France, **2014**;54–68.
4. Seca, A. M. L.; Pinto, D. C. G. A. *Int. J. Mol. Sci.* **2018**, *19*(1), 263. https://doi.org/10.3390/ijms19010263.
5. Nahar, L.; Sarker, S. D. *Chemistry for Pharmacy Students*, 2nd ed.; Wiley and Sons: Chichester, **2019**.
6. Kinghorn, A. D.; De Blanco, E. J. C.; Lucas, D. M.; Rakotondraibe, H. L.; Orjala, J.; Soejarto, D. D.; Oberlies, N. H.; Pearce, C. J.; Wani, M. C.; Stockwell, B. R.; Burdette, J. E.; Swanson, S. M.; Fuchs, J. R.; Phelps, M. A.; Xu, L.; Zhang, X.; Shen, Y. Y. *Anticancer Res.* **2016**, *36*, 5623–5638.
7. Paller, C. J.; Denmeade, S. R.; Carucci, M. A. *Clin. Adv. Hematol. Oncol.* **2016**, *14*(6), 446–455.
8. Newman, D. J.; Cragg, G. M. *J. Nat. Prod.* **2016**, *79*(3), 629–661.
9. Amaral, R. G.; dos Santos, S. A.; Andrade, L. N.; Severino, P.; Carvalho, A. A. *Clin. Oncol.* **2019**, *4,* 1562.
10. Jimenez, P. C.; Wilke, D. V.; Costa-Lotufo, L. V. *Clinics* **2018**, *73,* e82s.
11. Mangal, M.; Sagar, P.; Singh, H.; Raghava, G. P. S.; Agarwal, S. M. *Nucleic Acids Res.* **2013**, *42*(D1), D1124–D1129.
12. Fridlender, M.; Kapulnik, Y.; Koltai, H. *Plant Sci.* **2015**, *6,* 799.
13. Iqbal, J.; Abbasi, B. A.; Mahmood, T.; Kanwal, S.; Ali, B.; Shah, S. A.; Khalil, A. T. *Asian Pac. J. Trop. Biomed.* **2017**, 7(12), 1129–1150.
14. Rayan, A.; Raiyn, J.; Falah, M. *PLoS One* **2017**, *12*(11), e0187925.
15. Lichota, A.; Gwozdzinski, K. *Int. J. Mol. Sci.* **2018**, *19,* 3533. https://doi.org/10.3390/ijms19113533.
16. Khalifa, S. A. M.; Elias, N.; Farag, M. A.; Chen, L.; Saeed, A.; Hegazy, M. F.; Moustafa, M. S.; Chang, F.; Iwasaki, A.; Suenaga, K.; Alajlani, M.; Goransson, U.; El-Seedi, H. R. *Mar. Drugs* **2019**, *17,* 491. https://doi.org/10.3390/md17090491.
17. Howe, M.-J. R. *Lancet Oncol.* **2018**, *19*(3), 293–294.
18. Grothaus, P. G.; Cragg, G. M.; Newman, D. J. *Curr. Org. Chem.* **2010**, *14*(16), 1781–1791.
19. Cragg, G. M.; Pezzuto, J. M. *Med. Princ. Pract.* **2016**, *25*, 41–59.
20. Waksman, S. A.; Tishler, M. *J. Biol. Chem.* **1942**, *142*, 519–528.
21. Umezawa, H.; Maeda, K.; Takeuchi, T.; Okami, Y. *J. Antibiot.* **1966**, *19*(5), 200–209.
22. Tomasz, M. *Chem. Biol.* **1995**, *2*(9), 575–579.
23. Paz, M. M.; Zhang, X.; Lu, J.; Holmgren, A. *Chem. Res. Toxicol.* **2012**, *25*(7), 1502–1511.

24. Patel, A. G.; Kaufmann, S. H. *eLife* **2012**, *1,* e00387. https://doi.org/10.7554/eLife.00387.
25. Nicolaou, K. C.; Smith, A. L.; Yue, E. W. *Proc. Natl. Acad. Sci. U. S. A.* **1993**, *9013*, 5881–5888.
26. Van Lanen, S. G.; Shen, B. *Curr. Top. Med. Chem.* **2008**, *8*(6), 448–459.
27. Maiese, W. M.; Lechevalier, M. P.; Lechevalier, H. A.; Korshalla, J.; Kuck, N.; Fantini, A.; Wildey, M. J.; Thomas, J.; Greenstein, M. *J. Antibiot.* **1989**, *42*(4), 558–563.
28. Watanabe, C. M.; Supekova, L.; Schultz, P. G. *Chem. Biol.* **2002**, *9*(2), 245–251.
29. Shen, B.; Hindra, A.; Yan, X.; Huang, T.; Ge, H.; Yang, D.; Teng, Q.; Rudolf, J. D.; Lohman, J. R. *Bioorg. Med. Chem. Lett.* **2015**, *25*(1), 9–15.
30. Forli, S. *Curr. Top. Med. Chem.* **2014**, *14*(20), 2312–2321.
31. Galmarini, C. M.; D'Incalci, M.; Allavena, P. *Mar. Drugs* **2014**, *2*(2), 719–733.
32. Cancer Research UK. https://www.cancerresearchuk.org/about-cancer/cancer-in-general/treatment/cancer-drugs/drugs/trabectedin; **2019**, accessed on 13.11.2019.
33. Rangel, M.; Falkenberg, M. *J. Coast. Life Med.* **2015**, *3*(6), 421–428.
34. Cineli, M. A. *Med. Res. Rev.* **2019**, *39*(4), 1294–1337.
35. Lowe, J.; Li, H.; Downing, K. H.; Nogales, E. *J. Mol. Biol.* **2001**, *313*(5), 1045–1057.
36. Jordan, M. A.; Wilson, L. *Nat. Rev. Cancer* **2004**, *4*(4), 253–265.
37. Brito, D. A.; Yang, Z.; Rieder, C. L. *J. Cell Biol.* **2008**, *182*(4), 623–629.
38. Pommier, Y.; Leo, E.; Zhang, H.; Marchand, C. *Chem. Biol.* **2010**, *17*(5), 421–433.
39. Tacar, O.; Sriamornsak, P.; Dass, C. R. *J. Pharm. Pharmacol.* **2013**, *65*(2), 157–170.
40. Ulukan, H.; Swaan, P. W. *Drugs* **2002**, *62*(2), 2039–2057.
41. Nirmala, M.; Joyce, A.; Samundeeswari, P.; Sankar, D. *Res. Plant Biol.* **2011**, *1*(3), 1–14.
42. Pommier, Y. *Nat. Rev. Cancer* **2006**, *6*(10), 789–802.
43. Pang, B.; Qiao, X.; Janssen, L.; Velds, A.; Groothuis, T.; Kerkhoven, R.; Nieuwland, M.; Ovaa, H.; Rottenberg, S.; van Tellingen, O.; Janssen, J.; Huijgens, P.; Zwart, W.; Neefjes, J. *Nat. Commun.* **2013**, *4,* 1908. https://doi.org/10.1038/ncomms2921.
44. Koba, M.; Konopa, J. *Postep. Hig. Med. Dosw.* **2005**, *59*, 290–298.
45. Smith, A. L.; Nicolaou, K. C. *J. Med. Chem.* **1996**, *39*(11), 2103–2117.
46. Chen, Y.; Yin, M.; Horsman, G. P.; Shen, B. *J. Nat. Prod.* **2011**, *74*(3), 420–424.
47. Walker, S.; Landovitz, R.; Ding, W. D.; Ellestad, G. A.; Kahne, D. *Proc. Natl. Acad. Sci. U. S. A.* **1992**, *89*(10), 4608–4612.
48. Sarker, S. D.; Nahar, L. *Computational Phytochemistry.* Elsevier: United States, **2018**.
49. Viana, J. D.; Felix, M. B.; Maia, M. S.; Serafim, V. L.; Scotti, L.; Scotti, M. T. *Braz. J. Pharm. Sci.* **2018**, *54,* e01010.
50. Cavalcanti, E. B. V. S.; Felix, M. B.; Scotti, L.; Scotti, M. T. *Anticancer Agents Med. Chem.* **2019**, *19*(2), 154–171.
51. Shen, J.; Xu, X.; Cheng, F.; Liu, H.; Luo, X.; Shen, J.; Chen, K.; Zhao, W.; Shen, X.; Jiang, H. *Curr. Med. Chem.* **2003**, *10*, 2327–2342.
52. Alam, S.; Khan, F. *Sci. Rep.* **2018**, *8,* 5524.
53. Gogoi, B.; Gogoi, D.; Silla, Y.; Kakoti, B. B.; Singh, B. *Mol. Biosyst.* **2017**, *13*(2), 406–416.
54. Park, W. S.; Son, E. D.; Nam, G. W.; Kim, S. H.; Noh, M. S.; Lee, B. G.; Jang, I. S.; Jang, I. S.; Kim, S. E.; Lee, J. J.; Lee, C. H. *Planta Med.* **2003**, *69*(5), 459–461.
55. Li, X.; Zhang, X.-X.; Lin, Y.-X.; Xu, X.-M.; Li, L.; Yang, J.-B. *Chem. Biodivers.* **2019**, *16,* e1900170.
56. Floresta, G.; Amata, E.; Barbaraci, C.; Gentile, D.; Turnaturi, R.; Marazzo, A.; Rescifina, A. *Mar. Drugs* **2018**, *16*, 384. https://doi.org/10.3390/md16100384.
57. Dai, S.-X.; Li, W.-X.; Han, F.-F.; Guo, Y.-C.; Zheng, J.-J.; Liu, J.-Q.; Wang, Q.; Gao, Y.-D.; Li, G.-H.; Huang, J.-F. *Sci. Rep.* **2016**, *6,* 25462. https://doi.org/10.1038/srep25462.

58. ClinicalTrials.gov, NIH US National Library of Medicine. https://clinicaltrials.gov/ct2/show/NCT00003882; **2011**, accessed on 12.11.2019.
59. Greene, L. M.; Meegan, M. J.; Zisterer, D. N. *J. Pharmacol. Exp. Ther.* **2015**, *355*, 212–227.
60. Pettit, G. R.; Singh, S. B.; Niven, M. L.; Hamel, E.; Schmidt, J. M. *J. Nat. Prod.* **1987**, *50*(1), 119–131.
61. Hewlings, S. J.; Kalman, D. S. *Foods* **2017**, *6,* 92. https://doi.org/10.3390/foods6100092.
62. ClinicalTrials.gov, NIH US National Library of Medicine. https://clinicaltrials.gov/ct2/show/NCT00094445; **2014**, accessed on 12.11.2019.
63. ClinicalTrials.gov, NIH US National Library of Medicine. https://clinicaltrials.gov/ct2/show/NCT03072992; **2017**, accessed on 12.11.2019.
64. NIH National Cancer Institute. https://www.cancer.gov/about-cancer/treatment/clinical-trials/intervention/curcumin; **2019**, accessed on 12.11.2019.
65. Alchin, D. R. *Dermatol. Ther.* **2014**, *4*, 157–164.
66. Inxight: Drugs, NIH National Center for Advancing Translational Sciences. https://drugs.ncats.io/substance/V86L8P74GI; **2019**, accessed on 12.11.2019.
67. Beer, T. M.; Ryan, C.; Alumkal, J.; Ryan, C. W.; Sun, J.; Eilers, K. M. *Anticancer Drugs* **2010**, *21*(4), 33–438.
68. Galic, V. L.; Wright, J. D.; Lewin, S. N.; Herzog, T. J. *J. Expert Opin. Investig. Drugs* **2011**, *20*(6), 813–821.
69. DrugBank. https://www.drugbank.ca/drugs/DB13809; **2017**, accessed on 13.11.2019.
70. Meng, L. H.; Ding, J. *Acta Pharmacol. Sin.* **2007**, *28*(9), 1460–1465.
71. Pan, L.; Chai, H.; Kinghorn, A. D. *Phytochem Lett.* **2010**, *3*(1), 1–8.
72. Noel, P.; Von Hoff, D. D.; Saluja, A. K.; Velagapudi, M.; Borazanci, E.; Han, H. *Trends Pharmacol. Sci.* **2019**, *40*(5), 327–341.
73. ClinicalTrials.gov, NIH US National Library of Medicine. https://clinicaltrials.gov/ct2/show/NCT03117920; **2019**, accessed on 13.11.2019.
74. Shah, U.; Shah, R.; Acharya, S.; Acharya, N. *Chin. J. Nat. Med.* **2013**, *11*(1), 16–23.
75. Perreira, R. B.; Evdokimov, N. M.; Lefranc, F.; Valentao, P.; Kornienko, A.; Perreira, D. M.; Andrade, P. B.; Gomes, N. G. M. *Mar. Drugs* **2019**, *17*(6), 329. https://doi.org/10.3390/md17060329.
76. McCourt, C. K.; Deng, W.; Dizon, D. S.; Lankes, H. A.; Birrer, M. J.; Lomme, M. M.; Powell, M. A.; Kendrick, J. E.; Saltzman, J. N.; Warshall, D.; Tenney, M. E.; Kushner, D. M.; Aghajanian, C. *Gynecol. Oncol.* **2017**, *144*(1), 101–106.
77. Pham, J. V.; Yilma, M. A.; Feliz, A.; Majid, M. T.; Maffetone, N.; Walker, J. R.; Kim, E.; Cho, H. J.; Reynolds, J. M.; Song, M. C.; Park, S. R.; Yoon, Y. J. *Front. Microbiol.* **2019**, *10,* 1404.

CHAPTER THREE

Antimicrobial natural products

Mukhlesur Rahman[a,*], Satyajit D. Sarker[b]
[a]Medicines Research Group (MRG), School of Health, Sport and Bioscience, University of East London, London, United Kingdom
[b]Centre for Natural Products Discovery (CNPD), School of Pharmacy and Biomolecular Sciences, Liverpool John Moores University, Liverpool, United Kingdom
*Corresponding author: e-mail address: m.rahman@uel.ac.uk

Contents

1. Introduction

Infectious diseases are one of the major problems in today's healthcare system. Antibiotics have been used since the second world war for the treatment of various types of infections. The term antibiotic is originated from "antibiosis" that simply describes the interaction of two or more organisms having at least one being detrimental to other(s). Therefore, antibiosis is the process, where one organism in the presence of other organism(s) is capable of producing harmful effects to the later organism(s). "Antibiotics" literally describes a class of chemical compounds that are responsible for the treatment of infections by inhibiting the growth of bacteria or killing bacteria with minimum or no harm to the host. However, a more broad term "antimicrobial agents" is used to cover the treatment of infections caused

Annual Reports in Medicinal Chemistry, Volume 55
ISSN 0065-7743
https://doi.org/10.1016/bs.armc.2020.06.001

by various organisms including bacteria, fungi, protozoa and virus. Moreover, another term "anti-infective agent" has also been adopted worldwide in the research of antimicrobial drug discovery.

Even though the fortuitous discovery of penicillin in 1928 started the golden era of antibiotics, the empirical knowledge of treating wounds with the indigenous herbs and/or alternative approaches dates back to the ancient times. Without having any scientific knowledge, wounds were treated with spider's webs, cheese mold or moldy bread ingestion,[1] moldy soybean curd and honey.[2] The ancient Egyptians used moldy bread for the treatment of infected wounds, while the ancient Chinese used to apply moldy soybean curd to heal boils and managed foot infections by wearing sandals furried with mold.[2] In the middle ages, honey was used for the treatment of post-arrow wounds.[3] During this prehistorical era of antibiotics, such treatments were given without any knowledge of bacteria or other organisms that caused the infections.

The search for an effective agent to win in the battle against the bacteria or other organisms causing infections started in late 19th century. During an experiment with *Anthrax bacilli* in 1877, Pasteur and Joubert noticed that *Anthrax bacilli* were killed, while contaminated with other bacteria. Another experiment in 1901 revealed that a liquid culture of *Pseudomonas aeruginosa* injected to infected rabbits with anthrax recovered them from anthrax. Such experimental findings led to the conclusion that the metabolite(s) produced by one organism revealed their capabilities to inhibit the growth of other organisms, which supported the concept of antibiosis.[2] Later in 1910, a Nobel laureate Paul Ehrlich discovered an arsenic containing compound known as salvarsan or arsphenamine (**1**) that became the choice of medicine for the treatment of infections including syphilis and trypanosomiasis[3] until it was replaced by first antibiotic penicillin in 1945.

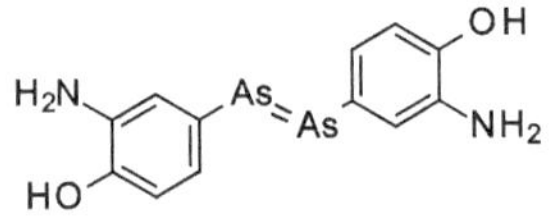

1

Alexander Fleming, a Scottish Physician and Microbiologist, serendipitously discovered penicillin G (**2**) in 1928. He was working on *Staphylococci* and stacked some petri-dishes of *Staphylococci* on a bench in a corner of his laboratory just before he went for holidays in August in 1928. Upon his return from holidays, he noticed that a fungus had contaminated a culture plate of *Staphylococcus* bacterium he had accidentally left uncovered. He also noticed that the fungus had inhibited the growth of *Staphylococcus* bacterium. Fleming grew the mold in a pure culture and noticed that it produced a metabolite that responded dramatically to treat a number of bacterial infections. He identified the mold as *Penicillium notatum* and named the metabolite as "penicillin," which became the choice of drug to fight bacterial infections during that period. Alexander Fleming shared the 1945 Nobel Prize in Medicine with Howard Florey and Ernst Chain for their contribution in mass production of the first mass-produced antibiotic. Considering the number of lives saved during the end of World War II, penicillin was nicknamed "the wonder drug."[1–3]

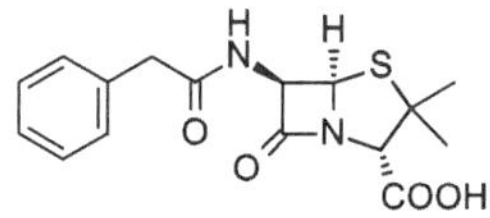

2

During the World War II, two more drugs proflavine (**3**) and prontosil (**4**) were introduced. Proflavine introduced in 1934 was effective in treating infections with deep surface wounds, however it was highly toxic. The discovery of prontosil, a sulfur-containing prodrug, in 1935 was a real breakthrough for the treatment of systematic infections until the availability of penicillin in early 1940s.[3]

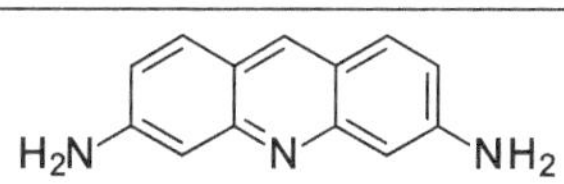

3

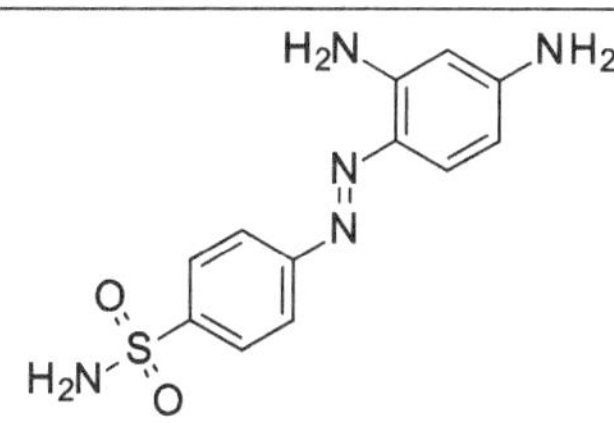

4

2. Antimicrobial resistance

Since the discovery of penicillin, there have been hundreds of antibiotics discovered that are in clinical use to treat infections. Considering the availability of hundreds of antibiotics for clinical use, one may raise question why we need new antibiotics. The answer is simple; the pathological organisms (bacteria, fungi, etc) have developed resistance to various antibiotics that have been designed to kill them (bugs) through a number of mechanisms (Section 2.2). For example, in 1962, *Staphylococcus aureus* developed resistance against methicillin, which was discovered in 1960. Similarly, because of developed resistance, penicillin became ineffective against *Streptococcus pneumoniae*, and vancomycin against *Enterococcus faecium*, 16 years after its discovery (Table 1).[4]

Antibiotic resistance (ABR) or more widely used term antimicrobial resistance (AMR) has become an increasing threat to today's healthcare system because of the unavailability of new and safe antibiotics to respond to the demand of antibiotics to treat increased number of life threatening infections. AMR infections currently counts more than 67,000 infections in Europe with approximately 33,000 death[5] costing around €1.1 billion, while there are more than 2.8 millions infections and more than 35,000 AMR related death certificates currently issued in the United State.[6]

Table 1 Antibiotics timeline with dates of discovery and development of resistance.[4]

Antibiotics	Year discovered	Resistant organisms	Year developed
Penicillin	1941	Penicillin-resistant *Staphylococcus aureus*	1942
		Penicillin-resistant *Streptococcus pneumoniae*	1967
		Penicillinase-producing *Neisseria gonorrhoeae*	1976
Tetracycline	1950	Tetracycline resistant *Shigella*	1959
Amphotericin B	1959	Amphotericin B-resistant *Candida auris*	2016
Methicillin	1960	Methicillin-resistant *Staphylococcus aureus*	1962
Vancomycin	1972	Plasmid-mediated vancomycin-resistant *Enterococcus faecium*	1988
		Vancomycin-resistant *Staphylococcus aureus*	2002
Azithromycin	1980	Azithromycin-resistant *Neisseria gonorrhoeae*	2011
Ciprofloxacin	1987	Ciprofloxacin-resistant *Neisseria gonorrhoeae*	2007
Fluconazole	1990	Fluconazole-resistant *Candida*	1988
Daptomycin	2003	Daptomycin-resistant methicillin-resistant *Staphylococcus aureus*	2004
Ceftazidime-avibactam	2015	Ceftazidime-avibactam-resistant KPC-producing *Klebsiella pneumoniae*	2015

Methicillin-resistant *Staphylococcus aureus* (MRSA) is a big concern as this is a major cause of both healthcare- and community-associated infections around the world. It has been estimated there are more than 150,000 patients due to MRSA infections in the Europe, which costs approximately €380 million for EU healthcare systems.[7] Pan-European surveillance data on bloodstream infections showed that more than 10% *Staphylococcus aureus* infections in 15 European countries are due to MRSA with some of these countries seeing such resistance closer to 50%.[8] If no action is taken on time, it has been estimated that AMR will be the leading healthcare problem counting in excess of 10 million deaths per year by 2050 costing the world an extra $100 trillion.[9]

2.1 Antimicrobial resistance: Its causes and ways it spreads

Antimicrobial resistance has become a catastrophic problem for public health in the recent years. Widespread uses of antibiotics in medicine, veterinary, agriculture and poultry have contributed a lot toward the development of bacterial resistance. The various causes beyond the development of resistance to the antibiotics are briefly summarized below.

2.1.1 Genetic modification of organisms

Antimicrobial resistance can happen naturally over time, usually through genetic changes in the bugs. Within our body, there are lots germs-some considered as good bacteria (protect the body from infections), other as bad bacteria (cause illness through infections) and only few as drug resistant bacteria. When antibiotics are taken, they usually kill both bad and good bacteria but cannot do any harm to the drug resistant bacteria which in turn are allowed to take over and grow. Some drug resistant bacteria give their resistance gene to other bacteria making the latter group of bugs resistant and causing the problem worse. Thus, antibacterial resistance occurs naturally through genetic modification.[10]

2.1.2 Over prescription of antibiotics

Antibiotics have been misused and/or overused in humans and animals, which has accelerated dramatically the problem of antimicrobial resistance. Sometimes antibiotics are given for viral infections like cold and flu which do not require antibiotics at all. Again, antibiotics are unreasonably given to animals for growth promotion and/or prevention of diseases in healthy animals. Antibiotics are also used unreasonably in agriculture and for the preservation of poultry products. Such misuses and/or overuses of antibiotics cause AMR.

2.1.3 Gratuitous prescription of broad spectrum antibiotics

Sometimes broad spectrum antibiotics are prescribed for the conditions, which could be treated with a narrow spectrum antibiotics. Such unjustified prescriptions also contributes toward the development of antimicrobial resistance.

2.1.4 Over-the-counter access to antibiotics and self-medication

Unrestricted access to antibiotics and self-medication have made AMR situation even worse in developing countries. Patients in some developing countries can buy antibiotics without any prescription which enable them to get hold of antibiotics whenever they wish to do so. Thus, they take antibiotics without consulting their physicians for conditions, where they should not require antibiotics at all. Sometimes they start taking antibiotics but stop before completing the course. Such irrational uses of antibiotics are an important route of developing AMR in some developing countries. However, AMR awareness activities have recently motivated the policy makers in developing countries to think carefully to take initiative to stop antibiotics without prescriptions.

2.1.5 Antibiotics trafficking

Because of easily availability, some people buy antibiotics when they travel to developing countries and carry antibiotics with them when they come back. Upon return they might consider taking antibiotics for some conditions even through the physicians have not prescribed antibiotics. Such ignorance hastens the problems of antimicrobial resistance.

2.1.6 Antimicrobial manufacturing contaminations

During the manufacturing antibiotics (both raw materials and finished products) in the Pharmaceutical Industries, antibiotics manufacturing wastes and effluents contaminate the environments (air, soil and water) and become a major concern for antimicrobial resistance as well.

Now the next concern regarding AMR is how it spreads. There are two ways of spreading AMR- (i) human to human and (ii) animals (poultry) to human. If a person takes antibiotics for the treatment of infection, but unfortunately develops resistant bugs, then the person may stay at home or gets care in hospital, care home or other inpatient facilities. If s/he stays at home, then AMR may spreads to family members and/or friends and ultimately to the community. At hospital or inpatient facilities, the infected person spreads AMR to doctors, nurses and other health professionals and thereby to the community.

When animals (common poultries) are given antibiotics but developed resistant bacteria in their guts, they spread these resistant gene to people when the meats from animals with drug-resistant bacteria are not handled properly and/or cooked properly. Sometimes, fertilizers or water containing animal feces and drug-resistant bacteria are used to grow vegetables and crops and thereby, such drug-resistant bacteria can be transferred to vegetables and crops. Ultimately, such drug-resistant bacteria spreads to humans when these vegetables and crops are consumed.

2.2 Mechanisms of antimicrobial resistance

Although several classes of antibiotics are in clinical uses and act in different modes of action, however the organisms have developed resistance to existing antibiotics using one or more of the following mechanisms.

2.2.1 Enzymatic inactivation of the antibiotics

This is the most common mechanism of developing resistance to several antibiotics such as penicillins, cephalosporins, chloramphenicol and aminoglycosides. This is simply happens when bacteria (both Gram-positive and Gram-negative) are capable of producing enzymes that inactivate or destroy the antibiotics. For example, β-lactamases enzymes produced by *Staphylococci* and also, sometime Gram-negative bacteria, cleave the β-lactam ring of penicillins and cephalosporins.[11,12] Some of these β-lactamases encoded by transposons may transfer resistance genes to other antibiotics as well. Another enzyme, chloramphenicol acetyltransferase, produced by both Gram-positive and Gram-negative bacteria inactivate chloramphenicol.[11] Similarly, kinases and other enzymes produced by both Gram-positive and Gram-negative bacteria can inactivate aminoglycosides by enzymatic reactions like phosphorylation, adenylation or acetylation.[11,13]

2.2.2 Alteration of antibiotic target sites

A large number of target sites alterations mechanisms are also found in clinical isolates. Chromosomal mutations can alter 30S subunit of ribosome which is the binding site protein for aminoglycoside and thereby, inactivate the antibiotic. Similarly, erythromycin may be inactivated through a plasmid-led changes on the 50S subunit of ribosome. Rifampicin resistance was reported to be related to the DNA-dependent RNA polymerase alteration.[11,13] Besides the enzymatic cleavages, resistance to β-lactam antibiotics is also conferred by target modification through a mutated chromosomal gene as seen in MRSA strains where its exogenous penicillin binding protein (PBP; its transpeptidase domain) is insensitive to the action of several different β-lactams.[11,13]

2.2.3 Decreased intracellular accumulation and/or efflux of antibiotics

This is considered as second common modes of developing antibiotic resistance in clinical strains. Decreased permeability is prominent in Gram-negative bacteria because of the presence of the additional outer membrane favoring a permeability barrier and offering an essential mechanism for protection against hydrophilic antibiotics like vancomycin.[14] Plasmid-determined inhibition of the porin genes and/or changes in their expression have been evident to further impact the susceptibility of Gram-negative bacteria to hydrophilic antibiotics.[15] Both Gram-positive and Gram-negative bacteria have also been reported to exhibit different types of active efflux pumps belonging to one of the five families: ABC, MFS, RND (Resistance-Nodulation-Division), MATE (Multidrug and Toxin Extrusion), and SMR (Small Multidrug Resistance).[16] The emergence of resistance to fluoroquinolone in *Streptococcus pneumoniae* has been evident by active efflux pumps.[17]

2.2.4 Modification of metabolic pathways or bypassing antibiotic inhibiting reactions

Plasmid mediated sulfonamide resistance in many bacteria appears due to the production of dihydropteroate synthetase with low affinity for antibiotic but no alteration in the affinity for PABA.[11] So sulfonamide resistance bacteria do not use PABA to synthesize nucleic acid and folic acid.[18]

3. Antimicrobial natural products

Since the dawn of human civilization, human beings have depended on natural resources, mostly plants and molds, for the treatment of various types of diseases including infections. Although the majority of the today's antibiotics are derived from microbial sources or their semisynthetic analogues, scientists around the globe have continued to carry out systematic research on microbes as well as on medicinal plants, marine organisms and animals for the discovery of new antibiotics. This section highlight key antimicrobial compounds isolated from natural sources covering microbes, medicinal plants, marine organisms and animals.

3.1 Antimicrobial natural products from microorganisms

Penicillin G (benzylpenicillin, **2**) and its semisynthetic analogue penicillin V (phenoxymethylpenicillin, **5**) belong to the class of β-lactam antibiotics that act by inhibiting the synthesis of the bacterial cell wall peptidoglycan. These are active against a wide range of organisms and considered as drugs of first choice for various infections. However, the poor absorption in the gastrointestinal tract and susceptibility to the enzyme β-lactamase are their main shortcomings.

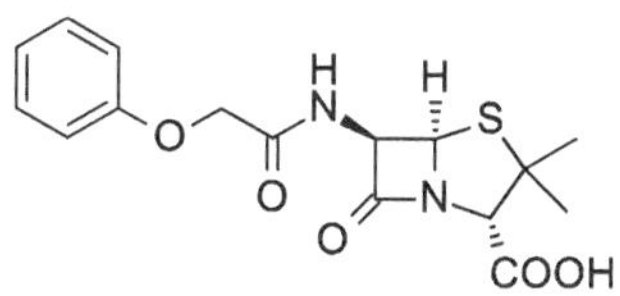

5

A number of further semi-synthetic penicillins such as methicillin, flucloxacillin, ampicillin and amoxycillin were developed by simple modification of side chains attached to the penicillin nucleus (R in Fig. 1). Among these, methicillin and flucloxacillin are β-lactamase resistant penicillins, while ampicillin and amoxycillin provide broad spectrum activity.[19]

Cephalosporins also belonging to β-lactam antibiotics were originated from the fungus *Cephalosporium*. Discovered in 1945 by an Italian pharmacologist Giuseppe Brotzu, cephalosporins are broad spectrum antibiotics used for the treatment of a number of infections including septicemia, pneumonia, meningitis, biliary-tract infections, peritonitis, and urinary-tract infections. Cephamycins, structurally similar to cephalosporins, are also classified as β-lactam antibiotics having similar mechanism of action as penicillins which were isolated from the microorganism *Streptomyces*, the largest genus of Actinobacteria. Various semisynthetic cephalosporins were

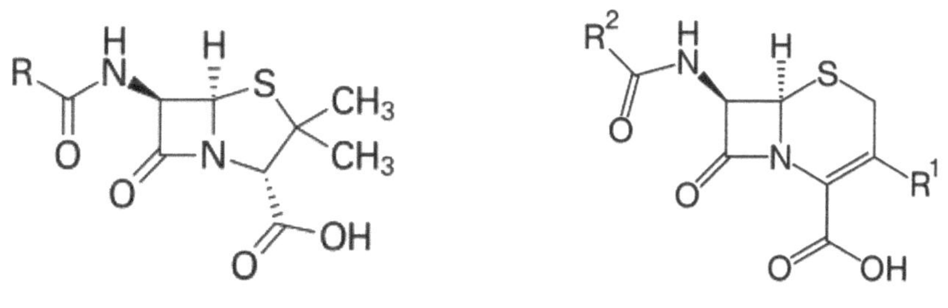

6 **7**

Fig. 1 General structures of penicillins (**6**) and cephalosporins (**7**).

developed by modifying the two side chains (R_1 and R_2 in Fig. 1) of cephalosporin nucleus.[19,20]

Vancomycin, a glycopeptide antibiotic with seven peptides present in the molecule, was discovered a Pharmaceutical company Eli Lilly in 1952 from a soil bacterium, *Streptomyces orientalis*.[21] It works by inhibiting the bacterial cell wall synthesis in the last stage through binding to the terminal D-alanyl-D-alanine of peptidoglycan precursors. This is used for the treatment of MRSA and some other bacteria resistant to β-lactam antibiotics. Its clinical uses also include the treatment of infections related to *Pseudomonas colitis*.[21]

Tetracyclines (**8**) discovered in the 1940s are a class of broad-spectrum antibiotics, which following entry into the organisms act by interfering the bacterial protein synthesis. They consist of four six-membered rings (designated as A, B, C and D) fused linearly to form tetracycline (four cyclic) nucleus to which various functional groups and/substituents are attached. As they contain a number of keto groups in the molecules they are also considered a class of compounds known as polyketides. Members belonging to this antibiotic class include tetracycline, oxytetracycline, chlortetracycline, demeclocycline, minocycline, doxycycline, methacycline and lymecycline. Among these, chlortetracycline and oxytetracycline are the first two members of natural tetracyclines both discovered in the 1940s from the filamentous bacteria *Streptomyces aureofaciens* and *S. rimosus*, respectively. Other naturally occurring antibiotics in the class such as tetracycline were isolated from *S. aureofaciens*, *S. rimosus*, and *S. viridofaciens* and demethylchlortetracycline from *S. aureofaciens*.[22]

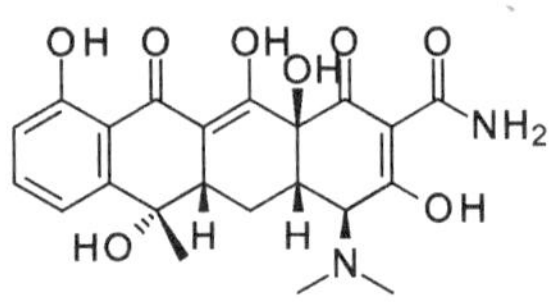

8

Semi-synthetic tetracyclines such as doxycycline, minocycline, methacycline and lymecycline are produced from natural tetracyclines by simple modifications on the functional groups and/or substituents. Tetracyclines are used for the treatment of a wide range of microorganisms including both Gram-positive and Gram-negative bacteria, *Chlamydia*, *Rickettsia*, *Mycoplasma* and protozoa.[22] Because of their capabilities of reacting with chelating ions such as calcium, magnesium, aluminum and iron forming nonabsorbable complexes, their absorption is reduced significantly in the presence of milk, iron preparations and antacids. Accordingly, patients are advised not to take oral tetracyclines with milk, antacids and iron preparations to avoid the consequence of forming non-absorbable complexes.

Chloramphenicol (**9**), an amphenicol, was isolated from a soil-dwelling bacterium, *Streptomyces venezuelae* in 1947 and its chemical structure was confirmed in 1949.[23] Following its discovery, it was manufactured in large scale through synthetic route and is considered as the first synthetic antibiotic as well. It acts by interfering the protein synthesis of bacteria by binding to the 50S ribosomes. It has a wide range of antimicrobial activity including Gram-positive and Gram-negative bacteria as well as rickettsia. Chloramphenicol ointment is widely used for the treatment of eye infections.[24] Together with antibiotics such as amphotericin B (**10**), griseofulvin (**11**) and streptomycin (**12**), chloramphenicol (**9**) is in the World Health Organisation's List of Essential Medicines, the safest and most effective medicines needed in a health system.[25] However, its most common side effect is idiosyncratic depression of bone marrow. It also causes grey baby syndrome in young children so it should administered to new-borns with great care by regular monitoring its plasma level.[24]

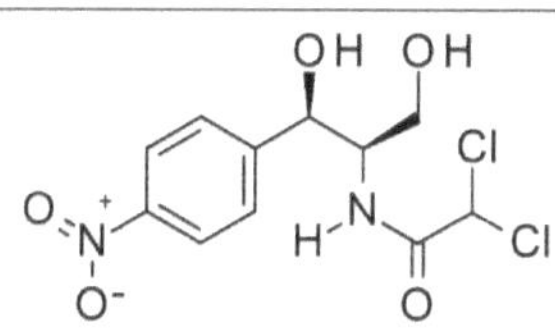

9

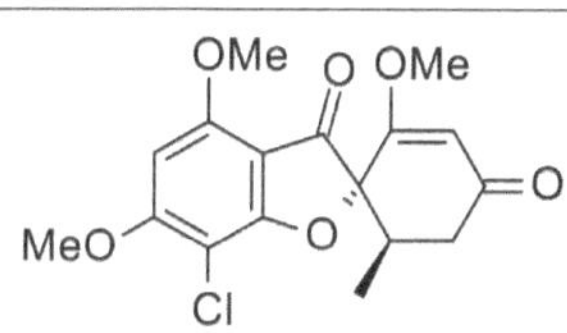

11

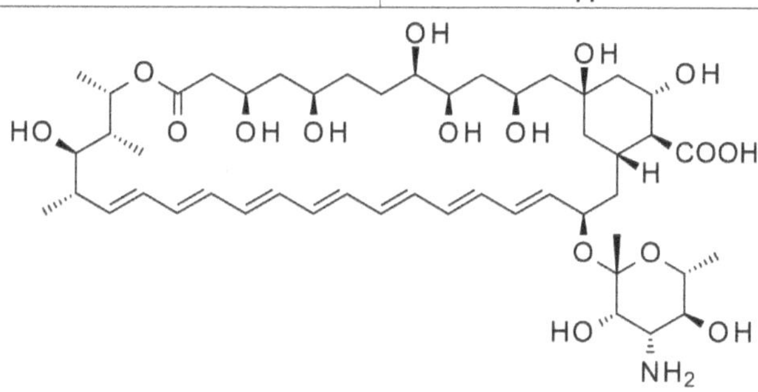

10

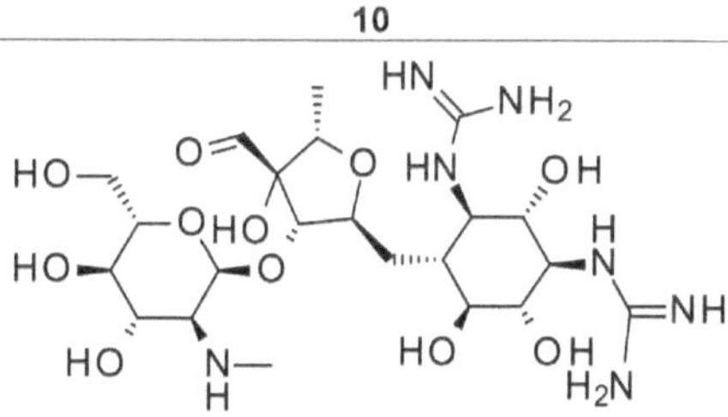

12

Amphotericin B (**10**) is an antifungal agent derived from the culture of *Streptomyces nodosus* in 1955.[26] This is a polyene macrolide that acts on the fungal cell membrane and binds with its cell membrane component ergosterol forming large pores leading to the gross disturbance in monovalent ion balance with ultimate leakage of intracellular K^+, Na^+, H^+ and Cl^- resulting in fungal cell death. This is a preferred antibiotic for the treatment for disseminated infections caused several fungi and yeast including *Aspergillus* and *Candida*.[27]

Griseofulvin (**11**), discovered in 1939 from a culture of *Penicillium griseofulvum*,[28] is a narrow spectrum antifungal antibiotic that works by binding to fungal microtubules and thus inferring fungal mitosis. It is administered orally for the treatment of dermatophytosis including the fungal infections of skins, scalp and nails when the local/topical antifungal agents become ineffective.

Streptomycin (**12**) belongs to the aminoglycoside class of antibiotics, and was discovered in 1943 from a culture of soil actinomycete, *Streptomyces griseus*.[29] Kanamycin (also known as kanamycin A) is another aminoglycoside antibiotic which was isolated in 1957 from the soil actinomycete, *Streptomyces kanamyceticus*.[29] Chemically, these aminoglycosides contain aminated carbohydrate rings connected to dibasic cyclitol through glycosidic linkage. Upon uptake to susceptible organisms, these antibiotics bind to the bacterial 30S ribosomal subunit and thereby disrupt the initiation and elongation steps in protein synthesis. As broad spectrum antibiotics, these are used for the treatment of several life threatening infections including tuberculosis, sepsis, endocarditis, brucellosis, and severe urinary tract infections.[30]

Echinocandin B (**13**) is an antifungal agent which is composed of a ring of six amino acids connected to a long-chain lipophilic side chain. It was discovered in 1974 from a culture of *Aspergillus nidulans*.[28,31] It acts by inhibiting the synthesis of 1,3-β-glucan, an essential component of fungal cell wall structure. It is fungicidal against some yeasts, mostly *Candida* species. Bacitracin is another similar polypeptide antibiotic composed of 11 amino acids (seven as part of ring and four in the side chain) was discovered in 1945 from a culture of the licheniformis group of Gram-positive bacterium *Bacillus subtilis*.[32] It is a narrow-spectrum antibiotic used for the treatment of infections caused by Gram-positive bacteria, especially those that cause skin infections specially those caused by small cuts, scrapes, or burns. It stops the growth of certain bacteria by inhibiting the bacterial cell and peptidoglycan synthesis.[32]

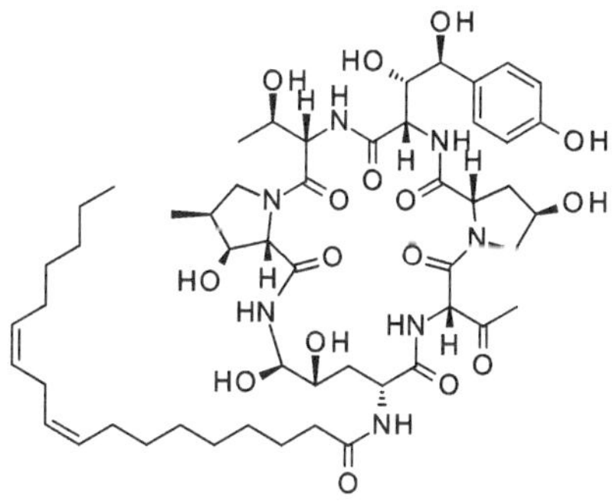

13

Polymyxins are cationic polypeptides composed of a cyclic heptapeptide plus a tripeptide side chain acylated at the *N* terminus by a fatty acid tail. Polymyxin B was first isolated in 1947 from a culture of a Gram-positive bacterium, *Bacillus polymyxa.*[33] Colistin (also known as polymyxin E) is a polypeptide antibiotic that was originally isolated in 1947 from the soil bacterium *Paenibacillus polymyxa* subsp. *colistinus.*[34] Colistin and polymyxin B are used for the treatment of infections caused by Gram-negative bacteria and act by breaking down the cytoplasmic membrane causing the ultimate death of bacterial cell.[33,34]

Neomycins (*e.g.*, neomycin B, **14** and neomycin C, **15**) are amino glycoside antibiotics comprising amino sugars connected through glycosidic linkages. The first member of this class was discovered in 1949 by the microbiologists Waksman and Lechevalier from a culture of the bacterium, *Streptomyces fradiae.*[35] This antibiotic is available in a number of topical preparations such as creams, ointments, and eyedrops. It acts by inhibiting the protein synthesis through binding with 30S ribosome causing genetic disruption. It also act by interfering the bacterial enzyme DNA polymerase.[35]

Nystatin (**16**), structurally related to amphotericin B, is an antifungal agent, which was isolated from the culture of actinomycetes, *Streptomyces noursei*, found in the soil of a dairy farm in USA.[36] Like other antifungal agent such as amphotericin B, it acts by binding with fungal cell membrane component, ergosterol, forming large pores in the membrane leading the leakage of K+ and ultimate death of fungal cell. It is given for the treatment of fungal infections caused mostly by Candia including oesophageal candidiasis, thrush and vaginal infections.[37]

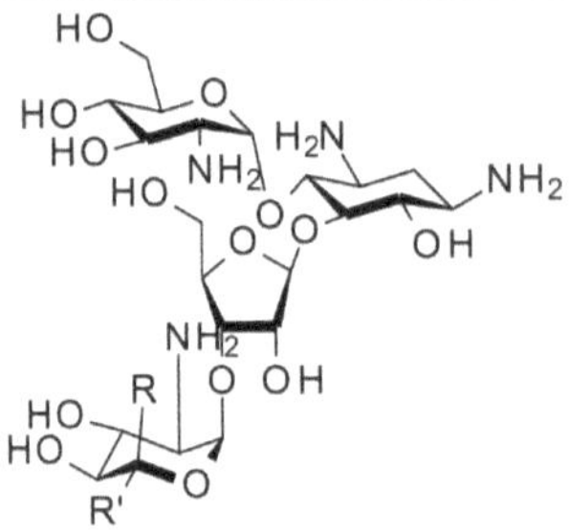

14 R = CH_2NH_2; R' = H
15 R = H; R' = CH_2NH_2

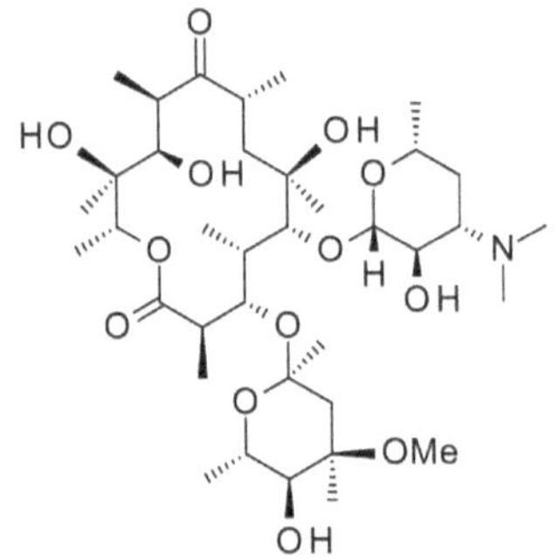

17

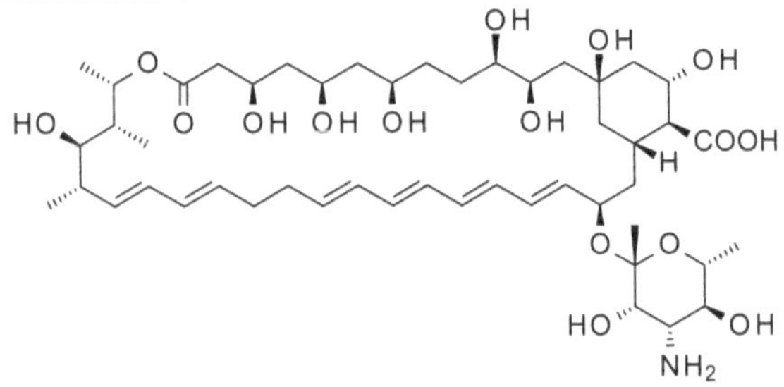

16

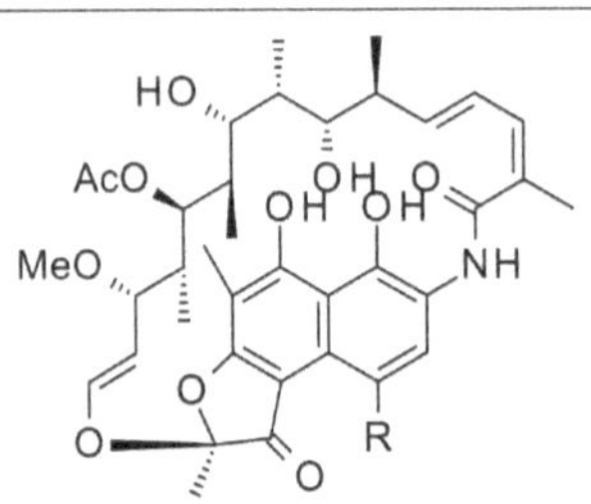

18 R = H
19 R = CH_2COOH

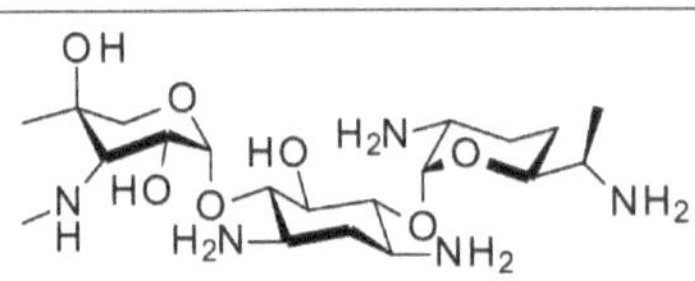

20

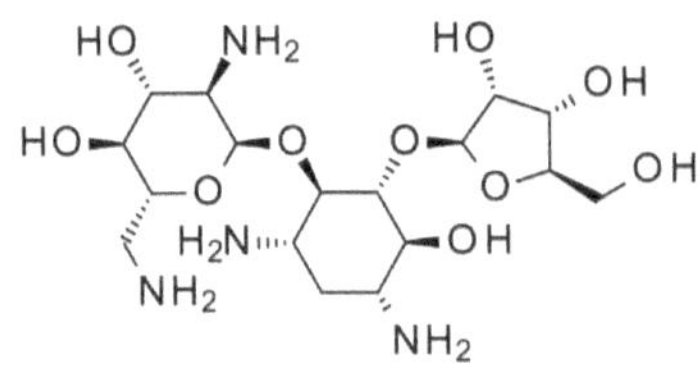

21

Erythromycin (**17**), a macrolide antibiotic, was discovered in 1952 from the culture of bacterium *Saccharopolyspora erythraea*.[38] It offers bacteriostatic activity by inhibiting the protein synthesis through binding with bacterial 50S subunit of rRNA, and is widely used for the treatment of chest infections (pneumonia), skin diseases (acne) and sexually transmitted diseases.[38]

Rifamycins (A–E), a group of structurally related secondary microbial metabolites, were discovered in 1957 from a product of fermentation from the Gram-positive bacterium *Amycolatopsis mediterranei* (also known as *Streptomyces mediterranei*).[39] Among these metabolites, rifamycin B (**18**) was isolated in pure form but with poor activity, but could easily be oxidized to more active form, rifamycin S, which was further modified to produce clinically relevant rifamycin SV (**19**), the first antibiotic used intravenously for the treatment of tuberculosis. Because of its high affinity to the prokaryotic enzyme, RNA polymerase, rifamycins act by inhibiting the bacterial DNA-dependent RNA synthesis.[40]

Like neomycins, gentamycin (**20**) and ribostamycin (**21**) are two other popular aminoglycoside antibiotics. Gentamicin (**20**) was identified in 1962 from the fermentation broth of a bacterium, *Micromonospora purpurea*.[41] This bactericidal antibiotic acts by binding the 30S subunit of the bacterial ribosome and thereby, inhibiting bacterial protein synthesis. It is active against a wide range of bacterial infections including urinary tract infections, respiratory tract infections, blood, bone and soft tissue infections.[41] On the other hand, ribostamycin (**21**) was first isolated in Japan in 1970 from the fermentation broth of a soil bacterium, *Streptomyces ribosidificus*.[42] Like other aminoglycosides, it also acts by inhibiting protein synthesis through binding with 30S subunit of bacterial ribosome.[43]

Mupirocin (**22**) was initially isolated in 1971 from the culture of a rod-shaped Gram-negative bacterium, *Pseudomonas fluorescens*.[44] It works by reversibly binding to the isoleucyl t-RNA synthetase in *Staphylococcus aureus* and *Streptococcus* and thereby inhibiting the protein synthesis. This topical antibiotic is used for the treatment of skin infections such as impetigo and folliculitis as well as for MRSA.[44]

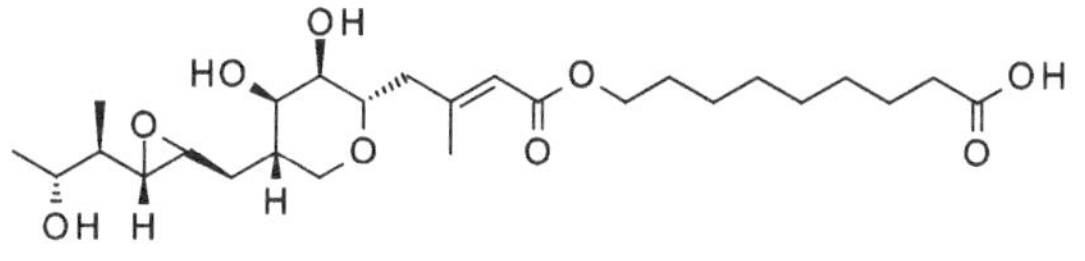

22

Among the recently introduced antibiotics, daptomycin, a lipopeptide antibiotic, was isolated from the soil bacterium, *Streptomyces roseosporus*.[45] It was approved in 2003 by the FDA for the treatment of complicated skin and soft tissue infections. It acts by disrupting multiple aspects of bacterial cell membrane function. Upon ingestion, it attacks the cell membrane (phosphatidylglycerol) forming holes that leak ions like K^+ and ultimate death of bacterial cell.[45] Platensimycin (**23**) and platencin are relatively recent novel antibiotics which were isolated from a bacterium *Streptomyces platensis* by using an antisense whole-cell differential sensitivity assay, where control organisms were compared to cells expressing fabF antisense RNA.[46] Both compounds a 3-amino-2,4-dihydroxy-benzoic acid and a different unusual diterpene (tetracyclic enone acid in platensimycin, while a tricyclic enone acid in platencin). They are potent and non-toxic natural products with potent activity against Gram-positive pathogens, including antibiotic-resistant strains and *Mycobacterium tuberculosis*.[47] Because of their unique structural features and promising antibacterial activity, they have been a breakthrough in the searches for novel antibiotics. Platensimycin was first isolated by the Merck group.

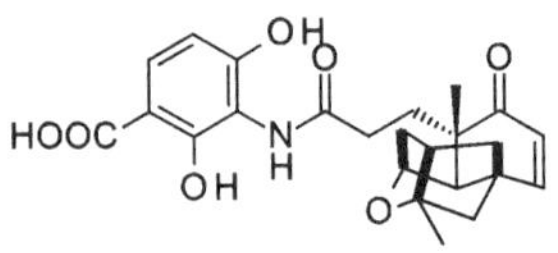

23

Lactic acid bacteria are good gut bacteria that produce secondary metabolites with potential antimicrobial activity. For example, nisin, which is a polycyclic antibacterial peptide composed of 34 amino acids in the molecule, was produced from the culture of a Gram-positive bacterium *Lactococcus lactis* that has been has been extensively used in the production of buttermilk and cheese. It is used as a food preservative because of its bactericidal activities against Gram-positive as well as spore forming food-borne bacteria including *S. aureus* and *Listeria monocytogenes*.[48] In the research laboratories, it is very useful as a selective agent in microbiological media for the isolation of gram-negative bacteria, yeast, and molds. Reuterin (3-hydroxypropionaldehyde; **24**),[49] reutericin 6[50] and reutericyclin[51] are produced by *Lactobacillus reuteri* and have a broad spectrum activity against food-borne pathogens and spoilage organisms. Reuterin (**24**), a simple aldehyde, in combination with nisin offer synergistic antimicrobial activity and reduce significantly the growth of Gram-positive *Staphylococcus aureus*, *Lactobacillus monocytogenes*, and Gram-negative *E. coli* and *S. typhimurium*. Reuterin inhibits the growth of some harmful bacteria (both Gram-positive and Gram-negative) as well as some fungi, yeasts and protozoa. *L. reuteri* secreting sufficient amounts of reuterin to achieve the desired antimicrobial effects is capable of eliminating gut invaders without harming other gut microbiota.[49]

24

The above examples show the history and progression of development of antibiotics from microbial sources throughout the past several decades, but in the recent past, there are hardly any new antibiotics reported from microbial sources, which have become commercially available for the treatment of infections. However, this does not mean the search for new antibiotics or antimicrobial agents has stopped. In fact, there are several papers published in recent years describing various antimicrobial compounds from microbial sources with different levels and spectrum of activities. A cultural broth of *Pseudomonas* sp. Ki19. was reported to produce four dialkyl resorcinols, 2-butyl-5-propylresorcinol (**25**), 2-hexyl-5-methylresorcinol (**26**), 2-hexyl-5-propylresorcinol (**27**), and 2-hexyl-5-pentylresorcinol (**28**) with antimicrobial activity against *S. aureus* (MIC 10 μg/mL), *Aspergillus fumigatus* (MIC = 50 μg/mL) and *Fusarium culmorum* (MIC = 50 μg/mL).[52]

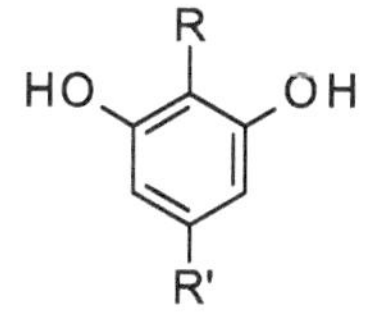

25 R = Butyl; R' = Propyl
26 R = Hexyl; R' = Methyl
27 R = Hexyl; R' = Propyl
28 R = Hexyl; R' = Pentyl

A fungus belonging to the genus *Phoma* produced three acetylenic acids, phomallenic acids A–C, which exhibited potent antibacterial activity against wild-type *S. aureus* with MICs in the range 3.9–7.8 μg/mL, with phomallenic acid C being the most active one, superior to commonly used FabF inhibitors such as cerulenin and thiolactomycin.[53] Ten-membered macrolides, phomolides A (**29**) and B (**30**), were isolated from the cultural broth of *Phomopsis* sp. hzla01-1 which revealed significant antimicrobial activities against *Escherichia coli* CMCC44103, *Candida albicans* AS2.538 and *Saccharomyces cerevisiae* ATCC9763 (MICs 5–10 μg/mL).[54]

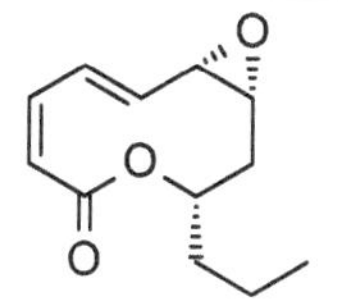

29

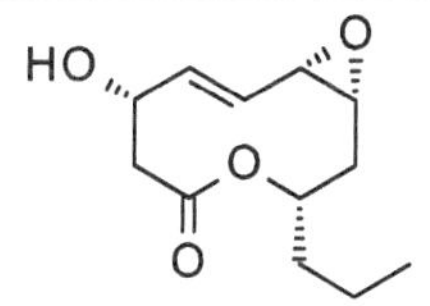

30

A culture broth of *Lysobacter* sp. produced antimicrobial polypeptides (AMPs), tripropeptins A-E and Z (MICs 0.39–12.5 μg/mL). Among these polypeptides, tripropeptins C and D displayed strong activities against Gram-positive bacteria, including both MRSA and vancomycin-resistant *Enterococcus*.[55] The culture broth of a cyanobacterium belonging to the genus *Hassallia* yielded broad-spectrum antifungal glycosylated lipopeptide hassallidins A and B (MICs of 8–16 μg/mL against 10 species of *Candida*).[56,57] The isolation of an unusual depsipeptide, teixobactin (**31**), from a soil bacterium is the most recent breakthrough in the search for antimicrobial drugs because of its intrinsic antibacterial activity, structural novelty and method of identification involving the culture production in natural soil environment. The hypothesis for thus drug discovery was based on that fact that the uncultured bacteria making up approximately 99% of all species in external environments could be an untapped source of new antibiotics. Accordingly, teixobactin (**31**) has been discovered through a screen of uncultured bacteria using a new device, i-Chip sealed with semi-permeable membranes.[58] It showed potent *in vitro* antimicrobial activity against Gram-positive bacterial strains including *Staphylococcus aureus*, *Bacillus anthracis*, *Enterococci* species and *Clostridium difficile* as well as *Mycobacterium tuberculosis*. It was also found effective *in vivo* against methicillin-resistant *Staphylococcus aureus* and *Streptomyces pneumoniae* in mice model. This compound was demonstrated to act by inhibiting the bacterial cell wall synthesis by binding at both lipid II (precursor of peptidoglycan) and lipid III (precursor of cell wall teichoic acid).[58]

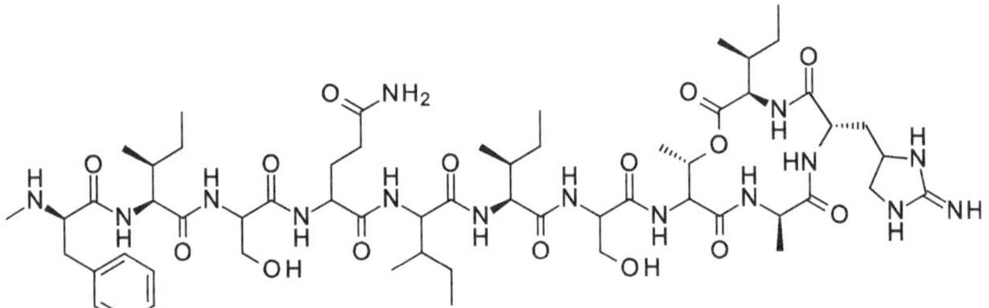

31

The first synthetic analogue of natural teixobactin was synthesized by substituting the L-allo-enduracididine residue with the naturally occurring L-arginine which exhibited the antibacterial activity against Gram-positive bacteria similar to that of teixobactin.[59]

Six-membered cyclic depsipeptides, enniatins [A (**32**), A_1, B, B_1 and B_2], were isolated from the methanolic extract of the endophyte *Fusarium tricinctum* Corda. The methanol extract of *F. tricinctum* displayed mild antibacterial and antileishmanial activities as well as inhibition of the activity of thioredoxin reductase enzyme of *Plasmodium falciparum*.[60]

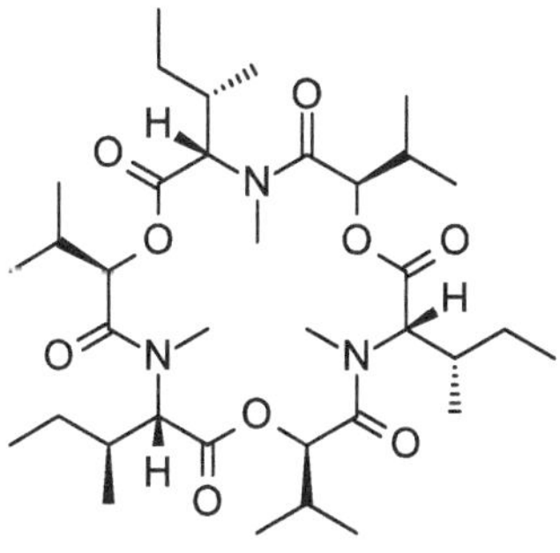

32

A co-cultivated fungus, *Coprinopsis cinereal*, produced a olypeptide antibiotic, copsin, which revealed its bactericidal property against a diversity of Gram-positive bacteria, including human pathogens such as *Enterococcus faecium* and *Listeria monocytogenes*.[61] Similarly, albicidin, a unique polyaromatic oligopeptide, mainly composed of *p*-aminobenzoic acids, was reported from the sugarcane invading bacterium *Xanthomonas albilineans*. This is a potent DNA gyrase inhibitor against a range of both Gram-positive and Gram-negative bacterial strains.[62] Three highly potent novel antibacterial compounds, the myxobacteria-derived cystobactamids 1–3, were isolated from *Cystobacter* sp.[63] These are inhibitors of bacterial DNA gyrase (type II topoisomerase) and revealed activity against *E. coli*, *A. baumannii*, *E. faecalis*, *S. aureus* and *S. pneumonia* with very lows MICs.[63] Despite promising antimicrobial activities of recently reported compounds, none of them has entered any proper clinical trials or extensive *in vivo* studies involving various animal models.

3.2 Plant derived antimicrobial natural products

Plants are main natural remedies which have been used for centuries for the treatment of various human ailments including infections. They are well known for the production of biologically active compounds and have played key role in modern drug discovery (see chapter "Medicinal natural products—An introduction" by Nahar and Sarker). A huge number of drugs in clinical uses today are either derived from plants or synthetic analogues of plant derived secondary metabolites. Some examples of such plant derived medicines include anticancer drugs like taxol, vinblastine, vincristine; antimalarial drugs like quinine, quinoline, artemisinin; analgesic drugs like morphine, codeine, eugenol; and cardioactive drugs like digoxin, digitoxin, lactoside C; CNS stimulants like caffeine; laxatives like sennosides and so many other drugs.[64] However, medicinal plants have been underexploited to some extent as a source of antimicrobial lead compounds. Nevertheless, there has been a significant body of ongoing research involving medicinal plants that have traditionally been used for the treatment of various infections in the traditional medicines including the Ayurvedic system[65] and the Traditional Chinese Medicine (TCM).[66] The antimicrobial property of plants is associated with and a part of their defense mechanisms against microbial attacks. Extensive phytochemical and/or bioassay directed studies on various medicinal plants have led to the characterization of a wide range of antimicrobial compounds. This section covers key plant-derived compounds with potential antimicrobial activity.

Aromatic medicinal plants, such as cinnamon, clove, cilantro, coriander, fennel, oregano, peppermint, rosemary, thyme etc. are good sources of essential oils, which have been well documented for their abilities to inhibit the growth of a variety of microorganisms.[67] Such naturally occurring essential oils act as preservatives (inhibit the growth of microorganisms) and flavoring agents and are sometimes incorporated in the food products. Chemically, these essential oils are terpenes, predominately mono- and sesquiterpenes. For example, the major compounds present in cinnamon, *Cinnamomum zeylanicum*, and clove, *Syzgium aromaticum*, are *trans*-cinnamaldehyde (**33**) and eugenol (**34**), respectively; both are simple monoterpenes. Such essential oils can inhibit the growth of molds, yeasts, and bacteria. Cinnamon and clove oils added at a concentration of 2% in potato dextrose agar (PDA) displayed complete inhibition of the growth of mycotoxigenic molds such as *Aspergillus flavus*, *A. parasiticus*, *A. ochraceus*, *Penicillium* sp. M46, *P. roqueforti*, *P. patulum*, and *P. citrinum* for a period of up to 21 days.[68] Both oils have also been documented for their ability to inhibit the growth of many microorganisms including *Lactobacillus* sp., *Bacillus thermoacidurans*, *Salmonella* sp., *Corynebacterium michiganense*, *Pseudomonas striafaciens*, *Clostridium botulinum*, *Aspergillus* sp., *Canninghamella* sp., *Fusarium* sp., and *Penicillium* sp.[69] The cinnamon oil was reported to inhibit the growth of *A. flavus*, *A. parasiticus*, *A. ochraceus*, and *Fusarium moniliforme* on PDA at very low concentration (500 ppm).[70] Eugenol, the main constituent of clove oil, has been used widely in perfumeries as flavoring agents, and also as an analgesic, local anesthetic, anti-inflammatory, and antibacterial agent.[71] It is used as an important constituent in the formulation of a paste or mixture as dental cement, filler, and restorative material local antiseptic and anesthetic in dentistry. Clove has also been reported to have strong antibacterial activity against *S. aureus*, *Escherichia coli* and *Listeria monocytogenes* with MICs in the range of 0.3–2.5 μg/mL.[72]

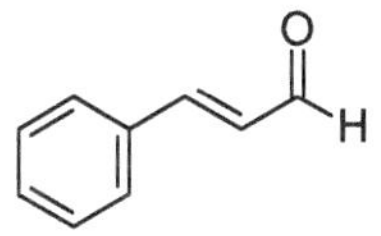

33

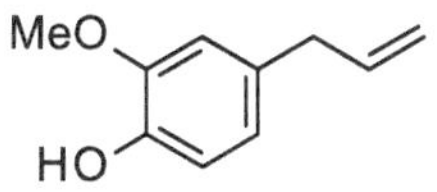

34

An antibacterial metabolite known as hypericin (**35**) has been isolated from St John's Wort (*Hypericum perforatum*), a herb widely used in the Western Herbal Medicine for the treatment of depression. This compound exhibited highly promising antibacterial activity against methicillin-resistant *Staphylococcus aureus* (MRSA) and penicillin-resistant variants reported with MICs of 0.1 μg/mL.[73] Medicinal Phytochemistry group at the UCL School of Pharmacy led by Professor Simon Gibbons[74] carried out extensive bioassay directed investigation on the genus *Hypericum* to explore antibacterial compounds with potential activity against a number of clinical isolates of methicillin resistant *S. aureus* (MRSA) strains. Bioassay-guided investigations into several *Hypericum* species led to the isolation of new acylphloroglucinols,[74–76] norlignans[77] and xanthone[78] with significant activity against *S. aureus*. Hyperenone A (**36**), a constituent of *H. acmosepalum*,[74] showed antibacterial activity against *Staphylococcus aureus* and *M. tuberculosis* as well as the inhibition of the adenosine triphosphate-dependent MurE ligase of *M. tuberculosis*, a crucial enzyme for peptidoglycan biosynthesis. An acylphloroglucinol, (*S*,*E*)-1-(2-((3, 7-dimethylocta-2,6-dien-1-yl)oxy)-4,6-dihydroxyphenyl)-2-methylbutan-1-one (trivial name, olympicin A) was isolated from *H. Olympicum* with promising (MICs of 1 μg/mL) activity against a panel of clinical isolates of multidrug-resistant (MDR) and methicillin resistant *Staphylococcus aureus*.[73] Such high antibacterial activity inspired the group to carry out the total synthesis of olympicin A as well as to make its analogues to fit into structure activity relationship (SAR) study. Olympicin A was synthesized in large scale and a number of its analogs were prepared by simply modifying two substituents. Among the analogues, the synthetic compound prepared by substituting gerenyl side chain with octyl group revealed same or twofold activity compared to natural olympicin A.[79]

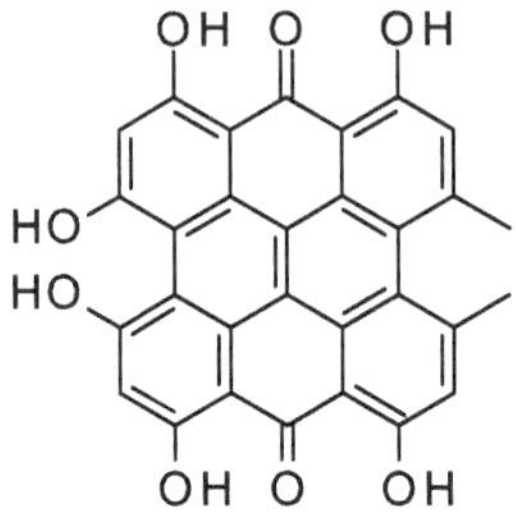

35

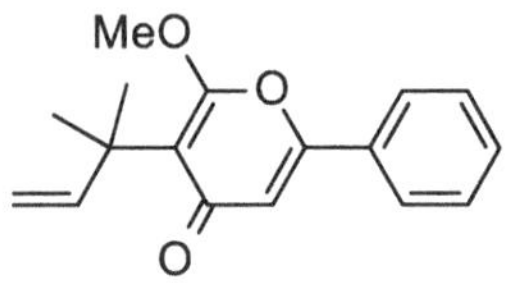

36

Like phloroglucinols, the antimicrobial activities of phenolic compounds have also been extensively studied. For example, eupomatenoid-3 (**37**), eupomatenoid-5 (**38**), eupomatenoid-6 (**39**) and conocarpan (**40**), isolated from *Piper regnellii* showed antibacterial activity. Among them, compounds **38** and **39** exhibited significant antibacterial activity against *Staphylococcus aureus*, *Bacillus subtilis*, and *B. subtilis* with MICs of 1.56–6.25 μg/mL.[80] Similarly, several antimicrobial isoflavanones including 5,7-dihydroxy-2′-methoxy-3′,4′-methylenedioxyisoflavanone and 4′,5-dihydroxy-2′,3′-dimethoxy-7-(5-hydroxyoxychromen-7yl)-isoflavanone were reported from a perennial herb, *Uraria picta* Desv., which has been traditionally used as an antidote to the venom of a dangerous Indian snake, *Echis carinata*.[81] These isoflavanones showed the antimicrobial activity against bacteria and fungi with MICs in the range of 12.5–100 μg/mL.[81] *Flemingia paniculata* several antimicrobial compounds including a simple salicylic acid derivative, 2-carboxy-3-(2-hydroxypropanyl)phenol,3-hydroxy-4-methoxycinnamaldehyde and several isoflavones including 5,7,4′-trihydroxy-8-(1,1-dimethyl-prop-2-enyl)-isoflavone, 5,7,2′,4′-tetra hydroxy-8-(1,1-dimethyl-prop-2-enyl)-isoflavone and 5,2′,4′-trihydroxy-4″,4″,5″(ζ)-trimethyl-4″,5″-dihydrofurano-(7,6,2″,3″)-isoflavone[82] with significant activities (MICs 1.57–200 μg/mL).[83] The highest potency (MIC 1.57 μg/mL; 0.005 mmol) was exhibited by 5,7,4′-trihydroxy-8-(1,1-dimethyl-prop-2-enyl)-isoflavone (**41**), against *S. aureus*.[83]

Quinones are well documented for their biological properties including antimicrobial activity. Bouldiaquinone, 2-acetylfuro-1,4-naphthoquinone, 2-hydroxy-3-methoxy-9,10-dioxo-9,10-dihydroanthracene-1-carbaldehyde and lapachol (**42**) were reported from the root bark of *Newbouldia laevis* with broad-spectrum *in vitro* antimicrobial activity against six Gram-positive and twelve Gram-negative bacterial species, as well as *Candida* strains with MIC values in the range 0.076–9.76 μg/mL.[84] Aerial part of *Saprosma fragrans* was reported to produce 3,4-dihydroxy-1-methoxyanthraquinone-2-carboxaldehyde and damnacanthal (**43**), which exhibited antifungal activities against *Trichophyton mentagrophytes*, *Cryptococcus neoformans*, *Sporothrix schenckii* and *T. mentagrophytes* (MIC = 1.56–12.5 μg/mL).[85]

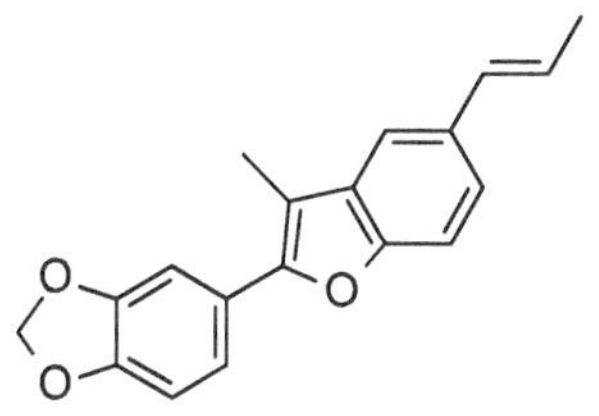

37

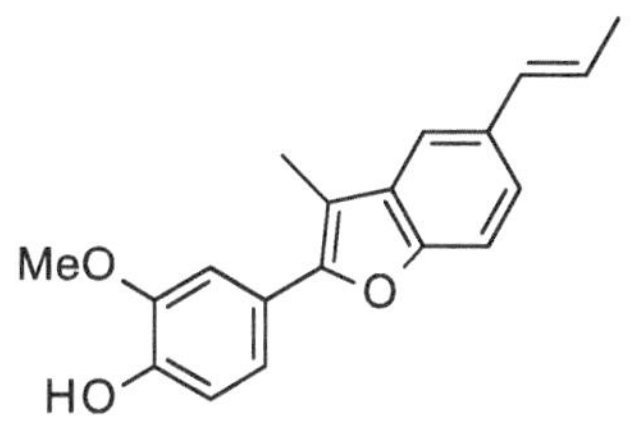

38 R = OMe
39 R = H

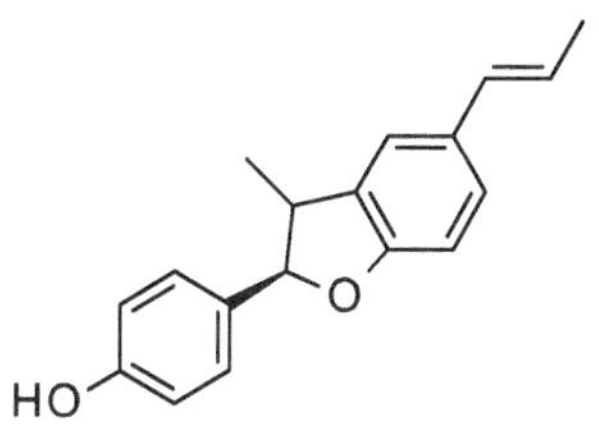

40

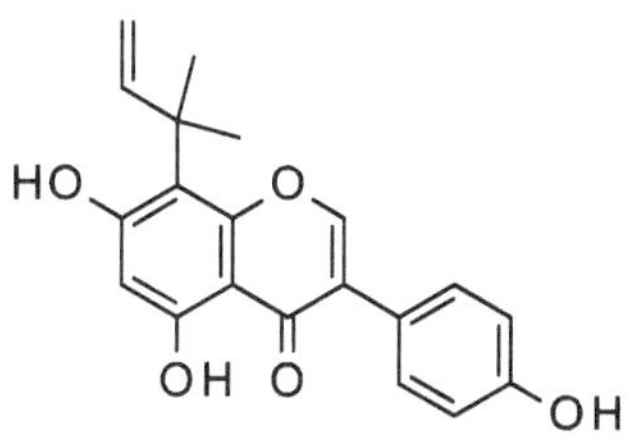

41

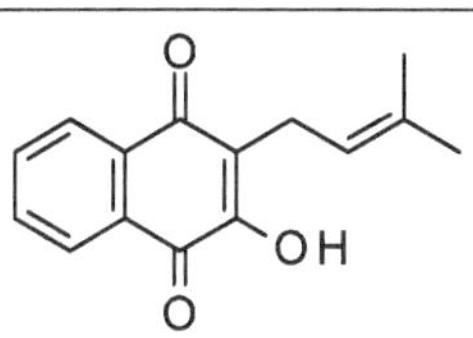

42

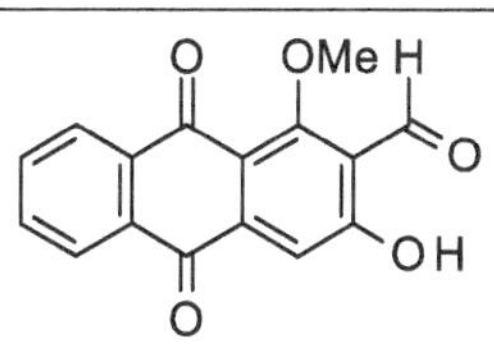

43

Many plant extracts containing steroidal saponins possess antimicrobial property. For example, the flower of *Allium leucanthum* produced aginoside and (25*R*)-5α-spirostan-3β,6β-diol-3-*O*-{β-D-glucopyranosyl-(1 → 2)-*O*-[β-D-xylopyranosyl-(1 → 3)]-*O*-β-D-glucopyranosyl-(1 → 4)-β-D-galactopyranoside}, which showed *in vitro* antifungal activity against seven *Candida* strains with MFCs in the range of 6.25–12.5 μg/mL.[86] Tigogenin 3-*O*-β-D-xylopyranosyl-(1 → 2)-[β-D-xylopyranosyl(1 → 3)]-β-D-glucopyranosyl-(1 → 4)-[α-L-rhamnopyranosyl-(1 → 2)]-β-D-galactopyranoside and tigogenin 3-*O*-β-D-glucopyranosyl-(1 → 2)-[β-D-xylopyranosyl-

(1 → 3)]-β-D-glucopyranosyl-(1 → 4)-β-D-galactopyranoside were isolated from *Tribulus terrestris*, and showed significant activity against *C. albicans* (MIC = 10–2.3 μg/mL) and *Cryptococcus neoformans* (MIC = 1.7–6.7 μg/mL).[87] Another steroidal saponin, dioscin, isolated from the rhizomes of *Smilacina atropurpurea*, showed antifungal against *C. albicans* and *C. glabrata* (MFCs ≤ 5.0 μg/mL).[88]

These are just a few examples of plant-derived antimicrobial compounds from several that have been reported in recent years. However, their *in vitro* antimicrobial efficacy is nowhere near any well-known antimicrobial agents obtained from microbial sources, and none of these compounds have been tested *in vivo*.

3.3 Antimicrobial natural products from marine organisms

The sections above have documented some examples of antimicrobial compounds predominantly from terrestrial organisms. However, various organisms from marine origin, have recently been shown as potential sources of new compounds with various therapeutic applications, including efficacy against infections. During last few decades there have been an incredible amount of research carried out on marine organisms and phytoplankton to explore the bioactive compounds to be considered as lead compounds in drug discovery including antimicrobial compounds. The marine environment is a rich source for unique actinomycete bacteria, which have produced thousands of metabolites with significant biological activity.[89] Many of the marine-derived actinomycetes have been produced structurally diverse secondary metabolites with promising anticancer properties.[90] With regards to the discovery of antibiotics, marine actinomycetes are less developed, however the isolation of promising antibiotics such as anthracimycin has inspired natural product chemists to explore this source further.[91] Anthracimycin (**44**) is a polyketide antibiotic isolated first in 2013 from a marine actinomycete of the genus *Streptomyces* (strain CNH365) collected off the shore of Santa Barbara at USA.[91] Another strain of *Streptomyces* (strain T676) isolated off the coast of St. John's Island, Singapore, also produced anthracimycin.[92] However, the research on marine actinomycetes for bioactive lead compounds started far earlier. Five structurally unique depsipeptides, salinamides A–E, were reported from a marine-derived *Streptomyces* sp. (strain CNB-091), which showed significant antibacterial and anti-inflammatory properties.[93] Subsequently, salinamide A exhibited significant inhibitory activity against RNA polymerase (RNAP) from Gram-positive and Gram-negative bacteria.[94] Further research was carried out on the *Streptomyces* sp. CNB-091 resulting in the isolation of salinamide F, from the ethyl acetate fraction, which also revealed significant RNAP-inhibitory activity against *S. aureus* (IC_{50} 4 μM) and *E. coli* (IC_{50} 4 μM) as well as antibacterial activity against *Enterococcus faecalis* (MIC 12.5 μg/mL), *S. aureus* (MIC100 μg/mL), *H. influenzae* (MIC 12.5 μg/mL), *Neisseria gonorrhoeae* (MIC 25 μg/mL), *Enterobacter cloacae* (MIC 50 μg/mL) and *E. coli* (MIC 0.20 μg/mL).[95]

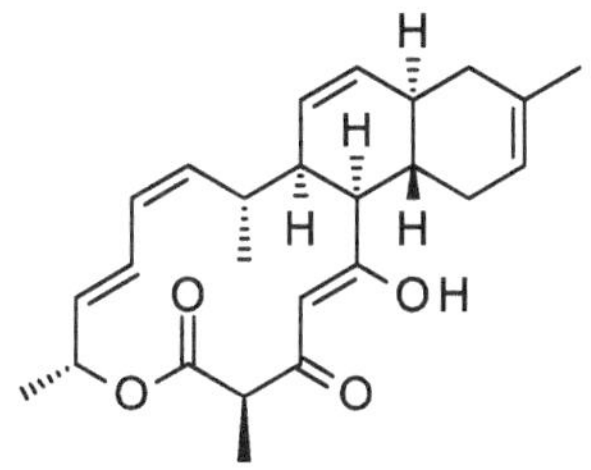

44

Cyclic depsipeptides, ulleungamides A and B,[96] were isolated from cultures of *Streptomyces* sp. which showed limited activity against *Staphylococcus aureus* and *Salmonella typhimurium* but no cytotoxicity. The antibacterial activity was tested using disc diffusion method. Zones of inhibitions of these two compounds against the above two organisms were found in the range of 9–16 mm compared to 20–25 mm of a positive standard, tetracycline. Eeudesmene-type sesquiterpenes, kandenols A–E (**45–49**)[97] from the culture broth of *Streptomyces* sp. (strain HKI0595) showed weak to moderate antimicrobial activities against *Bacillus subtilis* ATCC 6633 and *Mycobacterium vaccae*.[97]

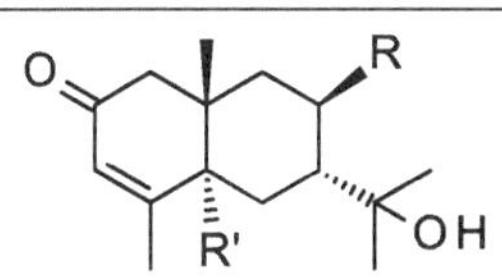

45 R = OH, R' = H
46 R = OH, R' = OH
47 R = OH, R' = O-OH
48 R = H, R' = O-OH

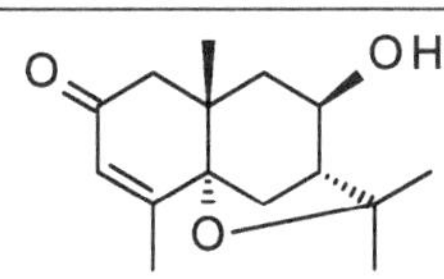

49

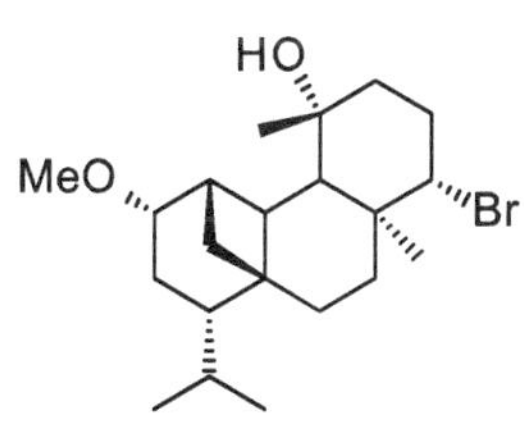

50

Naturally occurring brominated diterpenes with tetracyclic skeletons, ioniols I (**50**) and II,[98–100] were isolated from *Sphaerococcus coronopifolius*, a marine sponge collected from the rocky coasts of Corfu island in the Ionian Sea. These metabolites were evaluated for their antibacterial activity against a panel of clinical isolates of multidrug-resistant (MDR) and methicillin-resistant *Staphylococcus aureus* (MICs 0.5–128 μg/mL). Novel C_{15} eight-membered cyclic ethers[101] with a characteristic terminal *cis* eneeyne moiety were isolated from the red alga, *Laurencia glandulifera*, collected from the Crete island in South Greece. These compounds exhibited antibacterial activity against MRSA strains with MICs of 8–256 μg/mL.[101] A total of 17 diterpenes[102] featuring the dolabellane skeleton were isolated from the organic extracts of the brown alga, *Dilophus spiralis*. Some of these showed good antibacterial activity against six strains of *S. aureus*, including multidrug-resistant and methicillin-resistant variants.[102]

Neomaclafungins A-I,[103] 26-membered macrolides of the oligomycin subfamily, were isolated from *Actinoalloteichus* sp. NPS702, a marine sediment collected from USA Bay, Kochi Prefecture, Japan. These macrolides exhibited significant antifungal activity *in vitro* against *Trichophyton mentagrophytes* (ATCC 9533) with MICs of 1–3 μg/mL.[103]

Polycyclic secondary metabolites, citreamicins A and B, citreaglycon A and dehydrocitreaglycon A,[104] isolated from marine-derived *Streptomyces caelestis* exhibited antibacterial activity against *Staphylococcus haemolyticus*, *S. aureus*, and *B. subtillis*. Citreamicin A, citreamicin B, and citreaglycon A revealed strong activities against MRSA ATCC 43300 with MIC values of0.25, 0.25, and 8.0 μg/mL, respectively.[104] Isorhodoptilometrin-1-methyl ether, emodin, 1-methyl emodin, siderin, arugosin C, and variculanol obtained from the marine fungus *A. versicolor* were reported to exhibit antimicrobial activity, anticancer activity, and inhibition of Hepatitis C virus protease.[105] A marine-derived fungus *Nigrospora*, produced anthraquinone derivative, 3-acetoxy-4-deoxybostrycin, which exhibited promising activity against *Bacillus cereus* (MIC 48.8 nM).[106]C-glycosylated benz[α] anthraquinone derivatives, urdamycinone E, urdamycinone G, dehydroxyaquayamycin, isolated from the marine *Streptomycetes* sp. displayed potent activity against *M. tuberculosis* with MICs of 3.13–12.50 μg/mL.[107]

Trichodermaquinone and trichodermaxanthone from the marine-derived fungus *Trichoderma aureoviride* PSU-F95 demonstrated significant antibacterial activity against MRSA with MIC values of 8 and μg/mL, respectively.[108] Alkanoyl imidazoles, bulbimidazoles A–C,[109] isolated from the culture extract of the gammaproteobacterium *Microbulbifer* sp. DC3–6 collected from a stony coral of the genus *Tubastraea* displayed broad-spectrum antimicrobial activity (MICs ranging from 0.78 to 12.5 μg/mL).[109] Microketides A and B, a pair of C-11 epimeric polyketides, from the gorgonian-derived fungus *Microsphaeropsis* sp. RA10–14 collected from the South China Sea showed strong activity against *Pseudomonas aeruginosa*, *Nocardia brasiliensis*, *Kocuria rhizophila*, and *Bacillus anthraci* with the same MIC value as ciprofloxacin (0.19 μg/mL).[110] The liquid culture of a *Streptomyces* sp. (strain BD21–2) collected from a shallow-water sediment sample from Kailua Beach, Oahu, Hawaii produced bonactin (**51**), which showed antimicrobial activity against both Gram-positive and Gram-negative bacterial as well as fungal strains.[111]

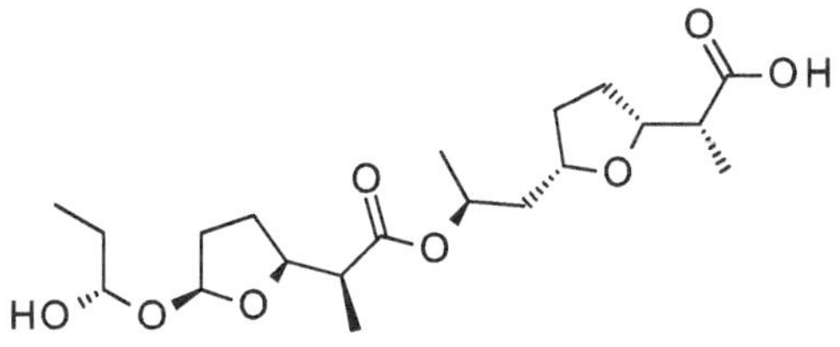

51

3.4 Antimicrobial natural products from animals

Animals have also been reported to secrete compounds of biological interests including those with significant antimicrobial properties. These animals include amphibians species, mammals and similar animals capable of producing metabolites with promising biological activities. The skin glands of amphibians species have been documented to produce antimicrobial peptides (AMPs) that are crucial in the first line of defense against microbial invasion. The vast majority of AMPs isolated from the frog skin are reported to exert potent activity against antibiotic-resistant bacteria, protozoa, yeasts, and fungi by permeating and destroying the plasma membrane and inactivating intracellular targets. Importantly, since they do not bind to a specific receptor, AMPs are less likely to induce resistance mechanisms. Although most of the AMPs have common characteristics, they differ in amino acids sequences which produce a wide range bioactivity with varying degrees of efficacy.

European fire-bellied toad, *Bombina* sps., is known to produce skin secretions with peptides, such as bombinins and bombinins H, with potential antimicrobial activity.[112] Bombinins and bombinins H, which are quite large peptides, were isolated from common precursors containing one or two bombinin copies at the amino and a single bombinin H at the carboxyl end. Bombinins have showed activity against Gram-positive and Gram-negative bacteria and fungi but virtually inactive in hemolysis assays. However, bombinins H showed lower bactericidal activities but lyse erythrocytes.[112] Bombinins, identified from the ancient toad belonging to the genus *Bombina*, are a group of amphibian-derived peptides with broad-spectrum antimicrobial activities. A novel bombinin precursor encoded a bombinin-like peptide (BLP-7) and a novel bombinin H-type peptide (Bombinin H-BO) were identified from the skin secretion of Oriental fire-bellied toad, *Bombina orientalis*.[113] In the antimicrobial experiment, the synthetic replicate of BLP-7 exhibited more potency than Bombinin H-BO against Gram-positive and Gram-negative bacteria and yeast.

A total of 11 AMPs was reported from *Pleurodema somuncurence* (Anura: Leptodactylidae: Leiuperinae). Three [Somuncurin-1 (FIIWPLRYRK), somuncurin-2 (FILKRSYPQYY), and thaulin-3 (NLVGSLLGGILKK)][114] inhibited the growth of *Escherichia coli* but only Somuncurin-1 showed antibacterial activity against *Staphylococcus aureus*. In a biophysical membrane model, this peptide showed a greater permeation effect in prokaryotic-like membranes and capability to restructure liposomes, suggesting fusogenic activity, which ultimate could cause cell aggregation and disruption of cell morphology. Eight new peptides isolated from the skin secretion of the frog, *Leptodactylus pustulatus*,[115] revealed structural similarities between them and other antimicrobial peptides reported from the same genus. Among these peptides, ocellatins-PT1 to -PT5 (25 amino acid residues) are amidated at the C-terminus, while ocellatins-PT6 to -PT8 (32 amino acid residues) have free carboxylates. All peptides, except for ocellatin-PT2, showed antibacterial activity against at least one Gram-negative strain. Ocellatin-PT8 inhibited the growth of *Staphylococcus aureus*, *Escherichia coli*, *Klebsiella pneumoniae*, and *Salmonella choleraesuis* strains with MICs of 60–240 μM. An LC-MS-MS to analysis on the skin samples of *Hypsiboas pulchellus*, an Argentinian wild frog, has identified antimicrobial peptides with molecular mass within 1000–2000 Da.[116] Out of 23 novel sequences identified by MS three (named P1-Hp-1971, P2-Hp-1935, and P3-Hp-1891) were synthesized. These three AMPs inhibited the growth of Gram-positive *Staphylococcus aureus* (MICs of P1-Hp-1971, P2-Hp-1935, and P3-Hp-1891: 8, 66, and 17 μM, respectively) and Gram-negative *E. coli* (MICs of P1-Hp-1971, P2-Hp-1935, and P3-Hp-1891: 16, 33, and

17 μM, respectively) revealing that P1-Hp-1971 and P3-Hp-1891 were the most active peptides.[116]

Chitosan, a polycationic biopolymer naturally present in the exoskeletons of crustaceans and arthropods, has been used as food preservative because of their ability to suppress fungal colony growth and inhibit fungal spore germination at a 0.01% (w/v) concentration.[117] Its antibacterial activity evaluated against several Gram-negative (*E. coli*, *P. fluorescens*, *S. typhimurium*, *Vibrio parahaemolyticus*) and Gram-positive bacteria (*L. monocytogenes*, *Bacillus megaterium*, *B. cereus*, *S. aureus*)[118] revealed that chitosan inhibited the growth of most of the tested bacteria showing stronger bactericidal effects against Gram-positive bacteria than Gram-negative bacteria at a concentration of 0.1% in agar medium. An investigation on the antimicrobial activity of a chitosan against various food poisoning and food spoilage bacteria revealed that the chitosan mixture retarded the growth of *Salmonella* spp. and reduced the population of *Staphylococcus* spp. in raw milk.[119] A study conducted to determine the shelf-life of oysters stored at 5 ± 1 °C established that chitosan at a concentration of 5.0 mg/mL extended the shelf-life of oysters from 8 to 9 days to 14–15 days which indicated the great potential of chitosan for seafood preservation.[120]

Lactoferrin, an iron-binding glycoprotein present in the milk of human and mammals, has shown to have significant antimicrobial activity against a wide range of bacteria and viruses.[121] Its antimicrobial activity has been reported against foodborne microorganisms including *Carnobacterium*, *L. monocytogenes*, *E. coli*, and *Klebsiella*[122] with a reduction of 4 log CFU/mL of *Cronobacter* species in the presence of 2.5 mg/mL lactoferrin in 0.2% peptone water within 4 h incubation at 37 °C. In combination with nisin (0.1 mg/g), lactoferrin (0.2 mg/g) displayed a significant reduction in spoilage bacterial counts (total aerobic bacteria, coli- form, *E. coli*, total psychrophilic bacteria, *Pseudomonas* species, yeast and molds) and extend shelf-life of up to 10 days in meatballs.[123] Milk-derived casein and whey proteins are known to possess various biological activities including antimicrobial activities.[124,125] Casocidin, a peptide produced by hydrolysis of a_{S2}-casein by chymosin, revealed antibacterial activity against *Staphylococcus* spp., *Sarcina* spp., *B. subtilis*, *Diplococcus pneumoniae*, and *Streptococcus pyogenes*.[124] Isracidin, another casein derived peptide, showed antibacterial activity against *S. aureus*, *L. monocytogenes*, and *C. albicans*.[125] Peptides such as casein A and B at 0.05 mM and 0.22 mM respectively, were shown to inhibit the growth of pathogenic *Enterobacter sakazakii*.[126] Peptides generated from a_{S2}- casein,

a_{S1}-casein, and k-casein have shown antibacterial effects against *E. coli* and *B. subtilis*.[127] The whey protein of bovine milk is composed of mainly of β-lactoglobulin and α-lactalbumin. The peptides released during the digestion of β-lactoglobulin with trypsin has shown to possess antimicrobial activity against food- borne pathogens such as *S. aureus*, *L. monocytogenes*, *Salmonella* spp. and *E. coli* O157 at concentration of 10–20 mg/mL.[124,128]

4. Antimicrobial compounds in preclinical and clinical trials

There are over 50 new antibiotics which are currently at different stages in clinical trials with the potential to treat serious bacterial infections including those which developed resistance to the existing antibiotics. A majority of these are of synthetic and/or semi-synthetic origin. Among the antimicrobial agents originated from natural sources, which are undergoing preclinical and clinical development stages are mostly antimicrobial peptides (AMPs)[129] and are usually derived from microorganisms or animal sources. Some of these AMPs with potential antimicrobial activity and proven mechanism actions have been outlined below.

MU1140, a lantibiotic peptide isolated from *Streptococcus mutans*, is currently undertaking preclinical development by Oragenics, Inc. It has been proven to exert activity against all gram-positive bacteria, especially methicillin-resistant *S. aureus* (MRSA). It acts through not only membrane disruption, but also inhibition of cell wall biosynthesis.[130]

Arenicin (AP139) isolated from lugworm *Arenicola marina* consists of 21-residue peptide, which has a positively charged amphipathic β-hairpin structure linked with one disulfide bond. This is undergoing through preclinical trial by Adenium Biotech. It reveals bactericidal activity against multidrug-resistant Gram-negative bacterial infections by membrane pore formation.[131]

EA-230, developed by Exponential Biotherapies, is a linear tetrapeptide derived from the β-chain of the human chorionic gonadotropin hormone (β-hCG). The mode of action is known to be immunomodulation, such as the release of pro-inflammatory cytokines and a reduction of neutrophil influx. It has demonstrated anti-inflammatory activity when injected intravenously. Clinical trials for acute systemic inflammation (*e.g.*, sepsis) and inflammation caused by organ dysfunction (*e.g.*, acute kidney injury) are undergoing.[132]

PAC113, developed by Pacgen Biopharmaceuticals, is a 12-amino acid linear peptide derived from histatin 5, a human salivary α-helical peptide. It exerts antimicrobial activity *via* membrane disruption, as well as immunomodulation. The phase II clinical trial has been completed treatment for oral candidiasis in HIV patients through mouth rinse.[133]

Novexatin (NP213), developed by Novabiotics, is a cyclic cationic peptide consisting of seven arginine residues. It is going through the phase IIb clinical trial for the treatment of fungal nail infections with topical administration. It exerts its action through membrane disruption.[133]

PXL01, a macrocyclic peptide comprised of 25 amino acids connected through a disulfide bond, is derived from human lactoferrin which is currently being evaluated by ProMore Pharma. It showed antimicrobial activity for the topical treatment of postsurgical adhesions and scar prevention. It acts through an immunomodulatory mechanism (*e.g.*, inhibition of the release of pro-inflammatory cytokines).[132]

Ramoplanin (NTI-851), developed by Nano-therapeutics, is a macrocyclic glycolipodepsipeptide produced by *Actinoplanes* spp. It reveals bactericidal activity by blocking the cell wall peptidoglycan synthesis of Gram-positive bacteria. Its phase III clinical study of the peptide was initiated for the oral treatment of vancomycin-resistant *enterococcus* (VRE) colonization and also its phase II trial against *Clostridium difficile.*[134]

5. Conclusions

Nature continues to produce metabolites of chemical diversity with a large array of potential biological activities, which have contributed significantly in drug discovery including the development of new antibiotics. A vast majority of today's antibiotics available for clinical uses are either derived directly from natural sources or their semisynthetic analogues. However, the global crisis of antimicrobial resistance urges the development of new antibiotics, which could fight against the so-called "superbugs" escaping the lethal actions of existing antibiotics. Such superbugs raising alarming situation of AMR have been grouped together and acronymically dubbed as ESKAPE pathogens including multidrug resistance *Enterococcus faecium, Staphylococcus aureus, Klebsiella pneumonia, Acinetobacter baumannii, Pseudomonas aeruginosa and Enterobacter species.*[135] These ESKAPE pathogens require immediate attention of antimicrobial drug discovery groups around the world to find new and safe antibiotics in order to control the situation of global healthcare threat. Collaborative research approaches among both academic and industrial scientists including natural product chemists, medicinal chemists, microbiologist and biotechnologists are also much needed.

Besides the discovery of new antibiotics, it is the responsibility of healthcare professionals and general public to preserve the efficacy of currently available antibiotics by raising public awareness on the appropriate uses of antibiotics. Although there has been a significant amount of work carried out through antibiotics stewardship, further work is necessary to encourage the reduction of over prescriptions and misuses of antibiotics. AMR situation is even much worse in the developing countries, where such situation could be improved by educating the general public for appropriate uses of antibiotics and also alerting the physicians to be more careful in prescribing antibiotics.

References

1. Ravina, E. *The Evolution of Drug Discovery*. Wiley-VCH Verlag GmbH &Co: Weinheim, 2011.
2. Pelczar, M. J.; Chan, E. C. S.; Krieg, N. R. *Microbiology*, 5th ed.; McGraw-Hill Book Company: New York, 1986.
3. Patrick, G. *An Introduction to Medicinal Chemistry*, 5th ed.; Oxford Publishing Press: Oxford, 2013.
4. Centres for Disease Control and Prevention. 2020, Available at https://www.cdc.gov/drugresistance/about.html (accessed on 10th April, 2020).
5. ECDC Surveillance Antimicrobial resistance in Europe. https://www.ecdc.europa.eu/sites/default/files/documents/surveillance-antimicrobial-resistance-Europe-2018.pdf; 2018.
6. CDC Antibiotic resistance threats in the United States. https://www.cdc.gov/drugresistance/pdf/threats-report/2019-ar-threats-report-508.pdf; 2019.
7. Köck, R.; Becker, K.; Cookson, B.; van Gemert-Pijnen, J. E.; Harbarth, S.; Kluytmans, J.; Mielke, M.; Peters, G.; Skov, R. L.; Struelens, M. J.; Tacconelli, E.; Torné, A. N.; Witte, W.; Friedrich, A. W. *Euro Surveill.* **2010**, *15*(41), 19688.
8. ECDC Antimicrobial Resistance Surveillance in Europe. https://www.ecdc.europa.eu/en/publications-data/antimicrobial-resistance-surveillance-europe-2013; 2013.
9. O'Neill, J. *Antimicrobial Resistance: Tackling a Crisis for the Health and Wealth of Nations*. https://amr-review.org/sites/default/files/AMR%20Review%20Paper%20-%20Tackling%20a%20crisis%20for%20the%20health%20and%20wealth%20of%20nations_1.pdf; 2014.
10. WHO, 2018, Antimicrobial Resistance, https://www.who.int/news-room/fact-sheets/detail/antimicrobial-resistance
11. Rang, H. P.; Dale, M. M.; Ritter, J. M.; Flower, R. J.; Henderson, G. *Rang and Dale's Pharmacology*, 7th ed.; Elsevier Churchill Livingstone: Edinburgh, London, New York, 2012.
12. Jacoby, G. A.; Munoz-Price, L. S. *N. Engl. J. Med.* **2005**, *352*(4), 380–391.
13. Sefton, A. M. *Drugs* **2002**, *62*(4), 557–566.
14. Nikaido, H. *Microbiol. Mol. Biol. Rev.* **2003**, *67*, 593–656. https://doi.org/10.1128/MMBR.67.4.593-656.2003.

15. Li, H.; Luo, Y. F.; Williams, B. J.; Blackwell, T. S.; Xie, C. M. *Int. J. Med. Microbiol.* **2012**, *302*, 63–68. https://doi.org/10.1016/j.ijmm.2011.10.001.
16. Schindler, B. D.; Kaatz, G. W. *Drug Resist. Updat.* **2016**, *27*, 1–13. https://doi.org/10.1016/j.drup.2016.04.003.
17. Jumbe, N. L.; Louie, A.; Miller, M. H.; Liu, W.; Deziel, M. R.; Tam, V. H.; Bachhawat, R.; Drusano, G. L. *Antimicrob. Agents Chemother.* **2006**, *50*(1), 310–317.
18. Sköld, O. *Drug Resist. Updat.* **2000**, *3*(3), 155–160.
19. Kardos, N.; Demain, A. L. *Appl. Microbiol. Biotechnol.* **2011**, *92*(4), 677–687.
20. Bergeron, M. G.; Brusch, J. L.; Barza, M.; Weinstein, L. *Antimicrob. Agents Chemother.* **1973**, *4*(4), 396–401.
21. Levine, D. P. *Clin. Infect. Dis.* **2006**, *42*(1), S5–12.
22. Chopra, I.; Roberts, L. *Microbiol. Mol. Biol. Rev.* **2001**, *65*(2), 232–260. https://doi.org/10.1128/MMBR.65.2.232-260.2001.
23. Pongs, O. Chapter 3: Chloramphenicol. In: *Mechanism of Action of Antibacterial Agents*; Hahn, F. E. Ed.; Antibiotics Volume V Part 1;Springer Berlin Heidelberg: Berlin, Heidelberg, 1979 pp 26–42.
24. Berendsen, B.; Stolker, L.; de Jong, J.; Nielen, M.; Tserendorj, E.; Sodnomdarjaa, R.; Cannavan, A.; Elliott, C. *Anal. Bioanal. Chem.* **2010**, *397*(5), 1955–1963. https://doi.org/10.1007/s00216-010-3724-6.
25. World Health Organisation *World Health Organization Model List of Essential Medicines: 21st List 2019.* World Health Organization: Geneva, 2019.
26. Caffrey, P.; Lynch, S.; Flood, E.; Finnan, S.; Oliynyk, M. *Chem. Biol.* **2001**, *8*(7), 713–723. https://doi.org/10.1016/S1074-5521(01)00046-1.
27. O'Keeffe, J.; Doyle, S.; Kavanagh, K. *J. Pharm. Pharmacol.* **2003**, *55*(12), 1629–1633. https://doi.org/10.1211/0022357022359.
28. Gupta, A. K.; Tomas, E. *Dermatol. Clin.* **2003**, *21*(3), 565–576.
29. Mandell, G.; Bennett, J.; Dolin, R. *Principles and Practice of Infectious Diseases*, 4th ed.; Churchill Livingstone: New York, 1995.
30. Mingeot-Leclercq, M.-P.; Glupczynski, Y.; Tulkens, P. M. *Antimicrob. Agents Chemother.* **1999**, *43*(4), 727–737.
31. Denning, D. W. *J. Antimicrob. Chemother.* **1997**, *40*, 611–614.
32. Trookman, N. S.; Rizer, R. L.; Weber, T. *J. Am. Acad. Dermatol.* **2011**, *64*(3 Suppl), S8–15. https://doi.org/10.1016/j.jaad.2010.11.011.
33. Benedict, R. G.; Langlykke, A. F. *J. Bacteriol.* **1947**, *54*, 24.
34. Falagas, M. E.; Rafailidis, P. I.; Matthaiou, D. K. *Drug Resist. Updat.* **2010**, *13*, 132–138. https://doi.org/10.1016/j.drup.2010.05.002.
35. Waksman, S. A.; Lechevalier, H. A. *Science* **1949**, *109*, 305–307, (March 1949).
36. Gupte, M.; Kulkarni, P.; Ganguli, B. N. *Appl. Microbiol. Biotechnol.* **2002**, *58*(1), 46–57. https://doi.org/10.1007/s002530100822.
37. Hammond, S. M. *Prog. Med. Chem.* **1977**, *14*, 105–179.
38. Liu, W.-B.; Shi, Y.; Yao, L.-L.; Zhou, Y.; Ye, B.-C. *PLoS One* **2013**, *8*(11), e80676, (2013-01-01).
39. Sensi, P. *Rev. Infect. Dis.* **1983**, *5*(Suppl. 3), S402–S406.
40. Calvori, C.; Frontali, L.; Leoni, L.; Tecce, G. *Nature* **1965**, *207*(995), 417–818. https://doi.org/10.1038/207417a0.
41. Fischer, J.; Ganellin, C. R. *Analogue-Based Drug Discovery*. John Wiley & Sons, 2006.
42. Shomura, T.; Ezaki, N.; Tsuruoka, T.; Niwa, T.; Akita, E.; Niida, T. *J. Antibiot.* **1970**, *23*(3), 155–161.
43. Zárate, S. G.; De la Cruz Claure, M. L.; Benito-Arenas, R.; Revuelta, J.; Santana, A. G.; Bastida, A. *Molecules* **2018**, *23*(2), 284 https://doi.org/10.3390/molecules23020284.
44. Pogliano, J.; Pogliano, N.; Silverman, J. A. *J. Bacteriol.* **2012**, *194*(17), 4494–4504. https://doi.org/10.1128/JB.00011-12.

45. Guay, D. R. *Consult. Pharm.* **2004**, *19*(7), 614–628.
46. Wang, J.; Soisson, S. M.; Young, K.; Shoop, W.; Kodali, S.; Galgoci, A.; Painter, R.; Parthasarathy, G.; Tang, Y. S.; Cummings, R.; Ha, S.; Dorso, K.; Motyl, M.; Jayasuriya, H.; Ondeyka, J.; Herath, K.; Zhang, C.; Hernandez, L.; Allocco, J.; Basilio, A.; Tormo, J. R.; Genilloud, O.; Vicente, F.; Pelaez, F.; Colwell, L.; Lee, S. H.; Michael, B.; Felcetto, T.; Gill, C.; Silver, L. L.; Hermes, J. D.; Bartizal, K.; Barrett, J.; Schmatz, D.; Becker, J. W.; Cully, D.; Singh, S. B. *Nature* **2006**, *441*, 358–361.
47. Martens, E.; Demain, A. L. *J. Antibiot.* **2011**, *64*(11), 705–710, doi: 10.1038.
48. Arqués, J. L.; Rodríguez, E.; Nuñez, M.; Medina, M. *J. Dairy Sci.* **2008**, *91*, 70–75.
49. Talarico, T. L.; Casas, I. A.; Chung, T. C.; Dobrogosz, W. *J. Antimicrob. Agents Chemother.* **1988**, *32*(12), 1854–1858.
50. Kabuki, T.; Saito, T.; Kawai, Y.; Uemura, J.; Itoh, T. *Int. J. Food Microbiol.* **1997**, *34*(2), 145–156. https://doi.org/10.1016/s0168-1605(96)01180-4.
51. Gänzle, M. G.; Höltzel, A.; Walter, J.; Jung, G.; Hammes, W. P. *Appl. Environ. Microbiol.* **2000**, *66*(10), 4325–4333.
52. Pohanka, A.; Levenfors, J.; Broberg, A. Antimicrobial dialkyl resorcinols from *Pseudomonas* sp. Ki19. *J. Nat. Prod.* **2010**, *73*, 825–830.
53. Ondeyka, J. G.; Zink, D. L.; Young, K.; Painter, R.; Kodali, S.; Galgoci, A.; Collado, J.; Tormo, J. R.; Basilio, A.; Vicente, F.; Wang, J.; Singh, S. B. *J. Nat. Prod.* **2006**, *69*, 377–380.
54. Du, X.; Lu, C.; Li, Y.; Zheng, Z.; Su, W.; Shen, Y. *J. Antibiot.* **2008**, *61*, 250–253.
55. Hashizume, H.; Igarashi, M.; Hattori, S.; Hori, M.; Hamada, M.; Takeuchi, T. *J. Antibiot.* **2001**, *54*, 1054–1059.
56. Neuhof, T.; Schmieder, P.; Preussel, K.; Dieckmann, R.; Pham, H.; Bartl, F.; von Döhren, H. *J. Nat. Prod.* **2005**, *68*, 695–700.
57. Neuhof, T.; Schmieder, P.; Seibold, M.; Preussel, K.; von Döhren, H. *Bioorg. Med. Chem. Lett.* **2006**, *16*, 4220–4222.
58. Ling, L. L.; Schneider, T.; Peoples, A. J.; Spoering, A. L.; Engels, I.; Conlon, B. P.; Mueller, A.; Schäberle, T. F.; Hughes, D. E.; Epstein, S.; Jones, M.; Lazarides, L.; Steadman, V. A.; Cohen, D. R.; Felix, C. R.; Fetterman, K. A.; Millett, W. P.; Nitti, A. G.; Zullo, A. M.; Chen, C.; Lewis, K. *Nature* **2015**, *517*(7535), 455–459. https://doi.org/10.1038/nature14098.
59. Jad, Y. E.; Acosta, G. A.; Naicker, T.; Ramtahal, M.; El-Faham, A.; Govender, T.; Kruger, H. G.; de la Torre, B. G.; Albericio, F. *Org. Lett.* **2015**, *17*, 6182–6185.
60. Zaher, A. M.; Makboul, M. A.; Moharram, A. M.; Tekwani, B. L.; Calderón, A. I. *J. Antibiot.* **2015**, *68*, 197–200.
61. Essig, A.; Hofmann, D.; Münch, D.; Gayathri, S.; Künzler, M.; Kallio, P. T.; Sahl, H. G.; Wider, G.; Schneider, T.; Aebi, M. *J. Biol. Chem.* **2014**, *289*, 34953–34964.
62. Cociancich, S.; Pesic, A.; Petras, D.; Uhlmann, S.; Kretz, J.; Schubert, V.; Vieweg, L.; Duplan, S.; Marguerettaz, M.; Noëll, J.; Pieretti, I.; Hügelland, M.; Kemper, S.; Mainz, A.; Rott, P.; Royer, M.; Süssmuth, R. D. *Nat. Chem. Biol.* **2015**, *11*, 195–197. https://doi.org/10.1038/nchembio.1734.
63. Baumann, S.; Herrmann, J.; Raju, R.; Steinmetz, H.; Mohr, K. I.; Hüttel, S.; Harmrolfs, K.; Stadler, M.; Müller, R. *Angew. Chem. Int. Ed. Engl.* **2014**, *53*, 14605–14609.
64. Heinrich, M.; Barnes, J.; Prieto-Garcia, J. M.; Gibbons, S.; Williamson, E. *Fundamentals of Pharmacognosy and Phytotherapy*, 3rd ed.; Elsevier: London, 2018.
65. Gautam, R.; Saklani, A.; Jachak, S. M. *J. Ethnopharmacol.* **2007**, *110*, 200–234.
66. Wang, Z.; Yu, P.; Zhang, G.; Xu, L.; Wang, D.; Wang, L.; Zeng, X.; Wang, Y. *Bioorg. Med. Chem.* **2010**, *18*, 4269–4274.
67. Burt, S. *Int. J. Food Microbiol.* **2004**, *94*, 223–253.

68. Azzouz, M. A.; Bullerman, L. B. *J. Food Prot.* **1982**, *45*, 1298–1301.
69. Conner, D. E.; Beuchat, L. R. *J. Food Sci.* **1984**, *49*, 429–434.
70. Soliman, K. M.; Badeaa, R. I. *Food Chem. Toxicol.* **2002**, *40*, 1669–1675.
71. Pinto, E.; Vale-Silva, L.; Cavaleiro, C.; Salgueiro, L. *J. Med. Microbiol.* **2009**, *58*, 1454–1462.
72. Farag, D. S.; Daw, Z. Y.; Hewedi, F. M.; El-Baroty, G. S. *J. Food Prot.* **1989**, *52*, 665–667.
73. Schempp, C. M.; Pelz, K.; Wittmer, A.; Schopf, E.; Simon, J. C. *Lancet* **1999**, *353*, 2129.
74. Gibbons, S.; Ohlendorf, B.; Johnsen, I. *Fitoterapia* **2002**, *73*, 300–304.
75. Shiu, W. K. P.; Gibbons, S. *Phytochemistry* **2006**, *67*, 2568–2572.
76. Shiu, W. K. P.; Rahman, M. M.; Curry, J.; Stapleton, P.; Zloh, M.; Malkinson, J. P.; Gibbons, S. *J. Nat. Prod.* **2012**, *75*, 336–343.
77. Osman, K.; Evangelopoulos, D.; Basavannacharya, C.; Gupta, A.; McHugh, T. D.; Bhakta, S.; et al. *Int. J. Antimicrob. Agents* **2012**, *39*, 124–129.
78. Wang, W.; Zeng, Y. H.; Osman, K.; Shinde, K.; Rahman, M. M.; Gibbons, S.; Mu, Q. *J. Nat. Prod.* **2010**, *73*, 1815–1820.
79. Rahman, M. M.; Shiu, W. K. P.; Gibbons, S.; Malkinson, J. P. *J. Nat. Prod.* **2012**, *75*, 336–343.
80. Pessini, G. L.; Dias Filho, B. P.; Nakamura, C. V.; Cortez, D. A. G. *Mem. Inst. Oswaldo Cruz* **2003**, *98*, 1115–1120, Rio de Janeiro.
81. Rahman, M. M.; Gibbons, S.; Gray, A. I. *Phytochemistry* **2007**, *68*, 1692–1697.
82. Rahman, M. M.; Sarker, S. D.; Byres, M.; Gray, A. I. *J. Nat. Prod.* **2004**, *67*, 402–406.
83. Rahman, M. M.; Khondkar, P.; Gray, A. I.; Sarker, S. D. *Pharm. Biol.* **2008**, *46*, 356–359.
84. Sathiamoorthy, B.; Gupta, P.; Kumar, M.; Chaturvedi, A. K.; Shukla, P. K.; Maurya, R. *Bioorg. Med. Chem. Lett.* **2007**, *17*, 239–242.
85. Singh, D. N.; Verma, N.; Raghuwanshi, S.; Shukla, P. K.; Kulshreshtha, D. K. *Bioorg. Med. Chem. Lett.* **2006**, *16*, 4512–4514.
86. Mskhiladze, L.; Kutchukhidze, J.; Chincharadze, D.; Delmas, F.; Elias, R.; Favel, A. *Georgian Med. News* **2008**, *154*, 39–43.
87. Zhang, J.-D.; Xu, Z.; Cao, Y.-B.; Chen, H.-S.; Yan, L.; An, M.-M.; Gao, P.-H.; Wang, Y.; Jia, X.-M.; Jiang, Y.-Y. *J. Ethnopharmacol.* **2006**, *103*, 76–84.
88. Zhang, Y.; Li, H.-Z.; Zhan, Y.-J.; Jacob, M. R.; Khan, S. I.; Li, X.-C.; Yang, C.-R. *Steroids* **2006**, *71*, 712–719.
89. Jensen, P. R.; Fenical, W. *Nat. Chem. Biol.* **2006**, *2*, 666–673.
90. Lam, K. S. Discovery of novel metabolites from marine actinomycetes. *Curr. Opin. Microbiol.* **2006**, *9*, 245–251.
91. Jang, K. H.; Nam, S. J.; Locke, J. B.; Kauffman, C. A.; Beatty, D. S.; Paul, L. A.; Fenical, W. *Angew. Chem. Int. Ed. Engl.* **2013**, *52*, 7822–7824.
92. Alt, S.; Wilkinson, B. *ACS Chem. Biol.* **2015**, *10*(11), 2468–2479.
93. Moore, B. S.; Trischman, J. A.; Seng, D.; Kho, D.; Jensen, P. R.; Fenical, W. *J. Org. Chem.* **1999**, *64*, 1145–1150.
94. Miao, S.; Anstee, M. R.; LaMarco, K.; Matthew, J.; Huang, L. H. T.; Brasseur, M. M. *J. Nat. Prod.* **1997**, *60*, 858–861.
95. Hassan, H. M.; Degen, D.; Jang, K. H.; Ebright, R. H.; Fenical, W. *J. Antibiot.* **2015**, *68*(3), 206–209.
96. Son, S.; Ko, S. K.; Jang, M.; Lee, J. K.; Ryoo, I. J.; Lee, J. S.; Lee, K. H.; Soung, N. K.; Oh, H.; Hong, Y. S.; Kim, B. Y.; Jang, J. H.; Ahn, J. S. *Org. Lett.* **2015**, *17*(16), 4046–4049.
97. Ding, L.; Maier, A.; Fiebig, H. H.; Lin, W. H.; Peschel, G.; Hertweck, C. *J. Nat. Prod.* **2012**, *75*, 2223–2237.

98. Smyrniotopoulos, V.; Vagias, C.; Rahman, M. M.; Gibbons, S.; Roussis, V. *Chem. Biodivers.* **2010**, 7, 666–676.
99. Smyrniotopoulos, V.; Vagias, C.; Rahman, M. M.; Gibbons, S.; Roussis, V. *J. Nat. Prod.* **2008**, *71*, 1386–1392.
100. Smyrniotopoulos, V.; Vagias, C.; Rahman, M. M.; Gibbons, S.; Roussis, V. *Chem. Biodivers.* **2010**, 7, 186–195.
101. Kladi, M.; Vagias, C.; Stavri, M.; Rahman, M. M.; Gibbons, S.; Roussis, V. *Phytochem. Lett.* **2008**, *1*, 31–36.
102. Ioannou, E.; Quesada, A.; Rahman, M. M.; Gibbons, S.; Vagias, C.; Roussis, V. *J. Nat. Prod.* **2011**, *74*, 213–222.
103. Sato, S.; Iwata, F.; Yamada, S.; Katayama, M. *J. Nat. Prod.* **2012**, *75*, 1974–1982.
104. Liu, L. L.; Xu, Y.; Han, Z.; Li, Y. X.; Lu, L.; Lai, P. Y.; Zhong, J. L.; Guo, X. R.; Zhang, X. X.; Qian, P. Y. *Mar. Drugs* **2012**, *10*, 2571–2583.
105. Hawas, U. W.; El-Beih, A. A.; El-Halawany, A. M. *Arch. Pharm. Res.* **2012**, *35*, 1749–1756.
106. Yang, K. L.; Wei, M. Y.; Shao, C. L.; Fu, X. M.; Guo, Z. Y.; Xu, R. F.; Zheng, C. J.; She, Z. G.; Lin, Y. C.; Wang, C. Y. *J. Nat. Prod.* **2012**, *75*, 935–941.
107. Supong, K.; Thawai, C.; Suwanborirux, K.; Choowong, W.; Supothina, S.; Pittayakhajonwut, P. *Phytochem. Lett.* **2012**, *5*, 651–656.
108. Khamthong, N.; Rukachaisirikul, V.; Tadpetch, K.; Kaewpet, M.; Phongpaichit, S.; Preedanon, S.; Sakayaroj, J. *Arch. Pharm. Res.* **2012**, *35*, 461–468.
109. Karim, M. R. U.; Harunari, E.; Oku, N.; Akasaka, K.; Igarashi, Y. *J. Nat. Prod.* **2020**, *83*, 1295–1299.
110. Liu, Y. F.; Zhang, Y. H.; Shao, C. L.; Cao, F.; Wang, C. Y. *J. Nat. Prod.* **2020**, *83*(4), 1300–1304.
111. Schumacher, R. W.; Talmage, S. C.; Miller, S. A.; Sarris, K. E.; Davidson, B. S.; Goldberg, A. *J. Nat Prod.* **2003**, *66*(9), 1291–1293.
112. Simmaco, M.; Kreil, G.; Barra, D. *Biochim. Biophys. Acta* **2009**, *788*(8), 1551–1555.
113. Zhou, C.; Wang, Z.; Peng, X.; Liu, Y.; Lin, Y.; Zhang, Z.; Qiu, Y.; Jin, M.; Wang, R.; Kong, D. *Chem. Biol. Drug Des.* **2018**, *91*(1), 50–61.
114. Cancelarich, N. L.; Wilke, N.; Fanani, M. A. L.; Moreira, D. C.; Pérez, L. O.; Alves Barbosa, E.; Plácido, A.; Socodato, R.; Portugal, C. C.; Relvas, J. B.; de la Torre, B. G.; Albericio, F.; Basso, N. G.; Leite, J. R.; Marani, M. M. *J. Nat. Prod.* **2020**, *83*(4), 972–984. https://doi.org/10.1021/acs.jnatprod.9b00906.
115. Marani, M. M.; Dourado, F. S.; Quelemes, P. V.; de Araujo, A. R.; Perfeito, M. L.; Barbosa, E. A.; Véras, L. M.; Coelho, A. L.; Andrade, E. B.; Eaton, P.; Longo, J. P.; Azevedo, R. B.; Delerue-Matos, C.; Leite, J. R. *J. Nat. Prod.* **2015**, *78*(7), 1495–1504. https://doi.org/10.1021/np500907t.
116. Siano, A.; Húmpola, M. V.; de Oliveira, E.; Albericio, F.; Simonetta, A. C.; Lajmanovich, R.; Tonarelli, G. G. *J. Nat. Prod.* **2014**, 77(4), 831–841. https://doi.org/10.1021/np4009317.
117. Tikhonov, V. E.; Stepnova, E. A.; Babak, V. G.; Yamskov, I. A.; Palma-Guerrero, J.; Jansson, H.-B.; Lopez-Llorca, L. V.; Gerasimenko, D. V.; Avdienko, I. D.; Varlamov, V. P. *Carbohydr. Polym.* **2006**, *64*, 66–72.
118. No, H. K.; Young Park, N.; Ho Lee, S.; Meyers, S. P. *Int. J. Food Microbiol.* **2002**, *74*(1), 65–72.
119. Tsai, G.-J.; Wu, Z.-Y.; Su, W.-H. *J. Food Prot.* **2000**, *63*(6), 747–752.
120. Cao, R.; Xue, C.-H.; Liu, Q. *Int. J Food Microbiol.* **2009**, *131*(2), 272–276.
121. Lönnerdal, B. *Nestle Nutr. Workshop Ser. Pediatr. Program.* **2011**, *2011*(67), 41–54.
122. Al-Nabulsi, A. A.; Holley, R. A. *Food Microbiol.* **2005**, *22*(2), 179–187.
123. Colak, H.; Hampikyan, H.; Bingol, E. B.; Aksu, H. *J. Food Saf.* **2008**, *28*(3), 355–375.

124. Szwajkowska, M.; Wolanciuk, A.; Barłowska, J.; Krol, J.; Litwinczuk, Z. *Anim. Sci. Paper Rep.* **2011**, *29*(4), 269–280.
125. Korhonen, H. J.; Rokka, S. Properties and applications of antimicrobial proteins and peptides from milk and eggs. In *Bioactive Food Proteins and Peptides: Applications in Human Health*; Hettiarachchy, N. S., Sato, K., Marshal, M. R., Kannan, A., Eds.; 2012.
126. Hayes, M.; Ross, R., Fitzgerald, G.; Hill, C.; Stanton, C. *Appl. Environ. Microbiol.* **2006**, *72*(3), 2260–2264.
127. Elbarbary, H. A.; Abdou, A. M.; Nakamura, Y.; Park, E. Y.; Mohamed, H. A.; Sato, K. *Biofactors* **2012**, *38*(4), 309–315.
128. Demers-Mathieu, V.; Gauthier, S. F.; Britten, M.; Fliss, I.; Robitaille, G.; Jean, J. *Int. Dairy J.* **2012**, *28*(2), 94–101.
129. Koo, H. B.; Seo, J. *Pept. Sci.* **2019**, *111, e24122.*
130. Chen, S.; Wilson-Stanford, S.; Cromwell, W.; Hillman, J. D.; Guerrero, A.; Allen, C. A.; Smith, L. *Appl. Environ. Microbiol.* **2013**, *79*, 4015–4023.
131. Shenkarev, Z. O.; Balandin, S. V.; Trunov, K. I.; Paramonov, A. S.; Sukhanov, S. V.; Barsukov, L. I.; Ovchinnikova, T. V. *Biochemistry* **2011**, *50*, 6255–6256.
132. Wiig, M.; Olmarker, K.; Hakansson, J.; Ekstrom, L.; Nilsson, E.; Mahlapuu, M. *J. Hand Surg. Eur. Vol.* **2011**, *36*, 656–662.
133. Greber, K. E.; Dawgul, M. *Curr. Top. Med. Chem.* **2017**, *17*, 620–628.
134. Res, Y.; Shin, D.; Hwang, I.; Boger, D. L. *J. Am. Chem. Soc.* **2004**, *126*, 1041–1043.
135. Rice, L. B. *J. Infect. Dis.* **2008**, *197*(8), 1079–1081.

CHAPTER FOUR

Antiparasitic natural products

Fyaz M.D. Ismail[a,*], Lutfun Nahar[a,b], Kalin Y. Zhang[c], Satyajit D. Sarker[a]

[a]Centre for Natural Products Discovery (CNPD), School of Pharmacy and Biomolecular Sciences, Liverpool John Moores University, Liverpool, United Kingdom
[b]Laboratory of Growth Regulators, Institute of Experimental Botany ASCR & Palacký University, Olomouc, Czech Republic
[c]School of Chinese Medicine, The University of Hong Kong, Pokfulam, Hong Kong SAR, China
[*]Corresponding author: e-mail address: fyaz.ismail@gmail.com

Contents

1. Introduction

An organism that lives on, or within, a host organism and obtains its food from, or at the expense of its host is known as a parasite. Antiparasitics constitute a class of medications, which are utilized for the treatment of widespread parasitic diseases/infections, usually prescribed by a physician

Annual Reports in Medicinal Chemistry, Volume 55
ISSN 0065-7743
https://doi.org/10.1016/bs.armc.2020.03.001

in the developed world or herbal practitioner, healer, medicine-man or shaman in areas, where modern medicine is unavailable. These individuals are an important source of herbal medicaments and knowledge, since commercially produced and regulated medications are unaffordable by the vast majority of the developing world. Moreover, some "synthetic" medications may be considered untrustworthy by some communities. Consequently, culturally acceptable traditional treatments that have not undergone rigorous modern tests for efficacy and/or safety are often used locally. Modernization is causing loss of indigenous knowledge and loss of habit due to urbanization, deforestation and climate change are exacerbating the situation.

Currently, there are three main classes of parasites, which inflict disease in humans: protozoa, helminths and ectoparasites. For protozoans, they include single-celled organisms, for instance *Plasmodia,* which are transmitted by the mosquito insect vector. Notably, protozoa can only multiply, divide and grow within the infected host erythrocytes causing malaria. Helminths (Greek for worms) are found in mummies and coprolites dating back several thousands of years.[1] Ancient documents describe characteristic clinical features of helminth infections from the writings of Chinese scrolls, Egyptian medical papyri, Hippocrates as well as the Ayurveda. For instance, evidence from Ancient Egyptian medical papyri or Assyrian medical texts reported both signs and symptoms indicative of schistosomiasis; also, various biblical passages describe an epidemic (depicted as a "curse"), which appears to show schistosomiasis spread in ancient Mesopotamia.[2,3] Direct observation of parasites required the invention of the microscope, which was instrumental for linking the vectors to the parasites they transmitted, especially for leishmania and malaria.[4]

Helminths are worm like parasites such as those that cause schistosomiasis. Other helminths include flukes, round worm, pinworm and *Trichina spiralis*. Those that reside upon the host are called ecotoparasites and include various fleas, lice and ticks.[5] Such observations were known to some of the oldest systems of medicine. For instance, a pertinent discourse on pathogenic organisms appears in chapters of the *Caraka Samahita,* an seminal text on general medicine in the Ayurveda. The text classifies organisms into normal (sahaja) and the pathogenic or invaders (agantu). Pathogenic parasites are subdivided into those that grow on the external surface or internal environment of the body. The external parasites include lice. Internal parasites thrive on fecal matter, mucus and blood.[6] Perhaps, the most extreme example of parasitism, is *Cymothoa exigua*, or the tongue-eating louse, which severs the blood vessels in a fish's tongue, causing the tongue to atrophy, fall off, and then the parasite replaces this organ and lives within the mouth of the living fish![7]

There are two major phyla of helminths. The nematodes (also known as roundworms) include the major intestinal worms (also known as soil-transmitted helminths) and the filarial worms that cause lymphatic filariasis (LF) and onchocerciasis, whereas the platyhelminths (also known as flatworms) include the flukes (also known as trematodes), such as the schistosomes, and the tapeworms (also known as the cestodes), such as the pork tapeworm that causes cysticercosis. The most prevalent helminthiases are intestinal helminths including ascariasis, trichuriasis, and hookworm, followed by schistosomiasis. Currently, hundreds of millions of people are chronically infected with parasitic worms, and as far back as 1947, Stoll in his seminal paper—"This wormy world," attempted to estimate the number of infections.[8] In an optimistic note looking forward Loker wrote in 2013 about "this de-wormed world,"[9] but currently, individuals in tropical regions still harbor single or multiple infections and it all requires the type of pharmaceutical philanthropy shown by Merk for treatment of onchocerca (river blindness).

Shortcomings in current antiparasitic chemotherapy, whether produced by pharmaceutical companies or from herbal sources, are evident by the lack of safe, general broad spectrum therapeutics. Given the reduced clinical success rates of current therapies, parasite resistance, toxicity and/or cost of current drugs, new, effective and safe alternatives are urgently required but funding remains a critical factor and cross-subsidy from the existing funding models/pharmaceutical sales should be considered. Advances in biochemistry, molecular biology and analytical techniques has allowed detailed examination of parasite biochemical pathways. Most commonly process unique to the invader allows off-target effects to be minimized. Many natural products are able to do so but why such secondary metabolites are present in plants but are active against parasites remains a matter of conjecture. What the methods are by which humans have associated beneficial effect for alleviating or curing a particular ailment, much local knowledge is available which due to rapid cultural, geographical and climate change may allow such knowledge to be lost. When a scientific basis for a particular therapy can be validated using an accepted scientific model of disease, or even clinical use, such compounds can be further considered as a putative treatment option providing issue of availability, safety and economics can be addressed. Pre-clinical studies should include evaluation in robust animal models representative of the different forms of parasitic diseases and precautions put in place that also identify pro-drugs. General criteria for efficacy evaluation on pre-clinical studies should be commonly agreed and employ a standardized model so that cross laboratory data can be readily compared. Resistance to antiparasitics has been a growing concern, especially in veterinary medicine. The egg hatch assay can be used to determine whether a parasite causing infection has become resistant to standard drug treatments.[10] For instance, *Setaria digitata*, a filarial parasite can causes fatal cerebrospinal nematodiasis in mammals such as goats, sheep and horses in the tropics, resulting in serious economic loss. Its close resemblance to *Wuchereria bancrofti* allows its use as a representative model organism to study human lymphatic filariasis.[11]

As a general class, many polyphenols of diverse structure possess promising anthelmintic effects and can be employed as additional or alternative treatment to current synthetic anthelmintics. They are also useful as novel lead structures for drug discovery in this neglected and underfunded area of human and veterinary medicine. Whether the mechanism(s) of action involves interactions with proline-rich surface-proteins or internal targets remains unknown. However, nematode species and their different developmental stages have variable sensitivity to tannins (as the protein composition in the respective target can vary between the species and the different development stages).[12]

A variety of reviews appear periodically on the use of plants as antiparasitic medications, the majority of which are for treating veterinary infections unless a specific disease affecting humans is being discussed.[13–18] A list of plants used for antihelminthic treatment of both animals and humans has been compiled by Veerakumari.[19] This chapter discusses principally the literature from the last decade, but still includes some older relevant materials as long as the molecules are active *in vivo* or, in the opinion of the authors, capable of modification to achieve clinical relevance. Here, the natural drug prototypes are classified by the type of parasitic infections rather than structural types.

2. Babesiosis

The four known *Babesia* species known to infect humans are *B. microti*, *B. divergens*, *B. duncani* and *B. venatorum*. Human babesiosis is a zoonotic infection caused by *Babesia microti* parasites transmitted by the bite of ixodid ticks crossing species from those infesting natural mammalian hosts including deer, domesticated cattle and rodents. However, the overall life cycle within humans shares the same trajectory as in other mammals. *Babesia* parasites are intracellular obligates that target erythrocytes, and the parasite's ability to first recognize, invade and deconstruct such cells is central to the disease pathology, which can be compared to some aspects of malaria infections.[20] Worryingly, reports of human infections have increased. Human babesiosis is prevalent in the Northeastern and Midwest USA and *Babesia divergens* in Western Europe with sporadic occurrence throughout the rest of the globe. It is associated with warm climates.[21] Since part of the life cycle involves erythrocyte invasion, it leads to symptoms of anemia, low hematocrit and circulating toxic metabolites released into the host's circulatory system but this can be resolved by exchange transfusion. Screening blood for infections avoids secondary transmission with contaminated blood for this (and any other blood borne parasite). Finally, infections can occur congenitally during pregnancy.[20,21] After *Trypanosoma*, *Babesia* is considered to be the second-most common blood parasite infecting mammals. In addition, domestic animals can also be infected in areas lacking severe winters; its prevalence is set to increase with climate change. Within cattle, it goes by the common names Texas cattle fever, redwater, or piroplasmosis.[22]

Until recently, babesiosis treatment for animals was based either on imidocarb or diminazene aceturate (Berenil). Parasite infection of cattle (bovine babesiosis) have been motivated in order to prevent economic losses. The efficacy of treating haemosporidan infections of cattle with the total alkaloid extract of *Peganum harmala* is known.[23] Out of 82 cattle naturally infected by haemosporidians, the cure rate rose above 85%. Harmala alkaloids from *Peganum harmala*, as well as tobacco leaves, include harmine, harmaline and harmol (Fig. 1). Such alkaloids have been in use for centuries in Amazonian shamanism (derived from other plants), where they function as monoamine oxidase inhibitors (MAOIs) and are controlled substances considered too toxic as patient medications. In experimental studies, the treatment proved effective in animals with single infections ascribed to either *B. bigemina* or *T. sergenti*. Unsurprisingly, mixed infections responded poorly to harmaline alkaloids therapy.

Mild babesiosis in humans responds to clindamycin treatment for 7–10 days in *B. divergens* and *B. venatorum* infections.[24,25] *B. duncani* requires

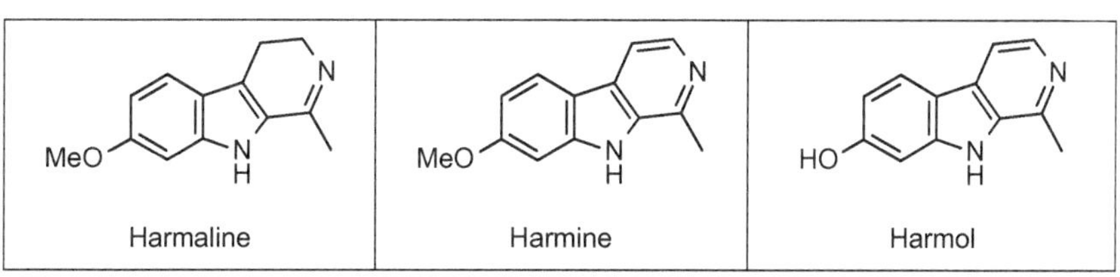

Fig. 1 Harmal alkaloids.

the supplementation with either quinidine or quinine. Since adverse reactions from quinine are thoroughly documented,[26] intravenously administered quinidine is a recommended alternative.[25,27] In mild *B. microti* infections, a 7–10 days course atovaquone plus azithromycin is superior to clindamycin/quinine combination, since it is just as effective, but with fewer side effects.[28]

Several natural products[20] have shown promise including artesunate, epoxomicin, gossypol and nerolidol (Fig. 2) and are discussed below. Since gossypol is an irreversible contraceptive in over 20% of men, it should be reserved as an agent of last resort.[29] The semisynthetic, artesunate was particularly effective against *B. microti* infected mice infected at doses equal to, or greater than 10 mg/kg of body weight on days 8–10 post-infection without notable side effects, indicating that artesunate could be used for treating *Babesia* infection.[30]

Many essential oils show activity against *Babesia in vitro*; for instance, nerolidol (Fig. 2). The IC_{50} values of nerolidol on the third day of culture for the growth were: *B. bovis* (21 ± 1), *B. caballi* (23.1 ± 1), *B. bigemina* (29.6 ± 3), *B. ovata* (26.9 ± 2), in micro-aerophilous stationary phase (MASP) cultures, but was ineffective against mice inoculated with *B. microti*.[31] The mechanism of inhibition of *Babesia* parasites is currently

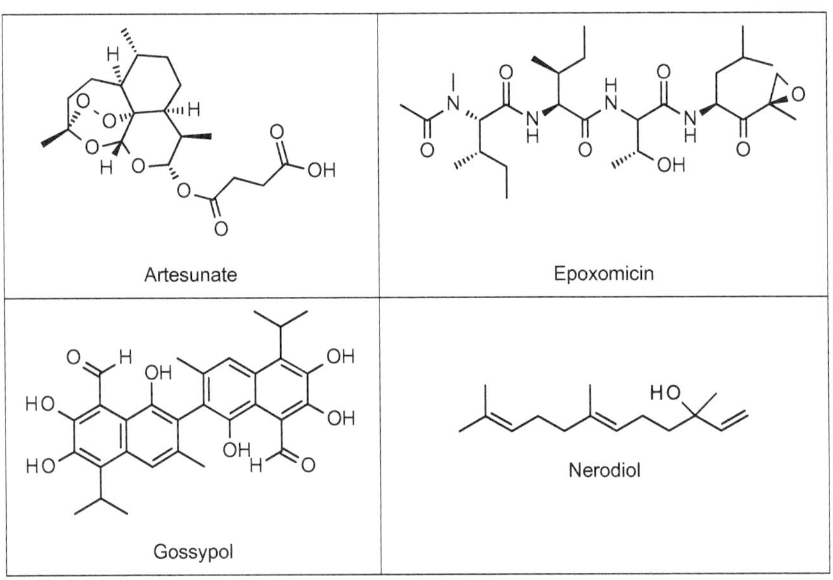

Fig. 2 Structures of artesunate, epoxomicin, gossypol and nerolidol.

unknown; however, *B. bovis* possesses an active isoprenoid pathway within its apicoplast, sharing similarities with both *P. falciparum* and *T. parva*.[32]

The *Babesia gibsoni* culture system can be used to screen antibabeisal activities of promising plant extracts.[33] For instance, of the 45 plant extracts selected from Central Kalimantan, Indonesia, appreciable antibabesial activity was observed within *Arcangelisia flava, Curcuma zedoaria, Garcinia benthamiana, Lansium domesticum* and *Peronema canescens*, with IC_{50} values ranging from 5.3 to 49.3 µg/mL without acute toxicity in mice at the intraperitoneal dose of 0.7 g/kg of body weight.[33] Other investigators demonstrated that aqueous extracts of *Achillea millefolium, Baeckea frutenscens, Brucea javanica, Curcuma xanthorrhizha, Strychnos lucida* and *Swietenia macrophylla* possessed notable antibabesial properties.[34] After screening the boiled extracts from 28 Indonesian medicinal plants, the fruit extract of *B. javanica*, known for its activity against *Plasmodia*, proved to be a potent antibabesial component against *B. gibsoni* (10 µg/mL). *Brucea javanica*, locally known as "buah makasar," is used in "Jamu" traditional folk medicine in Indonesia, to treat malaria, dysentery and cancer. Bioassay-guided fractionation of the fruit extract of *B. javanica* afforded two quassinoids, bruceantinol B and bruceine J. The structures were elucidated both using spectroscopic data and by chemical transformation to known compounds. In addition, the known quassinoids bruceines A to D, bruceantinol, and yadanziolide A were isolated. Antibabesial activities were determined *in vitro*, whereby bruceine A and bruceantinol (Fig. 3) proved more potent than diminazene aceturate, the veterinary drug used clinically against *B. gibsoni*, with IC_{50} values of 4 and 12 ng/mL, respectively.[35] It has only been used once in humans against *Babesia* infection but the patient did not survive; this drug has been used off-label successfully but against human African trypanosomiasis.[36,37]

Following on from the earlier studies on the active antibabesial quassinoids from *B. javanica*, brucein A (IC_{50} =7.7 nM) and bruceantinol

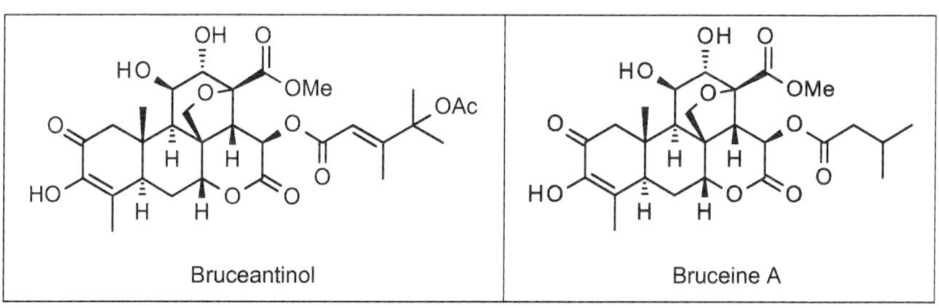

Fig. 3 Antibabesial components, bruceantinol and bruceine A, from *Brucea javanica*.

(IC_{50} = 19.8 nM) (Fig. 3) were identified as more potent antibabesial compounds compared to diminazene aceturate (IC_{50} = 172.6 nM); bruceine A at a concentration of 25 nM killed the babesial parasites within 24 h.[38] Oral administration of bruceine A at 6.4 mg/kg/day for 6 days effectively treated two parasite-infected dogs, without causing any apparent side effects.

A boiled extract of *Arcangelisia flava* provided a series of related protoberberine alkaloids (palmatine, berberine, jatrorrhizine and dihydorberberine) (Fig. 4) which inhibited the growth of *B. gibsoni* in culture, at concentrations between 100 and 1 μg/mL.[39] Many phytochemicals isolated from plants were tested for their antibabeisal activity, using a *Babeisa gibsoni* culture system. For instance, the antibabesial activity of 1-O-galloyl-6-O-luteoyl-α-D-glucose isolated from the boiled extract of *Phyllanthus niruri* has activity (IC_{50} value = 4.7 ± 0.8 μg/mL) within one log dose of diminazene (IC_{50} = 0.6 ± 0.04 μg/mL).[40]

In another study, the antibabesial activities of four compounds: 3′-demethoxy cyclocurcumin from *Curcuma xanthorrhiza*, *p*-hydroxybenaldehyde and cleomiscosin A from *B. javanica* and (+)-epiloliolide (Fig. 5) from *Excoecaria cochinchinensis* was evaluated with diminazene as the control.[41] The IC_{50} values of 3′-demethoxycyclocurcumin, *p*-hydroxybenaldehyde, cleomiscosin A and (+)-epiloliolide were 16.6, 7.6, 15.6 and 10 μg/mL, respectively, in comparison to diminazene aceturate (IC_{50} = 0.6 μg/mL). The authors suggested that

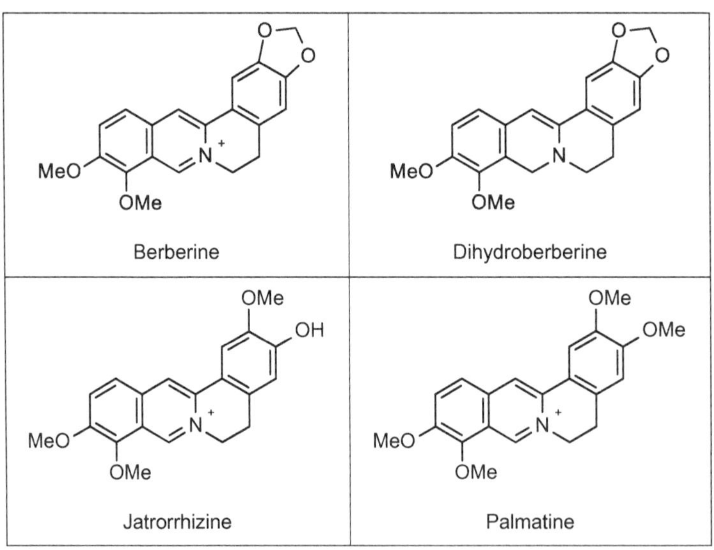

Fig. 4 Antibabesial protoberberine alkaloids from *Arcangelisia flava*.

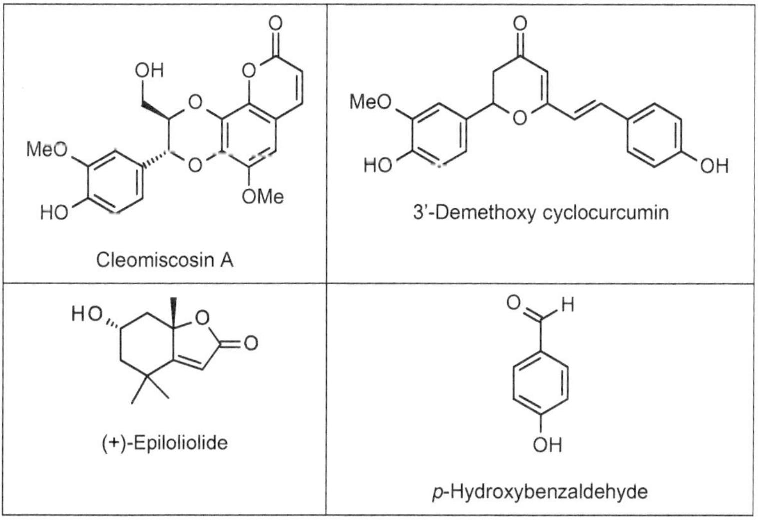

Fig. 5 Antibabesial compounds from *Brucea javanica, Curcuma xanthorrhiza* and *Excoecaria cochinchinensis.*

these could serve as prototypes for further development, since none were superior to the synthetic compound.

(−)-Epigalllocatechin-3-gallate (the major tea catechin) may prove useful for chemotherapy of *Babesia* infections. The IC_{50} values of 18 and 25 μM for *B. bovis* and *B. bigemina in vitro* were then evaluated in a mouse model innoculated with *B. microti in vivo* at dose of 5 and 10 mg/kg body weight.[42] The parasites were cleared by day 14 and 16 days post-inoculation. Similarly, the growth inhibitory assays of the same compound on equine *Babesia in vitro* demonstrated IC_{50} values of 4.42 and 3.64 μM against *B. caballi* and *B. equi*, respectively.

Clindamycin, a semisynthetic derivative of linomycin, a natural antibiotic expressed by actinobacterium *Stromyces lincolnensis*, obtained by 7(*S*)-chloro-substitution of the 7(*R*)-hydroxyl group of lincomycin. The synthesis of clindamycin was first announced by Magerlein et al.[43] and was marketed 2 years later.[44,45]

Epoxomicin (Fig. 6), originally isolated from isolated from an *Actinomycetes* strain, is a α′, β′-epoxyketone natural product. It binds the LMP7, X, Z and MECL1 catalytic *β* subunits of the proteasome, resulting in inhibition of three of its known proteolytic activities causing cell death.[46,47] It is a potent proteasome inhibitor, targeting a prototypic T1 threonine-peptidase multi-subunit complex, which plays a key role

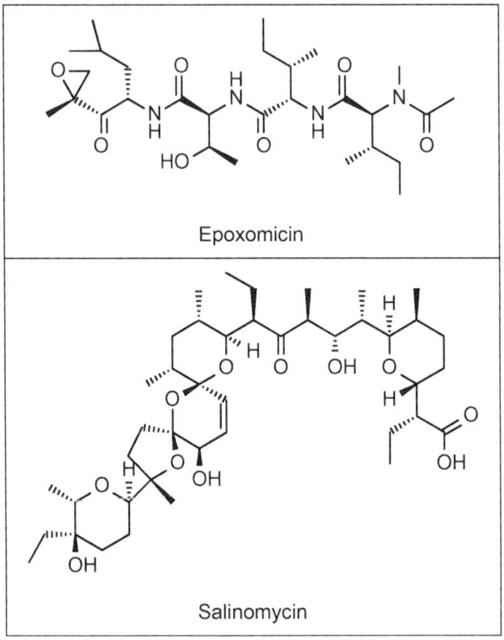

Fig. 6 Epoxomicin and salinomycin—two microbial antibabesial agents.

in endogenous protein turnover. Additionally, epoxomicin inhibits *Plasmodium falciparum* throughout the asexual, sexual, and mosquito midgut stages were destroyed.[48,49] *B. bovis* was inhibited at 10 nM epoxomicin, while *B. bigemina*, *B. ovata*, *B. caballi*, and *Theileria equi* required while only 5 nM epoxomicin treatment. However, *B. bigemina* and *T. equi* required 50 nM epoxomicin, to completely suppress growth. However, for a similar effect on *B. bovis*, *B. ovata*, and *B. caballi,* a much higher concentration was required (100 nM).[50]

Salinomycin (Fig. 7) belongs to a smaller subgroup of polyether antibiotics, known as carboxylic ionophores, that possess open-chained structure containing oxygenated heterocyclic rings and a single terminal carboxyl group capable of winding around group one cations such as Na^+. It is claimed to be a broad spectrum antibiotic also capable of attacking parasites such as the canine pathogen *Babesia gibsoni*.[51] Salinomycin then leads to both the activation of Na^+/K^+-ATPase and increased consumption of intracellular ATP, which ultimately result in the hemolysis of the parasite-infected erythrocytes. Unusually, the 20-deoxy salinomycin derivative, which is considerable less active as a coccodiostat, still crystalizes from hexane while still bound to a cation.[52,53]

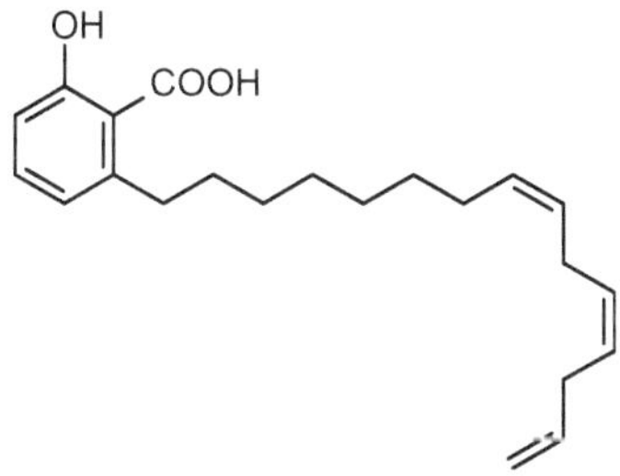

Fig. 7 Anacardic acid—a natural anti-Echinococcosis agent.

2.1 *Babesia* drug resistance

Currently, the combination of the semisynthetic clindamycin/quinine or atovaquone plus azithromycin remains effective, the speed of therapeutic response is poor and parasite resistance has arrived.[54,55] Worryingly, three incidents of drug resistance to atovaquone/azithromycin (defined by parasite relapse after more than 28 days of antibabesial treatment) in immunocompromised patients with *B. microti* were reported.[56] Recently, studies with artemether and lumefantrine *in vitro* showed a positive synergistic interaction of these compounds against *B. gibsoni*.[57] Compounds that directly target the pathways of parasite DNA and RNA synthesis of *B. gibsoni* and *B. bovis*, such as mycophenolate mofetil, mizoribine, ribavirin, and 7-nitroindole, directly inhibit propagations of these parasites *in vitro*.

2.2 Future prospects for *Babesia* rational drug design

All the aforementioned molecules offer novel options to control the disease. With the complete sequence of the *Babesia bovis* genome in hand and the *B. bigemina* genome project in progress,[58] the post-genomic era may allow rational drug design against novel targets. Meetei et al.[59] constructed a molecular model of babesipain-1, a cysteine protease from the bovine piroplasm *Babesia bigemina*, and virtually screened a chemical library of both natural and synthetic compounds, to identify putative inhibitors. The interactions between babesipain1 and the candidate lead compounds were then further examined using molecular dynamics (MD) simulation. Nine of the top hits in both synthetic and natural products were identified for further investigation.

3. Echinococcosis

Echinococcosis, a zoonotic infection caused by cestode species of the genus *Echinococcus*, has rather limited treatment options and new antihydatid agents are urgently required. Anacardic acid (AA) (Fig. 7), a natural product isolate from the Brazilian cashew-nut (*Anacardium* spp.) shell liquid, proved highly active against metacestodes of *Echinococcus multilocularis* (*E. multilocularis*) and *Echinococcus granulosus sensu stricto* (*E. granulosus* s.s.) both *in vitro* and *in vivo*.[60] AA exerted superior efficacy against *E. granulosus* s.s. *protoscoleces* and *E. multilocularis* metacestodes than either albendazole or dihydroartemisinin *in vitro*, and an inhibition on the growth of *Echinococcus* metacestode as effective as albendazole *in vivo*. The mechanisms of AA against *Echinococcus* may involve suppression of angiogenesis on/in the metacestode mass (through inhibiting vascular endothelial growth factor (VEGF)-induced signaling pathways). Consequently, AA is a putative potential candidate drug for echinococcosis treatment. However, related molecules are skin allergens (c.f. poison ivy oils), hence, detailed toxicological and immunological studies are warranted before widespread use.[60]

4. Cysticercosis

Cysticercosis is a parasitic disease that frequently involves the human central nervous system (CNS), and current treatment options are limited. Oxfendazole has revealed notable activity against the tissue stages of *Taenia solium* and has promise as an effective therapy for neurocysticercosis.[60] Bizhani details a number of herbal treatments including of wormicidal plant extracts of *Melia azedarach* Linn. (Meliaceae), which were superior to piperazine phosphate against *T. solium*.[61,62] *Ruta graveolens* (Rutaceae) is used as a popular crude herb infusion, but in crude form, is an unsafe natural antiparasitic agent. Although it presented a mild antiparasitic effect in mice, symptoms of natural-products-induced-liver-disease confirmed that its self-medication should be avoided.[63]

5. Balantidiasis

Balantidium coli, a single-cell parasite usually infects pigs or camels but can, in rare cases, jump hosts and cause intestinal infection in humans. Direct contact with pigs is the usual transmission mode or even by drinking contaminated water, usually in tropical regions. Three medications are used most often to treat *Balantidium coli*: tetracycline (Fig. 8), metronidazole, and iodoquinol. However, human cases are underreported and a detailed review of the literature finds at least 246 medical reports from 1945 onward. Some cases are mild while others can be fatal, especially if the patient is

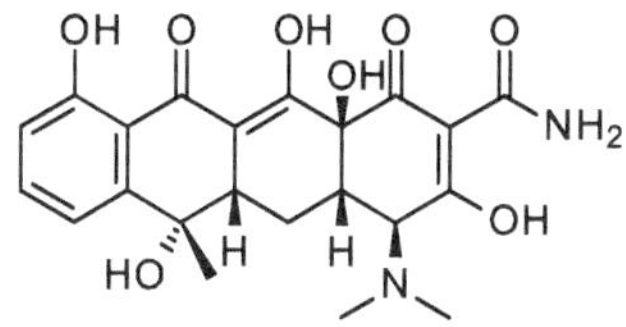

Fig. 8 Tetracycline—a first-line treatment for infections caused by *Baluntidium coli.*

immune-compromised. Several countries suspect that infections within human are more widespread than commonly reported.[64] For individuals with early, localized infection, oral administration of doxycycline remains a first-line treatment, since is effective against not only *Borrelia bacteria* but other infections transmitted by ticks. However, doxycycline is contraindicated in children less than 8 years of age as well as pregnant women or those breastfeeding children and can be photo-toxic.

Alternatives to doxycycline include amoxicillin, cefuroxime axetil, and azithromycin, a potent semisynthetic derived from erythromycin. The semisynthetic, azithromycin is recommended only in cases of intolerance to the other antibiotics. Ticks transmit the human strain of *babesiosis*, so it often co-occurs with other tick-borne illnesses such as Lyme disease.[24] When doubts arise in diagnosis between either Lyme disease or cellulitis, treatments are used that are effective against both infections: cefuroxime or amoxicillin/clavulanic acid.[65] Extracts were prepared from elder berry (*Sambucus nigra* L.) f. and are claimed to be useful against *Balantidium coli.*[66]

6. Blastocystosis

The blastocystis parasite enters humans through the fecal-oral route. Eating food or drink contaminated with human or animal feces can infect humans. *Blastocystis* is a protozoal, single-celled parasite that inhabits the gastrointestinal tracts of humans and other animals. Response to medication *for Blastocystis hominis* varies greatly between individuals and is often misdiagnosed. Many different types of *Blastocystis* are known, and they can infect humans, farm animals, birds, rodents, amphibians, reptiles, fish, as well as cockroaches. Blastocystosis may be a possible risk factor for development of irritable bowel syndrome. A number of studies found that military personnel suffered high rates of infection. One of the earliest accounts involves British troops in Egypt around 1916 who recovered following treatment with emetine.[67] Following up on reports in 1916,[68] modern investigations showed emetine was more effective than either metronidazole or

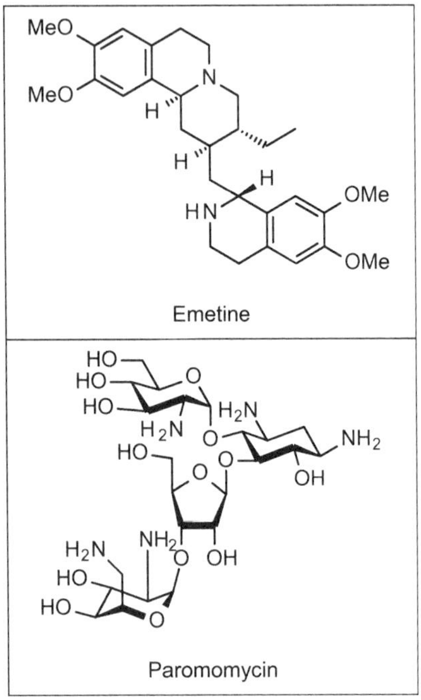

Fig. 9 Natural products against blastocystosis.

furazolidone when tested *in vitro*.[69] Emetine (Fig. 9) is available in the USA through special arrangement with the Center for disease control. Clioquinol (entero-vioform) was noted as successful in treatment of *Blastocystis* infection, but was withdrawn due to an adverse event in Japan.[70] Fourneau's acetarsol and narsenol, two arsenic-based antiprotozoals, were reported to be effective against the infection.[71] A 1990 study concluded symptomatic infection was more common in non USA nationals, immunocompromised individuals, as well as children.[72] Potential medications for treating *Blastocystis hominis* include various antibiotics, such as metronidazole or tinidazole. Combination medications, such as sulfamethoxazole and trimethoprim, nitazoxanide or paromomycin. Notably, paromomycin (Fig. 9) was isolated from an asteromyces strain in the 1950s and marketed in 1960. Interestingly, this antimicrobial can treat intestinal parasitic infections such as cryptosporidiosis and amoebiasis as well as leishmaniasis (cutaneous-1960s and much later, visceral-1990s in the former USSR). Paromomycin demonstrates superior performance when compared to both metronidazole and clioquinol and should be the agent of choice in treating *Blastocystis*.[73,74]

7. Coccidiosis

Coccidia are passed on through the fecal-oral route world wide and also affects human pets such as dogs and cats. Numerous therapeutic agents trialed, including bismuth, iodide, quinine, gentian violet, emetine and anthelmintics have failed to identify useful agents. Avian coccidiosis is hard to eradicate due to the tough oocyst wall and also the sporulated oocyst can survive harsh environmental conditions. Several parasitic protozoa that can cause enteric infections in humans, especially *Cryptosporidium* spp., *Cyclospora cayetanensis, Giardia duodenalis* (syn. *Giardia lamblia, Giardia intestinalis*), and *Cystoisospora belli* (previously *Isospora belli*).[75,76] Infections are typified by gastrointestinal clinical manifestations that may include diarrhea, vomiting, abdominal cramps, and general malaise.[77] A review of 68 plants used for anticoccidial actions was published in 2016, but for use in the poultry industry.[78] However, many compounds already in use for treatment of human *Plasmodial* infections were reported and of these should be tested against human case of cryptosporidium species including artemisinin, allicin and febrifugine (Fig. 10).

Dichroa febrifuga Lour., also known as Chang Shan, is a Chinese medicinal herb for protozoan diseases. Zhang and coworkers showed that crude extract of *D. febrifuga* was effective against *Eimeria tenella* infection in chickens.[79] The alkaloidal component, febrifugine, was halogenated to make an anticoccidial halogenated derivative, halofuginone. However, the strong emetic

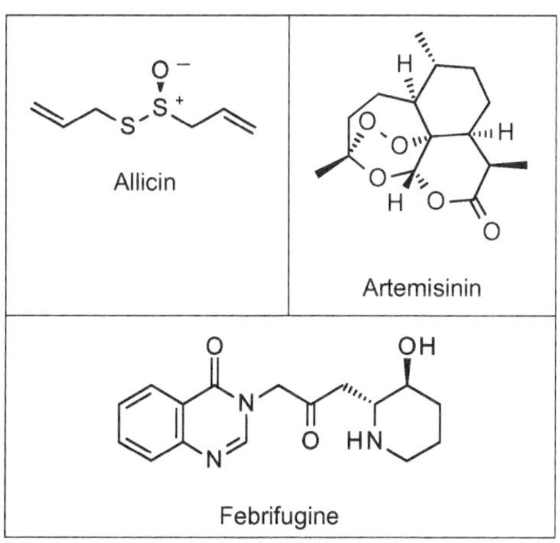

Fig. 10 Examples of anticoccidial natural products.

properties of this compound makes patient compliance problematic.[78] Three of these protozoa, *Cryptosporidium, Cyclospora*, and *Cystoisospora*, were previously classified as coccidian parasites (as such parasite infect enterocytes) with a complex life cycle which included asexual (meronts) and sexual (microgametocytes and macrogametocytes) reproductive stages. In contrast, *Giardia* is a flagellate, which fails to invade epithelial cells, reproduces by asexually, binary fission (see below). Since such infections are becoming more important in immunocompromised individuals (HIV, organ transplantation).

8. Amoebiasis

The parasite *Entamoeba histolytica* affects the intestines; due to long-term treatment toxicity and clinical resistance to drugs commonly used against this pathogen, new, safer and affordable drugs against amoebiasis are urgently needed. It is prevalent in tropical regions especially in areas of high population density and/or poor sanitation/hygiene practices and, therefore, transmitted through the fecal-oral route. In order to minimize dependency on toxic synthetic excipients, Sharma et al.[80] used a natural polysaccharide-based drug delivery system to deliver metronidazole (MNZ) into the colonic region for localized treatment of amoebiasis. They concluded that *Trigonella foenum-graecum* (fenugreek) polysaccharide-based double compression tablets were promising colon-specific drug carriers with reduced toxic effects associated with other commonly employed synthetic excipients. In terms of the poultry industry, berberine (Fig. 4) has been useful for synergystic treatment of severe diarrhea, amoebiasis and intestinal infections.[81]

The effects of epicatechin on *Entamoeba histolytica* morphology was analyzed by electronic microscopy. *E. histolytica* trophozoites were incubated for 48 h at 37 °C in the presence of 1.9 μg/mL epicatechin and analyzed using transmission electron microscopy. Epicatechin (Fig. 11) induced both nuclear and cytoplasmic changes in the treated trophozoites. These alterations appeared identical to the cellular changes seen in by *E. histolytica* trophozoites undergoing programmed cell death (PCD), suggesting that epicatechin could be an alternative compound to treat amoebiasis. However, caution is warranted since, depending on dose consumed, catechins (and their metabolites) can bind to erythrocytes and possibly induce release of autoantibodies, resulting in hemolytic anemia and renal failure.[82] This has resulted in the withdrawal of the catechin-containing drug Catergen [(+)-cyanidanol-3], used to treat viral hepatitis, from the

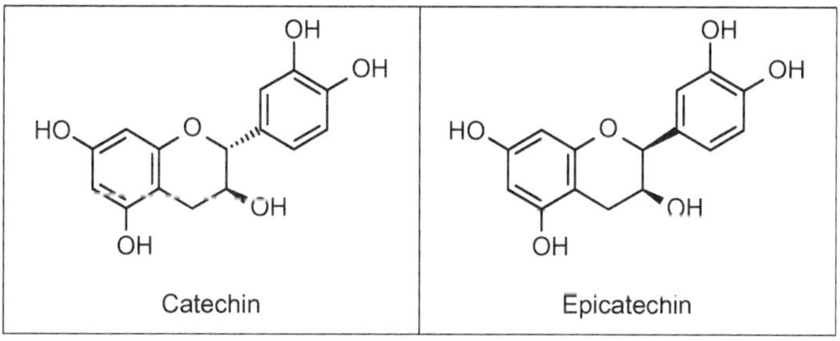

Fig. 11 Catechin and epicatechin—two natural antiamoebic agents.

European market in 1985.[83] The flavan-3-ol, (−)-epicatechin has been previously identified as the most important antiamoebic compound among the extracts from two medicinal plants: *Rubus coriifolius* and *Geranium mexicanum*.[84,85]

Castela texana (T. & G.) Rose (syn. *C. tortuosa* Liebm, *C. Nicholsonii* Hook. p *texana* T. & G.; Simaroubaceae) is a small shrub distributed in SouthWestern Texas and Northeastern Mexico. This plant has been employed in folk medicine to treat human dysentery,[86] despite the stem bark being intensely bitter. Although *C. texana* has long been used in traditional medicine in teas (or capsules containing dried plant-material) to treat patients with amoebic dysentery, little is known about its mutagenic, toxic or antigenotoxic effects. *C. texana* belongs to the *Simaroubaceae* family of which some members contain active principles, such as a potent antileukemia quassinoid from *Brucea antidysenterica*,[87] apart from antiplasmodial, antitrichomonic, antiamoebic and cytotoxic activities as reported in various conference papers, but not as full papers in scientific journal.

The treatment of amoebic infection of the cornea is extremely difficult and protracted in nature with current drugs *Acanthamoeba* keratitis. The main and most frequent disease caused by amoeba infections have risen sharply (85–88%) due to the popularity of wearing contact lenses.[88] The majority of cases (88%) are associated with wearing hydrogel lenses while rigid lenses are less affected (12%). The annual probability of contracting Acanthamoeba keratitis (AK) for individuals wearing contact lenses is around 1:30,000. AK is usually diagnosed in the advanced stage of the disease. Eye diseases caused by amoebae from the genus *Acanthamoeba* are normally chronic and severe, and their treatment is prolonged and often ineffective. A report by Hadas et al.[89] has shown that semi *in vivo* studies of tea tree oil (*Melaleuca alternifolia*) simulating eyeball infection demonstrated 100% effectiveness in the case of both trophozoites and cysts of amoebae from the genus *Acanthamoeba*. In

Brazilian folk medicine, *Tetradenia riparia* (Hochst.) Codd. (Lamiaceae) is used to treat toothaches and dental abscesses and diseases induced by worms, bacteria, or fungi. The action of tea tree oil (*Melaleuca alternifolia*) indicates that this is the first substance with a potential ability to quickly and effectively remove the amoebae from the eye. Tea tree oil has the ability to penetrate tissues, which allows it to destroy amoebae in both the shallow and deep layers of the cornea.

9. Giardiasis

Giardia, commonly called "beaver fever" affects the lumen of the small intestine. Consumption of food and/or water contaminated with feces, dormant cysts may go on to produce infection. The betel (*Piper betle*) is a vine belonging to the Piperaceae family, which includes both pepper and kava. Betel leaf is mostly chewed in Asia, and elsewhere in the world by some Asian emigrants, as betel quid or in paan, with Areca nut often with added tobacco. *Piqueria trinervia Cav.* (Asteraceae) is a plant species with a long history in traditional medicine to cure diarrhea and other digestive disorders. Its active constituents, trinervinol and piquerol (Fig. 12) showed antigiardial activity with an $IC_{50} = 2.03$ and 2.42 μg/mL, and $IC_{90} = 13.03$ and 8.74 μg/mL, respectively.[90] The concentrations of trinervinol employed ($CC_{50} = 590$ μg/mL) and piquerol ($CC_{50} = 501$ μg/mL) did not show cytotoxicity against human fibroblasts. Consequently, antigiardial activity may be enhanced with suitable chemical modification.

An ayurvedic herbal medicine, prepared from *Piper longum* (common name: Pippali) and *Butea monosperma* (Bengali name: Palash), has significant activity against giardiasis. *Butea monosperma* popularly known as Flame of the Forest, Dhak, or "Bastard teak," which has immense potential and appears to have a broad spectrum of activity on several ailments.

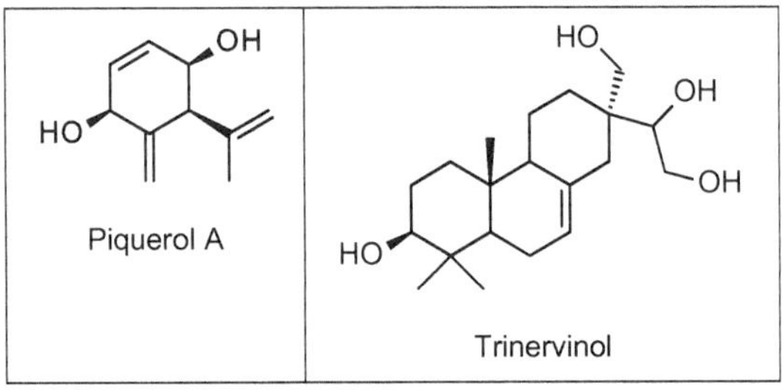

Fig. 12 Structures of piquerol and trinervinol.

Previous phytochemical examinations of this plant identified several constituents, including cajanin, isoformononetin, stigmasterol, butin, isobutrin (5-{(1*E*)-3-[4-(beta-D-glucopyranosyloxy)-2-hydroxyphenyl]-3-oxoprop-1-en-1-yl}-2-hydroxyphenyl beta-D-glucopyranoside) and the less active butrin; free sugars and free amino acids and (−)-3-hydroxy-9-methoxypterocarpan[(−)-medicarpin] (Fig. 13).[91] Notably its reported pharmacological properties include anthelmintic properties. The methanol extract of *Butea monosperma* seeds, tested *in vitro*, showed significant anthelmintic activity.

10. Plasmodia

A detailed exposition of malaria and its treatment can be found in a variety of recent reviews as well as the volume preceding this one.[92] Malaria is by far the most studied of all parasitic diseases given its widespread prevalence and association with military operations and conquest from colonization of tropical counties.[93] Unsurprisingly, many military organizations have spearheaded the effort including the Walter Reed Army Research Institute as well as selected pharmaceutical countries. The recent intervention of the Bill Gates and Melinda Gates organization is simply the latest manifestation. Compounds from natural sources were comprehensively discussed recently by Ismail.[94] Subsequently, a comprehensive listing of antiplasmodials appeared documenting all compounds active against *Plasmodia* (drug sensitive and resistant) *in vitro* possessing $IC_{50} \leq 3.0$ μM.[95] In contrast, this chapter critically reviews only compounds showing activity *in vivo*, since then.

Cassia siamea Lamk (Caesalpiniaceae) is a small to medium-sized tropical tree, growing up to 15–20 m tall, with a short bole and low branching high crown, native of Southeast Asia, and is well regarded for its medicinal, nutritional (vegetable-leaves and flowers), and economic importance. It is used in Burmese as well as Thai cuisine. In traditional medicine, it is used against fever, malaria, diabetes, hypertension, asthma, constipation, diuresis, CNS disorders, jaundice, abdominal pain and menstrual pain. Dosing is normally through an oral route using a decoction of leaves and flowers. The major phytochemical constituents present include anthraquinones, coumarins, alkaloids, flavonoids, glycosides, triterpenoids, sterols and other polyphenols which are analgesic, antioxidant, anti-inflammatory, antimalarial, vasorelaxant, antitumor, antiarrhythmic, antiandrogen, anti-HIV and anxiolytic activity in various *in vitro* and *in vivo* studies.[96] Cassiarin A (Fig. 13), isolated

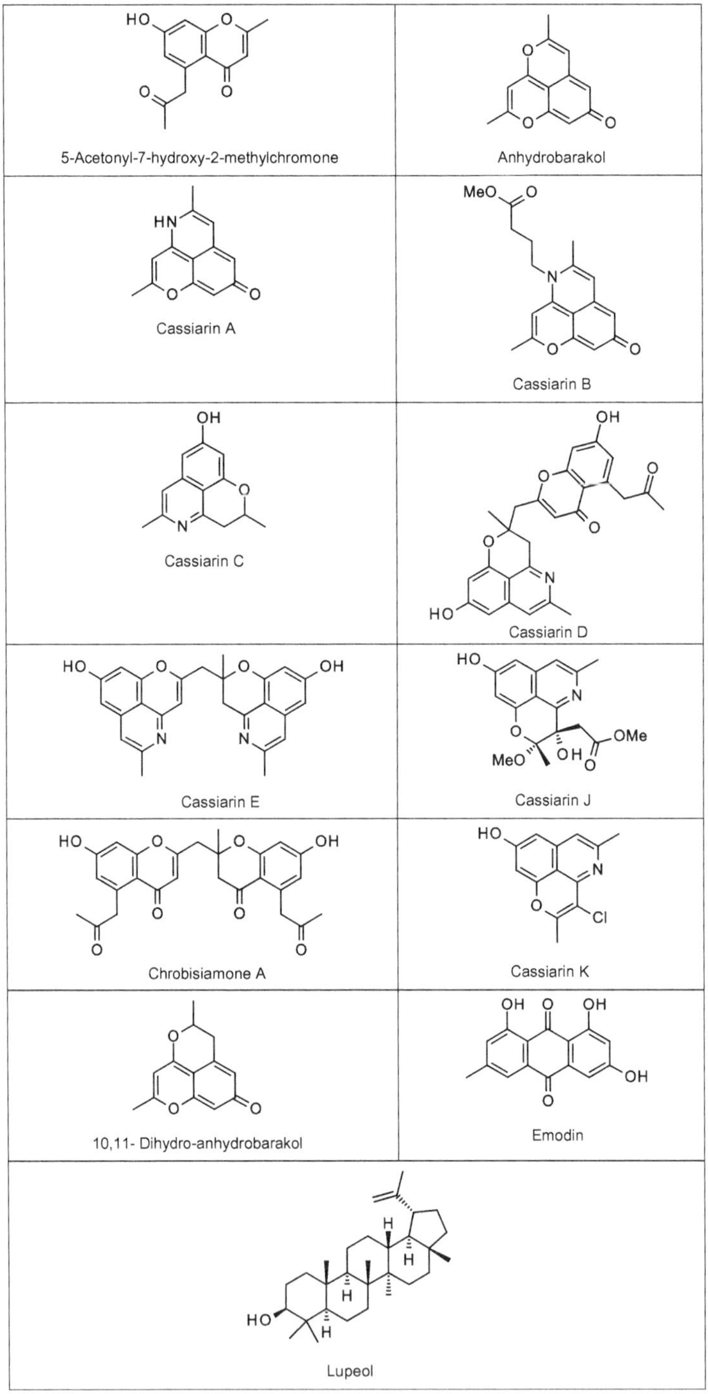

Fig. 13 Antimalarial compounds from *Cassia siamea*.

from the leaves of *C. siamea*, showed promising antimalarial activities. The ethanol and chloroform extracts, alkaloid fraction, cassiarin A, and chloroquine as a positive control exhibited the inhibition against *Plasmodium falciparum* (IC_{50}: 7.06, 2.41, 0.24, 0.005, and 0.006 μg/mL, respectively). The antimalarial activity *in vivo* was assessed using the 4-day suppressive Peter's Test. The ethanol, chloroform extracts, and alkaloid fraction were found to give an ED_{50} 34.7, 19.59, and 0.47 mg/kg, respectively (p.o.). The ED_{50} values of cassiarin A and chloroquine were 0.17 and 0.21 mg/kg, respectively (i.p.) in *P. berghei*. The plant contains other antimalarial constituents with lower efficacy: cassiarin A, cassiarin B, emodin, lupeol, chrobisiamone A, 10,11-dihydroanhydrobarakol (Fig. 13) and in addition the EtOAc extract of stem bark also possess activity. Cassiarin A was effective against *P. falciparum* (IC_{50} 0.005 μL/mL), while cassiarin B, emodin, lupeol and chrobisiamone A showed moderate activity (IC_{50} 6.9 μL/mL).

Three tricyclic isoquinoline alkaloids were isolated from the leaf extract of *Cassia siamea* (Fabaceae), a plant traditionally used to treat periodic fever and malaria in Indonesia.[97] Among the isolated compounds, cassiarin J and the first halogenated cassiarin congener, cassiarin K, inhibited the *in vitro* growth of 3D7 *P. falciparum*. However, cassiarin J was less active than the simpler, more potent cassiarin A, suggesting that the role of the pyran ring of the cassiarins in antiplasmodial activity should be further explored in structure-activity relationship (SAR) investigations.[98] The result showed that cassiarin A deserved further investigation toward development of a promising antimalarial drug.[99] After screening 1047 protein structures (belonging to 147 different proteins) from the protein data bank, they identified 16 protein targets for cassiarins. In a few cases of identified protein targets, the binding site was poorly studied, which encouraged them to perform comparative sequence and structural studies with their homologous proteins, like as in case of Kelch motif associated protein.

Cassiarins C–E and 10,11-dihydroanhydrobarakol (Fig. 13) showed moderate antiplasmodial activity *in vitro* against *Plasmodium falciparum* 3D7 (IC_{50} 1: 24.2 μM; 2: 3.6 μM; 3: 7.3 μM; 4: 2.3 μM) and no cytotoxicity against HL-60 (human blood premyelocytic leukemia) cells ($IC_{50} > 100$ μM). Anhydrobarakol (Fig. 13) and 5-acetonyl-7-hydroxy-2-methylchromone also showed moderate antiplasmodial activity (IC_{50} 5: 4.7 μM; 6: 8.6 μM) and no cytotoxicity against HL-60 cells. Cassiarins D and E were less active than cassiarin A (IC_{50} 0.023 μM), although they contained the cassiarin A-type skeleton. 10,11-Dihydroanhydrobarakol showed activity similar to that of anhydrobarakol.[100]

Most adults in West Africa treat acute febrile illnesses with local herbs, and patterns of herbs used for malaria were recently investigated in Sierra Leone.[101] Around 55% of participants surveyed reported taking one or more of seven herbs to treat symptoms of malaria. Among herb users, the most commonly used antimalarial herbs were *Moringa oleifera* (moringa, 52%) and *Sarcocephalus latifolius* (yumbuyambay, 50%). Other herbs used included *Senna siamea* (shekutoure, 18%), *Cassia sieberiana* (gbangba, 18%), *Uvaria afzelii* (gone-botai, 14%), *Morinda chrysorhiza* (njasui, 14%), and *Craterispermum laurinum* (nyelleh, 7%). Interestingly, combination herbal therapy was common, with 37% of herb users taking two or more herbs together when if malaria infections were suspected.

Quinine (Fig. 14), from *Cinchona succirubra* (Rubiaceae), remains an important antimalarial drug almost 400 years after its effectiveness was first documented. However, its continued use is challenged by its poor tolerability, poor compliance with complex dosing regimens, and the availability of more efficacious antimalarial drugs. This article reviews the historical role of quinine, considers its current usage and provides insight into its appropriate future use in the treatment of malaria. In light of recent research findings intravenous artesunate should be the first-line drug for severe malaria, with quinine as an alternative. The role of rectal quinine as pre-referral treatment for severe malaria has not been fully explored, but it remains a promising intervention. In pregnancy, quinine continues to play a critical role in the management of malaria, especially in the first trimester, and it will remain a mainstay of treatment until safer alternatives become available. For uncomplicated malaria, artemisinin-based combination therapy (ACT) offers a better option than quinine though the difficulty of maintaining a steady supply of ACT in resource-limited settings renders the rapid withdrawal of quinine for uncomplicated malaria cases risky. In HIV and TB infected populations, concerns about potential interactions between quinine and antiretroviral and anti-tuberculosis drugs exist, and these will need further research and pharmacovigilance.

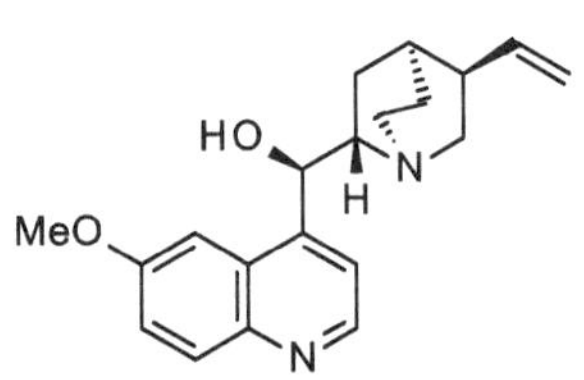

Fig. 14 Structure of the first antimalarial phytochemical, quinine.

Arguments have often been presented that isolation of a specific, active, component may differs in action when compared to the use of whole dried plant extracts, since efficacy of most natural medicines could lie upon the synergy or additivity of diverse pharmacological agents, rather than originating from a single compound.[102–104] Notably, it has been discovered that dried whole-plant *Artemisia annua* slows evolution of malaria drug resistance and overcomes resistance to artemisinin in rodent models of malarial infections.[105] Although the small trial involved only 18 patients, the authors suggest that dried whole-plant *Artemisia* should be rapidly incorporated into an antimalarial regimen for Africa and, possibly, wherever else ACT resistance has emerged.

Blighia sapida is a woody, perennial, multi-purpose, fruit tree species native to the Guinean forests of West Africa. The fruit aril known by various name (Ghana: Akye fufo/Akye) is the main constituent of the Jamaican National Dish Ackee with saltfish. Traditionally, around 22 diseases are treated with Ackee, most commonly for the treatment of: fever, malaria, dental decay, internal hemorrhage, dysentery, burns, eyes inflammation, yellow fever, constipation, cutaneous infections, whitlow (nail or digital herpes simplex infection) as well as head lice.[106] An ethanol extract of *Blighia sapida* (Sapindaceae) in chloroquine (CQ)-resistant *Plasmodium berghei* (ANKA strain)-infected mice (in prophylactic, suppressive, and curative models) with graded doses of the plant extracts with artemether-lumefantrine (Coartem) as control, showed a remarkable activity but only in the chemosuppression of *P. berghei* parasites ($P < 0.01$). This ranged from 57% to 36.5% at doses of 200–800 mg/kg, respectively, whereas co-artem (10 mg/kg) produced 62.1% chemosuppression, supporting the traditional use of Ackee in the prevention of malaria.[107] Although its phytochemical constituents are well studied, the exact constituent responsible for antiplasmodial activity remains unknown and only 19 compounds have been reportedly isolated from the parts of *B. sapida*.[108] Nonetheless, quinine remains an affordable and trusted antimalarial in developing countries when all else fails, despite its known side effects.[109,110]

11. Leishamaniasis

A detailed tabular description of natural compounds with potential therapeutic value in the treatment of Leishmaniasis has appeared in a useful review on Leishmaniasis and Chagas's disease.[111] Several other earlier review articles also covered plant-derived natural products in the treatment of this parasitic disease, Leishmaniasis.[112–115] Examples of a few of the antileishmanial phytochemicals are presented in Fig. 15.

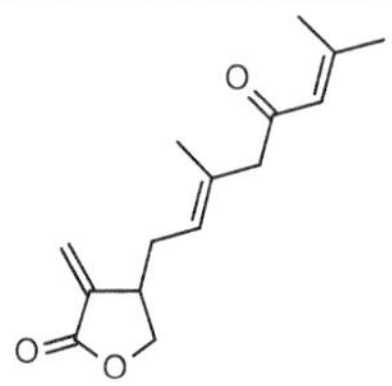

Anthecotulide

Source: *Anthemis cotula* (Asteraceae)

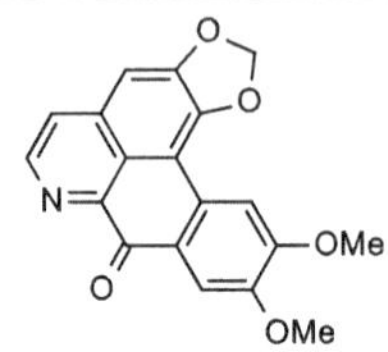

Dicentrinone

Source: *Duguetia furfuracea* (Annonaceae)

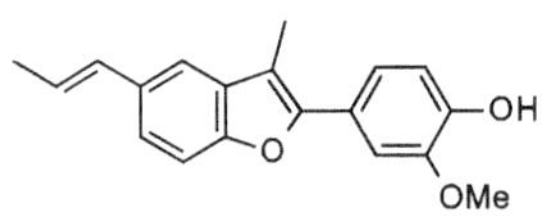

Eupomatenoid-5

Source: *Piper regnellii* var. *pallescens* (Piperaceae)

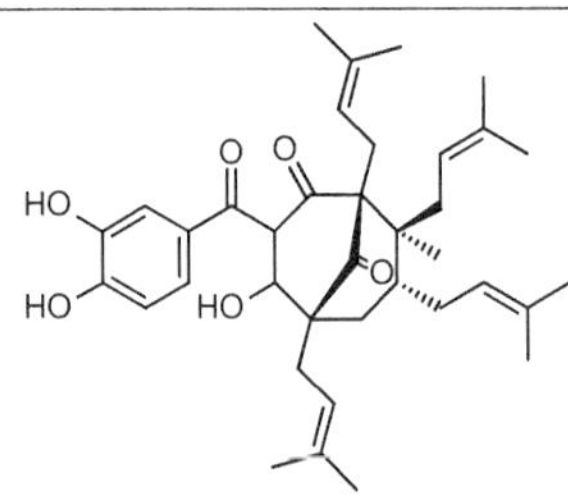

Guttiferone A

Source: *Garcinia livingstonei* (Clusiaceae)

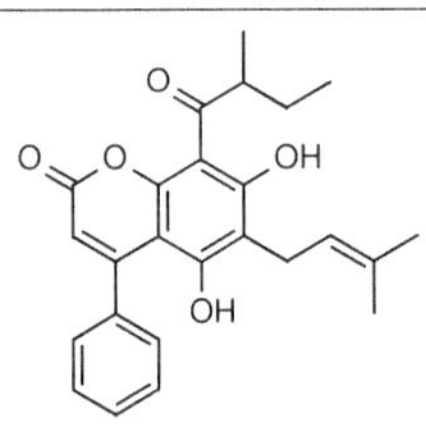

Mammea A/BB

Source: Mammea Americana (Calophyllaceae)

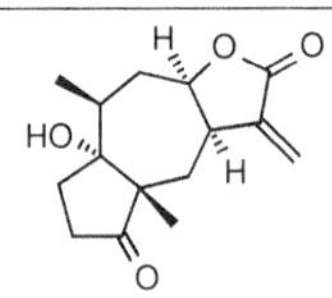

Peruvin

Source: *Lepidium meyenii* (Brassicaceae)

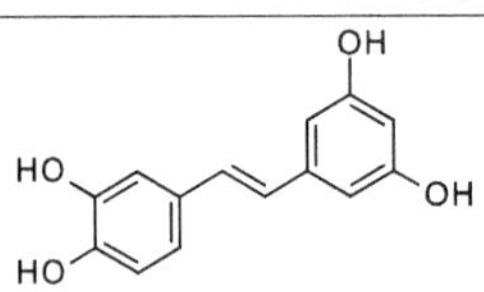

Piceatannol

Source: *Picea abies* (Pinaceae)

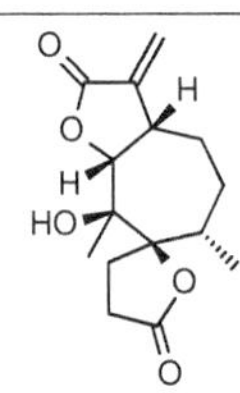

Psilostachyn

Source: *Ambrosia psilostachya* (Asteraceae)

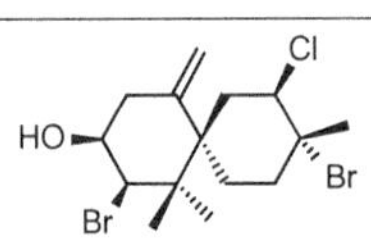

Obtusol

Source: *Laurencia dendroidea* (Rhodomelaceae)

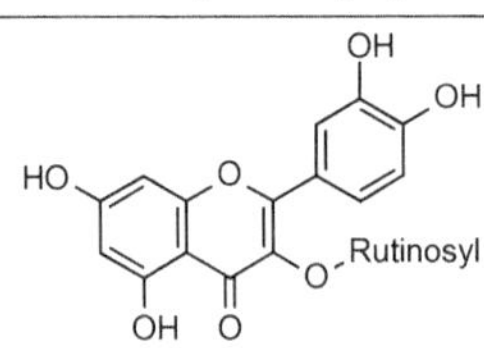

Rutin

Source: *Ruta graveolens* (Rutaceae)

Fig. 15 Examples of a few of the antileishmanial phytochemicals.

12. Schistosomiasis

Schistosomiasis is known as bilharzia (or bilharziosis), so named after the German physician Theodor Bilharz, who first described the cause of urinary schistosomiasis in 1851, is a disease caused by parasitic worms. The parasite is most commonly found throughout Africa, but also lives in parts of South America, the Caribbean, the Middle East and Asia. So far only a single drug, praziquantel, has been truly effective against this disease but its efficacy is decreasing.

In terms of antischistosomal activity of natural products *in vivo*, Utzinger et al.[116] reported that artemether (Fig. 16), the methyl ether derivative of artemisinin, exhibited antischistosomal properties by oral doses of 6 mg/kg in randomized controlled clinical trials. Since Schistosomiasis (bilharzia) has few affordable treatments, two Artemisia wormwood species, *A. annua* and *A. afra*, were compared with the current standard. Praziquantel (PZQ) treatment in a 2-month 800 patient clinical trial between August and November of 2015. The double blind, randomized, superiority clinical trial had three treatment arms: 400 for PZQ, 200 for *A. annua*, and 200 for *A. afra*. Artemisia-treated patients received 1 L/day of dry leaf/twig tea infusions divided into three aliquots daily, for 7 days with 28-day follow-up. Of 800 enrolled patients having an average of >700 *Schistosoma mansoni* eggs per fecal sample, 780 completed the trial. In all Artemisia-treated patients, eggs in fecal smears were undetectable within 14 days of treatment; a result sustained 28 days post treatment. Eggs in fecal smears of PZQ-treated patients were undetectable after day 21. Both adults and pediatric patients treated with *A. annua* responded better compared to PZQ treatment. Both *A. annua* and *A. afra* provided faster effective treatment of schistosomiasis and the authors suggested it should be considered for implementation on a global scale.

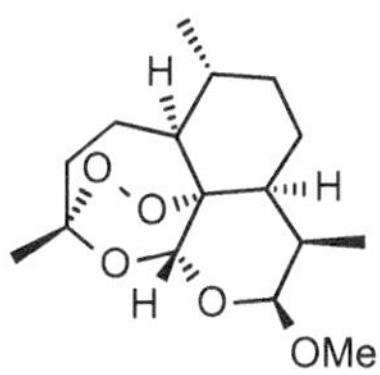

Fig. 16 Structure of artemether.

Balanites aegyptiaca fruit mesocarp in a dose of 200 mg/kg body weight in mice (p.o.) infected with Sudanese strain of *S. mansoni* showed a significant reduction in egg count per gram of feces.[117] Published literature shows that *Commiphora molmol* extract (Myrrh capsules) given to patients at a dose of 10 mg/kg of body weight per day for 3 or 6 consecutive days induced a curative rate higher than 90%.[118,119] However, contrary to these results, Botros et al.[120] and Barakat et al.[121] did not recommend the use of this plant drug in treating human cases of schistosomiasis based on negative results from several experiments performed on patients and infected animals.

In the screening of Egyptian native and cultivated plants for schistosomicidal activity using bioassay, *in vitro* screening of 346 methanol extracts originated from 281 native and cultivated plant species growing in Egypt, and related to 81 families, was carried out for schistosomicidal activity. The extracts were bioassayed at 100 μg/mL on viable *Schistosoma mansoni* mature worms in culture medium. Viability of worms was examined after exposure for 24 h, and mortality determined. Negative (DMSO) and positive (praziquantel) controls were used. Of the plant extracts evaluated, 72 were found to possess reproducible antischistosomal activity *in vitro*. Strong antischistosomal activity was found in the extracts of 15 species (possessing $LC_{50} \leq 15$ μg/mL), *viz. Agave americana.* L. var. *marginata* Trel. (Agavaceae), *A. lophantha* Schiede (Agavaceae), *Furcraea selloa* C. Koch. (Agavaceae), *Calotropis procera* (Aiton) W.T.Aiton (Asclepiadaceae), *Pergularia tomentosa* L. (Asclepiadaceae), *Asclepias sinaica* (Boiss.) Muschl. (Asclepiadaceae), *Alkanna orientalis* (L.) Boiss. (Boraginaceae), *Khaya grandifoliola* DC. (Meliaceae), *Swietenia mahogani* (L.) Jacq. (Meliaceae), *Pimenta racemosa* (Mill.) J.W.Moore (Myrtaceae), *Pinus canariensis* C.Sm. (Pinaceae), *Verbascum sinaiticum* L. (Scrophulariaceae), *Solanum elaeagnifolium* Cav. (Solanaceae), *S. nigrum* L. (Solanaceae), and *Brachychiton rupestris* (Lindl.) K.Schum. (Sterculiaceae). The authors suggested 15 species could represent promising bioactive sources that deserve further investigation and could result in discovering novel antischistosomal agents.[122]

The schistosomicidal activity of phytol (Fig. 17) against *S. mansoni* both *in vitro* and *in vivo* was reported.[123] Subsequently it was tested against *S. haematobium*. It showed interesting antischistosomal properties against *Schistosoma mansoni* adult worms both *in vitro* and *in vivo*. Specifically, the antischistosomal activity of phytol against *Schistosoma haematobium* juvenile was investigated Eraky in 2016.[124] Adult worms show alterations on the tegumental surface by SEM. The assay, which was carried out with six concentrations (25, 50, 75, 100, 125, and 150 μg/mL) of phytol, indicated dose and

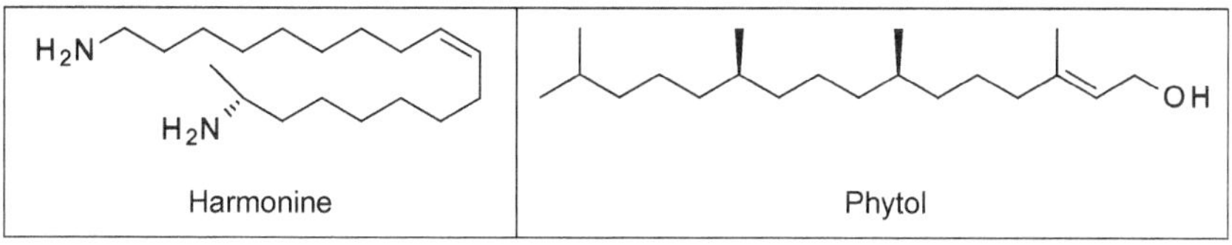

Fig. 17 Harmonine and phytol—two schistosomicidal agents.

time-dependent toxicity to the worms at 48 h after they had been exposed to phytol in the concentration of 150 μg/mL. Decline in worm motility followed by 100% mortality was found with Male worms especially susceptible. On an ultrastructural level, phytol also induced tegumental peeling, disintegration of tubercles and spines in addition to morphological disfiguring of both the oral and ventral suckers. Consequently, this is an affordable promising natural product that could be used for development of a new schistosomicidal agent.

Unusually, an insect-derived compound, harmonine (Fig. 17), an antimicrobial alkaloid from the harlequin ladybird *Harmonia axyridis*, was tested for antischistosomal activity. Pleiotropic effects of harmonine on physiological, cellular, and molecular processes in both adult male and female *Schistosoma mansoni* at concentrations as low as 5 μM *in vitro* was observed. Apart from tegumental damage, gut dilatation, dysplasia of gonads, caused increased production of abnormally shaped eggs at 5 μM and a complete stop of egg production occurred at 10 μM. Motility was reduced with an EC_{50} of 8.8 μM and lethal effects occurred at 10–20 μM within 3 days of culture. Enzyme inhibition assays revealed acetylcholinesterase (AChE) as one potential target of harmonine.[125] In addition, *P. falciparum* was inhibited by harmonine at IC_{50} values of 4.8 and 7.6 μM for the chloroquine-sensitive strain 3D7 and the chloroquine-resistant strain Dd2, respectively. Chloroquine treatment resulted in parasite growth inhibition with IC_{50} values of 24 (3D7) and 158 nM (Dd2), indicating the absence of cross resistance between harmonine and chloroquine. The pronounced activity of harmonine against mycobacteria together with activity against an MRSA strain is indicative for a novel mode of action, which might be exploited by the development of derivatives less toxic to human cells.

13. Trypanosomiasis

Trypanosomiasis or trypanosomosis is a name given to several diseases in vertebrates caused by parasitic protozoan trypanosomes of the genus

Trypanosoma. In humans, this includes both African trypanosomiasis and Chagas disease. Estimates suggest that 30,000 people in 36 countries within sub-Saharan Africa have African trypanosomiasis, which is caused by either *Trypanosoma brucei rhodesiense* or *Trypanosoma brucei gambiense*. The fly bite erupts into a red chancre sore and within a few weeks, the individual demonstrates fever, swollen lymph glands, aching muscles and joints, headaches and irritability and importantly, blood in urine. In the first phase, the patient suffers from intermittent bouts of fever (with lymphadenopathy) together with other, non-specific, signs and symptoms. The second stage of the disease involves the central nervous system, with extensive neurological effects, especially changes in personality, confusion, slurred speech, seizures and difficulty in walking and talking as well as alteration of the biological clock (the circadian rhythm). These problems may develop over many years and, if left untreated, can result in death. Chagas disease, also known as American trypanosomiasis, causes 21,000 deaths per year mainly in Latin America and is caused by the protist *Trypanosoma cruzi*. It is spread mostly by insects known as Triatominae, or "kissing bugs" or vampire bugs *T. brucei*.

Crude extracts of both *A. nilotica* and *B. buonopozense* revealed antitrypanosomal (*T. brucei*) effects at 200 and 300 mg/kg *in vivo*. In contrast, crude extracts of *T. avicennioides* and *Z. zanthoxyloides* could fail to clear parasitaemia, implying trypanostatic action rather than trypanocidal action.[126] *Crossopteryx febrifuga* and *Aristolochia gigantea* are two most cited plant species in relation to antitrypanosomal activity.[127]

In a recent study, ursolic acid and theviridoside (Fig. 18), isolated from *Bigonia binate* and *B. crucigera* of the family Bigoniaceae, have displayed antitrypanosomal activity with IC_{50} values of 29.12 and 17.80 μM, respectively, against *Trypanosoma brucei* S427.[128] C-benzylated dihydrochalcone, uvaretin has recently been isolated as a potential antitrypanosomal agent from the methanol extract of the root bark of *Uvaria tanzaniae* Verdc. (Annonaceae), showing significant activity against *T. brucei brucei* with a percentage inhibition of 97.3%.[129] Earlier, Lenz et al.[130] showed that the natural 4,15-isoatriplicolide type sesquiterpene lactones could inhibit trypanothione reductase; particularly, 4,15-isoatriplicolide tiglate (Fig. 18) exhibited exceptionally potent activity against *T. brucei rhodesiense* with an IC_{50} value of 15 nM. Similarly, dibenzylbutane neolignans from *Saururus cernuus* L. (Saururaceae) were also shown to possess antitrypanosomal activity; the neolignan *threo*-austrobailignan-6 (Fig. 18) was found to be the most active one against trypomastigote forms of *T. cruzi*.[131] It was shown that the activity was mediated through activations in the mitochondrial membrane

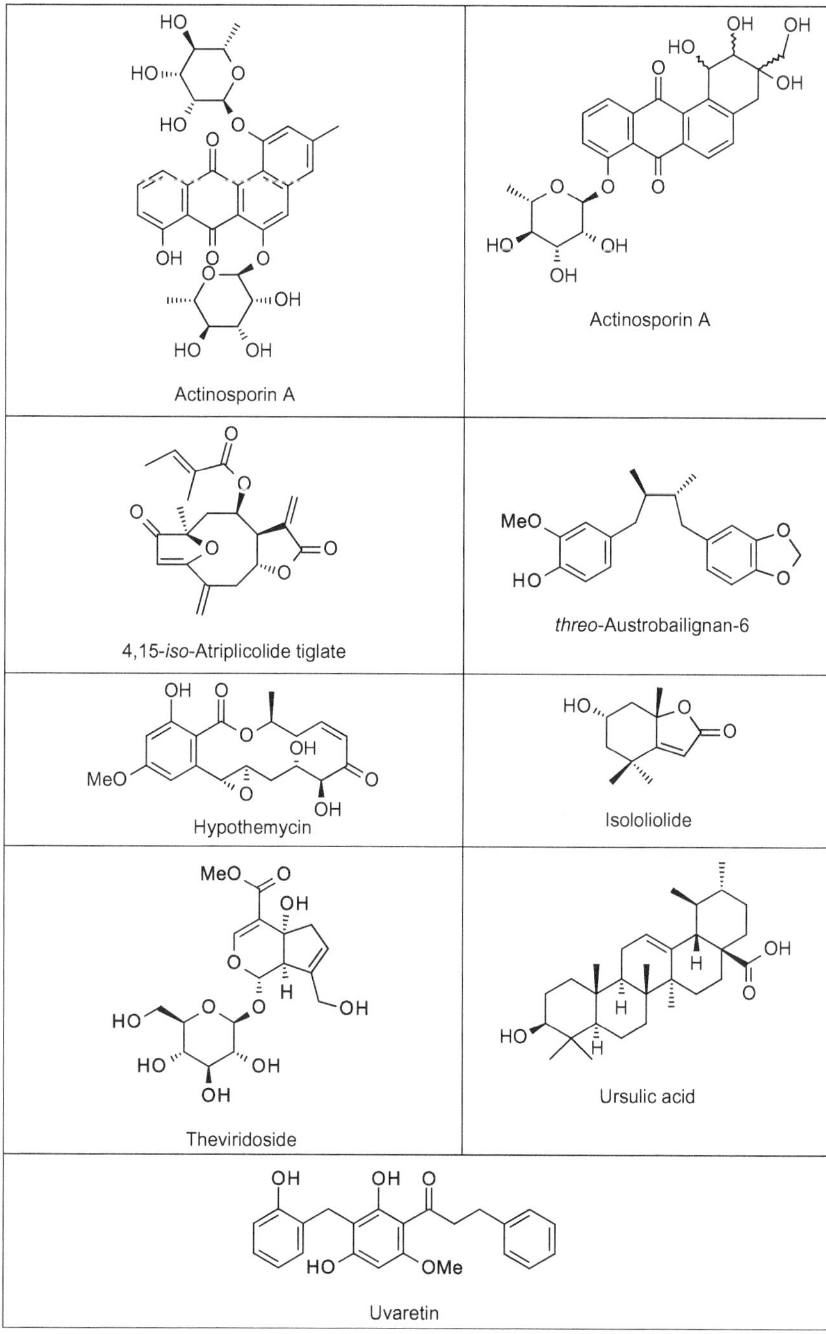

Fig. 18 Examples of a few antitrypanosomal natural products.

potential. Isololiolide (Fig. 18), isolated from the marine hydroid *Macrorhynchia philippina*, showed *in vitro* activity against both trypomastigote and intracellular amastigotes of *T. cruzi*, with IC_{50}, values of 32 and 40 μM, respectively, with no noticeable mammalian cytotoxicity (>200 μM).[132] The plausible mechanism of action of this compound was shown to be associated with disruption of the plasma membrane integrity and a strong depolarization of the mitochondrial membrane potential. However, most of these initial findings based on assays involving parasites, have not gone through any appropriate clinical trials involving human volunteers, and also there are only a handful of reports on the use of any animal models to establish antitrypanosomal activity of natural products.

Hypothemycin (Fig. 18) is macrolide that is isolated from the cultured broth of *Hypomyces subiculosus* and shows antifungal activity and inhibits the growth of some human cancer cells. This semisynthetic killed *T. brucei* cells cultured *in vitro* and also killed *T. brucei* inoculated into mice *in vivo*, but only one third of animals were cleared of parasitaemia with high doses of hypothemycin causing side effects. Importantly, employing a chemical biology approach coupled with quantitative mass spectrometry, Nishino et al.[133] reported the principal target of hypothemycin as a novel kinase essential for *T. brucei* survival. Despite the poor therapeutic index, hypothemycin identification of the drug target makes it a suitable as a prototype for a rational drug design medicinal chemistry campaign. Other non-classical targets have also been identified. For instance, during cell division, certain multi-protein complexes associate with the centromere of a chromosome called kinetochores which are disk-shaped protein structures associated with duplicated chromatids in eukaryotic cells where the spindle fibers attach during cell division to pull sister chromatids apart ensuring accurate transmission of genetic material to daughter cells. Although most eukaryotic kinetochores possess the same canonical kinetochore machinery, kinetochores of *T. brucei*, are an exception making it an attractive drug target.[134,135]

Strategies for the isolation of natural products continue to evolve. For instance, dereplication strategies for targeted isolation of new prototypes for drugs are exemplified in a paper in which antitrypanosomal actinosporins A and B were isolated from a marine sponge associated-Actinokineospora sp. EG49.[136]

14. Haemonchosis

Haemonchosis is one of the most significant parasitic diseases of livestock worldwide, affecting hundreds of millions of ruminants such as sheep and goats, causing billion dollar losses to the livestock industry annually.[137]

The causative agent, *Haemonchus contortus* (barber's pole worm), feeds on blood in the stomach (abomasum) and the host suffers gastritis, anemia and associated complications, leading to morbidity and mortality in severely affected animals. Picria fel-terrae has been used to treat helminthiasis in Indonesia activity but exact active components remain unknown. Consequently, eight extracts from the plants *Picria fel-terrae* Lour., *Linariantha bicolor*, *Lansium domesticum* and *Tetracera akara* were tested for nematocidal activity against seven strains of the free-living nematode *Caenorhabditis elegans* (one wild-type and six strains with GFP-tagged stress response pathways), and the stress responses caused by these extracts was determined.[138] These plants are widely distributed throughout Asia and have been used by indigenous Malaysian healers to treat worm infections and gastrointestinal disorders in humans.[138–140] Five of the eight plant extracts possessed significant nematocidal activity against both larval and adult stages of *C. elegans*. The most effective extracts were from *P. fel-terrae* and triggered stress response pathways that were distinct from commercially available anthelmintics (doramectin and levamisole) and so the mechanism of action may be novel.[141]

15. Conclusion

Most of the parasitic diseases are considered as "neglected disease," and investment in research in search for new molecules from natural sources for drug development for such diseases is much less than in other major health concerns like cancers, cardiovascular diseases and diabetes. Among parasitic diseases, much efforts have been put in the area of antimalarial drug development, especially after the introduction of the first antimalarial drug, quinine, and then the Noble Prize winning discovery of artemisinin as the last line of defense against malaria; however, there are reports on the emergence of artemisinin-resistant malarial parasites. A large body of published literature mainly covers antiparasitic work involving crude extracts or fractions from various natural matrices, much less on isolated compounds. Even the reports that deal with antiparasitic compounds from natural sources, predominantly relied on the results from indirect *in vitro* antiparasitic assays or evaluation on live parasites, but hardly any robust pre-clinical studies involving animal models, and/or any clinical trials involving human volunteers. Although there have been a few promising molecules discovered showing antiparasitic potentials and drug-like properties, further research is needed to develop any of those molecules as modern antiparasitic drugs.

Acknowledgments

L. Nahar gratefully acknowledges the financial support of the European Regional Development Fund—Project ENOCH (No. CZ.02.1.01/0.0/0.0/16_019/0000868). This review is dedicated to our friend and mentor, Wallace Peters (April 1, 1924–December 2018).

References

1. Amacho, M.; Araújo, A.; Morrow, J.; Buikstra, J.; Reinhard, K. *Parasit. Vectors* **2018**, *11*(1), 248. https://doi.org/10.1186/s13071-018-2729-4.
2. Hoeppli, R. *Exp. Parasitol.* **1956**, *5*(4), 398–419.
3. Hoeppli, R. *Parasites and Parasitic Infections in Early Medicine and Science.* University of Malaya Press: Singapore, **1959**.
4. Florentino, P. T. V.; Real, F.; Bonfim-Melo, A.; Orikaza, C. M.; Ferreira, E. R.; Pessoa, C. C.; Lima, B. R.; Sasso, G. R. S.; Mortara, R. A. *Biomed. Res. Int.* **2014**, *2014*, 565291. https://doi.org/10.1155/2014/565291.
5. Nabarro, L.; Morris-Jones, S.; Moore, D. *Peters' Atlas of Tropical Medicine and Parasitology*, 7th ed.; Elsevier, **2018**.
6. Manoha, P. R. *Indian J. Hist. Sci.* **2012**, *47*(4), 545–559.
7. Brusca, R. C.; Gilligan, M. R. *Copeia* **1983**, *3*(3), 813–816.
8. Stoll, N. R. *J. Parasitol.* **1999**, *85*(3), 392–396.
9. Loker, E. S. *J. Parasitol.* **2013**, *99*(6), 933–942.
10. Montresor, A. *Bench Aids for the Diagnosis of Intestinal Parasites*, 2nd ed.; World Health Organization (WHO), **2019.**
11. Perumal, A. N.; Gunawardene, Y. I.; Dassanayake, R. S. *J. Helminthol.* **2016**, *90*(2), 129–138.
12. Spiegler, V.; Liebau, E.; Hensel, A. *Nat. Prod. Rep.* **2017**, *34*(6), 627–643.
13. Tagboto, S.; Townson, S. *Adv. Parasitol.* **2001**, *50*, 199–295.
14. Kappagoda, S.; Singh, U.; Blackburn, B. G. *Mayo Clin. Proc.* **2011**, *86*(6), 561–583.
15. Garcia-Bustos, J. F.; Sleebs, B. E.; Gasser, R. B. *Parasit. Vectors* **2019**, *12*(1), 306. https://doi.org/10.1186/s13071-019-3537-1.
16. Zhou, L.; Wang, J.; Wang, K.; Xu, J.; Zhao, J.; Shan, T. Secondary metabolites with antinematodal activity from higher plants. In: *Studies in Natural Products Chemistry*; Atta-ur-Rahman, Ed.; Vol. 37, Elsevier: Amsterdam, **2012**; pp 67–114.
17. Kayser, O.; Kiderlen, A. F.; Croft, S. L. *Stud. Nat. Prod. Chem.* **2002**, *26*, 779–848.
18. Ghisalberti, E. L. *Stud. Nat. Prod. Chem.* **2002**, *26*, 425–506.
19. Veerakumari, L. *Asian J. Sci. Technol.* **2015**, *6*, 1881–1894.
20. Ord, R. L.; Lobo, C. A. *Curr. Clin. Microbiol. Rep.* **2015**, *2*(4), 173–181.
21. Richter, D.; Matuschka, F. R.; Spielman, A.; Mahadevan, L. *Proc. Biol. Sci.* **2013**, *280*(1773), 20131758. https://doi.org/10.1098/rspb.2013.1758.
22. Hunfeld, K. P.; Hildebrandt, A.; Gray, J. S. *Int. J. Parasitol.* **2006**, *38*(11), 1219–1237.
23. Hu, T.; Fan, B.; Liang, J.; Zhao, S.; Dang, P.; Gao, F.; Dong, M. *Tropl. Anim. Health Prod.* **1997**, *29*(4), 72S–76S.
24. Vannier, E.; Krause, P. J. *N. Engl. J. Med.* **2012**, *366*(25), 2397–2407.
25. Hildebrandt, A.; Gray, J. S.; Hunfeld, K. P. *Infection* **2013**, *41*(6), 1057–1072.
26. Brasseur, P.; Lecoublet, S.; Kapel, N.; Favennec, L.; Ballet, J. J. *Eur. J. Clin. Microbiol. Infect. Dis.* **1996**, *15*(10), 840–841.
27. Gray, J.; Zintl, A.; Hildebrandt, A.; Hunfeld, K. P.; Weiss, L. *Ticks Tick Borne Dis.* **2010**, *1*(1), 3–10.
28. Krause, P. J.; Lepore, T.; Sikand, V. K.; Gadbaw, J., Jr.; Burke, G.; Telford, S. R., 3rd. *N. Engl. J. Med.* **2000**, *343*(20), 1454–1458.

29. Coutinho, E. M. *Contraception* **2002**, *65*(4), 259–263.
30. Goo, Y. K.; Terkawi, M. A.; Jia, H.; Aboge, G. O.; Ooka, H.; Nelson, B.; Kim, S.; Sunaga, F.; Namikawa, K.; Igarashi, I.; Nishikawa, Y.; Xuan, X. *Parasitol. Int.* **2010**, *59*(3), 481–486.
31. AbouLaila, M.; Sivakumar, T.; Yokoyama, N.; Igarashi, I. *Parasitol. Int.* **2010**, *59*(2), 278–282.
32. Brayton, K. A.; Lau, A. O.; Herndon, D. R.; Hannick, L.; Kappmeyer, L. S.; Berens, S. J.; Bidwell, S. L.; Brown, W. C.; Crabtree, J.; Fadrosh, D.; Feldblum, T.; Forberger, H. A.; Haas, B. J.; Howell, J. M.; Khouri, H.; Koo, H.; Mann, D. J.; Norimine, J.; Paulsen, I. T.; Radune, D.; Ren, Q.; Smith, R. K.; Suarez, C. E.; White, O.; Wortman, J. R.; Knowles, D. P.; McElwain, T. F.; Nene, V. M. *PLoS Pathog.* **2007**, *3*(10), 1401–1413.
33. Subeki, M. H.; Yamasaki, M.; Yamato, O.; Maede, Y.; Katakura, K.; Suzuki, M.; Trimurningsih, C.; Yoshihara, T. *J. Vet. Med. Sci.* **2004**, *66*(7), 871–874.
34. Murnigsih, T.; Subeki, M. H.; Takahashi, K.; Yamasaki, M.; Yamato, O.; Maede, Y.; Katakura, K.; Suzuki, M.; Kobayashi, S.; Yoshihara, T. *J. Vet. Med. Sci.* **2005**, *67*(8), 829–831.
35. Peregrine, A. S.; Mamman, M. *Acta Trop.* **1993**, *54*(3–4), 185–203.
36. Pépin, J.; Milord, F. *Adv. Parasitol.* **1994**, *33*, 1–47.
37. Ubeki, M. H.; Takahashi, K.; Nabeta, K.; Yamasaki, M.; Maede, Y.; Katakura, K. *J. Nat. Prod.* **2007**, *70*(10), 1654–1657.
38. Nakao, R.; Mizukami, C.; Kawamura, Y.; Subeki, B. S.; Yamasaki, M.; Maede, Y.; Matsuura, H.; Nabeta, K.; Nonaka, N.; Oku, Y.; Katakura, K. *J. Vet. Med. Sci.* **2009**, *71*(1), 33–41.
39. Subeki, M. H.; Takahashi, K.; Yamasaki, M.; Yamato, O.; Maede, Y.; Katakura, K.; Suzuki, M.; Trimurningsih, C.; Yoshihara, T. *J. Vet. Med. Sci.* **2005**, *67*(2), 223–227.
40. Subeki, M. H.; Takahashi, K.; Yamasaki, M.; Yamato, O.; Maede, Y.; Katakura, K.; Kobayashi, S.; Trimurningsih, T.; Chairul, C.; Yoshihara, T. *J. Nat. Prod.* **2005**, *68*(4), 537–539.
41. Amada, K.; Subeki, N. K.; Yamasaki, M.; Katakura, K.; Matsuura, H. *Biosci. Biotechnol. Biochem.* **2009**, *73*, 776–780.
42. Aboulaila, M.; Yokoyama, N.; Igarashi, I. *Parasitology* **2010**, *137*(5), 785–791.
43. Magerlein, B. J.; Birkenmeyer, R. D.; Kagan, F. *Antimicrob. Agents Chemother.* **1966**, *6*, 727–736.
44. Birkenmeyer, R. D.; Kagan, F. *J. Med. Chem.* **1970**, *13*(4), 616–619.
45. Meyers, B. R.; Kaplan, K.; Weinstein, L. *Appl. Microbiol.* **1969**, *17*(5), 653–657.
46. Sin, N.; Kim, K. B.; Elofsson, M.; Meng, L.; Auth, H.; Kwok, B. H.; Crews, C. M. *Bioorg. Med. Chem. Lett.* **1999**, *9*(15), 2283–2288.
47. Meng, L.; Mohan, R.; Kwok, B. H.; Elofsson, M.; Sin, N.; Crews, C. M. *Proc. Natl. Acad. Sci. U. S. A.* **1999**, *96*(18), 10403–10408.
48. Czesny, B.; Goshu, S.; Cook, J. L.; Williamson, K. C. *Antimicrob. Agents Chemother.* **2009**, *53*(10), 4080–4085.
49. Kreidenweiss, A.; Kremsner, P. G.; Mordmuller, B. *Malar. J.* **2008**, 7, 187. https://doi.org/10.1186/1475-2875-7-187.
50. Aboulaila, M.; Nakamura, K.; Govind, Y.; Yokoyama, N.; Igarashi, I. *Vet. Parasitol.* **2010**, *167*(1), 19–27.
51. Yamasaki, M.; Nakamura, K.; Tamura, N.; Hwang, S. J.; Yoshikawa, M.; Sasaki, N.; Ohta, H.; Yamato, O.; Maede, Y.; Takiguchi, M. *J. Parasitol.* **2009**, *95*(6), 1532–1538.
52. Paulus, E.; Vertesy, L. *Z. Kristallogr. New Cryst. Struct.* **2003**, *218*, 609–611.
53. Antoszczak, M.; Huczyński, A. *Eur. J. Med. Chem.* **2019**, *176*, 208–227.
54. Haselbarth, K.; Tenter, A. M.; Brade, V.; Krieger, G.; Hunfeld, K. P. *Int. J. Med. Microbiol.* **2007**, *297*(3), 197–204.

55. Krause, P. J.; Spielman, A.; Telford, S. R.; Sikand, V. K.; McKay, K.; Christianson, D.; Pollack, R. J.; Brassard, P.; Magera, J.; Ryan, R.; Persing, D. H. *N. Engl. J. Med.* **1998**, *339*(3), 160–165.
56. Wormser, G. P.; Prasad, A.; Neuhaus, E.; Joshi, S.; Nowakowski, J.; Nelson, J.; Mittleman, A.; Aguero-Rosenfeld, M.; Topal, J.; Krause, P. J. *Clin. Infect. Dis.* **2010**, *50*(3), 381–386.
57. Iguchi, A.; Matsuu, A.; Matsuyama, K.; Hikasa, Y. *Parasitol. Int.* **2015**, *64*(2), 190–193.
58. Zintl, A.; Mulcahy, G.; Skerrett, H. E.; Taylor, S. M.; Gray, J. S. *Clin. Microbiol. Rev.* **2003**, *16*(4), 622–636.
59. Meetei, P. A.; Rathore, R. S.; Prabhu, N. P.; Vindal, V. *J. Mol. Model.* **2016**, *22*(4), 71. https://doi.org/10.1007/s00894-016-2945-9.
60. An, G.; Murry, D. J.; Gajurel, K.; Bach, T.; Deye, G.; Stebounova, L. V.; Codd, E. E.; Horton, J.; Gonzalez, A. E.; Garcia, H. H.; Ince, D.; Hodgson-Zingman, D.; Nomicos, E. Y. H.; Conrad, T.; Kennedy, J.; Jones, W.; Gilman, R. H.; Winokur, P. *Antimicrob. Agents Chemother.* **2019**, *63*(4), e02255–18.
61. Bizhani, N.; Iran, J. *Public Health* **2015**, *44*(11), 1555–1556.
62. Szewczuk, V. D.; Mongelli, E. R.; Pomilio, A. B. *Mol. Med. Chem.* **2003**, *1*, 54–57.
63. Freire, R. B.; Borba, H. R.; Coelho, C. D. *Indian J. Pharmacol.* **2010**, *42*(6), 345–350.
64. Sandoval, N. R.; Ríos, N.; Mena, A.; Fernández, R.; Perea, M.; Manzano-Román, R.; Santa-Quiteria, J. A.; Hernández-Gonzalez, A.; Siles-Lucas, M. *Acta Trop.* **2015**, *147*, 54–63.
65. Oysa, C.; Murtagh, R.; Lambert, J. S. *Antibiotics* **2019**, *8*(4). https://doi.org/10.3390/antibiotics8040223. pii: E223.
66. *Antiinfective proanthocyanidin compounds and methods of use thereof.* **2009**. Patent: WO 2009/026179.
67. Wenyon, C. M.; O'Connor, F. W. *J. R. Army Med. Corps* **1917**, *28*, 346–367.
68. Low, G. C. *J. Trop. Med. Hyg.* **1916**, *19*, 29–34.
69. Swan, J. C.; Hosseini, J. *J. Protozool.* **1983**, *30*(2), 332–334.
70. Zierdt, C. H. *Clin. Microbiol. Rev.* **1991**, *4*(1), 61–79.
71. da Costa, S. F. G. *Arch. Int. Pharmacodyn. Ther.* **1932**, *41*, 443–460.
72. Wilson, K. W.; Winget, D.; Wilks, S. *Mil. Med.* **1990**, *155*(9), 394–396.
73. van Hellemond, J. J.; Molhoek, N.; Koelewijn, R.; Wismans, P. J.; van Genderen, P. J. *J. Infect. Chemother.* **2013**, *19*(3), 545–548.
74. Neal, R. A.; Murphy, A. G.; Olliaro, P.; Croft, S. L. *Trans. R. Soc. Trop. Med. Hyg.* **1994**, *88*(2), 223–225.
75. Bern, C.; Hernandez, B.; Lopez, M. B.; Arrowood, M. J.; de Merida, A. M.; Klein, R. E. *Am. J. Trop. Med. Hyg.* **2000**, *63*(5–6), 231–235.
76. Ortega, Y. R.; Sterling, C. R.; Gilman, R. H.; Cama, V. A.; Diaz, F. *N. Engl. J. Med.* **1993**, *328*(18), 1308–1312.
77. Cama, V. A.; Mathison, B. A. *Clin. Lab. Med.* **2015**, *35*(2), 423–444.
78. Muthamilselvan, T.; Kuo, T. F.; Wu, Y. C.; Yang, W. C. *Evid. Based Complement. Alternat. Med.* **2016**, *2016*, 2657981. https://doi.org/10.1155/2016/2657981.
79. Zhang, D.-F.; Sun, B.-B.; Yue, Y.-Y.; Zhou, Q.-J.; Du, A.-F. *Parasitol. Res.* **2012**, *111*(6), 2229–2233.
80. Sharma, N.; Sharma, A.; Nishad, D. K.; Khanna, K.; Sharma, B. G.; Kakkar, D.; Bhatnagar, A. *AAPS PharmSciTech* **2018**, *19*(6), 2564–2571.
81. Malik, T. A.; Kamili, A. N.; Chishti, M. Z.; Tanveer, S.; Ahad, S.; Johri, R. K. *Microb. Pathog.* **2016**, *93*, 56–62.
82. Martinez, S. E.; Davies, N. M.; Reynolds, J. K. *"Toxicology and Safety of Flavonoids". Methods of Analysis, Preclinical and Clinical Pharmacokinetics, Safety, and Toxicology.* John Wiley & Son, **2013**;257.

83. Bode, J. C. *Assessment and Management of Hepatobiliary Disease. 371. Springer-Verlag: Berlin,* **1987**.
84. Soto, J.; Gómez, C.; Calzada, F.; Ramírez, M. E. *Planta Med.* **2010**, *76*(6), 611–612.
85. Reyes-López, M.; Villa-Treviño, S.; Arriaga-Alba, M.; Alemán-Lazarini, L.; Rodríguez-Mendiola, M.; Arias-Castro, C.; Fattel-Fazenda, S.; de la Garza, M. *Toxicol. In Vitro* **2005**, *19*(1), 91–97.
86. Calzado-Flores, C. C.; Segura, J. J.; Rodríguez, V. M.; Domínguez, X. A. *Proc. West. Pharmacol. Soc.* **1983**, *26*, 431–434.
87. Kupchan, S. M.; Britton, R. W.; Lacadie, J. A.; Ziegler, M. F.; Sigel, C. W. *J. Org. Chem.* **1975**, *40*(5), 648–654.
88. Dart, J. K. G.; Saw, V. P. J.; Kilvington, S. *Am. J. Opthalmol.* **2009**, *148*, 487–499.
89. Hadaś, E.; Derda, M.; Cholewiński, M. *Parasitol. Res.* **2017**, *116*(3), 997–1001.
90. Rufino-González, Y.; Ponce-Macotela, M.; Jiménez-Estrada, M.; Jiménez-Fragoso, C. N.; Palencia, G.; Sansón-Romero, G.; Anzo-Osorio, A.; Martínez-Gordillo, M. N. *Pharm. Biol.* **2017**, *55*(1), 1787–1791.
91. Prashanth, D.; Asha, M. K.; Amit, A.; Padmaja, R. *Fitoterapia* **2001**, *72*(4), 421–422.
92. Chibale, K. *Medicinal Chemistry Approaches to Malaria and Other Tropical Diseases.* Vol. 53, Elsevier, **2019**; Vol. 532–217?
93. Dascombe, M. J.; Drew, M. G. B.; Evans, P. G.; Ismail, F. M. D. Rational design strategies for the development of synthetic quinoline and acridine based antimalarials. In *Frontiers in Drug Design & Discovery: Structure-Based Drug Design in the 21st Century*, 3, Bentham Science Publishers, **2007**, pp 559–609.
94. Ismail, F. M. D. Nature's armamentarium against malaria: Antimalarials and their semi-synthetic derivatives. In: *Nutraceuticals and Natural Product Derivatives: Disease Prevention & Drug Discovery*; Ullah, M. F., Ahmad, A., Eds.; **2019**; pp 333–374.
95. Tajuddeen, N.; Van Heerden, F. R. *Malar. J.* **2019**, *18*(1), 404. https://doi.org/10.1186/s12936-019-3026-1.
96. Kumar, D.; Jain, A.; Verma, A. *Nat. Prod. J.* **2017**, 7(4), 255–266.
97. Deguchi, J.; Hirahara, T.; Hirasawa, Y.; Ekasari, W.; Widyawaruyanti, A.; Shirota, O.; Shiro, M.; Morita, H. *Chem. Pharm. Bull.* **2012**, *60*, 219–222.
98. Morita, H.; Oshimi, S.; Hirasawa, Y.; Koyama, K.; Honda, T.; Ekasari, W.; Indrayanto, G.; Zaini, N. C. *Org. Lett.* **2007**, *9*, 3691–3693.
99. Ekasari, W.; Wahyuni, T. S.; Arwaty, H.; Putri, N. T. *Afr. J. Infect. Dis.* **2018**, *12*(1), 110–115.
100. Oshimi, S.; Deguchi, J.; Hirasawa, Y.; Ekasari, W.; Widyawaruyanti, A.; Wahyuni, T. S.; Zaini, N. C.; Shirota, O.; Morita, H. *J. Nat. Prod.* **2009**, *72*(10), 1899–1901.
101. Ranasinghe, S.; Jacobsen, K. H.; Ansumana, R.; Lamin, J. M.; Bockarie, A. S.; Bangura, U.; Buanie, J. A. G.; Stenger, D. A. *J. Ethnopharmacol.* **2015**, *166*, 200–204.
102. Ginsburg, H.; Deharo, E. A. *Malar. J.* **2011**, *10*(1), S1. https://doi.org/10.1186/1475-2875-10-S1-S1.
103. Williamson, E. M. *Phytomedicine* **2001**, *8*, 401–409.
104. Houghton, P. J.; Howes, M. J.; Lee, C. C.; Steventon, G. *J. Ethnopharmacol.* **2007**, *110*, 391–400.
105. Elfawal, M. A.; Towler, M. J.; Reich, N. G.; Weathers, P. J.; Rich, S. M. *Proc. Natl. Acad. Sci. U. S. A.* **2015**, *112*(3), 821–826.
106. Ekué, M. R.; Sinsin, B.; Eyog-Matig, O.; Finkeldey, R. *J. Ethnobiol. Ethnomed.* **2010**, *6*, 12. https://doi.org/10.1186/1746-4269-6-12.
107. Otegbade, O. O.; Ojo, J. A.; Adefokun, D. I.; Abiodun, O. O.; Thomas, B. N.; Ojurongbe, O. *Drug Target Insights* **2017**, *28*, 11. https://doi.org/10.1177/1177392817728725.

108. Sinmisola, A.; Oluwasesan, B. M.; Chukwuemeka, A. P. *J. Ethnopharmacol.* **2019**, *235*, 446–459.
109. Achan, J.; Talisuna, A. O.; Erhart, A.; Yeka, A.; Tibenderana, J. K.; Baliraine, F. N.; Rosenthal, P. J.; D'Alessandro, U. *Malar. J.* **2011**, *10*, 144. https://doi.org/10.1186/1475-2875-10-144.
110. Elfawal, M. A.; Towler, M. J.; Reich, N. G.; Golenbock, D.; Weathers, P. J.; Rich, S. M. *PLoS One* **2012**, 7(12), e52746. https://doi.org/10.1371/journal.pone.0052746.
111. Souto, E. B.; Dias-Ferreira, J.; Craveiro, S. A.; Severino, P.; Sanchez-Lopez, E.; Garcia, M. L.; Silva, A. M.; Souto, S. B.; Mahant, S. *Pathogens* **2019**, *8*(3), 119. https://doi.org/10.3390/pathogens8030119.
112. Rohloff, J.; Hymete, A.; Tariku, Y. Plant-derived natural products for the treatment of Leishmaniasis. In: *Studies in Natural Products Chemistry*; Atta-ur-Rahman, Ed.; Vol. 39, ; Elsevier, **2013**; pp 381–429.
113. Scotti, M. T.; Scotti, L.; Ishiki, H.; Ribeiro, F. F.; Cruz, R. M.; Oliveira, M. P.; Mendonca, F. J. *Comb. Chem. High Throughput Screen.* **2016**, *19*(7), 537–553.
114. Oryan, A.; Iran, J. *Vet. Res.* **2015**, *16*(1), 1–19.
115. Rodrigues, I. A.; Mazotto, A. M.; Cardoso, V.; Alves, R. L.; Amaral, A. C.; Silva, J. R.; Pinheiro, A. S.; Vermelho, A. B. *Mediators Inflamm.* **2015**, 835910. https://doi.org/10.1155/2015/835910.
116. Utzinger, J.; Xiao, S.; N'Goran, E. K.; Bergquist, R.; Tanner, M. *Int. J. Parasitol.* **2001**, *31*, 1549–1562.
117. Koko, W. S.; Abdalla, H. S.; Galal, M.; Khalid, H. S. *Fitoterapia* **2005**, *76*, 30–34.
118. Massoud, A.; Sheir, Z.; Nasr, A. A.; Salama, O.; Badra, G. A.; El-Shennawy, H.; Hassan, N.; Hammad, S. M. *Am. J. Trop. Med. Hyg.* **2001**, *65*, 700–704.
119. Abo-Madyan, A. A.; Morsy, T. A.; Motawea, S. *J. Egypt. Soc. Parasitol.* **2004**, *34*, 423–446.
120. Botros, S.; William, S.; Ebeid, F.; Cioli, D.; Katz, N.; Day, T. A.; Bennett, J. L. *Am. J. Trop. Med. Hyg.* **2004**, 7, 206–210.
121. Barakat, R.; El-Morshedy, H.; Fenwick, A. *Am. J. Trop. Med. Hyg.* **2005**, *73*, 365–367.
122. Yousif, F.; Hifnawy, M. S.; Soliman, G.; Boulos, L.; Labib, T.; Mahmoud, S.; Ramzy, F.; Yousif, M.; Hassan, I.; Mahmoud, K.; El-Hallouty, S. M.; El-Gendy, M.; Gohar, L.; El-Manawaty, M.; Fayyad, W.; El-Menshawi, B. S. *Pharm. Biol.* **2007**, *45*(6), 501–510.
123. de Moraes, J.; de Oliveira, R. N.; Costa, J. P.; Junior, A. L. G.; de Sousa, D. P.; Freitas, R. M.; Allegretti, S. M.; Pinto, P. L. *PLoS Negl. Trop. Dis.* **2014**, *8, e2617.*
124. Eraky, M. A.; Aly, N. S. M.; Selem, R. F.; El-Monem, A.; El-Kholy, A.; Abd, E. -R. R. *Korean J. Parasitol.* **2016**, *54*(4), 477–484.
125. Kellershohn, J.; Thomas, L.; Hahnel, H. R.; Grunweller, A.; Hartmann, R. K.; Hardt, M.; Vilcinskas, A.; Grevelding, C. G.; Haeberlein, S. *PLoS Negl. Trop. Dis.* **2019**, *13*(3), e0007240.
126. Mann, A.; Ifarajimi, O. R.; Adewoye, A. T.; Ukam, C.; Udeme, E. E.; Okorie, I. I.; Sakpe, M. S.; Ibrahim, D. R.; Yahaya, Y. A.; Kabir, A. Y.; Ogbadoyi, E. O. *Afr. J. Tradit. Complement. Altern. Med.* **2011**, *8*(1), 15–21.
127. Vahekeni, N.; Neto, P. M.; Kayimbo, M. K.; Mäser, P.; Josenando, T.; da Costa, E.; Falquet, J.; van Eeuwijk, P. *J. Ethnopharmacol.* **2020**, *256*, 112382. https://doi.org/10.1016/j.jep.2019.112382.
128. Mahmoud, B. K.; Samy, M. N.; Hamed, A. N. E.; Abdelmohsen, U. R.; Hajar, D.; Yamano, Y.; Sugimoto, S.; Matsunami, K.; Kamel, M. S. *Phytochem. Lett.* **2020**, *35*, 200–205.
129. Christopher, R.; Mgani, Q. A.; Nyandoro, S. S.; Rousseau, A. L.; Isaacs, M.; Hoppe, H. C. *Nat. Prod. Res.* **2020**, 1–8. https://doi.org/10.1080/14786419.2019.1710705 (epub ahead of print).

130. Lenz, M.; Krauth-Siegel, R. L.; Schmidt, T. J. *Molecules* **2019**, *24*(20), 3737. https://doi.org/10.3390/molecules24203737.
131. Brito, J. R.; da Costa-Silva, T. A.; Tempone, A. G.; Ferreira, E. A.; Lago, J. H. G. *Fitoterapia* **2019**, *137*, 104251. https://doi.org/10.1016/j.fitote.2019.104251.
132. Lima, M. L.; Romanelli, M. M.; Borborema, S. E. T.; Johns, D. M.; Migotto, A. E.; Lago, J. H. G.; Tempone, A. G. *Biorg. Chem.* **2019**, *89*, 103002. https://doi.org/10.1016/j.bioorg.2019.103002.
133. Nishino, M.; Choy, J. W.; Gushwa, N. N.; Oses-Prieto, J. A.; Koupparis, K.; Burlingame, A. L.; Renslo, A. R.; McKerrow, J. H.; Taunton, J. *Elife* **2013**, *2, e00712*.
134. Nerusheva, O. O.; Ludzia, P.; Akiyoshi, B. *Open Biol.* **2019**, *9*(12), 190236. https://doi.org/10.1098/rsob.190236.
135. Gehrig, S.; Efferth, T. *Int. J. Mol. Med.* **2008**, *22*(4), 411–419.
136. Abdelmohsen, U. R.; Cheng, C.; Viegelmann, C.; Zhang, T.; Grkovic, T.; Ahmed, S.; Quinn, R. J.; Hentschel, U.; Edrada-Ebel, R. *Mar. Drugs* **2014**, *12*(3), 1220–1244.
137. Kumarasingha, R.; Preston, S.; Yeo, T. C.; Lim, D. S.; Tu, C. L.; Palombo, E. A.; Shaw, J. M.; Gasser, R. B.; Boag, P. R. *Parasit. Vectors* **2016**, *9*, 187. https://doi.org/10.1186/s13071-016-1458-1459.
138. Kumarasingha, R.; Palombo, E. A.; Bhave, M.; Yeo, T. C.; Lim, D. S. L.; Tu, C. L.; Shaw, J. M.; Boag, P. R. *Int. J. Parasitol.* **2014**, *44*(5), 291–298.
139. Chai, P. P. K. *Medicinal Plants of Sarawak.* Paul Chai P.K., **2006**.
140. Christensen, H. *Ethnobotany of the Iban & the Kelabit.* Nepcon. Malaysia: University of Aarhus; Forest Department Sarawak, **2002**.
141. Atilaya, P.; Husori, D.; Daulay, L. *Asian J. Pharma. Clin. Res.* **2018**, *11*, 59. https://doi.org/10.22159/ajpcr.2018.v11s1.26568.

CHAPTER FIVE

Anti-inflammatory natural products

Olumayokun A. Olajide[a,*], Satyajit D. Sarker[b]
[a]Department of Pharmacy, School of Applied Sciences, University of Huddersfield, Huddersfield, United Kingdom
[b]Centre for Natural Products Discovery (CNPD), School of Pharmacy and Biomolecular Sciences, Liverpool John Moores University, Liverpool, United Kingdom
*Corresponding author: e-mail address: o.a.olajide@hud.ac.uk

Contents

1. Introduction

Published literature suggests that inflammatory diseases and conditions with underlying pathology related to chronic inflammation contribute significantly to global mortality and morbidity statistics. In spite of their proven efficacy, currently available anti-inflammatory drugs have major limitations related to serious side effects and high cost of treatment. Natural products provide viable alternatives to these drugs because of their unique chemical

Annual Reports in Medicinal Chemistry, Volume 55
ISSN 0065-7743
https://doi.org/10.1016/bs.armc.2020.02.002

diversity, thus providing potentials for the identification of novel lead compounds for drug development for treating inflammatory diseases. This chapter highlights significant research on anti-inflammatory natural products, and covers basic studies on roles of natural products against inflammatory diseases.

2. Overview of inflammation

Inflammation, originated from the Latin word "inflammatio," is a biological process that is activated following a disruption to tissue structures as a result of physical, chemical or biological stimuli. The inflammatory response is an essential component of the immune system and triggered by infection and tissue injury in order to achieve homeostatic regulation. It is thus an adaptive response to the diverse noxious stimuli that frequently comes in contact with the human body.[1]

Inflammation can be described as being either acute or chronic, depending on the nature of the response and the ability of the response to achieve resolution through the clearing of the foreign agent or damaged tissues. Acute inflammation is a short-lived response characterized by events involving leakage of water, salt, and proteins from the vascular compartment. Other features of the acute inflammatory response includes activation of endothelial cells, recruitment of leukocytes, activation of tissue macrophages, platelet activation and aggregation, and the release of proteases and oxidants from phagocytic cells.[2] Regardless of the inflammatory stimulus, the earliest manifestation of the acute inflammatory response is the adhesion of polymorphonuclear leukocytes to the vascular endothelium.[2] On the other hand, chronic inflammation is defined as persistent acute inflammation due to failure to bring about resolution. This can be as a result of the continued presence of the inflammatory stimulus, persistent leukocyte migration and the resulting production of pro-inflammatory cytokines and reactive oxygen species (ROS), which cause tissue damage and remodeling, or processes, which cause the leukocytes to persist at the inflammatory site.[3]

3. Inflammatory diseases

Several diseases have been associated with acute inflammation and include sepsis and trauma. However, chronic inflammation is now widely accepted to contribute to diseases such as asthma, cancer, cardiovascular diseases (atherosclerosis), inflammatory bowel diseases (ulcerative colitis,

Crohn's disease), metabolic syndrome (diabetes, obesity), neurodegenerative diseases (Alzheimer's and Parkinson's diseases) and rheumatic conditions. An understanding of the roles played by inflammation in the pathogenesis of some of these diseases is critical in the discovery and development of medicinal natural products for treating them.

3.1 Rheumatoid arthritis

Rheumatoid arthritis (RA) is an inflammatory condition in the joints, resulting from auto-immune (inflammatory responses) targeted at the joints. In RA, there is accumulation of inflammatory cells and pro-inflammatory mediators within joints, resulting in bone loss, cartilage erosion, inflamed (swollen) synovium and swollen joint capsule. RA is initiated by immune cells (lymphocytes, neutrophils, and macrophages) which secrete pro-inflammatory cytokines which in turn promote cell migration.[4,5] Cytokines such as tumor necrosis factor-α (TNF-α), interleukin-1 (IL-1), and interleukin-6 (IL-6), which are released by macrophages, as well as cytokine interleukin-17 (IL-17), released by lymphocytes contribute to the inflammatory response and subsequent cartilage destruction.[5] Studies have also shown that pro-inflammatory cytokines interact with receptor activator of nuclear factor kappa-β ligand (RANKL) to induce osteoclast activity, and the resulting bone erosion.[6]

3.2 Role of inflammation in cardiovascular diseases

Cardiovascular diseases refer to conditions affecting the physiology of the heart and blood vessels. These conditions include hypertension, heart failure, angina pectoris, and myocardial ischemia. It has now been established that inflammation is central to the development and complications of cardiovascular diseases.[7,8] Damage to the heart during myocardial ischemia has been linked to the release of pro-inflammatory cytokines, which mediate leukocyte activation, neutrophil migration to the myocardium and the subsequent production of reactive oxidation species (ROS), proteases, and more cytokines.[9,10] These events can activate innate immune pathways, such as the Toll-like receptors (TLRs), and subsequent activation of various inflammatory signaling pathways such as the nuclear factor-kappa B (NF-κB) and the mitogen activated protein (MAP) kinases. Interestingly, modulation of NF-κB has been reported to be a therapeutic strategy in hypertension,[11] myocardial infraction,[12] and arteriosclerosis.[13]

3.3 Inflammation and cancer

The link between chronic inflammation and cancer was first proposed by Rudolf Virchow in 1863. Since then, many studies have reported the roles played by inflammatory cells, pro-inflammatory cytokines, and chemokines in cancer. Several studies have provided evidence that the key inflammatory transcription factor, NF-κB is an oncogene, which links inflammation and cancer.[14] It is noteworthy that activation of NF-κB has been reported in colon, gastric, pancreatic, ovarian, hepatocellular, breast, head and neck carcinomas, melanoma, lymphomas, leukemias.[14,15]

3.4 Neuroinflammation and neurodegenerative diseases

Neuro-inflammation is normally an innate immune response of cells of the CNS to block infection and eliminate pathogens, cell debris, and misfolded proteins.[16] Initially, neuroinflammation plays a role in the central nervous system (CNS) innate immunity to bring about resolution. However, in chronic neurological diseases, neuroinflammation becomes persistent with the subsequent damage to neuronal cells. Several studies have suggested that neuroinflammation is a principal pathology in neurodegenerative and other CNS diseases. The major cellular component of neuroinflammation are the microglia, which are CNS-resident immune cells. The microglia continuously carry out surveillance of the CNS microenvironment even in the resting state.[17] However, they are activated in the presence of hazardous stimulations or pathological states of the CNS.[18] Activated microglia secret pro-inflammatory mediators (TNFα, IL-6, IL-1β, IL-18, IL-23, IFN-γ, NO, and ROS). These mediators secreted by activated microglia contribute to the death of neurons in neurodegenerative diseases.

3.5 Inflammatory bowel disease

Inflammatory bowel disease (IBD) is an umbrella term for ulcerative colitis and Crohn's disease, and is characterized by abdominal pain, diarrhea, and weight loss. In IBD, the breakdown of the intestinal barrier results in bacterial passage, which in turn may damage the mucosa leading to the increased production of ROS.[19] A principal pro-inflammatory cytokine associated with IBD is TNFα, which is produced mainly by activated macrophages, monocytes and T cells. The intestinal inflammation caused by the increased, persistent secretion of TNFα causes reduced function of the tight junctions as well as apoptosis of intestinal epithelial cells, resulting in eventual damage to the intestinal barrier and increased barrier permeability.[20,21]

Increased secretions of TNFα further induces the production of other pro-inflammatory cytokines such as IL-1, IL-6 which further exacerbate inflammation.[22]

4. NF-κB activation pathway as a molecular drug target of anti-inflammatory activity

The NF-κB transcription factor regulates the expression of many genes associated with innate and adaptive immune responses as well as inflammation. Continuous stimulation of the cellular signaling pathway resulting in the activation of NF-κB is now known to be directly linked to many types of inflammatory and autoimmune diseases.[23] On the one hand, NF-κB is critical to the co-ordination of the host defense responses to stress, injury and infection.[24] On the other hand, NF-κB contributes to the pathogenesis of chronic inflammatory diseases.[25] Consequently, the NF-κB pathway has been consistently explored as an attractive molecular target for drug discovery.

5. Major anti-inflammatory natural products from plants

Several natural products from plants have been shown to produce potent anti-inflammatory activity through various mechanisms in animal and *in vitro* models of inflammatory diseases. This chapter will focus on the main classes of natural products with the most promising therapeutic potentials for treating inflammatory conditions. In this context, anti-inflammatory property of main alkaloids, flavonoids and terpenes will be highlighted.

5.1 Alkaloids

Alkaloids are pharmacologically active secondary metabolites mainly consisting of nitrogen, often as an integral part of the ring.[26] They are either proto-alkaloids (nitrogen-containing but not heterocyclic in structure) or the true alkaloids (nitrogen-containing heterocyclic compounds).[27] An increasing number of articles have reported pharmacological activities of different classes of alkaloids in human, animal and *in vitro* models of diseases. The anti-inflammatory activity of some major bioactive alkaloids[28–47] in models of inflammatory conditions is summarized in Table 1.

Table 1 Examples of some anti-inflammatory natural alkaloids.

Inflammatory disorders	**Alkaloids**	**Plant sources**	**Pharmacological activities**
Neurodegenerative disorders	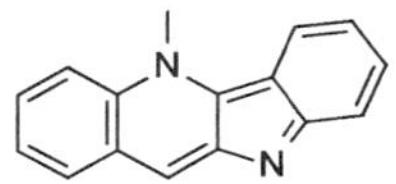 Cryptolepine	*Cryptolepis sanguinolenta* (Apocynaceae)	Reduction of TNFα, IL-6, IL-1β, NO, and PGE_2 levels in LPS-stimulated rat microglia. Protein and mRNA of COX-2 and iNOS were reduced. Inhibited activation of the activation of NF-κB, p38 MAPK[28]
	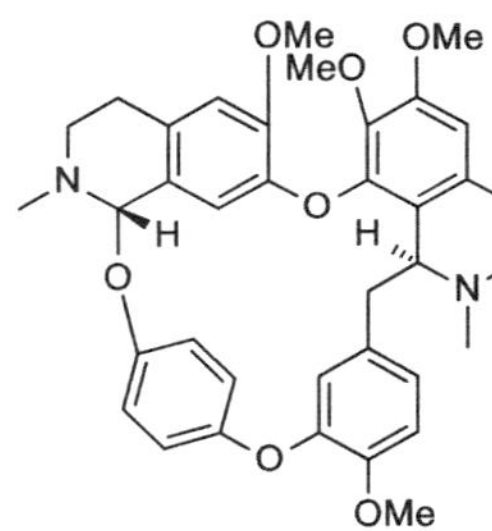 Tetrandrine	*Stephania tetrandra* (Menispermaceae)	Prevention of Aβ-induced spatial learning and memory disorder and NF-κB activity and the inhibition of IL-1β and TNFα in a rat model of Alzheimer's disease[29]
	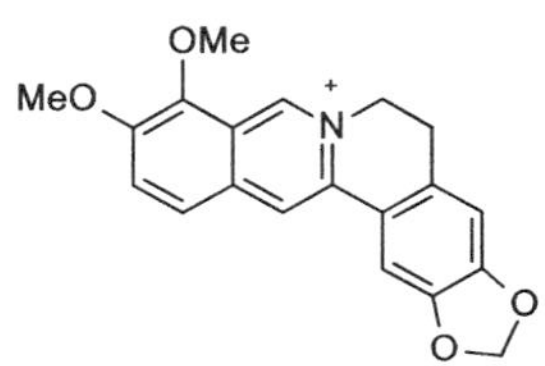 Berberine	*Hydrastis canadensis* (Ranunculaceae)	Inhibited the phosphorylation of IκBα in Aβ-activated microglial cells[30]

	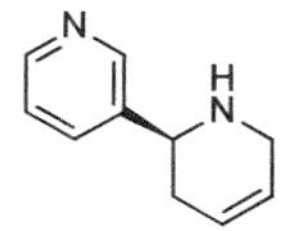 Anatabine	*Nicotiana tabacum* (Solanaceae)	Reduced NF-κB activation by inhibiting Aβ production *in vitro*. Reduced neuro-inflammation in microglia and SH-SY5Y cells, and *in vivo* in Tg APPsw mice[31,32]
Inflammatory bowel disease	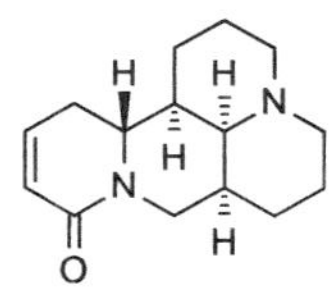 Sophocarpine 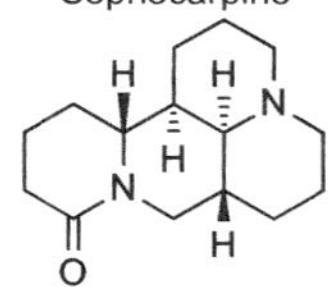 Sophoridine	*Sophora alopecuroides* (Fabaceae)	Reduction in DSS-induced colitis, which was associated with a reduction of serum IL-1β and IL-6 levels
	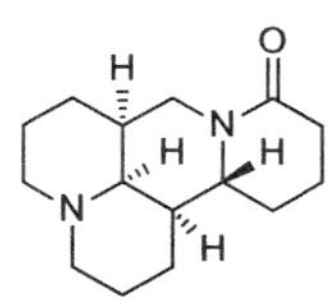 Matrine		Matrine induced significantly improved TNBS-induced colitis by reducing production of colonic TNF-α

Continued

Table 1 Examples of some anti-inflammatory natural alkaloids.—Cont'd

Inflammatory disorders	Alkaloids	Plant sources	Pharmacological activities
	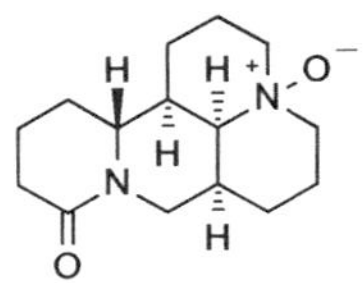 Oxymatrine		Oxymatrine prevented colonic injury of TNBS-induced colitis, and a decrease in the level of colonic IL-2 and increase the level of colonic IL-10. Oxymatrine also prevented NF-κB activation in acute intestinal inflammation[33–36]
	Berberine	*Hydrastis canadensis* (Ranunculaceae)	Berberine inhibited DSS-induced intestinal injury and colitis and reduced the release of pro-inflammatory cytokine (TNF-α, IFN-γ, and IL-17) in colonic macrophages by targeting NF-κB and mitogen-activated protein kinase (MAPK) pathways[37]
	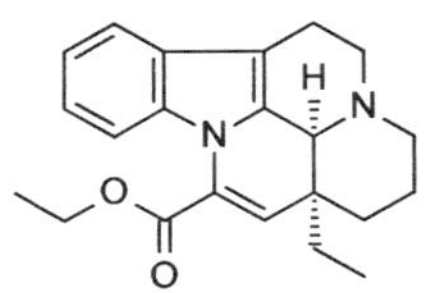 Vinpocetine	*Vinca minor* (Apocynaceae)	Vinpocetine reduced production of pro-inflammatory cytokines IL-1β, TNF-α, and IL-33 and inhibited activation of NF-κB in the colon in acetic acid-induced colitis in mice[38]
	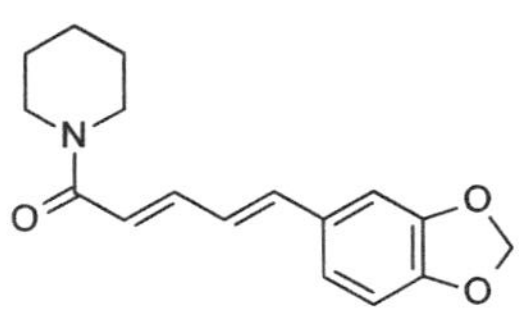 Piperine	*Piper longum* and *Piper nigrum* (Piperaceae)	Piperine prevented acetic acid-induced weight loss, diarrhoea, gross bleeding and infiltration of immune cells and reduced secretion of nitric oxide (NO), TNFα in acetic acid-induced colitis in mice[39]

Rheumatoid arthritis	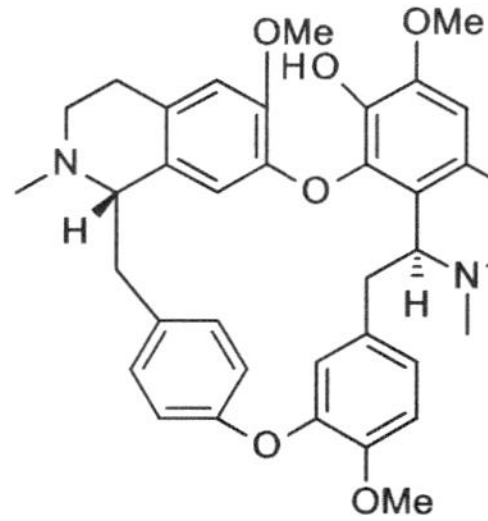 Fangchinoline	*Stephania tetrandra* (Menispermaceae)	ROS, TNF-α, and IL-6 levels were reduced in the chondrocytes of rheumatoid arthritis-induced rats[40]
	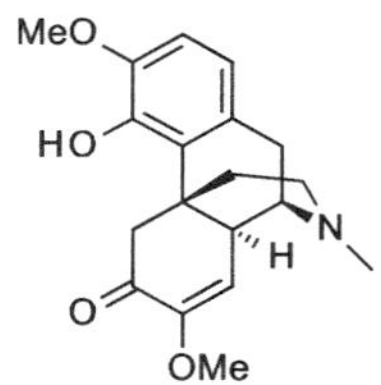 Sinomenine	*Sinomenium acutum* (Menispermaceae)	Suppression of inflammation-induced joint destructive progression and arthritis symptoms in adjuvant-induced arthritis rats. Sinomenine also decreased PGE_2 levels through inhibition of mPGES-1 expression in collagen-II induced arthritis in mice. Anti-inflammatory activity in synoviocytes was demonstrated to be through inhibition of NF-κB pathway[41–44]
	Tetrandrine	*Stephania tetrandra* (Menispermaceae)	Tetrandrine inhibits bone erosion in collagen-induced arthritis rats through mechanisms involving inhibition of translocation of NF-κB-p65, NFATc1 and tyrosine kinase[45]
	Oxymatrine	*Sophora alopecuroides* (Fabaceae)	Reduction in IL-6 and IL-8 expression and NF-κB activation in rheumatoid fibroblast-like synoviocytes[46]
	Sophocarpine		Suppression of pro-inflammatory cytokines, MAPK and NF-κB in human fibroblast-like synoviocytes and in mice with collagen-induced arthritis[47]

5.2 Flavonoids

Flavonoids are naturally occurring phenolic compounds with a 15 carbon skeleton consisting of two aromatic rings and a cyclohexane ring fused with one of those aromatic rings, and can be structurally classified as anthocyanins, catechins, flavones, flavonols, flavanols, flavanones, and isoflavonoids. They are found in abundance in flowers, fruits, barks, roots, stems, tea, wine and vegetables.[48] The anti-inflammatory activity of flavonoids has been shown to be mediated through various mechanisms including inhibition of the release of several pro-inflammatory cytokines, antioxidant mechanisms, as well as modulatory activities on transcription factors such as NF-κB, nuclear factor erythroid 2-related factor 2 (Nrf2), and activator protein 1 (AP-1).[49–51] The activities of some anti-inflammatory flavonoids[52–84] of natural origins in models of inflammatory diseases are summarized in Table 2.

5.3 Terpenes

Terpenoids are a large and structurally diverse group of compounds formed biosynthetically from a combination of two or more *isoprene* units (a five carbon unit, chemically known as 2-methyl-1,3-butadiene).[48] Terpenoids generally arise from *head-to-tail* joining of isoprene units. Terpenoids are found in all parts of higher plants and occur in mosses, liverworts, algae and lichens. Terpenoids of insect and microbial origins have also been found. About 60% of known natural products are terpenoids, and many of them possess various pharmacological properties. They can be classified into monoterpenes, sesquiterpenes, diterpenes, triterpenes and tetraterpenes, depending on the number of isoprene units. Many reports suggest that terpenes produce pharmacological actions in inflammation-associated disorders through mechanisms involving modulation of NF-κB signaling pathway.[85] Sesquiterpene lactones, such as parthenolide (Fig. 1), are especially potent inhibitors of NF-κB signaling.[86]

5.3.1 Monoterpenes

The most investigated anti-inflammatory monoterpene is limonene and the modified monoterpene, also known as iridoids, is aucubin (Fig. 2).

Rufino et al.[87] reported that limonene inhibited IL-1β-induced nitric oxide/iNOS production, and inhibited IL-1β-induced NF-κB, JNK and p38 activation in human chondrocytes, suggesting a potential therapeutic effect in rheumatoid- and osteo-arthritis. A further evidence of the anti-inflammatory property of limonene was reported by Yoon et al.,[88] who

Table 2 Examples of some anti-inflammatory natural flavonoids.

Inflammatory disorders	Flavonoid classes	Flavonoid	Pharmacological activities
Neurodegenerative disorders	Flavonols	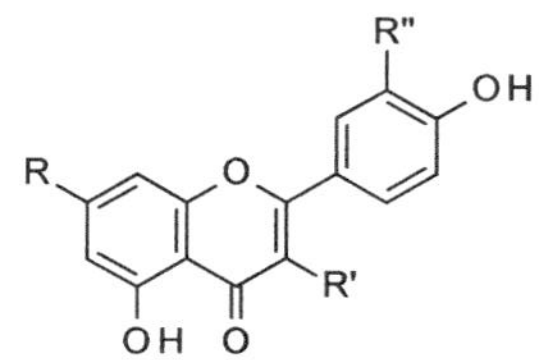	
		Quercetin R = R′=R″=OH	Reduction in 1-methyl-4-phenylpyridinium (MPP +)-induced production of IL-6, IL-1β and TNFα in a glial-neuronal system. Reduction in the expression of IL-6, TNF-α, IL-1β and COX-2 and prevention of neural damage in mouse hippocampus[52,53]
		Rutin R = R″ = OH; R′=Rutinosyloxy	Prevention of cognitive impairments by reducing oxidative stress and neuroinflammation in rat model of sporadic dementia[54]
		Kaempferol R = R′=OH; R″=H	Reduction in LPS-induced production of pro-inflammatory mediators in BV2 microglial cells through mechanisms involving TLR-4, NF-κB, p38-MAPK, JNK and AKT[55]
		Rhamnetin R = OMe; R′=R″=OH	Anti-inflammatory effects in BV-2 microglia and rat hippocampal slices[56,57]
	Flavones	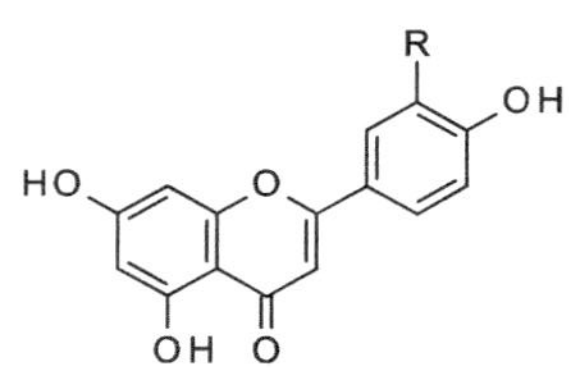	

Continued

Table 2 Examples of some anti-inflammatory natural flavonoids.—Cont'd

Inflammatory disorders	Flavonoid classes	Flavonoid	Pharmacological activities
		Luteolin R=OH	Reduction in the production of pro-inflammatory cytokines, NO and PGE2, as well as down-regulation of iNOS and COX-2 in LPS-activated BV2 microglia and LPS-treated adult mice. Inhibition of AP-1, NF-κB, p38, JNK, MAPK and AKT activation have been suggested as mechanisms involved in the inhibition of microglia activation[58–61]
		Apigenin R=H	Decreased release of the pro-inflammatory cytokines TNF-α and IL-6 in cultured N9 microglia and murine-derived primary microglia[62]
	Flavanols	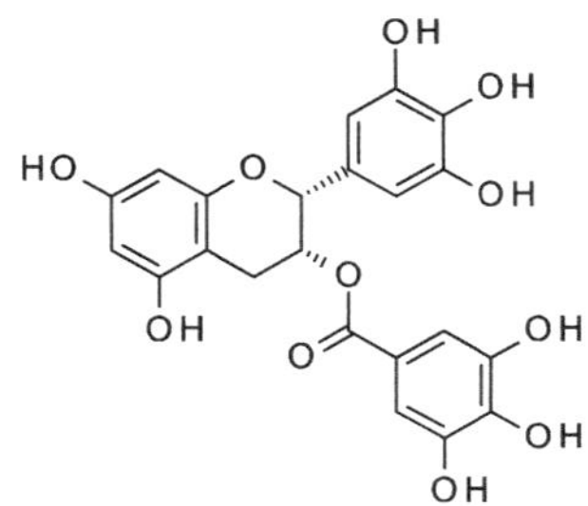 Epigallocatechin-3-gallate (EGCG)	Prevention of memory impairment, inhibition of reduction in the levels of Aβ generation and neurotoxicity induced by intraperitoneal injection of LPS in mice. Reduction in TNFα, IL-1β, IL-6, iNOS levels in Aβ-stimulated EOC13.31 microglia through mechanisms involving NF-κB. Microglia-mediated neurotoxicity was also inhibited[63–65]
	Flavanones	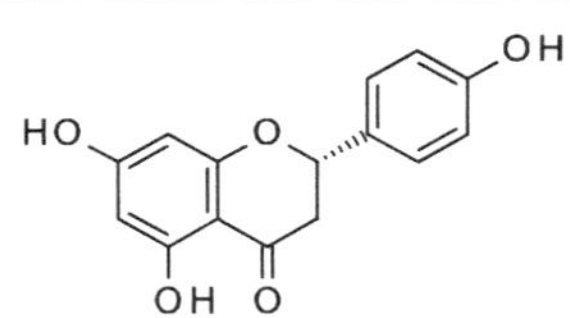 Naringenin	Inhibition LPS-induced neuroinflammation through modulation of NF-κB and MAPK activation in glial cells[66–68]

Inflammatory bowel disease	Flavonols	Rutin	Prevention of DSS-induced colitis in mice through mechanisms involving IL-1β and iNOS in enterocytes[69]
		Kaempferol	Reduction in DSS-induced colitis in mice through down-regulation of TNF-α, IL-6, IL-1β, iNOS, and COX-2 mRNA[70]
	Flavones	Luteolin	Suppression of macrophage and IFN-γ producing CD4+ T cell infiltration into the colon[71]
		Apigenin	Reduction in dextran sulfate sodium (DSS)-induced colitis through inhibition of inflammasome and of IL-1β production, as well as down-regulation of iNOS and COX-2[72]
	Flavanols	Epigallocatechin-3-gallate (EGCG)	Inhibition of IL-8, PGE2 and MIP-3α production in human colon epithelial cells stimulated with TNF-α[73]
Rheumatoid arthritis	Flavonols	Quercetin	Reduction in zymosan-induced hyperalgesia, articular edema and the recruitment of neutrophils to the knee joint cavity., accompanied by NF-κB-mediated reduction in TNFα and IL-1β production, and gp91phox, and cyclooxygenase-2 mRNA expression[74] Reduction in joint inflammation in collagen-induced arthritis, accompanied by decreased levels of TNF-α, IL-1β, IL-17, and MCP-1[75]
		Rutin	Produced improvement in inflammatory cells infiltration, synovial hyperplasia, pannus formation and cartilage and bone erosion, as well as a reduction in the levels of TNF-α, IL-1β in adjuvant arthritic rats[76] Produced a reduction in the arthritic score, NO/iNOS levels, as well as reduction in NF-κB activation in collagen-induced arthritis in rats[77]

Continued

Table 2 Examples of some anti-inflammatory natural flavonoids.—Cont'd

Inflammatory disorders	Flavonoid classes	Flavonoid	Pharmacological activities
		Kaempferol	Kaempferol produced a reduction in the mRNA and protein levels of MMP-1, MMP-3, COX-2 and PGE_2 in IL-1β-stimulated rheumatoid arthritis synovial fibroblasts. There was also inhibition of IL-1β-induced phosphorylation of ERK-1/2, p38 and JNK, as well as the activation of NF-κB[78]
	Flavones	Luteolin	Decreased the secretion of matrix metalloprotease-1 and -3 and the production of IL-6, IL-8, IL-15, and TGF-beta, as well as inactivating MAPK/ERKs and PI3K-Akt pathways in collagen-induced arthritic rats[79] Suppression of IL-1β-induced pro-inflammatory cytokines and matrix metalloproteinases production through mechanisms involving p38, JNK, NF-κB and AP-1 activation in human synovial sarcoma cells[80]
	Flavanols	Epigallocatechin-3-gallate (EGCG)	Suppression of TNFα-induced production of MMP-1 and MMP-3 in rheumatoid arthritis synovial fibroblasts; ERK1/2, p38, JNK and AP-1 activation were also inhibited. Inhibition of IL-1β-induced iNOS/NO and PGE_2/COX-2 production through NF-κB mechanisms in human chondrocytes[81–83]
	Flavanones	Naringenin	Reduction in the production of pro-inflammatory cytokines and chemokines, as well as inactivation of MAPK and NF-κB in a mouse model of collagen-induced arthritis[84]

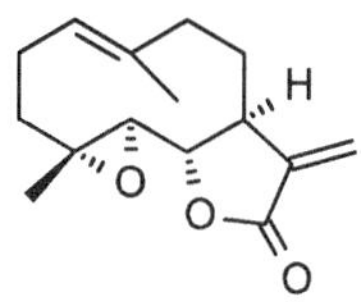

Fig. 1 Parthenolide, a naturally occurring sesquiterpene lactone of the germacranolide class found in feverfew (*Tanacetum parthenium*), of the family Asteraceae.

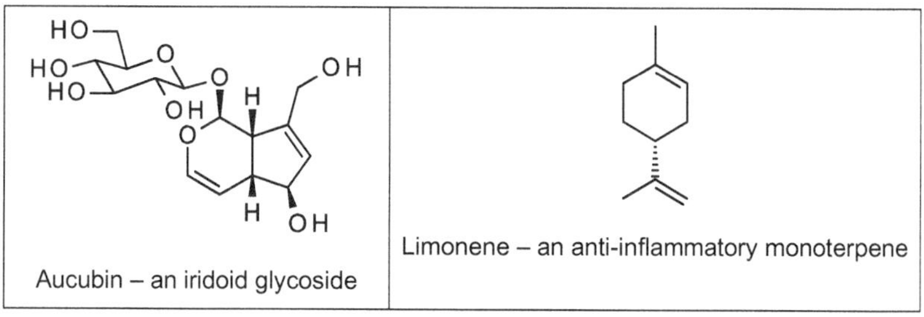

Fig. 2 Aucubin and limonene—two most studied anti-inflammatory terpenes.

showed that limonene produced a reduction in the levels of iNOS and COX-2/PGE_2, the pro-inflammatory cytokines TNF-α, IL-1β, and IL-6 in LPS-stimulated RAW264.7 macrophages. Considering the roles played by macrophages in the pathogenesis of inflammatory diseases, these observations are a significant in the therapeutics of limonene.

A study reported by Jeong et al.[89] suggested that aucubin could have therapeutic potentials in blocking inflammatory component of allergic conditions such as asthma through its inhibitory actions on TNFα and IL-6 production and NF-κB activation in mast cells. Similar NF-κB-mediated anti-inflammatory effects have been reported for aucubin in articular chondrocytes[90] and 3T3-L1 adipocytes,[91] demonstrating potentials in arthritis and metabolic syndrome/insulin resistance, respectively.

5.3.2 Sesquiterpenes

Sesquiterpene lactones, 15 carbon containing terpenoids, are a large group of bioactive secondary plant metabolites mostly found in the Asteraceae family.[48] Parthenolide (Fig. 1) is one of the most widely investigated biologically-active sesquiterpene lactones, and is present in *Tanacetum parthenium* (Asteraceae). Parthenolide has been reported to produce anti-inflammatory activity in many *in vivo* and *in vitro* models of inflammatory conditions. The compound was effective in reducing inflammation in

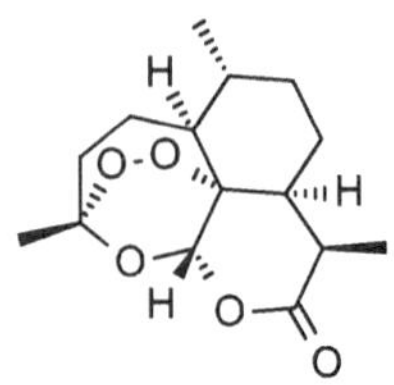

Fig. 3 Artemisinin from *Artemisia annua*.

obesity and obesity-induced inflammatory responses through the reduction of inflammatory cytokines, and inhibition of NF-κB and MAPK activation in 3T3-L1 adipocytes, RAW264.7, as well as in high-fat diet-fed mice.[92] The anti-inflammatory potential of parthenolide in rheumatoid arthritis was demonstrated by Liu et al.,[93] who reported that the compound suppressed LPS- and TNFα-induced increases in IL-1β, MMP-1, MMP-3, and iNOS mRNA in chondrocytes. They further demonstrated that parthenolide improved the symptoms of collagen-induced arthritis by reducing synovitis, inflammation, and pannus formation scores. The potential therapeutic benefits of parthenolide in inflammatory conditions has also been demonstrated in neuro-inflammation/neurodegeneration and colitis.[94,95]

Artemisinin (Fig. 3) is a sesquiterpene lactone found in the Chinese herb *Artemisia annua* (Qinghao) of the family Asteraceae. Artemisinin and its synthetic analogues were originally developed for the treatment of multi-drug resistant malaria. However, this compound is now receiving attention as a result of its reputed pharmacological properties beyond malaria. These include anti-inflammatory, anticancer and antiviral effects. Artemisinin has been studied in many models of inflammatory disease mainly due to its pharmacological action on multiple signaling pathways.[96] In neuroinflammation, it could inhibit LPS-induced release of TNFα, IL-6, MCP-1 and NO in BV-2 microglia through mechanisms that involve modulatory effects on NF-κB signaling.[97] These effects have been confirmed in an Alzheimer's disease (AD) mouse model, which showed that artemisinin B improved cognitive function and reduced levels of inflammatory cytokines in the hippocampus and the cortex.[98] Artemisinin has also been shown to produce promising anti-inflammatory effects in the cardiovascular system,[99,100] and in arthritis.[101] These observations suggest that artemisinin is a major source of drug repurposing to treat inflammatory conditions.

5.3.3 Diterpenes

Carnosol and carnosic acid (Fig. 4) are diterpene constituents (20 carbon containing terpenoids) found in abundance in the extracts of *Rosmarinus*

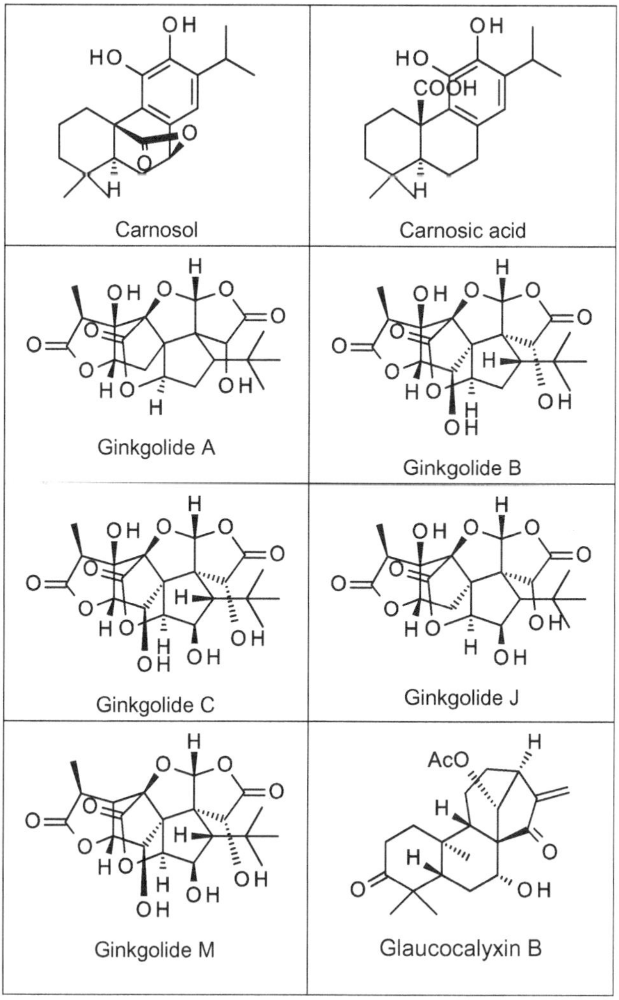

Fig. 4 Anti-inflammatory naturally occurring diterpenes.

officinalis (Rosemary) of the family Lamiaceae, and have been investigated as brain-permeable neuroprotective natural products, with inhibition of neuroinflammation accounting for part of their mechanisms of action.[102] These compounds have also been reported to produce anti-inflammatory activity in models of rheumatoid arthritis through reduction in the release of pro-inflammatory and catabolic mediators of cartilage breakdown in human chondrocytes.[103,104]

Ginkgolides are pharmacologically-active diterpenes found in *Ginkgo biloba* (Ginkgoaceae), which have been widely reported to have several pharmacological properties, including anti-inflammatory effects; for example, Ginkgolides A–C, J and M (Fig. 4). Anti-inflammatory effects have been reported for these ginkgolides in models of inflammatory conditions, including LPS-induced inflammation in chondrocytes,[105] LPS-induced neuroinflammation in BV-2 microglia,[106] LPS-induced inflammatory response in human coronary artery endothelial cells[107] and in asthma.[108] Inhibition of inflammation has also been reported to contribute to the neuroprotective actions of ginkgolides in models of cerebral ischemia and reperfusion injury.[109,110] Glaucocalyxin B (Fig. 4), a modified diterpene from *Rabdosia japonica* (Lamiaceae), was found to reduce the levels of NO, TNF-α, IL-1, ROS, COX-2 and iNOS in LPS-activated microglia cells, and also to inhibit activation of NF-κB, p38 MAPK.[111]

5.3.4 Triterpenes

One of the most investigated group of triterpenes (30 carbon containing terpenoids) are the boswellic acids, found in gum resins of *Boswellia* species. More than 12 different boswellic acids have been identified in the resin, but only 11-keto-β-boswellic acid and acetyl-11-keto-β-boswellic acid (Fig. 5) have been widely investigated for pharmacological activities.[112] Boswellic acids have been widely reported to produce anti-inflammatory activity in several models of human disease. They have been shown for their therapeutic potentials in chronic inflammatory diseases, such as rheumatoid arthritis,[113,114] atherosclerosis,[115] Parkinson's disease.[116] The anti-inflammatory activity of these compounds is thought to be mediated through their ability to target cellular signaling involving the NF-κB transcription factor.[117]

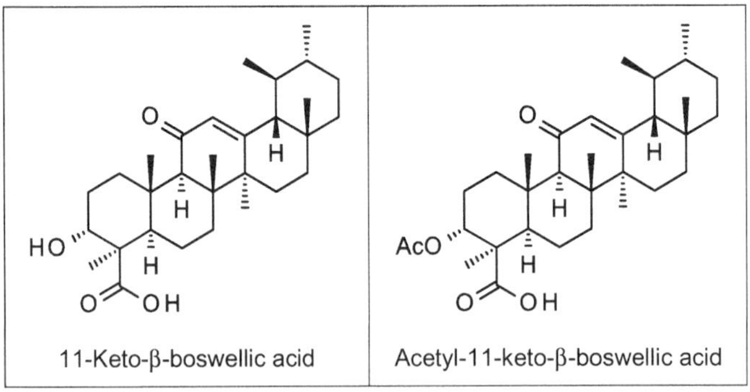

Fig. 5 Examples of anti-inflammatory naturally occurring triterpenes.

6. Anti-inflammatory marine natural products

Accumulating evidence suggests that natural products from marine sources are gaining attention as a source of pharmacologically active molecules for treating inflammatory disorders. In fact, the biological and chemical diversity of marine habitats present a rich reservoir of novel compounds; alkaloids, fatty acids, polysaccharides, proteins, steroids, terpenoids, and other chemical compounds, isolated from marine organisms have been found to exhibit anti-inflammatory activity.[118] Some of these compounds and their synthetic analogues are currently in pre-clinical and clinical trials. Gonzalez et al.[119] reviewed anti-inflammatory diterpenoids from marine sources, and reported that those diterpenes mainly belong to the classes of briarane, cembrane and eunicellane diterpenes.[48]

One of the most investigated marine natural products is astaxanthin (3,3′-dihydroxy-β, β′-carotene-4,4′-dione) (Fig. 6), which is a blood-red xanthophyll keto-carotenoid produced mainly by the freshwater microalgae *Haematococcus pluvialis*, *Chlorella zofingiensis*, *Chlorococcum* spp. and the yeast fungus *Phaffia rhodozyma* (also known as *Xanthophyllomyces dendrorhous*).[120] Investigations by Che et al.[121] revealed that astaxanthin could prevent cognitive disorders by reducing pathological features including neuroinflammation in APP/PS1 mice model, suggesting that the compound might be beneficial in Alzheimer's disease. It can be noted that APP/PS1 are double transgenic mice expressing a chimeric mouse/human amyloid precursor protein (Mo/HuAPP695swe) and a mutant human presenilin 1 (PS1-dE9), both directed to CNS neurons. Other studies suggested that astaxanthin could offer neuroprotection by inhibiting neuroinflammation.[122–124] Studies have also shown that astaxanthin may be beneficial in UV-exposure-induced skin damage through anti-inflammatory mechanisms. It was reported to reduce the levels of iNOS and COX-2 following UV irradiation in human keratinocytes through mechanisms

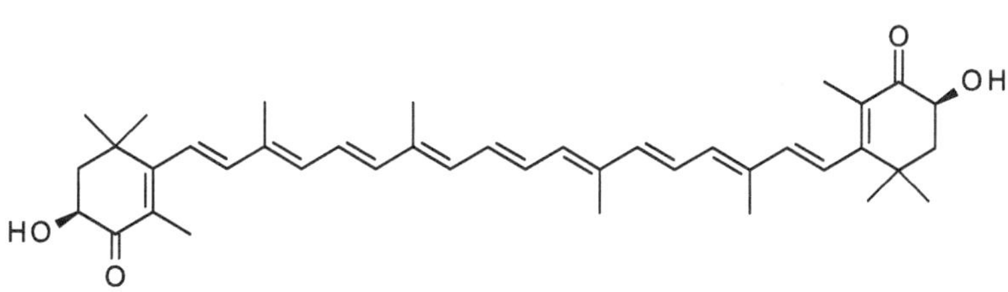

Fig. 6 Astaxanthin—a natural anti-inflammatory agent from marine organisms.

involving NF-κB activation.[125,126] Interestingly, several clinical studies have proven the benefits of astaxanthin in dermatological disorders.[127] The benefits of astaxanthin in cardiovascular diseases have also been established to be due in part to the anti-inflammatory activity of the compound in a variety of *in vitro* and *in vivo* models.[128,129]

Pharmacological evidence further demonstrated that astaxanthin attenuated DSS-induced colitis by blocking the activation of NF-κB, AP-1 and MAPK pathways,[130] suggesting the potential of the compound in the treatment of inflammatory bowel disease.

7. Conclusions and future perspectives

Due to the limitations (side effects and high cost) associated with existing anti-inflammatory therapies, increasing research efforts have focused on anti-inflammatory natural products. In fact, a great deal of effort in recent years has focused on using cutting-edge *in vivo* and *in vitro* techniques to evaluate anti-inflammatory activity of natural products. Plant-derived natural products are an important source of new drugs for treating conditions associated with chronic inflammation. Due to their chemical diversity and established mechanisms of anti-inflammatory actions, further investigations on these natural products may result in the development of novel potent anti-inflammatory agents with minimal side effects. Artemisinin, which is already approved by regulatory agencies for treating malaria is an example of natural products which can be re-purposed as an anti-inflammatory drug. Many anti-inflammatory natural products have not been tested in rigorous clinical trials in patients. Also, in spite of the strides made in identifying natural products which inhibit neuroinflammation and have potentials in treating AD, it is not yet clear if these compounds are able to cross the blood-brain barrier and achieve effective concentrations in the brain. Future research efforts on anti-inflammatory natural products need to focus on human trials to determine efficacy and safety.

References

1. Varela, M. L.; Mogildea, M.; Moreno, I.; Lopes, A. *Inflammation* **2018**, *41*, 1115–1127.
2. Ward, P. A. In: *Fundamentals of Inflammation*; Serhan, C. N., Ward, P. A., Gilroy, D. W., Eds.; Cambridge University Press: New York, 2010.
3. Roda, J. M.; Eubank, T. D. In: *Chronic Inflammation: Molecular Pathophysiology, Nutritional and Therapeutic Interventions*; Roy, S., Bagchi, D., Raychaudhuri, S. P., Eds.; CRC Press LLC: Baton Rouge, 2012.
4. Choy, E. *Rheumatology* **2012**, *51*, 3–11.

5. Alwarith, J.; Kahleova, H.; Rembert, E.; Yonas, W.; Dort, S.; Calcagno, M.; Burgess, N.; Crosby, L.; Barnard, N. D. *Front. Nutr.* **2019**, *6*, 141.
6. Smolen, J. S.; Aletaha, D.; Barton, A.; Burmester, G. R.; Emery, P.; Firestein, G. S.; Kavanaugh, A.; McInnes, I. B.; Solomon, D. H.; Strand, V.; Yamamoto, K. *Nat. Rev. Dis. Prim.* **2018**, *4*, 18001.
7. Libby, P. *Am. J. Clin. Nutr.* **2006**, *83*, 456S–460S.
8. Ruparelia, N.; Chai, J. T.; Fisher, E. A.; Choudhury, R. P. *Nat. Rev. Cardiol.* **2017**, *14*, 133–144.
9. von Hundelshausen, P.; Weber, C. *Circ. Res.* **2007**, *100*, 27–40.
10. Vinten-Johansen, J. *Cardiovasc. Res.* **2004**, *61*, 481–497.
11. Koeners, M. P.; Wesseling, S.; Sánchez, M.; Braam, B.; Joles, J. A. *Am. J. Hypertens.* **2016**, *29*, 123–131.
12. Zhao, H.; Shen, R.; Dong, X.; Shen, Y. *Inflammation* **2017**, *40*, 232–239.
13. Wang, X.; Chen, Q.; Pu, H.; Wei, Q.; Duan, M.; Zhang, C.; Jiang, T.; Shou, X.; Zhang, J.; Yang, Y. *Lipids Health Dis.* **2016**, *15*, 33.
14. DiDonato, J. A.; Mercurio, F.; Karin, M. *Immunol. Rev.* **2012**, *246*, 379–400.
15. Karin, M.; Cao, Y.; Greten, F. R.; Li, Z. W. *Nat. Rev. Cancer.* **2002**, *2*, 301–310.
16. Yang, Q. Q.; Zhou, J. W. *Glia* **2019**, *67*, 1017–1035.
17. Nimmerjahn, A.; Kirchhoff, F.; Helmchen, F. *Science* **2005**, *308*, 1314–1318.
18. Davalos, D.; Grutzendler, J.; Yang, G.; Kim, J. V.; Zuo, Y.; Jung, S.; Gan, W. B. *Nat. Neurosci.* **2005**, *8*, 752–758.
19. Schoultz, I.; Söderholm, J. D.; McKay, D. M. *Inflamm. Bowel Dis.* **2011**, *17*, 2008–2018.
20. Horiuchi, T.; Mitoma, H.; Harashima, S.; Tsukamoto, H.; Shimoda, T. *Rheumatology* **2010**, *49*, 1215–1228.
21. Reimund, J. M.; Wittersheim, C.; Dumont, S.; Muller, C. D.; Kenney, J. S.; Baumann, R.; Poindron, P.; Duclos, B. *Gut* **1996**, *39*, 684–689.
22. Schoultz, I.; Keita, A. V. *Cells* **2019**, *8*, 193, pii: E193.
23. Freitas, R. H. C. N.; Fraga, C. A. M. *Curr. Drug Targets* **2018**, *19*, 1933–1942.
24. Hayden, M. S.; Ghosh, S. *Genes Dev.* **2012**, *26*, 203–234.
25. Bennett, J.; Capece, D.; Begalli, F.; Verzella, D.; D'Andrea, D.; Tornatore, L.; Franzoso, G. *Int. J. Biochem. Cell Biol.* **2018**, *95*, 108–112.
26. Ziegler, J.; Facchini, P. J. *Annu. Rev. Plant Biol.* **2008**, *59*, 735–769.
27. Rosa, E. A. S.; Bennett, R. N.; Aires, A. In: *Handbook of Organic Food Safety and Quality*; Cooper, J., Niggli, U., Leifert, C., Eds.; Woodhead Publising, 2007.
28. Olajide, O. A.; Bhatia, H. S.; de Oliveira, A. C.; Wright, C. W.; Fiebich, B. L. *Evid. Based Compl. Alt. Med.* **2013**, *2013*, 459723.
29. He, F. Q.; Qiu, B. Y.; Zhang, X. H.; Li, T. K.; Xie, Q.; Cui, D. J.; Huang, X. L.; Gan, H. T. *Brain Res.* **2011**, *1384*, 89–96.
30. Jia, L.; Liu, J.; Song, Z.; Pan, X.; Chen, L.; Cui, X.; Wang, M. *J. Pharm. Pharmacol.* **2012**, *64*, 1510–1521.
31. Paris, D.; Beaulieu-Abdelahad, D.; Bachmeier, C.; Reed, J.; Ait-Ghezala, G.; Bishop, A.; Chao, J.; Mathura, V.; Crawford, F.; Mullan, M. *Eur. J. Pharmacol.* **2011**, *670*, 384–391.
32. Paris, D.; Beaulieu-Abdelahad, D.; Abdullah, L.; Bachmeier, C.; Ait-Ghezala, G.; Reed, J.; Verma, M.; Crawford, F.; Mullan, M. *Eur. J. Pharmacol.* **2013**, *698*, 145–153.
33. Wang, X. J.; Deng, H. Z.; Jiang, B.; Yao, H. *Int. J. Colorectal. Dis.* **2012**, *27*, 575–581.
34. Cheng, H.; Xia, B.; Zhang, L.; Zhou, F.; Zhang, Y. X.; Ye, M.; Hu, Z. G.; Li, J.; Li, J.; Wang, Z. L.; Li, C.; Guo, Q. S. *Pharmacol. Res.* **2006**, *53*, 202–208.
35. Fan, H.; Chen, R.; Shen, L.; Lv, J.; Xiong, P.; Shou, Z.; Zhuang, X. *J. Huazhong Univ. Sci. Technol. Med. Sci.* **2008**, *28*, 415–420.
36. Guzman, J. R.; Koo, J. S.; Goldsmith, J. R.; Mühlbauer, M.; Narula, A.; Jobin, C. *Sci. Rep.* **2013**, *3*, 1629.

37. Yan, F.; Wang, L.; Shi, Y.; Cao, H.; Liu, L.; Washington, M. K.; Chaturvedi, R.; Israel, D. A.; Cao, H.; Wang, B.; Peek, R. M., Jr.; Wilson, K. T.; Polk, D. B. *Am. J. Physiol. Gastrointest. Liver Physiol.* **2012**, *302*, G504–G514.
38. Colombo, B. B.; Fattori, V.; Guazelli, C. F. S.; Zaninelli, T. H.; Carvalho, T. T.; Ferraz, C. R.; Bussmann, A. J. C.; Ruiz-Miyazawa, K. W.; Baracat, M. M.; Casagrande, R.; Verri, W. A., Jr. *Inflammation* **2018**, *41*, 1276–1289.
39. Gupta, R. A.; Motiwala, M. N.; Dumore, N. G.; Danao, K. R.; Ganjare, A. B. *J. Ethnopharmacol.* **2015**, *164*, 239–246.
40. Shan, L.; Tong, L.; Hang, L.; Fan, H. *Biomed. Pharmacother.* **2019**, *111*, 142–150.
41. Mu, H.; Yao, R. B.; Zhao, L. J.; Shen, S. Y.; Zhao, Z. M.; Cai, H. *Inflammation* **2013**, *36*, 1136–1144.
42. Zhou, H.; Liu, J. X.; Luo, J. F.; Cheng, C. S.; Leung, E. L.; Li, Y.; Su, X.-H.; Liu, Z.-Q.; Chen, T.-B.; Duan, F.-G.; Dong, Y.; Zuo, Y.-H.; Li, C.; Lio, C. K.; Li, T.; Luo, P.; Xie, Y.; Yao, X.-J.; Wang, P.-X.; Liu, L. *Biochem. Pharmacol.* **2017**, *142*, 133–144.
43. Wang, Y.; Fang, Y.; Huang, W.; Zhou, X.; Wang, M.; Zhong, B.; Peng, D. *J. Ethnopharmacol.* **2005**, *98*, 37–43.
44. Yao, R. B.; Zhao, Z. M.; Zhao, L. J.; Cai, H. *Pharmazie* **2017**, *72*, 355–360.
45. Jia, Y.; Miao, Y.; Yue, M.; Shu, M.; Wei, Z.; Dai, Y. *FASEB J.* **2018**, *32*, 3398–3410.
46. Liang, J.; Chang, B.; Huang, M.; Huang, W.; Ma, W.; Liu, Y.; Tai, W.; Long, Y.; Lu, Y. *Int. Immunopharmacol.* **2018**, *55*, 105–111.
47. Zhu, L.; Zhu, L. *Eur. Cytokine Network* **2017**, *28*, 120–126.
48. Nahar, L.; Sarker, S. D. *Chemistry for Pharmacy Students*, 2nd ed.; Wiley and Sons: United Kingdom, 2019.
49. Rathee, P.; Chaudhary, H.; Rathee, S.; Rathee, D.; Kumar, V.; Kohli, K. *Inflamm. Allergy Drug Targets* **2009**, *8*, 229–235.
50. García-Lafuente, A.; Guillamón, E.; Villares, A.; Rostagno, M. A.; Martínez, J. A. *Inflamm. Res.* **2009**, *58*, 537–552.
51. Bharti, S.; Rani, N.; Krishnamurthy, B.; Arya, D. S. *Planta Med.* **2014**, *80*, 437–451.
52. Bournival, J.; Plouffe, M.; Renaud, J.; Provencher, C.; Martinoli, M. G. *Oxid. Med. Cell Longev.* **2012**, *2012*, 921941.
53. Mehta, V.; Parashar, A.; Udayabanu, M. *Physiol. Behav.* **2017**, *171*, 69–78.
54. Javed, H.; Khan, M. M.; Ahmad, A.; Vaibhav, K.; Ahmad, M. E.; Khan, A.; Ashafaq, M.; Islam, F.; Siddiqui, M. S.; Safhi, M. M.; Islam, F. *Neuroscience* **2012**, *210*, 340–352.
55. Park, S. E.; Sapkota, K.; Kim, S.; Kim, H.; Kim, S. J. *Br. J. Pharmacol.* **2011**, *164*, 1008–1025.
56. Lutz, J. A.; Kulshrestha, M.; Rogers, D. T.; Littleton, J. M. *Fitoterapia* **2014**, *98*, 11–21.
57. Lutz, J. A.; Carter, M.; Fields, L.; Barron, S.; Littleton, J. M. *Alcohol Clin. Exp. Res.* **2015**, *39*, 2345–2353.
58. Zhu, L. H.; Bi, W.; Qi, R. B.; Wang, H. D.; Wang, Z. G.; Zeng, Q.; Zhao, Y. R.; Lu, D. X. *Neurol. Res.* **2011**, *33*, 927–934.
59. Burton, M. D.; Rytych, J. L.; Amin, R.; Johnson, R. W. *Rejuvenation Res.* **2016**, *19*, 286–292.
60. Jang, S.; Kelley, K. W.; Johnson, R. W. *Proc. Natl. Acad. Sci. U. S. A* **2008**, *105*, 7534–7539.
61. Zhu, L.; Bi, W.; Lu, D.; Zhang, C.; Shu, X.; Lu, D. *Exp. Ther. Med.* **2014**, *7*, 1065–1070.
62. Rezai-Zadeh, K.; Ehrhart, J.; Bai, Y.; Sanberg, P. R.; Bickford, P.; Tan, J.; Shytle, R. D. *J. Neuroinflamm.* **2008**, *5*, 41.
63. Lee, Y. J.; Choi, D. Y.; Yun, Y. P.; Han, S. B.; Oh, K. W.; Hong, J. T. *J. Nutr. Biochem.* **2013**, *24*, 298–310.

64. Lee, Y. K.; Yuk, D. Y.; Lee, J. W.; Lee, S. Y.; Ha, T. Y.; Oh, K. W.; Yun, Y. P.; Hong, J. T. *Brain Res.* **2009**, *1250*, 164–174.
65. Cheng-Chung Wei, J.; Huang, H. C.; Chen, W. J.; Huang, C. N.; Peng, C. H.; Lin, C. L. *Eur. J. Pharmacol.* **2016**, *770*, 16–24.
66. Vafeiadou, K.; Vauzour, D.; Lee, H. Y.; Rodriguez-Mateos, A.; Williams, R. J.; Spencer, J. P. *Arch. Biochem. Biophys.* **2009**, *484*, 100–109
67. Raza, S. S.; Khan, M. M.; Ahmad, A.; Ashafaq, M.; Islam, F.; Wagner, A. P.; Safhi, M. M.; Islam, F. *Neuroscience* **2013**, *230*, 157–171.
68. Wu, L. H.; Lin, C.; Lin, H. Y.; Liu, Y. S.; Wu, C. Y.; Tsai, C. F.; Chang, P. C.; Yeh, W. L.; Lu, D. Y. *Mol. Neurobiol.* **2016**, *53*, 1080–1091.
69. Kwon, K. H.; Murakami, A.; Tanaka, T.; Ohigashi, H. *Biochem. Pharmacol.* **2005**, *69*, 395–406.
70. Park, M. Y.; Ji, G. E.; Sung, M. K. *Dig. Dis. Sci.* **2012**, *57*, 355–363.
71. Nishitani, Y.; Yamamoto, K.; Yoshida, M.; Azuma, T.; Kanazawa, K.; Hashimoto, T.; Mizuno, M. *BioFactors* **2013**, *39*, 522–533.
72. Márquez-Flores, Y. K.; Villegas, I.; Cárdeno, A.; Rosillo, M. Á.; Alarcón-de-la-Lastra, C. *J. Nutr. Biochem.* **2016**, *30*, 143–152.
73. Porath, D.; Riegger, C.; Drewe, J.; Schwager, J. *J. Pharmacol. Exp. Ther.* **2005**, *315*, 1172–1180.
74. Guazelli, C. F. S.; Staurengo-Ferrari, L.; Zarpelon, A. C.; Pinho-Ribeiro, F. A.; Ruiz-Miyazawa, K. W.; Vicentini, F. T. M. C.; Vignoli, J. A.; Camilios-Neto, D.; Georgetti, S. R.; Baracat, M. M.; Casagrande, R.; Verri, W. A., Jr. *Biomed. Pharmacother.* **2018**, *102*, 175–184.
75. Haleagrahara, N.; Miranda-Hernandez, S.; Alim, M. A.; Hayes, L.; Bird, G.; Ketheesan, N. *Biomed. Pharmacother.* **2017**, *90*, 38–46.
76. Sun, C. L.; Wei, J.; Bi, L. Q. *Pharmacology* **2017**, *100*, 40–49.
77. Gul, A.; Kunwar, B.; Mazhar, M.; Faizi, S.; Ahmed, D.; Shah, M. R.; Simjee, S. U. *Int. Immunopharmacol.* **2018**, *59*, 310–317.
78. Yoon, H. Y.; Lee, E. G.; Lee, H.; Cho, I. J.; Choi, Y. J.; Sung, M. S.; Yoo, H. G.; Yoo, W. H. *Int. J. Mol. Med.* **2013**, *32*, 971–977.
79. Hou, Y.; Wu, J.; Huang, Q.; Guo, L. *Cell Biol. Int.* **2009**, *33*, 135–147.
80. Choi, E. M.; Lee, Y. S. *Food Chem. Toxicol.* **2010**, *48*, 2607–2611.
81. Yun, H. J.; Yoo, W. H.; Han, M. K.; Lee, Y. R.; Kim, J. S.; Lee, S. I. *Rheumatol. Int.* **2008**, *29*, 23–29.
82. Ahmed, S.; Rahman, A.; Hasnain, A.; Lalonde, M.; Goldberg, V. M.; Haqqi, T. M. *Free Radical Biol. Med.* **2002**, *33*, 1097–1105.
83. Singh, R.; Ahmed, S.; Islam, N.; Goldberg, V. M.; Haqqi, T. M. *Arthritis Rheum.* **2002**, *46*, 2079–2086.
84. Li, Y. R.; Chen, D. Y.; Chu, C. L.; Li, S.; Chen, Y. K.; Wu, C. L.; Lin, C. C. *J. Nutr. Biochem.* **2015**, *26*, 1467–1478.
85. Salminen, A.; Lehtonen, M.; Suuronen, T.; Kaarniranta, K.; Huuskonen, J. *Cell. Mol. Life Sci.* **2008**, *65*, 2979–2999.
86. Souza, M. T.; Almeida, J. R.; Araujo, A. A.; Duarte, M. C.; Gelain, D. P.; Moreira, J. C.; dos Santos, M. R.; Quintans-Júnior, L. J. *Basic Clin. Pharmacol. Toxicol.* **2014**, *115*, 244–256.
87. Rufino, A. T.; Ribeiro, M.; Sousa, C.; Judas, F.; Salgueiro, L.; Cavaleiro, C.; Mendes, A. F. *Eur. J. Pharmacol.* **2015**, *750*, 141–150.
88. Yoon, W. J.; Lee, N. H.; Hyun, C. G. *J. Oleo Sci.* **2010**, *59*, 415–421.
89. Jeong, H. J.; Koo, H. N.; Na, H. J.; Kim, M. S.; Hong, S. H.; Eom, J. W.; Kim, K. S.; Shin, T. Y.; Kim, H. M. *Cytokine* **2002**, *18*, 252–259.
90. Wang, S. N.; Xie, G. P.; Qin, C. H.; Chen, Y. R.; Zhang, K. R.; Li, X.; Wu, Q.; Dong, W. Q.; Yang, J.; Yu, B. *Int. Immunopharmacol.* **2015**, *24*, 408–415.

91. Park, K. S. *Cytokine* **2013**, *62*, 407–412.
92. Syrovets, T.; Buchele, B.; Krauss, C.; Laumonnier, Y.; Simmet, T. J. *Immunology* **2005**, *174*, 498–506.
93. Liu, Q.; Zhao, J.; Tan, R.; Zhou, H.; Lin, Z.; Zheng, M.; Romas, E.; Xu, J.; Sims, N. A. *Scand. J. Rheumatol.* **2015**, *44*(3), 182–191.
94. Khare, P.; Datusalia, A. K.; Sharma, S. S. *Neuromolecular Med.* **2017**, *19*(1), 101–112.
95. Zhao, Z. J.; Xiang, J. Y.; Liu, L.; Huang, X. L.; Gan, H. T. *Int. Immunopharmacol.* **2012**, *12*(1), 169–174.
96. Shakir, M.; Hussain, A.; Javeed, M.; Ashraf, A.; Riaz, A. *Eur. J. Pharmacol.* **2011**, *668*, 6–14.
97. Zhu, C.; Xiong, Z.; Chen, X.; Peng, F.; Hu, X.; Chen, Y.; Wang, Q. *PLoS One* **2012**, 7(4), e35125.
98. Qiang, W.; Cai, W.; Yang, Q.; Yang, L.; Dai, Y.; Zhao, Z.; Yin, J.; Li, Y.; Li, Q.; Wang, Y.; Weng, X.; Zhang, D.; Chen, Y.; Zhu, X. *Neuroscience* **2018**, *395*, 1–12.
99. Jiang, Y.; Du, H.; Liu, X.; Fu, X.; Li, X.; Cao, Q. *J. Drug Target.* **2020**, *28*(1), 70–79.
100. Cao, Q.; Jiang, Y.; Shi, J.; Xu, C.; Liu, X.; Yang, T.; Fu, P.; Niu, T. *J. Surg. Res.* **2015**, *194*(2), 667–678.
101. Zhong, G.; Liang, R.; Yao, J.; Li, J.; Jiang, T.; Liu, J.; Le, Y.; Shen, C.; Long, H.; Lu, H.; Ma, S.; Luo, J.; Miao, Z.; Su, W.; Zheng, L.; Zhao, J. *Cell Physiol. Biochem.* **2018**, *51*(6), 2575–2590.
102. de Oliveira, M. R. *Mol. Neurobiol.* **2016**, *53*, 6155–6168.
103. Sanchez, C.; Horcajada, M. N.; Membrez Scalfo, F.; Ameye, L.; Offord, E.; Henrotin, Y. *PLoS One* **2015**, *10,* e0136118.
104. Schwager, J.; Richard, N.; Fowler, A.; Seifert, N.; Raederstorff, D. *Molecules* **2016**, *21*, 465.
105. Hu, H.; Li, Y.; Xin, Z.; Zhanga, X. *Adv. Clin. Exp. Med.* **2018**, *27*, 913–920.
106. Zhou, J. M.; Gu, S. S.; Mei, W. H.; Zhou, J.; Wang, Z. Z.; Xiao, W. *Cell Stress Chaperones* **2016**, *21*, 1037–1053.
107. Zhaocheng, J.; Jinfeng, L.; Luchang, Y.; Yequan, S.; Feng, L.; Kai, W. *Pharmazie* **2016**, *71*, 588–591.
108. Chu, X.; Ci, X.; He, J.; Wei, M.; Yang, X.; Cao, Q.; Li, H.; Guan, S.; Deng, Y.; Pang, D.; Deng, X. *Molecules* **2011**, *16*, 7634–7648.
109. Jiang, M.; Li, J.; Peng, Q.; Liu, Y.; Liu, W.; Luo, C.; Peng, J.; Li, J.; Yung, K. K.; Mo, Z. *J. Neuroinflamm.* **2014**, *11*, 167.
110. Gu, J. H.; Ge, J. B.; Li, M.; Wu, F.; Zhang, W.; Qin, Z. H. *Eur. J. Pharm. Sci.* **2012**, *47*, 652–660.
111. Gan, P.; Zhang, L.; Chen, Y.; Zhang, Y.; Zhang, F.; Zhou, X.; Zhang, X.; Gao, B.; Zhen, X.; Zhang, J.; Zheng, L. T. *J. Pharmacol. Sci.* **2015**, *128*, 35–46.
112. Ammon, H. P. T. In: *Anti-Inflammatory Nutraceuticals and Chronic Diseases*; Gupta, S., Prasad, S., Aggarwal, B., Eds.; Advances in Experimental Medicine and Biology.; Vol. 928, Springer: Cham, 2016.
113. Khanna, D.; Sethi, G.; Ahn, K. S.; Pandey, M. K.; Kunnu-Makkara, A. B.; Sung, B.; Aggarwal, A.; Aggarwal, B. B. *Curr. Opin. Pharmacol.* **2007**, 7, 344–351.
114. Singh, S.; Khajuria, A.; Taneja, S. C.; Khajuria, R. K.; Singh, J.; Qazi, G. N. *Bioorg. Med. Chem. Lett.* **2007**, *17*, 3706–3711.
115. Cuaz-Perolin, C.; Billiet, L.; Bauge, E.; Copin, C.; Scott-Algara, D.; Genze, F.; Buchele, B.; Syrovets, T.; Simmet, T.; Rouis, M. *Arterioscler. Thromb. Vasc. Biol.* **2008**, *28*, 272–277.
116. Ameen, A. M.; Elkazaz, A. Y.; Mohammad, H. M. F.; Barakat, B. M. *Can. J. Physiol. Pharmacol.* **2017**, *95*(7), 819–829.
117. Kim, C. Y.; Kang, B.; Suh, H. J.; Choi, H. S. *Pharmacol. Res.* **2019**, *145*, 104259.

118. Cheung, R. C.; Ng, T. B.; Wong, J. H.; Chen, Y.; Chan, W. Y. *Appl. Microbiol. Biotechnol.* **2016**, *100*(4), 1645–1666.
119. Gonzalez, Y.; Torres-Mendoza, D.; Jones, G. E.; Fernandez, P. L. *Mediators Inflamm.* **2015**, *2015*, 263543. https://doi.org/10.1155/2015/263543.
120. Ambati, R.; Phang, S.-M.; Ravi, S.; Aswathanarayana, R. *Mar. Drugs* **2014**, *12*, 128–152.
121. Che, H.; Li, Q.; Zhang, T.; Wang, D.; Yang, L.; Xu, J.; Yanagita, T.; Xue, C.; Chang, Y.; Wang, Y. *J. Agric. Food Chem.* **2018**, *66*, 4948–4957.
122. Fakhri, S.; Abbaszadeh, F.; Dargahi, L.; Jorjani, M. *Pharmacol. Res.* **2018**, *136*, 1–20.
123. Grimmig, B.; Kim, S. H.; Nash, K.; Bickford, P. C.; Douglas Shytle, R. *Geroscience* **2017**, *39*, 19–32.
124. Wu, H.; Niu, H.; Shao, A.; Wu, C.; Dixon, B. J.; Zhang, J.; Yang, S.; Wang, Y. *Mar. Drugs.* **2015**, *13*, 5750–5766.
125. Yoshihisa, Y.; Rehman, M. U.; Shimizu, T. *Exp. Dermatol.* **2014**, *23*, 178–183.
126. Terazawa, S.; Nakajima, H.; Shingo, M.; Niwano, T.; Imokawa, G. *Exp. Dermatol.* **2012**, *21*, 11–17.
127. Singh, K. N.; Patil, S.; Barkate, H. *J. Cosmet. Dermatol.* **2019**, 1–6.
128. Visioli, F.; Artaria, C. *Food Funct.* **2017**, *8*, 39–63.
129. Fassett, R. G.; Coombes, J. S. *Future Cardiol.* **2009**, *5*, 333–342.
130. Sakai, S.; Nishida, A.; Ohno, M.; Inatomi, O.; Bamba, S.; Sugimoto, M.; Kawahara, M.; Andoh, A. *J. Clin. Biochem. Nutr.* **2019**, *64*, 66–72.

CHAPTER SIX

Neuroprotective natural products

Subrata Das[a], Lutfun Nahar[b,c], Rajat Nath[a], Deepa Nath[d], Satyajit D. Sarker[b], Anupam Das Talukdar[a,*]

[a]Department of Life Science and Bioinformatics, Assam University, Silchar, Assam, India
[b]Centre for Natural Products Discovery (CNPD), School of Pharmacy and Biomolecular Sciences, Liverpool John Moores University, Liverpool, United Kingdom
[c]Laboratory of Growth Regulators, Institute of Experimental Botany ASCR & Palacký University, Olomouc, Czech Republic
[d]Department of Botany, Gurucharan College, Silchar, Assam, India
[*]Corresponding author: e-mail address: anupam@bioinfoaus.ac.in

Contents

1. Introduction

The term "neuroprotection" is defined as a strategy or method for safeguarding the central nervous system (CNS) from neuronal damages caused by chronic or acute neurodegenerative diseases (NDs).[1] NDs are the result of collapse and deterioration of neurons of the CNS and often affect the mental as well as the intellectual abilities of the sufferers. NDs are characterized by progressive loss of structure and/or function of neurons and incorporate a wide variety of conditions from Alzheimer's, Parkinson's, and Huntington's disease to glaucoma. The symptoms of NDs are usually progressive after the onset and include loss of memory, mostly short term, difficulty in learning, motor coordination, and many other functional loses.[2] Aging, a complex physiological process involving both biochemical and

Annual Reports in Medicinal Chemistry, Volume 55
ISSN 0065-7743
https://doi.org/10.1016/bs.armc.2020.02.009

morphological changes that gradually unfold as we get older,[3] has been found to be strictly linked with NDs. Aging denotes a major risk factor among the other etiological factors of NDs, including hypertension, genetic factors, environmental factors, and infections. With increasing age, proteins aggregation, inflammation, oxidative stress, and loss of neurotransmitters, are common to the pathology of NDs.[3,4]

In order to achieve neuroprotection and to prevent CNS diseases, the use of phytomedicine and phytofood represents one of the popular options, in association with a healthy lifestyle as well as moderate physical activity. In complementary and alternative therapy, herbal medicine or phytomedicine means the use of plant organs (leaves, flowers, stems, roots, fruits and seeds) for their healing properties,[3] often contributed by the presence of complex mixtures of active phytochemicals.[5,6] Considering the fact that NDs epitomize chronic and incurable conditions, bioactive compounds extracted and isolated from medicinal plants have become a viable choice to prevent and alleviate neurological disorders[7] owing to their diverse biological properties, which also include neuroprotective effects, direct uptake of free-radicals and modulation of enzymes associated with oxidative stress.[8] Phytochemicals of vegetables and fruits are known to reduce the risk of major health complications including cancers, cardiovascular diseases as well as NDs. Consequently, people who eat plenty of vegetables and fruits may be at a lower risk for diseases caused by neuronal dysfunction.[9,10]

Herbal therapeutics have long been used to treat neural symptoms. Even though the exact mechanisms of action of plant drugs have yet to be confirmed, many of them have been shown to exert antioxidant or anti-inflammatory effects in a variety of peripheral systems. Evidence suggests that anti-inflammatory herbal medicine and its elements are effective neuroprotectors against various brain pathologies. Chemical diversity that exists in medicinal plants makes them an important source of novel lead compounds for various therapeutic targets that are newly identified by proteomics, genomics, and high-throughput screening.[11]

Plant based therapeutics have also become popular in the Western societies resulting in a surge in their sales.[12] In the developing countries around 80% of people rely on plant-based medicine for primary healthcare.[13] In ethnic practices of medicine, numerous plants have been identified and used to cure cognitive disorders, including neurodegenerative diseases such as Alzheimer's disease (AD), Parkinson's disease and other memory associated disorders. In Asian countries there are more than 120 traditional therapeutics used for the treatment of CNS disorders.[14] In the Ayurvedic system of medicine the following medicinal plants are just a few examples that have been depicted to possess promising neuropharmacological actions: *Acorus calamus, Allium sativum, Angelica sinensis, Bacopa monnierae, Centella asiatica, Celastrus paniculatus, Crocus sativus, Curcuma longa, Enhydra fluctuans, Ginkgo biloba, Glycyrrhiza glabra, Huperiza serrata, Hypericum perforatum, Nicotiana tabaccum, Physostigma venosum, Ricinus communis, Salvia officinalis, Terminalia chebula, Uncaria tomentosa, Valeriana wallichii* and *Withania somnifera.*[15]

Not only the terrestrial plants that offer neuroprotection and neuroprotective compounds, several marine organisms, especially macroalgae, are also know to produce neuroprotective compounds.[16] During the period, 1999–2015, approximately 100 pure compounds with potential neuroprotective properties as identified from a variety of *in vitro* assays, e.g., inhibition of beta amyloid protein, inhibition of cholinesterases, reduction in oxidative stress, anti-inflammatory activity, inhibition of kinases, enhancement of neurite outgrowth and the reduction in dopaminergic neurotoxicit, were reported from marine sources, predominantly from Phaeophyceae (58%), followed by Rhodophyta (28%) and Chlorophyta (14%).[16]

This chapter aims to present an overview on neuroprotective natural products, predominantly from plant origins, their plausible mechanisms of actions, and *in silico* approaches to identify natural products leads with potential neuroprotective and therapeutic applications in CNS disorders.

2. Commonly used neuroprotective plants and their bioactive components

There are several plants mentioned in various traditional medicinal systems, e.g., the Ayurveda, to possess neuroprotective properties, and different compounds isolated and identified from those plants were also demonstrated to have significant neuroprotective actions. For example, *Bacopa monniera* "Brahmi," has been used in the Ayurvedic system of medicine for centuries. Traditionally, it is used as a brain tonic to enhance memory development, learning and to provide relief to patients with anxiety or epileptic disorders. The constituents responsible for its cognitive effects are bacosides A (**71**) and B (**72**).[17] The bacosides aid in repair of damaged neurons by enhancing kinase activity, neuronal synthesis, and restoration of synaptic activity, and ultimately nerve impulse transmission.[18] *Centella asiatica* (Apiaceae) is a psychoactive medicinal plant being used from centuries in the Ayurvedic system of medicine as a medhya rasayna.[19,20]

It has been shown to decrease the oxidative stress parameters. *Curcuma longa* (Zingiberaceae), commonly known as turmeric, is a common dietary supplement in the Asian cuisine. Neuroprotective effect of curcumin (**54**), the major bioactive compound of turmeric, has been reported by several researchers; treatment with curcumin (**54**) can protect neurons against ischemic cell death in experimental animal studies.[21] While an *in vitro* cell culture experiment showed that curcumin (**54**) could prevent or slow down age-related neurodegenerative diseases, it could also increase the expression of brain derived neurotrophic factor (BDNF). *Ginkgo biloba* (Ginkgoaceae), known as maiden hair tree or kew tree, is indigenous to East Asia, and has been used for the improvement of memory loss connected with irregularities in the blood circulation.[22] The leaf extract shows memory enhancing action by increasing oxygen supply, and removing free-radicals. *Panax ginseng* (Araliaceae) growing in north-eastern Asia, is used as anti-aging herb and as a tonic and revitalizing agent.[23] Ginseng root, characterized by the presence of ginsenosides (**75**), is considered an adaptogenic herb, able to increase the body's resistance to stress, trauma, anxiety and fatigue by modulating the immune function, and may provide protection against neurodegeneration by multiple mechanisms. *Hypericum perforatum* (Hypericaceae), commonly known as "St. John's Wort", possesses antidepressant activity and the mechanism of action was stated to be mediated through inhibition of mono amine oxidase (MAO).[24] The dried powder of *Picrorhiza scrophulariiflora* (Scrophulariaceae) has long been used to treat neurodegenerative diseases in Southeast Asia[25]; picrosides I (**39**) and II (**40**) isolated from this plant were shown to possess neuritogenic activity. The mixture of picrosides showed a substantial neurite outgrowth in PC12D cells by intracellular Mitogen-Activated Protein Kinase (MAPK)-dependent signaling pathway.[26] Table 1 lists 37 of such plant species and their major bioactive compounds (Fig. 1).[27–56]

3. Mechanisms of neuroprotective action of natural products

Phytochemicals play an important role in upregulation of gene response resulting from the cellular stress-response pathways. Experimental evidence suggests that neuroprotective effects of natural products are mainly govern by inhibiting the apoptotic pathways.[57] Nearly 50 neurotransmitters of varied chemical natures have been recognized in the brain. Acetylcholine is one of the most important targets because of its deficiency or inefficiency has been thought to be responsible for dementia and other neurodegenerative diseases,

Table 1 Examples of some neuroprotective medicinal plants and their major neuroprotective compounds.

Plant species	Family	Major neuroprotective compounds
Aconitum sp.	Ranunculaceae	Aconitine (**1**), heteratisine (**13**), lappaconitine (**18**) and mesaconitine (**19**)[27]
Acorus calamus	Acoraceae	Asarone (**86**)[28]
Angelica sinensis	Apiaceae	Ferulic acid (**45**)[29]
Areca catechu	Arecaceae	Arecaidine (**2**), arecoline (**3**) and guvacine (**4**)[30]
Bacopa monniera	Scrophularaceae	Brahmine (**6**), bacosides A (**71**) and B (**72**) herpestine, hersaponin, D-mannitol (**87**) and monnierin[17]
Biota orientalis	Cupressaceae	Pinusolide (**82**)[31]
Cassia siamea	Caesalpinaceae	Barakol (**52**)[32]
Centella asiatica	Apiaceae	Asiatic acid (**68**), asiaticoside (**69**) and madecassoside (**70**)[33]
Corydalis ternata	Papaveraceae	Protropine (**22**)[14]
Crocus sativus	Liliaceae	Safranal (**88**)[34]
Curcuma longa	Zingiberaceae	Curcumin (**54**)[35]
Emblica officinalis	Phyllanthaceae	Phyllemblin (**59**)[36]
Evolvulus alsinoides	Convolvulaceae	Betaine (**5**), evolvine and sankhapushpine[37]
Fraxinus griffithii	Oleaceae	Ligustrosid (**58**)[38]
Galanthus nivalis	Amaryllidaceae	Galanthamine (**12**)[39]
Gastrodia elata	Orchidaceae	Gastrodin (**53**) and vanillin (**66**)[40]
Ginkgo biloba	Ginkgoaceae	Bilobalide (**73**), ginkgolide B (**74**), isorhamnetin (**27**), kaempferol (**26**) and quercetin (**28**)[22]
Glycyrrhiza glabra	Fabaceae	Glycyrrhizin (**77**)[41]
Huperzia serrata	Huperziaceae	Huperzines A (**14**) and B (**15**)[42]

Continued

Table 1 Examples of some neuroprotective medicinal plants and their major neuroprotective compounds.—Cont'd

Plant species	Family	Major neuroprotective compounds
Hypericum perforatum	Hypericaceae	Hyperforin (**78**)[43]
Ilex paraguariensis	Aquifoliaceae	Caffeine (**7**), chlorogenic acid (**46**), kaempferol (**26**), quercetin (**28**) theobromine (**8**) and theophylline (**9**)[44]
Magnolia officinalis	Magnoliaceae	Honokiol (**41**), magnolol (**43**) and 4-*O*-methyl honokiol (**42**)[45]
Paeonia lactiflora	Paeoniaceae	Paeoniflorin (**81**)[46]
Panax ginseng	Araliaceae	Ginsenoside (**75**)[23]
Passiflora incarnate	Passifloraceae	Chrysin (**25**)[47]
Perilla frutescens	Lamiaceae	Apigenin (**24**), caffeic acid (**44**) and rosmarinic acid (**50**)[48]
Picrorhiza scrophulariiflora	Scrophulariaceae	Picrosides I (**39**) and II (**40**)[26]
Piper methysticum	Piperaceae	Kavain (**47**), methysticin (**51**) and yangonin (**48**)[49]
Piper nigrum	Piperaceae	Piperine (**21**)[50]
Salvia miltiorrhiza	Lamiaceae	Lithospermate B (**49**)[41]
Scutellaria baicalensis	Labiatae	Baicalein (**29**), baicalin (**30**) and wogonin (**37**)[51]
Tilia tomentsa	Tiliaceae	Chrysin (**25**) and kaempferol (**26**)[52]
Uncaria rhynchophylla	Rubiaceae	Corynoxeine (**10**), isocorynoxeine (**16**), isorhynchophylline (**17**) and rhynchophylline (**23**)[45]
Withania somnifera	Solanaceae	Sitoindoside IX (**83**), sitoindoside X (**84**), withanolide (**85**) and withanol[53]
Ziziphus jujube	Rhamnaceae	Jujuboside A (**79**), jujuboside B (**80**) and spinosin (**38**)[54]
Zingiber officinale	Zingiberaceae	Gingerol (**56**), shogaol (**64**) and zingerone (**67**)[55]
Ziziphus spinosa	Rhamnaceae	Jujuboside A (**79**) and jujuboside B (**80**)[56]

Alkaloids	
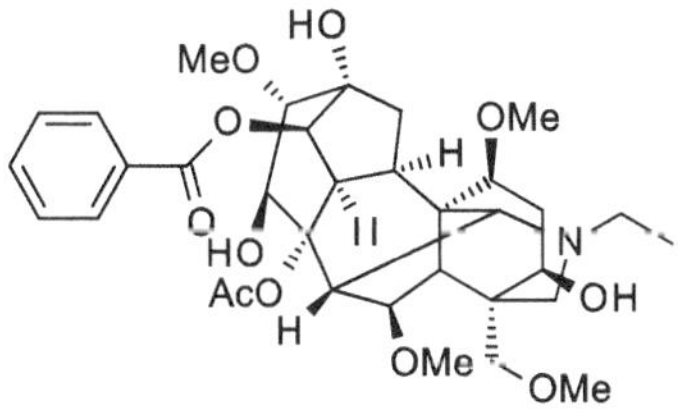 Aconotine (**1**)	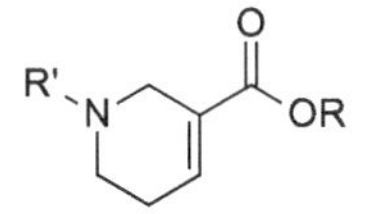 Arecaidine (**2**) R = H; R' = Me Arecoline (**3**) R = R' = Me Guvacine (**4**) R = R' = H
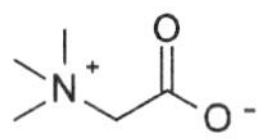 Betaine (**5**)	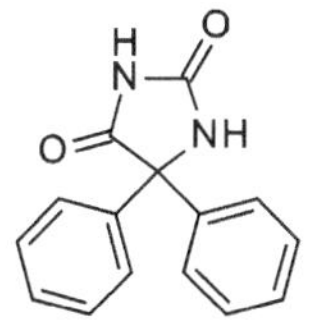 Brahmine (**6**)
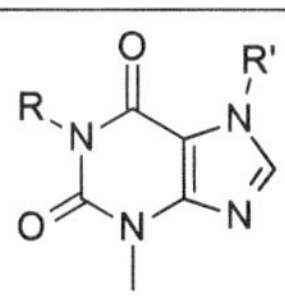 Caffeine (**7**) R = R' = Me Theobromine (**8**) R = H; R' = Me Theophylline (**9**) R = Me; R' = H	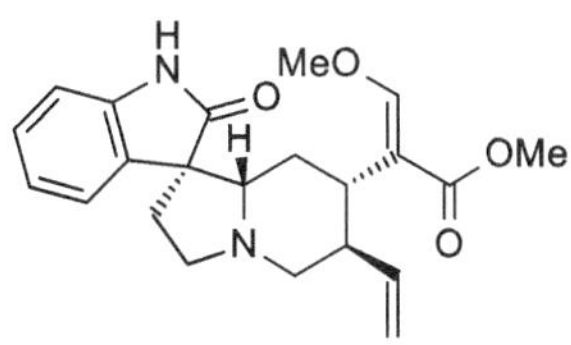 Corynoxeine (**10**)
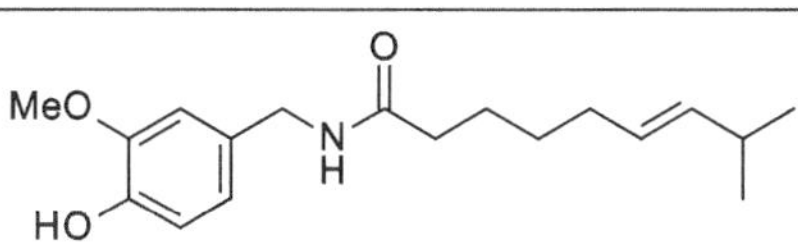 Capsaicin (**11**)	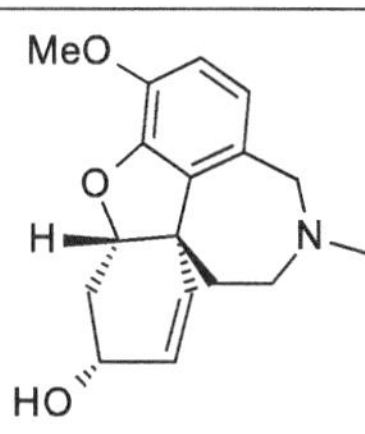 Galanthamine (**12**)
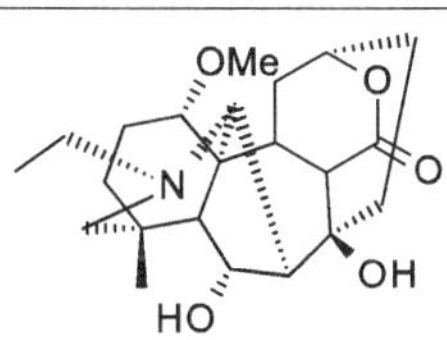 Heteratisine (**13**)	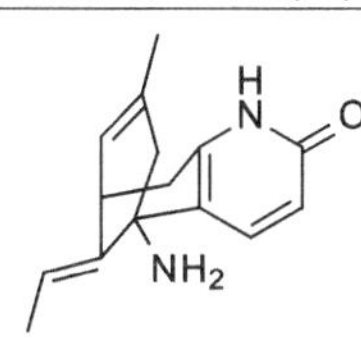 Huperzine A (**14**)
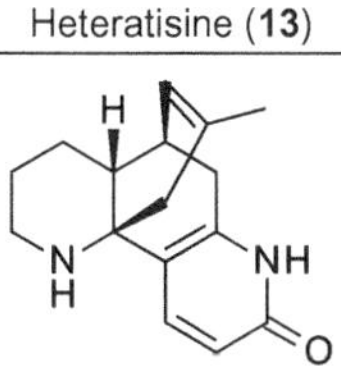 Huperzine B (**15**)	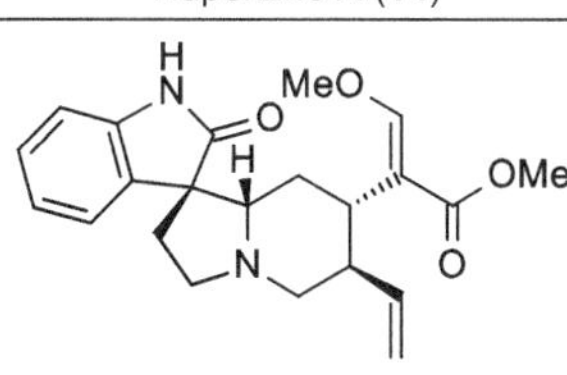 Isocorynoxeine (**16**)

Fig. 1 Structures of some major bioactive compounds from neuroprotective plants.

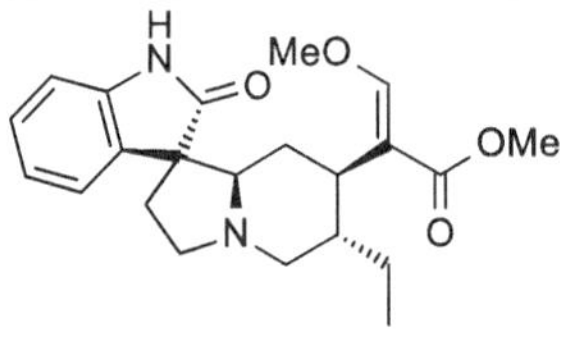

Isorhynchophylline (**17**)

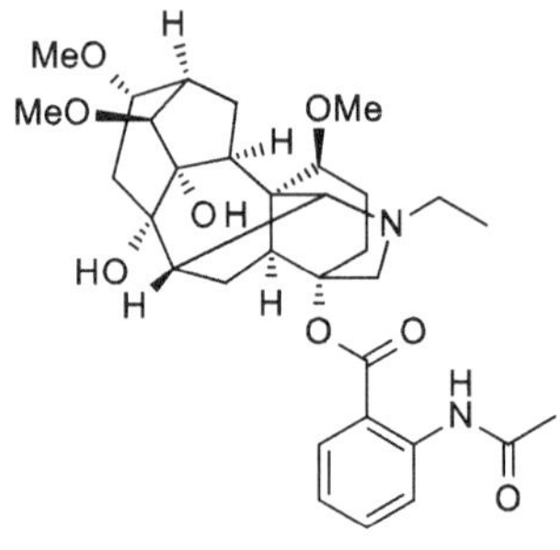

Lappaconitine (**18**)

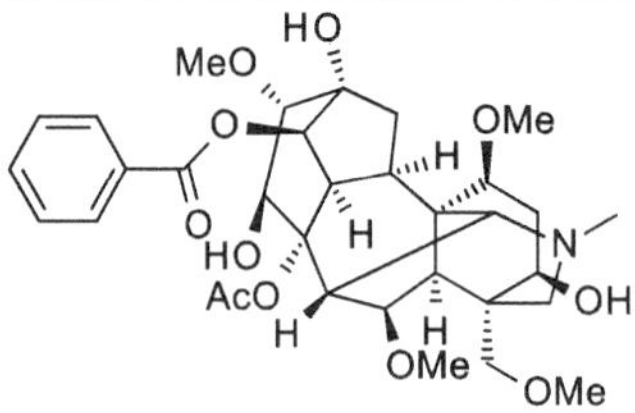

Mesaconitine (**19**)

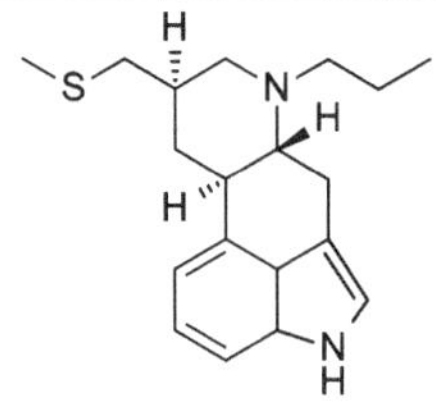

Pergolide (**20**)

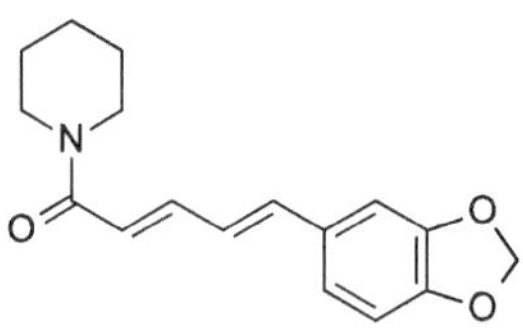

Piperine (**21**)

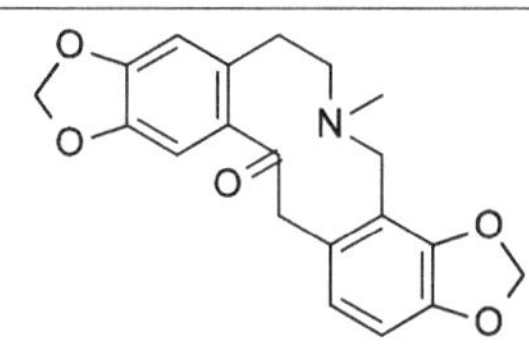

Protropine (**22**)

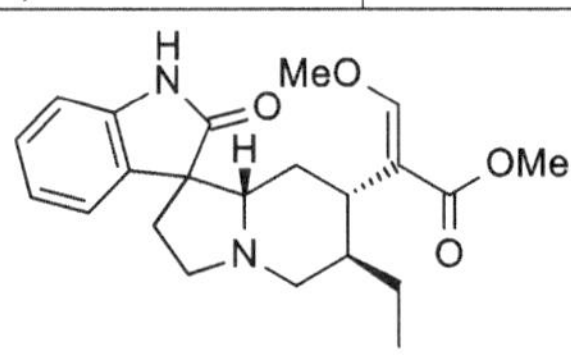

Rhynchophylline (**23**)

Flavonoids

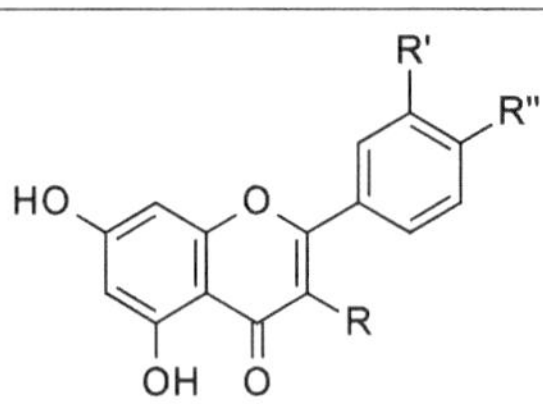

Apigenin (**24**) R = R' = H; R'' = OH
Chrysin (**25**) R = R' = R'' = H
Kaempferol (**26**) R = R'' = OH; R' = H
Isorhamnetin (**27**) R = R'' = OH; R' = OMe
Quercetin (**28**) R = R' = R'' = OH

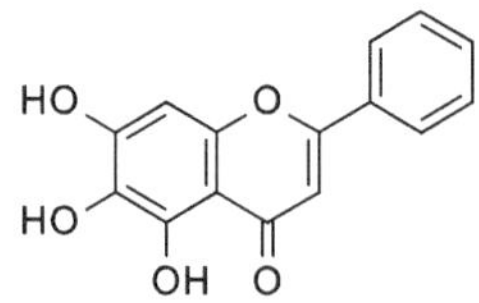

Baicalein (**29**)

Fig. 1—Cont'd

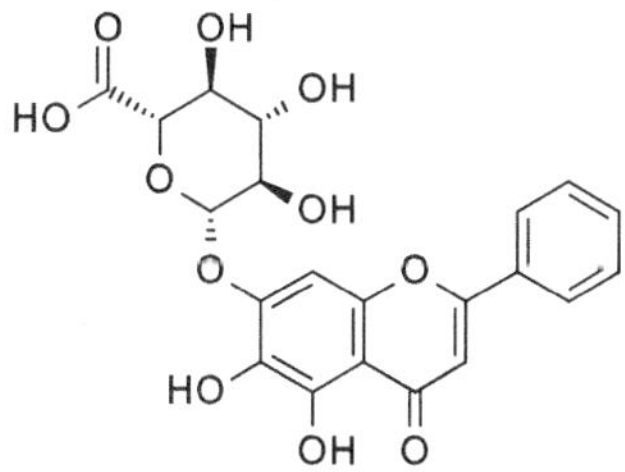

Baicalin (**30**)

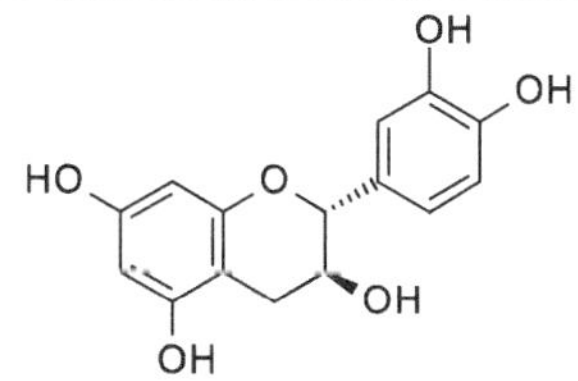

Catechin (**31**)

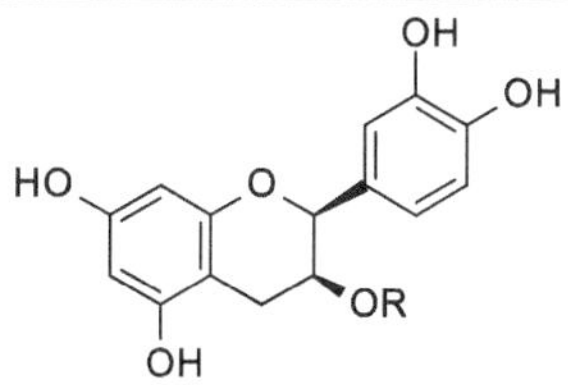

Epicatechin (**32**) R = H
Epicatechin gallate (**33**) R = Galloyl

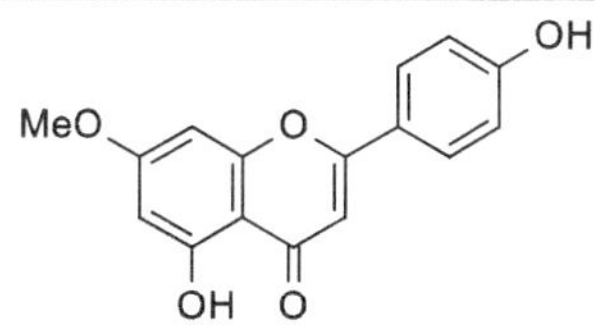

Genkwanin (**34**)

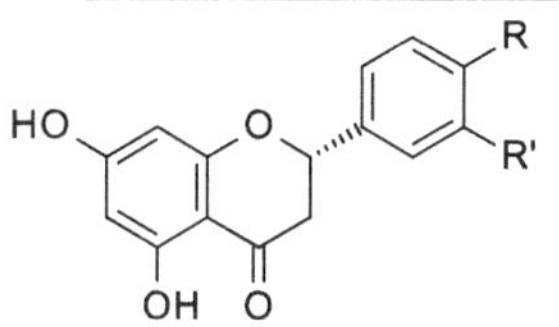

Hesperetin (**35**) R = OMe; R' = OH
Naringenin (**36**) R = OH; R' = H

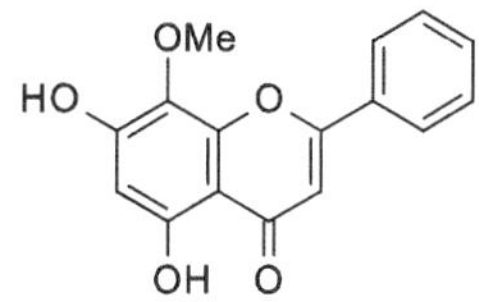

Wogonin (**37**)

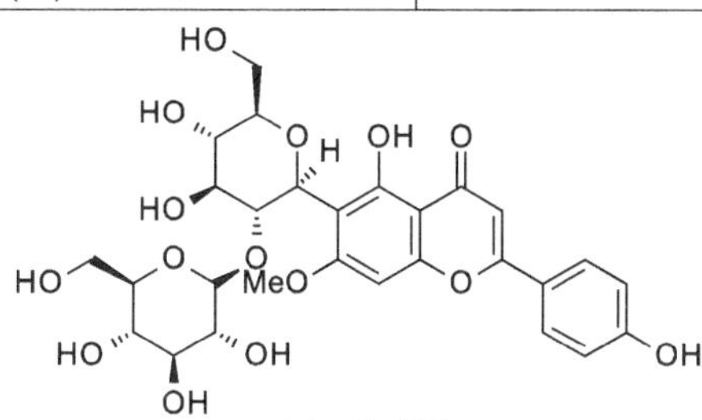

Spinosin (**38**)

Iridoids

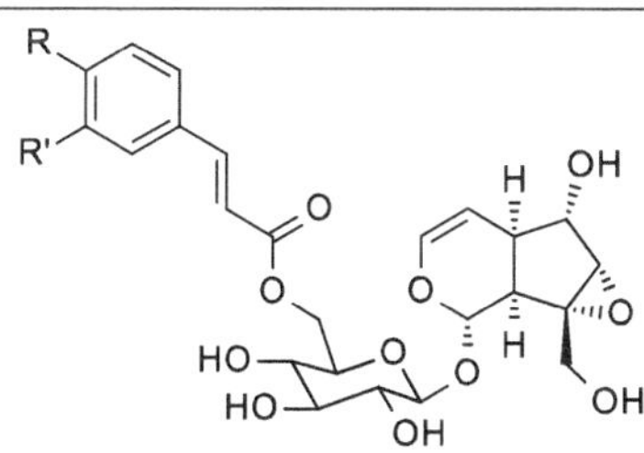

Picroside (**39**) I R = R' = H
Picroside II (**40**) R = OH; R' = OMe

Fig. 1—Cont'd

Neolignans

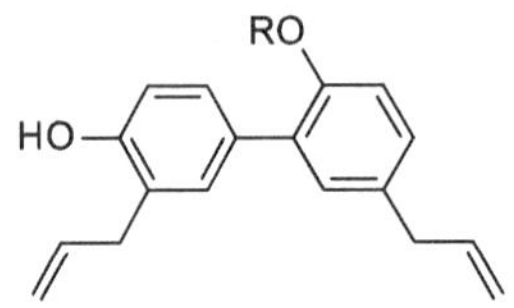

Honokiol (**41**) R = H
4-*O*-Methyl-honokiol (**42**) R = Me

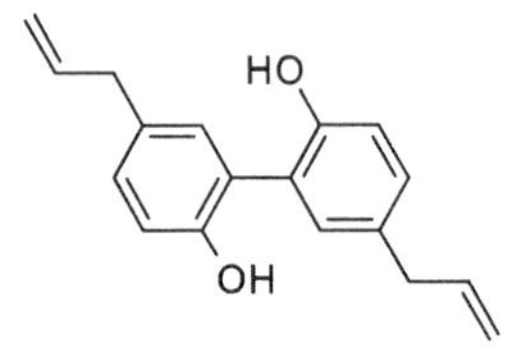

Magnolol (**43**)

Phenylpropanoids

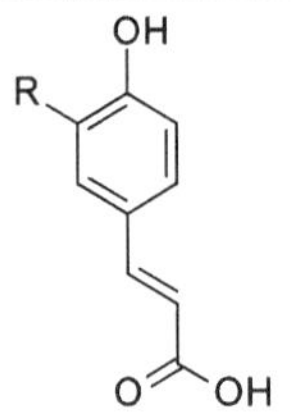

Caffeic acid (**44**) R = OH
Ferulic acid (**45**) R = OMe

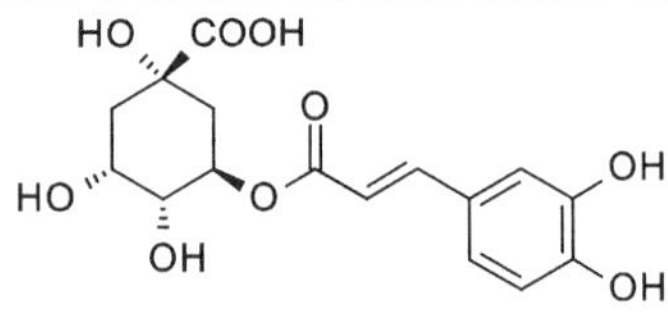

Chlorogenic acid (**46**)

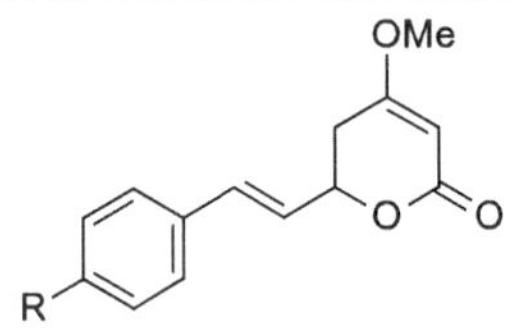

Kavain (**47**) R = H
Yangonin (**48**) R = OMe

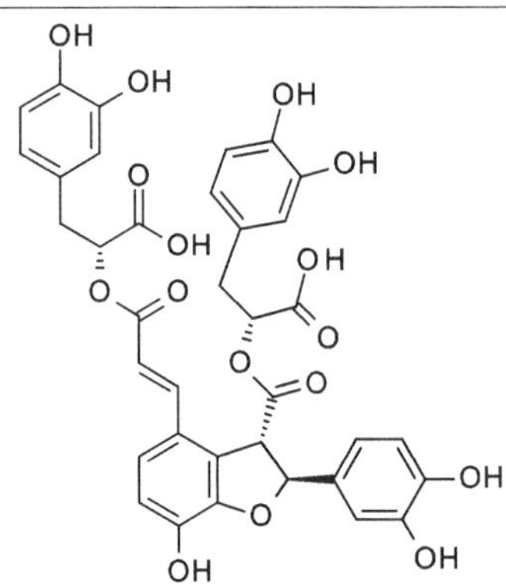

Lithospermate B (**49**)

Fig. 1—Cont'd

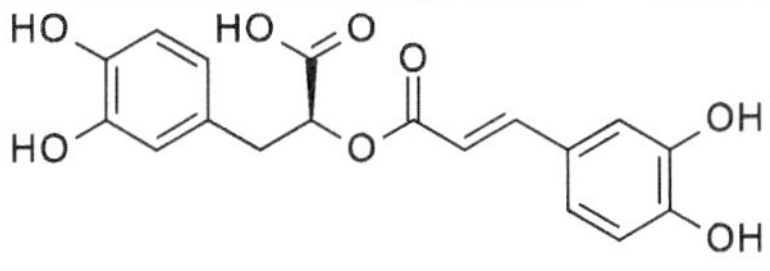

Rosmarinic acid (**50**)

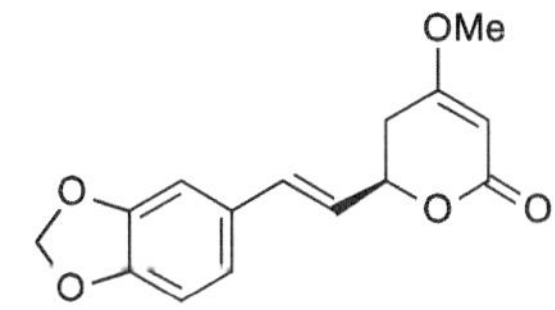

Methysticin (**51**)

Other phenolics

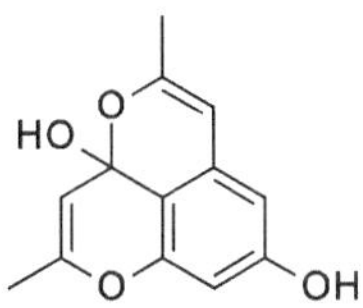

Barakol (**52**)

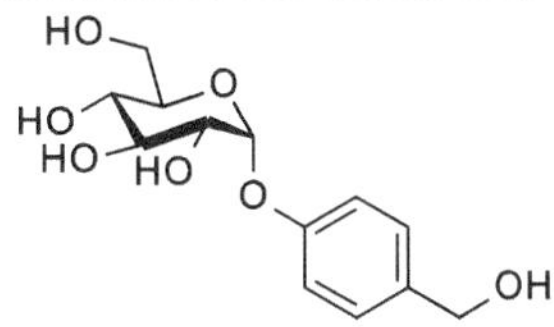

Gastrodin (**53**)

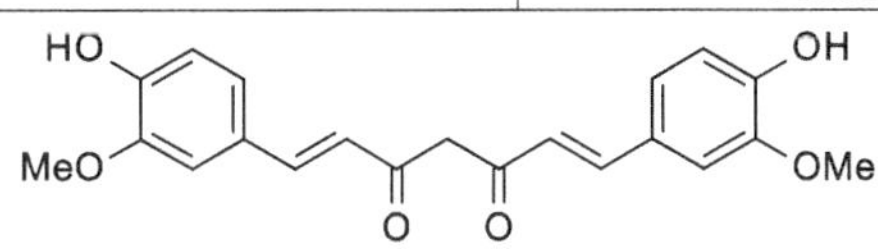

Curcumin (**54**)

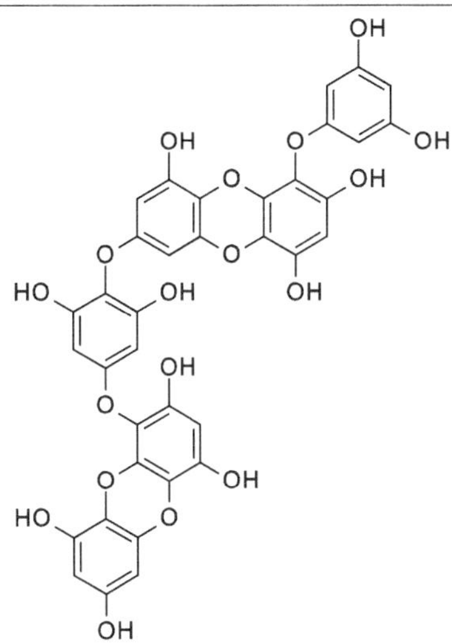

Dieckol (**55**)

Fig. 1—Cont'd

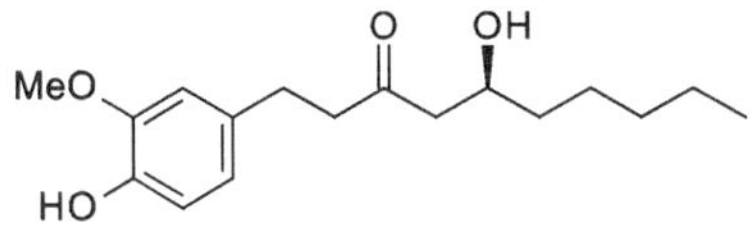
Gingerol (**56**)

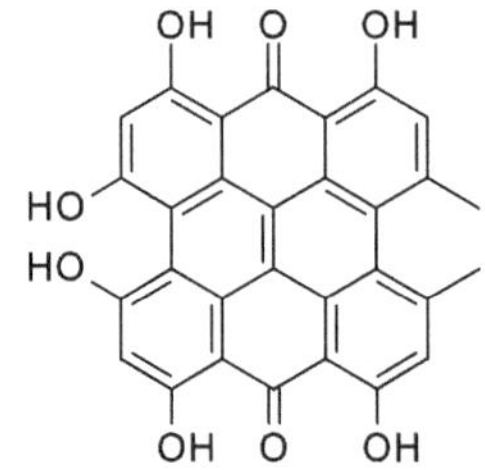
Hypericin (**57**)

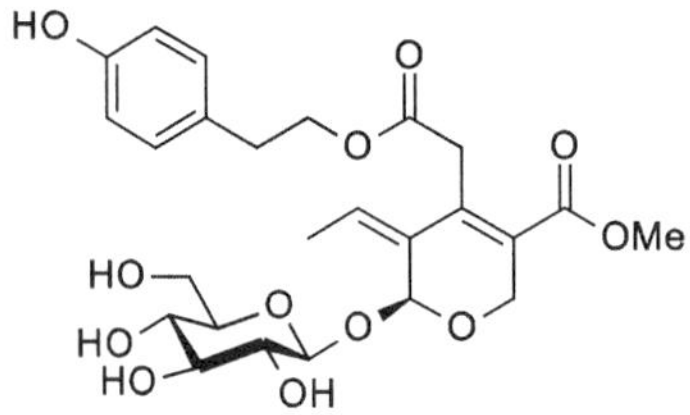
Ligustrosid (**58**)

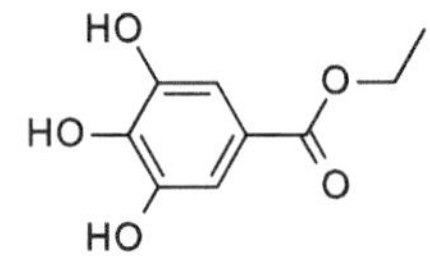
Phyllemblin (**59**)

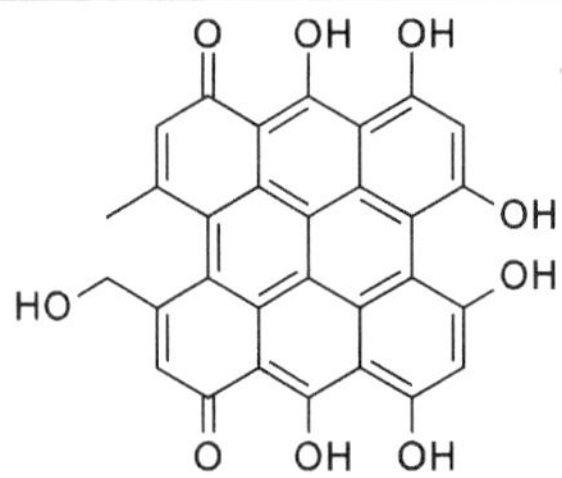
Pseudohypericin (**60**)

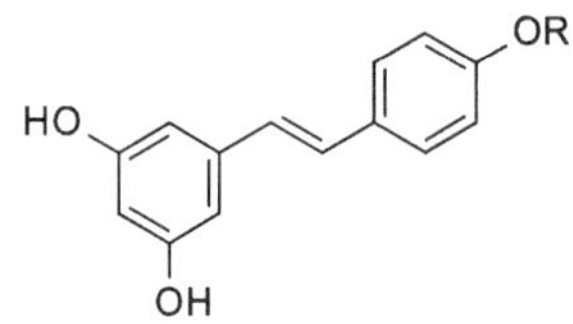
Resveratrol (**61**) R = H
Resveratroloside (**62**) R = Glucosyl

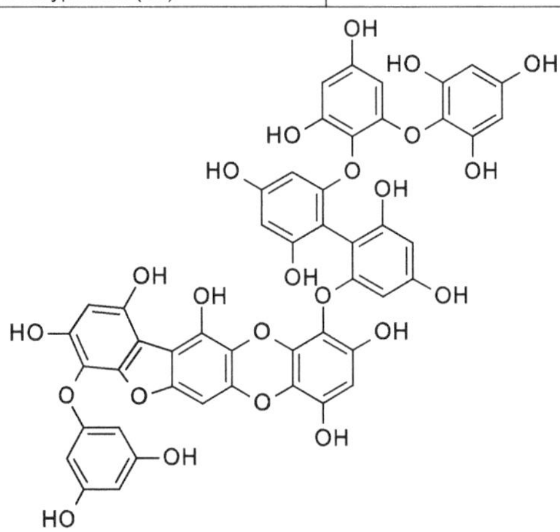
Phlorofucofluoroeckol (**63**)

Fig. 1—Cont'd

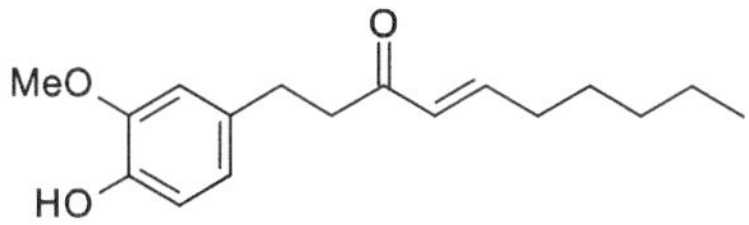 Shogaol (**64**)	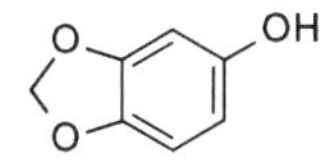 Sesamol (**65**)
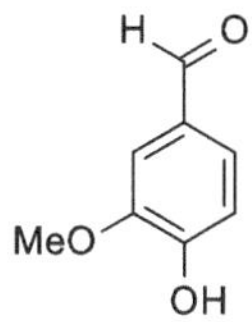 Vanillin (**66**)	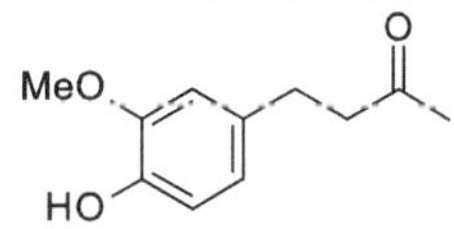 Zingerone (**67**)
Terpenoids	
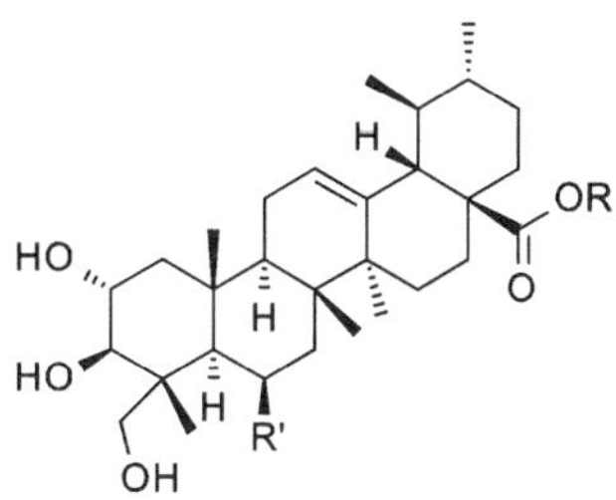 Asiatic acid (**68**) R = R' = H Asiaticoside (**69**) R = 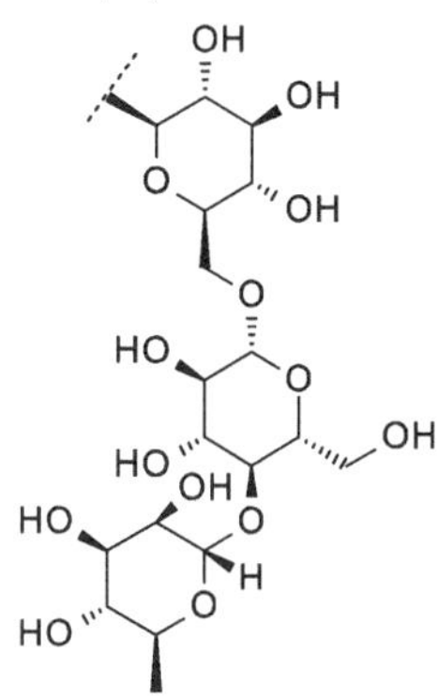 ; R' = H Madecassoside (**70**) R = 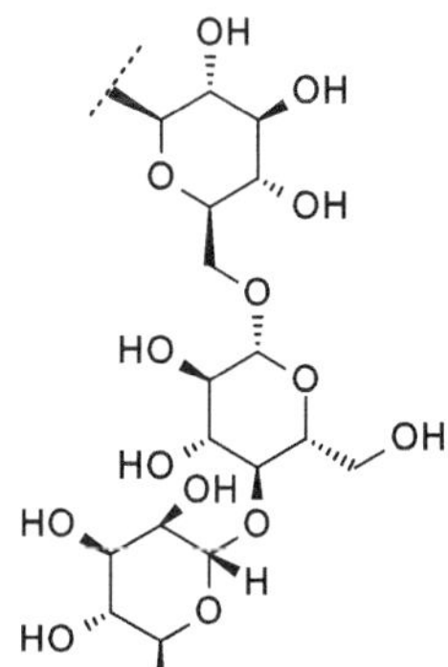 ; R' = OH	

Fig. 1—Cont'd

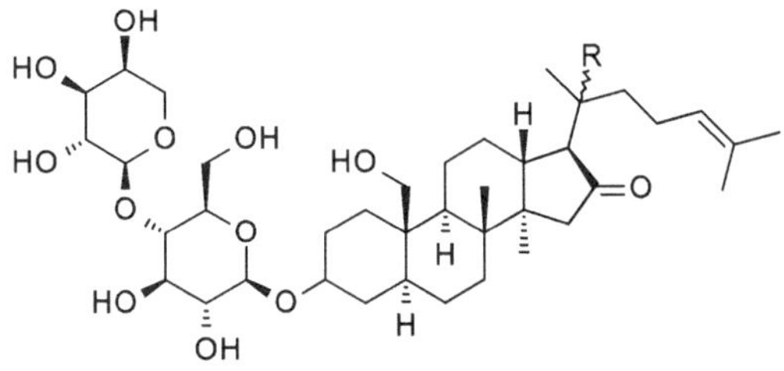

Bacoside A (**71**) R = α-OH
Bacoside B (**72**) R = β-OH

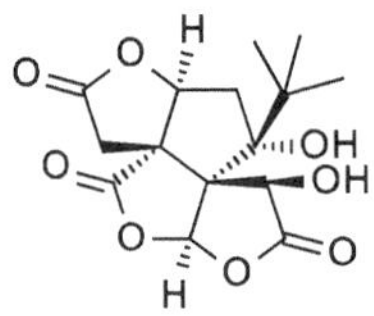

Bilobalide (**73**)

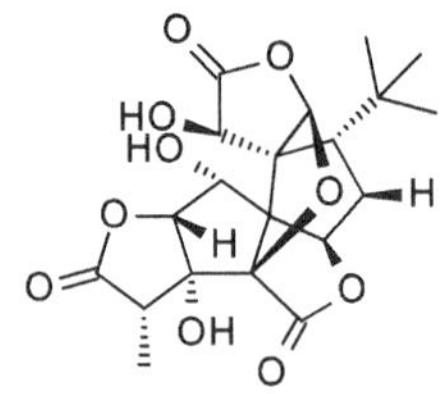

Ginkgolide B (**74**)

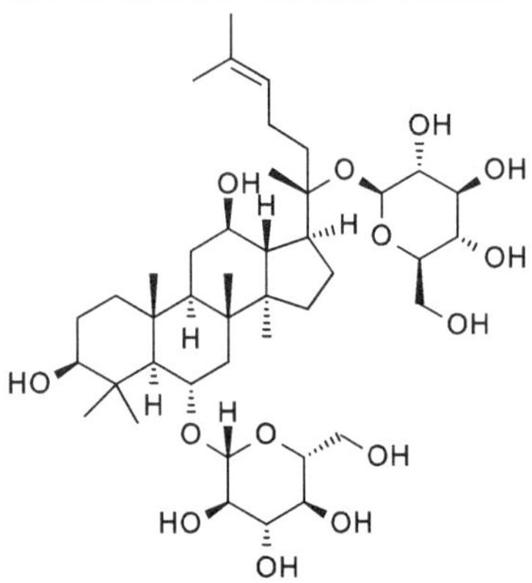

Ginsenoside (**75**)

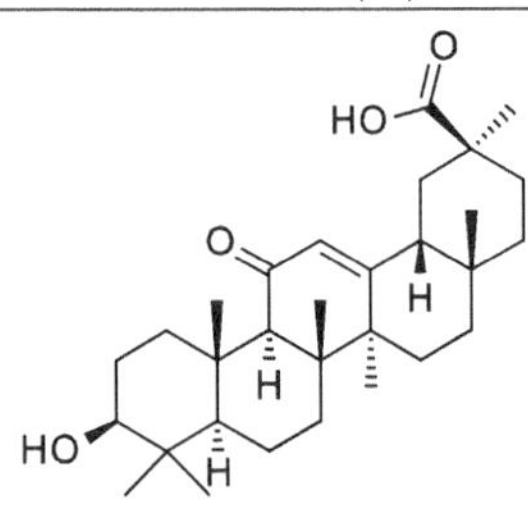

Glycyrrhetinic acid (**76**)

Fig. 1—Cont'd

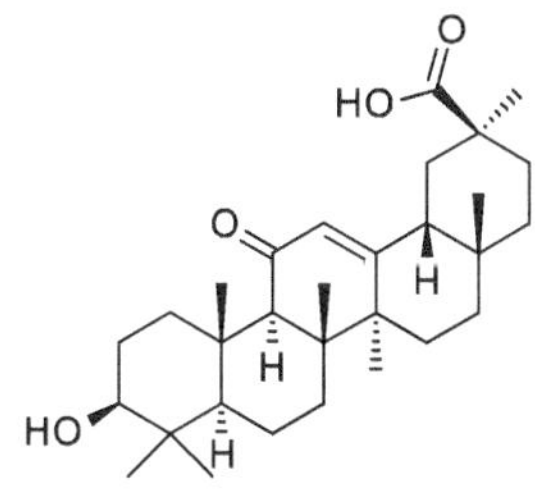

Glycyrrhetinic acid (**76**)

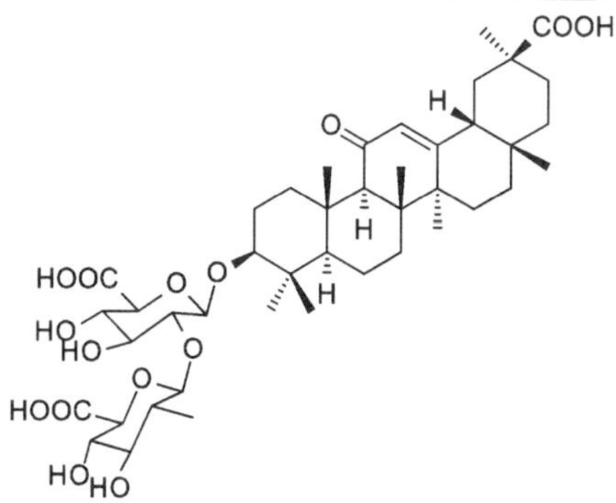

Glycyrrhizin (**77**)

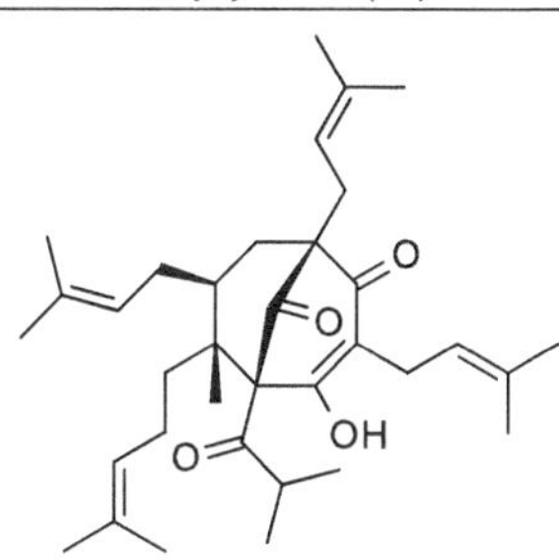

Hyperforin (**78**)

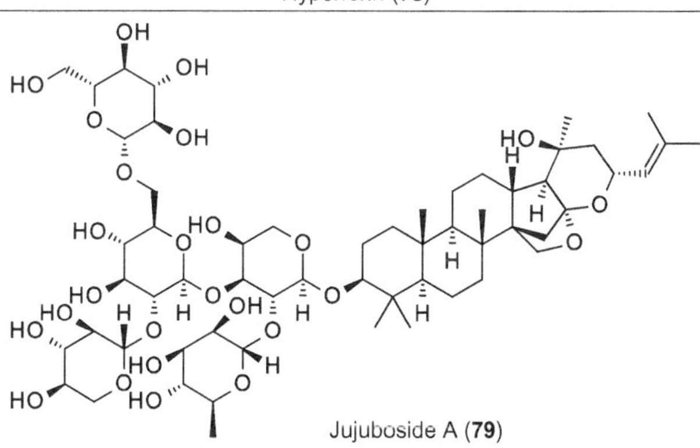

Jujuboside A (**79**)

Fig. 1—Cont'd

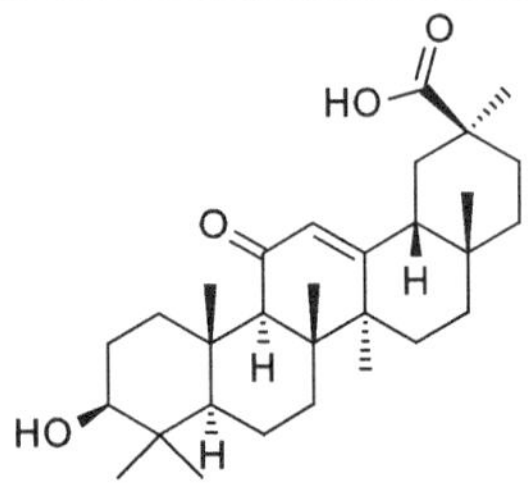

Glycyrrhetinic acid (**76**)

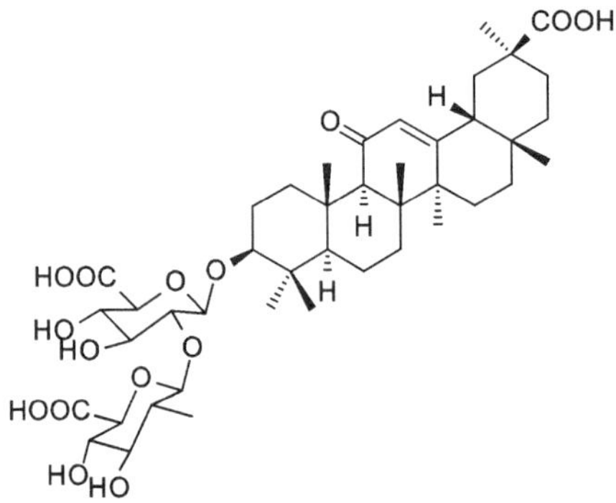

Glycyrrhizin (**77**)

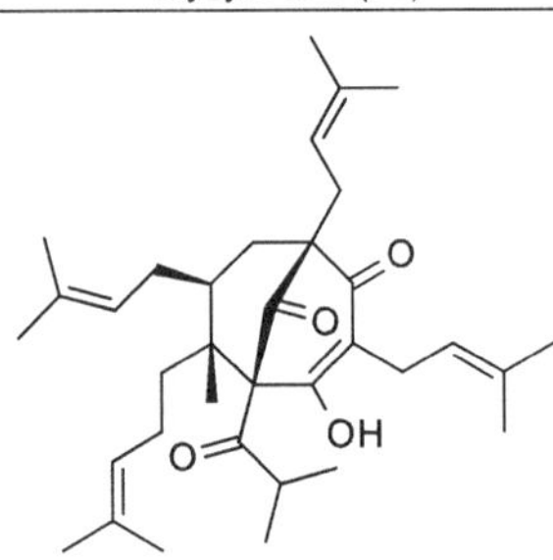

Hyperforin (**78**)

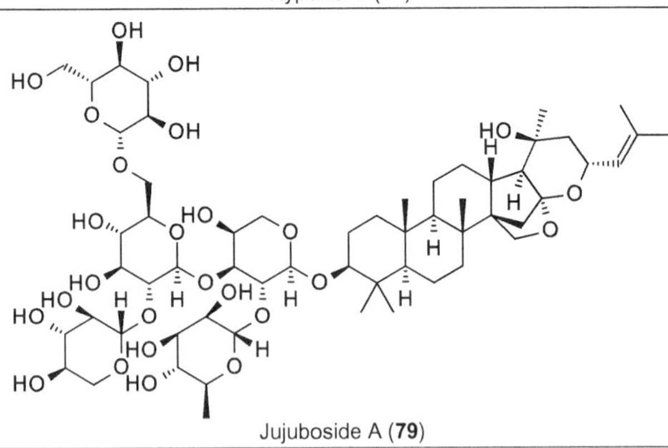

Jujuboside A (**79**)

Fig. 1—Cont'd

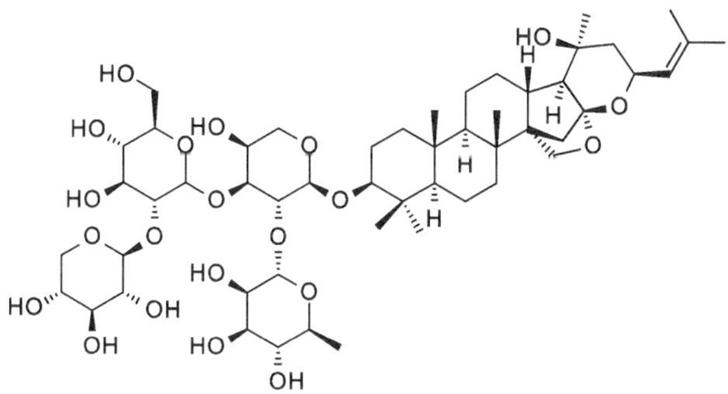

Jujuboside B (**80**)

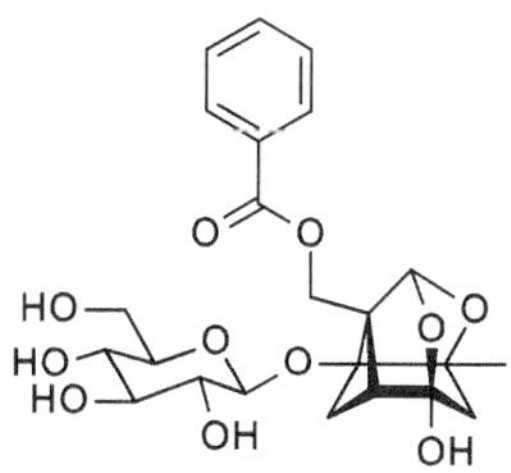

Paeoniflorin (**81**)

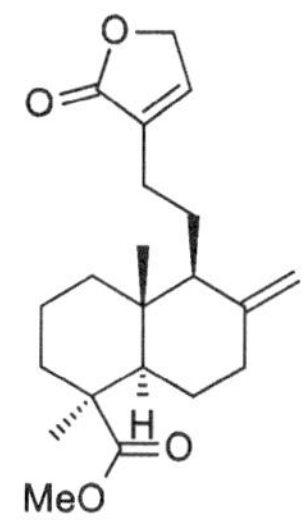

Pinusolide (**82**)

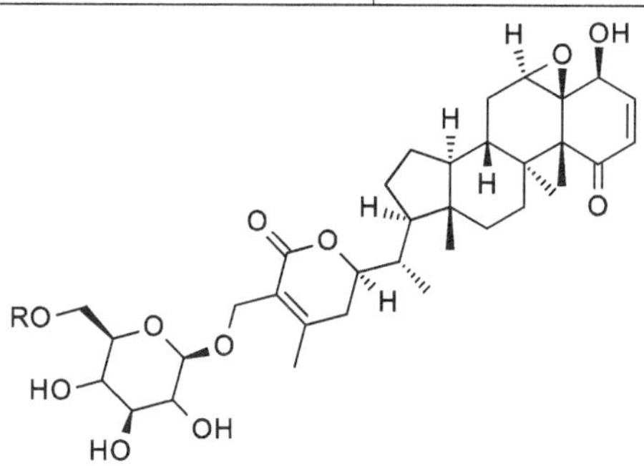

Sitoindoside IX (**83**) R = H
Sitiindoside X (**84**) R = Palmitoyl

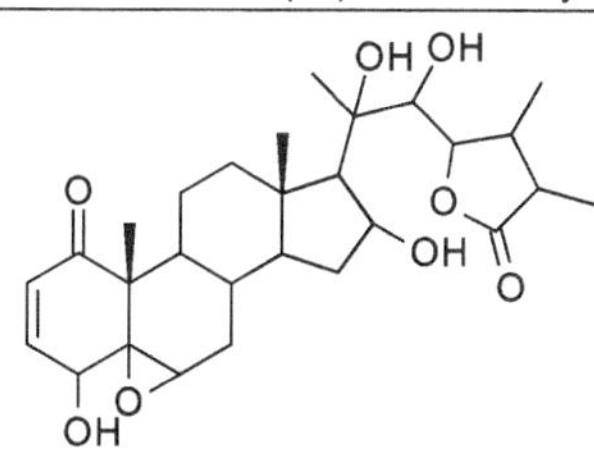

Withanolide (**85**)

Fig. 1—Cont'd

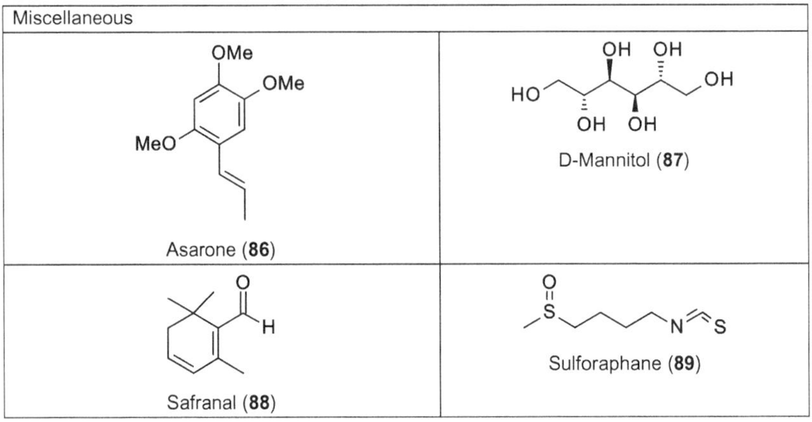

Fig. 1—Cont'd

especially Alzheimer's disease.[58] Phytochemicals can induce the expression of antioxidant enzymes and Bcl-2 by stimulating the transcription factor NF-κB, and can stimulate multiple signaling processes by acting as ligands, which bind to several receptors like growth factor receptors (GFR), G-protein-coupled receptors (GPCR) and insulin receptors (IR). Activated receptors in turn activate various kinase cascades that involve mitogen-activated protein kinases (MAPK), protein kinase C (PKC) and phosphatidylinositol-3-kinase (PI3K).

Flavonoids play a vital role in activation of ERE-CREB pathway and the P13 kinase-mTOR cascade, which lead to changes in synaptic plasticity.[24] Plant secondary metabolites like terpenes, isothiocyanates, insoles, curcuminoids, diallyl sulfides and cytokines have been shown to stimulate pathways upstream of nuclear factor erythroid 2-related factor 2 (NRF2). Activation of transient receptor potential (TRP) ion channels in the plasma membrane of neurons by ellison results in Ca^{2+} influx, and which in turn activates neuroprotective kinase cascades via camp-response-element-binding protein (CREB).[59] CREB induces the expression of brain-derived neurotrophic factor (BDNF). A major neurotrophic factor is BDNF, which can initiate PI-3K/Akt and MAPK/ERK pathways, and thus activate the downstream molecules, which promote neurogenesis and cell survival.[60]

Flavanones, like hesperetin (**35**) and its metabolite, 5-nitrohesperetin, have been recognized to inhibit neuronal apoptosis that occurs due to oxidant induced via a machinery involving the activation/phosphorylation of important signaling proteins in the prosurvival pathways. Capsaicin (**11**) (Fig. 1) can activate specific Ca^{2+} channels. Flavonoids, terpenes and other phytochemicals are known to affect the function of ionotropic receptors for GABA, the major inhibitory neurotransmitter in the brain.[61] The GABA modulators are available in vegetables, fruits, various beverages, and herbal extracts such as Ginseng and *Ginkgo biloba*. The pharmacological effects of many plant derived compounds and herbal extracts are known to be involved in the mechanisms for neuroregeneration.[24] A summary of possible mechanism of neuroprotective actions of natural products in Alzheimer's pathway is presented in Fig. 2.

4. Virtual screening of neuroprotective plant-derived biomolecules

Nowadays, natural products drug discovery processes involve concerted efforts combining *in vitro*, *ex vivo*, *in vivo* and *in silico* methodologies and approaches.[62] Fig. 3 captures the steps involved in an *in silico* lead discovery process. Plant metabolite is an important source of neuroprotective agents.

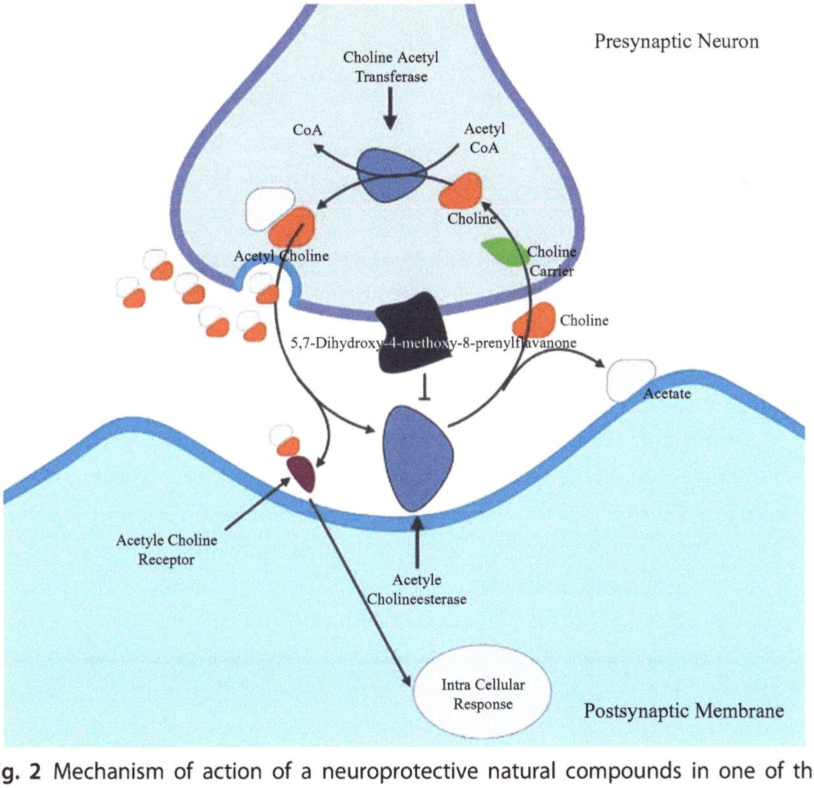

Fig. 2 Mechanism of action of a neuroprotective natural compounds in one of the Alzheimer's pathway.

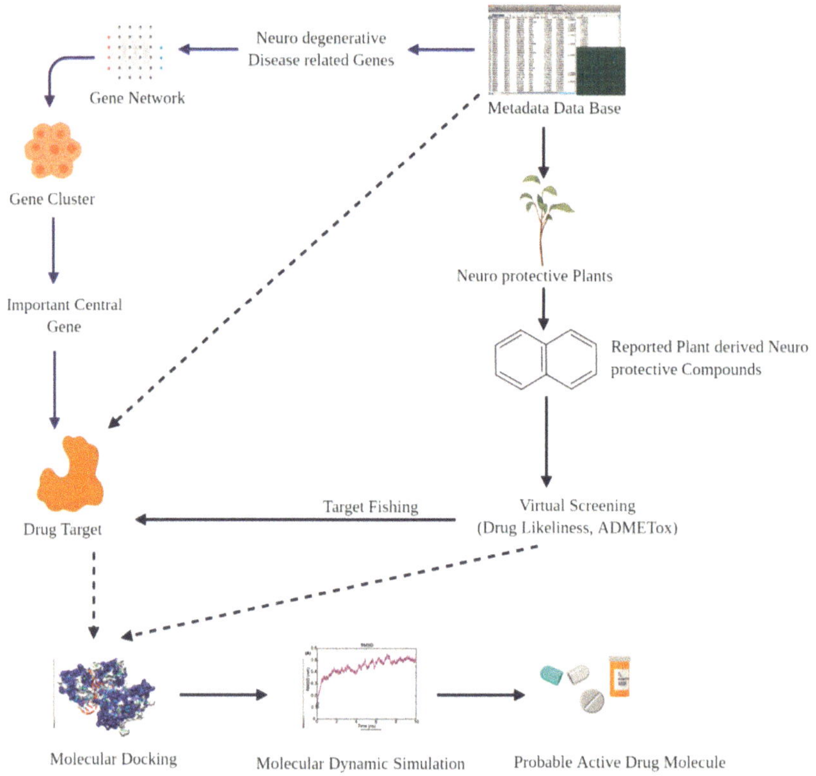

Fig. 3 Natural products drug discovery—an in silico approach.

The selection and development of novel neuroprotective leads from the pool of phytochemicals is exceptionally challenging. Virtual screening is one of the best suited strategies to meet this challenge and to minimize the time and cost involved in the discovery process. It has shown an exceptional potential in drug discovery in digging out lead phytochemical.[63]

With the aid of computer knowledge, the screening of phytochemicals is undertaken to find out the leads based on various suited criteria. *In silico* approach is the key method under the virtual screening strategy for drug development. Researchers use computational methods to assess virtual libraries or databases against virtual targets or receptor aiming at minimizing the time required for drug development process. In general, virtual screening is planned for searching large number of chemical structures or virtual libraries in hypothetical databases by using computational analysis and for picking up a restricted number of drug molecules possibly to be effective against a selected biological receptor.[62,64] Therefore, virtual screening may be described as a logical extension of 3D pharmacophore-based database searching (PBDS) capable of automatically assessing very large databases of compounds.

4.1 *In silico* approach for designing potent neuroprotective biomolecules

Computer aided drug designing (CADD) is the widely used approach in *in silico* drug designing. To develop and design a neuroprotective drug or biomolecule *in silico*, two main approaches have to be followed; they are (a) Ligand-based drug design and (b) Target-based drug design.

4.1.1 Ligand-based drug design

In ligand-based drug design approaches, a known structure of a ligand or small molecule is taken from the meta data of phytochemicals available in the database like, Dr. Duke's Phytochemical and Ethnobotanical Databases. Based on the structure of the ligand, target fishing is the first and foremost step to go forward. Before target fishing it is to be kept in mind that the structure of ligand should have drug likeliness characteristic, which is an important criterion for a compound to act as drug. There are subsequent steps which are described below.

Target fishing: In target fishing, target database is screened with ligand and based on probable interaction, suitable target(s) for the particular (neurodegenerative) diseases is selected. In this process use of PDB to recognize potent hits from the databases carried out when 3D structures of the targets are unknown. Various online servers are available for target fishing such as SuperPred database, PharmMapper (Fig. 4), Swiss Target Prediction etc.[65] With citing an example,[58] worked on ligand-based drug designing for Alzheimer's disease and they come up with an active neuroprotective compound 5,7-dihydroxy-4-methoxy-8-prenylflavanone against acetylcholinesterase (Fig. 5).

Molecular docking: It is performed between ligand and the target along with some known inhibitors of the target available on the database like Binding DB, Drug bank etc. In this approach, the ligand is allowed to interact with the specific target and it is the basic step of the ligand-based drug design method.[66] The affinity of the ligands toward the target is checked by the molecular docking and the affinity of the ligand is compared with the known inhibitors of the target. There are several software's available for performing molecular docking. A few software's are Flex X, Auto Dock, Gold, FlexAID, HYBRID, idock etc. Fig. 5 shows that

Result of 150630150457

Top 300 targets ranked by fit score in descending order

Ligand: NoName

	Rank	PDB ID	Target Name	Number of Feature	Fit Score	Normalized Fit Score	z'-score
+	1	1AQU	Estrogen sulfotransferase, testis isoform	8	5.014	0.6267	2.42093
+	2	1GP6	Leucoanthocyanidin dioxygenase	8	4.828	0.6036	6.32509
+	3	1I52	2-C-methyl-D-erythritol 4-phosphate cytidylyltransferase	16	4.65	0.2907	0.443475
+	4	1RY0	Aldo-keto reductase family 1 member C3	7	4.611	0.6588	1.62738
+	5	1DB1	Vitamin D3 receptor	13	4.557	0.3505	1.23347
+	6	1CBS	Cellular retinoic acid-binding protein 2	10	4.548	0.4548	0.501957
+	7	1HBJ	Acetylcholinesterase	9	4.538	0.5042	0.170555
+	8	1EPB	Epididymal-specific lipocalin-5	9	4.462	0.4957	1.0609

Fig. 4 PharmMapper—an online database for target fishing.

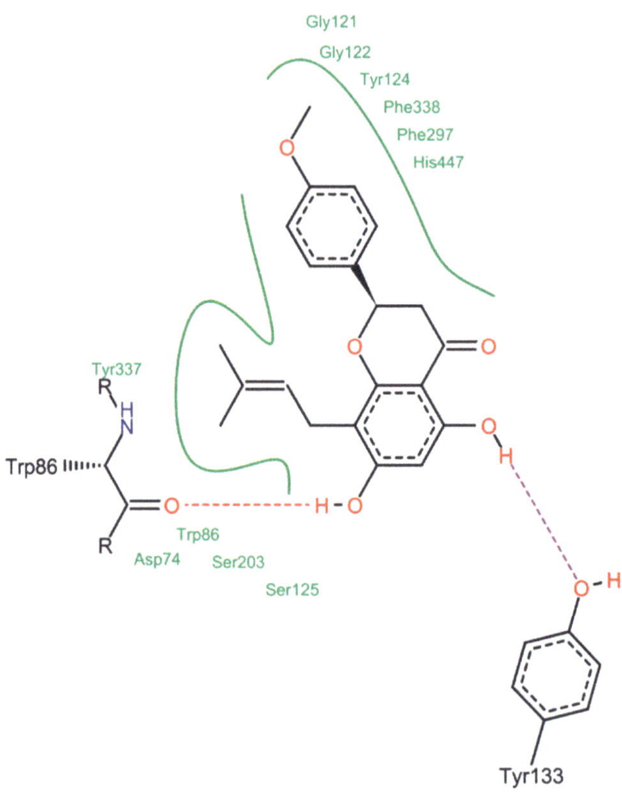

Fig. 5 Binding of 5,7-dihydroxy-4-methoxy-8-prenylflavanone with the target acetylcholinesterase.

5,7-dihydroxy-4-methoxy-8-prenylflavanone is bound with the target acetylcholinesterase with 3 strong hydrogen bond with the gln527, ala526 and Gly523 amino acids of the target enzyme.

Quantitative structure activity relationship (QSAR): This is an important parameter to understand the bioactivity of a ligand. Based on the best possible interaction between a ligand and a receptor, a 3D pharmacophore structure is constructed by dot plot analysis following the structure activity relationship (SAR) to get an idea about the bioactivity of the ligand. KNIME, easyQSAR, etc. are some software useful in the calculation of IC_{50} of the compound from the QSAR analysis. Example, from the QSAR equation the IC_{50} value of 5,7-dihydroxy-4-methoxy-8-prenylflavanone was predicted as 1659.59 nM (Fig. 6).[58]

Molecular dynamic simulation: If all the checked parameters, provide satisfactory result, then to know the physical movements of atoms and molecules in the biological system, molecular dynamics simulation is performed. It is a computer simulation method for analyzing the physico-chemical movements of atoms and molecules of ligands and target in virtual biological

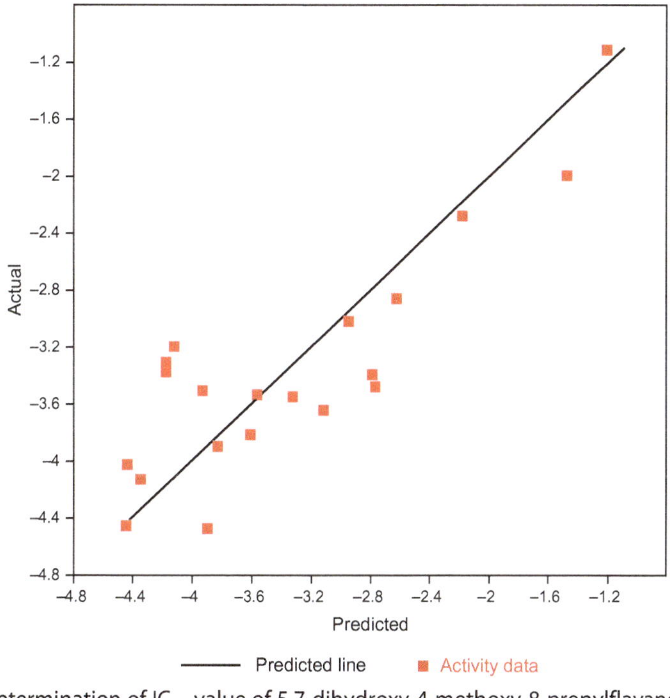

Fig. 6 Determination of IC_{50} value of 5,7-dihydroxy-4-methoxy-8-prenylflavanone from the QSAR equation.

system. The atoms and molecules are allowed to interact for a fixed period of time, giving a view of the dynamic "evolution" of the system.[67] Example, Growmacs, SimScale, Simul 8, Simulink etc. are some of software useful for Molecular Dynamic Simulation.

4.1.2 Target-based drug design

In target-based drug design, the drug targets of various diseases are known and in that case ligand search is taken into action to find out the best suited molecule for the particular target protein or receptors. Researchers perform gene network analysis to identify the target gene of interest involved in a particular disease for which drug designing is required.

Gene network analysis: In gene network analysis all the reported genes of a particular disease (neurodegenerative disease) are retrieved from the available database like, NCBI gene database, Disgenet database etc. The protein–protein interactions of the related genes are analyze by STRING online server to get a gene–gene interaction network.

Gene cluster analysis: Form gene–gene interaction network clustering analysis is performed with software available in the Cytoscape platform. Cytoscape is a visualizing and analyzing platform of gene network constructed in the STRING online server. For cluster analysis of the genes involved in the networks Mcode, ClusterONE, Animo etc. are the cytoscape plugin used by following different algorithms. The gene clusters identified from the network consist of a group of genes which are important in the network.

Gene enrichment analysis: From the clustered gene, the gene enrichment is achieved by which we can identify the central gene of the cluster which will be treated as the possible drug target for the particular disease. Various parameters are involved to identify the central gene like, average shortest path length, degree of freedom, betweenness centrality, closeness centrality, clustering coefficient etc.

Gene ontology study: For validation of importance of selected target genes in the selected disease, gene ontology is performed by several software as well as by various online server such as David online gene ontology server, in which molecular process, cellular process and disease pathway are studied. Links to a disease are retrieve from the databases and through cluster analysis from the network, target gene has been identified. Target may be a protein or enzyme which is having an important role in the disease causing pathway. An example of target based drug designing, a target monoamine oxidase-B (MAO-B) was identified an important drug target for Parkinson's disease and which was inhibited by a phytochemical garcinol (**00**) isolated from *Garcinia* sp.[68]

Generally, these two approaches are in use simultaneously because the earlier can filter out the potent ligand quickly and the latter can screen the ligand-target or ligand-receptor binding more precisely. In the above two approaches, target-based and ligand-based drug design, a few *in silico* steps are common; ligand in both the approaches should have passed some of the crucial filters such as Lipinski drug likeliness and ADMETox screening. In Lipinski drug likeliness, the Lipinski rule of 5 for the ligand is checked whether it is following or not. The rule says, to become a drug, a ligand should have molecular weight less than 500 Da, log P value should be less than 5, hydrogen bond acceptor should be less than 10 and hydrogen bond donor should be less than 5. In ADMETox screening the absorption, distribution, metabolism, excretion and toxicity of the ligand is calculated. If both the parameters qualify by the ligand than it may consider that the ligand may act as drug.[58]

5. Development of neuroprotective agents from natural products

Several studies with natural products have been performed to date to discover neuroprotective biomolecules, that can be developed as drugs; many of these studies are based on virtual screening and/or a combination of lab based and *in silico* studies. Plant-derived natural products that have been reported to possess neuroprotective property belong to several classes of secondary metabolites, but most of them are alkaloids, phenolics and terpenoids (Fig. 1). Among the phenolics, flavonoids were quite extensively studied, probably because of their presence in edible fruits and vegetables. Despite several structurally related flavonoids have been studied, no conclusive structure activity relationships among those flavonoids have been established to date. However, because of the phenolic nature of those compounds, they possess strong antioxidant property which is thought to be one of the driving forces for them to act as neuroprotecting agents, by mitigating oxidative stress which often causes neural damage. On Parkinson's disease (PD), several *in silico* studies, phytochemicals such as naringenin (**36**), quercetin (**28**), resveratrol (**61**) and sesamol (**65**) (Fig. 1) were screened, and quercetin (**28**) showed the satisfactory result.[69,70] worked on PD and glycyrrhetinic acid (**76**), resveratroloside (**62**), genkwanin (**34**), protropine (**22**), naringenin (**36**), pergolide (**20**) and magnolol (**43**) (Fig. 1) were screened against dopamine receptor D3 and found glycyrrhetinic acid (**76**) and resveratroloside (**62**) were potent neuroprotective agents. Catechin (**31**), epicatechin (**32**) and epicatechin gallate (**33**) (Fig. 1) have been found to suppress neuroinflammation and can attenuate and inhibit stimulation of microglia.[71] Allicin, a sulfur containing compound from garlic and onion, has been shown to possess neuroprotective and free-radical scavenging activities.[59]

Sulforaphane (**89**) could protect cultured neurons in high oxidative stress and dopaminergic neurons from mitochondrial toxins,[72] while resveratrol (**61**), showed antioxidant activity, and peripheral administration of resveratrol (**61**) could protect neurons in brain and spinal cord against ischemic injury.[73]

Hypericin (**570**) and pseudohypericin (**60**) (Fig. 1), major constituents of *Hypericum perforatum,* have antidepressant property.[74] Dieckol (**55**) and phlorofucofluoroeckol (**63**), isolated from a marine algae displayed memory enhancing and acetyl choline esterase (AChE) inhibitory activity to prevent the Alzheimer's disease.[75]

6. Conclusion

Plant secondary metabolites or phytochemicals play a pivotal role in protecting and improving human health. Various phytochemicals have recently been reported to have neuroprotective effects in several experimental models of neurological disorders as well as in *in silico* studies. Mechanisms of action of natural neuroprotective agents is mainly based on their inherent antioxidant and anti-inflammatory properties.

Acknowledgment

L. Nahar gratefully acknowledges the financial support of the European Regional Development Fund—Project ENOCH (No. CZ.02.1.01/0.0/0.0/16_019/0000868).

References

1. Elufioye, T. O.; Berida, T. I.; Habtemariam, S. *Evid. Based Complement. Alternat. Med.* **2017**, 3574012. https://doi.org/10.1155/2017/3574012.
2. Nieoullon, A. *J. Appl. Biomed.* **2011**, *9*(4), 173–183.
3. Iriti, M.; Vitalini, S.; Fico, G.; Faoro, F. *Molecules* **2010**, *15*(5), 3517–3555.
4. Reeve, A.; Simcox, E.; Turnbull, D. *Ageing Res. Rev.* **2014**, *14*, 19–30.
5. McClatchey, W. C.; Mahady, G. B.; Bennett, B. C.; Shiels, L.; Savo, V. *Pharmacol. Ther.* **2009**, *123*(2), 239–254.
6. Suk, K. *Neurosignals* **2005**, *14*(1–2), 23–33.
7. Abdulwanis Mohamed, Z.; Mohamed Eliaser, E.; Mazzon, E.; Rollin, P.; Cheng Lian Ee, G.; Abdull Razis, A. F. *Molecules* **2019**, *24*(17), 3109. https://doi.org/10.3390/molecules24173109.
8. Pérez-Hernández, J.; Zaldívar-Machorro, V. J.; Villanueva-Porras, D.; Vega-Ávila, E.; Chavarría, A. *Oxid. Med. Cell. Longev.* **2016**, 8378613. https://doi.org/10.1155/2016/8378613.
9. Lobo, V.; Patil, A.; Phatak, A.; Chandra, N. *Pharmacogn. Rev.* **2010**, *4*(8), 118–126.

10. Selvam, A. *Pharmacogn. Rev.* **2008**, *2*(3), 61–94.
11. Uriarte, P. I.; Calvo, M. *Pharmacogn. Mag.* **2009**, *5*(20), 287–290.
12. Capasso, R.; Izzo, A. A.; Pinto, L.; Bifulco, T.; Vitobello, C.; Mascolo, N. *Fitoterapia* **2000**, *71*, S58–S65.
13. Harvey, A. L. *Trends Pharmacol. Sci.* **1999**, *20*(5), 196–198.
14. Kumar, V. *Phytother. Res.* **2006**, *20*(12), 1023–1035.
15. Gong, X.; Sucher, N. J. *Trends Pharmacol. Sci.* **1999**, *20*(5), 191–196.
16. Alghazwi, M.; Kan, Y. Q.; Zhang, W.; Gai, W. P.; Garson, M. J.; Smis, S. *J. Appl. Phycol.* **2016**, *28*(6), 3599–3616.
17. Singh, H.; Dhawan, B. *Ind. J. Pharmacol.* **1997**, *29*(5), S359.
18. Russo, A.; Borrelli, F. *Phytomedicine* **2005**, *12*(4), 305–317.
19. Anand, T.; Mahadeva, N.; Phani, K. G.; Farhath, K. *Pharmacogn. J.* **2010**, *2*(17), 53–58.
20. Nalini, K.; Aroor, A.; Rao, A.; Karanth, K. *Fitoterapia* **1992**, *63*(3), 231–238.
21. Xu, Y.; Ku, B.; Cui, L.; Li, X.; Barish, P. A.; Foster, T. C.; Ogle, W. O. *Brain Res.* **2007**, *1162*, 9–18.
22. Chandrasekaran, K.; Mehrabian, Z.; Spinnewyn, B.; Drieu, K.; Fiskum, G. *Brain Res.* **2001**, *922*(2), 282–292.
23. Nocerino, E.; Amato, M.; Izzo, A. A. *Fitoterapia* **2000**, *71*, S1–S5.
24. Kumar, G. P.; Anilakumar, K.; Naveen, S. *Pharmacogn. J.* **2015**, 7(1), 1–17. https://doi.org/10.5530/pj.2015.1.1.
25. Zou, L.-C.; Zhu, T.-F.; Xiang, H.; Yu, L.; Yan, Z.-H.; Gan, S.-C.; Wang, D.-C.; Zeng, S.; Deng, X.-M. *Molecules* **2008**, *13*(9), 2049–2057.
26. Ansari, J.; Siraj, A.; Inamdar, N. *Int. J. Pharmacol.* **2010**, *6*(5), 584–590.
27. Ameri, A.; Gleitz, J.; Peters, T. *Brain Res.* **1996**, *738*(1), 154–157.
28. Zanoli, P.; Avallone, R.; Baraldi, M. *Phytother. Res.* **1998**, *12*(S1), S114–S116.
29. Yan, J. J.; Cho, J. Y.; Kim, H. S.; Kim, K. L.; Jung, J. S.; Huh, S. O.; Suh, H. W.; Kim, Y. H.; Song, D. K. *Brit. J. Pharmacol.* **2001**, *133*(1), 89–96.
30. Dar, A.; Khatoon, S.; Rahman, G. *Phytomedicine* **1997**, *4*(1), 41–45.
31. Koo, K. A.; Sung, S. H.; Kim, Y. C. *Chem. Pharm. Bull.* **2002**, *50*(6), 834–836.
32. Sukma, M.; Chaichantipyuth, C.; Murakami, Y.; Tohda, M.; Matsumoto, K.; Watanabe, H. *J. Ethnopharmacol.* **2002**, *83*(1–2), 87–94.
33. Kumar, M. V.; Gupta, Y. *J. Ethnopharmacol.* **2002**, *79*(2), 253–260.
34. Hosseinzadeh, H.; Khosravan, V. *Arch. Iran. Med.* **2002**, *5*(1), 44–47.
35. Park, S.-Y.; Kim, D. S. *J. Nat. Prod.* **2002**, *65*(9), 1227–1231.
36. Joseph, J. A.; Shukitt-Hale, B.; Casadesus, G. *Am. J. Clin. Nutr.* **2005**, *81*(1), 313S–316S.
37. Gupta, P.; Siripurapu, K. B.; Ahmad, A.; Palit, G.; Arora, A.; Maurya, R. *Chem. Pharm. Bull.* **2007**, *55*(5), 771–775.
38. Basori, A. *Folia Med. Indonesia.* **2004**, *40*(3), 103–107.
39. Woodruff-Pak, D.; Vogel, R., III; Wenk, G. *Neuroscience* **2003**, *117*(2), 439–447.
40. An, H.; Kim, I. S.; Koppula, S.; Kim, B. W.; Park, P. J.; Lim, B. O.; Choi, W. S.; Lee, K. H.; Choi, D. K. *J. Ethnopharmacol.* **2010**, *130*(2), 290–298.
41. Dhingra, D.; Parle, M.; Kulkarni, S. *J. Ethnopharmacol.* **2004**, *91*(2–3), 361–365.
42. Wang, L. S.; Zhou, J.; Shao, X. M.; Tang, X. C. *Brain Res.* **2002**, *949*(1–2), 162–170.
43. Klusa, V.; Germane, S.; Nöldner, M.; Chatterjee, S. *Pharmacopsychiatry* **2001**, *34*, 61–69.
44. Prediger, R. D.; Fernandes, M. S.; Rial, D.; Wopereis, S.; Pereira, V. S.; Bosse, T. S.; Da Silva, C. B.; Carradore, R. S.; Machado, M. S.; Cechinel-Filho, V. *J. Ethnopharmacol.* **2008**, *120*(3), 465–473.
45. Lee, H.; Kim, Y. O.; Kim, H.; Kim, S. Y.; Noh, H. S.; Kang, S. S.; Cho, G. J.; Choi, W. S.; Suk, K. *FASEB J.* **2003**, *17*(13), 1943–1944.
46. Ohta, H.; Matsumoto, K.; Watanabe, H.; Shimizu, M. *Jpn. J. Pharmacol.* **1993**, *62*(4), 345–349.
47. Dhawan, K.; Kumar, S.; Sharma, A. *J. Ethnopharmacol.* **2001**, *78*(2–3), 165–170.

48. Takeda, H.; Tsuji, M.; Matsumiya, T.; Kubo, M. *Jpn. J. Psychopharmacol.* **2002**, *22*(1), 15–22.
49. Feltenstein, M. W.; Lambdin, L. C.; Ganzera, M.; Ranjith, H.; Dharmaratne, W.; Nanayakkara, N. D.; Khan, I. A.; Sufka, K. J. *Phytother. Res.* **2003**, *17*(3), 210–216.
50. D'Hooge, R.; Pei, Y.; Raes, A.; Lebrun, P. *Arzneimittelforschung* **1996**, *46*(6), 557–560.
51. Liao, J.-F.; Hung, W.-Y.; Chen, C.-F. *Eur. J. Pharmacol.* **2003**, *464*(2–3), 141–146.
52. Viola, H.; Wolfman, C.; De Stein, M. L.; Wasowski, C.; Pena, C.; Medina, J.; Paladini, A. *J. Ethnopharmacol.* **1994**, *44*(1), 47–53.
53. Bhattacharya, S.; Bhattacharya, A.; Kumar, A.; Ghosal, S. *Phytother. Res.* **2002**, *14*(3), 174–179.
54. Shou, C.; Feng, Z.; Wang, J.; Zheng, X. *Planta Med.* **2002**, *68*(09), 799–803.
55. Topic, B.; Tani, E.; Tsiakitzis, K.; Kourounakis, P.; Dere, E.; Hasenöhrl, R.; Häcker, R.; Mattern, C.; Huston, J. *Neurobiol. Aging* **2002**, *23*(1), 135–143.
56. Ying-Jun, L.; Jun, Z.; Shao-Min, Z.; Heng-Yi, Z.; Xiao-Xiang, Z. *J. Zhejiang Univ. Sci. B.* **2005**, *6*(4), 265–271.
57. Haddad, J. *J. Cell. Signalling* **2002**, *14*(11), 879–897.
58. Das, S.; Laskar, M. A.; Sarker, S. D.; Choudhury, M. D.; Choudhury, P. R.; Mitra, A.; Jamil, S.; Lathiff, S. M. A.; Abdullah, S. A.; Basar, N. *Phytochem. Anal.* **2017**, *28*(4), 324–331.
59. Oi, Y.; Kawada, T.; Shishido, C.; Wada, K.; Kominato, Y.; Nishimura, S.; Ariga, T.; Iwai, K. *J. Nutr.* **1999**, *129*(2), 336–342.
60. Cheng, J.; Kuang, P.; Wu, W.; Zhang, F. *J. Tradit. Chin. Med.* **1999**, *19*(2), 135–140.
61. Curtis, D.; Duggan, A.; Felix, D.; Johnston, G. *Nature* **1970**, *228*(5272), 676–677.
62. Sarker, S. D.; Nahar, L. *Computaional Phytochemistry*, 1st ed.; Elsevier: USA, **2018**.
63. Shen, J.; Xu, X.; Cheng, F.; Liu, H.; Luo, X.; Shen, J.; Chen, K.; Zhao, W.; Shen, X.; Jiang, H. *Curr. Med. Chem.* **2003**, *10*(21), 2327–2342.
64. Waszkowycz, B.; Perkins, T. D. J.; Sykes, R. A.; Li, J. *IBM Syst. J.* **2001**, *40*(2), 360–376.
65. Pan, Q.; Zhou, R.; Su, M.; Li, R. *Med. Sci. Monit.* **2019**, *25*, 4648–4654.
66. Acharya, C.; Coop, A.; Polli, J. E.; MacKerell, A. D. *Curr. Comput. Aided Drug Des.* **2011**, 7, 10–22.
67. Durrant, J. D.; McCammon, J. A. *BMC Biol.* **2011**, *9*(1) 71 http://www.biomedcentral.com/1741-7007/9/71.
68. Mazumder, M. K.; Paul, R.; Phukan, B. C.; Dutta, A.; Chakrabarty, J.; Bhattacharya, P.; Borah, A. *Med. Hypotheses* **2018**, *117*, 54–58.
69. Jha, S. K.; Kumar, P. *Int. J. Pharm. Pharm. Sci.* **2017**, *9*(9), 149–155. https://doi.org/10.22159/ijpps.2017v9i9.20445.
70. Mirza, M. U.; Mirza, A. H.; Noor-Ul-Huda Ghori, S. F. *Drug Des. Devel. Ther.* **2015**, *9*, 187.
71. Wu, D.; Guo, Z.; Ren, Z.; Guo, W.; Meydani, S. N. *Free Rad. Biol. Med.* **2009**, *47*(5), 636–643.
72. Han, J. M.; Lee, Y. J.; Lee, S. Y.; Kim, E. M.; Moon, Y.; Kim, H. W.; Hwang, O. *J. Pharmacol. Exp. Ther.* **2017**, *321*(1), 249–256.
73. Huang, S. S.; Tsai, M. C.; Chih, C. L.; Hung, L. M.; Tsai, S. K. *Life Sci.* **2001**, *69*(9), 1057–1065.
74. Chatterjee, S.; Bhattacharya, S.; Wonnemann, M.; Singer, A.; Müller, W. *Life Sci.* **1998**, *63*(6), 499–510.
75. Myung, C.-S.; Shin, H.-C.; Bao, H. Y.; Yeo, S. J.; Lee, B. H.; Kang, J. S. *Arch. Pharmacal Res.* **2005**, *28*(6), 691–698.

CHAPTER SEVEN

Hepatoprotective natural products

Sitesh C. Bachar[a,*], Ritesh Bachar[b], Khoshnur Jannat[c], Rownak Jahan[c], Mohammed Rahmatullah[c]
[a]Department of Pharmacy, Faculty of Pharmacy, University of Dhaka, Dhaka, Bangladesh
[b]Department of Pharmacy, University of Information Technology and Sciences, Dhaka, Bangladesh
[c]Department of Biotechnology & Genetic Engineering, University of Development Alternative, Dhaka, Bangladesh
[*]Corresponding author: e-mail address: bacharsc@du.ac.bd

Contents

Annual Reports in Medicinal Chemistry, Volume 55
ISSN 0065-7743
https://doi.org/10.1016/bs.armc.2020.06.003

1. Introduction

One of the most largest and important organs of the body is liver. The maintenance, performance, regulation of homeostasis, secretions of bile, storage of vitamin,[1] energy production, metabolism and reproduction[2] and detoxification to control almost all systems of the body is done by this metabolic "engine room." The major functions of liver include carbohydrate, protein, fat metabolism; detoxification of endogenous (waste metabolites) and/or exogenous (toxic compounds) substances; detoxification of several drugs and xenobiotics coming from food, drinks, drugs and environment. The hepatic reticuloendothelial system clears activated clotting factors, proteolytic enzyme/inhibitor complexes, and fibrin and fibrinogen degradation products.[3–6] It plays vital role in all the biochemical pathways to growth, immune system, nutrient supply, energy provision and reproduction.[2]

The liver usually filters all the blood from the digestive tract, before passing it to the rest of the body to avoid entering toxins in the other body systems.[7] Liver also synthesizes many complement systems and proteins for supporting immune system.[8] Therefore, appropriate functioning of liver is indispensable for the healthy living of an individual.

2. Hepatic problems

Many risk factors influence an individual to hepatic drug injury such as pre-existing liver disease, aging, female sex, and genetics. However, liver diseases are the most serious among all the liver disorders. These can be classified as acute or chronic hepatitis (inflammatory liver diseases), hepatosis (non-inflammatory liver diseases) and cirrhosis (degenerative disorder) causing fibrosis of the liver. Toxic chemicals such as certain antibiotics, chemotherapeutics, peroxidised oil, aflatoxin, carbon tetrachloride, acetaminophen, chlorinated hydrocarbons, etc., are responsible for liver toxicities and damages. Others are excess consumption of alcohol and auto-immune disorders.[9–12] Liver microsomal metabolism of ethanol may result in hepatitis and cirrhosis due to enhanced lipid peroxidation. It has been found that viruses are responsible for 90% acute hepatitis.[11]

Hepatic diseases lead to liver damage. A major contributory factor for liver damage is excess alcohol intake in developed countries.[13] Starvation, blood deficiency, communicable diseases, unsupervised drug use and accessibility of over the counter-hepatotoxic drugs are the most recurrent factors of liver cell injures in under developed and developing countries according to a 1992 World Health Organization (WHO) Bulletin.

It is evident from the reports of the WHO in 2016 that non-communicable diseases were the cause of 68% of all global deaths in 2012, rising from 60% in 2000. Hepatic complications have turned out to be multifactorial diseases that affected a population of around almost 600 million in 2014, and it is likely to amplify by about 33% over the next two decades.[14,15]

In England, 2% of total deaths between 2001 and 2009 were due to liver diseases.[16,17] The report states that, in the same period, alcoholic liver disease accounted for well over a third (37%) of liver disease deaths. In the United States, over 25 million Americans are afflicted with liver disease and more than 27,000 die from cirrhosis, the seventh leading cause of death. Additionally, 43,000 die of liver disease, of which about 50% are alcohol related.[18] Moreover, liver diseases such as fibrosis are a major worldwide health problem, with high prevalence in developing countries where hundreds of millions are afflicted.[19] In fact, with the global obesity epidemic and the developing of non-alcoholic steatohepatitis (NASH), fibrosis, cirrhosis and hepatocellular carcinoma have become a worldwide health concern for all ages and ethnicities.[20–22]

Liver diseases can be non-inflammatory, inflammatory and degenerative. High levels of plasma total cholesterol (LDL-C) and triacylglycerols (TGs) are associated with high risk of atherosclerosis and cardiovascular disease owing to the hepatic insufficiency.[23,24]

Acute renal failure[3]—This is a general term applied to the rapid development of hepatic synthetic dysfunction associated with significant coagulopathy, usually defined by a prothrombin time or factor V level less than 50% of normal. The most common cause of acute renal failure is drugs and viral hepatitis.

Acute and chronic viral hepatitis[3]—The agents of acute viral hepatitis can be broadly classified into two groups: Enterically transmitted agents like Hepatitis A virus, Hepatitis E virus and blood borne agents like Hepatitis B virus, Hepatitis D virus and Hepatitis C virus. Chronic viral hepatitis describes persistent inflammation of the liver for 6 months or more after initial exposure and/or initial detection of liver disease. The primary cause of chronic hepatitis is viral infection.

Drug induced hepatotoxicity[3]—Hepatotoxicity may occur as an unexpected idiosyncratic reaction to a medication's therapeutic dose or as an expected consequence of the agent's intrinsic toxicity. Serum alanine and aspartate aminotransferase and lactate dehydrogenase levels may be elevated 10–100 times in acute hepatocellular injury, while alkaline phosphatase levels are usually less than three times the upper limit of normal. The serum bilirubin may be elevated or within the normal range.

Cirrhosis[3]—The diffuse process characterized by fibrosis and the conversion of normal liver architecture into structurally abnormal nodules that lack normal lobular organization. Structural changes in the liver may cause impairment of hepatic function manifested as jaundice, ascites, hepatorenal syndrome, hepatic encephalopathy, spontaneous bacterial peritonitis.

Portal hypertension[3]—It is defined as an increase in the portal venous pressure gradient and is a function of portal venous blood flow and hepatic and portocollateral resistance.

Alcohol consumption—Alcohol hinders the functional aspects of various tissue components and hepatocytes in particular. Alcohol diffuses, crossing the membrane barrier, and is distributed throughout the cell and tissue system, interacting with the major proteins and cellular component present in it.[25] Development of toxic molecules like reactive oxygen species (ROS) is another negative upshot of alcohol. In addition to ROS, it also produces acetaldehyde and nitric oxides, which are extremely reactive and toxic by-products that chip into tissue damage.[26,27] Under conditions of alcohol mediated hepatic complications, mitochondrial respiration is hindered and, in turn, hypoxia occurs.

Fatty liver and non-alcoholic steatohepatitis[3]—Non-alcoholic fatty liver disease includes a spectrum of abnormalities from hepatic steatosis with associated necroinflammatory changes and varying degrees of fibrosis. Non-alcoholic steatohepatitis is a clinical pathological syndrome of steatosis and associated hepatic necroinflammatory changes that are diagnosed only by liver biopsy.

Budd-Chiari syndrome (BCS)[3]—It results from obstruction to hepatic venous outflow and may result from either thrombotic or nonthrombotic occlusion.

Hepatic veno-occlusive disease (VOD)[3]—It is also referred to as sinusoidal obstruction syndrome. VOD is most often seen in an acute form following bone marrow transplantation; thought to be due to toxicity from the preparative regimen of high dose cytoreductive therapy with or without hepatic irradiation.

3. Rationale for the hepatoprotective natural products

Physicians and patients need effective therapeutic agent with minimum incidence of toxic effects. In recent years, researchers have examined the effects of plants used traditionally by indigenous healers and herbalists to support liver function and treat diseases of the liver. Mono and polyherbal preparations have been used in various liver disorders. According to one estimate, more than 700 mono and polyherbal preparations in the form of decoction, tincture, tablets and capsules from more than 100 plants are in clinical use. In most cases, research has borne out the traditional experience and wisdom by discovering the mechanisms and modes of action of these plants, as well as confirming the therapeutic effectiveness of certain plants or plant extracts in clinical studies.[3]

A large number of plants and formulations have been claimed to have hepatoprotective activity. In spite of the tremendous advances made, no significant and safe hepatoprotective agent is available in modern therapeutics. Therefore, due importance has been given globally to develop plant-based hepatoprotective drugs or agents, effective against a variety of liver disorders. Several hundred plants have been examined for use in a wide variety of liver disorders.[28]

Hepatoprotectives are a class of therapeutic agents that includes many synthetic and natural products used to protect against hepatic damage induced by various toxins. Natural resources such as plants are always considered and used in the search for new molecules to be used as drugs. Numerous medicinal plants and their formulations are used for liver disorders in ethnomedical practice as well as traditional system of medicine in India.[29] Plant-based therapeutic agents like silybin, silymarin from *Silybum marianum* (milk thistle) are accepted and used worldwide as hepatoprotective.[30] Silymarin is a flavonolignan that has been introduced recently as a hepatoprotective agent. Silymarin is extracted from the seeds and fruits of *Silybum marianum* which is a mixture of three structural components: Silibinin, Silydianine and Silychristine.[29] In India, about 40 polyherbal commercial formulations reputed to have hepatoprotective action are being used. It has been reported that 160 phytoconstituents from 101 plants have hepatoprotective activity.[31,32] Liver protective herbal drugs contain a variety of chemical constituents like phenols, coumarins, lignans, essential oil, monoterpenes, carotenoids, glycosides, flavanoids, organic acids, lipids, alkaloids and xanthines.[31,33] Some of these well researched plants including *Silybum marianum* L. (milk thistle), *Eucommia ulmoides* Oliver, *Gymnema sylvestre* R.Br., *Microcos paniculata* L., *Allium sativum* L., *Zingiber officinalis* Roscoe, *Curcuma longa* L., *Phyllanthus niruri* L., *Lawsonia inermis* L. and others have been discussed in this review chapter (Table 1).

4. Purpose of the review article

This review chapter emphasizes the literature about purified or semi-purified chemicals ofherbal origins (i.e., chemically defined molecules) with reported experimental and/or clinical hepatoprotective activity. The findings are likely to help further to investigate hepatoprotective agents at both the experimental and the clinical levels. For this a literature search using mainly PubMed and Google search engine was performed. An extensive number of articles which address many plant constituents with reported experimental (*in vitro* or *in vivo*) effects are summarized in various reviews.[99–104]

The proper uses of some of these plants and their constituents of different fruits have played fundamental roles in human health care. Approximately 80% of the world's population has employed traditional medicine for health care, which is based predominantly on plant materials.[105,106]

Despite huge advances in contemporary medicine, there are no entirely useful drugs that stimulate hepatic function, suggesting absolute safeguard to the organ, or aid in regenerating hepatic cells. It is necessary to identify, accumulate and collect information on alternative pharmaceuticals for the treatment of hepatic diseases, with the aim of these agents or constituents being more effective, more useful and less toxic.

Table 1 List of plants including active phytoconstituents with proposed hepatoprotective effect.

Plant name	Phytochemicals	Reference
Eucommia ulmoides Oliy, *Aucuba japonica* Thunb., *Plantago asiatica* L.	Aucubin	34
Gymnema sylvestre	Gurmarin, gymnemic acid, gymnemasaponins, gymnemanol, gymnemoside, gymmestrogenin, gymnemasin	35
Silybum marianum L.	Silymarin, silibinin A&B, isosilybin A&B, silychristin, silydianin	36
Piper longum	Piperine, pierlongumine, flavonoids, essential oils, steroids	37
Microcos paniculata	Apigenin, kaempferol, quercetin, epicatechin, vitexin, isovitexin	38
Securidaca inappendiculata Hassk	(7*R*,8*S*)-4-hydroxy-3,3′-dimethoxy-8,4′-oxyneoligna-7,9,9′-triol-4-*O*-β-D-glucopyranosyl-(1 → 4)-β-D-glucopyranoside, (7*R*,8*S*)-4-hydroxy-3,5′-dimethoxy-4′,7-epoxy-8,3′-neoligna-9,9′-diol-9′-*O*-β-D-glucopyranosyl-4-*O*-[β-D-glucopyranosyl-(1 → 4)]-β-D-glucopyranoside, (7*R*,8*S*)-4-hydroxy-3,5,5′-trimethoxy-4′,7-epoxy-8,3′-neoligna-9,9′-diol-9′-*O*-β-D-glucopyranosyl-4-*O*-[β-D-glucopyranosyl-(1 → 4)]-β-D-glucopyranoside, phenolic glycoside, securiphenoside B, hemiterpene glycosides, securiterpenoside E-F (5, 6)	39
Boswellia carterii Birdw	Eighteen new cembrane-type diterpenoids, named as Boscartins P-AG (1–18), incensole, incensole oxide acetate, incensole oxide, 1,4-epoxy-8,13-cembrandien-5,12-diol, 4,8-epoxy-8,12-cembrandien-5,12-diol	40
Annona crassiflora Mart.	Phenolic compounds, alkaloids, annonaceous acetogenins, tocols, carotenoids, phytosterols, dietary fiber, vitamins, minerals and essential oils	41

Continued

Table 1 List of plants including active phytoconstituents with proposed hepatoprotective effect.—cont'd

Plant name	Phytochemicals	Reference
Allium sativum L.	Allicin, Allin, diallyl sulfide, diallyl disulfide, diallyl trisulfide, ajoene, S-allyl-cysteine	42
Solanum surattense	Beta sitosterol	43
Phylanthus amarus	Gallic acid, ellagic acid, corilagin, geraniin, niranthin, phyllanthin, hypophyllanthin, phyltetralin, isonirtetralin	44
Lepidium sativum L.	Chlorogenic acid, quercetin, kaempferol, ferulic acid, sinapic acid, protocatechuic acid, vanillic acid, caffeic acid, P-coumaric acid	45
Berry	Anthocyanines, phenolic acids	46
Cynanchum otophyllum	Cynanchin A–G	47
Muntingia calabura	Gallic acid, ferulic acid, quercetin, genistin	48
Leea asiatica	−epicatechin, −epiafzelechin, juglanin, mearnsetin-3-*O*-arabinopyranoside, myricitrin, afzelin, quercitrin	49
Ammannia baccifera	Chlorogenic acid, quercetin, protocatchuic acid, lamioside, crocetin, khayasinc	50
Zingiber officinalis	Chlorogenic acid, hespiridin, curcumene, linalool	51
Malas domestica (apple)	Phloretin	52
Canarium tramdenum	Alpha & beta amyrins	53
Cyclea barbata	Phytoecdysones 3,5,9,6	54
Caesalpinia gilliesii (Hook.) D. Dietr.	Gossypetin-3-*O*-β-D-robinobioside	55
Ilex chinensis	Ilexchinenosides J, L, N, P	56
Malvaviscus arboreus Cav.	β-Resorcylic, caffeic, protocatechuic, 4-hydroxyphenylacetic acids, trifolin, astragalin	57
Origanum vulgare ssp. vulgare	Chlorogenic, rosmarinic acids	58
Capsicum annum L.	Capsaicin, 2,3-dihydroxy 1,4-dioxane, 2,6-dimethylnapthol [1,2-b] furan-4,5-dione,	59

Table 1 List of plants including active phytoconstituents with proposed hepatoprotective effect.—cont'd

Plant name	Phytochemicals	Reference
	4-[5-(methylsulfinyl)-1,3,4 thiadiazole-2-yl] pyridine	
Hygrophila auriculata (Schum.)	Apigenin, luteolin, ellagic acid, gallic acid, quercetin, asteracanthine asteracanthicine, lupeol, lupenone, hentricontane betulin, stigmasterol, asterol	60
Aloe vera	Anthraquinones, naphthalenones	61
Artemisia vulgaris	Sesquiterpene lactones, coumarins, acetylenes, monoterpenes, sesquiterpene	62
Aegle marmelos Linn.	Aegeline, aegelenine, aegelinosides, marmelin, marmelosin, malondialdehyde, anhydromarmeline, marmelide, umbelliferone β-D-galactopyranoside, lupeol, halfordinol, butyl *p*-tolyl sulfide, 6-methyl-4-chromanone, butylated hydroxyanisole, imperatorin, xanthorrhizol, xanthoarnol, 1-hydroxy-5,7-dimethoxy-2-naphthalene-carboxaldehyde, 1-methyl2-(3′-methyl-but-2′-enyloxy)-anthraquinone	63
Croton bonplandianus Baill.	α-Amyrin	64
Iris sibirica	Palmitic acid, linoleic acids, α-linolenic acids	65
Quercus dilatata Lindl.	Gallic acid, quercetin, ascorbic acid	66
Moringa oleifera Lam.	Isothiocyanates	67
Coffea arabica	Chlorogenic acid	68
Muntingia calabura L.	Gallic acid, quercetin	69
Yucca aloifolia	Hesperidin, kaemp-3-(2-*p*-coumaroyl), gallic acid, protocatechuic acid	70
Turraea fischeri	Cinchonains	71
Acacia catechu	4-Hydroxybenzoic acid, kaempferol, quercetin, 3,4,7-trihydroxyl-3,5-dimethoxyflavone, catechin, rutin, isorhamnetin, epicatechin, afzelechin,	72

Continued

Table 1 List of plants including active phytoconstituents with proposed hepatoprotective effect.—cont'd

Plant name	Phytochemicals	Reference
	epiafzelechin, mesquitol, ophioglonin, aromadendrin	
Lycium europaeum	Caffeic acid	73
Stevia rebaudiana bert.	Stevioside	74
Neanotis wightiana	Ursolic acid, oleanolic acid, neanoside B	75
Phyllanthus Niruri	Ellagic acid, phyllanthin	76
Cortex Moutan	Monoterpene glycosides	77
Butea monosperma	Quercetin, buteaspermin B, ononin	78
Phyllanthus emblica L.	Vitamin C, mucic acid, ellagic acid	79
Lagerstroemia speciosa (L.) Pers.	γ-Sitosterol, 1,2,3-Benzenetriol	80
Matrcaria recuitta chamomile	Aligning, apigenin-7-*O*-glucoside, caffeine acid, cholorogenic acid, luteolin, luteolin-7-*O*-glucoside, Terpene disability, farnesene, chamazulene, quercetin, platuletin and coumarin	81
Asphodeline lutea (L.)	Anthraquinones, caffeic acid, naphthalenes	82
Amaranthus spinosus	Gallic acid, catechin, caffeic acid, epicatechin, vanillic acid, protocatechuic acid, ferulic acid, coumaric acid, cinnamic acid, rutin, quercetin, isorhamentin, kaempferol	83
Nerium oleander (L.)	Apocynin, tocopherol, squalene, amyrin, lupeol	84
Aquilaria malaccensis	Sesquiterpener, 2(-2-phenylethyl)-4*H*-chromen-4-one derivatives, genkwanins, mangiferins, iriflophenones, cucurbitacins	85
Rosmarinus officinalis (L.)	Carnosic acid, rosmerinic acid, betulinic acid, carnosol, rosmanol, ursolic acid	86
Hypericum androsaemum	Shikimic acid, chlorogenic acid	87
Symplocos recemosa Roxb	Symplocoside, betulinic acid, acetyloleanolic acid, oleanolic acid, quercetin	88

Table 1 List of plants including active phytoconstituents with proposed hepatoprotective effect.—cont'd

Plant name	Phytochemicals	Reference
Withania somnifera L.	Withanoside, withaferin A,1,2 deoxywithastramonolide, withanone, withanolide A, withanolide B	89
Cichorium intybus L.	Sesquiterpenes	90
Physalis peruviana	15-Desacetylphysabubenolide, betuline	91
Tiliacora racemosa Colebr.	Tiliamosine, tiliacorine, tiliarine, *N*-methyl-tiliamosine, tiliaresine, tiliacorinine, acorine, mosine	92
Lawsonia inermis L.	Lawsone	93
Glycyrrhiza glabra	Glycyrrhizin, flavonoids, 18-β-glycyrrhetinic acid, two glucuronic acid (18β-glycyrrhetinic acid-3-*O*-β-D-glucuronopyranosyl-(1 → 2)-beta-D-glucuronide), coumarin, alkaloids, polysaccharides, sitosterol, and amino acids	94–96
Vitis vinifera L.	Resveratrol, *trans-*&*cis*-resveratrol, viniferins, lycopene, quercetin, melatonin, catechins, procyanidins, flavonols, flavonoids, anthocyanins and others	97,98

5. Potential plant constituents for hepatic diseases and protection

Phytochemicals are naturally occurring compounds in plants which can protect liver by providing medicinal value or nutrients. Worldwide for ages, researchers are working to find out the chemical compounds from plants that can either prevent liver diseases or cure them. Below is given a list of plants and phytochemicals and their reported hepatoprotective effects (Table 1).

Plants that are listed above have been used as hepatoprotective either in ethnomedicine or reported in pharmacological studies. Researchers have discovered the phytochemicals in various plants responsible particularly for protecting liver, their action in combating liver diseases can be same or different. Different types of compounds such as alkaloid, flavonoid, anthocyanine, terpene give hepatoprotective effect.

6. Pharmacological evaluation of herbal active constituents

Natural compounds, for their diverse chemical structures, exhibit various biological or pharmacological properties, and thus also show hepatoprotective effects. They offer hepatic protections against oxidative damages by donating hydrogen or electron to free radicals, promoting regeneration of hepatocytes, inhibiting fibrogenesis, lowering lipid peroxidation and GSH levels, ameliorating serum glutamic pyruvic transaminase (GPT), glutamic oxaloacetic transaminase (GOT), glutathione peroxidase (GSH-Px), IL-6, and hepatic nuclear NF-κB, and nitric oxide synthase (iNOS), modulating reactive oxygen species (ROS)/MAPKs/apoptosis axis and nuclear factor erythroid 2 (Nrf-2)-related factor, suppressing lipid peroxidation, accumulating and activating SOD gene expression, etc. The pharmacological or biological actions of such natural compounds are discussed in this chapter in order to offer health benefits especially on hepatic protection.

6.1 Aucubin

Aucubin (Scheme 1) is an iridoid glycoside (iridoid known as defensive compound),[107] which has been reported as hepatoprotective.[108] It can protect liver from poisoning caused by α-amanitin,[109] and showed a strong preventive effect from carbontetrachloride or CCl_4-induced hepatic damage by shortening the duration of hexobarbital sodium salt-induced sleep, decreasing serum glutamic oxalacetic transaminase and glutamic pyruvate transaminase activity in CCl_4 treated mice.[110] In a study, aucubin reduced activity of aucubigenin on CYP450 (cytochromes P450) and also showed inhibitory effect on hepatic RNA and protein synthesis which clearly indicates that this

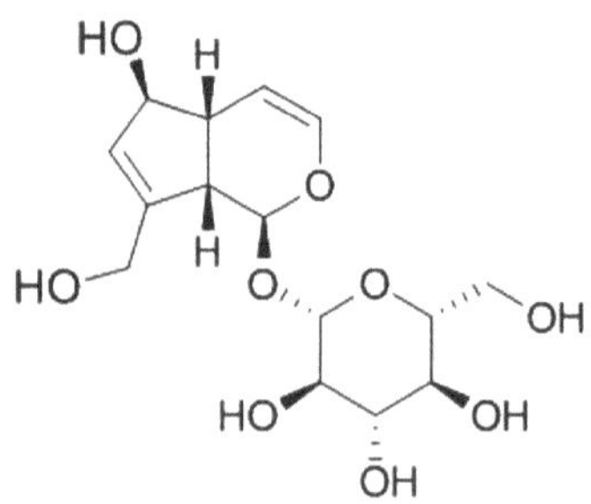

Scheme 1 Aucubin

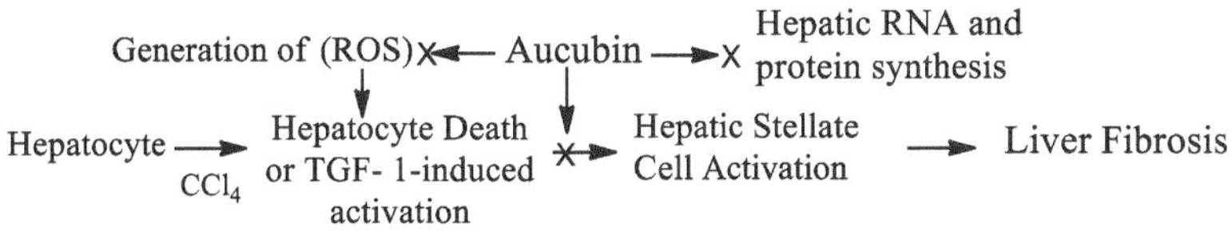

Fig. 1 Schematic diagram for possible mechanism of action of Aucubin in hepatic protection in CCl_4 induced liver fibrosis and other mediators.

compound has a liver protective effect.[111] Suppression of hepatic lipid accumulation and cell death (induced by palmitate) by lysosomal activity,[112,113] blocking TGF-β1-induced activation of human hepatic stellate cells and decreasing generation of reactive oxygen species (ROS) by aucubin (Fig. 1) were also reported.[114]

6.2 Lawsone

Lawsone (Scheme 2) is a napthoquinone which showed hepatoprotective effect by lowering rifampicin-isoniazid induced cytotoxicity to HepG2 cells and lowering transaminase leakage from HepG2 cells as well.[93]

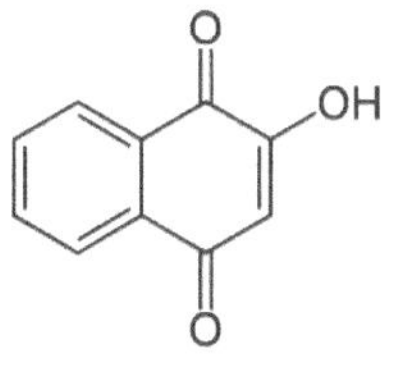

Scheme 2 Lawsone

6.3 Tiliamosine

Non-alcoholic steatohepatitis (NASH) is an aggressive form of non-alcoholic fatty liver disease (NAFLD) which eventually causes cirrhosis and hepatocellular carcinoma (HCC). Presence of steatosis, inflammation and ballooning injury in hepatocytes are the main characteristics of NASH.[115] Tiliamosine (Scheme 3) is one of the bisbenzyl isoquinoline alkaloids of *Tiliacora racemosa*, which showed significant hepatoprotective effect at different sites. For *in vivo* study male Wistar rats were used. There was an increase amount of T_g accumulation and lipotoxicity in Po-BSA induced steatotic HepG2 cells, but treatment with tiliamosine exerted a significant reduction in T_g (triglyceride)-lipotoxicity level. HFD-DEN (high fat diet-diethylnitrosamine) accelerated fibrosis and cirrhosis progression and

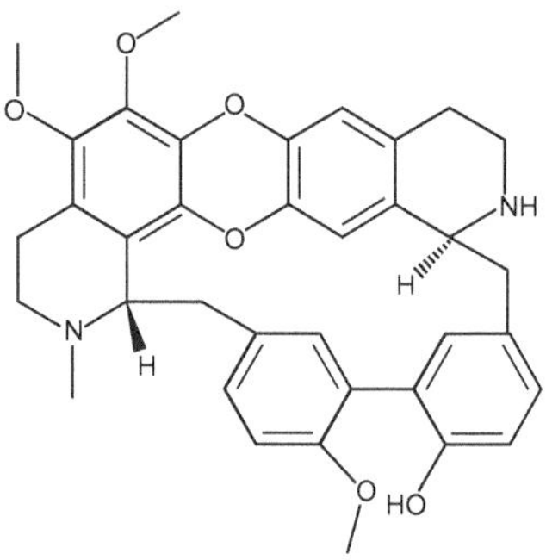

Scheme 3 Tiliamosine

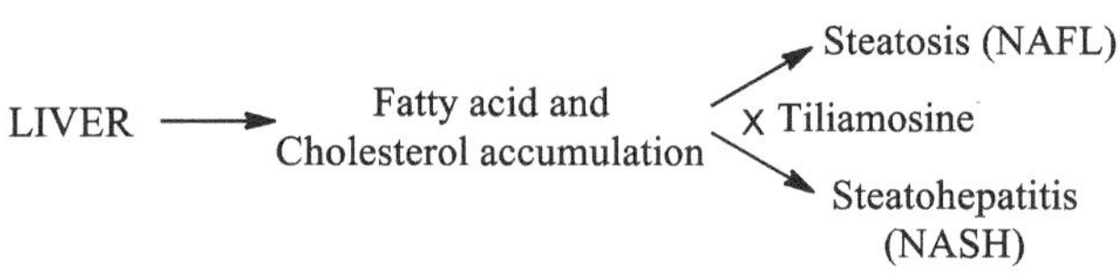

Fig. 2 Schematic diagram for possible mechanism of action of tiliamosine in steatosis (NAFL) and steatohepatitis (NASH).

other NASH characteristics, treatment with tiliamosine showed (Fig. 2) improved lipid level as well as markers of hepatic cell damage and inflammation. Tiliamosine also increased albumin level which indicates the recovery of hepatic cell function. Tiliamosine also increased tumor necrosis factor α (TNF-α) expression along with FXR (Farnesoid X receptor, which regulates bile, lipid, glucose metabolism in liver fibrosis, hepatocyte generation, inflammation, cholestiasis). Tiliamosine was docked (molecular docking) with FXR; the result showed that it strongly binds with the active site of the receptor which denotes a good bioavailability.[92]

6.4 Silymarin

Fruits and seeds of *Silybum marianum* have been used to treat liver disease for centuries. It contains flavonolignans silymarin (Scheme 4) which gave promising hepatoprotective effect by promoting regeneration of hepatocytes, inhibit oxidative stress, fibrogenesis inhibiting and reducing liver steatosis (Fig. 3).[116,117]

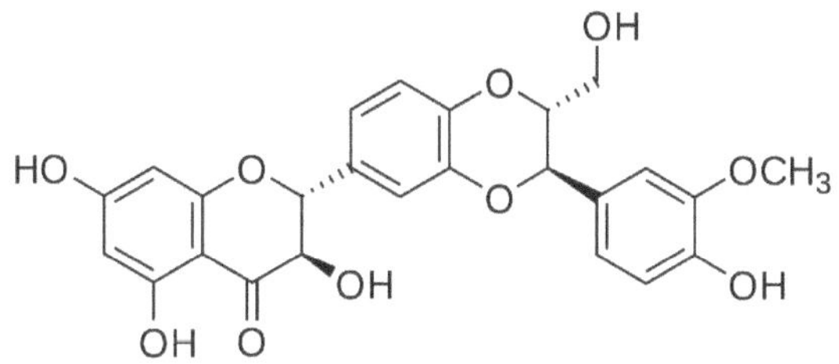

Scheme 4 Silymarin

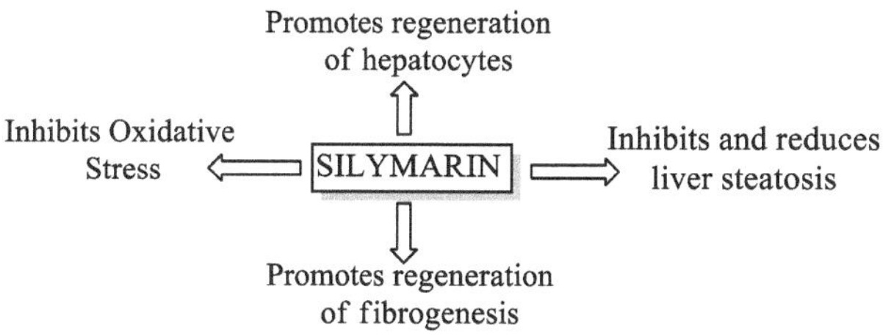

Fig. 3 Performance of silymarin in hepatoprotection.

6.5 Piperine

Piperine (Scheme 5), an alkaloidal chemical compound found from the extract of *Piper longum* was tested to evaluate its hepatoprotective effect. This compound gave a significant protection against tert-butyl hydroperoxide and carbon tetrachloride hepatotoxicity by reducing both *in vitro* and *in vivo* lipid peroxidation, enzymatic leakage of GPT (liver blood test) and alkaline phosphate, and by preventing the reduction of GSH and total thiols in intoxicated mice. However, piperine was found to be less active than silymarin.[118] In another study, piperine (100 mg/kg) and silymarine (50 mg/kg) combination was studied against D-galactosamine induced liver damage in albino rats. This combination significantly protected liver from damage by reducing serum enzymes alanine aminotransferase (ALT), aspartate aminotransferase (AST), alkaline phosphatase (ALP) and total bilirubin.[119]

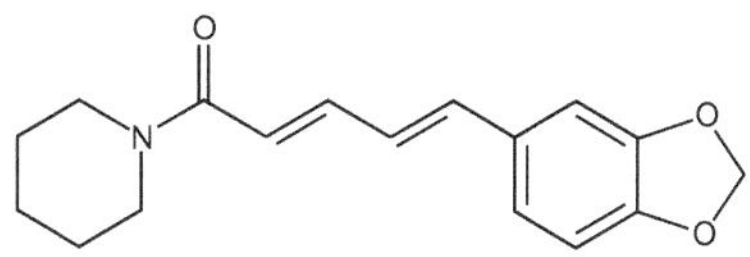

Scheme 5 Piperine

6.6 Catechin

A polyphenol enriched plant *Microcos paniculata* contains catechin (Scheme 6), quercetin (Scheme 28) and similar polyphenolic compounds. Leaf extract of this plant is highly enriched with such polyphenols which were tested for hepatoprotective effects against oxidative stress and acetaminophen-induced hepatotoxicity. The polyphenols gave hepatoprotective effect mainly through modulating reactive oxygen species (ROS)/MAPKs/apoptosis axis (Fig. 4) and nuclear factor erythroid 2 (Nrf-2)-related factor mediated antioxidant response.[120]

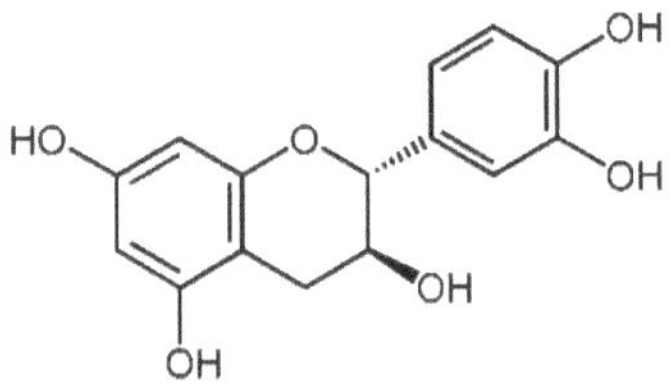

Scheme 6 Catechin

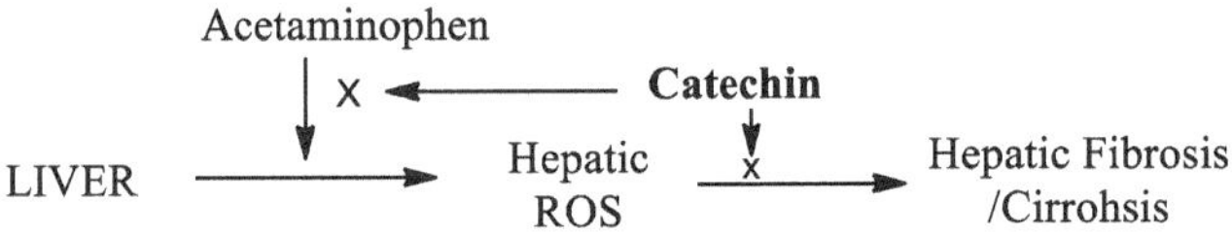

Fig. 4 Schematic diagram for possible mechanism of action of catechin in acetaminophen-induced hepatotoxicity.

6.7 Neoligna-β-D-glucopyranoside and incensole

Among six new compound from *Securidaca inappendiculata*, (7*R*,8*S*)-4-hydroxy-3,5′-dimethoxy-4′,7-epoxy-8,3′-neoligna-9,9′-diol-9′-*O*-β-D-glucopyranosyl-4-*O*-[β-D-glucopyranosyl-(1 → 4)]-β-D-glucopyranoside gave a moderate hepatoprotective effect against acetaminophen-induced hepatic cell injury.[6] *Boswellia carterii* contains cembrane-type diterpenoids and a total 26 compound were isolated. Among all of these Boscartin P, Boscartin U-Y, Boscartin AA-AB, Boscartin AE-AF, incensole (Scheme 7), incensole oxide acetate, incensole oxide, 1,4-epoxy-8,13-cembrandien-5,12-diol, 4,8-epoxy-8,12-cembrandien-5,12-diol were found to be moderately hepatoprotective. They were examined against D-galactosamine-induced toxicity in HL-7702 cells and exerted a significant inhibition in toxicity compared to control and showed enhanced survival rate.[40]

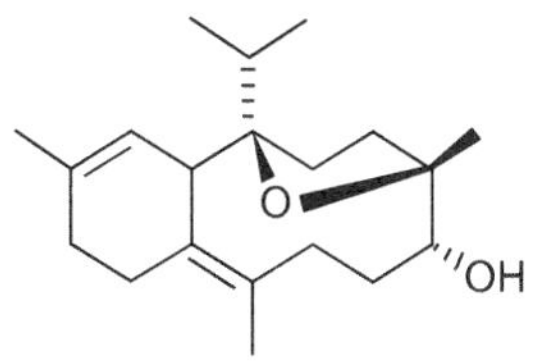

Scheme 7 Incensole

6.8 Caffeic acid, ferulic acid and rutin

Annona crassiflora acts as hepatoprotective for its phenolic compounds. Phenolic compounds, such as caffeic acid (Scheme 8), ferulic acid (Scheme 9), xanthoxylin, caffeoyltartaric acid, caffeoylglucose, quercetin derivative and rutin (Scheme 10) were found to be hepatoprotective when fruit peel and seed (hydroethanolic extract) extract was tested against CCl_4

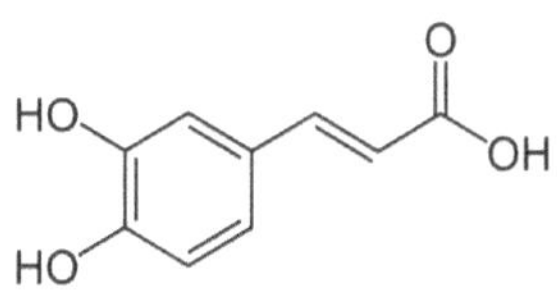

Scheme 8 Caffeic acid

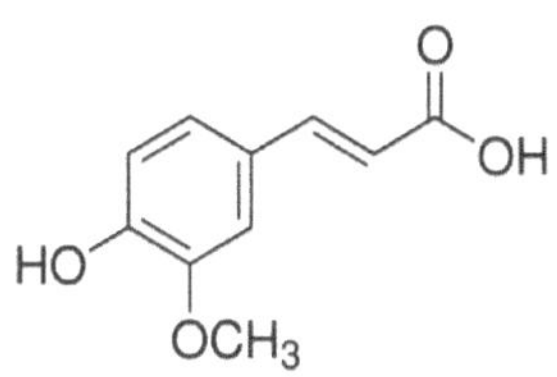

Scheme 9 Ferulic acid

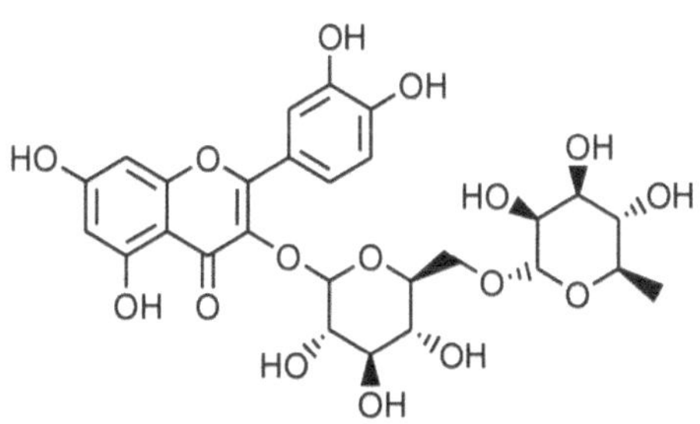

Scheme 10 Rutin

induced liver damage in male Wistar rats. The compounds showed this activity by lowering lipid peroxidation and GSH levels as well as enhancement or maintaining of hepatic antioxidant enzymes activities.[121]

6.9 Allicin

Allium sativum or garlic is a very common plant which have been used for various ailments along with hepatic disorders. Organosulfur compounds such as diallyl sulfide, allicin (Scheme 11; diallyl disulfide), *S*-methyl-l-cysteine were found to be positive in curing acute and chronic ethanol-induced liver damage. Allicin (diallyl disulfide) also worked against non-alcoholic fatty liver diseases. It was very effective in inhibiting the expression of CYTP450 2E1 (Fig. 5) thus enhancing antioxidant activity and significantly reduced pro-inflammatory cytokines release in liver.[122,123]

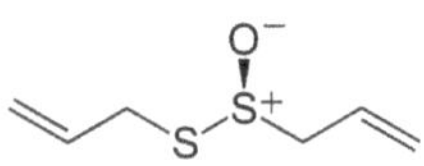

Scheme 11 Allicin

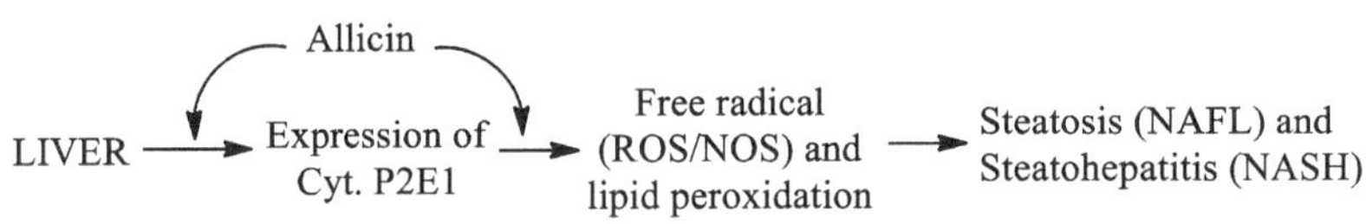

Fig. 5 Schematic diagram for possible mechanism of action of allicin in non-alcoholic fatty liver diseases.

6.10 Withanolides (A & B)

Withanolides from *Withania somnifera* also gave significant hepatoprotective effect. Withanolides (A & B) (Schemes 12 and 13) rich extract was given to acetaminophen-induced hepatic damaged male Wistar albino rats. The

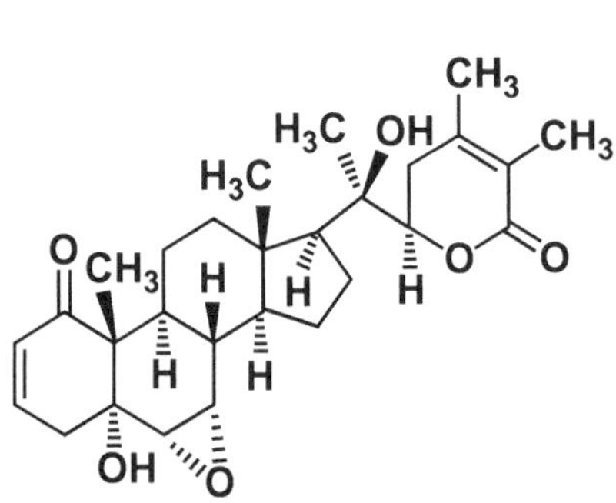

Scheme 12 Withanolide A

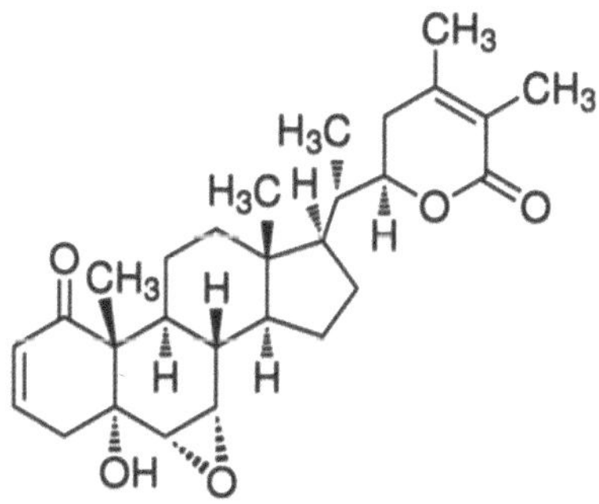

Scheme 13 Withanolide B

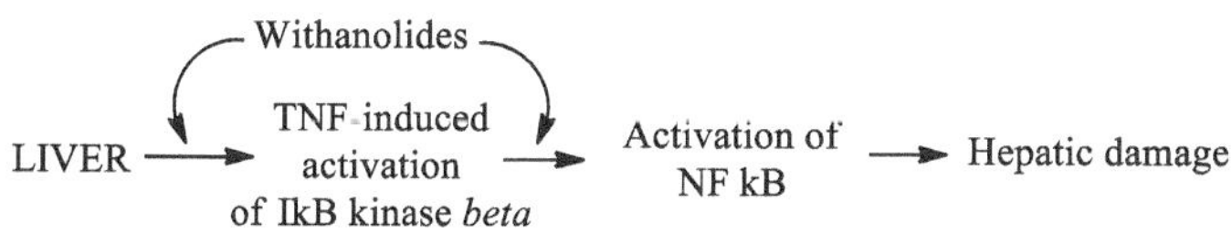

Fig. 6 Schematic diagram for possible mechanism of action of allicin in non-alcoholic fatty liver diseases.

compounds gave hepatoprotective effects by their antioxidant effect, decreasing serum bilirubin, ALP, AST, ALT level and also by inhibiting the expression of inflammatory markers TNF-α, NF-κB, COX-II and IKKb (Fig. 6).[89]

6.11 4-*O*-Caffeoylquinic acid

Phyllanthus niruri has been widely used for liver diseases in traditional medicine. Phytochemicals of this plant were found promising in curing hepatic diseases. From ethanolic extract, 4-*O*-caffeoylquinic acid (Scheme 14) and quercetin-3-*O*-rhamnoside (Scheme 15) were isolated which mainly produced the hepatoprotective effect. The extract was given to thioacetamide induced liver cirrhosis in male Sprague Dawley rats and it resulted in a successful prevention of high synthesis of extracellular matrix as well as reduced transforming growth factor β, collagen a1 and metalloproteinase

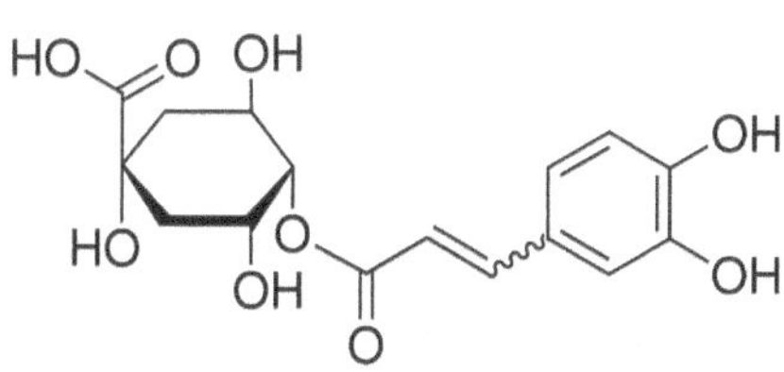

Scheme 14 4-*O*-caffeoylquinic acid

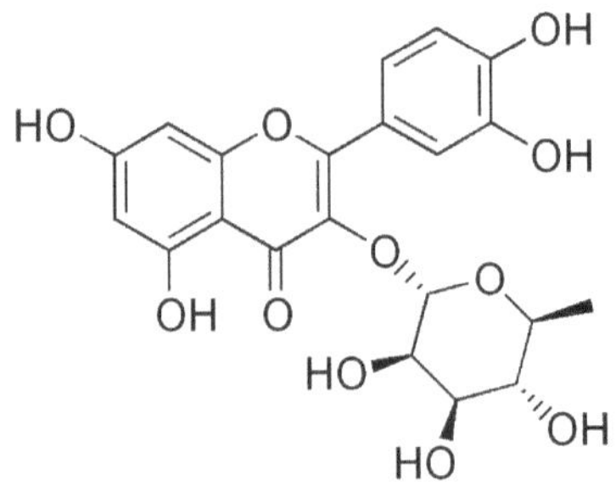

Scheme 15 Quercetin-3-*O*-rhamnoside

2 expression. Phyllanthin (Scheme 16) and hypophyllanthin protected liver from CCl_4 induced liver injury in rats. These two compounds significantly reduced liver injury markers produced by CCl_4 such as glutamate oxaloacetate transaminase and glutamate pyruvate transaminase. Some other phytochemicals from this plant such as protocatechuic acid (Scheme 17), niruriflavone, pectolinarin showed hepatoprotective effect as well.[124]

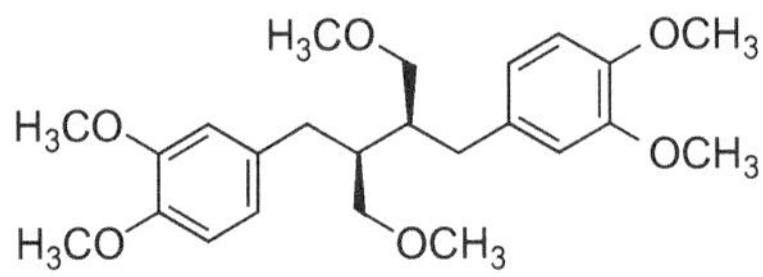

Scheme 16 Phyllanthin

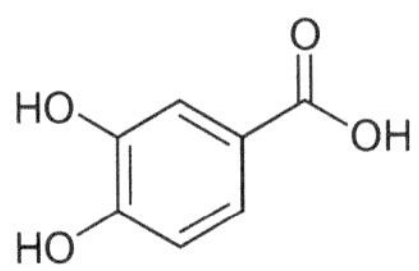

Scheme 17 Protocatechuic acid

6.12 Withaferin A

Withaferin A (Scheme 18) showed protective role against bromobenzene-induced liver and kidney damage in mice which was administered orally (10 mg/kg) to the mice. This experiment showed that the levels of liver and kidney functional markers—lipid peroxidation, and cytokines (TNF-α and IL-1β) were increased and there was a decrease in antioxidant activity in the bromobenzene-treated group of mice. The results validated that bromobenzene caused liver and kidney damage via mitochondrial

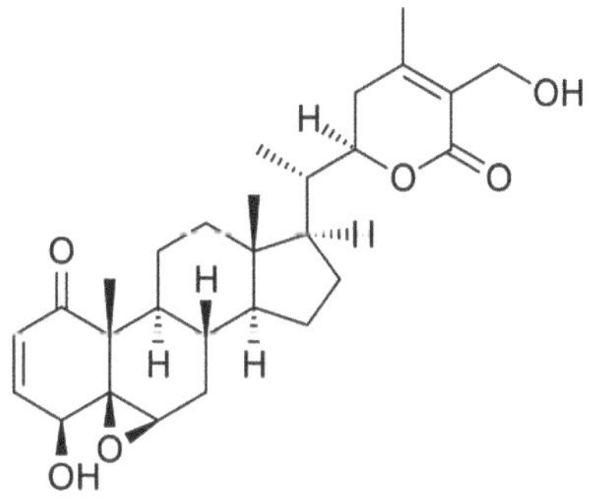

Scheme 18 Withaferin A

pathway and withaferin A provided significant protection against it. Thus, withaferin A may have possible usage in clinical liver and kidney diseases in which oxidative stress and mitochondrial dysfunction may be existent.[125]

6.13 Silibinin A & B

Silibinin A and Silibinin B are called diastereoisomers. Both silibinin A & B (Scheme 19) inhibited Hepatitis C Virus infection in cell culture by variably blocking viral entry, viral fusion, viral RNA and protein synthesis, HCV N5SB RNA dependent RNA polymerase activity, and virus transmission showing protection in hepatic cells.[126,127]

6.14 Vitexin

Administration of vitexin (Scheme 20) decreased the liver levels of ALT and TC in mice with liver injury and reduced the release amounts of 4% dextran

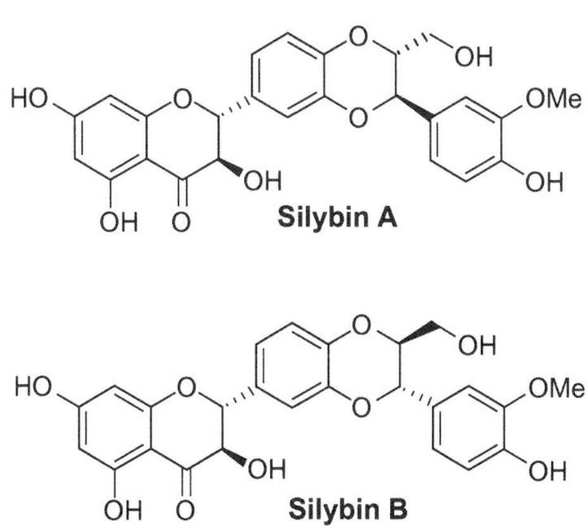

Scheme 19 Silibinin A&B

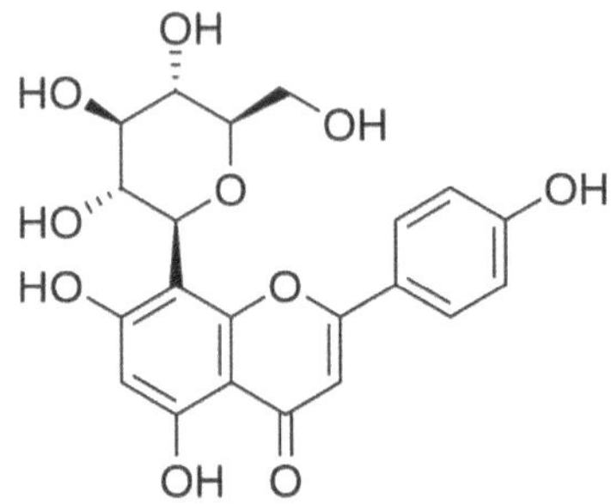

Scheme 20 Vitexin

sodium sulfate (DSS)-induced pro-inflammatory cytokines TNF-α, IL-6 and IL-1β. In the study, vitexin showed inhibitory activity of TLR4/NF-κB signaling pathway induced by DSS (Fig. 7). Vitexin demonstrated hepatoprotective activities against colitis-induced liver injury; it has potential application prospects in the treatment of liver injury induced by ulcerative colitis.[128]

6.15 Glycyrrhizin

Glycyrrhizin (Scheme 21), found in *Glycyrrhiza glabra* has been shown to be effective on viral hepatitis (in particular chronic viral hepatitis) in a

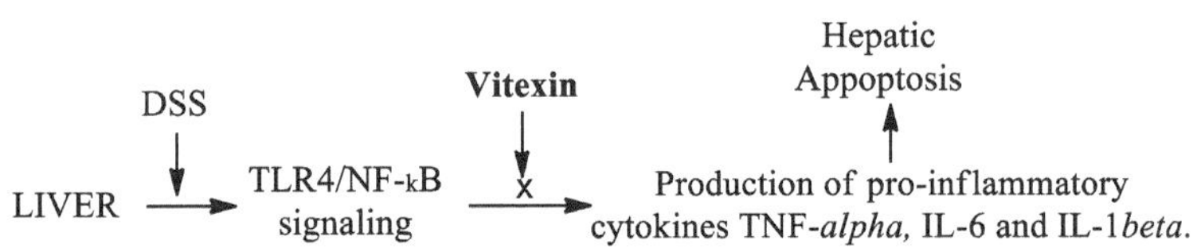

Fig. 7 Schematic diagram for possible mechanism of action of vitexin in dextran sodium sulfate (DSS)-induced liver damage.

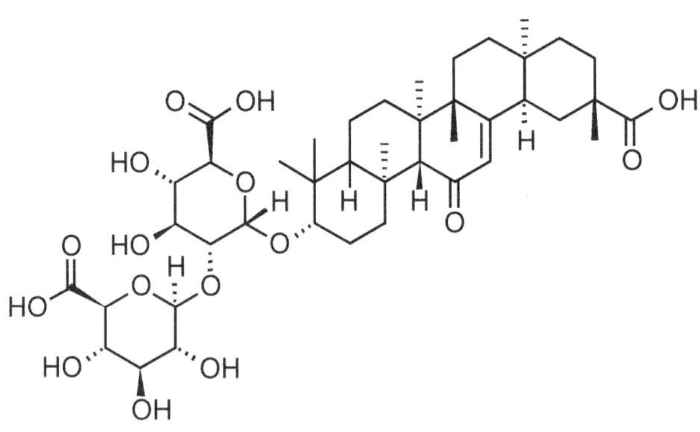

Scheme 21 Glycyrrhizin

double blind study. It was shown that *G. glabra* can stimulate endogenous interferon production.[129] Through *in vitro* studies with human hepatoma cells, Crance et al. showed that glycyrrhizin inhibited penetration of hepatitis A virus likely by altering cell membrane fluidity.[130] In addition, glycyrrhizin has been shown to decrease the release of AST via inhibiting the activation of phospholipase A2, as well as via prevention of changes in the hepatocyte membrane permeability. Also, glycyrrhizin was found to suppress hepatitis B surface antigen (HBsAg) production.[131]

6.16 α- and β-Amyrin

Two triterpenes α- and β-amyrin (Scheme 22) isolated from *Canarium tramdenum* demonstrated hepatoprotective potential against toxic liver injury and it has been suggested that the diminution in oxidative stress and toxic metabolite formation are likely mechanisms involved in their hepatoprotection.[132]

6.17 Curcumin

Curcumin (Scheme 23) is a major compound, which plays a role in the hepatoprotective activity. It is able to inhibit GST and cytochrome P450 activities.[133,134] Mechanisms of the preventive effect of curcumin may be related to its suppressing of pro-inflammatory cytokines, lipid peroxidation products, hepatic stellate cells, and Akt activation (Fig. 8). Curcumin

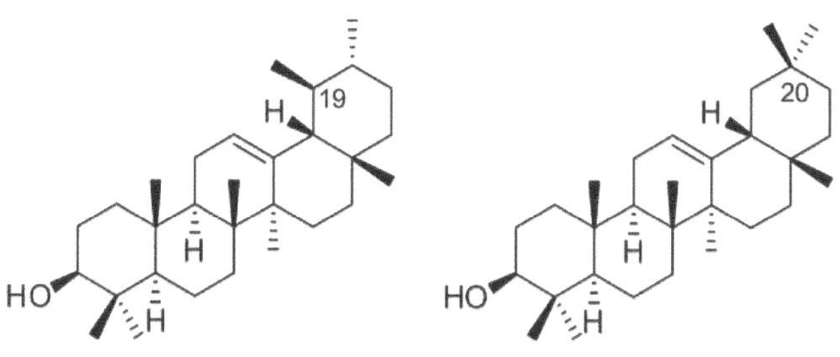

α-amyrin (**1a**) β-amyrin (**1b**)

Scheme 22 α- & β-Amyrins

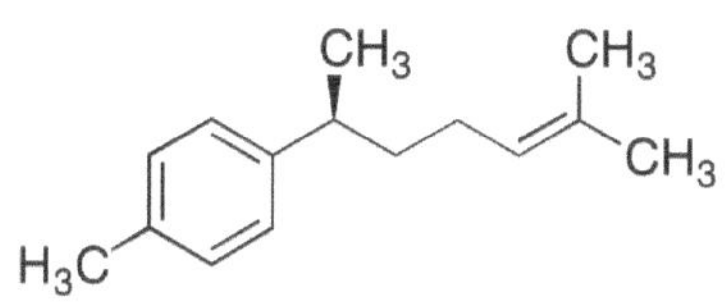

Scheme 23 Curcumin

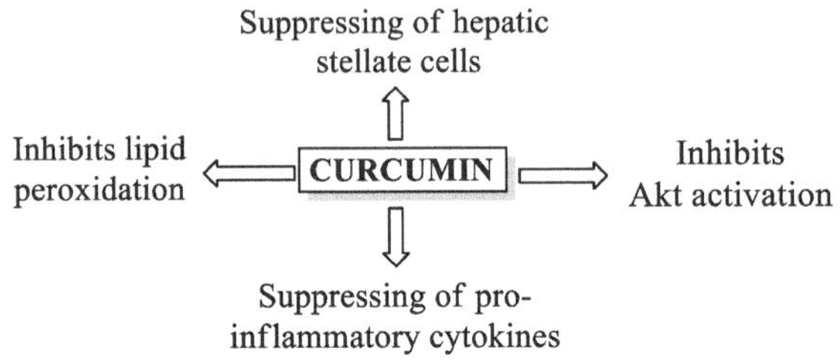

Fig. 8 Proposed preventive effect of curcumin in liver damage.

restored Bcl-2/Bax ratio, thus reducing the PCT-induced hepatocyte apoptosis. Curcumin downregulated the mRNA expression of Bax and upregulated the mRNA expression of Bcl-2.[135]

6.18 Apigenin

Apigenin (Scheme 24) also prevented paracetamol-induced increase in malondialdehyde (MDA) level in animal model. The activities of both CAT (catalase) and GR (glutathione reductase) enzymes following a toxic dose of paracetamol were significantly increased in liver homogenates, which were reversed by apigenin to near control values. So it indicates that apigenin inhibits the level of lipid peroxidation and significantly increases the enzyme antioxidant defense mechanisms in paracetamol-induced hepatotoxicity.[136]

6.19 Kaempferol

The hepatoprotective[86] properties of kaempferol (Scheme 25) against alcohol-induced liver injury were exhibited by attenuating the activity and expression of CYP2E1 by enhancing the protective role of anti-oxidative defense system.[137] Kaempferol and its glycosides induced liver protective activation through STAT3 pathway by inhibiting TNF-α and activating IL-10.[138]

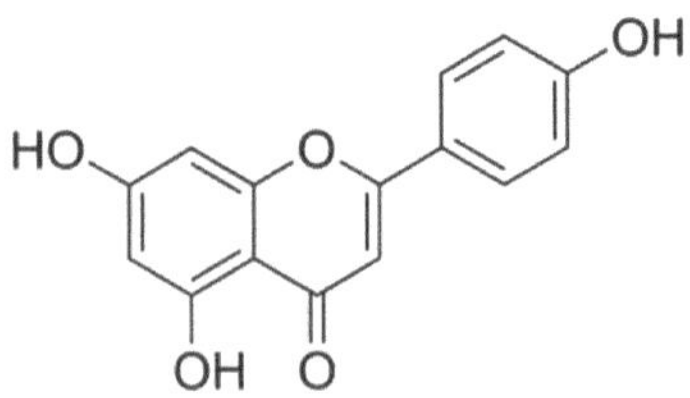

Scheme 24 Apigenin

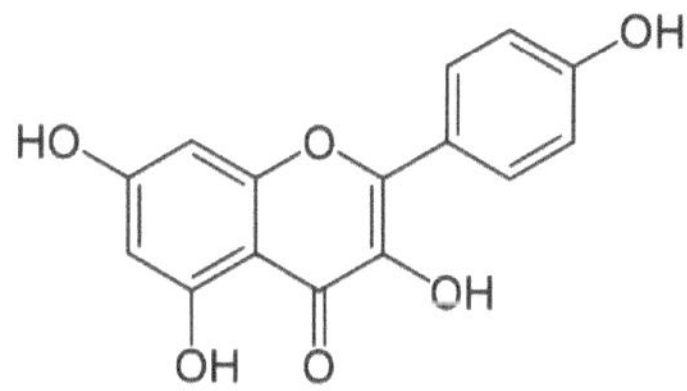

Scheme 25 Kaempferol

6.20 (−)-Epicatechin

(−)-Epicatechin (Scheme 26) decreased the MDA levels in hepatitis rats. The pretreatment of it decreased the production of free radicals in hepatitis rats; this caused downregulation of MDA levels. Thus, (−)-epicatechin seems a potent chain breaking agent, which may prevent reaction between free radical and cell membranes interactions and protect the liver tissue under D-GalN intoxication.[139]

6.21 Quercetin

It has been reported that quercetin (Scheme 27) preserves the structural integrity of the hepatocellular membrane against lethal dose of paracetamol in mice as well as significantly reduces the paracetamol-induced rise in serum GOT and GPT levels.[140] Drugs with calcium channel blocking,[141] antioxidant and free radical scavenging[142] properties are considered useful in liver damage, and hepatoprotective activity of quercetin found in this study may

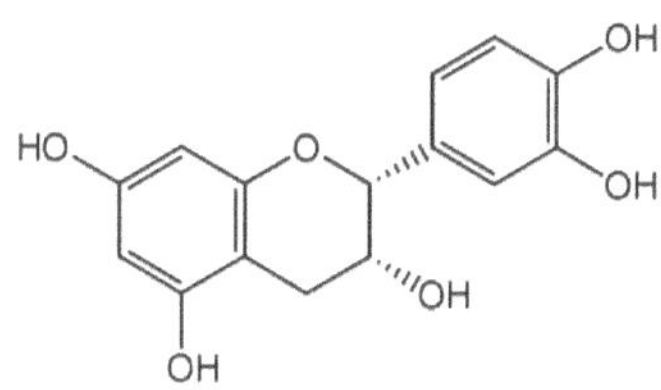

Scheme 26 Epicatechin

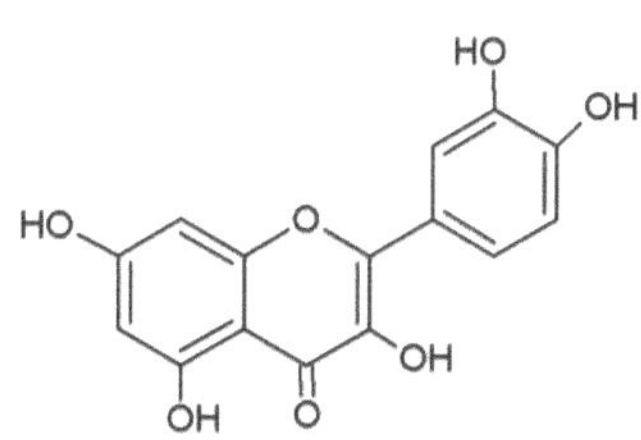

Scheme 27 Quercetin

be attributed to the activities of quercetin reported earlier.[143] Such results indicate that quercetin possesses hepatoprotective activity and support the folkloric use of the plant containing the compound in liver damage.

The impact of quercetin on various types of liver dysfunction has collectively revealed that it modulates the different hallmarks of liver diseases, notably cell death, inflammation and oxidative stress mechanisms. Moreover, the protective mechanism of quercetin has been shown to prevent apoptosis in normal cells through various mechanisms including NF-κB, p53, PPAR-α and PI3K/AKT downregulation (Fig. 9).[144]

6.22 Niranthin

Niranthin (Scheme 28) is a lignin, which significantly decreased the secretion of HBsAg and HBeAg after treatment with niranthin. In DHBV-infected ducklings, it significantly reduced the serum DHBV DNA, HBsAg, HBeAg, ALT and AST. The analysis of the liver pathological changes confirmed the hepatoprotective effect of niranthin.[145]

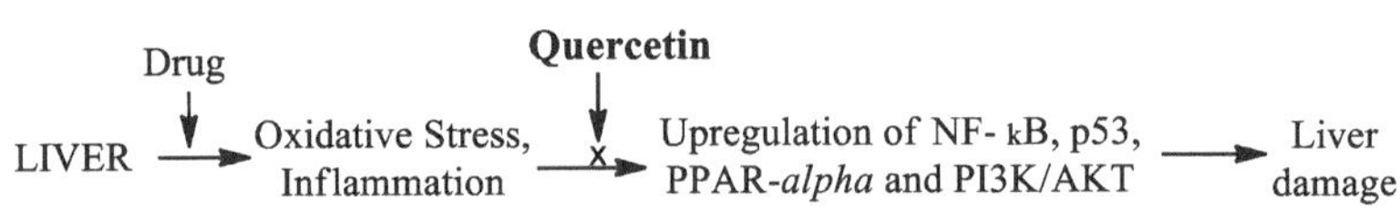

Fig. 9 Schematic diagram for possible mechanism of action of quercetin in liver diseases.

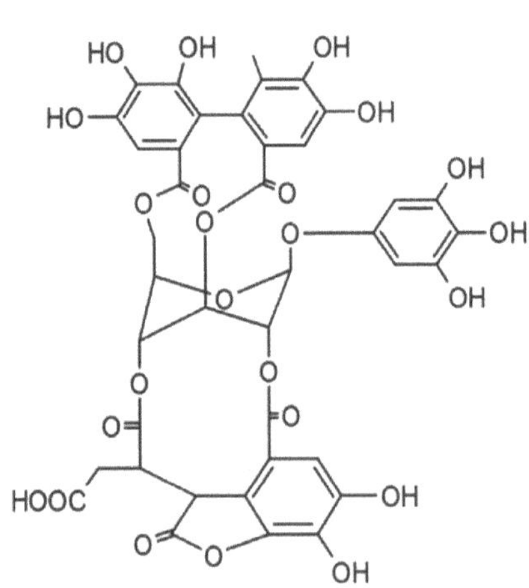

Scheme 28 Geraniin

6.23 Geraniin

It significantly reduced CCl_4 induced lipid peroxidation, increase in amount of glutathione, glutathione reductase and Heme oxygenase-1 (HO-1). On the other hand, it inhibited significant reduction in catalase activity and expression caused by CCl_4 administration. Pretreatment with geraniin (Scheme 29) reduced phosphorylation of translation initiation factor eIF2α at serine 51, caused by CCl_4 exposure and reduced elevated expression of its upstream kinase, Heme-regulated inhibitor (HRI). These results clearly demonstrate hepatoprotective activity of geraniin against CCl_4-induced acute hepatotoxicity via its free radical scavenging and antioxidant activities.[146]

6.24 Gossypetin-3-*O*-β-D-robinobioside

Polyoxygenated flavonol robinobioside such as gossypetin-3-*O*-β-D-robinobioside (Scheme 30) was isolated from the leaves of *Caesalpinia gilliesii* along with 13 known phenolic secondary metabolites. The isolated compounds were characterized using spectroscopic analysis, including 1D and

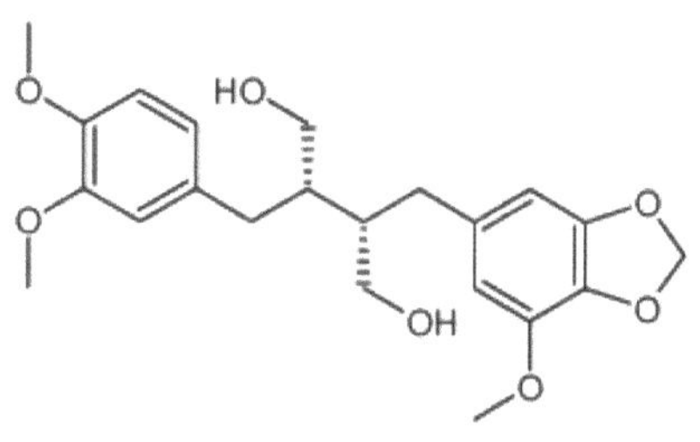

Scheme 29 Niranthin

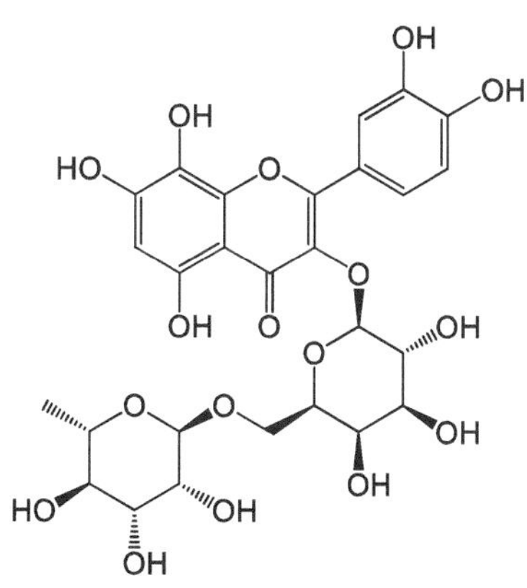

Scheme 30 Gossypetin-3-*O*-b-D-robinobioside

2D NMR and mass spectrometry (MS) analyses. The extract reduced the level of liver damage in CCl_4-induced liver injury in rats. A decrease of the liver biomarkers—aspartate aminotransferase (AST) and alanine aminotransferase (ALT) and an increase of total antioxidant capacity (TAC) levels—were observed similar to the liver protecting drug silymarin.[147]

6.25 Rosmarinic acid

Rosmarinic acid (Scheme 31) demonstrated hepatoprotective effect against lipopolysaccharide (LPS) and D-galactosamine (DGalN)-sensitized mice. The mechanism of rosmarinic acid protection was confirmed by investigating the effects of anti-TNF antibody, SOD, and aminoguanidine. Treatment with rosmarinic acid significantly reduced the elevation of plasma aspartate aminotransferase (AST) levels, as well as had anti-TNF and SOD effect as confirmed by histology. It has been suggested that rosmarinic acid's liver protective effect might be due to scavenging or reducing activities of superoxide or peroxynitirite.[148]

In CCl_4 induced liver toxicity, transforming growth factor-β1 (TGF-β1) drives toward fibrosis, and TNF-α appear to direct toward activation of caspases and apoptosis.[149] CCl_4 destroys not only plasma membrane but also phospholipid bilayer in mitochondria and this triggers caspase-3-dependent apoptosis.[150] Rosmarinic acid exhibits (Fig. 10) *in vitro* and *in vivo* antifibrotic effects on experimental liver fibrosis induced by CCl_4. In the study it has been shown that rosmarinic acid inhibited hepatic stellate cells (HSCs) proliferation, TGF-β1, connective transforming growth factor (CTGF) and α-smooth muscle actin (α-SMA) expression in cultured HSCs (see Fig. 11).[151]

6.26 Capsaicin

Capsaicin (Scheme 32) pretreatment protected mice from Concanavalin A (Con A)-induced hepatic damage and was partially involved in inhibiting

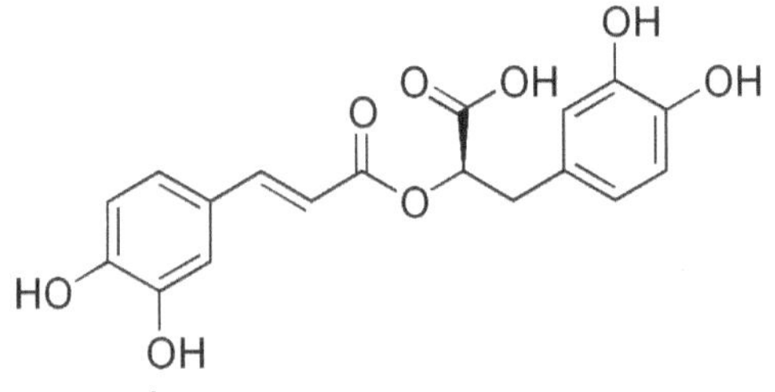

Scheme 31 Rosmarinic acid

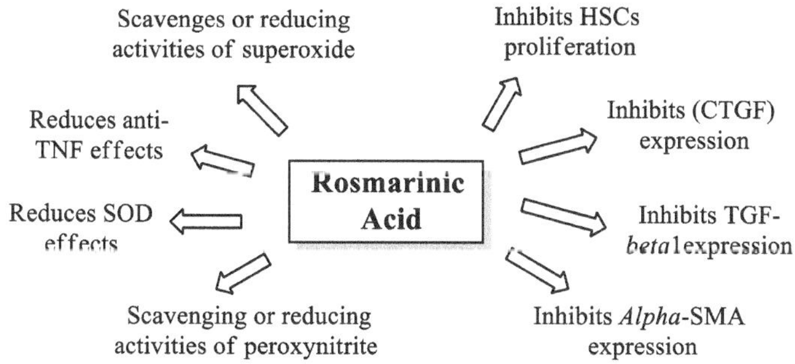

Fig. 10 Schematic diagram for possible mechanism of action of *Rosmarinic acid* in LPS and DGalN-sensitized mice and CCl_4 induced liver toxicity.

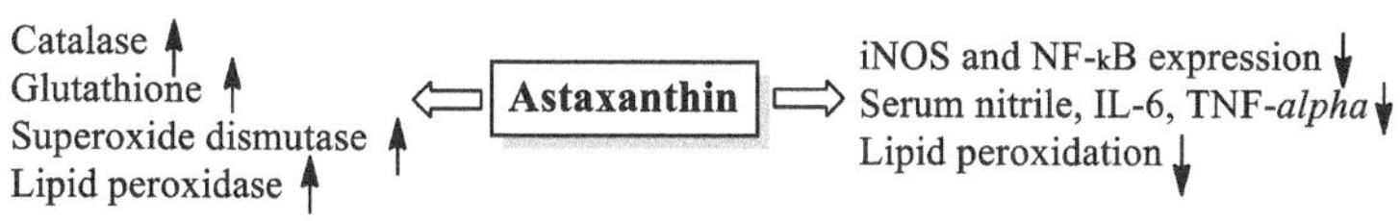

Fig. 11 Schematic diagram for possible mechanism of action of Astaxanthin against CCl_4 induced hepatic damage.

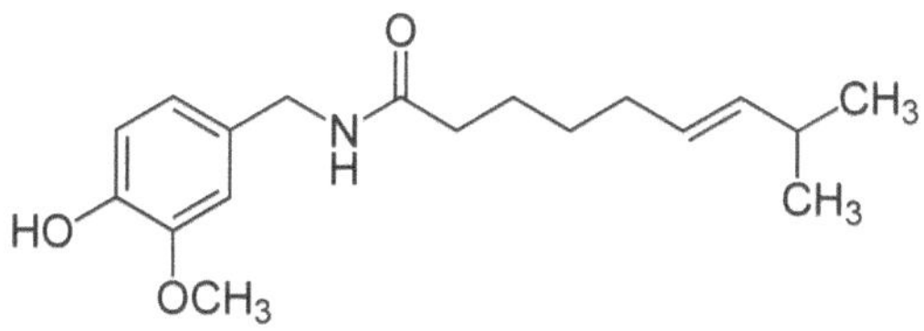

Scheme 32 Capsaicin

hepatocyte apoptosis, oxidative stress, and inflammatory mediators as well as regulating activation and recruitment of intrahepatic leukocytes.[152] Capsaicin confers an hepatoprotective effect, which might be explained partially through its diminishing the generation of MDA, induction of antioxidant systems and inhibition of active caspase-3.[153]

6.27 Apigenin-7-*O*-glucoside

The hepatoprotective activity of apigenin-7-O-glucoside (Scheme 33) is possibly due to its antioxidant properties, acting as scavengers of ROS. The results suggest that apigenin-7-O-glucoside has protective effects against hepatic oxidative injury induced by CCl_4.[154]

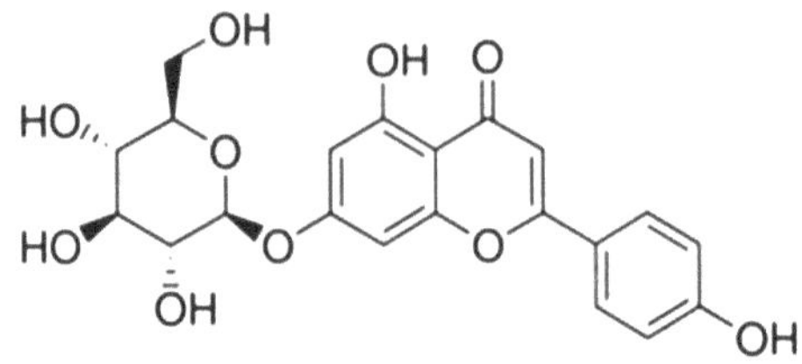

Scheme 33 Apigenin-7-*O*-glucoside

6.28 Ellagic acid

Pretreatment with ellagic acid (Scheme 34) decreases the damage of the liver tissue, significantly decreases the activated Bax and caspase-3-positive hepatocytes and, decreases the biochemical marker levels in D-GalN-induced liver toxicity in rats. It showed better hepatoprotective results in terms of histopathological and biochemical values. It can be concluded that ellagic acid might play an important role in repairing D-GalN-induced liver damage both as a protective and a therapeutic agent.[155] An administration of CCl_4 to normal rats and consequent oral feeding with ellagic acid (50 mg/kg) provided a significant protection against the biochemical alterations in serum and liver produced by CCl_4.[156]

6.29 Astaxanthin

Effect of astaxanthin (Scheme 35) was evaluated for its hepatoprotective activity against CCl_4 induced liver damage. The antioxidant enzymes

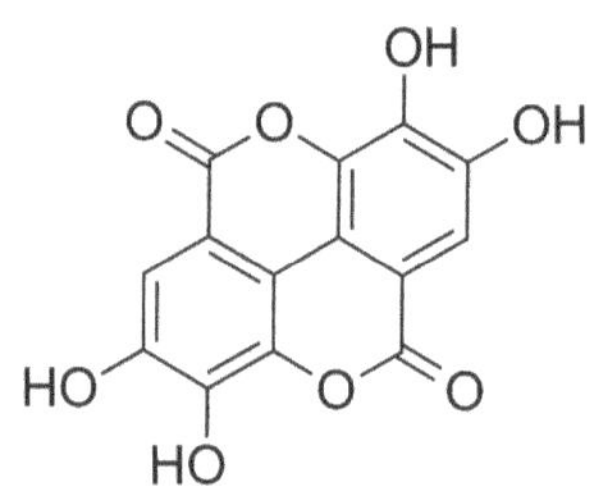

Scheme 34 Ellagic acid

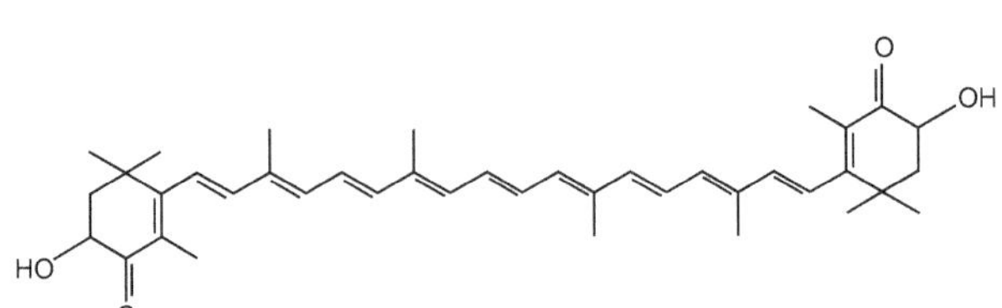

Scheme 35 Astaxanthin (Asteracanthine)

catalase, glutathione, superoxide dismutase and lipid peroxidase levels as well as lipid peroxidation was restored by astaxanthin. The result indicates that it has better hepatoprotection and can be used in pharmaceutical and nutraceutical applications.[157] The liposome-encapsulated astaxanthin exhibited significant amelioration of serum glutamic pyruvic transaminase (GPT), glutamic oxaloacetic transaminase (GOT), blood urea nitrogen (BUN), creatinine (CRE), hepatic malondialdehyde (MDA) and glutathione peroxidase (GSH-Px), IL-6, and hepatic nuclear NF-κB and inducible nitric oxide synthase (iNOS), suggesting that at a 10 mg/kg-day dosage it renders superior hepatoprotective effects in comparison to *N*-acetylcysteine. It indicated that hydrophobic astaxanthin after being properly encapsulated by liposomes improves bioavailability and can also function as potential drug delivery system in treating hepatotoxicity.[158]

6.30 Hesperidin

Hesperidin (Scheme 36) significantly prevented thioacetamide (TAA)-activated Wnt3a/β-catenin and Wnt5a gene expressions pathways of hepatocellular carcinoma (HCC) in rats. It exerted hepatoprotective effect by significantly improving the oxidative imbalance, inflammation and liver function parameters, serum ALT, AST activities, and albumin level. Hesperidin had a prophylactic effect against hepatocarcinogenesis via preventing the induction of both canonical and non-canonical Wnt pathways.[159]

6.31 Resveratrol

Resveratrol (Scheme 37) can serve an essential role in mitigating ethanol-induced hepatic oxidative stress *in vivo* and *in vitro* through protective effects resulting from suppression of lipid peroxidation and accumulation, and the activation of SOD gene expression. It was suggested that PPARγ activation was involved in resveratrol-enhanced SOD gene regulation (Fig. 12).

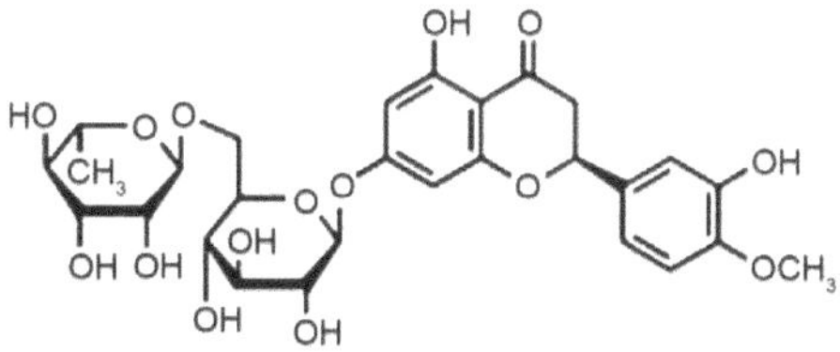

Scheme 36 Hesperidin

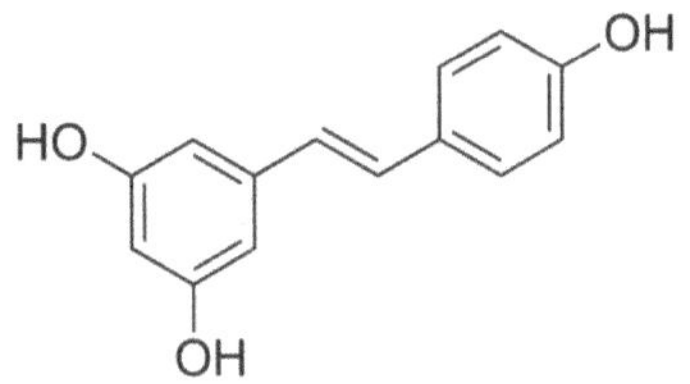

Scheme 37 Resveratrol

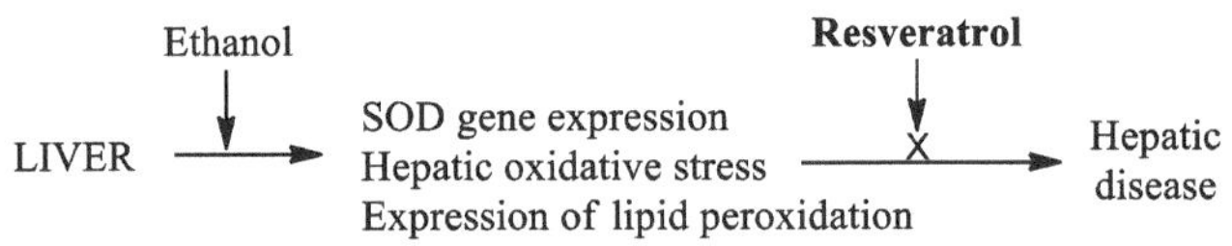

Fig. 12 Schematic diagram for possible mechanism of action of resveratrol in ethanol-induced hepatic damage and liver disease.

The results identify the novel mechanisms underlying the protective actions of resveratrol and provide a novel insight into the prevention of ethanol-induced hepatic damage and liver disease.[160]

7. Conclusion

It can be concluded from the whole literature survey that plant-derived herbal medicines can serve as effective candidates against hepatic disorders.[3] Natural products have been used as medicine in Ayurveda for a long time. Some medicinal plants are hepatoprotective/hepatogenic agents against hepatotoxicity caused by hepatotoxicants. Many researches have been done and more remain to be done on their safe and effective use. Natural products and specially phytoconstituents have become a promising therapeutic alternative and prospective replacement of conventional marketed drug in practice due to their effectiveness, minimal side effects, and protective properties. Furthermore, their dietary nature and availability is a bonus, and gives all the more reason to decline those generally available drugs that also cause toxicity to cells. Remarkable plant constituents like silymarin, silibinin A &B, Piperine, piperlongumine, flavonoids, lawsone, sesquiterpenes, withanoside, withaferin A and so on are pharmacologically tested hepatoprotective agents against prevention of hepatic disorders. Thus far, the natural hepatoprotective products or medicines have been considered non-toxic; however this hypothesis needs to be scientifically evaluated. As a result, appropriate selection of the natural based plants constituents as drug is an utmost obligation. All the important phytoconstituents and their derivatives, though studied and well reported for, have not yet been fully analyzed for their immense therapeutic usage, as there are not enough studies available regarding them. Phytochemicals in plant(s) vary with temperature and cultivation process. Furthermore, variation in the physicochemical properties could result from different modes of production of such plants, including agricultural and environmental factors. Understanding biological function, composition, and therapeutic effects could help prevent adverse effects from long-term administration of such compounds, and develop health promoting properties. It is envisaged from this present review that plant-based constituents can not only reduce the risk of hepatopathy, but can also be a safe substitute that can be used for various hepatotoxicity mediated diseases.

References

1. Ahsan, M. R.; Islam, K. M.; Bulbul, I. J. Hepatoprotective Activity of Methanol Extract of Some Medicinal Plants Against Carbon Tetrachloride-Induced Hepatotoxicity in Rats. *Eur. J. Sci. Res.* **2009**, *37*, 302–310.
2. Ward, F. M.; Daly, M. J. Hepatic disease. In: *Clinical Pharmacy and Therapeutics*; Walker, R., Edwards, C., Eds.; Churchill Livingstone: New York, 1999, pp 195–212.
3. Jharna, B.; Nitin, K.; Rishabha, M.; Pramod, K. S. Hepatoprotective Models and Various Natural Product Used in Hepatoprotective Agents: A Review. *Pharmacogn. Commun.* **2014**, *4*(3), 2–30. https://doi.org/10.5530/pc.2014.3.2.
4. Subramoniam, A.; Pushpangadan, P. Development of Phytomedicine for Liver Diseases. *Indian J. Pharmacol.* **1999**, *31*, 166–175.
5. Shanani, S. Evaluation of Hepatoprotective Efficacy of APCL-A Polyherbal Formulation *In Vivo* In Rats. *Indian Drugs* **1999**, *36*, 628–631.
6. Bodakhe, S. H.; Ram, A. Hepatoprotective Properties of *Bauhinia variegata* Bark Extract. *Yakugaku Zasshi* **2007**, *127*, 1503–1507.
7. Dienstag, J. L.; Isselbacger, K. J. Toxic and drug induced hepatitis. In: *Harrison's Principles of Internal Medicine*, 15th ed.; Kaspe, D. L., Fauci, A. S., Hauser, S. L., Longo, D. L., Jameson, J. L., Loscalzo, J., Eds.; The McGraw-Hill Companies, Inc., 2001;Vol. 2, pp 737–1742. Chapter 296.
8. Dey, P.; Saha, M. R.; Sen, A. Hepatotoxicity and the Present Herbal Hepatoprotective Scenario. *Int. J. Gr. Pharm.* **2013**, 7, 265–273.
9. Pandey, G. Medicinal Plants Against Liver Diseases. *Int. Res. J. Pharm.* **2011**, *2*(5), 115–121.
10. Pandey, G. P. *Pharmacological Studies of Livol® With Special Reference to Liver Function.* MVSc & AH. thesis, JNKVV: Jabalpur, MP, India, 1980;.
11. Pandey, G. P. *Hepatogenic effect of some indigenous drugs on experimental liver damage.* PhD. Thesis, JNKVV: Jabalpur, 1990;.
12. Kumar, C. H.; Ramesh, A.; Suresh Kumar, J. N.; Mohammed, I. B. A Review on Hepatoprotective Activity of Medicinal Plants. *Int. J. Pharm. Sci. Res.* **2011**, *2*(3), 501–515.
13. Nadeem, M.; Dangiya, P. C.; Pasha, K. V.; Imara, M.; Balani, D. K.; Vohora, S. B. Hepatoprotective Activity of *Solanum nigrum* Fruits. *Fitoterapia* **1997**, *58*, 245–254.

14. Saha P, Das A, Talukdar RN, Sarker SD, Nahar L, Sahu J. Dutta MC. (2019). Role of Natural Phenolics in Hepatoprotection: A Mechanistic Review and Analysis of Regulatory Network of Associated Genes. Front. Pharmacol.; 10: 509. doi:https://doi.org/10.3389/fphar.2019.00509.
15. Finkelstein, E. A.; Khavjo, O. A.; Thompson, H.; Trogdon, J. G.; Pan, L.; Sherry, B.; Dietz, W. Obesity and Severe Obesity Forecasts Through 2030. *Am. J. Prevent. Med.* **2012**, *42*, 563–570. https://doi.org/10.1016/j.amepre.2011.10.026.
16. Blachier, M.; Leleu, H.; Peck-Radosavljevic, M.; Valla, D. C.; Roudot-Thoraval, F. The Burden of Liver Disease in Europe: A Review of Available Epidemiological Data. *J. Hepatol.* **2013**, *58*(3), 593–608.
17. Effong, K.; Osinowo, A.; Pring, A.; Verne, J. *Death From Liver Disease: Implication for End of Life Care in England.* National End of Life Care Intelligence Network: Leicester, 2012. http://ncbi.nih.gov/pmc/articles/PMC5369469/pdf. access on 12 April 2020.
18. Farghali, H.; Canová, N. K.; Zakhari, S. Hepatoprotective properties of extensively studied medicinal plant active constituents: possible common mechanisms. *Pharm. Biol.* **2015**, *53*(6), 781–791.
19. Sanchez-Valle, V.; Chavez-Tapia, N. C.; Uribe, M.; Mendez-Sanchez, N. Role of Oxidative Stress and Molecular Changes in Liver Fibrosis: A Review. *Curr. Med. Chem.* **2012**, *19*, 4850–4860.
20. Alonso, F. T.; Garmendia, M. L.; de Aguirre, M.; Searle, J. Mortality Trend From Liver Cirrhosis in Chile From 1990 to 2007. *Rev. Med. Chile* **2010**, *138*, 1253–1258.
21. Nobili, V.; Carter-Kent, C.; Feldstein, A. E. The Role of Lifestyle Changes in the Management of Chronic Liver Disease. *BMC Med.* **2011**, *9*, 70. https://doi.org/10.1186/1741-7015-9-70.
22. Torres, D. M.; Williams, C. D.; Harrison, S. A. Features, Diagnosis, and Treatment of Nonalcoholic Fatty Liver Disease. *Clin. Gastroenterol. Hepatol.* **2012**, *10*, 837–858.
23. Dominiczak, M. H. Lipids, lipoproteins. In: *Medical Biochemistry*; Baynes, J. W., Dominiczak, M. H., Eds.; Elsevier Mosby: Philadelphia, 2005 pp 234–242.
24. Ekaidem, I. S.; Akpan, H. D.; Usoh, I. F.; Etim, O. E.; Ebong, P. E. Effects of Ethanolic Extract of Azadirachta indica Leaves on Lipid Peroxidation and Serum Lipids of Diabetic Wistar Rats. *Acta Biol. Szegeden.* **2007**, *51*, 17–20.
25. Li, M.; He, Y.; Zhou, Z.; Ramirez, T.; Gao, Y.; Gao, Y.; Ross, R. A.; Cao, H.; Cai, Y.; Xu, M.; Feng, D.; Zhang, P.; Liangpunsakul, S.; Gao, B. MicroRNA-223 Ameliorates Alcoholic Liver Injury by Inhibiting the IL-6–p47phox–Oxidative Stress Pathway in Neutrophils. *Gut* **2016**, *66*, 705–715. https://doi.org/10.1136/gutjnl-2016-311861.
26. Madrigal-Santillán, E.; Bautista, M.; Gayosso-De-Lucio, J. A.; Reyes-Rosales, Y.; Posadas-Mondragón, A.; Morales-González, Á.; Soriano-Ursúa, M. A.; García-Machorro, J.; Madrigal-Bujaidar, E.; Álvarez-González, I.; Morales-González, J. A. Hepatoprotective Effect of Geranium Schiedeanum Against Ethanol Toxicity During Liver Regeneration. *World J. Gastroenterol.* **2015**, *21*, 7718. https://doi.org/10.3748/wjg.v21.i25.7718.
27. Marshall, J. Alcohol: pharmacokinetics and pharmacodynamics. In: *The SAGE Handbook of Drug & Alcohol Studies: Biological Approaches*; Wolff, K., White, J., Karch, S., Eds.; SAGE Publications Ltd., 2016; p 63. https://doi.org/10.4135/9781473922143
28. Srivastava, R.; Srivastava, P. Hepatotoxicity and the Role of Some Herbal Hepatoprotective Plants in Present Scenario. *Int. J. Dig. Dis.* **2018**, *4*, 2. https://doi.org/10.4172/2472-1891.100034 (2:3).
29. Maity, T. K.; Veerendra, Y.; Nayak, T. G.; Rajahrigam, D.; Sengupta, P.; Maiti, B. C.; Deepak, K. D. Evaluation of Hepatoprotective and Antioxidant Activity of Ichnocarpus Frutescens (Linn.) R.Br. on Paracetamol-Induced Hepatotoxicity in Rats. *Trop. J. Pharm. Res.* **2007**, *6*(3), 755–765.

30. Fraschini, F.; Demartini, G.; Esposti, D. Pharmacology of Silymarin. *Clin. Drug Investig.* **2002**, *22*, 51–65.
31. Handa, S. S.; Sharma, A.; Chakraborti, K. K. Natural Products and Plants as Liver Protecting Drug. *Fitoterapia* **1986**, *57*, 307–351.
32. Jannu, V.; Baddam, P. G.; Boorgula, A. K.; Jambula, S. R. A Review on Hepatoprotective Plants. *Int. J. Drug Dev. Res.* **2012**, *4*, 1–8.
33. Dhilleswara Rao, V.; Dattatreya, A.; Dan, M. M.; Sarangi, T.; Sasidhar, K.; Rahul, J. Translational Approach in Emerging Infectious Disease Treatment: An Update. *Biomed. Res.* **2017**, *28*, 5678–5686.
34. Zeng, X.; Guo, F.; Ouyang, D. A Review of the Pharmacology and Toxicology of Aucubin. *Fitoterapia* **2020**, *140*, 104443. https://doi.org/10.1016/j.fitote.2019.104443.
35. Khan, F.; Sarker, M. M. R.; Ming, L. C.; Mohamed, I. N.; Zhao, C.; Sheikh, B. Y.; Tsong, H. F.; Rashid, M. A. Comprehensive Review on Phytochemicals, Pharmacological and Clinical Potentials of *Gymnema sylvestre*. *Front. Pharmacol.* **2019**, *10*, 1223. https://doi.org/10.3389/fphar.2019.01223.
36. Francini-Pesenti, F.; Spinella, P.; Calò, L. A. Potential Role of Phytochemicals in Metabolic Syndrome Prevention and Therapy. *Diabetes Metab. Syndr. Obes.* **2019**, *12*, 1987–2002. https://doi.org/10.2147/DMSO.S214550.
37. Yadav, V.; Krishnan, A.; Vohora, D. A Systematic Review on *Piper longum* L.: Bridging Traditional Knowledge and Pharmacological Evidence for Future Translational Research. *J. Ethnopharmacol.* **2020**, *247*, 112255. https://doi.org/10.1016/j.jep.2019.112255.
38. Jiang, Y. Q.; Liu, E. H. Microcos Paniculata: A Review on Its Botany, Traditional Uses, Phytochemistry and Pharmacology. *Chin. J. Nat. Med.* **2019**, *17*(8), 561–574. https://doi.org/10.1016/S1875-5364(19)30058-5.
39. Ji, J.; Wang, Q.; Wang, M.; Chen, J.; Li, X. Chemical Constituents From the Stems of *Securidaca inappendiculata* Hassk. *Fitoterapia* **2019**, *137*, 104271. https://doi.org/10.1016/j.fitote.2019.104271.
40. Wang, Y. G.; Ren, J.; Ma, J.; Yang, J. B.; Ji, T.; Wang, A. G. Bioactive Cembrane-Type Diterpenoids From the Gum-Resin of *Boswellia carterii*. *Fitoterapia* **2019**, *137*, 104263. https://doi.org/10.1016/j.fitote.2019.104263.
41. Arruda, H. S.; Pastore, G. M. Araticum (*Annona crassiflora* Mart.) as a Source of Nutrients and Bioactive Compounds for Food and Non-food Purposes: A Comprehensive Review. *Food Res. Int.* **2019**, *123*, 450–480.
42. Shang, A.; Cao, S. Y.; Xu, X. Y.; Gan, R. Y.; Tang, G. Y.; Corke, H. Bioactive Compounds and Biological Functions of Garlic (*Allium sativum* L.). *Foods* **2019**, *8*(7), 246. https://doi.org/10.3390/foods8070246.
43. Parvez, M. K.; Al-Dosari, M. S.; Arbab, A. H.; Alam, P.; Alsaid, M. S.; Khan, A. A. Hepatoprotective Effect of *Solanum surattense* Leaf Extract Against Chemical-Induced Oxidative and Apoptotic Injury in Rats. *BMC Complement. Altern. Med.* **2019**, *19*(1), 154. https://doi.org/10.1186/s12906-019-2553-1.
44. Alagan, A.; Jantan, I.; Kumolosasi, E.; Ogawa, S.; Abdullah, M. A.; Azmi, N. Protective Effects of *Phyllanthus amarus* Against Lipopolysaccharide-Induced Neuroinflammation and Cognitive Impairment in Rats. *Front. Pharmacol.* **2019**, *10*, 632.
45. Ullah, M. A.; Tungmunnithum, D.; Garros, L.; Hano, C.; Abbasi, B. H. Monochromatic Lights-Induced Trends in Antioxidant and Antidiabetic Polyphenol Accumulation in In Vitro Callus Cultures of Lepidium Sativum L. *J. Photochem. Photobiol. B* **2019**, *196*, 111505.
46. Tsakiroglou, P.; VandenAkker, N. E.; Del Bo', C.; Riso, P.; Klimis-Zacas, D. Role of Berry Anthocyanins and Phenolic Acids on Cell Migration and Angiogenesis: An Updated Overview. *Nutrients* **2019**, *11*(5), 1075.

47. Dong, J.; Peng, X.; Lu, S.; Zhou, L.; Qiu, M. Hepatoprotective Steroids From Roots of *Cynanchum otophyllum*. *Fitoterapia* **2019**, *136*, 104171.
48. Zakaria, Z. A.; Mahmood, N. D.; Omar, M. H.; Taher, M.; Basir, R. Methanol Extract of Muntingia Calabura Leaves Attenuates CCl_4-Induced Liver Injury: Possible Synergistic Action of Flavonoids and Volatile Bioactive Compounds on Endogenous Defense System. *Pharm. Biol.* **2019**, *57*(1), 335–344.
49. Kil, H. W.; Rho, T.; Yoon, K. D. Phytochemical Study of Aerial Parts of *Leea asiatica*. *Molecules* **2019**, *24*(9), 1733.
50. Goodla, L.; Manubolu, M.; Pathakoti, K.; Jayakumar, T.; Sheu, J. R.; Fraker, M.; Tchounwou, P. B.; Poondamalli, P. R. Protective Effects of *Ammannia baccifera* Against CCl_4-Induced Oxidative Stress in Rats. *Int. J. Environ. Res. Public Health* **2019**, *16*(8), 1440.
51. Fahmi, A.; Hassanen, N.; Abdur-Rahman, M.; Shams-Eldin, E. Phytochemicals, Antioxidant Activity and Hepatoprotective Effect of Ginger (*Zingiber officinale*) on Diethylnitrosamine Toxicity in Rats. *Biomarkers* **2019**, *24*(5), 436–447.
52. Mariadoss, A. V. A.; Vinyagam, R.; Rajamanickam, V.; Sankaran, V.; Venkatesan, S.; David, E. Pharmacological Aspects and Potential Use of Phloretin: A Systemic Review. *Mini Rev. Med. Chem.* **2019**, *19*(13), 1060–1067.
53. Quan, N. V.; Xuan, T. D.; Tran, H. D.; Thuy, N. T. D.; Trang, L. T.; Huong, C. T.; Andriana, Y.; Tuyen, P. T. Antioxidant, α-Amylase and α-Glucosidase Inhibitory Activities and Potential Constituents of *Canarium tramdenum* Bark. *Molecules* **2019**, *24*(3), 605.
54. Wang, X. J.; Zhang, Q.; Peng, Y. R.; Li, L.; Qu, J.; Liu, Y. B.; Xu, S.; Ma, S. G.; Li, Y.; Zou, Z. M.; Wang, R. B.; Yu, S. S. Two Azafluoranthene Alkaloids and a Phytoecdysone From the Stems of *Cyclea barbata*. *J. Asian Nat. Prod. Res.* **2019**, *21*(3), 217–226.
55. Emam, M.; El Raey, M. A.; El-Haddad, A. E.; El Awdan, S. A.; Rabie, A. M.; El-Ansari, M. A.; Sobeh, M.; Osman, S. M.; Wink, M. A New Polyoxygenated Flavonol Gossypetin-3-*O*-*β*-d-Robinobioside From *Caesalpinia gilliesii* (Hook.) D. Dietr. And in Vivo Hepatoprotective, Anti-Inflammatory, and Anti-Ulcer Activities of the Leaf Methanol Extract. *Molecules* **2018**, *24*(1), 138.
56. Lin, H. Y.; Chang, T. C.; Chang, S. T. A Review of Antioxidant and Pharmacological Properties of Phenolic Compounds in *Acacia confusa*. *J. Tradit. Complement. Med.* **2018**, *8*(4), 443–450.
57. Abdelhafez, O. H.; Fawzy, M. A.; Fahim, J. R.; Desoukey, S. Y.; Krischke, M.; Mueller, M. J.; Abdelmohsen, U. R. Hepatoprotective Potential of Malvaviscus Arboreus Against Carbon Tetrachloride-Induced Liver Injury in Rats. *PLoS One* **2018**, *13*(8), e0202362.
58. Oniga, I.; Puscas, C.; Silaghi-Dumitrescu, R.; Olah, N. K.; Sevastre, B.; Marica, R.; Marcus, I.; Sevastre-Berghian, A. C.; Benedec, D.; Pop, C. E.; Hanganu, D. *Origanum vulgare* ssp. *vulgare*: Chemical Composition and Biological Studies. *Molecules* **2018**, *23*(8), 2077.
59. Das, M.; Basu, S.; Banarjee, B.; Sen, A.; Jana, K.; Datta, G. Hepatoprotective Effects of Green Capsicum Annum Against Ethanol Induced Oxidative Stress, Inflammation and Apoptosis in Rats. *J. Ethnopharmacol.* **2018**, *227*, 69–81.
60. Sethiya, N. K.; Ahmed, N. M.; Shekh, R. M.; Kumar, V.; Kumar Singh, P.; Kumar, V. Ethnomedicinal, Phytochemical and Pharmacological Updates on Hygrophila Auriculata (Schum.) Hiene: An Overview. *J. Integr. Med.* **2018**, *16*(5), 299–311.
61. Gao, Y.; Kuok, K.; Jin, Y.; Wang, R. Biomedical Applications of *Aloe vera*. *Crit. Rev. Food Sci. Nutr.* **2010**, *59*(Suppl. 1), S244–S256.

62. Abiri, R.; Silva, A. L. M.; de Mesquita, L. S. S.; de Mesquita, J. W. C.; Atabaki, N.; de Almeida, E. B., Jr.; Shaharuddin, N. A.; Malik, S. Towards a Better Understanding of Artemisia Vulgaris: Botany, Phytochemistry, Pharmacological and Biotechnological Potential. *Food Res. Int.* **2018**, *109*, 403–415.
63. Dutta, S.; Chakrabarty, A. K.; Dey, P.; Kar, P.; Guha, P.; Sen, S.; Kumar, A.; Sen, A.; Chaudhuri, T. K. Amelioration of CCl_4 Induced Liver Injury in Swiss Albino Mice by Antioxidant Rich Leaf Extract of *Croton bonplandianus* Baill. *PLoS One* **2018**, *13*(4), e0196411.
64. Mykhailenko, O.; Kovalyov, V.; Kovalyov, S.; Toryanik, E.; Osolodchenko, T.; Buidin, Y. Fatty Acid Composition of Lipids of *Iris sibirica*. *Ceska Slov. Farm. Spring* **2018**, *66*(5), 220–226.
65. Manandhar, B.; Paudel, K. R.; Sharma, B.; Karki, R. J. Phytochemical Profile and Pharmacological Activity of *Aegle marmelos* Linn. *Integr. Med.* **2018**, *16*(3), 153–163.
66. Kazmi, S. T. B.; Majid, M.; Maryam, S.; Rahat, A.; Ahmed, M.; Khan, M. R.; Haq, I. U. Quercus Dilatata Lindl. Ex Royle Ameliorates BPA Induced Hepatotoxicity in Sprague Dawley Rats. *Biomed. Pharmacother.* **2018**, *102*, 728–738.
67. Kou, X.; Li, B.; Olayanju, J. B.; Drake, J. M.; Chen, N. Nutraceutical or Pharmacological Potential of *Moringa oleifera* Lam. *Nutrients* **2018**, *10*(3), 343.
68. Islam, M. T.; Tabrez, S.; Jabir, N. R.; Ali, M.; Kamal, M. A.; da Silva, A. L.; De Oleveira Santos, J. V.; Da Mata, A. M. O. F.; De Aguiar, R. P. S.; de Carvalho Melo Cavalcante, A. A. An Insight Into the Therapeutic Potential of Major Coffee Components. *Curr. Drug Metab.* **2018**, *19*(6), 544–556.
69. Zakaria, Z. A.; Mahmood, N. D.; Mamat, S. S.; Nasir, N.; Omar, M. H. Endogenous Antioxidant and LOX-Mediated Systems Contribute to the Hepatoprotective Activity of Aqueous Partition of Methanol Extract of Muntingia calabura L. Leaves against Paracetamol Intoxication. *Front. Pharmacol.* **2018**, *8*, 982.
70. El Hawary, S.; El Sayed, A.; Helmy, M. W.; El Naggar, E. M. B.; Marzouk, H. S.; Bassam, S. M. DNA Fingerprinting, Biological and Chemical Investigation of Certain *Yucca* Species. *Nat. Prod. Res.* **2018**, *32*(21), 2617–2620.
71. Sobeh, M.; Mahmoud, M. F.; Sabry, O. M.; Abel, R.; Dmirieh, M.; El-Shazly, A. M.; Wink, M. HPLC-PDA-MS/MS Characterization of Bioactive Secondary Metabolites From *Turraea fischeri* Bark Extract and its Antioxidant and Hepatoprotective Activities In Vivo. *Molecules* **2017**, *22*(12), 2089.
72. Lakshmi, T.; Ezhilarasan, D.; Nagaich, U.; Vijayaragavan, R. *Acacia catechu* Ethanolic Seed Extract Triggers Apoptosis of SCC-25 Cells. *Pharmacogn. Mag.* **2017**, *13*(Suppl. 3), S405–S411.
73. Rjeibi, I.; Feriani, A.; Ben Saad, A.; Ncib, S.; Sdayria, J.; Saidi, I.; Souid, S.; Hfaiedh, N.; Allagui, M. S. Phytochemical Characterization and Bioactivity of Lycium Europaeum: A Focus on Antioxidant, Antinociceptive, Hepatoprotective and Nephroprotective Effects. *Biomed. Pharmacother.* **2017**, *95*, 1441–1450.
74. SL Chaudhary, S RSR. (2017). Hydroalcoholic Extract of *Stevia rebaudiana* Bert. Leaves and Stevioside Ameliorates Lipopolysaccharide Induced Acute Liver Injury in Rats. Biomed. Pharmacother.; 95: 1040–1050.
75. Das, N.; Atanasov, A. G.; Deb, P. K.; Mocan, A.; Nabavi, S. M.; Ghosh, R.; Dinda, B. Hepatoprotective Naphthalene Diglucoside From *Neanotis wightiana* Aerial Parts. *Phytomedicine* **2017**, *33*, 14–20.
76. Al Zarzour, R. H.; Ahmaed, M.; Asmawi, M. Z.; Kaur, G.; Saeed, M. A. A.; Al-Mansoub, M. A.; Saghir, S. A. M.; Usman, N. S.; Al-Dulaimi, D. W.; Yam, M. F. Phyllanthus Niruri Standardized Extract Alleviates the Progression of Non-Alcoholic Fatty Liver Disease and Decreases Atherosclerotic Risk in Sprague-Dawley Rats. *Nutrients* **2017**, *9*(7), 766.

77. Wang, Z.; He, C.; Peng, Y.; Chen, F.; Xiao, P. Origins, Phytochemistry, Pharmacology, Analytical Methods and Safety of Cortex Moutan (Paeonia Suffruticosa Andrew): A Systematic Review. *Molecules* **2017**, *22*(6), 946.
78. Kaur, V.; Kumar, M.; Kaur, P.; Kaur, S.; Singh, A. P.; Kaur, S. Hepatoprotective Activity of *Butea monosperma* Bark Against Thioacetamide-Induced Liver Injury in Rats. *Biomed. Pharmacother.* **2017**, *89*, 332–341.
79. Huang, C. Z.; Tung, Y. T.; Hsia, S. M.; Wu, C. H.; Yen, G. C. The Hepatoprotective Effect of Phyllanthus emblica L. Fruit on High Fat Diet-Induced Non-alcoholic Fatty Liver Disease (NAFLD) in SD Rats. *Food Funct.* **2017**, *8*(2), 842–850.
80. Tiwary, B. K.; Dutta, S.; Dey, P.; Hossain, M.; Kumar, A.; Bihani, S.; Nanda, A. K.; Chaudhuri, T. K.; Chakraborty, R. Radical Scavenging Activities of *Lagerstroemia speciosa* (L.) Pers. Petal Extracts and Its Hepato-Protection in CCl4-Intoxicated Mice. *BMC Complement. Altern. Med.* **2017**, *17*(1), 55.
81. Miraj, S.; Alesaeidi, S. A Systematic Review Study of Therapeutic Effects of *Matricaria recuitta chamomile* (Chamomile). *Electron Physician* **2016**, *8*(9), 3024–3031.
82. Lazarova, I.; Simeonova, R.; Vitcheva, V.; Kondeva-Burdina, M.; Gevrenova, R.; Zheleva-Dimitrova, D.; Zengin, G.; Danchev, N. D. Hepatoprotective and Antioxidant Potential of *Asphodeline lutea* (L.) Rchb. Roots Extract in Experimental Models In Vitro/In Vivo. *Biomed. Pharmacother.* **2016**, *83*, 70–78.
83. Rjeibi, I.; Ben Saad, A.; Hfaiedh, N. Oxidative Damage and Hepatotoxicity Associated With Deltamethrin in Rats: The Protective Effects of *Amaranthus spinosus* Seed Extract. *Biomed. Pharmacother.* **2016**, *84*, 853–860.
84. Dey, P.; Dutta, S.; Biswas-Raha, A.; Sarkar, M. P.; Chaudhuri, T. K. Haloalkane Induced Hepatic Insult in Murine Model: Amelioration by Oleander Through Antioxidant and Anti-Inflammatory Activities, an In Vitro and In Vivo Study. *BMC Complement. Altern. Med.* **2016**, *16*(1), 280.
85. Hashim, Y. Z.; Kerr, P. G.; Abbas, P.; Mohd, S. H. *Aquilaria* spp. (Agarwood) as Source of Health Beneficial Compounds: A Review of Traditional Use, Phytochemistry and Pharmacology. *J. Ethnopharmacol.* **2016**, *189*, 331–360.
86. Hassani, F. V.; Shikrani, K.; Hosseinzadeh, H. Rosemary (*Rosmarinus officinalis*) as a Potential Therapeutic Plant in Metabolic Syndrome: A Review. *Naunyn-Schmiedeberg's Arch. Pharmacol.* **2016**, *389*(9), 931–949.
87. Caprioli, G.; Alunno, A.; Beghelli, D.; Bianco, A.; Bramucci, M.; Frezza, C.; Iannarelli, R.; Papa, F.; Quassinti, L.; Sagratini, G.; Tirillini, B.; Venditti, A.; Vittori, S.; Maggi, F. Polar Constituents and Biological Activity of the Berry-Like Fruits from *Hypericum androsaemum* L. *Front. Plant Sci.* **2016**, 7, 232.
88. Acharya, N.; Acharya, S.; Shah, U.; Shah, R.; Hingorani, L. A Comprehensive Analysis on *Symplocos racemosa* Roxb.: Traditional Uses, Botany, Phytochemistry and Pharmacological Activities. *J. Ethnopharmacol.* **2016**, *181*, 236–251.
89. Devkar, S. T.; Kandhare, A. D.; Zanwar, A. A.; Jagtap, S. D.; Katyare, S. S.; Bodhankar, S. L.; Hegde, M. V. Hepatoprotective Effect of Withanolide-Rich Fraction in Acetaminophen-Intoxicated Rat: Decisive Role of TNF-α, IL-1β, COX-II and iNOS. *Pharm. Biol.* **2016**, *54*(11), 2394–2403.
90. Elgengaihi, S.; Mossa, A. T.; Refaie, A. A.; Abouubaker, D. Hepatoprotective Efficacy of *Cichorium intybus* L. Extract Against Carbon Tetrachloride-Induced Liver Damage in Rats. *J. Diet. Suppl.* **2016**, *13*(5), 570–584.
91. Khalaf-Allah, A.-R.; El-Gengaihi, S. E.; Hamed, M. A.; Zahran, H. G.; Mohammed, M. A. Chemical Composition of Golden Berry Leaves Against Hepato-Renal Fibrosis. *J. Diet. Suppl.* **2016**, *13*(4), 378–392.
92. Darvin, S. S.; Toppo, E.; Esakkimuthu, S.; Krishna, T. P. A.; Ceasar, S. A.; Stalin, A.; Balakrishna, K.; Muniappan, N.; Pazhanivel, N.; Mahaprabhu, R.; Paulraj, M. G.; Pandikumar, P.; Ignacimuthu, S.; Al-Dhabi, N. A. Hepatoprotective Effect of

Bisbenzylisoquinoline Alkaloid Tiliamosine From *Tiliacora racemosa* in High-Fat Diet/Diethylnitrosamine-Induced Non-alcoholic Steatohepatitis. *Biomed. Pharmacother.* **2018**, *108*, 963–973.
93. Darvin, S. S.; Esakkimuthu, S.; Toppo, E.; Balakrishna, K.; Muniappan, N.; Paulraj, M. G.; Pandikumar, P.; Ignacimuthu, S.; Al-Dhabi, N. A. Hepatoprotective Effect of Lawsone on Rifampicin-Isoniazid Induced Hepatotoxicity in In Vitro and In Vivo Models. *Environ. Toxicol. Pharmacol.* **2018**, *61*, 87–94.
94. Chen, J. F.; Xu, P.; Zhu, S. Y.; Hu, Y. J. Advance in Research for Pharmacokinetics and Drug Interactions of Licorice. *Chin. J. Clin. Pharmacol.* **2010**, *15*(10), 1174–1182.
95. Matsui, S.; Matsumoto, H.; Sonoda, Y.; Ando, K.; Aizu-Yokoto, E.; Sato, T.; Kasahara, T. Glycyrrhizin and Related Compounds Down-Regulate Production of Inflammatory Chemokines IL-8 and Eotaxin 1 in a Human Lung Fibroblast Cell Line. *Int. Immunopharmacol.* **2004**, *4*(13), 1633–1644.
96. Park, H.-Y.; Park, S.-H.; Yoon, H.-K.; Myung, J. H.; Kim, D.-H. Anti-Allergic Activity of 18-β-Glycyrrhetinic Acid-3-O-β-D-Glucuronide. *Arch. Pharmacal Res.* **2004**, *27*(1), 57–60.
97. Singh, C. K.; Liu, X.; Ahmad, N. Resveratrol, in Its Natural Combination in Whole Grape, for Health Promotion and Disease Management. *Ann. N. Y. Acad. Sci.* **2015**, *1348*(1), 150–160. https://doi.org/10.1111/nyas.12798.
98. Higdon, J.; Drake, V. J.; Delage, B.; Espín, J. C. *Resveratrol: Newsletters (2015 Updated)*. Linus Pauling Institute, Oregon State University, 307 Linus Pauling Science Center: Corvallis, Oregon, USA, 2015.
99. Alvari, A.; Mehrnaz, S. O.; Ahmad, F. J.; Abdin, M. Z. Contemporary Overview on Clinical Trials and Future Prospects of Hepato-Protective Herbal Medicines. *Rev. Recent Clin. Trials* **2012**, 7, 214–223.
100. Dhiman, A.; Nanda, A.; Ahmad, S. A Recent Update in Research on the Antihepatotoxic Potential of Medicinal Plants. *J Chin. Integr. Med.* **2012**, *10*, 117–127.
101. Ding, R. B.; Tian, K.; Huang, L. L.; He, C. W.; Jiang, Y.; Wang, Y. T.; Wan, J. B. Herbal Medicines for the Prevention of Alcoholic Liver Disease: A Review. *J. Ethnopharmacol.* **2012**, *144*, 457–465.
102. Girish, C.; Pradhan, S. C. Indian Herbal Medicines in the Treatment of Liver Diseases: Problems and Promises. *Fund. Clin. Pharmacol.* **2012**, *26*, 180–189.
103. Muriel, P.; Rivera-Espinoza, Y. Beneficial Drugs for Liver Diseases. *J. Appl. Toxicol.* **2008**, *28*, 93–103.
104. Zhang, A.; Sun, H.; Wang, X. Recent Advances in Natural Products From Plants for Treatment of Liver Diseases. *Eur. J. Med. Chem.* **2013**, *63*, 570–577.
105. Adewusi, E. A.; Afolayan, A. J. A Review of Natural Products With Hepatoprotective Activity. *J. Med. Plants Res.* **2010**, *4*, 1318–1334. https://doi.org/10.5897/JMPR09.472.
106. Deshwal, N.; Sharma, A. K.; Sharma, P. Review on Hepatoprotective Plants. *Int. J. Pharm. Sci. Rev. Res.* **2011**, 7, 15–26.
107. Nieminen, M.; Soumi, J.; Van Nouhuys, S. Effect of Iridoid Glycoside Content on Oviposition Host Plant Choice and Parasitim in a Specialist Herbivore. *J. Chem. Ecol.* **2003**, *29*(4), 823–843.
108. Cobbina, E.; Akhlaghi, F. Non-alcoholic Fatty Liver Disease (NAFLD)—Pathogenesis, Classification, and Effect on Drug Metabolizing Enzymes and Transporters. *Drug Metab. Rev.* **2017**, *49*(2), 197–211.
109. Chang, L. M.; Yun, H. S.; Kim, Y. S.; Ahn, J. W. Aucubin: Potential Antidote for Alpha-Amanitin Poisoning. *J. Toxicol. Clin. Toxicol.* **1984**, *22*(1), 77–85.
110. Chang, I. M.; Ryu, J. C.; Park, Y. C.; Yun, H. S.; Yang, K. H. Protective Activities of Aucubin Against Carbon Tetrachloride-Induced Liver Damage in Mice. *Drug Chem. Toxicol.* **1983**, *6*(5), 443–453.

111. Bartholomaeus, A.; Ahokas, J. Inhibition of P-450 by Aucubin: Is the Biological Activity of Aucubin Due to Its Glutaraldehyde-Like Aglycone?*J. Toxicol. Lett.* **1995**, *80*(1–3), 75–83.
112. Lee, H. Y.; Lee, G. H.; Lee, M. R.; Kim, H. K.; Kim, N. Y.; Kim, S. H.; Lee, Y. C.; Kim, H. R.; Chae, H. J. Eucommia ulmoides Oliver Extract, Aucubin, and Geniposide Enhance Lysosomal Activity to Regulate ER Stress and Hepatic Lipid Accumulation. *PLoS One* **2013**, *8*(12), e81349.
113. Lee, G. H.; Lee, M. R.; Lee, H. Y.; Kim, S. H.; Kim, H. K.; Kim, H. R.; Chae, H. J. Eucommia Ulmoides Cortex, Geniposide and Aucubin Regulate Lipotoxicity Through the Inhibition of Lysosomal BAX. *PLoS One* **2014**, *9*(2), e88017.
114. Lv, P. Y.; Feng, H.; Huang, W. H.; Tian, Y. Y.; Wang, Y. Q.; Qin, Y. H.; Li, X. H.; Hu, K.; Zhou, H. H.; Ouyang, D. S. Aucubin and its Hydrolytic Derivative Attenuate Activation of Hepatic Stellate Cells Via Modulation of TGF-β Stimulation. *Environ. Toxicol. Pharmacol.* **2017**, *50*, 234–239.
115. Sharma, M.; Mitnala, S.; Vishnubhotla, R. K.; Reddy, D. N.; Rao, P. N. The Riddle of Nonalcoholic Fatty Liver Disease: Progression From Nonalcoholic Fatty Liver to Nonalcoholic Steatohepatitis. *J. Clin. Exp. Hepatol.* **2015**, *5*, 147–158.
116. Fehér, J.; Lengyel, G. Silymarin in the Prevention and Treatment of Liver Diseases and Primary Liver Cancer. *Curr. Pharm. Biotechnol.* **2012**, *13*, 210–217.
117. Tajmohammadi, A.; Rajavi, B. M.; Hosseinzadeh, H. Silybum Marianum (Milk Thistle) and Its Main Constituent, Silymarin, as a Potential Therapeutic Plant in Metabolic Syndrome: A Review. *Phytother. Res.* **2018**, *32*, 1933–1949.
118. Koul, I. B.; Kapil, A. Evaluation of the Liver Protective Potential of Piperine, an Active Principle of Black and Long Peppers. *Planta Med.* **1993**, *59*(5), 413–417.
119. Shukla, R.; Surana, S. J.; Tatiya, A. U.; Das, S. K. Investigation of Hepatoprotective Effects of Piperine and Silymarin on D-Galactosamine Induced Hepatotoxicity in Rats. *Res. J. Pharm. Biol. Chem. Sci.* **2011**, *2*(3), 975–982.
120. Wu, H.; Zhang, G.; Huang, L.; Pang, H.; Zhang, N.; Chen, Y.; Wang, G. Hepatoprotective Effect of Polyphenol-Enriched Fraction From *Folium Microcos* on Oxidative Stress and Apoptosis in Acetaminophen-Induced Liver Injury in Mice. *Oxid. Med. Cell. Longevity* **2017**, *2017*, 3631565.
121. Roesler, R. Effect of Extracts from Araticum (*Annona crassiflora)* on CCl_4-Induced Liver Damage in Rats. *Food Sci. Tech.* **2011**, *31*(1), 93–100.
122. Lai, Y. S.; Chen, W. C.; Ho, C. T.; Lu, K. H.; Lin, S. H.; Tseng, H. C.; Lin, S. Y.; Sheen, L. Y. Ginger Essential Oil Ameliorates Hepatic Injury and Lipid Accumulation in High Fat Diet-Induced Nonalcoholic Fatty Liver Disease. *J. Agric. Food Chem.* **2014**, *62*, 5897–5906.
123. Guan, M. J.; Zhao, N.; Xie, K. Q.; Zeng, T. Hepatoprotective Effects of Garlic Against Ethanol-Induced Liver Injury: A Mini-Review. *Food Chem. Toxicol.* **2018**, *111*, 467–473.
124. Kaur, N.; Kaur, B.; Sirhindi, G. Phytochemistry and Pharmacology of Phyllanthus Niruri L.: A Review. *Phytother. Res.* **2017**, *31*(7), 980–1004.
125. Vedi, M.; Sabina, E. P. Assessment of Hepatoprotective and Nephroprotective Potential of withaferinA on Bromobenzene-Induced Injury in Swiss Albino Mice: Possible Involvement of Mitochondrial Dysfunction and Inflammation. *Cell Biol. Toxicol.* **2016**, *32*(5), 373–390. https://doi.org/10.1007/s10565-016-9340-2.
126. Wagoner, J.; Morishima, C.; Graf, T. N.; Oberlies, N. H.; Teissier, E.; Pecheur, E. I.; Tavis, J. E.; Polyak, S. J. Differential In Vitro Effects of Intravenous Versus Oral Formulations of Silibinin on the HCV Life Cycle and Inflammation. *PLoS One* **2011**, *6*(1), e16464.
127. Ahmed-Belkacem, A.; Ahnou, N.; Barbotte, L.; Wychowski, C.; Pallier, C.; Brillet, R.; Pohl, R. T.; et al. Silibinin and Related Compounds Are Direct Inhibitors of Hepatitis C Virus RNA-Dependent RNA Polymerase. *Gastroenterology* **2010**, *138*, 1112–1122.

128. Duan, S.; Du, X.; Chen, S.; Liang, J.; Huang, S.; Hou, S.; Gao, J.; Ding, P. Effect of Vitexin on Alleviating Liver Inflammation in a Dextran Sulfate Sodium (DSS)-Induced Colitis Model. *Biomed. Pharmacother.* **2020**, *121*, 109683. https://doi.org/10.1016/j.biopha.2019.109683.
129. Luper, S. A Review of Plants Used in the Treatment of Liver Disease: Part 1. *Altern. Med. Rev.* **1998**, *3*, 410–421.
130. Crance, J. M.; Leveque, F.; Biziagos, E.; van Cuyck-Gandre, H.; Jouan, A.; Deloince, R. Studies on Mechanism of Action of Glycyrrhizin against Hepatitis A Virus Replication In Vitro. *Antiviral Res.* **1994**, *23*, 63–76.
131. Levy, C.; Seeff, L. D.; Lindor, K. D. Use of Herbal Supplements for Chronic Liver Disease. *Clin. Gastroenterol. Hepatol.* **2004**, *2*, 947–956.
132. Oliveira, F. A.; Chaves, M. H.; Almeida, F. R. C.; Lima, R. C. P.; Silva, R. M.; Maia, J. L.; Brito, G. A. C.; Santos, F. A.; Rao, V. S. Protective Effect of α- and β-Amyrin, a Triterpene Mixture From Protiumheptaphyllum (Aubl.) March. Trunk Wood Resin, against Acetaminophen-Induced Liver Injury in Mice. *J. Ethnopharmacol.* **2005**, *98*, 103–108.
133. Martono, S.; Hakim, L.; Soemhoedi, M.; Commandeur, J. N.; Vermeulen, N. P. Effect of Curcumin and Its Derivates on Glutathione-S-Transferase Activity in Liver Cytosolic of β-Naphthoflavone-Treated Rats. In Proceedings of the International Symposium on Curcumin Pharmacochemistry (ISCP), Yogyakarta, Faculty of Pharmacy Universitas Gadjah Mada, 1995. 126-36.36.
134. Oetari, S.; Commandeur, J. N.; Suwaldi, M.; Sugiyanto, D.; Bijloo, G.; Samhoedi, M.; et al. Interaction Between Curcumin and Curcumin Analogues and Cytochrome P-450: Inhibition of Rat Hepatic Microsomal Ethoxyresorufin-O-Dealkylase and Their Structure-Activity Relationship. In Proceedings of the International Symposium on Curcumin Pharmacochemistry, Yogyakarta, Faculty of Pharmacy Universitas Gadjah Mada, 1995. 136-51.37.
135. Li, G.; Chen, J. B.; Wang, C.; Xu, Z.; Nie, H.; Qin XYChen XM, Gong Q. Curcumin Protects Against Acetaminophen-Induced Apoptosis in Hepatic Injury. *World J. Gastroenterol.* **2013**, *19*, 7440–7446.
136. Raskovic, A.; Gigov, S.; Capo, I.; PautKuaturica, M.; Milijasevic, B.; Kojic-Damjanov, S.; Martic, N. Antioxidative and Protective Actions of Apigenin in a Paracetamol-Induced Hepatotoxicity Rat Model. *Eur. J. Drug Metab. Pharmacokinet.* **2017**, *42*(5), 849–856. https://doi.org/10.1007/s13318-017-0407-0.
137. Wang, M.; Sun, J.; Jiang, Z.; Xie, W.; Zhang, X. Hepatoprotective Effect of Kaempferol Against Alcoholic Liver Injury in Mice. *Am. J. Chin. Med.* **2015**, *43*(2), 241–254. https://doi.org/10.1142/S0192415X15500160.
138. Nguyen Cong, T. T.; Ninh, T. S.; Nguyen, T. N.; Vu Thi, T. P.; Nguyen, T. C.; Do Thi, P.; Gilles, N. M. C.; Do, T. T. The Hepatoprotective Activity of a New Derivative Kaempferol Glycoside From the Leaves of Vietnamese Phyllanthus Acidus (L.) Skeels. *Med. Chem. Res.* **2017**, *26*, 2057–2064.
139. Shanmugam, B.; Shanmugam, K. R.; Ravi, S.; Subbaiah, G. V.; Ramakrishana, C.; Mallikarjuna, K.; Reddy, K. S. Exploratory Studies of (−)-Epicatechin, a Bioactive Compound of Phyllanthusniruri, on the Antioxidant Enzymes and Oxidative Stress Markers in D-Galactosamine-Induced Hepatitis in Rats: A Study With Reference to Clinical Prospective. *Pharmacog. Magaz.* **2017**, *13*, S56. https://doi.org/10.4103/0973-1296.203973.
140. Gillani, A. H.; Janbaz, K. H.; Shah, B. H. Quercetin Exhibits Hepatoprotective Activity in Rats. *Biochem. Soc. Trans.* **1997**, *25*, S619.
141. Landon, E. J.; Naukam, R. J.; Rama Sastry, B. V. Effects of Calcium Channel Blocking Agents on Calcium and Centrilobular Necrosis in the Liver of Rats Treated With Hepatotoxic Agents. *Biochem. Pharmacol.* **1986**, *35*, 697–705.

142. Liu, G. T.; Zhang, T. M.; Wang, B. E.; Wang, Y. W. Protective Action of Seven Natural Phenolic Compounds Against Peroxidative Damage to Biomembranes. *Biochem. Pharmacol.* **1992**, *43*, 147–152.
143. Morales, M. A.; Lozoya, X. Calcium-Antagonist Effect of Quercetin on Aortic Smooth Muscle. *Planta Med.* **1994**, *60*(4), 313–317.
144. Miltonprabu, S.; Tomczyk, M.; Skalicka-Woźniak, K.; Rastrelli, L.; Daglia, M.; Nabavi, S. F.; Alavian, S. M.; Nabavi, S. M. Hepatoprotective Effect of Quercetin: From Chemistry to Medicine. *Food Chem. Toxicol.* **2017**, *108*(B), 365–374. https://doi.org/10.1016/j.fct.2016.08.034.
145. Sheng, L.; Wanxing, W.; Kaichuang, S.; Xun, C.; Min, Z.; Zhiping, L. In Vitro and In Vivo Anti-Hepatitis B Virus Activities of the Lignanniranthin Isolated From *Phyllanthusniruri L. J. Ethnopharmacol.* **2014**, *155*(2), 1061–1067. https://doi.org/10.1016/j.jep.2014.05.064.
146. Aayadi, H.; Mittal, S. P. K.; Deshpande, A.; Gore, M.; Ghaskadbi, S. S. Protective Effect of Geraniin Against Carbon Tetrachloride Induced Acute Hepatotoxicity in Swiss Albino Mice. *Biochem. Biophys. Res. Commun.* **2017**, *487*(1), 62–67. https://doi.org/10.1016/j.bbrc.2017.04.013.
147. Emam, M.; El Raey, M. A.; El-Haddad, A. E.; El Awdan, S. A.; Rabie, A.-G. M.; El-Ansari, M. A.; Sobeh, M.; Osman, S. M.; Wink, M. A New Polyoxygenated Flavonol Gossypetin-3-O-β-d-Robinobioside From *Caesalpini agilliesii* (Hook.) D. Dietr. And in Vivo Hepatoprotective, Anti-Inflammatory, and Anti-Ulcer Activities of the Leaf Methanol Extract. *Molecules* **2019**, *24*(1), 138. https://doi.org/10.3390/molecules24010138.
148. Osakabe N, Yasuda A, Natsume M, Sanbongi C, Kato Y, Osawa T. Yoshikawa T. (2002). Rosmarinic Acid, a Major Polyphenolic Component of Perillafrutescens, Reduces Lipopolysaccharide (LPS)-Induced Liver Injury in D-Galactosamine (D-GalN)-Sensitized Mice. Free Radical Biol. Med.; 33 (6): 798–806
149. Weber, A.; Boll, L. W.; Stampfl, M. Hepatotoxicity and Mechanism of Action of Haloalkanes: Carbon Tetrachloride as a Toxicological Model. *Crit. Rev. Toxicol.* **2003**, *33*, 105–136.
150. Sun, S. Y.; Yue, P.; Zhou, J. Overexpression of Bcl2 Blocks TNF-Related Apoptosis-Inducing Ligand (TRAIL)-Induced Apoptosis in Human Lung Cancer Cells. *Biochem. Biophys. Res. Commun.* **2001**, *280*, 788–797.
151. Li, G.; Xie, Q.; Shi, Y.; Li, D.; Zhang, M.; Jiang, S.; Zhou, H.; Lu, H.; Jin, Y. Inhibition of Connective Tissue Growth Factor by siRNA Prevents Liver Fibrosis in Rats. *J. Gene Med.* **2006**, *8*, 889–900.
152. Zhang, H.; Bai, Y.; Gao, M.; Zhang, J.; Dong, G.; Yan, F.; Ma, Q.; Fu, X.; Zhang, Q.; Li, C.; Shi, H.; Ning, Z.; Dai, J.; Li, Z.; Ming, J.; Xue, Q.; Si, C.; Xiong, H. Hepatoprotective Effect of Capsaicin Against Concanavalin A-Induced Hepatic Injury Via Inhibiting Oxidative Stress and Inflammation. *Am. J. Transl. Res.* **2019**, *11*(5), 3029–3038.
153. Hassan, M. H.; Edfawy, M.; Mansour, A.; Hamed, A. A. Antioxidant and Antiapoptotic Effects of Capsaicin Against Carbon Tetrachloride-Induced Hepatotoxicity in Rats. *Toxicol. Ind. Health* **2012**, *28*(5), 428–438. https://doi.org/10.1177/0748233711413801.
154. Zheng, Q. S.; Sun, X. L.; Xu, B.; Li, G.; Song, M. Mechanisms of Apigenin-7-Glucoside as a Hepatoprotective Agent. *Biomed. Environ. Sci.* **2005**, *18*(1), 65–70.
155. Cengiz M, Ali JH, Kutlu HM, Vejselova D, Ayhanci A. (2017). Potential Recruiting and Hepatoprotective Effects of Ellagic Acid in D-Galactosamine-Induced Liver Damage in Rats. Pak. J. Zool.; 49(4): 1251–1259; doi:org/10.17582/journal.pzj/2017.49.4.1251.1259

156. Singh, K.; Khanna, A. K.; Chander, R. Hepatoprotective Activity of Ellagic Acid Against Carbon Tetrachloride Induced Hepatotoxicity in Rats. *Indian J. Exp. Biol.* **1999**, *37*(10), 1025–1026.
157. Ranga Rao, A.; Sarada, R.; Shylaja, M. D.; Ravishankar, G. A. Evaluation of Hepatoprotective and Antioxidant Activity of Astaxanthin and Astaxanthin Esters From Microalga *Haematococcuspluvialis*. *J. Food Sci. Technol.* **2015**, *52*(10), 6703–6710. https://doi.org/10.1007/s13197-015-1775-6.
158. Chiu, C.-H.; Chang, C.-C.; Lin, S.-T.; Chyau, C.-C.; Peng, R. Y. Improved Hepatoprotective Effect of Liposome-Encapsulated Astaxanthin in Lipopolysaccharide-Induced Acute Hepatotoxicity. *Int. J. Mol. Sci.* **2016**, *17*(7), 1128. https://doi.org/10.3390/ijms17071128.
159. Zaghloul, R. A.; Elsherbiny, N. M.; Kenawy, H. I.; El-Karef, A.; Eissa, L. A.; El-Shishtawy, M. M. Hepatoprotective Effect of Hesperidin in Hepatocellular Carcinoma: Involvement of Wnt Signaling Pathways. *Life Sci.* **2017**, *185*, 114–125. https://doi.org/10.1016/j.lfs.2017.07.026.
160. Chen, W. M.; Shaw, L. H.; Chang, P. J.; Tung, S. Y.; Chang, T. S.; Shen, C. H.; Hsieh, Y. Y.; Wei, K. L. Hepatoprotective Effect of Resveratrol Against Ethanol-Induced Oxidative Stress Through Induction of Superoxide Dismutase In Vivo and In Vitro. *Exp. Ther. Med.* **2016**, *11*(4), 1231–1238. https://doi.org/10.3892/etm.2016.3077.

CHAPTER EIGHT

Nephroprotective natural products

Emil Jivishov[a,*], Lutfun Nahar[b], Satyajit D. Sarker[b]

[a]College of Humanities and Sciences, ADA University, Baku, Azerbaijan
[b]Centre for Natural Products Discovery (CNPD), School of Pharmacy and Biomolecular Sciences, Liverpool John Moores University, Liverpool, United Kingdom
[*]Corresponding author: e-mail address: ejivishov@ada.edu.az

Contents

1. Introduction

The kidneys are vital organs that play an important role in removing waste and toxic materials from the blood, and maintain electrolyte balance, regulate homeostasis of the blood plasma, blood volume, blood pressure, and red blood cell genesis. Simply, to help maintain homeostasis, the kidneys take care of the removal of metabolic waste and toxins from the blood and dilute them with water and electrolytes. The kidneys maintain blood volume and pressure by secreting renin,[1] activating vitamin D,[2] secreting erythropoietin hormone that controls red blood cell production,[3] performs gluconeogenesis while fasting for long periods.[4]

The kidneys can be damaged or may become totally inactive by several factors such as oxidative stress, inflammation, ischemia/reperfusion injury, diabetes, and nephrotoxic agents, most often various modern drugs that are in use today. Chronic kidney disease (CKD) affects 15% of the

Annual Reports in Medicinal Chemistry, Volume 55
ISSN 0065-7743
https://doi.org/10.1016/bs.armc.2020.02.003

United States Medicare population aged 65 and older, and costs 24% fee-for-service Medicare spending. For Medicare population aged younger than 65 only 8% had, but still accounted for 18% of the spending. In 2016, total Medicare spending on both CKD and kidney failure (end-stage kidney disease, ESRD) was over $114 billion.[5]

Diabetic kidney disease (DKD), sometimes referred to as diabetic nephropathy, is the leading cause of the CKD and ESRD, which may cause cardiovascular failure as well.[6] Although DKD is not certainly classified as an inflammatory disease, genome-wide transcriptome analysis,[7,8] along with several *in vitro*, pre-clinical, clinical and epidemiological studies,[9–12] indicated the presence and involvement of inflammatory pathways in DKD.

Another major cause of kidney problems is obesity, which is a metabolic disorder and/or arise from excessive consumption of food and no or little physical activities to utilise the energy. Since 1975 the number of officially obese people has nearly tripled. In 2016, over 650 million people worldwide were obese.[13] It has been known that there is a strong relationship between obesity, hypertension and renal diseases.[14–17] Non-alcoholic fatty liver disease (NAFLD) is associated with obesity along with type 2 diabetes and hypertension[17] and thus has detrimental effect on the kidney health, the fact which has been supported by international meta-analysis and cohort studies.[18,19] Obesity and NAFLD are metabolic disorders with multiple pathways and mechanisms, which complicate their pathophysiology. Lifestyle modifications, such as diet modification, sustained weight loss, and increase in exercise are accepted first line of treatment for patients with a few pharmacological agents, such as vitamin E, metformin and statins, to be used as rather supporting remedy.[17] Obviously, controlling or preventing obesity, hypertension, cardiovascular diseases and NAFLD may prevent CKD associated with them.[20,21]

Nephrolithiasis (kidney stones), which often associated with the fluid intake, diet, metabolic disorders and obesity, is another common disease affecting the kidneys of about 1 in 10 adult men and less than that women.[22–24] Kidney stones were found to be linked to increased papillary renal cell carcinoma risk.[25]

Oxidative stress, as mentioned earlier, has been implicated to CKD[26] and diabetic nephropathy.[27] Disturbances in key endogenous oxidative stress sources [nicotinamide adenine dinucleotide phosphate (NADPH) oxidase, endothelium nitric oxide synthase (eNOS)] and enzymatic (superoxide dismutase, catalyses and peroxidases) and nonenzymatic antioxidants (glutathione) may be observed as results of many pathophysiological processes and can start in the early stages of CKD. Increased generation of reactive oxygen species (ROS) and reactive nitrogen species (RNS) has been observed in obesity and diabetes. Thus, natural compounds that can prevent or mitigate such oxidative stress, may eventually help protect the kidneys. Several phenolic and polyphenolic compounds, *e.g.*, flavonoids and isoflavonoids, from various natural sources, are commonly known as natural antioxidants, are effective nephroprotectors.[28]

As hypertension is considered as both cause and effect of CKD, antihypertensive drug treatment has shown to slow progress of CKD as well as reduce the damage of cardiovascular disease.[15] Renin-angiotensin is a hormone system involved in maintenance of renal function *via* regulation of blood pressure and fluid balance.[29] It does so by producing angiotensin II peptide, which acts on brain to control thirst, blood pressure, ventilation rate as well as participates in memory process. Angiotensin II is known for its ROS generation function, which exceeds endogenous antioxidant capacity in hypertension and leads to oxidative stress.

According to the National Health Service UK (NHS, UK), currently there is no cure available for CKD and only symptomatic treatment or disease progress control is available.[30] For early stage CKD, main recommendations include lifestyle changes, hypertension and high cholesterol control medicine such as angiotensin converting enzyme (ACE) inhibitor and statins, respectively. For ESRD, dialysis and kidney transplant are only available treatment choices.

Only a few clinical trials looked into possible effect of antioxidant interventions in CKD.[31] Low-cost choices, such as vitamin E, *N*-acetylcysteine, allopurinol (Fig. 1) have been assessed for their benefit on cardiovascular disease in end-stage renal disease patients. While vitamin E and *N*-acetylcysteine (NAC) did not show beneficial effects on all-cause mortality, allopurinol was reported to slow down progression of CKD by decreasing uric acid levels. However, despite the conflicting research outcomes regarding the effectiveness of NAC in kidney disease, the 2012 Kidney Disease: Improving Global Outcomes Guidelines recommend the use of oral NAC for the prevention of contrast-induced nephropathy in high-risk individuals, considering its plausible benefit, low possibility for adverse effects, and most importantly, low cost.

Atorvastatin (Fig. 1) might reduce serum inflammatory and oxidative markers while improving cholesterol levels.[32] More recent systematic review and meta-analysis of the literature suggest that while statin therapy does not reduce the risk for kidney failure events in adults not receiving

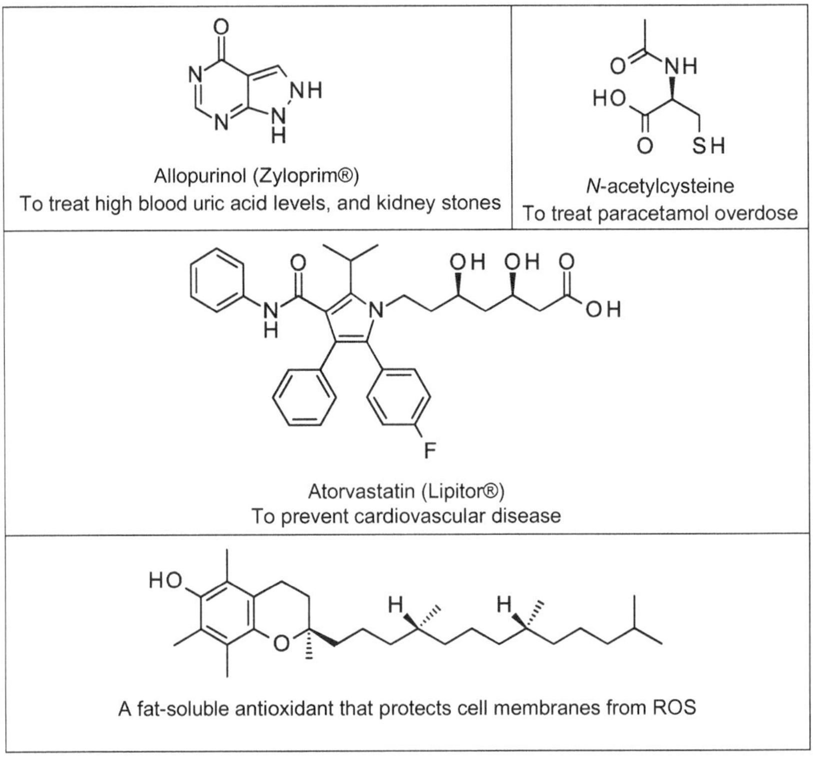

Fig. 1 Structures of allopurinol, *N*-acetylcysteine, atorvastatin and vitamin E.

dialysis for whom kidney disease outcomes were reported, but might modestly reduce proteinuria and rate of estimated glomerular filtration rate decline.[33] Statins, *e.g.*, atorvastatin, might have some beneficial effect on preventing the loss of kidney function and on reducing loss of protein in urine in people with cardiovascular diseases, but statins, before heart surgery, does not prevent acute kidney injury. Some tentative evidence suggests that atorvastatin may decrease the risk of contrast-induced nephropathy (CIN).[34]

Bardoxolone methyl (Fig. 2), a semi-synthetic triterpenoid based on scaffold of oleanolic acid, was used as antioxidant inflammation modulator and was promoted to Phase III clinical trial for CKD patients with type 2 diabetes mellitus, but due to serious adverse effects the trial was terminated.[35] This compound has been shown to be an activator of the KEAP1-Nrf2 pathway to inhibit the pro-inflammatory transcription factor NF-κB.

Several natural products possess significant antioxidant properties, which have been validated through various *in vitro* and *in vivo* studies, and thus have the potential to provide new leads for the development of new and effective

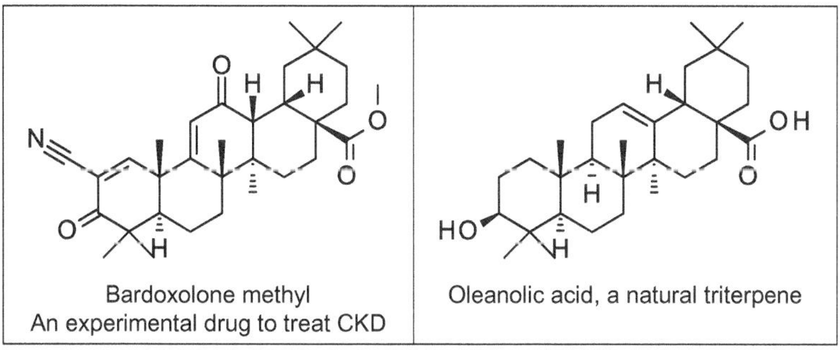

Fig. 2 Bardoxolone methyl, a semisynthetic triterpenoid based on oleanolic acid.

nephroprotective agents. This chapter primarily focuses on potential nephroprotective natural products that might have undergone preclinical and/or clinical trials. Additionally, this chapter provides a summary of information on possible mechanisms of actions of nephroprotective natural products.

2. Nephroprotective natural products

Natural products have played a significant role in providing several drug leads for the treatment of various human ailments, as well as in offering a number of molecules with definitive preventive roles against toxicities, and organ damages, for example, cancer chemoprevention by phytochemicals (*see* chapter "Cancer chemopreventive natural products" by Ritchie and Sarker). Most of the preventative actions of natural products are often associated with their antioxidant and/or anti-inflammatory properties. Several natural products, especially from higher plants, have been reported in the literature to possess nephroprotective properties; some of these compounds, for example, curcumin (Fig. 3), have been in the clinical trials for its role in nephroprotection.[36,37] Generally, polyphenolic compounds, exhibit considerable radical-scavenging and antioxidant activity and thus offer nephroprotective functions. However, most often, their poor bioavailability in humans prevents them from being considered as "drug leads."

2.1 Nephroprotective natural products from plant origin

Plants are easy to access, and thus, most earlier discoveries of medicinal and bioactive principles for human use were from plants. Plants are a good source of bioactive compounds, often with defined medicinal properties, and protective functions including nephroprotection. Plant polyphenols, by virtue of their significant antioxidant property, can offer

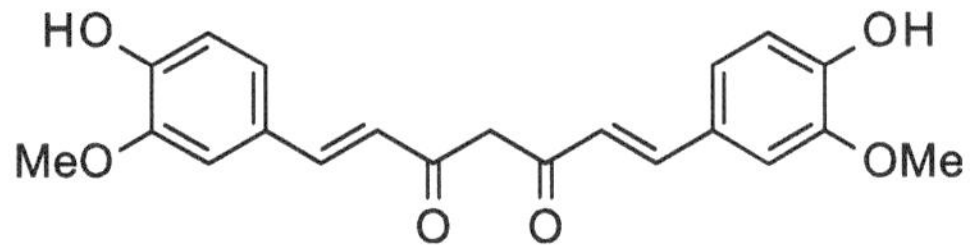

Fig. 3 Curcumin, a well-known nephroprotective phytochemical.

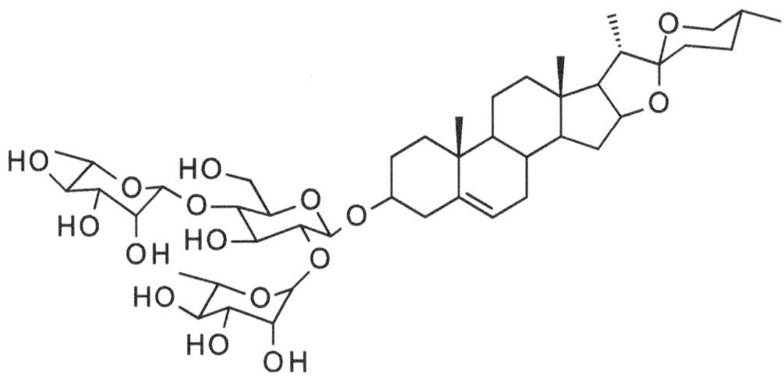

Fig. 4 Dioscin from *Dioscorea* spp. possesses nephroprotective property.

nephroprotection, a classic example being the green tea polyphenols.[38] Similarly, dioscin (Fig. 4), a natural steroidal saponin, isolated from *Dioscorea* spp. (Dioscoreaceae) has been shown to possess nephroprotective property in various *in vivo* animal models, and its plausible mechanism of action has been extensively studied.[39] Ferulic acid, a phenylpropanoid found in various plants, in combination with omega-3 rich fish oil, was found to offer nephroprotection in a gentamicin-induced nephrotoxicity rat model.[40]

In a recent animal study using a methotrexate-induced nephrotoxicity rat model, naringin and silymarin (Fig. 5), two naturally occurring flavonoids, were found to offer nephroprotection, evident from their ability to decrease levels of various parameters which were elevated by methotrexate.[41]

Steveoside (Fig. 6), a natural sweetener from *Stevia rebaudiana* (Asteraceae) was found to protect kidneys from cisplatin-induced injuries in mice.[42] Cisplatin-induced nephrotoxicity could also be ameliorated by quercetin, a well-known flavonol distributed widely in the plant kingdom.[43] Thymoquinone, the major constituent of the well-known medicinal plant *Nigella sativa* (Ranunculaceae), was found to offer nephroprotection in chemotherapy induced nephrotoxicity mediated through decreasing lipid peroxidation and increasing the activity of antioxidant enzymes in renal tissue.[44]

Vincamine (Fig. 7), a monoterpene indole alkaloids obtained from the leaves of *Vinca minor*, a relative of *Vinca rosea* that produces the well-known anticancer drugs, vinblastine and vincristine (*see* chapter "Anticancer natural products" by Sarker et al.), is sold in Europe as a prescription medicine for

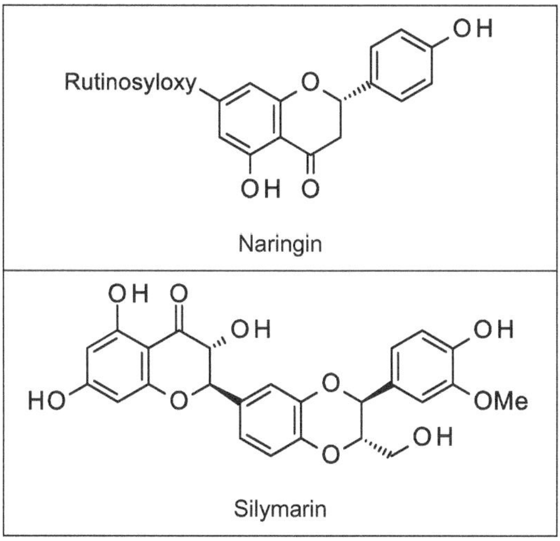

Fig. 5 Naringin and silymarin, two nephroprotective flavonoids of plant origin.

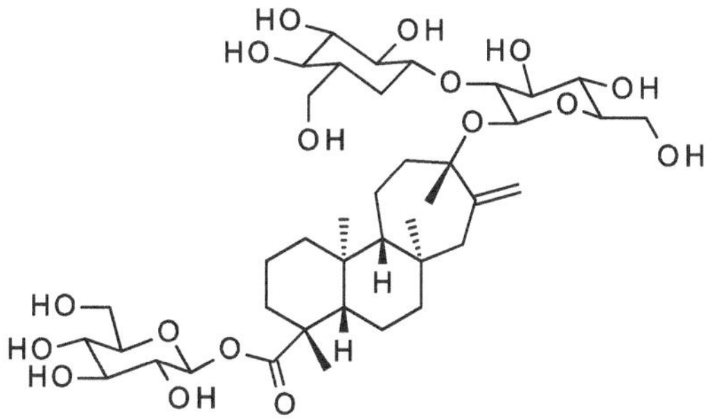

Fig. 6 Steveoside from *Stevia rebaudiana.*

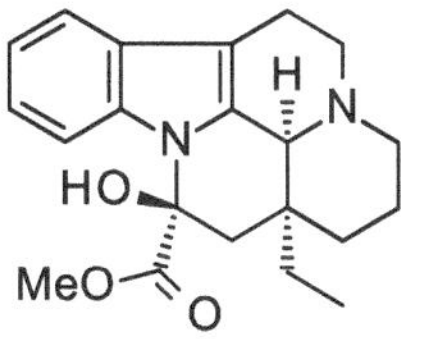

Fig. 7 Structure of vincamine, a monoterpene indole alkaloid from *Vinca minor.*

the treatment of primary degenerative and vascular dementia, but recently has been implicated to its probable role in protecting renal injury.[45]

Several other plant-derived nephroprotective agents,[46–74] as identified from both *in vitro* and *in vivo* animal studies are summarized in Table 1.

Table 1 Examples of some nephroprotective agents (in animal models) of plant origin.

Compounds	**Sources**	**Nephroprotective functions**
Alkaloids		
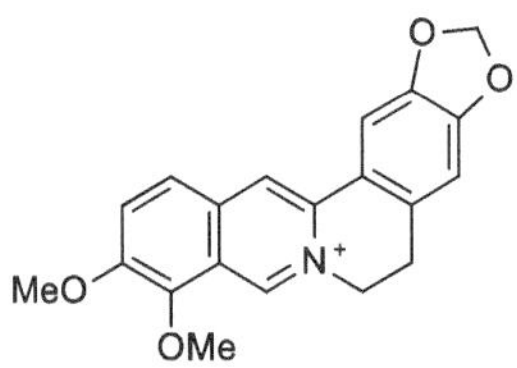 Berberine	*Berberis vulgaris* (Berberidaceae)	Attenuated inflammation responses in mesangial cells[46]; improved diabetes-induced glomerular pathological changes; increased protein expressions of nephrin and podocin.[47]
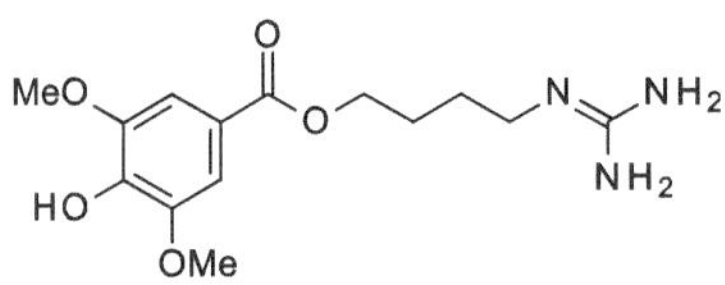 Leonurine	*Leonurus cardiac* (Lamiaceae)	Could attenuate renal fibrosis.[48]
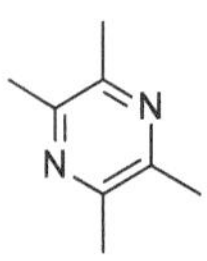 Ligustrazine	*Ligusticum wallichii* (Apiaceae)	Attenuated tubulointerstitial fibrosis.[49]
Anthraquinone		
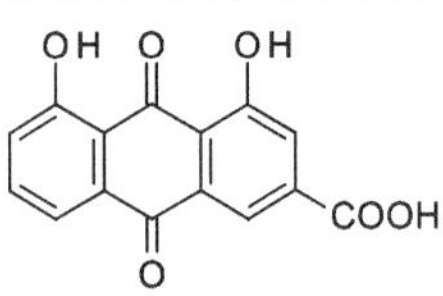 Rhein	*Rheum palmatum* (Polygonaceae)	Reduced drug-induced oxidative damage, increased the excretion of uric acid and reduced inflammation.[50,51]
Diarylheptanoids		
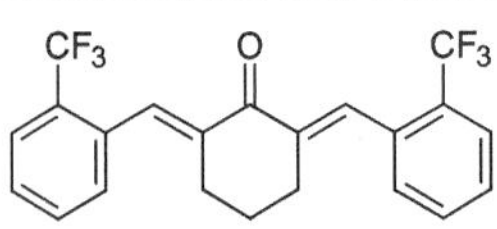 Curcumin derivative	Synthesis inspired by curcumin from *Curcuma longa* (Zingiberaceae)	Improved histological abnormalities and fibrosis of diabetic kidneys, inhibited renal inflammatory gene expression,[52] could attenuate oxidative stress, inflammation, and renal fibrosis,[53] ameliorate renal and mesangial matrix area expansion, reduction in blood urea nitrogen, creatinine and albuminuria.[54]

Table 1 Examples of some nephroprotective agents (in animal models) of plant origin.—cont'd

Compounds	Sources	Nephroprotective functions
Flavonoids		
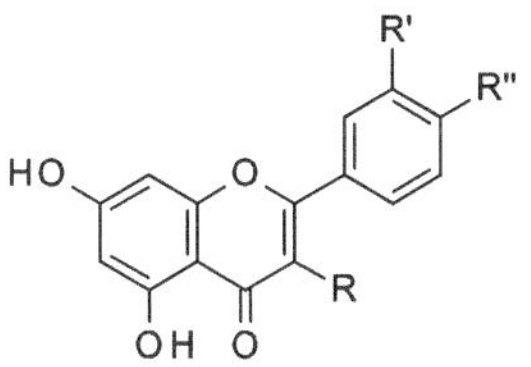 Chrysin R = R' = R'' = H	*Passiflora* spp. (Passifloraceae)	Reduced accumulation of myofibroblast-like cells and matrix proteins in diabetic kidneys.[55]
Kaempferol R = R'' = OH	Fruits and vegetables	Nephroprotective properties against mercury chloride induced nephrotoxicity.[56]
Quercetin R = R' = R'' = OH		Could protect the impairment of kidney functions, protect against oxidative injury,[57] ameliorate the pathological damage of kidneys, improve renal function, alleviate renal oxidative stress injury, restrict inflammatory cells infiltration,[58] reduce serum levels of fibroblast growth factor 23, and kidney inflammation.[59]
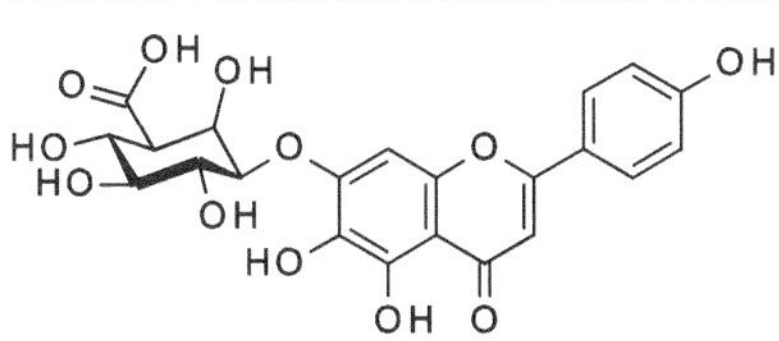 Scutellarin	*Erigeron breviscapus* (Asteraceae)	Exerted anti-inflammatory effect, hypoglycemic and renal protective effects.[60]
Phenylbutanoid		
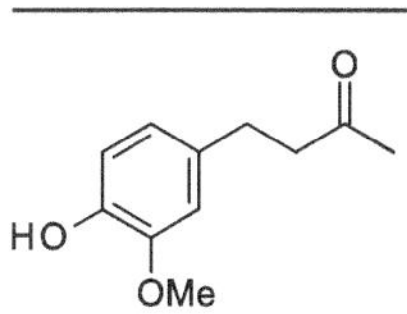 Zingerone	*Zingiber officinale* (Zingiberaceae)	Mitigated CCl_4-induced renal toxicity.[61]

Continued

Table 1 Examples of some nephroprotective agents (in animal models) of plant origin.—cont'd

Compounds	Sources	Nephroprotective functions
Phenylethanoid		
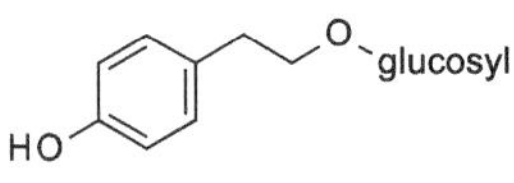 Salidroside	*Rhodiola rosea* (Crassulaceae)	Could attenuate diabetic renal injury by inhibiting albuminuria, serum creatinine and blood urea nitrogen.[62]
Stibenoids		
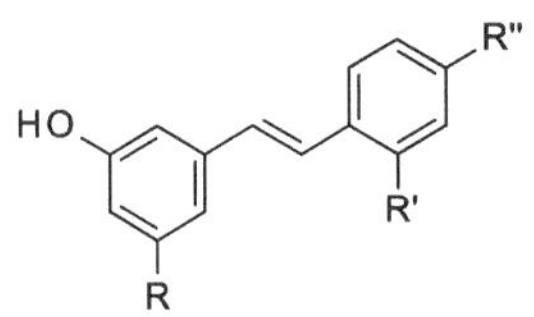 Mulberroside A R = R" = Glucosyloxy, R' = OH	*Morus alba* (Moraceae)	Uricosuric and nephroprotective effects.[63]
Resveratrol R=R″=OH, R'=H	*Vitis vinifera* (Vitaceae)	Anti-inflammatory and histoprotective effect in diabetic kidneys,[64–66] could prevent cisplastin-induced nephrotoxicity[67]; and diabetic renal fibrosis[68]
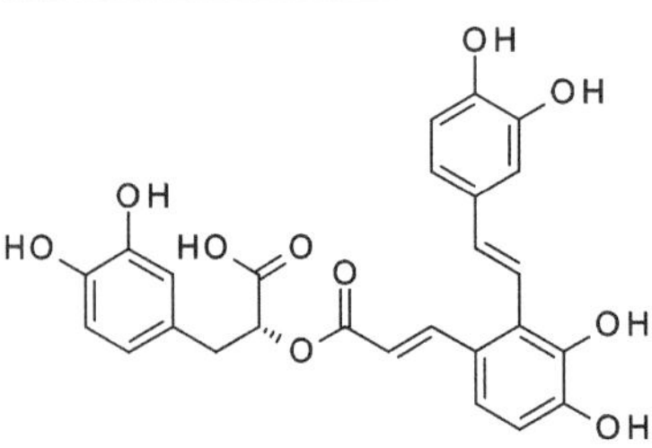 Salvianolic acid	*Salvia miltiorrhiza* (Lamiaceae)	Attenuated glomerular filtration dysfunction, protected against diabetic nephropathy, demonstrated antioxidative effect and ameliorated the inflammation and enhanced impaired autophagy.[69]
Terpenoids		
Crocin (Fig. 8)	*Crocus sativus* (Iridaceae)	Could decrease the high level of serum creatinine and blood urea nitrogen, inflammation, and tubular desquamation in diabetic kidneys,[70] reduce histological damage, oxidative stress and apoptosis, and improve renal function.[71]

Table 1 Examples of some nephroprotective agents (in animal models) of plant origin.—cont'd

Compounds	Sources	Nephroprotective functions
Miscellaneous		
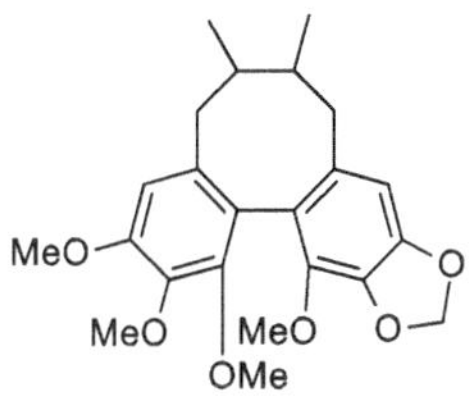 Schisandrin B	*Schisandra chinensis* (Schisandraceae)	Suppressed inflammatory response and oxidative stress in diabetic nephropathy.[72]
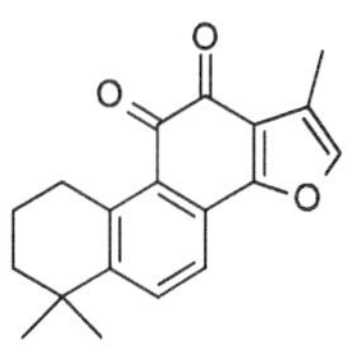 Tanshinone IIA	*Salvia miltiorrhiza* (Lamiaceae)	Improved creatinine, and urine albumin excretion, suppressed renal fibrosis and inflammation[73,74]

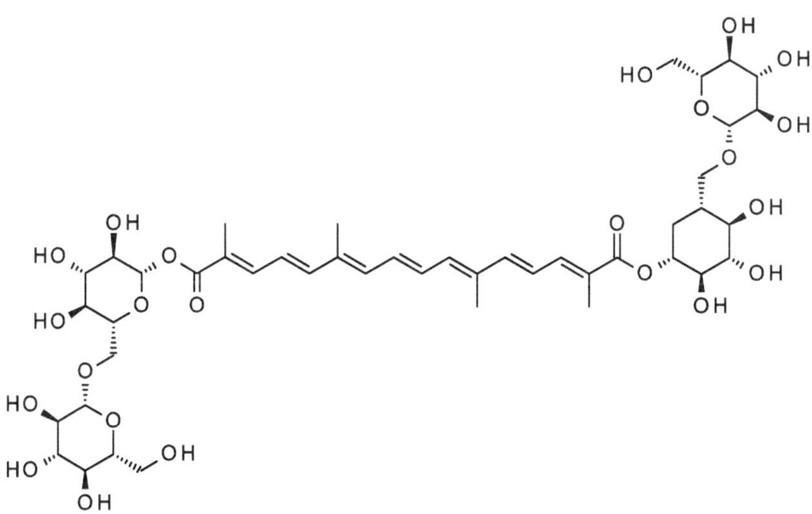

Fig. 8 Structure of crocin, from *Crocus sativus.*

2.2 Nephroprotective natural products of microbial origin

Although plants are the major sources of nephroprotective natural products, there have been a handful of nephroprotective compounds from microbial sources appeared in the literature.[75] FK506, also known as tacrolimus and fujimycin (Fig. 9), was discovered in 1987 from the fermentation broth of a Japanese soil sample that contained *Streptomyces tsukubaensis*, and has recently been shown to offer nephroprotective effects mediated through its anti-inflammatory properties.

Similarly, rapamycin (Fig. 10), which is a microbial secondary metabolite produced by *Streptomyces hygroscopicus*, can diminish the activation and proliferation of inflammatory cells and the expression of inflammatory cytokines, provide neuroprotection.[75] It also has immunosuppressant property in humans, which makes this compound particularly useful in preventing the rejection of kidney transplants.

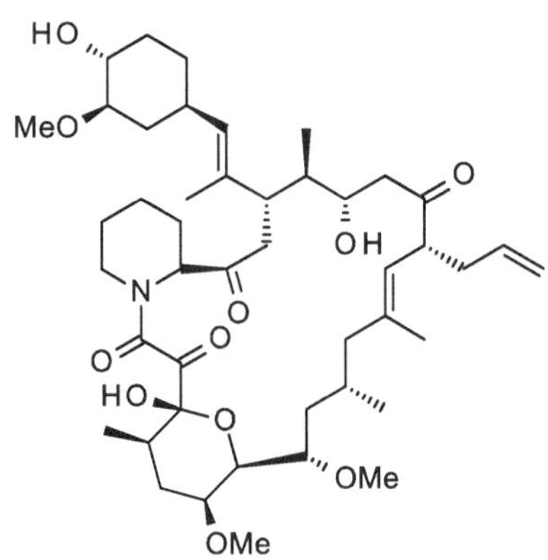

Fig. 9 FK506, a nephroprotective agent from *Streptomyces tsukubaensis*.

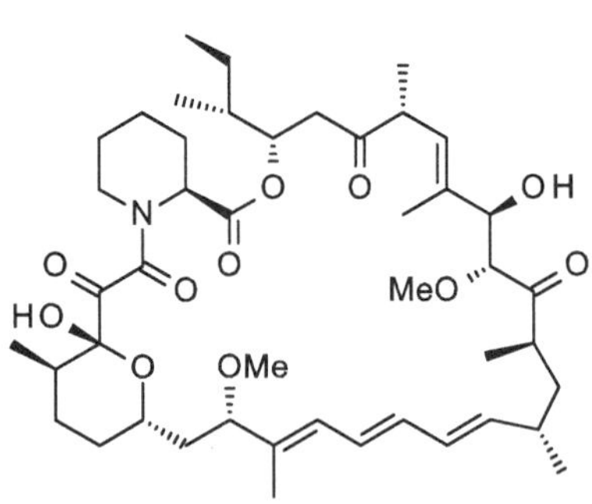

Fig. 10 Rapamycin, a nephroprotective agent from *Streptomyces hygroscopicus*.

3. Mechanisms of action of nephroprotective natural products

Nephroprotective actions of natural products are generally mediated by their antioxidant properties, and to some extent, anti-inflammatory functions. Natural products may ameliorate the advancement of diabetic nephropathy and improve renal function through targeting several pathological processes and inhibiting or depleting many pathways, *e.g.*, p38MAPK, JNK, ERK, TGF-beta, RhoA, NF-kB, Wnt, JAK-STAT, AMPK, mTOR, Akt and TXNIP.[76] Most of the mechanistic studies of nephroprotective natural products have been performed in animal models with induced hepatotoxicity. For example, fructose-induced renal injuries and degeneration was found to be rescued significantly by dioscin (Fig. 4) treatment.[39] In renal tissue of rat, it was observed that the inhibitory effects of dioscin against fibrosis, oxidative stress and abnormal lipid metabolism could be mediated though restoration of the normal expression of Sirt3. It can be noted that the reduced expression of Sirt3 could result in NF-κB activation and cell death. In inflammatory renal insult induced by lipopolysaccharide (LPS), dioscin could supress TLR4/MyD88 *via* upregulating miR-let-7i. Renal injury induced by doxorubicin could also be attenuated by this saponin through farnesoid X receptor mediated signalling. There are several other ways dioscin could protect the kidneys from damages.[39]

The nephroprotective effect of curcumin is mainly mediated by its strong antioxidant properties as well as through mitochondrial function maintenance.[77] It has been shown that the nephroprotective property of this compound could be related to its ability to regulate autophagic flow by classic signalling pathways, *e.g.*, suppression of protein kinase B (Akt)/mammalian target of rapamycin (nTOR) and/or by stimulation of adenosine monophosphate (AMP)-activated protein kinase (AMPK) and extracellular signal-dependent kinase (ERK) pathways.

A methotrexate-induced renal toxicity model using rats was used to study the mechanism of action of nephroprotective property of vincamine (Fig. 5), and it was shown that treatment with this indole alkaloid could ameliorate renal injury through increasing the expression of Nrf2 and HO-1 supressing oxidative stress, decreasing the expression of NF-κB and caspase-3 pathways and increasing ATP levels.[45] Ferulic acid, in combination with omega-3 rich fish oil, was found to offer nephroprotection in a gentamicin-induced nephrotoxicity rat model, and the possible mechanism of action was studied.[40] It was concluded that the nephroprotective action of ferulic acid in fish oil could be mediated through their peroxisome proliferator-activated receptor (PPAR)-γ agonist activity.

Steveoside (Fig. 6) was found to exerts its nephorprotective activity in cisplatin-induced nephrotoxicity by supressing oxidative stress, inflammation and apoptosis through mechanisms involving ERK1/1, STAT3 and NF-κB suppression.[42] In another similar study, quercetin was found to display its protection on cisplatin nephrotoxicity through its antioxidant, vascular, anti-inflammatory and anti-apoptotic effects.[43]

4. Virtual screening of nephroprotective natural products

Virtual screening of compounds, including various natural products, has become an effective tool for drug discovery, as it offers various advantages over traditional and preliminary *in vitro* or *in vivo* screening methods.[78] A recent example of virtual screening of natural products for nephroprotective activity can be observed in the paper published by Guo et al.[62] where they studied the nephroprotective property of salidroside using *in silico* method to identify the main therapeutic targets and pathways of this compound in diabetes with uninephrectomy (Table 1). In another recent *in silico* study has revealed that allicin (found in garlic) and losartan (Fig. 11) could have a common mechanism involving interaction with AT1 receptor to exerts its nephroprotective effect.[79]

The bromodomain-containing protein-4, Brd4, a family member of the BET proteins that regulate gene transcription by binding to acetylated

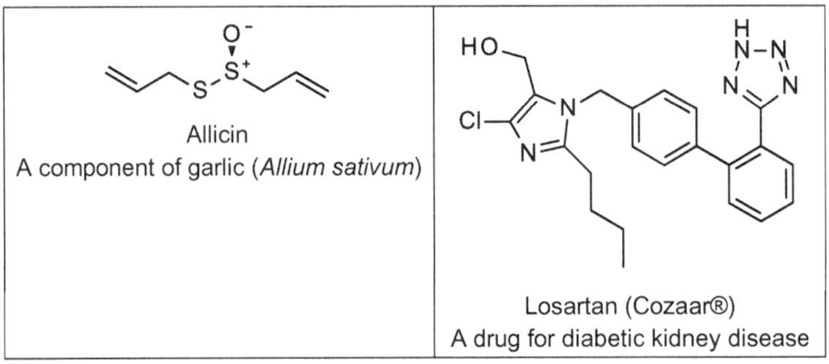

Fig. 11 Allicin and losartan.

lysine residues on histones. The Brd4 protein is known for regulating cell proliferation process and inhibition of the protein results in antifibrotic effects in the CKD.[80] Docking analysis of 10 rare flavonoids, 3-*O*-acetylpinobanksin, 3,7-*O*-diacetylpinobanksin, naringenin triacetate, kaempferol tetraacetate, jaceidin triacetate, artemetin acetate, 5-acetoxy-7-hydroxyflavone, 5,7-diacetoxyflavone, and 5-hydroxy-7-acetoxy-8-methoxy flavone, for their Brd4 inhibitory activity *in silico*, found them to occupy hydrophobic WPF shelf, with 3-*O*-acetylpinobanksin, 3,7-*O*-acetylpinobanksin, naringenin diacetate and naringenin triacetate and kaempferol tetraacetate (Fig. 12), having highest docking scores.[81]

Another study reported *in silico* screening and docking of 4500 flavonoids and extended flavonoids on BD1 of Brd4 receptor,[82] resulting in a kaempferol derivative as the top-ranked compound inhibiting Brd4 receptor, and highlighting that the compound by occupying acetylated lysine site of BD1 of Brd4 prevented binding of Brd4 to acetylated lysines on nucleosome histones, and thus inhibited RNA polymerase II mediated transcription elongation. Virtual screening of *Armillaria mellea* compounds,[83] namely daidzein and genistein (Fig. 13), Donnelly_1a and Donnelly_1c

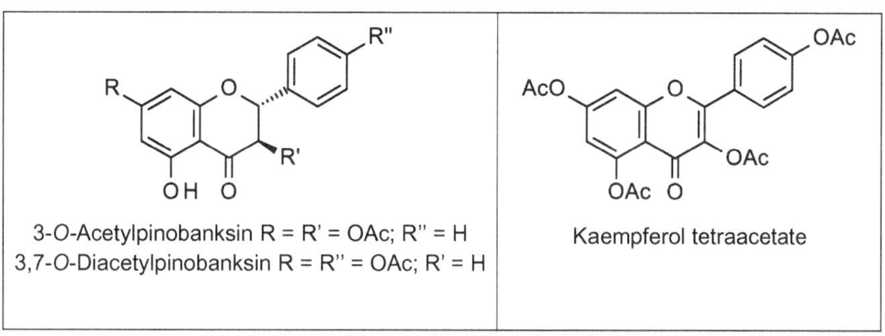

Fig. 12 Structures of 3-*O*-acetylpinobanksin, 3,7-*O*-diacetylpinobanksin and kaempferol tetraacetate.

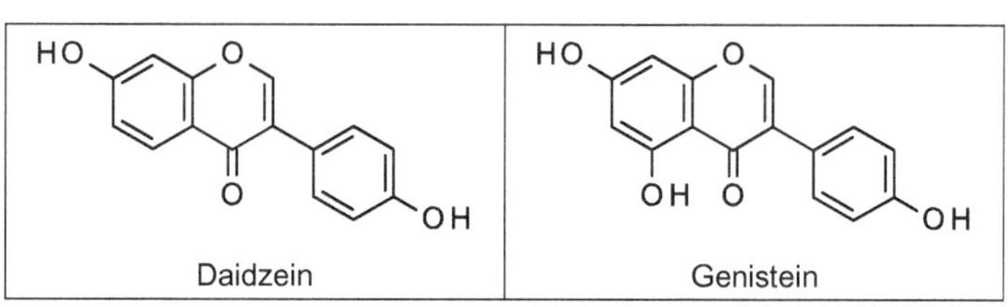

Fig. 13 Structures of daidzein and genistein.

was performed using molecular docking method to determine their mechanism of action. Donnelly_1a and Donnelly_1c had lowest interaction energies, −112.20 and −122.52 kcal/mol, respectively, compared to −40.95 and −41.20 kcal/mol interaction energies of daidzein and genistein, respectively. All compounds formed hydrogen bond, Pi–Pi and Pi-sigma interactions with various amino acid moieties located in the defined active pocket of XOD, thus suggesting that compounds might possess hypouricemic activity.

Known metabolites of *Agrocybe aegerita* were virtually screened using molecular docking method for their mechanism of action against XOD[84] and four compounds (Fig. 14) were selected as the most energetically favourable binding candidates as they were located in the hydrophobic cavity and interacted with the molybdenum atomic domain. Specifically, the compounds interacted with amino acid residues of XOD's active pocket by forming hydrogen, Pi–Pi and Pi-sigma bonding. Interaction energies of these compounds were calculated as −255.419, −150.1420, −163.64157 and −133.116 kcal/mol, respectively.

Taraxerol (Fig. 15), isolated from leaf of *Abroma augusta,* was subjected to molecular docking analysis for its possible interactions with proteins PKC σ, PKC ß, IRS, PI3K, Akt, AMPK, PARP, and NF-kB.[85]

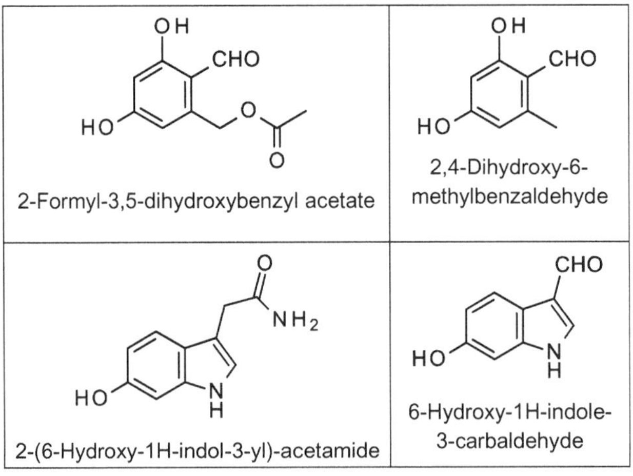

Fig. 14 Structures of four potentially nephroprotective compounds of *Agrocybe aegerita.*

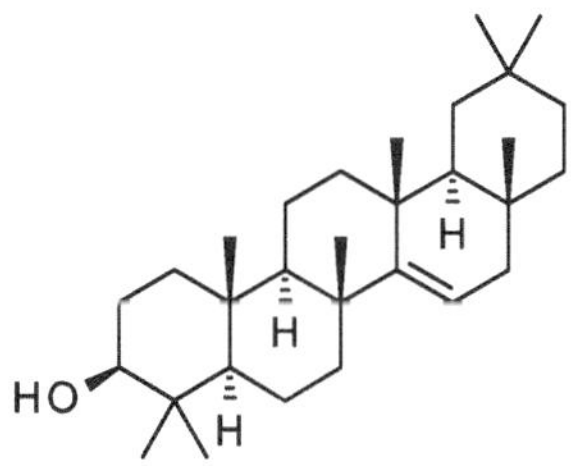

Fig. 15 Structure of taraxerol, isolated from leaf of *Abroma augusta*.

The compound was observed to interact with all tested proteins by either H-bond or hydrophobic interactions with amino acid residues, although, both, interactions and amino acids sometimes being different than the ones mentioned in proteins' PDB files.

5. Nephroprotective natural products in clinical trials

Curcumin, a diarylheptanoid from *Curcuma longa* (Zingiberaceae), is a well-known antioxidant,[86] offers protective functions against the liver, the heart, the central nervous system (CNS) and the kidneys, evident from a number of *in vitro* as well as *in vivo* animal studies.[37] This compound, and its nano-formulations are currently in the clinical trials for their nephroprotective potentials.[37]

Resveratrol (Fig. 16), a natural stilbenoid, found in grape skins and red wine, is a well-known antioxidant, and has recently been in a double blinded, *placebo*-controlled clinical trial involving 60 patients with type 2 diabetes to assess its nephroprotective potentials.[87] It was observed that this phenolic compound could be an effective adjunct to angiotensin receptor blockers for reducing urinary albumin excretion in patients with diabetic nephropathy.

6. Conclusions

Various plant extracts have long been used to treat kidney problems. In recent years, the focus has been given to purify and identify active components from those nephroprotective plant extracts, and to try to understand how these natural compounds could potentially offer protection against nephrotoxicity and renal injuries. Broadly, two main mechanisms of actions of these nephroprotective natural products have emerged through

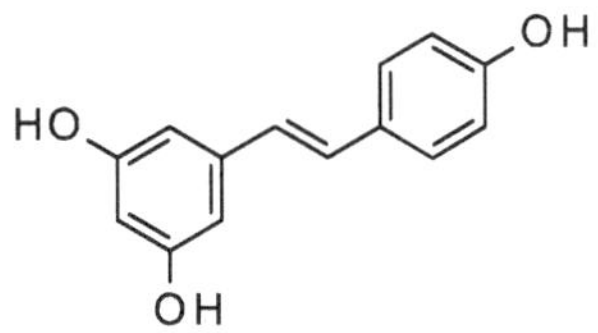

Resveratrol

Sources: Grapes (*Vitis* species), blueberries (*Vaccinium corymbosum*), raspberries (*Rubus idaeus*), mulberries (*Morus nigra*), and peanuts (*Arachis hypogaea*)

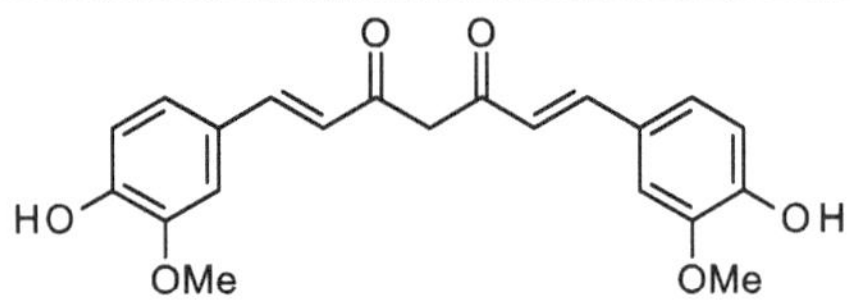

Curcumin

Source: *Curcuma longa* (Zingiberaceae)

Fig. 16 Examples of some nephroprotective natural products in clinical trials.

numerous *in vivo* animal studies, and they are antioxidant and anti-inflammatory mechanisms at cellular levels. However, there are hardly any of such molecules developed as a nephroprotective drugs, albeit, some of the are in clinical trials. Among the natural sources, plants have appeared to be almost an exclusive source of the natural nephroprotective agents.

References

1. Barbosa-Filho, J. M.; Martins, V. K. M.; Rabelo, L. A.; Moura, M. D.; Silva, M. S.; Cunha, E. V. L.; Souza, M. F. V.; Almeida, R. N.; Medeiros, I. A. *Rev. Bras* **2006**, *16*(3), 421–446.
2. Zehnder, D.; Quinkler, M.; Eardley, K. S.; Bland, R.; Lepenies, J.; Hughes, S. V.; Raymond, N. T.; Howie, A. J.; Cockwell, P.; Stewart, P. M.; Hewison, M. *Kidney Int.* **2008**, *74*(10), 1343–1353.
3. Donnelly, S. *Adv. Exp. Med. Biol.* **2003**, *543*(2), 73–87.
4. Kaneko, K.; Soty, M.; Zitoun, C.; Duchampt, A.; Silva, M.; Philippe, E.; Gautier-Stein, A.; Rajas, F.; Mithieux, G. *Mol. Metab.* **2018**, *16*, 203–212.
5. United States Renal Data System *USRDS Annual Data Report: Epidemiology of Kidney Disease in the United States*. National Institutes of Health, National Institute of Diabetes and Digestive and Kidney Diseases: Bethesda, MD, 2018; 2018. https://www.usrds.org/2018/view/Default.aspx (accessed on 15 September 2019).
6. Furuya, F.; Ishii, T.; Kitamura, K. *Contrib. Nephrol.* **2019**, *198*, 33–39.
7. Woroniecka, K. I.; Park, A. S. D.; Mohtat, D.; Thomas, D. B.; Pullman, J. M.; Susztak, K. *Diabetes* **2011**, *60*(9), 2354–2369.
8. Martini, S.; Eichinger, F.; Nair, V.; Kretzler, M. *Rev. Endocr. Metab. Disord.* **2008**, *9*(4), 267–274.
9. Mora, C.; Navarro, J. F. *Curr. Diab. Rep.* **2006**, *6*(6), 463–468.

10. Brosius, F. C.; Tuttle, K. R.; Kretzler, M. *Diabetologia* **2016**, *59*(8), 1624–1627.
11. Ruiz-Andres, O.; Sanchez-Niño, M. D.; Moreno, J. A.; Ruiz-Ortega, M.; Ramos, A. M.; Sanz, A. B.; Ortiz, A. *Am. J. Physiol. Ren. Physiol.* **2016**, *311*(6), F1329–F1340.
12. Starkey, J. M.; Haidacher, S. J.; LeJeune, W. S.; Zhang, X.; Tieu, B. C.; Choudhary, S.; Brasier, A. R.; Denner, L. A.; Tilton, R. G. *Diabetes* **2006**, *55*(5), 1252–1259.
13. World Health Organisation *Obesity and Overweight*. https://www.who.int/news-room/fact-sheets/detail/obesity-and-overweight; 2019 (accessed on 15 September 2019).
14. Hall, J. E.; Henegar, J. R.; Dwyer, T. M.; Liu, J.; Da Silva, A. A.; Kuo, J. J.; Tallam, L. *Adv. Ren. Replace. Ther.* **2004**, *11*(1), 41–54.
15. Hall, M. E.; do Carmo, J. M.; da Silva, A. A.; Juncos, L. A.; Wang, Z.; Hall, J. E. *Int. J. Nephrol. Renov. Dis.* **2014**, 7, 75–88.
16. Kopple, J. D.; Feroze, U. *J. Ren. Nutr.* **2011**, *21*(1), 66–71.
17. Sarwar, R.; Pierce, N.; Koppe, S. *Diabetes Metab Syndr Obes* **2018**, *11*, 533–542.
18. Musso, G.; Gambino, R.; Tabibian, J. H.; Ekstedt, M.; Kechagias, S.; Hamaguchi, M.; Hultcrantz, R.; Hagström, H.; Yoon, S. K.; Charatcharoenwitthaya, P.; George, J.; Barrera, F.; Haflioadottir, S.; Bjornsson, E. S.; Amstraong, M. J.; Hopkins, L. J.; Gao, X.; Francque, S.; Verriken, A.; Yilmaz, Y.; Lindor, K. D.; Charlton, M.; Haring, R.; Lerch, M. M.; Rettig, R.; Volzke, H.; Ryu, S.; Li, G.; Wong, L. L.; Machado, M.; Cortez-Pinto, H.; Yasui, K.; Cassader, M. *PLoS Med.* **2014**, *11*(7). e1001680. https://doi.org/10.1371/journal.pmed.1001680.
19. Sinn, D. H.; Kang, D.; Jang, H. R.; Gu, S.; Cho, S. J.; Paik, S. W.; Ryu, S.; Chang, Y.; Lazo, M.; Guallar, E.; Cho, J.; Gwak, G. Y. *J. Hepatol.* **2017**, *67*(6), 1274–1280.
20. Haynes, R.; Zhu, D.; Judge, P. K.; Herrington, W. G.; Kalra, P. A.; Baigent, C. *Nephrol. Dial. Transplant.* **2019**, 1–7. https://doi.org/10.1093/ndt/gfz058.
21. Imamura, Y.; Takahashi, Y.; Hayashi, T.; Iwamoto, M.; Nakamura, R.; Goto, M.; Takeba, K.; Shinohara, M.; Kubo, S.; Joki, N. *Clin. Exp. Nephrol.* **2019**, *23*(4), 484–492.
22. Antonelli, J. A.; Maalouf, N. M.; Pearle, M. S.; Lotan, Y. *Eur. Urol.* **2014**, *66*(4), 724–729.
23. Streeper, N. M.; Dubnansky, A.; Sanders, A. B.; Lehman, K.; Thomaz, E.; Conroy, D. E. *Urology* **2019**, *133*, 57–66. https://doi.org/10.1016/j.urology.2019.05.056.
24. Taylor, E. N.; Stampfer, M. J.; Curhan, G. C. *JAMA* **2005**, 293(*4*), 455–462.
25. van de Pol, J. A. A.; van den Brandt, P. A.; Schouten, L. J. *Br. J. Cancer* **2019**, *120*(3), 368–374.
26. Daenen, K.; Andries, A.; Mekahli, D.; Van Schepdael, A.; Jouret, F.; Bammens, B. *Pediatr. Nephrol.* **2019**, *34*(6), 975–991.
27. Stanton, R. C. *Curr. Diab. Rep.* **2011**, *11*(4), 330–336.
28. Nahar, L.; Sarker, S. D. *Chemistry for Pharmacy Students*; 2nd ed.; Wiley and Sons: Chichester, 2019.
29. Wong, M. K. S. *Chapter 29. Renin-Angiotensin System*; Elsevier Inc., 2015.
30. National Health Service *Chronic Kidney Disease-Treatment-NHS*. https://www.nhs.uk/conditions/kidney-disease/treatment/; 2019 (accessed on 15 October 2019).
31. Impellizzeri, D.; Esposito, E.; Attley, J.; Cuzzocrea, S. *Pharmacol. Res.* **2014**, *81*, 91–102.
32. Goicoechea, M.; De Vinuesa, S. G.; Lahera, V.; Cachofeiro, V.; Gómez-Campderá, F.; Vega, A.; Abad, S.; Luño, J. *J. Am. Soc. Nephrol.* **2006**, *17*, 231–235.
33. Su, X.; Zhang, L.; Lv, J.; Wang, J.; Hou, W.; Xie, X.; Zhang, H. *Am. J. Kidney Dis.* **2016**, *67*(6), 881–892.
34. Briasoulis, A.; Pala, M.; Telila, T.; Merid, O.; Akintoye, E.; Vogiatzi, G.; Oikonomou, E.; Tousoulis, D. *Curr. Pharm. Des.* **2017**, *23*(46), 7141–7148.

35. de Zeeuw, D.; Akizawa, T.; Audhya, P.; Bakris, G. L.; Chin, M.; Christ-Schmidt, H.; Goldsberry, A.; Houser, M.; Krauth, M.; Lambers Heerspink, H. J.; McMurray, J. J.; Meyer, C. J.; Parving, H. H.; Remuzzi, G.; Toto, R. D.; Vaziri, N. D.; Wanner, C.; Wittes, J.; Wrolstad, D.; Chertow, G. M. *N. Engl. J. Med.* **2013**, *369*(26), 2492–2503.
36. Ahmad, Q. Z.; Jahan, N.; Ahmad, G.; Tajuddin *J. Nephrol. Ther.* **2014**, 4(*4*) 1000170.
37. Ma, Z.; Wang, N.; He, H.; Tang, X. *J. Control. Release* **2019**, 316, 359–380.
38. Ben Saad, A.; Ncib, S.; Rjeibi, I.; Saidi, I.; Zouari, N. *Appl. Physiol. Nutr. Metab.* **2019**, *44*(11), 1134–1140.
39. Yang, L.; Ren, S.; Xu, F.; Ma, Z.; Liu, X.; Wang, L. *Biomed. Res. Int.* **2019**, *2019,* 5763602. https://doi.org/10.1155/2019/5763602.
40. El-Ashmawy, N. E.; Khedr, N. F.; El-Bahrawy, H. A.; Helal, S. A. *Biomed. Pharmacother.* **2018**, *99*, 504–510.
41. Kandemir, F. M.; Kucukler, S.; Caglayan, C.; Gur, C.; Batil, A. A.; Gulcin, I. *J. Food Biochem.* **2017**, *41*(5), e12398.
42. Potocnjak, I.; Broznic, D.; Kindl, M.; Kropek, M.; Vladimir-Knezevic, S.; Domitrovic, R. *Food Chem. Toxicol.* **2017**, *107*, 215–225.
43. Sanchez-Gonzalez, P. D.; Lopez-Hernandez, F. J.; Duenas, M.; Prieto, M.; Sanchez-Lopez, E.; Thomale, J.; Ruiz-Ortega, M.; Lopez-Novoa, J. M.; Morales, A. I. *Food Chem. Toxicol.* **2017**, *207*, 226–236.
44. Cascella, M.; Plma, G.; Barbieri, A.; Bimonte, S.; Amruthraj, N. J.; Muzio, M. R.; del Vecchio, V.; Rea, D.; Falco, M.; Luciano, A.; Arra, C.; Cuomo, A. *Nutrients* **2017**, *9*(6), 625. https://doi.org/10.3390/nu9060625.
45. Shalaby, Y. M.; Menze, E. T.; Azab, S. S.; Awad, A. S. *Arch. Toxicol.* **2019**, *93*(5), 1417–1431.
46. Jiang, Q.; Liu, P.; Wu, X.; Liu, W.; Shen, X.; Lan, T.; Xu, S.; Peng, J.; Xie, X.; Huang, H. *Mol. Cell. Endocrinol.* **2011**, *331*, 34–40.
47. Wu, D.; Wen, W.; Qi, C. L.; Zhao, R. X.; Lü, J. H.; Zhong, C. Y.; Chen, Y. Y. *Phytomedicine* **2012**, *19*, 712–718.
48. Cheng, H.; Bo, Y.; Shen, W.; Tan, J.; Jia, Z.; Xu, C.; Li, F. *Int. Immunopharmacol.* **2015**, *25*, 406–415.
49. Yuan, X. P.; Liu, L. S.; Fu, Q.; Wang, C. X. *Phytother. Res.* **2012**, *26*, 697–703.
50. Zhao, Y. L.; De Zhou, G.; Yang, H. B.; Wang, J. B.; Shan, L. M.; Li, R. S.; Xiao, X. H. *Food Chem. Toxicol.* **2011**, 49, 1705–1710.
51. Meng, Z.; Yan, Y.; Tang, Z.; Guo, C.; Li, N.; Huang, W.; Ding, G.; Wang, Z.; Xiao, W.; Yang, Z. *Planta Med.* **2015**, *81*, 279–285.
52. Pan, Y.; Wang, Y.; Cai, L.; Cai, Y.; Hu, J.; Yu, C.; Li, J.; Feng, Z.; Yang, S.; Li, X.; Liang, G. *Br. J. Pharmacol.* **2012**, *166*, 1169–1182.
53. Soetikno, V.; Sari, F. R.; Lakshmanan, A. P.; Arumugam, S.; Harima, M.; Suzuki, K.; Kawachi, H.; Watanabe, K. *Mol. Nutr. Food Res.* **2013**, *57*, 1649–1659.
54. Huang, J.; Huang, K.; Lan, T.; Xie, X.; Shen, X.; Liu, P.; Huang, H. *Mol. Cell. Endocrinol.* **2013**, *365*, 231–240.
55. Lee, E. J.; Kang, M. K.; Kim, D. Y.; Kim, Y. H.; Oh, H.; Kang, Y. H. *Nutrients* **2018**, *10*, 1–17.
56. Vijayaprakash, S.; Langeswaran, K.; Gowtham Kumar, S.; Revathy, R.; Balasubramanian, M. P. *Biomed. Aging Pathol.* **2013**, *3*, 119–124.
57. Yousef, M. I.; Omar, S. A. M.; El-Guendi, M. I.; Abdelmegid, L. A. *Food Chem. Toxicol.* **2012**, *48*, 3246–3261.
58. Tong, F.; Liu, S.; Yan, B.; Li, X.; Ruan, S.; Yang, S. *Int. J. Nanomed.* **2017**, *12*, 7799–7813.
59. Yang, H.; Song, Y.; Liang, Y. N.; Li, R. *Med. Sci. Monit.* **2018**, *24*, 4760–4766.
60. Liu, Y.; Wang, J.; Zhang, X.; Wang, L.; Hao, T.; Cheng, Y.; Wang, D. *Oxid. Med. Cell. Longev.* **2019**, https://doi.org/10.1155/2019/1354345
61. Safhi, M. M. *Oxid. Med. Cell. Longev.* **2018**, *2018,* 2474831. https://doi.org/10.1155/2018/2474831.

62. Guo, C. C.; Li, Y.; Zhang, R.; Zhang, Y. Q.; Zhang, J. Y.; Yao, J. M.; Sun, J.; Dong, J. J.; Liao, L. *Front. Pharmacol.* **2018**, 9, 1433. https://doi.org/10.3389/fphar.2018.01433.
63. Wang, C. P.; Wang, Y.; Wang, X.; Zhang, X.; Ye, J. F.; Hu, L. S.; Kong, L. D. *Planta Med.* **2011**, 77, 786–794.
64. Palsamy, P.; Subramanian, S. *Biochim. Biophys. Acta Mol. basis Dis.* **1812**, *2011*, 719–731.
65. Moridi, H.; Karimi, J.; Sheikh, N.; Goodarzi, M. T.; Saidijam, M.; Yadegarazari, R.; Khazaei, M.; Khodadadi, I.; Tavilani, H.; Piri, H.; Asadi, S.; Zarei, S.; Rezaei, A. *Int. J. Endocrinol. Metab.* **2015**, *13*, 8–13.
66. Jang, I. A.; Kim, E. N.; Lim, J. H.; Kim, M. Y.; Ban, T. H.; Yoon, H. E.; Park, C. W.; Chang, Y. S.; Choi, B. S. *Nutrients* **2018**, *10*, 1–15.
67. Osman, A. M. M.; Telity, S. A.; Damanhouri, Z. A.; Al-Harthy, S. E.; Al-Kreathy, H. M.; Ramadan, W. S.; Elshal, M. F.; Khan, L. M.; Kamel, F. *Cancer Cell Int.* **2015**, *15*, 1–8.
68. He, T.; Xiong, J.; Nie, L.; Yu, Y.; Guan, X.; Xu, X.; Xiao, T.; Yang, K.; Liu, L.; Zhang, D.; Huang, Y.; Zhang, J.; Wang, J.; Sharma, K.; Zhao, J. *J. Mol. Med.* **2016**, *94*, 1359–1371.
69. Hou, B.; Qiang, G.; Zhao, Y.; Yang, X.; Chen, X.; Yan, Y.; Wang, X.; Liu, C.; Zhang, L.; Du, G. *Cell. Physiol. Biochem.* **2018**, *44*, 2378–2394.
70. Altinoz, E.; Oner, Z.; Elbe, H.; Cigremis, Y.; Turkoz, Y. *Hum. Exp. Toxicol.* **2015**, *34*, 127–134.
71. Yaribeygi, H.; Mohammadi, M. T.; Rezaee, R.; Sahebkar, A. *J. Cell. Biochem.* **2018**, *119*, 6080–6093.
72. Mou, Z.; Feng, Z.; Xu, Z.; Zhuang, F.; Zheng, X.; Li, X.; Qian, J.; Liang, G. *Biochem. Biophys. Res. Commun.* **2019**, *508*, 243–249.
73. Jiang, C.; Shao, Q.; Jin, B.; Gong, R.; Zhang, M.; Xu, B. *Biomed. Res. Int.* **2015**, *2015*, 867140. https://doi.org/10.1155/2015/867140.
74. Wang, D. T.; Huang, R. H.; Cheng, X.; Zhang, Z. H.; Yang, Y. J.; Lin, X. *Int. Immunopharmacol.* **2015**, *26*, 4–12.
75. Pham, J. V.; Yilma, M. A.; Feliz, A.; Majid, M. T.; Maffetone, N.; Walker, J. R.; Kim, E.; Cho, H. J.; Reynolds, J. M.; Song, M. C.; Park, S. R.; Yoon, Y. *J. Front. Microbiol.* **2019**, https://doi.org/10.3389/fmicb.2019.01404.
76. Parveen, A.; Jin, M.; Kim, S. Y. *Phytomedicine* **2018**, *39*, 146–159.
77. Avila-Rojas, S. H.; Lira-Leon, A.; Aparicio-Trejo, O. E.; Reyes-Fermin, L. M.; Pedraza-Chaverri, J. *Medicina-Lithuania* **2019**, 55(7), 360. https://doi.org/10.3390/medicina55070360.
78. Sarker, S. D.; Nahar, L. *Computational Phytochemistry*; Elsevier: USA, 2018.
79. Trejo, E. M. A. G.; Buendia, A. S. A.; Reyes, O. S.; Arroyo, F. E. G.; Garcia, R. A.; Mendoza, M. L. L.; Tapia, E.; Lozada, L. G. S.; Alonso, H. O. *Int. J. Mol. Sci.* **2017**, 18(*9*), 1980. https://doi.org/10.3390/ijms18091980.
80. Morgado-Pascual, J. L.; Rayego-Mateos, S.; Tejedor, L.; Suarez-Alvarez, B.; Ruiz-Ortega, M. *Front. Pharmacol.* **2019**, *10*, 1–15.
81. Dhananjayan, K. *J.Cancer Res.* **2015**, *2015*, 1–15.
82. Spiegler, V.; Sendker, J.; Petereit, F.; Liebau, E.; Hensel, A. *Molecules* **2015**, *20*(8), 14810–14832.
83. Yong, T.; Chen, S.; Xie, Y.; Chen, D.; Su, J.; Shuai, O.; Hu, H.; Zuo, D.; Liang, D. *Am. J. Chin. Med.* **2018**, *46*(3), 585–599.
84. Yong, T.; Chen, S.; Xie, Y.; Shuai, O.; Li, X.; Chen, D.; Su, J.; Jiao, C.; Liang, Y. *Front. Pharmacol.* **2018**, *9*, 1–10.
85. Khanra, R.; Bhattacharjee, N.; Dua, T. K.; Nandy, A.; Saha, A.; Kalita, J.; Manna, P.; Dewanjee, S. *Biomed. Pharmacother.* **2017**, *94*, 726–741.
86. Hewlings, S. J.; Kalman, D. S. *Foods* **2017**, 6, 92. https://doi.org/10.3390/foods6100092.
87. Sattarinezhad, A.; Roozbeh, J.; Yeganeh, B. S.; Omrani, G. R.; Shams, M. *Diabetes Metab.* **2019**, *45*(1), 53–59.

CHAPTER NINE

Cancer chemopreventive natural products

Kenneth J. Ritchie*, Satyajit D. Sarker
Centre for Natural Products Discovery (CNPD), School of Pharmacy and Biomolecular Sciences, Liverpool John Moores University, Liverpool, United Kingdom
*Corresponding author: e-mail address: k.j.ritchie@ljmu.ac.uk

Contents

1. Introduction

Cancer has haunted mankind since time immemorial with early records of cancer incidence dating as far back as 3000 years.[1] Cancer

Annual Reports in Medicinal Chemistry, Volume 55
ISSN 0065-7743
https://doi.org/10.1016/bs.armc.2020.02.004

incidence during the twentieth century, however, seen the largest rise since records began, and continues to be a major cost for health care providers worldwide.[2] The latest statistics indicate that, in 2019, about 1.7 million US citizens are likely to be diagnosed with the diease.[2] The prevention of cancer is consequently a major priority for governments and health care provides globally and has resulted in large scale public awareness campaigns to raise awareness of the dangers of tobacco smoking (lung cancer) and exposure to sunlight (skin cancer). Such lifestyle choices are also reflected in dietary advice to prevent cancer with the well-known '5-a-day' campaign to increase the consumption of fruits and vegetables in an attempt to increase the general health of a population and as a consequence, reduce the incidence of major diseases such as cancer.[3] This general philosophy of natural products being beneficial to health has been expanded and solidified through the large body of research, which has investigated the ability of plant phytochemicals to prevent cancer.

2. Cancer chemoprevention

Cancer chemoprevention was first used as a term by Sporn et al. in the context vitamin A preventing cancer *in vivo*.[4] Since that point the term has become widely used and can be broken down into two distinct areas, which align closely with the generally accepted theory of carcinogenesis involving an initiation event (initial DNA damage), followed by a promotional series of events involving signal transduction pathways and associated hormones, inflammation, the avoidance of apoptosis and metastasis. Consequently, cancer chemopreventives are traditionally defined into two categories:

2.1 Blocking agents

These are chemicals that prevent DNA damage from occurring following exposure to DNA damaging agents of both internal (endogenous free radicals) and external (carcinogens and pro-carcinogens) origins. Such activity prevents not only initial DNA damage and hence the initiation of cancer, but also subsequent gain of additional mutations or 'hits,' which contribute to the increasing genomic instability characteristic of oncogenic development. The cellular mechanisms involved in such blocking strategies include the induction of cellular detoxification pathways (Phase II drug metabolism), the prevention of the metabolic activation of pro-carcinogens (Phase I drug metabolism) and the deceased uptake of such chemicals, the inhibition of inflammatory processes, the promotion of apoptosis and a reduction in the production of reactive oxygen species and nitrogen species within the cell.

2.2 Suppressing agents

These are agents generally aimed at reducing the progression of the initiated (mutated) cell to clinically defined cancer and/or the subsequent spread of the primary cancer to secondary sites. Inhibition of the promotional activity of hormones and signal transduction pathways such as testosterone in prostate cancer and estrogen in breast cancer are consequently important mechanisms of suppression. Signaling pathways associated with chronic inflammation, which is also an important promoter of cancer, the induction of cell defense genes that can dampen inflammation, the stimulation of apoptosis in transformed cells and the inhibition of metastasis are also central to the idea of suppressing agents in cancer chemoprevention.

In consideration of the above summary, it is clear that several mechanisms of cancer chemoprevention are common to both the blocking and suppression mechanisms, and can be grouped into four common areas: induction of cell defense genes, inhibition of inflammation, stimulation of apoptosis and inhibition of metastasis (Fig. 1).

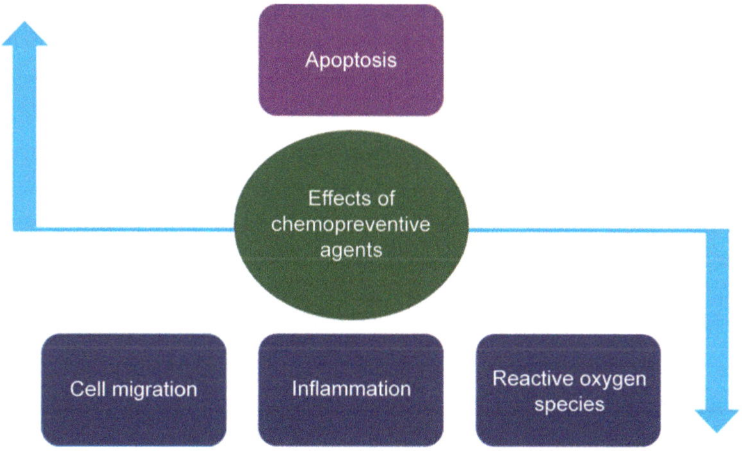

Fig. 1 Four common areas of chemoprevention.

3. Central tenets of cancer chemoprevention

3.1 Reduction of inflammation

Inflammation is a well-known driver of carcinogenesis. Several cancers are known to be directly linked to inflammatory processes. Long term chronic inflammation can act as a powerful promoter of oncogenic transformation and promotion, with more than one fifth of all human cancers being attributed to long term and dysregulated inflammation.[5] Important in the cellular machinery that is central to inflammation is the transcription factor NF-κB, which regulates the transcription of many genes involved in inflammation.[6] Natural products, predominantly phenolic compounds from higher plants, such as apigenin, β-carotene, capsaicin, curcumin, genistein, kaempferol, luteolin, lycopene, quercetin, resveratrol and silymarin supress NF-κB activation (Fig. 2).[7] Long term bacterial infection, such as that seen with *Helicobacter pylori*, is also directly linked to carcinogenesis and many natural products are noted to have antibacterial properties.[8] Similarly, the prevention of infection by viruses associated with cancer initiation, such as hepatitis B and C, HIV and HPV, has been linked to natural products.[9]

3.2 Induction of cellular defense mechanisms

Oxidative stress is the unavoidable consequence of utilizing cellular biochemistry, which depends upon molecular oxygen to produce Adenosine Triphosphate (ATP). The consequent leakage of reactive oxygen species (ROS) such as the superoxide anion from the mitochondria allows the production of a chain reaction of ROS with ultimate production of the hydroxyl ion. The hydroxyl ion is commonly accepted to be the most damaging ROS and can readily attack DNA causing mutations and increased cancer risk. The traditional view point of the ability of natural products to reduce such oxidative stress was centred on the ability of molecules such as vitamin C and vitamin E to quench and deactivate ROS. Although this is still the case in the last two decades, the ability of natural products to induce enzyme based cellular defense mechanisms to counter act the effects of ROS have been widely investigated with much of that research focused on the master regulator molecule of antioxidant gene induction, Nrf2 (nuclear erythroid 2-related factor).

3.3 Promotion of apoptosis

Apoptosis is cellular suicide. It is initiated by the cell itself, is energy dependent and requires the activation of genes and the production of proteins

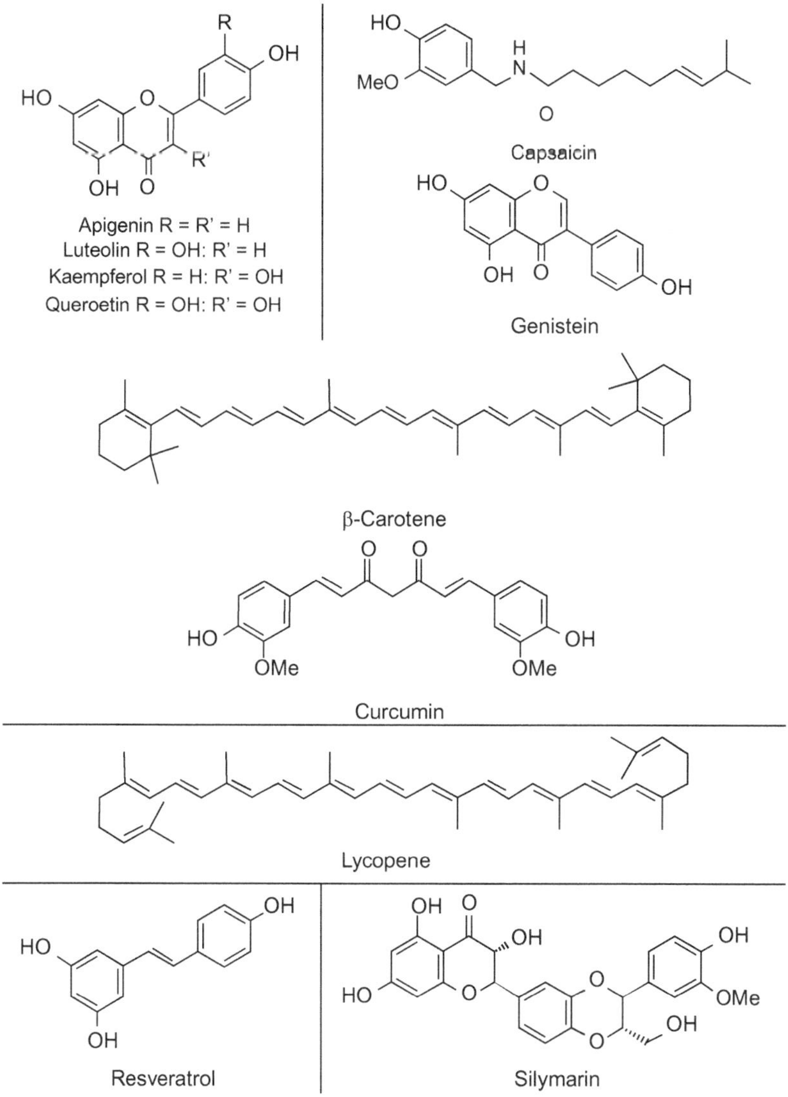

Fig. 2 Examples of some natural products that suppress NF-κB activation.

to deconstruct the cell in an orderly manner that does not cause inflammation to its surroundings. Many human cancers, however, lack the ability to undergo apoptosis due to mutations in the apoptotic transcription factor p53.[10] The ability to induce cell death before this key ability is lost is therefore central to tenets of chemoprevention. Many natural products have been shown to induce apoptosis including apigenin, curcumin, fisetin,

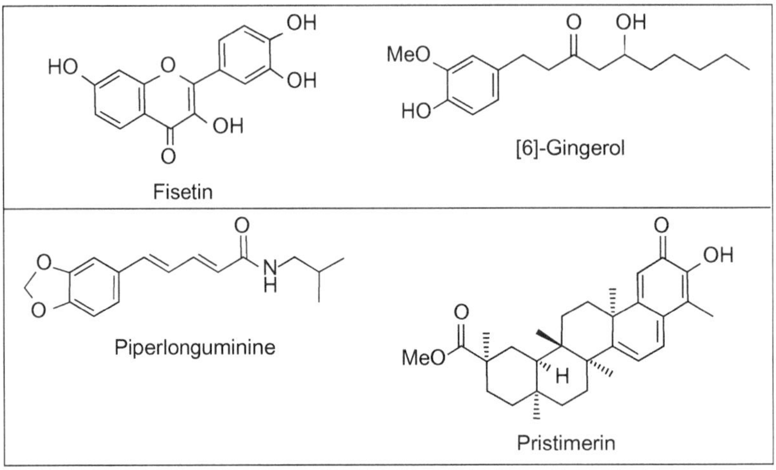

Fig. 3 Examples of some natural products that induce apoptosis.

[6]-gingerol, resveratrol, piperlonguminine, pristimerin and quercetin (Figs. 2 and 3). Recently it has also been reported that a companion mechanism to apoptosis, termed autophagy is also influenced by natural products. Autophagy occurs prior to apoptosis and involves the selective destruction of cellular components such as organelles by the lysosome and is also instigated by many of the previously mentioned natural products.[11]

3.4 Inhibition of metastasis

The ability of cancer to spread or metastasize is one of the key factors in determining the severity of the cancer as the degree of spread around the body is often used to grade cancer progression. Such movement of cancerous cells away from the primary tumor is dependent upon the cellular pathways that govern cell-to-cell interactions. E-cadherin is a molecule critically involved in cell-cell adhesion in epithelial tissues the down regulation of which is thought to critical to the ability of cancer cells to invade basal tissue, migrate in the circulatory system, seed into distant organs and begin growth into new metastasis.[12] Similarly transition of cancerous epithelial cells to a mesenchymal phenotype is also associated with cancer spread to areas distant from the primary site of tumorigenesis.[13] Importantly natural products such as curcumin, resveratrol (Fig. 2) and green tea polyphenols have been noted to both increase the expression of E-cadherin and decrease epithelial-mesenchymal transition (EMT) as detailed later in this chapter.

4. Cellular mechanisms of cancer chemoprevention

4.1 Induction of cellular defense mechanisms

Human physiology poses the remarkable ability to process and detoxify xenobiotic compounds, which are yet to be invented, indeed our metabolic biochemistry has not significantly changed in humans for millennia, yet today we regularly are exposed to hundreds of synthetic chemicals that we easily metabolize.[14] In this process of metabolism, the central aim is to take exogenous xenobiotic chemical entities, which are lipophilic and hence not readily excreted into more polar compounds that may be removed easily from our body. The enzymatic systems responsible for this important process are termed Phase I and Phase II drug metabolism. Phase I drug metabolism is principally concerned with the alteration of chemical structure through processes such as hydroxylation, dealkylation and dehalogenation (not exhaustive list) and is principally carried-out by enzymes of the cytochrome P450 family with the ultimate aim of producing a chemical moiety, which is amenable to further modification and detoxification by Phase II drug metabolism. Although generally considered to be part of the chemical detoxification pathway Phase I drug metabolism as a consequence of its alteration of chemical structures can in fact cause an increase the carcinogenicity of chemicals (activation of pro-carcinogens). Phase II drug metabolism is considered to be the 'true' detoxification pathway and sees the addition of an endogenous moiety to the products of Phase I drug metabolism hence increasing polarity, water solubility and excretion. Central to the induction and activation of Phase II drug metabolizing enzymes is the transcription factor protein Nrf2.

In the last two decades, the Nrf2 protein has been established as the gate keeper of Phase II drug metabolism gene induction.[15,16] Nrf2 is normally inactive in the cell due to its rapid polyubiquitination and hence degradation by the action of a E3 ubiquitin ligase complex which is formed between the proteins Keap1 and Cullin3.[17] The consequence of this degradation is the depression of Nrf2 gene induced activation. Crucially, however, the Keap1 protein acts as an electrophilic sensor due to reactive cysteine residues found within it, which are easily altered by electrophiles and oxidative stress resulting in the loss of interaction with Cullin3.[18–20] The consequence of this interaction is that newly synthesized Nrf2 is not immediately degraded and can translocate to the nucleus of the cell. Once inside the nucleus Nrf2 combines with a small Maf (sMAF) transcription factor forming a dimer, which then has the ability to bind to the antioxidant-responsive element

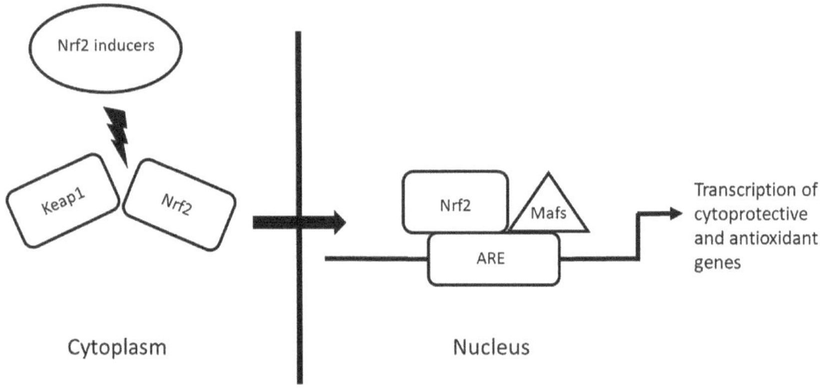

Fig. 4 Simplified diagram of Nrf2 activation.

(ARE) or the electrophile-responsive element (EpRE) found with the promoters of over 200 genes many of which are involved in the protection of the cell against electrophilic and oxidative damage.[21–23] (see Fig. 4.). This master regulator of cellular defense is, however, crucially not only activated by toxic substances that are potentially harmful to the cell, but also a wide variety of phytochemicals.[24]

Although an understanding of the cancer preventing potential of natural Nrf2 inducers was not realized until the early 2000's, as early as 1972 observations were beginning to be made regarding the potential of phenolic antioxidant butylated hydroxyanisole (BHA) to decrease the incidence of cancer.[25,26] Indeed the archetypal indicator gene of Nrf2 induction NAD(P)H:quinone reductase (NQ01), was one of the first genes to be associated with both BHA exposure and the prevention of cancer.[27] Common to the ability of such natural products to activate the Nrf2 system is the presence of olefinic bonds, which are made electrophilic upon reaction with electron withdrawing substrates.[28,29] Such intermediates may then react with the thiol groups present on the cysteine molecules of Keap1 this rendering it inactive and allowing nascent Nrf2 to be directed to the nucleus.[30] The action of such 'soft' electrophiles (non-genotoxic) consequently switches on Nrf2 activated cellular defenses in readiness for exposure of the cell to 'hard' electrophiles (genotoxic) is termed the 'electrophile counter attack response.'[28]

4.2 Inhibition of inflammation

Long term chronic inflammation has been implicated to the occurrence of cancer for at least 500 years as noted by the Ayurvedic medicine.[31]

Such inflammation may be induced by a variety of factors including the repetitive exposure to chemicals and allergens, infection by bacteria and viruses and obesity.[32] Consistent in the environment induced by chronic inflammation is the production of oxidative stress, which is present in inflammatory sites due to the presence of mast cells and leukocytes which produce large quantities of ROS during a process known as the 'respiratory burst.'[33] The resultant oxidative stress stimulates the continued recruitment of inflammatory cells into the inflammatory site, the release of further soluble mediators such as arachidonic acid, cytokines and chemokines to attract further immune cells continuing the cycle. Traditionally, the occurrence of oxidative stress and the production of ROS has been associated with the production of DNA mutation in proto-oncogenes, tumor suppressor genes, DNA strand breaks and point mutations.[34] Although such DNA damage is undoubtedly present in the context of inflammation-induced carcinogenesis the role of low level but constant oxidative stress has now, however, been expanded to include the activation of cell signaling pathways related to cell proliferation and cell survival such as such as activator protein-1 (AP-1), extracellular signal-regulated kinase/mitogen-activated protein kinase (ERK/MAPK), and phosphoinositide 3-kinase/AKT8 virus oncogene cellular homologue (PI3K/Akt).[32] However, in consideration of the focus of this chapter on natural products, the most common cell signaling pathway involved in inflammation that is also associated with cancer and can be modified by natural products is the pro-inflammatory transcription factor and master regulator of inflammation NF-κB nuclear factor kappa B (NF-κB) which is thought to regulated more than 500 oncogenes.[35]

Cancer has been famously defined as requiring six 'hallmarks' or capabilities that are required for the progression of a cancer.[36] A commonality between all six is that they have all been found to involve NF-κB activation.[37] Similar to Nrf2, NF-κB is held inactively in the cytoplasm of the cell by the inhibitory molecule IκB. Upon receiving relevant cell signaling stimulation IκB kinase is activated and phosphorylates IκB inhibitor allowing it to be ubiquitinated and degraded by the proteasome. NF-κB is then free to travel to the nucleus to bind to and activate target genes.[38] Notably there is a plethora of reports in the literature that denote the ability of various natural products to inhibit this pathway. The most commonly accepted mechanisms by which natural products are thought to mediate their effects are through inhibition of IκB kinase activation and subsequent reduction of IκB inhibitor phosphorylation, ubiquitination and proteasomal degradation.[35] Extensively reported phytochemicals, such as curcumin, epigallocatechin gallate (EGCG) and resveratrol, are thought to inhibit NF-κB activity using such mechanisms.[39–41]

4.3 Promotion of apoptosis

The ability of a cell to enter apoptosis is a key mechanism by which an initiated cell (which has sustained damage in a key proto-oncogene) is prevented from gaining further mutations and progressing to a fully transformed (cancerous) state. Accordingly the goal of chemoprevention in this case is to stimulate the occurrence of apoptosis. At a molecular level the transcription factor p53 is central to this process as exemplified by the higher rate of cancer incidence in individuals who inherit a mutant p53 allele and the increased cancer rates observed in mouse models of cancer that have been bred to mouse models with mutations or absence of the p53 gene.[42,43] Under normal cellular conditions p53 is prevented from binding to DNA and inducing the pro-apoptotic response by the E3 ubiquitin ligase Mdm2.[44] However, under conditions of oxidative stress, DNA damage, hypoxia, telomere shortening and oncogenic activation the interaction between Mdm2 and p53 is reduced and p53 is freed to translocate to the nucleus to induce a plethora of genes involved in a myriad of cellular functions which also includes the induction of genes associated with the regulation of oxidative stress.[45] Enzymes such as manganese superoxide dismutase (MnSOD) and glutathione peroxidase (GPx1) are subsequently induced by p53, which also regulates reduced glutathione levels.[46,47]

Much research has been carried-out to find natural products that inhibit Mdm2 and/or activate p53. Careful observation of this literature, however, reveals distinct categories of potential 'mechanism-of-action' which these natural products may be divided into. Natural products consequently are found to either inhibit Mdm2's expression and stability, directly interfere in the interaction between p53 and Mdm2 or inhibit the ubiquitin ligase activity of Mdm2 allowing the stabilization of p53.[48]

4.4 Inhibition of metastasis

Although cancer chemoprevention is commonly thought of as being the prevention of the initial DNA damaging event, it is important to realize that most cancer deaths do not occur as a consequence of the impact the primary tumor has on its initial location. Rather it is the movement or 'spread' of the primary cancer to a distant site through the process of metastasis which accounts for 90% of cancer-related deaths.[13] As a consequence the inhibition of metastasis is an important target for cancer chemoprevention.

Such metastasis is driven by a process known as epithelial-to-mesenchymal transition (EMT). During the EMT process epithelial cells of the primary tumor gain a mesenchymal phenotype and exhibit increased motility, the gain of a spindle-like morphology and loss of the typical apical-basolateral polarity normally observed in epithelial cells.[12] These carcinoma cells, which have undergone this transition then 'escape' from the primary tumor into the surrounding vasculature and proceed to seed distant organs in a process that is thought to occur in minutes and potentially before the primary tumor has been even detected.[49] This fact underlines the importance of chemoprevention both in the prevention of the initial DNA damage, which drives the initial cancer formation but also in the inhibition of early metastasis.

At a mechanistic level EMT is controlled by several transcription factors, the so called major EMT-TFs which are SNAI1, SNAI2, ZEB1, ZEB1, and TWIST1, which suppress the expression of the cell adhesion molecule E-cadherin at the level of transcription.[50] Recent research has, however, revealed that these EMT-TFs are themselves regulated and controlled by a myriad of complex cellular pathways such as miRNAs (MiR-200 and Mir-34), other transcription factors (including p53), epigenetic regulators (HDAC1/2) and ubiquitin-mediated proteasomal degradation among others.[13] Despite this multitude of regulatory factors EMT has only been found to be consistently affected by a relatively small amount of natural products. Natural products such as apigenin, curcumin, genistein, resveratrol (Fig. 2) and EGCG have been consequently identified as regulating EMT.[51–56]

5. Phytochemicals principally associated with chemoprevention

In consideration of the previously outlined factors that are proposed as being central to the action of a chemopreventive agent, the following selected natural products have been reported to poses all of the associated biological activities. It is important to note, however, that this is not an exhaustive list, but rather an attempt to provide an example of how by clearly highlighting the capability of the selected natural products to modulate all of the previously mentioned chemopreventive pathways a focus may be applied to the field of natural product chemoprevention.

5.1 Curcumin

Curcumin (Fig. 2) is the most studied natural product in relation to cancer and has been reported to affect a multitude of different cancers mostly through the ability to interfere with cellular signaling pathways (also see chapter "Anticancer natural products" by Sarker et al.). The phytochemical itself is derived from turmeric, the well-known yellow colored spice commonly associated with the Asian culinary dish of curry. Indeed the extremely low incidence of breast cancer incidence in India was one of the first rationales for investigating the chemopreventive ability of curcumin.[57] Turmeric is classically found in the rhizomes of the herb *Curcuma longa* Linn. and consists of several curcuminoids including curcumin (77%), demethoxycurcumin (17%), and bisdemethoxycurcumin (3%). Curcumin is a polyphenol [1,7-bis(4-hydroxy-3-methoxyphenyl)-1,6- heptadiene-3,5-dione] and also found in other *Curcuma* species such as *C. aromatic, C. mangga, C. phaeocaulis, C. xanthorrhiza, C. zedoaria,* as well as species of other genera of the Zingiberaceae family, for example, *Etlingera elatior* and *Zingiber cassumunar.*[58] Curcumin has been found to have an anticancer effect on almost all types of cancer cells and has been reported to inhibit cancer cells between the 1 and 100 μM concentration.[59] Although undoubtedly active against cancer the ability of curcumin as a drug is limited by its poor solubility and susceptibility to first pass metabolism, which combine to result in low oral bioavailability.[60] Of the central mechanisms of cancer prevention as outlined earlier, curcumin has been found to modify all. In consideration of the process of EMT curcumin has been found to supress a host of EMT-suppressive miRNAs (miR-34a, miR-101, miR-141, miR-200b, miR-200c and miR-429).[61] It has also been reported to induce apoptosis in a wide variety of cancers.[62] Interestingly, part of this ability to induce apoptosis comes from its ability to induce cellular oxidative stress *via* the production of ROS. Such activity has been noted in glioma cells, hepatocellular carcinoma cells, neuroblastoma cells, leukaemic cells and osteosarcoma cells.[63–67] Such activity (the production of ROS) is also likely to be responsible for the induction of Nrf2 by curcumin.[68–73] The downregulation of the transcription factor NF-κB and as a consequence inflammatory signaling pathways has also been reported consistently.[74–76] Clinically curcumin has been investigated in numerous clinical trials against a variety of cancers such as breast, cervical, colorectal, prostate and pancreatic cancer and has been expertly reviewed recently.[77]

5.2 Resveratrol

Resveratrol is perhaps one of the best known phytochemicals as a consequence of the general public being aware of the health benefits of drinking red wine. This knowledge arose due to the observation that the wine drinking nation of France had relatively low levels of cardiovascular disease despite eating a diet relatively high in saturated fat.[78] Resveratrol (3,4′,5-trihydroxy-*trans*-stilbene) is a polyphenol and is also classified as a phytoalexin. It is commonly found in various foods such as peanuts, berries and grapes in which it is believed to be involved in host defense against infection by the fungus *Botrytis cinerea* (gray mold). The biosynthesis of resveratrol is catalyzed by the enzyme resveratrol synthase the gene for which has been added into the genome of various plants (tomatoes, potatoes and rice) to confer resistance against the mold.[79] Resveratrol is composed of two phenolic rings connected by a styrene double bond and can occur in both the *trans*- and *cis*-isomers, although it is the *trans*-isomer that is the major isomer and has been subjected to the most studies.

Since the initial seminal report that resveratrol could inhibit the development of skin cancer in a mouse model, numerous studies have shown the ability of resveratrol to act as a chemopreventive agent.[80] This stilbene has consequently been shown to be excellent at reducing oxidative stress as a result of its intrinsic ability as a highly efficient scavenger of ROS such as hydroxyl and super oxide radicals and as a consequence preventing DNA damage.[81] Resveratrol has also been found to inhibit the aryl hydrocarbon receptor (AhR)-mediated activation of Phase I enzymes such as CYP1A1 and CYP1B1 and the conversion of the polycyclic aromatic hydrocarbon (PAH) benzo[*a*]pyrene (B[*a*]P) to its mutagenic metabolite.[82,83] Phase II enzymes such as NAD(P)H:quinone oxidoreductase-1 (NQO1) and heme oxygenase-1 (OH-1) are also found to be induced by resveratrol through Nrf2-mediated activation. Such activity has been found to prevent oxidative damage to lung epithelial cells following exposure to cigarette smoke, and estrogen induced mammary carcinogenesis.[84–86]

Consistent with the most common anticancer phytochemicals that are being reviewed here, resveratrol is also reported to induce both apoptosis and autophagy through various different pathways and mechanisms. Of the most recent findings resveratrol is reported to induce apoptosis in ovarian cancer cells by inducing the enhanced breakdown of galectin-3 (GAL-3), which is an anti-apoptotic lectin that is highly overexpressed in ovarian cancer and has been found to increase the efficacy of tumor necrosis factor-related apoptosis-inducing ligand (TRAIL) induced apoptosis in renal cell carcinoma.[87,88] The transcription factor NF-κB, which is involved in both inflammation (as previously noted) and the induction of apoptosis is also reported to be negatively regulated by resveratrol.[89,90] Numerous studies have shown that resveratrol is capable of inhibiting EMT.[91–94]

5.3 Epigallocatechin-gallate (EPCG)

After water itself, tea is estimated to be the second most consumed beverage in the world.[95] The dried leaves of *Camellia sinensis* are used to make the most common types of tea (black and green), although a wide range of different teas are available originating from many different plants, berries and fruits. Of the most common types, black tea and green tea are perhaps among the most widely researched, with black tea differing from green tea simply due to a fermentation process being involved in the preparation of the leaves of the former which encourages oxidation.[95] Due to the lack of processing involved in its preparation, green tea contains a higher amount of the flavanol group of polyphenols known as catechins, which may constitute up to 30% of the dry leaf weight.[95] The major polyphenolic component found in green tea is EGCG (Fig. 5), which accounts for at least 50% of the total catechin content in green tea leaves.[96] The molecule itself has been extensively researched in terms of cancer prevention and has been found to poses various properties associated with the prevention of cancer. Consistent with the majority of well-known anticancer phytochemicals EGCG has the capacity to reduce oxidative stress produced by ROS, but also possesses strong pro-oxidant activity.[97,98] EGCG is also widely reported to inhibit inflammation through the inhibition of NF-κB signaling.[99,100] The induction of apoptotic cell death by EGCG has also been noted in many cells *in vitro* including oral, esophageal and lung cancer cell lines.[101–103] EGCG is also found to inhibit EMT in thyroid and pancreatic cells *in vitro*.[104,105] The inhibition of EMT-mediated transformation by EGCG has also been reported to be induced by Nrf2 activation.[106] The induction of Phase II detoxification enzymes such as glutathione S-transferase, glutathione peroxidase, glutamate cysteine ligase, hemeoxygenase-1 has also been widely reported as consequence of exposure to EGCG and has been comprehensively reviewed elsewhere.[107]

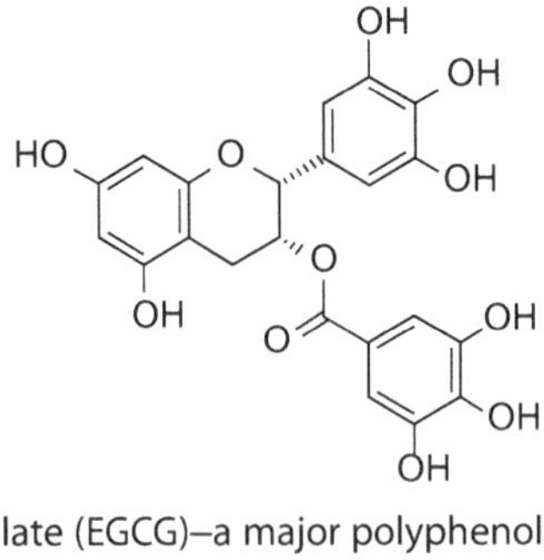

Fig. 5 Epigallocatechin gallate (EGCG)–a major polyphenolic compound present in tea (*Camellia sinensis*).

5.4 Quercetin

Quercetin (Fig. 2) is one of the most commonly consumed flavonoids in the Western diet due to it being widely found in both fruits and vegetables. It is estimated that average daily ingested amount is 25 mg with the consumption of onions being a particularly good source of the molecule.[108] The flavonoid itself is found in normally ingested in a glycosylated form, which is then processed in the intestine by β-glycosidases before absorption. Quercetin then undergoes complex biotransformation involving glucuronidation, sulfation, methylation, possible deglucuronidation and ring fission.[109] This flavonol, however, has limited water-solubility and as a consequence, has low bioavailability.[109] Nonetheless quercetin has been reported to possess many biological properties important in the prevention of cancer. The ability of quercetin to induce apoptosis is well recognized and has been shown in several cancers including ovarian, gastric, cervical and skin.[110–112] While inhibition of metastasis by down regulation of EMT is observed cells derived from prostate, colon, pancreatic and lung cancers.[113–117] A reduction in inflammation by modulation of NF-κB by quercetin is seen in many precancerous conditions such as hepatitis and skin damage by UV light.[118–121] The induction of cell protective enzymes by Nrf2 stimulation has also been observed following exposure to quercetin in many different cells including HaCaT, HepG2 and granulosa cells.[122–126]

5.5 Genistein

Genistein (Fig. 2) is an isoflavone, which is found in soy based foods such as soy cheeses and soy based drinks which are made from soy beans that contain on average of 81 mg/100 g of soy.[127] The molecule Genistein was first isolated from *Genista tinctoria* L. and subsequently named after it and is classified as a phytoestrogen due to its structural similarity to estradiol and its ability to bind to estrogen receptors.[128] Although oral bioavailability is poor (owing to its poor solubility in water) genistein is largely attributed as being responsible for the low rates of both breast and prostate cancer in Asian countries such as Japan and China where soy-rich diets are more common than that in western countries (USA and Europe).[129,130] At a molecular level genistein is reported to act as a caner chemopreventive agent through the modulation cell signaling pathways for apoptosis, inflammation, the inhibition of metastasis and induction of cytoprotective enzymes. Specifically, genistein has been found to induce apoptosis in cervical, hepatocellular, and retinoblastoma cancer cells among others.[131–134] Inhibition of the master inflammation transcription factor NF-κB has been recorded in breast, colon and leukemic cancer cells.[135–137] This isoflavone is also is found to inhibit EMT in hepatocellular, gastric and thyroid cancer cells lines, while it modulates all three molecular mechanisms (apoptosis, NF-κB and EMT) in colon cancer cells.[136,138–140] The induction of cell defense enzymes *via* the transcription factor Nrf2 is also observed following exposure of the colorectal cancer cell line CaCo-2 cells to genistein.[141]

5.6 Sulforaphane

Sulforaphane is found in cruciferous vegetables such as broccoli and brussels sprouts and was first isolated and its structure elucidated by Talalay and Zhang who were also the first to show the anticancer potential of the chemical as it was noted to prevent mammary cancer in rats following exposure to the carcinogen 9,10-dimethyl-1,2-benzanthracene (DMBA).[142,143] It is important to note, however, that very little sulforaphane is present in plants under normal unstressed conditions. It is only after the plant is mechanically injured that the precursor of sulforaphane, glucoraphanin is exposed to the enzyme myrosinase and the conversion to the reactive isothiocyanate sulforaphane is made. Crucially myrosinase is not present in mammalian cells but is found in the bacterial microflora of the gastrointestinal tract.[144] Such conversion can therefore be influenced by external factors that modify gut bacteria. Sulforaphane is perhaps the best known and potent natural product inducer of Nrf2 and hence Phase II cell defense enzymes such as NQO1, HO-1 and GSH among others.[145] The induction of such genes has been directly linked to an increase in the incidence of cancer in mice in which Nrf2 has been deleted. Nrf2 knock-out mice have consequently been reported to be susceptible to various cancers following exposure to carcinogens.[146–149] Similarly feeding of sulforaphane to mice has reduced the incidence of oral and skin cancer.[150,151] In addition sulforaphane also reduces inflammation *via* inhibition of NF-κB and also induces apoptosis by stimulation of pro-apoptotic pathways and the inhibition of anti-apoptotic pathways.[152–154] EMT has also been observed to be inhibited by sulforaphane in hepatocellular and lung cancer cells.[155,156]

6. Conclusion

In this chapter we have focused on the most established and accepted mechanisms by which natural products have been found to prevent cancer. In doing so an attempt has been made to cut through the large amount of studies present in the literature which report a spectrum of biological effects produced by phytochemicals to simply focus on the phytochemicals which have the most sustained evidence of significant anticancer effects. Crucially, we have distilled the chemopreventive action of these natural products down to their fundamental actions; the induction of cell defense genes and apoptosis and the inhibition of both inflammation and EMT. This is consequently an attempt to refocus the attention of researchers into taking a comprehensive and responsible approach to the definition of chemoprevention and leads to the proposal of a new term for any natural product that possess the four 'hallmarks of chemoprevention' to be defined as a 'complete chemopreventive.' Notably chemicals such as lycopene, although very good at reducing ROS levels, would not qualify under this proposal as being a 'complete chemopreventive' owing to the lack of suitable evidence suggesting it prevents EMT. Similarly synthetic chemicals, which are known to be active chemopreventives and clinical approved for human use would also not qualify under for this definition due to them generally possessing a highly defined activity against one particular biological pathway. To account for this we further suggest the classification of all chemopreventive agents (both natural and synthetic) into either class I or class II chemopreventive agents with the former having been shown to reduce cancer risk at clinical trial and the latter only being suspected of reducing cancer risk (similar to the definitions used for carcinogens). Such classification may consequently allow a focusing of research efforts on to the most promising chemicals and as such finally allow the progression of such potential into a quantifiable benefit for human health.

References

1. Hajdu, S. I. *Cancer* **2011**, *117*(5), 1097–1102.
2. Siegel, R. L.; Miller, K. D.; Jemal, A. Cancer Statistics, 2019. *CA Cancer J. Clin.* **2019**, *69*(1), 7–34.
3. International Agency for Research on Cancer (IARC) *Handbook of Cancer Prevention*. IARC Press: Lyon, 2003.
4. Sporn, M. B.; Dunlop, N. M.; Newton, D. L.; Smith, J. M. *Fed. Proc.* **1976**, 1332–1338.
5. Chai, E. Z. P.; Siveen, K. S.; Shanmugam, M. K.; Arfuso, F.; Sethi, G. *Biochem. J.* **2015**, *468*, 1–15.
6. Didonato, J. A.; Mercurio, F.; Karin, M. *Immunol. Rev.* **2012**, *246*(1), 379–400.
7. Bellik, Y.; Boukraâ, L.; Alzahrani, H. A.; Bakhotmah, B. A.; Abdellah, F.; Hammoudi, S. M.; Iguer-Ouada, M. *Molecules* **2013**, *18*, 322–353.
8. Moloney, M. G. *Trends Pharmacol. Sci.* **2016**, *37*(8), 689–701.
9. Dhama, K.; Karthik, K.; Khandia, R.; Munjal, A.; Tiwari, R.; Rana, R.; Khurana, S. K.; Ullah, S.; Khan, R. U.; Alagawany, M.; Mahmoud, F.; Dadar, M.; Joshi, M.; Kumar, S. *Curr. Drug Metab.* **2018**, *19*(3), 236–263.

10. Hainaut, P.; Pfeifer, G. P. *Cold Spring Harb. Perspect. Med.* **2016**, *6*(11), pii: a026179. https://doi.org/10.1101/cshperspect.a026179.
11. Mizushima, N.; Levine, B. *Nat. Cell Biol.* **2010**, *12*(9), 823–830.
12. Nieto, M. A.; Huang, R. Y.; Jackson, R. A.; Thiery, J. P. *Cell* **2016**, *166*, 21–45.
13. Zhang, Y.; Weinberg, R. A. *Front. Med.* **2018**, *12*, 361–373.
14. Lewis, D. F. V.; Watson, E.; Lake, B. G. *Mutat. Res. Rev. Mutat. Res.* **1998**, *410*(3), 245–270.
15. Cullinan, S. B.; Gordan, J. D.; Jin, J.; Harper, J. W.; Diehl, J. A. *Mol. Cell. Biol.* **2004**, *24*(19), 8477–8486.
16. Zhang, D. D.; Lo, S.-C.; Cross, J. V.; Templeton, D. J.; Hannink, M. *Mol. Cell. Biol.* **2004**, *24*(24), 10941–10953.
17. Kobayashi, A.; Kang, M.-I.; Okawa, H.; Ohtsuji, M.; Zenke, Y.; Chiba, T.; Igarashi, K.; Yamamoto, M. *Mol. Cell. Biol.* **2004**, *24*(16), 7130–7139.
18. Dinkova-Kostova, A. T.; Holtzclaw, W. D.; Wakabayashi, N. *Biochemistry* **2005**, *44*(18), 6889–6899.
19. Rachakonda, G.; Xiong, Y.; Sekhar, K. R.; Stamer, S. L.; Liebler, D. C.; Freeman, M. L. *Chem. Res. Toxicol.* **2008**, *21*(3), 705–710.
20. Ishii, T.; Itoh, K.; Takahashi, S.; Sato, H.; Yanagawa, T.; Katoh, Y.; Bannai, S.; Yamamoto, M. J. *Biol. Chem.* **2000**, *275*(21), 16023–16029.
21. Itoh, K.; Chiba, T.; Takahashi, S.; Ishii, T.; Igarashi, K.; Katoh, Y.; Oyake, T.; Hayashi, N.; Satoh, K.; Hatayama, I.; Yamamoto, M.; Nabeshima, Y. *Biochem. Biophys. Res. Commun.* **1997**, *236*(2), 313–322.
22. Rushmore, T. H.; Pickett, C. B. J. *Biol. Chem.* **1990**, *265*(24), 14648–14653.
23. Friling, R. S.; Bensimon, A.; Tichauer, Y.; Daniel, V. *Proc. Natl. Acad. Sci. U. S. A.* **1990**, *87*(16), 6258–6262.
24. Kumar, H.; Kim, I.-S.; More, S. V.; Kim, B.-W.; Choi, D.-K. *Nat. Prod. Rep.* **2014**, *31*(1), 109–139.
25. Wattenberg, L. W. *J. Natl. Cancer Inst.* **1972**, *48*(5), 1425–1430.
26. Wattenberg, L. W. *Food Chem. Toxicol.* **1986**, *24*(10 – 11), 1099–1102.
27. Benson, A. M.; Hunkeler, M. J.; Talalay, P. *Proc. Natl. Acad. Sci. U. S. A.* **1980**, 77(9 II), 5216–5220.
28. Prestera, T.; Zhang, Y.; Spencer, S. R.; Wilczak, C. A.; Talalay, P. *Adv. Enzyme Regul.* **1993**, *33*(C), 281–296.
29. Talalay, P.; De Long, M. J.; Prochaska, H. J. *Proc. Natl. Acad. Sci. U. S. A.* **1988**, *85*(21), 8261–8265.
30. Kobayashi, A.; Kang, M.-I.; Watai, Y.; Tong, K. I.; Shibata, T.; Uchida, K.; Yamamoto, M. *Mol. Cell. Biol.* **2006**, *26*(1), 221–229.
31. Garodia, P.; Ichikawa, H.; Malani, N.; Sethi, G.; Aggarwal, B. B. *J. Soc. Integr. Oncol.* **2007**, *5*, 25–37.
32. Reuter, S.; Gupta, S. C.; Chaturvedi, M. M.; Aggarwal, B. B. *Free Radic. Biol. Med.* **2010**, *49*(11), 1603–1616.
33. Coussens, L. M.; Werb, Z. *Nature* **2002**, *420*(6917), 860–867.
34. Hussain, S. P.; Hofseth, L. J.; Harris, C. C. *Nat. Rev. Cancer* **2003**, *3*(4), 276–285.
35. Gupta, S. C.; Kunnumakkara, A. B.; Aggarwal, S.; Aggarwal, B. B. *Front. Immunol.* **2018**, *9*, 2160.
36. Hanahan, D.; Weinberg, R. A. Hallmarks of Cancer: The Next Generation. *Cell* **2011**, *144*, 646–674.
37. Taniguchi, K.; Karin, M. *Nat. Rev. Immunol.* **2018**, *18*, 309–324.
38. Gilmore, T. D. *Oncogene* **2006**, *25*(51), 6680–6684.
39. Singh, S.; Aggarwal, B. B. J. *Biol. Chem.* **1995**, *270*(42), 24995–25000.
40. Banerjee, S.; Bueso-Ramos, C.; Aggarwal, B. B. *Cancer Res.* **2002**, *62*(17), 4945–4954.

41. Nomura, M.; Ma, W.; Chen, N.; Bode, A. M.; Dong, Z. *Carcinogenesis* **2000**, *21*(10), 1885–1890.
42. Guha, T.; Malkin, D. *Cold Spring Harb. Perspect. Med.* **2017**, 7(4), a026187.
43. Donehower, L. A.; Lozano, G. *Nat. Rev. Cancer* **2009**, 831–841.
44. Kubbutat, M. H. G.; Jones, S. N.; Vousden, K. H. *Nature* **1997**, *387*(6630), 299–303.
45. Labuschagne, C. F.; Zani, F.; Vousden, K. H. *Biochim. Biophys. Acta Rev. Cancer* **2018**, *1870*(1), 32–42.
46. Suzuki, S.; Tanaka, T.; Poyurovsky, M. V.; Nagano, H.; Mayama, T.; Ohkubo, S.; Lokshin, M.; Hosokawa, H.; Nakayama, T.; Suzuki, Y.; Sugano, S.; Sato, E.; Nagao, T.; Yokote, K.; Tatsuno, I.; Prives, C. *Proc. Natl. Acad. Sci. U. S. A.* **2010**, *107*(16), 7461–7466.
47. Hussain, S. P.; Amstad, P.; He, P.; Robles, A.; Lupold, S.; Kaneko, I.; Ichimiya, M.; Sengupta, S.; Mechanic, L.; Okamura, S.; Hofseth, L.; Moake, M.; Nagashima, M.; Forrester, K.; Harris, C. *Cancer Res.* **2004**, *64*(7), 2350–2356.
48. Qin, J.-J.; Li, X.; Hunt, C.; Wang, W.; Wang, H.; Zhang, R. *Genes Dis.* **2018**, *5*(3), 204–219.
49. Lambert, A. W.; Pattabiraman, D. R.; Weinberg, R. A. *Cell* **2017**, *168*, 670–691.
50. Lamouille, S.; Xu, J.; Derynck, R. *Nat. Rev. Mol. Cell Biol.* **2014**, *15*, 178–196.
51. Qin, Y.; Zhao, D.; Zhou, H.; Wang, X.; Zhong, W.; Chen, S.; Gu, W.; Wang, W.; Zhang, C.; Liu, Y.; Liu, H.; Zhang, Q.; Guo, Y.; Sun, T.; Yang, C. *Oncotarget* **2016**, 7(27), 41421–41431.
52. Huang, X.; Dai, S.; Dai, J.; Xiao, Y.; Bai, Y.; Chen, B.; Zhou, M. *Onco. Targets. Ther.* **2015**, *8*, 2989–3001.
53. Ko, H.; So, Y.; Jeon, H.; Jeong, M.-H.; Choi, H.-K.; Ryu, S.-H.; Lee, S.-W.; Yoon, H.-G.; Choi, K.-C. *Cancer Lett.* **2013**, *335*(1), 205–213.
54. Kim, Y.-S.; Choi, K.-C.; Hwang, K.-A. *Phytomedicine* **2015**, *22*(11), 993–999.
55. Li, W.; Ma, J.; Ma, Q.; Li, B.; Han, L.; Liu, J.; Xu, Q.; Duan, W.; Yu, S.; Wang, F.; Wu, E. *Curr. Med. Chem.* **2013**, *20*(33), 4185–4194.
56. Jiao, D.; Wang, J.; Lu, W.; Tang, X.; Chen, J.; Mou, H.; Chen, Q. Y. *Mol. Ther. Oncolytics* **2016**, *3*, 16018.
57. Kumar, P.; Kadakol, A.; Shasthrula, P.; Mundhe, N.; Jamdade, V.; Barua, C.; Gaikwad, A. *Anticancer Agents Med. Chem.* **2015**, *15*(5), 647–656.
58. Aggarwal, B. B.; Sundaram, C.; Malani, N.; Ichikawa, H. *Advances in Experimental Medicine and Biology*. Springer New York LLC, 2007.
59. Heger, M.; van Golen, R. F.; Broekgaarden, M.; Michel, M. C. *Pharmacol. Rev.* **2014**, *66*, 222–307.
60. Shen, L.; Ji, H. F. *Trends Mol. Med.* **2012**, *18*, 138–144.
61. Toden, S.; Okugawa, Y.; Jascur, T.; Wodarz, D.; Komarova, N. L.; Buhrmann, C.; Shakibaei, M.; Richard Boland, C.; Goel, A. *Carcinogenesis* **2014**, *36*(3), 355–367.
62. Mortezaee, K.; Salehi, E.; Mirtavoos-Mahyari, H.; Motevaseli, E.; Najafi, M.; Farhood, B.; Rosengren, R. J.; Sahebkar, A. J. *Cell. Physiol.* **2019**, *234*(8), 12537–12550.
63. Zhang, Y.; Tu, L.; Zhou, X.; Li, B. *Med. Sci. Monit. Basic Res.* **2018**, *24*, 216–224.
64. Zheng, R.; You, Z.; Jia, J.; Lin, S.; Han, S.; Liu, A.; Long, H.; Wang, S. *Mol. Med. Rep.* **2016**, *13*(2), 1570–1576.
65. Picone, P.; Nuzzo, D.; Caruana, L.; Messina, E.; Scafidi, V.; Di Carlo, M. *Free Radic. Res.* **2014**, *48*(12), 1397–1408.
66. Gopal, P. K.; Paul, M.; Paul, S. *Asian Pac. J. Cancer Prev.* **2014**, *15*(1), 93–100.
67. Chang, Z.; Xing, J.; Yu, X. *Tumor Biol.* **2014**, *35*(1), 753–758.
68. Shen, T.; Jiang, T.; Long, M.; Chen, J.; Ren, D. M.; Wong, P. K.; Chapman, E.; Zhou, B.; Zhang, D. D. *Antioxid. Redox Signal.* **2015**, *23*(8), 651–664.
69. Chen, B.; Zhang, Y.; Wang, Y.; Rao, J.; Jiang, X.; Xu, Z. *J. Steroid Biochem. Mol. Biol.* **2014**, *143*, 11–18.

70. Das, L.; Vinayak, M. *PLoS One* **2015**, *10*(4), e0124000. https://doi.org/10.1371/journal.pone.0124000.
71. Ben Yehuda Greenwald, M.; Frušić-Zlotkin, M.; Soroka, Y.; Ben Sasson, S.; Bitton, R.; Bianco-Peled, H.; Kohen, R. *Oxid. Med. Cell. Longev.* **2017**, *2017*, 1–17.
72. Xie, Y. L.; Chu, J. G.; Jian, X. M.; Dong, J. Z.; Wang, L. P.; Li, G. X.; Yang, N. B. *Biomed. Pharmacother.* **2017**, *91*, 70–77.
73. Tu, Z. S.; Wang, Q.; Sun, D. D.; Dai, F.; Zhou, B. *Eur. J. Med. Chem.* **2017**, *134*, 72–85.
74. Qiao, Q.; Jiang, Y.; Li, G. *J. Pharmacol. Sci.* **2013**, *121*(4), 247–256.
75. Vageli, D. P.; Doukas, S. G.; Spock, T.; Sasaki, C. T. *J. Cell. Mol. Med.* **2018**, *22*(9), 4209–4220.
76. Giordano, A.; Tommonaro, G. Curcumin and Cancer. *Nutrients* **2019**, *11*(10), 2376.
77. Kunnumakkara, A. B.; Bordoloi, D.; Padmavathi, G.; Monisha, J.; Roy, N. K.; Prasad, S.; Aggarwal, B. B. *Br. J. Pharmacol.* **2017**, *174*, 1325–1348.
78. Ko, J. H.; Sethi, G.; Um, J. Y.; Shanmugam, M. K.; Arfuso, F.; Kumar, A. P.; Bishayee, A.; Ahn, K. S. *Int. J. Mol. Sci.* **2017**, *18*, 2589.
79. Cucciolla, V.; Borriello, A.; Oliva, A.; Galletti, P.; Zappia, V.; Fulvio Ragione, D. *Cell Cycle* **2007**, *6*(20), 2495–2510.
80. Jang, M.; Cai, L.; Udeani, G. O.; Slowing, K. V.; Thomas, C. F.; Beecher, C. W. W.; Fong, H. H. S.; Farnsworth, N. R.; Kinghorn, A. D.; Mehta, R. G.; Moon, R.; Pezzuto, J. *Science* **1997**, *275*(5297), 218–220.
81. Leonard, S. S.; Xia, C.; Jiang, B. H.; Stinefelt, B.; Klandorf, H.; Harris, G. K.; Shi, X. *Biochem. Biophys. Res. Commun.* **2003**, *309*(4), 1017–1026.
82. Ciolino, H. P.; Yeh, G. C. *Mol. Pharmacol.* **1999**, *56*(4), 760–767.
83. Santostefano, M.; Merchant, M.; Arellano, L.; Morrison, V.; Denison, M. S.; Safe, S. *Mol. Pharmacol.* **1993**, *43*(2), 200–206.
84. Zhou, X.; Zhao, Y.; Wang, J.; Wang, X.; Chen, C.; Yin, D.; Zhao, F.; Yin, J.; Guo, M.; Zhang, L.; Du, L.; Zhang, B.; Yin, X. *Biochem. Pharmacol.* **2018**, *155*, 252–263.
85. Kode, A.; Rajendrasozhan, S.; Caito, S.; Yang, S.-R.; Megson, I. L.; Rahman, I. *Am. J. Physiol. Lung Cell. Mol. Physiol.* **2008**, *294*(3), L478–L488.
86. Hsieh, T.; Lu, X.; Wang, Z.; Wu, J. M. *Med. Chem.* **2006**, *2*(3), 275–285.
87. Zeng, Y.; Li, F.; Shi, C.-W.; Du, J.-L.; Xue, Y.-J.; Liu, X.-Y.; Cao, X.; Wei, N. *Cancer Gene Ther.* **2019**. https://doi.org/10.1038/s41417-019-0150-6.
88. El-Kott, A. F.; Shati, A. A.; Ali Al-Kahtani, M.; Alharbi, S. A. *J. Food Biochem.* **2019**, *43*, e13072.
89. Wu, F.; Cui, L. *Arch. Dermatol. Res.* **2017**, *309*(10), 823–831.
90. Rasheduzzaman, M.; Jeong, J. K.; Park, S. Y. *Life Sci.* **2018**, *208*, 208–220.
91. Jin, X.; Wei, Y.; Liu, Y.; Lu, X.; Ding, F.; Wang, J.; Yang, S. *Cancer Med.* **2019**, *8*(3), 1246–1257.
92. Song, Y.; Chen, Y.; Li, Y.; Lyu, X.; Cui, J.; Cheng, Y.; Zheng, T.; Zhao, L.; Zhao, G. *Biomed Res. Int.* **2019**, *2019*, 1321973. https://doi.org/10.1155/2019/1321973.
93. Chen, K.-Y.; Chen, C.-C.; Chang, Y.-C.; Chang, M.-C. *PLoS One* **2019**, *14*(7), e0219317. https://doi.org/10.1371/journal.pone.0219317.
94. Yuan, L.; Zhou, M.; Huang, D.; Wasan, H.; Zhang, K.; Sun, L.; Huang, H.; Ma, S.; Shen, M.; Ruan, S. *Mol. Med. Rep.* **2019**, *20*(3), 2783–2795.
95. Graham, H.; Prev, N. *Medicine (Baltimore)* **1992**, *21*(3), 334–350.
96. Khan, N.; Afaq, F.; Saleem, M.; Ahmad, N.; Mukhtar, H. *Cancer Res.* **2006**, *66*, 2500–2505.
97. Lambert, J. D.; Elias, R. J. *Arch. Biochem. Biophys.* **2010**, *501*, 65–72.
98. Ellinger, S.; Müller, N.; Stehle, P.; Ulrich-Merzenich, G. *Phytomedicine* **2011**, *18*(11), 903–915.
99. Fatemi, A.; Safa, M.; Kazemi, A. *Tumor Biol.* **2015**, *36*(11), 8425–8437.

100. Li, Y. J.; Wu, S. L.; Lu, S. M.; Chen, F.; Guo, Y.; Gan, S. M.; Shi, Y. L.; Liu, S.; Li, S. L. *Tumor Biol.* **2015**, *36*(4), 2747–2761.
101. Liu, L.; Zuo, J.; Wang, G. *Oncol. Lett.* **2017**, *14*(4), 4391–4395.
102. Huang, J.; Chen, S.; Shi, Y.; Li, C.-H.; Wang, X.-J.; Li, F.-J.; Wang, C.-H.; Meng, Q.-H.; Zhong, J.-N.; Liu, M.; Wang, Z. *J. BUON* **2017**, *22*(6), 1422–1427.
103. Yoshimura, H.; Yoshida, H.; Matsuda, S.; Ryoke, T.; Ohta, K.; Ohmori, M.; Yamamoto, S.; Kiyoshima, T.; Kobayashi, M.; Sano, K. *Mol. Med. Rep.* **2019**, *20*(2), 1139–1148.
104. Li, T.; Zhao, N.; Lu, J.; Zhu, Q.; Liu, X.; Hao, F.; Jiao, X. *Bioengineered* **2019**, *10*(1), 282–291.
105. Wei, R.; Cortez Penso, N.; Hackman, R.; Wang, Y.; Mackenzie, G. G. *Nutrients* **2019**, *11*(8), 1856.
106. Kanlaya, R.; Khamchun, S.; Kapincharanon, C.; Thongboonkerd, V. *Sci. Rep.* **2016**, *6*, 30233. https://doi.org/10.1038/srep30233.
107. Na, H. K.; Surh, Y. J. *Food Chem. Toxicol.* **2008**, *46*(4), 1271–1278.
108. Rothwell, J. A.; Perez-Jimenez, J.; Neveu, V.; Medina-Remón, A.; M'Hiri, N.; García-Lobato, P.; Manach, C.; Knox, C.; Eisner, R.; Wishart, D. S.; Scalbert, A. Database **2013**, *2013,* bat070. https://doi.org/10.1093/database/bat070.
109. Guo, Y.; Bruno, R. S. Endogenous and Exogenous Mediators of Quercetin Bioavailability. *J. Nutr. Biochem.* **2015**, *26*, 201–210.
110. Kim, S.-H.; Yoo, E.-S.; Woo, J.-S.; Han, S.-H.; Lee, J.-H.; Jung, S.-H.; Kim, H.-J.; Jung, J.-Y. *Eur. J. Pharmacol.* **2019**, *860*, 172568.
111. Kedhari Sundaram, M.; Raina, R.; Afroze, N.; Bajbouj, K.; Hamad, M.; Haque, S.; Hussain, A. *Biosci. Rep.* **2019**, *39*(8), 1.
112. Shen, X.; Si, Y.; Wang, Z.; Wang, J.; Guo, Y.; Zhang, X. *Int. J. Mol. Med.* **2016**, *38*(2), 619–626.
113. Feng, J.; Song, D.; Jiang, S. Y.; Yang, X. H.; Ding, T. T.; Zhang, H.; Luo, J.; Liao, J.; Yin, Q. *Biochem. Biophys. Res. Commun.* **2018**, *498*(1), 132–138.
114. Bhat, F. A.; Sharmila, G.; Balakrishnan, S.; Arunkumar, R.; Elumalai, P.; Suganya, S.; Raja Singh, P.; Srinivasan, N.; Arunakaran, J. *J. Nutr. Biochem.* **2014**, *25*(11), 1132–1139.
115. Baruah, M. M.; Khandwekar, A. P.; Sharma, N. *Tumor Biol.* **2016**, *37*(10), 14025–14034.
116. Chang, J. H.; Lai, S. L.; Chen, W. S.; Hung, W. Y.; Chow, J. M.; Hsiao, M.; Lee, W. J.; Chien, M. H. *Biochim. Biophys. Acta Mol. Cell Res.* **2017**, *1864*(10), 1746–1758.
117. Yu, D.; Ye, T.; Xiang, Y.; Shi, Z.; Zhang, J.; Lou, B.; Zhang, F.; Chen, B.; Zhou, M. *Onco. Targets. Ther.* **2017**, *10*, 4719–4729.
118. Shin, E. J.; Lee, J. S.; Hong, S.; Lim, T. G.; Byun, S. *Int. J. Mol. Sci.* **2019**, *20*(21), 5262.
119. Vicentini, F. T. M. C.; He, T.; Shao, Y.; Fonseca, M. J. V.; Verri, W. A.; Fisher, G. J.; Xu, Y. *J. Dermatol. Sci.* **2011**, *61*(3), 162–168.
120. Li, X.; Liu, H. C.; Yao, Q. Y.; Xu, B. L.; Zhang, S. C.; Tu, C. T. *Inflammation* **2016**, *39*(1), 96–106.
121. Granado-Serrano, A. B.; Martn, M. Á.; Bravo, L.; Goya, L.; Ramos, S. *Nutr. Cancer* **2012**, *64*(4), 588–598.
122. Schadich, E.; Hlaváč, J.; Volná, T.; Varanasi, L.; Hajdúch, M.; Džubák, P. *Biomed. Res. Int.* **2016**, *2016*, 2173275.
123. Rashidi, Z.; Aleyasin, A.; Eslami, M.; Nekoonam, S.; Zendedel, A.; Bahramrezaie, M.; Amidi, F. *Reprod. Biol.* **2019**, *19*(3), 245–254.
124. Ramyaa, P.; Krishnaswamy, R.; Padma, V. V. *Biochim. Biophys. Acta Gen. Subj.* **2014**, *1840*(1), 681–692.
125. Saw, C. L. L.; Guo, Y.; Yang, A. Y.; Paredes-Gonzalez, X.; Ramirez, C.; Pung, D.; Kong, A. N. T. *Food Chem. Toxicol.* **2014**, *72*, 303–311.

126. Lee, S.; Lee, J.; Lee, H.; Sung, J. *J. Food Biochem.* **2019**, *43*(11), e13002.
127. Spagnuolo, C.; Russo, G. L.; Orhan, I. E.; Habtemariam, S.; Daglia, M.; Sureda, A.; Nabavi, S. F.; Devi, K. P.; Loizzo, M. R.; Tundis, R.; Nabavi, R.; Mohammad, S. *Adv. Nutr.* **2015**, *6*(4), 408–419.
128. Yoon, K.; Kwack, S. J.; Kim, H. S.; Lee, B. M. *J. Toxicol. Environ. Health B Crit. Rev.* **2014**, *17*(3), 127–174.
129. Chen, M.; Rao, Y.; Zheng, Y.; Wei, S.; Li, Y.; Guo, T.; Yin, P. *PLoS One* **2014**, *9*(2), 0089288.
130. Applegate, C. C.; Rowles, J. L.; Ranard, K. M.; Jeon, S.; Erdman, J. W. *Nutrients* **2018**, *10*(1), 40.
131. Wei, D.; Yang, L.; Lv, B.; Chen, L. *Mol. Vis.* **2017**, *23*, 385–394.
132. Roh, T.; Kim, S. W.; Moon, S. H.; Nam, M. J. *Food Chem. Toxicol.* **2016**, *97*, 127–134.
133. Li, S.; Li, J.; Dai, W.; Zhang, Q.; Feng, J.; Wu, L.; Liu, T.; Yu, Q.; Xu, S.; Wang, W.; Lu, X.; Chen, K.; Xia, Y.; Lu, J.; Zhou, Y.; Fan, X.; Mo, W.; Xu, L. Guo, C. *Br. J. Cancer* **2017**, *117*(10), 1518–1528.
134. Yang, Y.-M.; Yang, Y.; Dai, W.-W.; Li, X.-M.; Ma, J.-Q.; Tang, L.-P. *Eur. Rev. Med. Pharmacol. Sci.* **2016**, *20*(15), 3292–3296.
135. Pan, H.; Zhou, W.; He, W.; Liu, X.; Ding, Q.; Ling, L.; Zha, X.; Wang, S. *Int. J. Mol. Med.* **2012**, *30*(2), 337–343.
136. Zhou, P.; Wang, C.; Hu, Z.; Chen, W.; Qi, W.; Li, A. *BMC Cancer* **2017**, *17*(1), 813. https://doi.org/10.1186/s12885-017-3829-9.
137. Yamasaki, M.; Mine, Y.; Nishimura, M.; Fujita, S.; Sakakibara, Y.; Suiko, M.; Morishita, K.; Nishiyama, K. *Cell Biol. Int.* **2013**, *37*(7), 742–747.
138. Dai, W.; Wang, F.; He, L.; Lin, C.; Wu, S.; Chen, P.; Zhang, Y.; Shen, M.; Wu, D.; Wang, C.; Lu, J.; Zhou, Y.; Xu, X.; Xu, L.; Guo, C. *Mol. Carcinog.* **2015**, *54*(4), 301–311.
139. Zhang, C.; Lv, B.; Yi, C.; Cui, X.; Sui, S.; Li, X.; Qi, M.; Hao, C.; Han, B.; Liu, Z. G. *J. Cancer* **2019**, *10*(3), 737–748.
140. Cao, X.; Ren, K.; Song, Z.; Li, D.; Quan, M.; Zheng, Y.; Cao, J.; Zeng, W.; Zou, H. *Oncol. Rep.* **2016**, *36*(2), 1157–1165.
141. Malavolta, M.; Bracci, M.; Santarelli, L.; Sayeed, M. A.; Pierpaoli, E.; Giacconi, R.; Costarelli, L.; Piacenza, F.; Basso, A.; Cardelli, M.; Provinciali, M. Mediators of Inflammation. Hindawi Limited, **2018**.
142. Zhang, Y.; Kensler, T. W.; Cho, C. G.; Posner, G. H.; Talalay, P. *Proc. Natl. Acad. Sci. U. S. A.* **1994**, *91*(8), 3147–3150.
143. Zhang, Y.; Talalay, P.; Cho, C. G.; Posner, G. H. *Proc. Natl. Acad. Sci. U. S. A.* **1992**, *89*(6), 2399–2403.
144. Yagishita, Y.; Fahey, J. W.; Dinkova-Kostova, A. T.; Kensler, T. W. *Molecules* **2019**, *24*(19), 3593.
145. Thimmulappa, R. K.; Mai, K. H.; Srisuma, S.; Kensler, T. W.; Yamamoto, M.; Biswal, S. *Cancer Res.* **2002**, *62*(18), 5196–5203.
146. Becks, L.; Prince, M.; Burson, H.; Christophe, C.; Broadway, M.; Itoh, K.; Yamamoto, M.; Mathis, M.; Orchard, E.; Shi, R.; McLarty, J.; Pruitt, K.; Zhang, S.; Kleiner-Hancock, H. *BMC Cancer* **2010**, *10*, 540. https://doi.org/10.1186/1471-2407-10-540.
147. Iida, K.; Itoh, K.; Kumagai, Y.; Oyasu, R.; Hattori, K.; Kawai, K.; Shimazui, T.; Akaza, H.; Yamamoto, M. *Cancer Res.* **2004**, *64*(18), 6424–6431.
148. Khor, T. O.; Huang, M. T.; Prawan, A.; Liu, Y.; Hao, X.; Yu, S.; Cheung, W. K. L.; Chan, J. Y.; Reddy, B. S.; Yang, C. S.; Kong, A.-N. *Cancer Prev. Res.* **2008**, *1*(3), 187–191.
149. Ramos-Gomez, M.; Kwak, M. K.; Dolan, P. M.; Itoh, K.; Yamamoto, M.; Talalay, P.; Kensler, T. W. *Proc. Natl. Acad. Sci. U. S. A.* **2001**, *98*(6), 3410–3415.

150. Xu, C.; Huang, M. T.; Shen, G.; Yuan, X.; Lin, W.; Khor, T. O.; Conney, A. H.; Kong, A. N. T. *Cancer Res.* **2006**, *66*(16), 8293–8296.
151. Lan, A.; Li, W.; Liu, Y.; Xiong, Z.; Zhang, X.; Zhou, S.; Palko, O.; Chen, H.; Kapita, M.; Prigge, J. R.; Schmidt, E.; Chen, X.; Sun, Z.; Chen, L. *Oncotarget* **2016**, 7(33), 53502–53514.
152. Jiang, X.; Liu, Y.; Ma, L.; Ji, R.; Qu, Y.; Xin, Y.; Lv, G. Drug Design, Development and Therapy. dove press, **2018**, 2905–2913.
153. Kim, H. N.; Kim, D. H.; Kim, E. H.; Lee, M. H.; Kundu, J. K.; Na, H. K.; Cha, Y. N.; Surh, Y. J. *Cancer Lett.* **2014**, *351*(1), 41–49.
154. Heiss, E.; Herhaus, C.; Klimo, K.; Bartsch, H.; Gerhäuser, C. *J. Biol. Chem.* **2001**, *276*(34), 32008–32015.
155. Wu, J.; Han, J.; Hou, B.; Deng, C.; Wu, H.; Shen, L. *Oncol. Rep.* **2016**, *35*(5), 2977–2983.
156. Wang, D. X.; Zou, Y. J.; Zhuang, X. B.; Chen, S. X.; Lin, Y.; Li, W. L.; Lin, J. J.; Lin, Z. Q. *Acta Pharmacol. Sin.* **2017**, *38*(2), 241–251.

CHAPTER TEN

Antipsoriatic natural products

Anca Miron[a,*], Lutfun Nahar[b,c], Elvira Gille[d], Satyajit D. Sarker[b]

[a]Department of Pharmacognosy, Faculty of Pharmacy, "Grigore T. Popa" University of Medicine and Pharmacy, Iasi, Romania
[b]Centre for Natural Products Discovery (CNPD), School of Pharmacy and Biomolecular Sciences, Liverpool John Moores University, Liverpool, United Kingdom
[c]Laboratory of Growth Regulators, Institute of Experimental Botany ASCR & Palacký University, Olomouc, Czech Republic
[d]National Institute of Research and Development for Biological Sciences Bucharest/"Stejarul" Biological Research Centre, Piatra Neamt, Romania
*Corresponding authors: e-mail addresses: anca.miron@umfiasi.ro; ancamiron@yahoo.com

Contents

1. Introduction

Psoriasis is a chronic inflammatory and hyperplastic skin disorder affecting 1–3% of the population worldwide.[1,2] It is characterized by inflamed, red plaques with silvery scales and thickened epidermis (acanthosis) usually appearing on the elbows, knees, scalp, trunk and nails. Psoriatic plaques are often itchy and painful. In severe cases, they may crack and bleed affecting the patients' quality of life.[3–5] The itching condition is suggested by the etymology of the term psoriasis; it derives from two Greek words: *psora* meaning itch and *sis* meaning action.[6] In brief, the main hallmark features of psoriatic lesions are keratinocyte abnormal differentiation and hyperproliferation, infiltration of inflammatory cells and increased vascularization.[5]

Annual Reports in Medicinal Chemistry, Volume 55
ISSN 0065-7743
https://doi.org/10.1016/bs.armc.2020.02.005

Psoriasis has a multifactorial etiology involving genetic, epigenetic and environmental factors. The increased incidence of psoriasis in relatives of the psoriatic patients' and the higher concordance rate between monozygotic twins in comparison with dizygotic ones (65–72% *vs.* 15–30%) support the genetic component of psoriasis.[7] There have been reported susceptibility loci associated with psoriasis, which are involved in the immune and inflammation responses (*HLA-C, IL-12B, TNFAIP3, IL13, TNIP1, IL23A, ZNF313*), epidermal differentiation and hyperplasia (*IL-23R, LCE3D*). The major psoriasis susceptibility locus is *PSORS1* accounting for 35–40% of the genetic susceptibility to psoriasis.[7,8] Epigenetic changes have also been observed in psoriasis (increase in genome-wide methylation in blood mononuclear cells not only in psoriatic lesions, but also in peripheral blood, decrease in acetylation level of histone H4).

Environmental factors such as certain drugs, smoking, alcohol intake, diet, mental stress and depression play a crucial role in the development of psoriasis.[7] Imiquimod, immune activator used in the topical treatment of genital and perianal warts, actinic keratoses and non-melanoma skin cancer (basal cell carcinoma), can induce and exacerbate psoriasis both locally and systemically. Being a Toll-like receptor 7/8 agonist, imiquimod activates type I interferon (IFN), an inductor of psoriatic lesions. IL-23/IL-17 pathway, involved in the pathogenesis of psoriasis, is also activated by imiquimod. Therefore, the latter is frequently used to induce psoriasis in animal models.[7,9] Other drugs such as beta-blockers, antidepressants (lithium), antimalarials (chloroquine and hydroxychloroquine), angiotensin-converting enzyme inhibitors, antibiotics (tetracyclines and penicillins), nonsteroidal anti-inflammatory (naproxen and indomethacin) and immunotherapeutic (pembrolizumab) agents might also provoke psoriasis.[7,10] Both smoking and alcohol consumption stimulate the production of pro-inflammatory cytokines, thus exacerbating psoriasis severity. Obese individuals are at high risk of developing psoriasis, especially severe forms of psoriasis. On the other hand, a gluten-free diet rich in vitamins D and B12, selenium, ω-3 fatty acids (fish oil) has benefits in psoriatic patients. Besides, mental stress and depression facilitate psoriasis onset and exacerbation.[7]

Psoriasis vulgaris, also known as plaque *psoriasis*, is associated with several comorbidities including psoriatic arthritis, cardiovascular disorders, type 2 diabetes mellitus, metabolic syndrome, Crohn's disease, lymphoma, depression and anxiety.[4,8]

Nowadays both topical and systemic therapies are used to ameliorate psoriatic symptoms. Topical drugs are recommended in mild to moderate forms and include calcineurin inhibitors (tacrolimus and pimecrolimus), corticosteroids, vitamin D_3 analogues (calcipotriol, calcitriol and tacalcitol), retinoids, keratolytics (salicylic acid, lactic acid and urea), dithranol and coal tar. Calcineurin inhibitors and corticosteroids have anti-inflammatory and immunosuppressive effects, whereas vitamin D_3 analogues, retinoids, dithranol and coal tar act primarily by reducing keratinocyte hyperproliferation. All these topical drugs cause local side effects such as redness, itching, burning sensation, tenderness. Long-term topical application of corticosteroids can lead to skin atrophy. Systemic therapy of psoriasis includes methotrexate, cyclosporine, acitretin, fumaric acid esters, apremilast and biological agents. The latter inhibit T cell activity (alefacept and efalizumab) and pathways mediated by tumor necrosis factor α (TNF-α) (etanercept, adalimumab, certolizumab, golimumab and infliximab), IL-17 (brodalumab, ixekizumab and secukinumab), IL-23 (guselkumab), IL-12 and IL-23 (ustekinumab and briakinumab) causing a significant immunosuppression.[4,8,11]

Phototherapy has been widely used in the treatment of psoriasis. Repeated exposure of psoriatic lesions to UVB or UVA irradiation in combination with psoralens (topical or oral) counteracts keratinocyte hyperproliferation. Exposure to UVA is effective only in combination with a photosensitizing agent such as a psoralen (PUVA therapy). Phototherapy is also associated with several side effects such as burning, itching and a higher risk to develop skin cancer.[4,12]

All these conventional therapies cause side effects. Moreover, they cannot cure psoriasis, they can only attenuate the symptoms of the disease. Therefore, there has been an increasing interest for natural product-based therapeutics with greater efficacy, safety and tolerability in the treatment of psoriasis.[4] It is worth mentioning that a natural product, namely 8-methoxypsoralen (8-MOP, also known as xanthotoxin) (Fig. 1), has revolutionized psoriasis treatment; 8-MOP (oral or topical) combined with UVA exposure (photochemotherapy, PUVA) is one of the oldest and most effective therapy for psoriasis.[4,13]

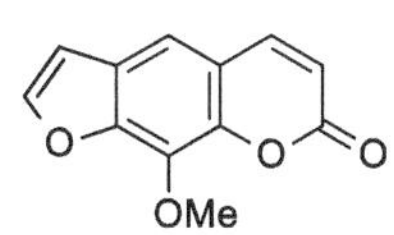

Fig. 1 Xanthotoxin (8-MOP), a plant-derived furanocoumarin, with antipsoriatic property.

2. Antipsoriatic natural products

During the past 15 years, several natural products and their analogues have been investigated for their antipsoriatic activity in different experimental models (cell-based, *ex vivo* and animal-based *in vivo* assays). Most of these natural products are plant constituents; only a few have been isolated from marine organisms or microbial sources.

2.1 Plant-derived antipsoriatic natural products

Plant constituents of various chemical classes: alkaloids, cyclopeptides, flavonoids, furanocoumarins, organic acid derivatives, polyphenols, sphingolipids and terpenoids were found to be "lead" compounds for the development of novel antipsoriatic drugs (Fig. 2).

In 1994 fumaric acid esters were licensed in Germany for the treatment of severe psoriasis under the name of Fumaderm® (dimethylfumarate and calcium, magnesium and zinc salts of monoethylfumarate).[14] Fumaric acid was initially detected in a fungus, *Boletus pseudoignarius* (Boletaceae) and further isolated from *Fumaria officinalis* L. (Papaveraceae, common fumitory).[15] In the 17th century, common fumitory was used to treat leprosy and other skin disorders. As at that time psoriasis was not distinguished from leprosy, it can be assumed that common fumitory was used in the treatment of psoriasis 400 years ago.[14] Nowadays fumaric acid is produced by chemical synthesis or microbial conversion of renewable resources.[16]

The standard treatment involves a gradual increase in dosage over 8 weeks from one tablet/day of Fumaderm Initial® (30 mg dimethylfumarate, 75 mg monoethylfumarate as salts) to maximum six tablets/day of Fumaderm® (120 mg dimethylfumarate, 95 mg monoethylfumarate as salts). The standard treatment ensures therapeutic benefits and good tolerability. Monomethylfumarate and dimethylfumarate (Fig. 2) are the main bioactive metabolites of fumaric acid esters.[14,17] Fumaric acid esters offered significant clinical improvements in Physician's Global Assessment and Psoriasis Area and Severity Index (PASI) in psoriatic patients, including those with severe recalcitrant psoriasis. Fumaric acid esters have a 4-year drug survival rate (overall measurement of treatment effectiveness) of 60%, whereas biological agents such as etanercept, adalimumab and infliximab have 4-year survival rates of 40% and 70%, respectively.[17–19] In addition, concomitant exposure to narrowband UVB (NB-UVB) light resulted in an accelerated therapeutic response

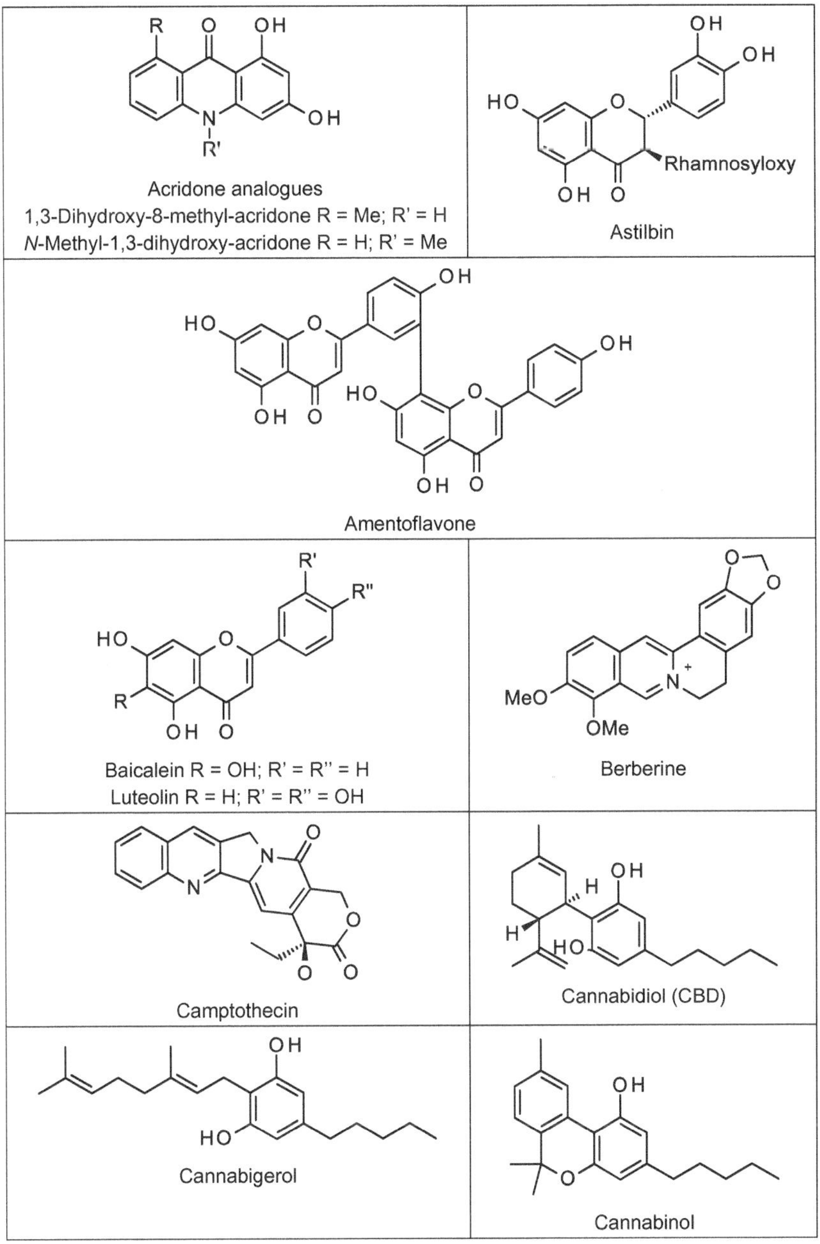

Fig. 2 Plant-derived antipsoriatic "lead" compounds.

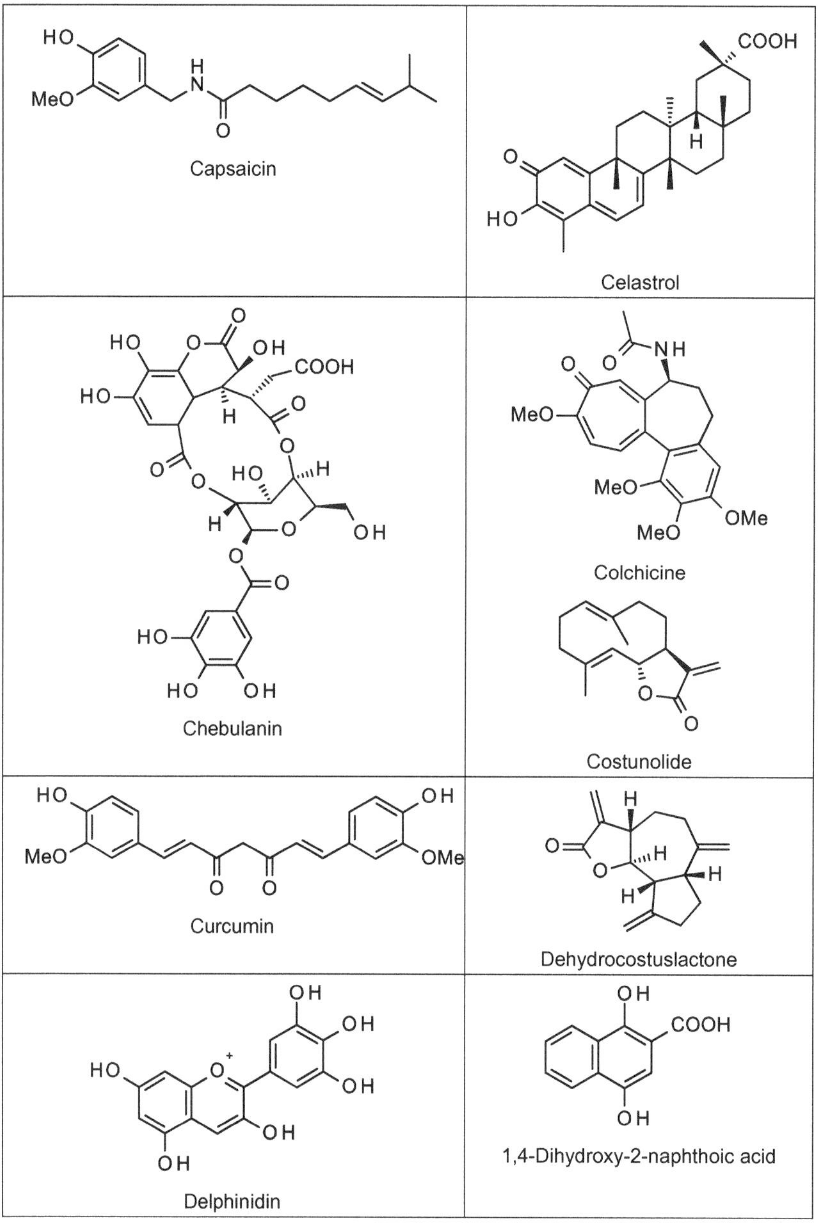
Capsaicin
Celastrol
Chebulanin
Colchicine
Costunolide
Curcumin
Dehydrocostuslactone
Delphinidin
1,4-Dihydroxy-2-naphthoic acid

Fig. 2—Cont'd

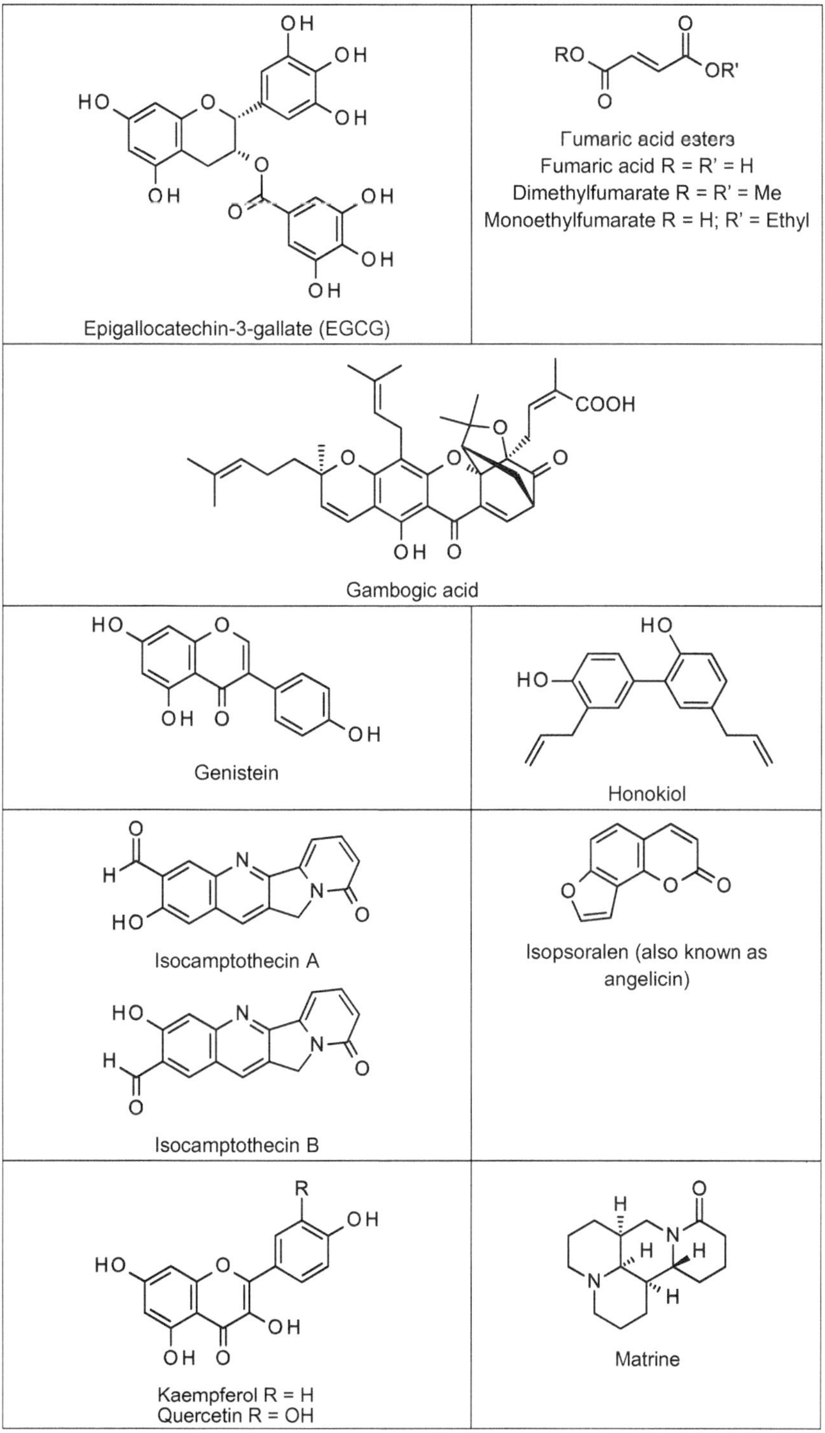
Epigallocatechin-3-gallate (EGCG)
Fumaric acid esters
Fumaric acid R = R' = H
Dimethylfumarate R = R' = Me
Monoethylfumarate R = H; R' = Ethyl
COOH
Gambogic acid
Genistein
Honokiol
Isocamptothecin A
Isocamptothecin B
Isopsoralen (also known as angelicin)
Kaempferol R = H
Quercetin R = OH
Matrine

Fig. 2—Cont'd

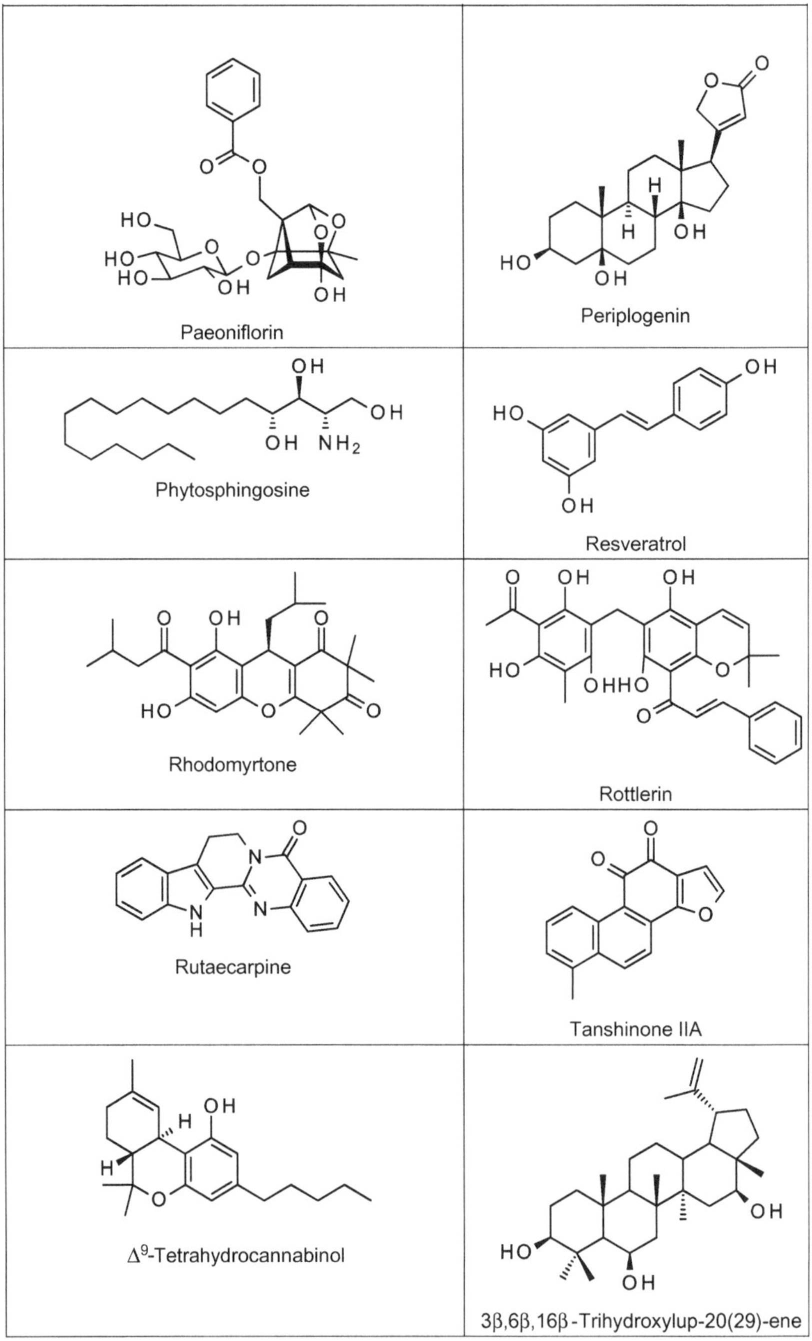
Paeoniflorin
Periplogenin
Phytosphingosine
Resveratrol
Rhodomyrtone
Rottlerin
Rutaecarpine
Tanshinone IIA
Δ9-Tetrahydrocannabinol
3β,6β,16β-Trihydroxylup-20(29)-ene

Fig. 2—Cont'd

in patients with moderate-to-severe plaque psoriasis. Six-week combination of NB-UVB exposure (250–700 mJ/cm^2 depending on skin phototype, increase in dose by 10% and 20% until a maximum single dose of 3 J/cm^2, three irradiations/week) with administration of gradually increasing doses of fumaric acid esters resulted in a more pronounced improvement in PASI, Psoriasis-Log-based Area and Severity Index (PLASI) and Dermatology Life Quality Index (DLQI).[20] Fumaric acid esters may cause gastrointestinal upset, hot flushes, lymphopenia and proteinuria, the latter two being reversible with dose reduction.[17] However, there are opinions standing that more clinical evidence is needed to support the efficacy and safety of fumaric acid esters in psoriasis treatment.[18]

1,4-Dihydroxy-2-naphthoic acid (Fig. 2), anthraquinone precursor in *Rubia cordifolia* L. (Rubiaceae), was proven to be a promising alternative for the treatment of psoriasis. In comparison with dithranol, the compound was less active in inhibiting the growth of HaCaT and NCTC 2544 human keratinocytes (approximately, 4-fold and 12-fold, respectively) but less toxic on Hs-68 fibroblasts and PIG1 melanocytes (approximately, 8-fold and 10-fold, respectively). Topical antipsoriatic agents in general and dithranol in particular cause skin irritation such as erythema, itching, edema and pain by triggering the release of IL-1α from keratinocytes followed by release of chemokines, vasodilation and immune cell infiltration.[21] Therefore, identification of antipsoriatic agents inducing less local irritation is of great interest for psoriasis treatment. The milder cytotoxicity of 1,4-dihydroxy-2-naphthoic acid in comparison with dithranol makes the former a good candidate for the development of novel topical treatments.

Paeoniflorin, a bioactive benzoyl monoterpene glucoside (Fig. 2) from *Paeonia lactiflora* Pall. and other *Paeonia* species (Paeoniaceae), significantly reduced skin lesions in imiquimod-induced psoriasis mouse model. Paeoniflorin (75–300 mg/kg/day for 8 and 10 days, intraperitoneal and intragastric administration) reduced skin erythema and thickening, scaly patches and smoothened the skin with no alteration in liver and kidney functions.[9,22] According to PASI scores, paeoniflorin (240 mg/kg/day, 8 day intragastric administration) was as active as methotrexate (1 mg/kg/day, 8 day intragastric administration) in reducing erythema and scaling, but more active than the latter in reducing skin thickness. Overall, the cumulative PASI score (erythema + scaling + thickness) was significantly lower in the paeoniflorin group than in the methotrexate group revealing a higher efficacy of the former in alleviating the psoriatic skin lesions.[9] A recent study revealed the anti-inflammatory potential of paeoniflorin supporting its use in topical treatment of psoriasis. Local application of 9% paeoniflorin emulsion (0.2 g/day, 1 week) decreased the Baker score by 37% in a guinea pig model of topical propranolol-induced psoriasis-like lesions. Paeoniflorin attenuated psoriasis-like lesions caused by propranolol such as hyperkeratinization, parakeratosis and disappearance of the granular layer.[23]

Furanocoumarins, plant metabolites with photosensitizing properties, have a narrow distribution in the plant kingdom, and are found mainly in plants from the families Apiaceae, Fabaceae, Moraceae, Rosaceae and Rutaceae.[8,13] Among furanocoumarins, as mentioned earlier, 8-MOP (Fig. 1), also known as methoxsalen and commercialized under various brand names (Oxsoralen® Ultra has better bioavailability compared to others), is used in combination with exposure to UVA (320–400 nm) in the treatment of psoriasis. The procedure, known as photochemotherapy (PUVA), was approved for the management of cutaneous psoriasis by the Food and Drug Administration (FDA) in 1982.[24–26] The main vegetal sources of 8-MOP are the fruits of *Ammi majus* L. (Apiaceae, bishop's weed) and seeds of *Psoralea coryfolia* L. (Fabaceae, babchi).[8,13] In contrast to other constituents in *P. coryfolia* seeds (psoralen, psoralidin, bakuchiol), 8-MOP and isopsoralen (also known as angelicin; Fig. 2) showed the highest absorption and penetration rates into and across animal skins.[24] It is interesting to note that while 8-MOP is a linear furanocoumarin, isopsoralen is an angular one.

The effectiveness of 8-MOP in combination with UVA exposure was proven in different experimental models. Simultaneous exposure to 8-MOP (0.25 μM) and UVA (365 nm, 15 J/cm^2) almost completely reduced the viability of HaCaT keratinocytes. A strong cytotoxicity was also achieved by pre-treatment of keratinocytes with 8-MOP (2 μM, 2 h) followed by UVA exposure (365 nm, 15 J/cm^2). In imiquimod-induced psoriasis mouse model, topical application of 8-MOP (0.2 g/day, 4 days) followed after 30 min by UVA exposure (2.5 J/cm^2) significantly reduced transepidermal water loss (from 55.5 to 33.3 g/m^2/h). The latter increases in psoriasis as a consequence of disruption of skin barrier. Besides, 8-MOP decreased epidermal thickening (from 117.1 to 61.5 μm) and showed good tolerance after 7 day administration (no skin barrier damage and no irritation).[24] In K5.hTGF-β1 transgenic mice (exhibit skin lesions similar to human psoriatic ones), topical application of 8-MOP (200 μL, 0.1 mg/mL in ethanol) followed by exposure to UVA (351–400 nm, peak 365 nm, 0.25 J/cm^2 for 29 s or 0.50 J/cm^2 for 59 s), twice a week for 4 weeks, suppressed acanthosis, hyperkeratosis/parakeratosis, inflammatory infiltrates and increased dermal blood vessels.[27] In psoriatic patients, PUVA treatment involves an initial clearing phase (0.4–0.6 mg 8-MOP/kg body weight and exposure to UVA after 75 min, 2–3 treatments/week for 9–15 weeks). The second phase, the maintenance phase (one treatment/month) is applied to patients having almost complete clearing of psoriasis (95%). In clearing phase, the initial UVA dose is 0.5–6 J/cm^2 and can be increased by 0.5–2.5 J/cm^2 whereas in the maintenance phase the UVA dose is held constant. Generalized and local immersions in 8-MOP solutions prior to UVA exposure are the most common topical PUVA treatments.[25] PUVA treatment might cause both short-term (burning, itching and skin pigmentation) and long-term (skin aging, non-melanoma and melanoma skin cancers) side effects. Photoprotective measures (avoidance of sunlight exposure and use of sunscreens) can minimize the adverse effects.[25,28]

Isopsoralen was found to be a potential candidate for PUVA therapy; it showed high cytotoxicity in HaCaT keratinocytes and markedly reduced transepidermal water loss and epidermal thickening in imiquimod-induced psoriasis-like lesions.[24] The effects were comparable to those of 8-MOP except the reduction in epidermal thickening which was more pronounced in the case of isopsoralen (from 117.1 to 25.9 μm).[24]

Several naturally occurring flavonoids exhibited antipsoriatic effects in different experimental models. Kaempferol and quercetin (Fig. 2), flavonols abundant in many plants, attenuated imiquimod-induced psoriatic skin inflammation in mice.[29,30] Kaempferol (50 and 100 mg/kg, 7 days, oral administration) reduced the severity of skin lesions (erythema, scaling and epidermal thickness) with significant reductions in PASI scores, more pronounced in the group treated with higher dose of kaempferol.[29] On the other hand, pre-treatment with quercetin (30, 60 and 120 mg/kg, 7 days, oral administration) dose-dependently alleviated the temperature (sign of inflammation) and improved the morphology of psoriasis-like lesions induced by imiquimod with a significant reduction in PASI score. Quercetin could also smoothen the epidermis and attenuated parakeratosis and epidermal thickening.[30] Recent preclinical findings support the development of delphinidin (Fig. 2), an anthocyanidin widely spread in fruits and vegetables, as antipsoriatic drug due to its anti-inflammatory, antiproliferative and prodifferentiation effects observed in different experimental models (3D reconstructed human psoriatic skin equivalent model, cell-based models and imiquimod-induced psoriasis-like dermatitis model).[31,32] Topical application of delphinidin (1 mg/cm^2, 10 days) on imiquimod-induced skin lesions reduced inflammation, scaling and epidermal thickness.[32] Another flavonoid that might be developed as antipsoriatic drug is the isoflavone genistein, which was first isolated from *Genista tinctoria* (Fabaceae) in 1899 (Fig. 2). The antipsoriatic potential of genistein appears to be related to its antiproliferative effects on keratinocytes and immunosuppressive properties, but also to its ability to modulate gene expression proven not only *in vitro* (normal and "psoriasis-like" keratinocytes), but also in psoriatic patients (75 or 150 mg genistein/day for 56 days).[2,33,34]

Baicalein (Fig. 2), the major bioactive constituent of *Scutellaria baicalensis* Georgi (Lamiaceae) roots, targeted not only the proliferation rate/viability of HaCaT keratinocytes, but also their differentiation. Keratinocyte hyperproliferation and abnormal differentiation are major hallmarks of psoriasis. Baicalein (10 μM) promoted keratinocyte differentiation with changes in cell morphology (appearance of dendritic horn-like structures).[35] Astilbin (Fig. 2), a flavanone glycoside, is the major bioactive constituent in the rhizomes of *Smilax glabra* Roxb. (Smilacaceae). Oral administration of astilbin (25 and 50 mg/kg/day, 7 days after imiquimod application) reduced keratinocyte abnormal differentiation and proliferation and inflammatory reactions in imiquimod-induced psoriasis-like inflammation. Luteolin (10–100 μM) (Fig. 2), a well-known flavone widely distributed in the plant kingdom, reduced TNF-activation of HaCaT and normal human epidermal keratinocytes, reduced the proliferation rate of the former and demonstrated anti-inflammatory properties in human psoriatic skin.[36]

The biflavonoid amentoflavone (Fig. 2) could reduce the proliferation rate of cytokine-stimulated HaCaT keratinocytes and demonstrated anti-inflammatory effects in imiquimod-induced psoriasis-like mouse model.[36] Another flavonoid of the flavanol group with promising benefits in psoriasis treatment is epigallocatechin-3-gallate (EGCG) (Fig. 2), the major catechin found in green tea (*Camellia sinensis* (L.) Kuntze, Theaceae). Both topical application (30 mg/mL in glycerine solvent, early and medium-term intervention) and gavage administration (150 and 300 mg/kg/day, 3 weeks) of EGCG attenuated the severity of imiquimod-induced psoriasis dermatitis in mice (reduction of PASI scores consisting of erythema, scales and leukocyte infiltration, reduction of parakeratosis, acanthosis and inflammatory cell infiltration). As the extraction of EGCG from green tea leaves is easy to perform and inexpensive, the compound is an attractive candidate for the development of novel antipsoriatic therapies.[37]

Chebulanin (Fig. 2), a bioactive polyphenol isolated from *Terminalia chebula* Retzius (Combretaceae), has been reported as a novel potential therapeutic option for psoriasis owing to its antiproliferative activity against keratinocytes and attenuation of imiquimod-induced psoriatic skin lesions in mice (erythema, scaling, epidermal, right ear fold and back skinfold thickness, acanthosis, hyperkeratosis and elongation of rete-like ridge in mice skin) after intragastrical administration (50 mg/kg/day, 7 days).[38]

Recent findings suggest that resveratrol (Fig. 2), a stilbenoid found in various plant species, can be an effective treatment for psoriasis lesions.[39] Resveratrol is abundant in grapes, berries and peanuts, but also in some plant species (*Polygonum cuspidatum* Siebold & Zucc., Polygonaceae); its *trans*-isomer is biologically active.[40] Resveratrol inhibited the proliferation of normal human epidermal keratinocytes (NHEKs) and induced cell death in HaCaT keratinocytes in a dose-dependent manner.[39,41] In addition, resveratrol (400 mg/kg/day, 7 days) attenuated the severity of imiquimod-induced psoriasis-like skin inflammation in mice (reduction in erythema, scaling, skin fold, ear and epidermal thickness).[42]

Curcumin (diferuloylmethane) (Fig. 2), a major bioactive diarylheptanoid found in the rhizomes of turmeric (*Curcuma longa* L., Zingiberaceae), can induce cell death in HaCaT keratinocytes (only in the presence of TNF-α, cytokine playing an important role in apoptosis, immune response and inflammation).[40,43] Visible light and UVA irradiation increased the susceptibility of both HaCaT and normal keratinocytes to curcumin. Curcumin pre-treatment (1 μg/mL, 1 h) followed by UVA exposure (1 J/cm^2) caused cell death (apoptosis) in more than 40% of treated HaCaT keratinocytes whereas curcumin itself (without UVA exposure) caused apoptosis in only 0.5% of HaCaT cells.[44] Similar effects (reduction in viability and proliferation rate, induction of apoptosis) were observed in TNF-α-activated HaCaT keratinocytes when pre-treated with low doses of curcumin (<5 μM) followed by exposure to a combination of red light (630 or 660 nm) and blue light (405 nm).[45] Curcumin, administered orally as Curcuminoid C3 Complex (500 mg capsules containing 95% curcuminoids: 450 mg curcumin, 40 mg demethoxycurcumin, 10 mg bisdemethoxycurcumin; 4.5 g/day for 12 weeks), showed limited efficacy in plaque psoriasis patients.[46,47] The study enrolled 12 patients of which only eight completed the treatment.[47] In contrast, a recent clinical study reported that co-administration of Meriva (curcumin-phosphatidylcholine phytosome complex, 2 g/day, 12 weeks) in patients with mild-to-moderate plaque psoriasis undergoing topical treatment with methylprednisolone aceponate 0.1% resulted in a more pronounced decrease in PASI values in comparison with topical methylprednisolone aceponate and placebo.[46,48] The antipsoriatic efficacy showed by oral curcumin in the latter study might be ascribed to the higher oral bioavailability of Meriva in comparison with Curcuminoid C3 Complex.

Honokiol (Fig. 2), a biphenolic neolignan isolated from *Magnolia* species (Magnoliaceae), is a promising antipsoriatic agent. Honokiol could normalize psoriatic phenotype (acanthosis, parakeratosis, rete pegs, inflammation and sub-epidermal layer thickness) in K14-VEGF transgenic mice which developed skin inflammation resembling psoriasis. According to Baker score system, honokiol cream (0.5% and 1%) showed better therapeutic efficacy than tretinoin ointment (0.025%); the treatment involved daily application of 0.3 g of cream/cm^2 on the back ears on both sides for 4 weeks.[49]

Gambogic acid (Fig. 2), a polyprenylated xanthone isolated from the resin of *Garcinia hanburyi* Hook f. (Clusiaceae) (traditional Chinese medicine), inhibited the proliferation of HaCaT keratinocytes and showed antipsoriatic effects in the K14-VEGF transgenic mice model, propranolol-induced psoriasis-like model of guinea pigs and mouse tail model. Administered as cream (0.1%, 0.25%, 0.5%) for 4 weeks, gambogic acid time- and dose-dependently reduced epidermal hyperplasia, hyperkeratosis, acanthosis, microvessel density and inflammatory infiltrates (in K14-VEGF transgenic mice), erythema, scaling, hyperkeratosis and parakeratosis (in propranolol treated guinea pigs) and induced an increase in epidermal thickness and orthokeratosis in mouse tail skin (in inter-hinge areas of scale).[50,51]

Rhodomyrtone (Fig. 2), an acylphloroglucinol isolated from *Rhodomyrtus tomentosa* (Aiton) Hassk. (Myrtaceae), inhibited the growth of HaCaT keratinocytes in a dose- and time-dependent manner.[52,53] The antiproliferative activity of rhodomyrtone was confirmed in a wound scratch assay, the delay in wound closure suggesting a decrease in migration rate of HaCaTs and consequently, an antipsoriatic activity. A rhodomyrtone formulation was further developed showing no skin irritation (erythema and edema) in a rabbit skin model.[52] Moreover, rhodomyrtone showed anti-inflammatory activity in an *ex vivo* skin inflammation model (human skin organ cultures stimulated with IL-17A and TNF-α) and a mouse model of imiquimod-induced skin inflammation.[53]

Rottlerin (Fig. 2), a polyphenolic compound isolated from *Mallotus philippinensis* Muell. (Euphorbiaceae), is another promising candidate for the development of novel antipsoriatic agents. It inhibited the proliferation and senescence of NHEKs and angiogenesis in human umbilical vein endothelial cells (HUVECs). Besides inflammation, angiogenesis plays an important role in the pathogenesis of psoriasis. Rottlerin was also investigated *in vivo*. According to PASI scores, oral administration of rottlerin (20 mg/kg/day, 6 days) reduced the psoriasiform lesions induced by topical application of imiquimod in mice. Reductions in hyperkeratosis, acanthosis, inflammatory infiltrates and dermal vascular density were also noticed.[54]

Several terpenoids exhibited promising antipsoriatic effects, which support their use in drug development. Dehydrocostuslactone and costunolide (Fig. 2), sesquiterpene lactones present in different plants (*Magnolia sieboldii* L., *Laurus nobilis* L. and *Saussurea costus* L.), regulated the proliferative and inflammatory responses of human keratinocytes to different pro-inflammatory cytokines (IL-22, interferon-γ (IFN-γ), TNF-α).[55] Tanshinone IIA (Fig. 2), a diterpene quinone, the major constituent of *Salvia miltiorrhiza* Bunge (Lamiaceae) roots (Danshen), displayed significant dose- and time-dependent antiproliferative effects in primary mouse keratinocytes (IC_{50} = 4.33, 1.89 and 1.14 μg/mL after 24, 48 and 72 h incubation, respectively), supporting the traditional use of Danshen for the treatment of psoriasis.[56]

Celastrol and 3β,6β,16β-trihydroxylup-20(29)-ene (TTHL) (Fig. 2) are triterpenoids isolated from *Celastrus orbiculatus* Thunb. (Celastraceae) and *Combretum leprosum* Mart. (Combretaceae), respectively. Both compounds showed antiproliferative effects in HaCaT keratinocytes; in addition, TTHL reduced 12-*O*-tetradecanoylphorbol 13-acetate (TPA)-induced psoriasis-like inflammation in mice ears.[57,58] In the case of single and multiple topical application of TPA, TTHL (1.313 and 0.656 μmoL/ear twice daily, respectively) showed similar effectiveness to dexamethasone (0.255 and 0.127 μmoL/ear twice daily, respectively) in inhibiting edema (96.9% *vs.* 98.4% and 62.4% *vs.* 64.7%, respectively). The increase in epidermis thickness induced by TPA was efficiently reduced by TTHL (59.8% *vs.* 60.1% for dexamethasone).[58]

Periplogenin, cardiac aglycone isolated from the roots of *Periploca sepium* Bunge. (Apocynaceae), a well-known traditional Chinese medicine, attenuated skin inflammation in mice induced by TPA and imiquimod and epidermal hyperplasia generated by the former (2 μg periplogenin/day, 11 days after induction of skin lesions).[5,59]

Cannabinoids, active constituents of *Cannabis sativa* L. (Cannabaceae), have potential application for psoriasis therapy as they are endowed with anti-inflammatory properties, cannabinoid receptors are present in human skin and the endocannabinoid anandamide was found to inhibit the differentiation of epidermal keratinocytes. Δ^9-Tetrahydrocannabinol, cannabinol, cannabidiol and cannabigerol (Fig. 2) inhibited the proliferation of hyper-proliferating human keratinocytes (HPV-16 E6/E7 transformed human skin keratinocytes) with IC_{50} values ranging between 2 and 2.7 μM and cannabidiol being the most active. There was found no relation between the activation of CB1 or CB2 receptors and antiproliferative activity.[60]

Alkaloid-rich plant extracts as well as isolated alkaloids have been extensively investigated for potential application in psoriasis treatment. Extracts of *Mahonia aquifolium* (Pursh.) Nutt. (Berberidaceae) having berberine (Fig. 2) as an active component, indigo naturalis (*Baphicacanthus cusia* (Nees) Bremek), camptothecin (source: *Camptotheca acuminata* L., Nyssaceae), capsaicin (source: *Capsicum* spp., Solanaceae) and colchicine (source: *Colchicum* spp., Colchicaceae) (Fig. 2) are only few examples.[4,8,61] The following are some examples of recent studies on potential antipsoriatic properties of various plant-derived alkaloids.

Rutaecarpine (or rutecarpine), an indolopyridoquinazolinone alkaloid isolated from *Evodia rutaecarpa* (Juss.) Benth. (Rutaceae), a plant commonly used in the traditional Chinese medicine, ameliorated imiquimod-induced psoriasis-like dermatitis in mice. Five-day topical application of rutaecarpine cream (0.5% and 1%) normalized the thickness of the epidermis, decreased the number of inflammatory cell infiltrates and significantly reduced the PASI scores.[62,63] Matrine, the major quinolizidine alkaloid in the roots of *Sophora flavescens* Ait. (Kushen) (Fabaceae), suppressed dose- and time-dependently the proliferation of HaCaT keratinocytes (24, 48 and 72 h incubation with 0.2–1.6 mg/mL matrine).[64,65] At a dose of 50 mg/kg/day, administered intraperitoneally in mice with imiquimod-induced psoriasis-like lesions, matrine reduced the PASI score (erythema, scaling and thickening). When matrine was combined with the retinoid acitretin (active metabolite of etretinate, used to treat moderate and severe psoriasis), both *in vitro* and *in vivo* effects were potentiated suggesting that the combination was more effective than each drug alone. These findings support the clinical efficacy of matrine glucose injection in combination with oral acitretin in plaque psoriasis; the combination showed a better tolerability profile than acitretin alone. It should be mentioned that matrine (injection or in combination with hyperbaric oxygen) is used to treat psoriasis in China.[65]

Several analogues of plant alkaloids, having stronger effects and lower toxicity than the parent compound, were developed. Some of them appear to be good candidates for the treatment of psoriasis. Isocamptothecin (mixture of isocamptothecin A and isocamptothecin B) (Fig. 2) is a camptothecin analogue biosynthesized from the suspension cell culture of *Camptotheca acuminata* L. In HaCaT keratinocytes, isocamptothecin showed similar antiproliferative activity (dose- and time-dependent) with camptothecin, but less cytotoxicity. More precisely, the concentration of isocamptothecin producing the largest inhibition on HaCaTs proliferation rate was 46-fold its least effective concentration (69 *vs.* 1.5 μg/mL); in the case of camptothecin, the concentration inducing the largest inhibition was 1000-fold its least effective concentration (16 *vs.* 0.016 μg/mL).[66]

Acridone alkaloids usually occur in species of the Rutaceae family.[4] From a structural point of view, they can be considered as 10-aza-analogues of anthralin (dithranol, 1,8-dihydroxy-10*H*-anthracen-9-one) (Fig. 3), a well-known topical antipsoriatic agent. But anthralin causes skin irritation, skin and cloth staining. These side effects are mainly attributed to the generation of oxygen radicals, the methylene moiety at C-10 being the site of radical generation.[67,68] Therefore, a series of analogues, namely *N*-unsubstituted acridones (*N*-unsubstituted 10*H*-acridin-9-ones), was synthesized to identify agents with enhanced antiproliferative activity in HaCaT keratinocytes and attenuated side effects. The analogue comprising a 1,3-dihydroxy-8-methyl substitution (Fig. 2) pattern was the most active in inhibiting HaCaTs proliferation with an IC_{50} value comparable to that of anthralin (0.8 *vs.* 0.7 μM). The analogue did not generate radicals (lack of interaction with 2,2-diphenyl-1-picrylhydrazyl (DPPH) radical) and was devoid of membrane damaging effects as shown by lactate dehydrogenase (LDH) assay in culture medium. Minor changes in the substitution pattern (methyl group at C-5 or C-6) resulted in loss of antiproliferative activity.[67] Another series of acridone analogues, namely 10-substituted acridones (10-substituted hydroxy-10*H*-acridin-9-ones), was further synthesized. The analogue having a *N*-methyl-1,3-dihydroxy skeleton (Fig. 2) was the most potent inhibitor of HaCaTs proliferation, being slightly more active than anthralin ($IC_{50}=0.6$ *vs.* 0.7 μM) (Fig. 3); the analogue did not produce any radicals and did not damage membrane integrity of keratinocytes as revealed by DPPH and LDH assays, respectively. Overall, the antipsoriatic potential resides within the 1,3-dihydroxy-substituted acridone skeleton.[68]

Phytosphingosine (Fig. 2) is found in plants such as rice (*Oryza sativa* L.), wheat (*Triticum aestivum* L.), potato (*Solanum tuberosum* L.) and sweet potato

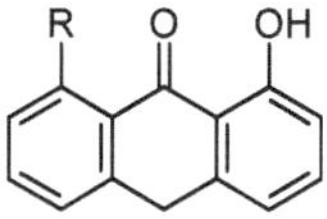

Fig. 3 Anthralin (also known as dithranol), an antipsoriatic drug.

(*Ipomoea batatas* (L.) Lam.), maize (*Zea mays* L.), soybean (*Glycine* max (L.) Merr.), kidney bean (*Phaseolus vulgaris* L.) and beet (*Beta vulgaris* L.), but also in fungi and mammalian epidermis.[69] In mammalian epidermis, it prevents water loss, regulates the growth and differentiation of epidermal cells, exerts antibacterial and anti-inflammatory activities. Despite these beneficial effects, the therapeutic use of phytosphingosine and its derivatives is limited by severe toxicity, especially at high doses. Two synthesized phytosphingosine derivatives, mYG-II-6 ((*Z*)-4-oxo-4-(((2*S*,3*S*,4*R*)-1,3,4-trihydroxyoctadecan-2-yl)amino)but-2-enoic acid) and fYG-II-6 ((*E*)-4-oxo-4-(((2*S*,3*S*,4*R*)-1,3,4-trihydroxyoctadecan-2-yl)amino)but-2-enoic acid), suppressed TPA-induced inflammation in mice skin. In addition, fYG-II-6 suppressed IL-23-induced inflammation in mice ears. Both derivatives were more potent and less toxic than phytosphingosine.[70]

2.2 Antipsoriatic natural products from marine sponges

Several marine constituents, potential leads for the development of novel antipsoriatic drugs, have been isolated (Fig. 4). Avarol, a sesquiterpenoid hydroquinone isolated from the sponge *Dysidea avara* (Dysideidae), exhibited antipsoriatic effects in different experimental models. It inhibited the psoriatic differentiation of human keratinocytes without affecting their growth at non-cytotoxic doses.[71] Avarol (Fig. 4) also showed anti-inflammatory effects in different assays (zymosan-stimulated human monocytes, HaCaT keratinocytes, zymosan-stimulated air pouch model, TPA-induced epidermal hyperplasia model). In mouse air pouch model, avarol (1% in ethanol) was co-administered with zymosan injection into pouches, whereas in TPA-induced hyperplasia model, it was topically applied (0.6 and 1.2 μmol/site daily) 1 h before TPA administration and 2 days after.[72] Several avarol derivatives (ester, 3′-thiosalicylate and amino derivatives) were developed. Similar to avarol, avarol-3′-thiosalicylate reduced inflammatory responses in cell-based and animal models.[73] In contrast to avarol, two benzylamine derivatives (benzylamine and *p*-methoxybenzylamine derivatives) (Fig. 4) inhibited human keratinocyte growth. The former showed an interesting pharmacological profile which will be detailed under next section and therefore it is considered a promising antipsoriatic drug candidate.[71]

Two cyclopeptides, perthamides C and E, were isolated from the tube sponge *Theonella swinhoei* (Theonellidae), a well-known source of marine bioactives. Both cyclopeptides demonstrated anti-inflammatory properties

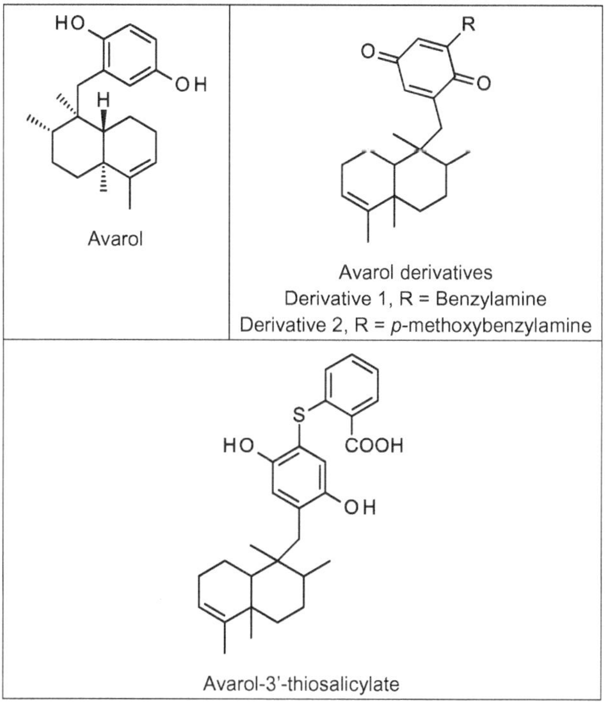

Fig. 4 Antipsoriatic natural products from marine sources.

in primary human keratinocytes stimulated with TPA with perthamide C showing a more promising pharmacological profile.[74]

2.3 Antipsoriatic microbial natural products

Cyclosporine (Fig. 5), microbial natural product, is an immunosuppressant drug, taken orally or by intravenous injection for the treatment of various medical conditions including psoriasis. The first cyclosporine was isolated in 1971 from the fungus *Tolypocladium inflatum* (Ophiocordycipitaceae). In 1997, the Food and Drug Administration (FDA) approved Neoral for adults with severe psoriasis and otherwise normal immune systems. Cyclosporine can effectively treat severe and disabling psoriasis, and it works quickly.[75] In clinical trials, 80–90% of patients who received cyclosporine for 12–16 weeks had rapid improvement.

Fumarates, as discussed in the earlier section, are also found in various fungal strains, and are useful in the treatment of psoriasis. Similarly, antipsoriatic drug phytosphingosine is also produced by several fungi.

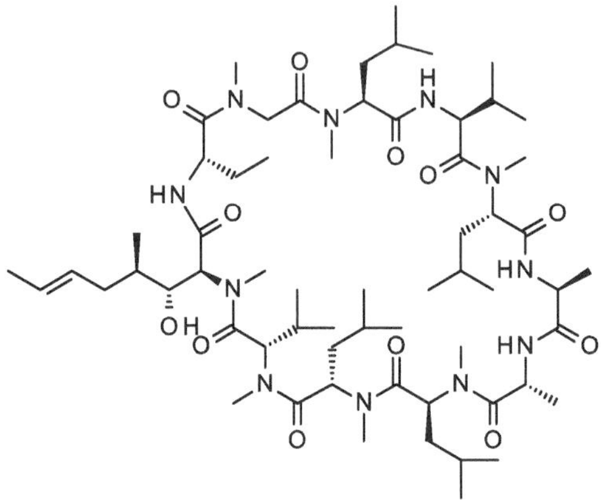

Fig. 5 Cyclosporine, a microbial antipsoriatic drug.

3. Mechanisms of action of antipsoriatic natural products

The pathogenesis of psoriasis is complex and not fully elucidated; nevertheless, several hallmark features can be highlighted: epidermal hyperplasia, abnormal keratinocyte differentiation, inflammatory cell infiltration and increase in dermal vascularization.[21,22] T-cell dysregulation (hyperactive Th1 cells, dysfunction of Th2 cells), activation of nuclear factor-κB (NF-κB) pathway, overexpression of vascular endothelial growth factor (VEGF) and vascular inflammatory markers proved to play a crucial role in inflammation and angiogenesis.[49] The mechanisms underlying the antipsoriatic activity of recently investigated natural products of plant and marine origins are systematized in Tables 1 and 2, respectively.

4. Virtual screening of antipsoriatic natural products

Virtual screening proved to be an important strategy in drug discovery.[76] This technique was used to predict natural products or their analogues as antipsoriatic agents based on their binding affinity to different targets. Fifteen constituents of *Cassia tora* L. (Fabaceae) were assessed for their ability to interact with 15 antipsoriatic protein targets. According to London dG docking scores, luteolin 7-*O*-β-glucopyranoside, quercetin 3-*O*-β-glucuronide and gluco-aurantioobtusin (Fig. 6) offered stronger binding affinities to 14 of 15 antipsoriatic targets (TNF-α, adenosine A2 receptor (AA2R), mitogen-activated protein kinase p38 (p38MAPK), sphingosine

Table 1 Antipsoriatic plant-derived products and analogues—mechanisms underlying antipsoriatic activity.

Compound	Antipsoriatic mechanisms ↑ ↓		
	Keratinocyte-based models	**Animal models**	**Human studies**
Fumaric acid esters	–	–	↑ GSH, Nrf2, GATA3, NR3C1 ↓ PTTG1, NFkBIZ[19]
1,4-Dihydroxy-2-naphtoic acid	↑ Apoptosis[21]	–	–
Paeoniflorin	–	↓ F4/80$^+$CD68$^+$ macrophages, CD11b$^+$ Gr-1$^+$ neutrophils, TNF-α, IL-1β, -6, -12, -17A, -17F, -21, -22, -23, IFN-γ, MIP-2, iNOS[9,22,23]	–
8-Methoxypsoralen (8-MOP)	Formation of covalent bonds with pyrimidine bases of DNA[24,27]	↓ IL-1β[24] ↓ IL-23/Th17, Th1, STAT3, RORγt[a28] ↑ Th2, CD4$^+$CD25$^+$Foxp3$^+$ T_{regs}[a28]	–
Isopsoralen		↓ IL-1β, -6[24]	–
Kaempferol	–	↓ CD3$^+$ T cells, IL-17A$^+$CD4$^+$ T cells, TNF-α, IL-6, -17A, NF-κB[29] ↑ CD4$^+$FoxP3$^+$ T_{regs}[29]	–
Quercetin	–	↓ TNF-α, IL-16, -17, NF-κB, IKKα, NIK, RelB[30] ↑ GSH, CAT, SOD, TRAF3[30]	–
Delphinidin	↓ Ki67, PCNA, iNOS, S100A7-psoriasin, S100A15-koebnerisin[31] ↑ Caspase-14, Filaggrin, Loricrin, Involucrin[31] ↓ PI3K/Akt/mTOR[32]	↓ PI3K/Akt/mTOR[32]	–

Continued

Table 1 Antipsoriatic plant-derived products and analogues—mechanisms underlying antipsoriatic activity.—cont'd

Compound	Antipsoriatic mechanisms ↑ ↓		
	Keratinocyte-based models	**Animal models**	**Human studies**
Genistein	↓ TNF-α, IL-1β, -6, -8, -20, -23, CCL2, VEGFA, MCP1, STAT3, NF-κB, ROS[2,33] Modulation of gene activity[33]	↓ TNF-α, IL-1β, -6, -17, -23 p40, MCP1, STAT3[2]	Modulation of gene activity[34]
Baicalein	↑ K1, K10, Ca^{2+} influx[35]	–	–
Astilbin	–	↓ TNF-α, IFN-γ, IL-2, -6, -17A[36]	–
Luteolin	↓ IL-6, -8, VEGF[36]	–	–
Amentoflavone	↓ IL-17A, -22, NF-κB[36] ↑ Apoptosis[36]	↓ TNF-α, IL-17A, -22, -23[36]	–
EGCG	–	↓ PCNA, IL-17A, -17F, -22, -23, MDA, $CD11c^+$ dendritic cells[37] ↑ caspase-14, SOD, CAT, $CD4^+$ T cells[37]	–
Chebulanin	↓ TNF-α, IL-17A, -23, ROS, NF-κB[38]	↓ TNF-α, IL-17A, -23, MMP-9, TBARS[38] ↑ GSH[38]	–
Resveratrol	↑ Sirt1[39] ↓ Aquaporin 3[41]	–	–
Curcumin	↑ Apoptosis (↑ TRAIL-R1/R2, Caspase-8 and -9[b]; ↓ IAP1, IAP2, Bcl-X_L, NF-κB[b], PKB/Akt[b], ERK1/2[b], EGF[b])[43,44] ↓ NF-κB, IL-6/IL-8[43]	–	↓ IL-22[48]
Honokiol	–	↓ Th1/Th2-expression $CD4^+$ T cells, TNF-α, IFN-γ, NF-κB, VEGFR-2[49]	–

Gambogic acid	–	↓ ICAM-1, *E*-selectin, VEGFR-2, IL-17, -22[51]	–
Rhodomyrtone	↑ Apoptosis[52] ↓ IL-36G, -36RN, MMP-13, S100A7, S100A8, S100A12, CXCL1, HBD2, MAPK, NF-κB[53]	–	–
Rottlerin	↓ NF-κB, ROS, miR-21, -31, Involucrin, Loricrin, TNF-α, IL-6, -23[54] ↑ Apoptosis, Autophagy[54]	↓ $CD3^+CD4^+$ T cells, $CD3^+$ $CD8^+$ T cells, $CD3^+$, $CD11b^+$, $CD11c^+$, TNF-α, IL-6, -23[54]	–
Dehydrocostuslactone	↓ STAT-3, STAT-1, CCL2, CXCL10, ICAM-1, SOCS3[55] ↑ Apoptosis, EGFR, EFK1/2[55]	–	–
Costulonide	↓ STAT-3, STAT-1, CCL2, CXCL10, ICAM-1, SOCS3[55] ↑ Apoptosis, EGFR, EFK1/2[55]	–	–
Tanshinone IIa	↑ Apoptosis (↑ Caspase-3, PARP, Cytochrome *c*)[55]	–	–
Celastrol	↑ Apoptosis (↑ Caspase-3, Bax; ↓ Bcl-2, NF-κB)[57]	–	–
3β,6β,16β-Trihydroxylup-20 (29)-ene (TTHL)	↑ Apoptosis[58]	↓ MPO, NAG, PCNA[58]	–
Periplogenin	↑ Necroptosis (↑ ROS)[5]	–	–
Rutaecarpine	–	↓ IFN-α, IL-17A, IL-23p19, NF-κB, TRL7[63]	–

Continued

Table 1 Antipsoriatic plant-derived products and analogues—mechanisms underlying antipsoriatic activity.—cont'd

Compound	Antipsoriatic mechanisms ↑ ↓		
	Keratinocyte-based models	**Animal models**	**Human studies**
Matrine	↑ G0/G1 phase arrest, p21, LC3II/I, Beclin I[65] ↓ Cyclin D1[65]	–	–
Isocamptothecine	↑ Apoptosis[66] ↓ Telomerase[66]	–	–
mYG-II-6	↓ NF-κB, JAK/STAT, MAPK[70] ↑ Apoptosis[70]	–	–
fYG-II-6	↓ NF-κB, JAK/STAT, MAPK[70] ↑ Apoptosis[70]	↓ IL-1α, -1β, -6, -17A, -22, TNF-α, IFN-γ, CXCL1, CCL17, CCL20, CCL27[70] ↑ IL-10[70]	–

[a]Effects developed by PUVA treatment.
[b]Effects developed in combination with light exposure.
CAT, catalase; EGF, epidermal growth factor; EGFR, epidermal growth factor receptor; ERK1/2, extracellular signal-regulated protein kinases 1 and 2; GATA3, GATA binding protein 3; GSH, glutathione; HBD2, human beta defensin 2; IAP1,2, inhibitor of apoptosis 1,2; ICAM-1, intercellular adhesion molecule 1; IFN-γ, interferon γ; IL, interleukin; iKKα, IkB kinase α; iNOS, inducible nitric oxide synthase; JAK/STAT, Janus kinase/signal transducers and activators of transcription pathway; K1/K10, keratins 1 and 10; MAPK, mitogen activated protein kinase; MDA, malondialdehyde; MCP-1, monocyte chemoattractant protein-1; MIP-2, macrophage inflammatory protein-2; MMP-9, matrix metalloproteinase-9; MPO, myeloperoxidase; NAG, *N*-acetylglucosaminidase; NF-κB, nuclear factor-κB; NIK, NF-κB-inducing kinase; NF-κBIZ, NF-κB inhibitor zeta; Nrf2, nuclear factor erythroid 2-related factor 2; NR3C1, nuclear receptor subfamily 3 group C member 1; PARP, poly (ADP-ribose) polymerase; PCNA, proliferating cell nuclear antigen; PI3K/Akt/mTOR, phosphatidylinositol-3-kinase and the mammalian target of rapamycin signaling pathways; PKB/Akt, protein kinase B/Akt pathway; PTTG1, pituitary tumor transforming gene 1; SOD, superoxide dismutase; STAT3, signal transducer and activator of transcription 3; TBARS, thiobarbituric acid reactive substances; Th, T helper; TNF-α, tumor necrosis factor-α; TRAF3, TNF receptor associated factor 3; RORγt, retinoic acid receptor-related orphan receptor gamma t; ROS, reactive oxygen species; Sirt1, sirtuin 1; VEGFA, vascular endothelial growth factor A; VEGFR, vascular endothelial growth factor receptor.

Table 2 Antipsoriatic marine natural products and analogues—mechanisms underlying antipsoriatic activity.

Compound	Antipsoriatic mechanisms		
	Keratinocyte-based models	Animal models	Human studies
Avarol	↓ NF-κB, TNF-α[72]	↓ NF-κB, TNF-α, IL-1β, -2, PGE_2, LTB_4, MPO[72]	–
Benzylamine derivative of avarol	↓ TNF-α, COX-2[71]	–	–
Avarol-3′-thiosalicylate	–	↓ TNF-α, IL-1β, -2, PGE_2, LTB_4, MPO, NF-κB[72]	–
Perthamides C and E	↓ TNF-α, IL-8[74]	–	–

COX-2, cyclooxygenase-2; IL, interleukin; LTB_4, leukotriene B_4; MPO, myeloperoxidase; NF-κB, nuclear factor-κB; PGE_2, prostaglandin E_2; TNF-α, tumor necrosis factor-α.

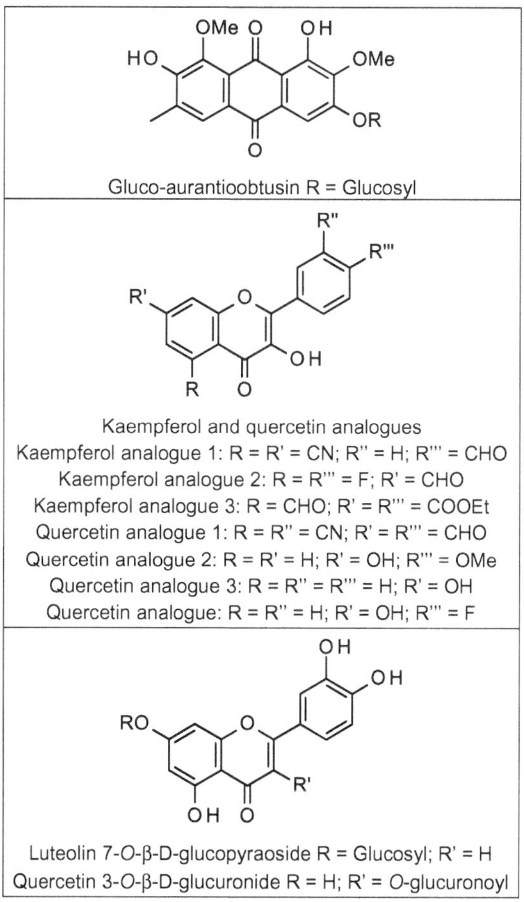

Fig. 6 Molecular docked analogues of natural products with antipsoriatic activity.

1-phosphate receptor (S1PR), protein kinase (PK), Bruton's tyrosine kinase (BTK), peptidyl arginine deiminases (PAD), spleen protein kinase (SPK), protein kinase C (PKC), IL-17A, phosphodiesterase-4 (PDE-4), cathepsin S (CCS), Janus kinase (JAK-3), Ras-related C3 botulinum toxin substrate 1 (RAC-1)). Several structural features such as sugar (especially uronic acid), pyranone and *ortho*-diphenolic moieties seem to be essential for binding to target proteins.[77]

Analogues of kaempferol and quercetin were designed using Schrödinger-Maestro 11 and further docked on their ability to bind to calcineurin using PyRx software. Calcineurin suppresses the activation and proliferation of memory $CD4^+$ T cells, events playing a crucial role in the pathophysiology of psoriasis. The analogues with the best predicted binding affinities are shown in Fig. 6.[78]

5. Conclusions

As prescribed antipsoriatic drugs are associated with severe side effects, there is a high interest in the development of novel agents having an improved pharmaco-toxicological profile. Recent studies reviewed in this chapter highlight that natural products derived mainly from plants and marine organisms contain structural scaffolds that can be exploited for the development of novel antipsoriatic drugs. Human clinical trials, assessing the pharmacokinetic profile, safety and therapeutic efficacy of natural products and their analogues, are still much needed.

Acknowledgment

L.N. gratefully acknowledges the financial support of the European Regional Development Fund—Project ENOCH (No. CZ.02.1.01/0.0/0.0/16_019/0000868).

References

1. Saelee, C.; Thongrakard, V.; Tencomnao, T. *Molecules* **2011**, *16*, 3908–3932.
2. Wang, A.; Wei, J.; Lu, C.; Chen, H.; Zhong, X.; Lu, Y.; Li, L.; Hung, H.; Dai, Z.; Han, L. *Int. Immunopharmacol.* **2019**, *69*, 270–278.
3. Rendon, A.; Schäkel, K. *Int. J. Mol. Sci.* **2019**, *20*, 1475.
4. Huang, T.-H.; Lin, C.-F.; Alalaiwe, A.; Yang, S.-C.; Fang, J.-Y. *Int. J. Mol. Sci.* **2019**, *20*, 2558.
5. Zhang, W.-J.; Song, Z.-B.; Bao, Y.-L.; Li, W.-I.; Yang, X.-G.; Wang, Q.; YU, C.-L.; Sun, L.-G.; Huang, Y.-X.; Li, Y.-X. *Biochem. Pharmacol.* **2016**, *105*, 66–79.
6. Singh, K. K.; Tripathy, S. *J. App. Pharm. Sci.* **2014**, *4*(11), 114–121.
7. Zeng, Z.; Luo, S.; Huang, Y.; Lu, Q. *J. Dermatol.* **2017**, *44*, 863–872.
8. Kaur, A.; Kumar, S. *Pharm. Biol.* **2012**, *50*(12), 1573–1591.

9. Zhao, J.; Di, T.; Wang, Y.; Wang, Y.; Liu, X.; Liang, D. *Eur. J. Pharmacol.* **2016**, *772*, 131–143.
10. Kim, G. K.; Del Rosso, J. Q. *J. Clin. Aesthet. Dermatol.* **2010**, *3*(1), 32–38.
11. Kim, W. B.; Jerome, D.; Yeung, J. *Can. Fam. Physician* **2017**, *63*, 278–285.
12. Wong, T.; Hsu, L.; Liao, W. *J. Cutan. Med. Surg.* **2013**, *17*(1), 6–12.
13. Christensen, L. P. Polyphenols and polyphenol-derived compounds from plants and contact dermatitis. In: *Polyphenols—Prevention and Treatment of Human Disease*, 2nd ed.; Watson, R. R., Preedy, V. R., Zibadi, S., Eds.; Elsevier Inc.: San Diego, 2018.
14. Harries, M. J.; Chalmers, R. J. G.; Griffiths, C. E. M. *Br. J. Dermatol.* **2005**, *153*, 549–551.
15. European Commission *Health and Consumer Protection Directorate-General.* https://ec.europa.eu/food/sites/food/files/safety/docs/animal-feed_additives_rules_scan-old_report_out112.pdf; 2003, accessed on 22 December 2019.
16. Sauer, M.; Mattanovich, D.; Marx, H. Microbial production of organic acids for use in food. In: *Microbial Production of Food Ingredients, Enzymes and Nutraceuticals*; McNeil, B., Archer, D., Giavasis, I., Harvey, L., Eds.; Woodhead Publishing, 2013.
17. Ismail, N.; Collins, P.; Rogers, S.; Kirby, B.; Lally, A. *Br. J. Dermatol.* **2014**, *171*, 397–402.
18. Atwan, A.; Ingram, J. R.; Abbott, R.; Kelson, M. J.; Pickles, T.; Bauer, A.; Piguet, V. *Br. J. Dermatol.* **2016**, *175*, 873–881.
19. Onderdijk, A. J.; Balak, D. M. W.; Baerveldt, E. M.; Florencia, E. F.; Kant, M.; Laman, J. D.; van Ijcken, W. F. J.; Racz, E.; de Ridder, D.; Thio, H. B.; Prens, E. P. *Br. J. Dermatol.* **2014**, *171*, 732–741.
20. Tzaneva, S.; Geroldinger, A.; Trattner, H.; Tanew, A. *Br. J. Dermatol.* **2018**, *178*, 682–688.
21. Mok, C.-F.; Xie, C.-M.; Sham, K. W.-Y.; Lin, Z.-X.; Cheng, C. H.-K. *Evid. Based Complement. Alternat. Med.* **2013**, *2013*, 792840.
22. Sun, Y.; Zhang, J.; Huo, R.; Zhai, T.; Li, H.; Wu, P.; Zhu, X.; Zhou, Z.; Shen, B.; Li, N. *Int. Immunopharmacol.* **2015**, *24*, 392–399.
23. Yu, Z.; Xiao, Z.; Zhao, R.; Lu, C.; Zhang, Y. *Life Sci.* **2017**, *180*, 17–22.
24. Alalaiwe, A.; Hung, C.-F.; Leu, Y.-L.; Tahara, K.; Chen, H.-H.; Hu, K.-Y.; Fang, J.-Y. *Eur. J. Pharm. Sci.* **2018**, *124*, 114–126.
25. Farahnik, B.; Nakamura, M.; Singh, R. K.; Abrouk, M.; Zhu, T. H.; Lee, K. M.; Jose, M. V.; DaLovisio, R.; Koo, J.; Bhutani, T.; Liao, W. *Dermatol. Ther.* **2016**, 6, 315–324.
26. Griffith, J. L.; Zarbo, A. J.; Lim, H. W. Phototherapy and photochemotherapy. In: *Evidence-Based Psoriasis, Updates in Clinical Dermatology*; Bhutani, T., Liao, W., Nakamura, M., Eds.; Springer International Publishing AG, 2018.
27. Singh, T. P.; Schön, M. P.; Wallbrecht, K.; Michaelis, K.; Rinner, B.; Mayer, G.; Schmidbauer, U.; Strohmaier, H.; Wang, X.-Y.; Wolf, P. *J. Immunology* **2010**, *184*, 7257–7267.
28. Valejo Coelho, M. M.; Apetato, M. *Clin. Dermatol.* **2016**, *34*, 556–562.
29. Liu, C.; Liu, H.; Lu, C.; Deng, J.; Yan, Y.; Chen, H.; Wang, Y.; Liang, C.-L.; Wei, J.; Han, L.; Dai, Z. *Clin. Exp. Immunol.* **2019**, *198*(3), 403–415.
30. Chen, H.; Lu, C.; Liu, H.; Wang, M.; Zhao, H.; Yan, Y.; Han, L. *Int. Immunopharmacol.* **2017**, *48*, 110–117.
31. Chamcheu, J. C.; Pal, H. C.; Siddiqui, I. A.; Adhami, V. M.; Ayehunie, S.; Boylan, B. T.; Noubissi, F. K.; Khan, N.; Syed, D. N.; Elmets, C. A.; Wood, G. S.; Afaq, F.; Mukhtar, H. *Skin Pharmacol. Physiol.* **2015**, *28*(4), 177–188.
32. Chamcheu, J. C.; Adhami, V. M.; Esnault, S.; Sechi, M.; Siddiqui, I. A.; Satyshur, K. A.; Syed, D. N.; Dodwad, S.-J. M.; Chavez-Rodriquez, M.-I.; Longley, B. J.; Wood, G. S.; Mukhtar, H. *Antioxid. Redox Signal.* **2017**, *26*, 49–69.

33. Smolińska, E.; Moskot, M.; Jakóbkiewicz-Banecka, J.; Węgrzyn, G.; Banecki, B.; Szczerkowska-Dobosz, A.; Purzycka-Bohdan, D.; Gabig-Cimińska, M. *PLoS One* **2018**, *13,* e0192297.
34. Smolińska, E.; Węgrzyn, G.; Gabig-Cimińska, M. *Acta Biochim. Pol.* **2019**, *66*(1), 101–110.
35. Huang, K.-F.; Ma, K.-H.; Liu, P.-S.; Chen, B.-W.; Chueh, S.-H. *Exp. Dermatol.* **2016**, *25*, 623–629.
36. Bonesi, M.; Loizzo, M. R.; Menichini, F.; Tundis, R. Flavonoids in treating psoriasis. In: *Immunity and Inflammation in Health and Disease*; Chatterjee, S., Jungraithmayr, W., Bagchi, D., Eds.; Academic Press, 2018.
37. Zhang, S.; Liu, X.; Mei, L.; Wang, H.; Fang, F. *BMC Complement. Altern. Med.* **2016**, *16*, 334.
38. An, J.; Li, T.; Dong, Y.; Li, Z.; Huo, J. *Cell. Physiol. Biochem.* **2016**, *39*, 531–543.
39. Lee, J.-H.; Kim, J.-S.; Park, S.-Y.; Lee, Y.-J. *Oncol. Rep.* **2016**, *35*, 524–529.
40. Luca, S. V.; Macovei, I.; Bujor, A.; Miron, A.; Skalicka-Woźniak, K.; Aprotosoaie, A. C.; Trifan, A. *Crit. Rev. Food Sci. Nutr.* **2020**, *60*(4), 626–659.
41. Wu, Z.; Uchi, H.; Morino-Koga, S.; Shi, W.; Furue, M. *J. Dermatol. Sci.* **2014**, *75*, 16–23.
42. Kjaer, T. N.; Thorsen, K.; Jessen, N.; Stenderup, K.; Pedersen, S. B. *PLoS One* **2015**, *10,* e0126599.
43. Sun, J.; Han, J.; Zhao, Y.; Zhu, Q.; Hu, J. *Int. Immunopharmacol.* **2012**, *13*, 170–174.
44. Dujic, J.; Kippenberger, S.; Hoffmann, S.; Ramirez-Bosca, A.; Miquel, J.; Diaz-Alperi, J.; Bereiter-Hahn, J.; Kaufmann, R.; Bernd, A. *J. Invest. Dermatol.* **2007**, *127*, 1992–2000.
45. Niu, T.; Tian, Y.; Cai, Q.; Ren, Q.; Wei, L. *PLoS One* **2015**, *10,* e0138754.
46. Trifan, A.; Aprotosoaie, A. C.; Miron, A. Curcumin in food. In: *Handbook of Dietary Phytochemicals*; Xiao, J., Sarker, S. D., Asakawa, Y., Eds.; Springer Nature Singapore Pte Ltd., 2019.
47. Kurd, S. K.; Smith, N.; VanVoorhees, A.; Troxel, A. B.; Badmaev, V.; Seykora, J. T.; Gelfand, J. M. *J. Am. Acad. Dermatol.* **2008**, *58*(4), 625–631.
48. Antiga, E.; Bonciolini, V.; Volpi, W.; Del Bianco, E.; Caproni, M. *Biomed. Res. Int.* **2015**, 283634.
49. Wen, J.; Wang, X.; Pei, H.; Xie, C.; Qiu, N.; Li, S.; Wang, W.; Cheng, X.; Chen, L. *J. Pharmacol. Sci.* **2015**, *128*, 116–124.
50. Yi, T.; Yi, Z.; Cho, S.-G.; Luo, J.; Pandey, M. K.; Aggarwal, B. B.; Liu, M. *Cancer Res.* **2008**, *68*, 1843–1850.
51. Wen, J.; Pei, H.; Wang, X.; Xie, C.; Li, S.; Huang, L.; Qiu, N.; Wang, W.; Cheng, X.; Chen, L. *J. Dermatol. Sci.* **2014**, *74*, 242–250.
52. Chorachoo, J.; Saeloh, D.; Srichana, T.; Amnuaikit, T.; Musthafa, K. S.; Sretrirutchai, S.; Voravuthikunchai, S. P. *Eur. J. Pharmacol.* **2016**, *772*, 144–151.
53. Chorachoo, J.; Lambert, S.; Furnholm, T.; Roberts, L.; Reingold, L.; Auepemkiate, S.; Voravuthikunchai, S. P.; Johnston, A. *PLoS One* **2018**, *13,* e0205340.
54. Min, M.; Yan, B.-X.; Wang, P.; Landeck, L.; Chen, J.-Q.; Li, W.; Cai, S.-Q.; Zheng, M.; Man, X.-Y. *PLoS One* **2017**, *12,* e0190051.
55. Scarponi, C.; Butturini, E.; Sestito, R.; Madonna, S.; Cavani, A.; Mariotto, S.; Albanesi, C. *PLoS One* **2014**, *9,* e107904.
56. Li, F.-L.; Xu, R.; Zeng, Q.-C.; Li, X.; Chen, J.; Wang, Y.-F.; Fan, B.; Geng, L.; Li, B. *Evid. Based Complement. Alternat. Med.* **2012**, 927658.
57. Zhou, L.-L.; Lin, Z.-X.; Fung, K.-P.; Cheng, C. H. K.; Che, C.-T.; Zhao, M.; Wu, S.-H.; Zuo, Z. *Eur. J. Pharmacol.* **2011**, *670*, 399–408.
58. Horinouchi, C. D.; Mendes, D. A.; Nolte, S.; Brito, P. S.; Soley, B. D.; Favero, G. M.; Facundo, V. A.; Santos, A. R.; Cabrini, D. A.; Otuki, M. F. *J. Ethnopharmacol.* **2017**, *195*, 298–308.

59. Zhang, Y.-W.; Bao, Y.-L.; Wu, Y.; Yu, C.-L.; Li, Y.-X. *Acta Crystallogr., Sect. E: Struct. Rep. Online* **2012**, *68*(Pt. 6), o1582–o1583.
60. Wilkinson, J. D.; Williamson, E. M. *J. Dermatol. Sci.* **2007**, *45*, 87–92.
61. Lin, W.; Huang, W.; Ning, S.; Wang, X.; Ye, Q.; Wei, D. *PLoS One* **2018**, *13*, e0199788.
62. Lee, S. H.; Son, J. K.; Jeong, B. S.; Jeong, T. C.; Chang, H. W.; Lee, E.-S.; Jahng, Y. Progress in the studies on rutaecarpine. *Molecules* **2008**, *13*, 272–300.
63. Li, Y.; Zhang, G.; Chen, M.; Tong, M.; Zhao, M.; Tang, F.; Xiao, R.; Wen, H. *Biomed. Pharmacother.* **2019**, *109*, 1876–1883.
64. Tang, C.; Qian, X.; Jia, R.; Wu, M.; Liang, Z. *Oncol. Rep.* **2013**, *30*, 2529–2535.
65. Jiang, W.-W.; Wang, Y.-M.; Wang, X.-Y.; Zhang, Q.; Zhu, S.-M.; Zhang, C.-L. *Chin Med J (Engl)* **2019**, 132, 2079–2088.
66. Lin, J.; Liu, X.; Bao, Y.; Hou, S.; An, L.; Lin, X. *Exp. Dermatol.* **2008**, *17*, 530–536.
67. Putic, A.; Stecher, L.; Prinz, H.; Müller, K. *Eur. J. Med. Chem.* **2010**, *45*, 3299–3310.
68. Putic, A.; Stecher, L.; Prinz, H.; Müller, K. *Eur. J. Med. Chem.* **2010**, *45*, 5345–5352.
69. Tessema, E. N.; Gebre-Mariam, T.; Neubert, R. H. H.; Wohlrab, J. *Skin Pharmacol. Physiol.* **2017**, *30*, 115–138.
70. Kim, B.-H.; Lee, J. M.; Jung, Y.-G.; Kim, S.; Kim, T.-Y. *J. Invest. Dermatol.* **2014**, *134*, 1023–1032.
71. Amigó, M.; Schalkwijk, J.; Olthuis, D.; De Rosa, S.; Payá, M.; Terencio, M. C.; Lamme, E. *Life Sci.* **2006**, *79*, 2395–2404.
72. Amigó, M.; Payá, M.; Braza-Boïls, A.; De Rosa, S.; Terencio, M. C. *Life Sci.* **2008**, *82*, 256–264.
73. Amigó, M.; Payá, M.; De Rosa, S.; Terencio, M. C. *Br. J. Pharmacol.* **2007**, *152*, 353–365.
74. Festa, C.; De Marino, S.; Sepe, V.; D'Auria, M. V.; Bifulco, G.; Andrés, R.; Terencio, M. C.; Payá, M.; Debitus, C.; Zampella, A. *Tetrahedron* **2011**, *67*, 7780–7786.
75. Colombo, M. D.; Cassano, N.; Bellia, G.; Vena, G. A. *ScientificWorldJournal* **2013**, *2013*, 805705.
76. Sarker, S. D.; Nahar, L. *Computational Phytochemistry*. Elsevier: USA, 2018.
77. Akachukwu, I.; Amara, E. E. *J. Appl. Pharm. Sci.* **2019**, *9*(04), 082–087.
78. Sharma, P.; Mishra, R.; Vakil, B. *Int. J. Curr. Res. Rev.* **2018**, *10*(3), 19–27.

CHAPTER ELEVEN

Medicinal natural products in osteoporosis

Didem Şöhretoğlu[a,*], **Gülin Renda**[b]
[a]Department of Pharmacognosy, Faculty of Pharmacy, Hacettepe University, Ankara, Turkey
[b]Department of Pharmacognosy, Faculty of Pharmacy, Karadeniz Technical University, Trabzon, Turkey
*Corresponding author: e-mail address: didems@hacettepe.edu.tr

Contents

Abbreviations

ALP alkaline phosphatase
BALP bone-specific alkaline phosphatase
BMC bone mineral content
BMD bone mineral density
BMP-2 bone morphogenetic protein-2
BMS bone marrow-derived mesenchymal stem
Col I type I collagen
E2 17β-estradiol
ER estrogen receptor
ERK the extracellular-signal-regulated kinase
JUN Jun N-terminal kinase

Annual Reports in Medicinal Chemistry, Volume 55
ISSN 0065-7743
https://doi.org/10.1016/bs.armc.2020.03.002

LPS	lipopolysaccharide
MAPK	mitogen-activated protein kinase
NF-κB	nuclear factor kappa B
OCN	osteocalcin
OPG	osteoprotegerin
OSCAR	osteoclast-associated receptor
OVX	ovariectomized
PTGS2	prostaglandin-endoperoxide synthase 2
RANKL	RANKL
RUNX2	runt-related transcription factor 2
TRAP	tartrate-resistant acid phosphatase

1. Introduction

Osteoporosis is characterized by reduction in bone mass and bone mineral density (BMD), as well as micro-structural deterioration of the bone tissue due to depletion of calcium and bone protein, associated with high risk of fracture and defined as a metabolic bone disorder resulting from an imbalance of bone remodeling, in which the rate of bone resorption is higher than the rate of bone formation.[1,2] Worldwide, osteoporosis is reported to cause more than 8.9 million fractures annually. Osteoporosis is estimated to affect 200 million women worldwide. It is estimated that by 2050, the worldwide incidence of hip fracture in men is projected to increase by 310% and 240% in women, compared to rates in 1990.[3]

Osteoporosis can be classified into primary (Type I) and secondary (Type II, senile) osteoporosis. Primary type osteoporosis occurs in women soon after menopause (postmenopausal osteoporosis) due to decreased estrogen levels and during and after middle-age in men due to loss of androgen. Secondary osteoporosis is due to medications or certain medical conditions, such as hypogonadism, hyperparathyroidism, or leukemia. Also prolonged use of some medications, such as oral or high-dose inhaled corticosteroids, thyroid hormone replacement, and aromatase inhibitors can lead to bone loss.[2] According to the World Health Organization (WHO), osteoporosis has been diagnosed by bone mineral density at the hip and/or the spine. The bone microarchitecture reflects the mechanical strength and bone quality. It is also used in *in vivo* experiments to reflect bone deterioration.[4–8]

There is a continuous remodeling process in bone tissue to grow, regulate calcium and phosphate metabolism as well as to repair any damage.

This process is carried out by the activities of two principal cell types, osteoblasts and osteoclasts. Osteoblasts are bone forming cells, they fill up the bone cavities with new tissues, and on the contrary, osteoclasts degrade bones by resorption. In osteoporosis, the balance between bone formation and resorption is impaired. There are several indicators used in clinical tests for the assessment of bone health and evaluation of therapeutic responses. Many of these indicators can also be applied to experimental studies using *in vitro* (cell-based) assays or *in vivo* (animal) models. Bone remodeling indicators include the measurements of bone mineral density, bone microarchitecture, and biochemical markers.[6–9]

Bone formation biochemical markers are osteocalcin (OCN), total alkaline phosphatase (ALP) and bone ALP, C-terminal propeptides of type I procollagen (PICP) and N-terminal propeptides of type I procollagen.[10,11]

ALP is produced in many tissues including liver, kidney and bone. Bone ALP is an enzyme attached to the external surface of the cell membrane of osteoblasts. It is partly released into the circulation. Bone ALP is a major regulator of bone mineralization and it plays role in osteoid formation. Total serum ALP, which is mainly hepatic, renal skeletal isoenzyme, is also a non-specific marker of bone formation.[11,12] OCN, produced by osteoblasts and odontoblasts, is the most abundant non-collagenous protein in bone. OCN is generally regarded as a marker of bone formation, whereas it is also involved in bone mineralization and calcium ion homeostasis rather than matrix production. Since the plasma concentration of OCN is the marker of bone formation, the circulating concentration generally well associated with bone formation rate.[13]

Procollagen I is precursor of type I collagen (Col I), which is the most abundant protein of bone (about 90%). Therefore, C-terminal propeptides of type I procollagen (PICP) and N-terminal propeptides of type I procollagen (PINP) are the convenient markers of bone formation.[11] Bone resorption markers are some collagen degradation products, osteopontin (OPN), osteoclastic enzymes like tartrate-resistant acid phosphatase (TRAP), cathepsin K and osteoclast activity markers, including receptor activator of nuclear factor kappa B ligand (RANKL), osteoprotegerin (OPG), Dickkopf-related protein 1, and sclerostin. The main component of the bone matrix is Col I. Therefore, during bone resorption, the degradation products such as telopeptides Col I, hydroxyproline, pyridinoline, deoxypyridinoline, are formed. OPN, which is a marker of osteoclast activity and bone resorption, has a critical role in the maintenance of bone, including inhibition of bone mineralization and ectopic calcification.[11,12,14]

Cathepsin K is a collagenase enzyme, secreted by osteoclasts. The secretion of cathepsin K results in bone degradation, primarily of type I collagen.[15] TRAP is a metalloenzyme produced by osteoclasts as well as different tissues. Hence, it is a non-specific marker of bone resorption. However, TRAP5b isoform is osteoclast specific and used as a specific biomarker of osteoclast activity.[12] Dickkopf-related protein and sclerostin are Wnt signaling pathway inhibitors. The serum Dickkopf-related protein level reflects the inhibition of bone formation. Sclerostin is secreted into circulation by osteocytes, and serum level shows inhibition of bone formation.[11,12]

RANKL is a transmembrane protein, released by osteoblasts. RANKL binds RANK and stimulates osteoclastic bone resorption. This binding activates different pathways like nuclear factor kappa light chain enhancer of activated B cells (NF-κB), mitogen-activated protein kinases (MAPKs) including ERK, JNK and p38.[12]

Osteoprotogenin (OPG) is also known as osteoclastogenesis inhibitory factor, produced by osteoblasts as well as many different tissues. OPG production is regulated by Wnt/β-catenin signaling in osteoblasts. OPG binds to RANKL and inhibits osteoclast differentiation.[16] Most of the treatments currently available for osteoporosis aim to inhibit bone resorption and reduce bone remodeling. Parathyroid hormone, and its analogue teriparatide, are the only anabolic therapies available to treat severe osteoporosis. Agents for the treatment of osteoporosis are classified as either antiresorptive or anabolic. Antiresorptive agents work by inhibiting the activity of osteoclasts and, therefore, reducing bone resorption. Currently available anti-osteoporosis agents include bisphosphonates, selective estrogen receptor modulators, calcitonin and estrogen. The current drug therapies have been reported to improve bone mineral density and reduce the risk of fracture, but long-term use may cause various side effects.[4–6]

Herbal treatments are considered safe alternatives in the current treatment of osteoporosis due to side effects and cost factors. In this respect, the role of several plant extracts in the treatment of osteoporosis was evaluated. Some of these studies have been resulted in the isolation of the bioactive molecules that are capable enough to induce bone formation with promising inhibition of osteoclast formation and/or differentiation.[17] The anti-osteoporotic effects of a large number of plants have been demonstrated by cellular and animal studies and their identified bioactive constituents are summarized in this chapter.

2. Anti-osteoporotic natural products

There are numerous plant extracts and compounds tested for their possible role for osteoporosis. Some of plants especially the ones used in Chinese medicine for the treatment of osteoporosis seems promising candidates to use in clinic. Anti-osteoporotic properties of selected plant extracts were given in Table 1. The roles of some natural compounds in osteoporosis are also investigated by mechanisms and they were given according to their secondary metabolite group:

2.1 Flavonoids

2.1.1 Flavanone

Hesperidin (5 g/kg) and 7 g/kg of α-glucosylhesperidin treatment for 4 weeks significantly ameliorated bone loss in ovariectomized (OVX) mices (Fig. 1). α-Glucosylhesperidin prevented the trabecular bone volume and trabecular thickness in the femoral distal metaphysis according to histomorphometric analyses. Moreover, hesperidin decreased the osteoclast number of the femoral metaphysis in OVX mice like 17β-estradiol (E2). Also, femur Ca, P and Zn concentrations in hesperidin and E2 groups were higher than that of OVX control group. Serum concentration of hesperetin, aglycon of both compounds was higher in α-glucosylhesperidin fed group due to higher solubility of α-glucosylhesperidin. Hence, protective effect of α-glucosylhesperidin was slightly higher than that of hesperidin.[76]

Pinostrobin increased the proliferation of murine pre-osteoblastic MC3T3-E1 cells at 5–80 μg/mL ($P < 0.05$ or $P < 0.01$), and caused a significant elevation of ALP activity, Col I content, and mineralization of osteoblasts at 10–40 μg/mL ($P < 0.05$ or $P < 0.01$), and upregulated the expression level of osteocalcin mRNA gene ($P < 0.01$). Moreover, pinostrobin ameliorated dexamethasone-induced inhibition of cell differentiation (Fig. 1).[77]

2.1.2 Flavanol

Chrysosplenetin time-dependently promoted proliferation and osteoblastogenesis of human-derived bone marrow stromal (BMS) cells reaching its maximal effects at a concentration of 10 μM (Fig. 1). 0–20 μM chrysosplenetin upregulated gene expression of the osteoblastic marker genes RUNX2, BGLAP, CTNNB1, and BMP2 in a non-dose-dependent at day 14 of culture. Moreover, it upregulated the expressions of target genes

Table 1 Anti-osteoporotic properties of selected plant extracts.

Phyto constituents	Plant part	Extract type	Result	Major active molecule	Potential mechanism
Alpinia officinarum (Hance) Farw.	–	98% EtOH	Decreases bone loss and increases bone formation	Galangin	Increases cell viability, differentiation, and mineralization in osteoblasts, increases osteopenia and bone formation rate, enhances the bone strength, and prevents the deterioration[18]
Areca catechu L.	Nut	Water	Protects against ovariectomy-induced osteoporosis		Inhibits the release of ROS and bone resorption[19]
Artemisia princeps Pamp.	–	EtOH	Prevents bone loss and osteoclast activity	–	Prevents the increase in bone turnover by controlling the serum Ca/P ratio, anti-inflammatory and antioxidant properties[20]
Aspalathus linearis (Burm.f.) R.Dahlgren	Leaf	Water	Inhibits osteoclast formation		Reduces osteoclast resorption and the associated gene expression, attenuates NF-κB activity[21]
Bixa orellana L.	Corn	Oil	Normalizes conditions associated with metabolic syndrome and exhibits both anabolic and anti-osteoporotic effects on bone	Tocotrienol	Regulation of hormone levels and inflammatory response[22]

Bombax ceiba L.	Stem bark	Petroleum ether and MeOH	Improves the bone porosity, restores the normal architecture of bone	Lupeol, gallic acid, β-sitosterol	Reverses the debilitating state of bone remodeling due to estrogenic modulation *in vitro* and *in vivo* and maintains calcium, phosphorus level, and bone integrity[23]
Bupleurum falcatum L.	Root	Water	Prevents estrogen deficiency-induced bone loss, anti-osteoclastogenic activity	–	Inhibits osteoclastogenic transcription factors c-Fos and nuclear factor of activated T cells c1 (NFATc1), downregulates expression of osteoclast marker genes, upregulates iNOS protein expression[24]
Camellia sinensis (L.) Kuntze	Leaf	Water	Ameliorate the symptoms of osteoporosis		Inhibits RANKL-induced osteoclast-specific gene and protein expression[25]
Cassia occidentalis L.	Stem	95% EtOH	Prevents glucocorticoid-induced osteopenia	Apigenin, 4′,7′-dihydroxyflavone, luteolin, 3′,4′,7′-trihydroxy flavone, emodin, apigenin-6-C-glucopyranoside	Reduces osteopenia and increases bone strength *via* the osteogenic as well as antiresorptive mechanisms[26]
Cichorium ṅtybusL.	Leaf	Water	Bone protection against glucocorticoid-induced osteoporosis	Flavonoids and inulin	Increases Ca absorption and results in a corresponding increase in BMD[27]

Continued

Table 1 Anti-osteoporotic properties of selected plant extracts.—Cont'd

Phyto constituents	Plant part	Extract type	Result	Major active molecule	Potential mechanism
Citrus bergamia Risso and Poit.	Fruit	Juice	Stimulates osteoblast differentiation and production of collagen	Polyphenols (neoeriocitrin, naringin, neohesperidin)	β-Catenin pathway, upregulates β-catenin expression ($P = 0.001$), osteoblast differentiation markers (RUNX2 and COL 1A), and downregulates RANKL ($P = 0.028$)[28]
Citrus junos Siebold ex Tanaka	Fruit peel	EtOH	Prevents OVX-induced osteopeni, promotes osteoblastogenesis	Flavanones	Activation of the BMP/p38/Smad/Runx2 pathways[29]
Cornus officinalis Siebold and Zucc.	Fructus	EtOH	Improves osteoporosis activity	Sweroside	Stimulates osteocalcin secretion. Increases cell proliferation, ALP activity and inhibits apoptotic cell death[30]
Corylopsis coreana Uyeki	Flower	80% EtOH	Inhibits osteoclast differentiation	Bergenin, quercetin, quercitrin	Inhibited mRNA expression levels of NFATc1, OSCAR, cathepsin K, TRAP, and the translational expression of NFATc1[31]
Cynanchum wilfordii (Maxim.) Hemsl.	Root	Water	Inhibits bone resorption and stimulates bone formation	–	Decreases the number of tartrate-resistant acid phosphatase-positive multinucleated osteoclasts, increases BMD and BMC[32]

Cynomorium songaricum Rupr.	Stem	80% EtOH	Restrains osteoclastogenesis, prevents bone loss		Inhibition of RANKL/RANK/TRAF6 and the subsequent NF-κB and PI3K/AKT signaling pathways[33]
Dalbergia sissoo DC.	Heartwood	EtOH	Promotes fracture healing and exhibits osteoprotective effect	–	Stimulation of osteogenic genes at callus region and osteoblast function[34]
Daucus carota L., *Vitis* spp., *Sasa veitchii* (Carrière) Rehder, *Euptelea polyandra Siebold and Zucc.*, *Sesamum indicum* L.	–	MeOH	Stimulates osteoblast differentiation	–	Promotes osteoblast differentiation while inhibiting osteoclast differentiation without cell damage[35]
Dimocarpus longan Lour.	Fruit	Water	Inhibits osteoclast differentiation	–	Repeals RANKL-induced down-regulation of inhibitor of NF-κB, reduces NF-κB and AP-1 pathway[36]
Dioscorea spongiosa J.Q.Xi, M.Mizuno and W.L.Zhao	Rhizome	90% EtOH fraction of water extract	Inhibits glucocorticoid-induced osteoporosis	–	Increases the bone mass by increasing its Ca content[37]

Continued

Table 1 Anti-osteoporotic properties of selected plant extracts.—Cont'd

Phyto constituents	Plant part	Extract type	Result	Major active molecule	Potential mechanism
Drynaria fortunei (Kunze ex Mett.) J. Sm.	Rhizome	–	Increases the proliferation and osteogenic differentiation of BMS cells	Naringin, neoeriocitrin	Activates the estrogen receptors selectively; replaces estrogen, stimulates BMP-2 and ALP, increases expression of upregulated Runx2 and osteocalcin[38]
Eleutherococcus senticosus (Rupr. and Maxim.) Maxim.	–	–	Inhibits RANKL-induced osteoclast formation	–	Attenuates the NF-κB and MAPKs signaling pathway[39]
Epimedium sp.	–	–	Prevents and treats osteoporosis	–	Estrogen-like mechanism, regulates 11 estrogen related targets and target proteins on estrogen related pathways[40]
Epimedium brevicornu Maxim., *Ligustrum lucidum* W.T.Aiton, *Psoralea corylifolia* L.	Aerial part, fruit	Water	Enhances bone formation, reduces bone reabsorption	–	Stimulates osteoblast proliferation (ED_{35}: 150 μg/mL), inhibits osteoclast formation (ED_{45}: 150 μg/mL), stimulates osteoblast differentiation (ED_{70}: 150 μg/mL)[41]
Eucommia ulmoides Oliv., *Dipsacus asperoides* C.Y. Cheng and T.M.Ai	Bark, radix	Water	Anti-osteoporotic effect mediated *via* phytoestrogenic effects, prevents bone loss		Reduces serum (TRACP-5b) levels, increases BMD and bone turnover marker OCN levels dose dependently[42]

Ficus carica L.	–	–	Modulate bone remodeling	–	Balance between bone resorption phase regulated by osteoclast and bone formation phase regulated by osteoblast, inhibits osteoclastogenesis in RANKL pathway and regulates expression of osteoblast-specific genes such as BMP-2, OPG and OCN[43]
Ficus deltoidea Jack	Leaf	50% aqueous EtOH	Stimulates bone formation and suppresses bone resorption		Downregulates pro-inflammatory NF-κB, TNF-α, and IL-6 mRNA expressions[44]
Ginkgo biloba L.	–	Water	Restores bone alkaline phosphatase and BMD		Reduces BMD and the biomechanical parameters of bone; reduces the apoptosis of osteoblasts[45]
Hemidesmus indicus (L.) R. Br.	Root	EtOH	Prevents bone loss in OVX-induced osteoporosis	–	Increases bone strength and restores bone turnover markers (ALP, tartrate-resistant acid phosphatase and hydroxyproline), improves femur parameters[46]

Continued

Table 1 Anti-osteoporotic properties of selected plant extracts.—Cont'd

Phyto constituents	Plant part	Extract type	Result	Major active molecule	Potential mechanism
Humulus lupulus L.	–	EtOH	Prevents ovariectomy-induced bone loss	–	Decreases the number of osteoclasts in the tibial metaphysis and prevents reduction of the trabecular thickness[47]
Hydrilla verticillata (L.f.) Royle	–	Water	Decreases BMD	–	–[48]
Hygrophila spinosa (Schumach.) Heine	Whole plant	95% EtOH	Restoration of normal bone architecture	–	Improves biomechanical parameters, phosphorous and calcium level in serum and urine, and bone turnover markers[49]
Labisia pumila (Blume) Mez	–	96% EtOH	Normalizes osteoblastic bone formation	–	Reduces serum levels of osteocalcin[50]
Labisia pumila	Leaf	50% aqueous EtOH	Reduces osteophytes and cartilage erosions	Myricetin and gallic acid	Upregulates the anti-inflammatory interleukin-10, collagen type II and downregulates pro-inflammatory PTGS2 mRNA expressions; reduces serum collagenases (MMP-1 and MMP-3) and collagen type II degradation biomarker (CTX-II) levels[51]

Lycopus lucidus Turcz. ex Benth.	Leaf, stem	Water	Inhibits osteoclastogenesis and bone resorption	–	Inhibits the formation of TRAP-positive osteoclasts, modulates the NFATc/c-Fos pathway[52]
Morinda citrifolia L.	Leaf	Water	Increases bone formation, decreases bone resorption, osteoclast density	–	Elevation of serum calcium level *via* the increased calcium absorption, the increase in both serum calcium and osteoblast density together with the elevation of osteocalcin and alkaline phosphatase gave rise to the increase in BMD[53]
	Fruit juice	–	Stimulates osteoblastogenesis	–	Enhances the proliferation rate of the BMSC and upregulates the osteogenic differentiation marker genes ALP and OCN, and Runx2[54]
Morinda officinalis F.C.How	Root	Water	Prevents bone loss	Monotropein and rubiadin-1-methyl ether	Interfering with arachidonic acid metabolism, antiglucocorticoid-induced osteoporosis effect[55]

Continued

Table 1 Anti-osteoporotic properties of selected plant extracts.—Cont'd

Phyto constituents	Plant part	Extract type	Result	Major active molecule	Potential mechanism
Morus alba L.	Leaf	Water	Develops bone quality through maintenance of calcium homeostasis	–	Regulates PTH/VDR/CaBP signaling, and eliminates of oxidative stress *via* regulating AGEs/RAGE/Nox4/NF-κB signaling[56]
Ocimum basilicum L.	leaf	Water	Bone protection against glucocorticoid-induced osteoporosis	Rosmarinic acid, phytoestrogens	Brings the bone MDA back to normal, anti-osteoporotic effect by stimulating osteoblastic activity[27]
Olea europaea L.	Fruit	–	Stimulates of bone formation	Oleuropein, hydroxytyrosol	Enhances osteoblast differentiation and proliferation[43]
Oryza sativa L.	Seed	70% EtOH	Induces osteoblast differentiation, prevents decreases in BMD and strength in OVX rats		Stimulates ALP activity in both C3H10T1/2 and primary bone marrow cells; selectively stimulates Wnt-driven luciferase activities[57]
			Anti-adipogenic activity	Anthocyanin (cyanidin-3-*O*-glucoside)	Inhibits lipid accumulation and induction of adipocyte markers during adipocyte differentiation[57]
				Anthocyanin (peonidin-3-*O*-glucoside)	Suppresses lipid accumulation and expression of PPARγ[57]

Passiflora foetida L.	Fruit	BuOH fraction of EtOH extract	Bone sparing/conserving effects (50 and 100 mg/kg/day)		Stimulates osteoblast function and inhibits osteoclast function[58]
Petroselinum crispum (Mill.) Nyman ex A.W. Hill	Leaf	Water	Bone protection against glucocorticoid-induced osteoporosis	Quercetin and diosmetin glycosides	Vitamin K positively affects calcium balance, quercetin induces apoptosis in mature osteoclasts and inhibits bone resorption and diosmetin induce osteoblastic differentiation[27]
Phlomis umbrosa Turcz.	–	Water	Osteogenic effect, recovers bone mineral loss	–	Upregulates Runx2[57]
Pholidota articulata Lindl.	Leaf, pseudo-bulb	95% EtOH	Fracture healing activity	Oxoflavidin	Enhances ALP activity, mineral nodule formation and mRNA levels of osteogenic markers like BMP-2, Col I, RUNX-2, osteocalcin[59]
Plastrum testudinis	–	EtOAc and water	Promotes BMSC proliferation and osteogenic differentiation	–	Positive regulation of *Let-7f-5p* expression and negative modulation of the TNFR2/PI3K/AKT signaling pathway[60]

Continued

Table 1 Anti-osteoporotic properties of selected plant extracts.—Cont'd

Phyto constituents	Plant part	Extract type	Result	Major active molecule	Potential mechanism
Plastrum testudinis	Back shell	EtOAc	Improves rBMSCs osteogenic differentiation and mineralization	–	Inhibits TRAF6 protein expression and promotes STE20, IGF1R, and phosphorylated p38 MAPK protein expression[61]
Polygonum multiflorum Thunb.	Radix	Water	Reduces osteoclast differentiation and stimulates osteoblast formation		Decreases the mRNA expression levels of NFATc1 target genes, enhances the differentiation of primary osteoblasts, and increases the mRNA, expression of runt-related transcription factor 2, ALP, osterix, osteocalcin[62]
Polygonum multiflorum		75% EtOH	Exerts protective effect on bone tissue in glucocorticoid-induced osteoporosis	Emodin and 2,3,5,4tetrahydroxystilbene-2-*O*-β-glucoside	Increases apoptosis of osteoblasts and mature osteocytes *via* activation of caspase 3, repression of BMP-2 and expression of core binding factor a1[63]
	Root	EtOH	Reduces glucocorticoid-induced osteoporosis	–	Direct upregulation of the canonical Wnt/β-catenin signaling pathway[64]

Psidium guajava L.	Fruit	EtOH	Osteoanabolic effect, stimulates osteoblast differentiation	Triterpenes (ursolic acid, 2α-hydroxy ursolic acid)	Induces osteogenic differentiation by activating Wnt/β-catenin pathway, activation of canonical Wnt signaling leading to osteoblast differentiation[65]
Punica granatum L.	Fresh fruit peel	70% EtOH	Improves BMD, modulates bone cell differentiation	–	Stimulates osteoblastic MC3T3-E1 ALP activity, mineralization, and the transcription level of osteogenic markers[66]
Punica granatum	Seed	n-Hexane	Upregulation of osteoblastogenesis and down-regulation of osteoclastogenesis	–	Improves bone architecture, BMD and bone markers in OVX rats, phytoestrogenic activity, elevation of serum E2 level[67]
Ramalina litoralis Asahina	Thallus	MeOH	Anti-osteoclastic effect	–	Inhibits the expression level of NFATcl through the RANKL signaling pathway[68]

Continued

Table 1 Anti-osteoporotic properties of selected plant extracts.—Cont'd

Phyto constituents	Plant part	Extract type	Result	Major active molecule	Potential mechanism
Rehmannia glutinosa (Gaertn.) DC.	Radix	Water	Decreases bone loss	Iridoid (catalpol)	Increases BMD in the femurs and lumbar and enhances integrin β1 mRNA expression, decreases the ALP level, urinary Ca^{2+} and urinary deoxypyridinoline/creatinine(H/C) without affecting the E2 level[69]
Rehmannia glutinosa	Rhizome	Water	Enhances BMD, improves the micro architecture of bone	–	Increases the cell proliferation and ALP activity, enhances the extracellular matrix mineralization level, improves the expression of RUNX2 and osteopontin[70]
Rhus javanica L.	Gall	Water, EtOH	Inhibits osteoclast differentiation and bone resorption selectively	–	Augments the inhibition of BMD and BMC, decreases bone turnover markers[71]

Salvia miltiorrhiza Bunge +*liquefied calcium supplement*	–	Water	Protection of estrogen-dependent bone loss	–	Inhibits tibial bone loss, sustains trabecular bone state, ameliorates bone biochemical markers, blocking expression of TRAF6 and NFTAc1 and produces cathepsin K and calcitonin receptor to develop osteoclast differentiation[72]
Schizonepeta tenuifolia (Benth.) Briq.	–	EtOH	Reduces osteoclast formation and activation *in vitro* and protects against lipopolysaccharide-induced bone loss		Inhibits phosphorylation of Akt and IkB at early stages of RANKL-induced osteoclastogenesis, blocks filamentous actin (F-actin) ring formation and bone resorbing activity of mature osteoclasts *in vitro*[73]
Scutellaria baicalensis Georgi	Root	–	Increases BMD of OVX osteoporotic rats	Baicalein	Reduced ALP activity, positive effects on the osteometabolic indicators of OVX osteoporotic rats, acted as a weak phytoestrogen[74]
Stauntonia hexaphylla Decne.	Leaf	MeOH	Inhibits osteoclastogenesis	–	Downregulated RANKL-associated c-Fos protein induction by suppressing its translation and suppression of NFATc1 induction[75]

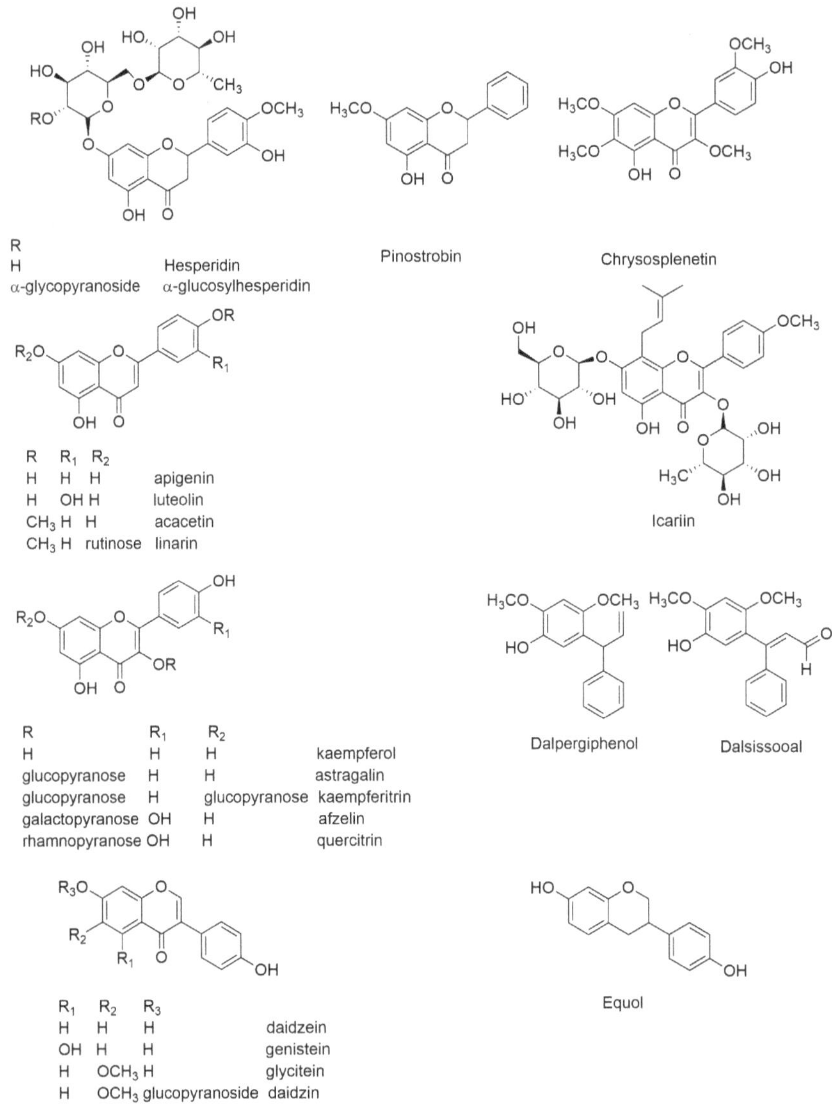

Fig. 1 Structures of selected flavonoids.

of Wnt/β-catenin pathway and it decreased phosphorylation of β-catenin, promoted nuclear translocation of β-catenin. It's effect on Wnt/β-catenin pathway was blocked by Wnt inhibitor Dickkopf-related protein 1 and bone morphogenetic protein 2 (BMP2) antagonist Noggin. Chrysosplenetin treatment at the dose of 3 mg/kg every 2 days for 6 weeks suppressed estrogen deficiency related bone loss in OVX mices.[78]

Icariin is the main active compound of Herba Epimedii, which is used as aphrodisiac and antirheumatic in Chinese medicine (Fig. 1). Effect of icariin on osteoporosis is well studied. 10^{-8}–10^{-6} M icariin increased MC3T3-E1 osteoblastic cell and rat BMS cells proliferation and reduced apoptosis, associated with increased mRNA levels of positive regulators of cell cycle gene, Cyclin E and proliferating cell nuclear antigen (PCNA), decreased mRNA level of negative regulator gene, Cyclin-dependent kinase 4 inhibitor B, and reduced Bax/Bcl-2 ratio and caspase-3 activity.[79,80] It also enhanced MC3T3-E1 osteblastic cell and rBMS cell differentiation and mineralization through up-regulations of the expression of differentiation markers, ALP and Col I, also increased Ca deposition the number and areas of calcified nodules.[79–81] Underlying mechanisms explained by its effect on mitogen-activated protein kinase (MAPK) signaling, inhibition of the extracellular-signal-regulated kinase (ERK) and Jun N-terminal kinase (JNK) activation. Treatment with specific inhibitors of MAPKs, U0126 (ERK inhibitor) and SP600125 (JNK inhibitor) reduced effect of icariin. Moreover, treatment of osteoblasts with estrogen receptor antagonist ICI182780, decreased icariin-induced proliferation and mineralization.[80] Icariin also increased nitric oxide synthase (NOS) activity, inducible NO synthase (iNOS) and endothelial increased nitric oxide (eNOS) expression, increased nitric oxide (NO) production, sGC and cGMP contents and PKG expression, as well as level of p-Akt, indicating PI3K-AKT-eNOS-NO-cGMP-PKG signal pathway also involves in anti-osteoporotic effect of icariin.[81] RhoA and its effector Rho-associated protein kinase (ROCK) regulate the cytoskeletal organization and activation of RhoA leads to osteogenic differentiation, while inactivation of RhoA causes adipogenic differentiation. Icariin at the concentration of 10^{-7} M upregulated RhoA protein expression and ROCK substrate molecule p-MYPT1 expression in adipose-derived stem cells. The RhoA inhibitor C3 treatment supressed the increased proliferation and osteogenic differentiation of rat adipose-derived stem cells (rASCs) induced by icariin. These results suggested that icariin promotes proliferation and osteogenic differentiation of rASCs by activating the RhoA signaling pathway.[82]

Icariin reduced the size of LPS-induced osteoclasts formation, and diminished their osteoclasts differentiation maker, TRAP and acid phosphatase (ACP) activity. Icariin also inhibited LPS-induced bone resorption and interleukin-6 (IL-6), and tumor necrosis factor-α (TNF-α) expression. It upregulated gene expression of OPG and downregulated the RANKL. Icariin also inhibited the synthesis of cyclooxygenase type-2 (COX-2) and prostaglandin E2 (PGE2). Moreover, icariin reduced LPS-induced hypoxia inducible factor-1α (HIF-1α) expression of osteoclasts. On osteoclasts, icariin suppresses LPS-mediated activation of the p38 and JNK; while on the osteoblasts, icariin reduced the LPS-induced activation of ERK1/2 and I-kappa-B-alpha (IκB-α) but promoted the activation of p38.[83] Icariin (100 mg/kg) could protect iron overload-induced osteoporosis through attenuating iron deposition in bone marrow and its antioxidative effect. It also suppressed excess iron ions-induced oxidative damage in osteoblast cells by increasing cell viability, inhibiting cell apoptosis, reducing mitochondrial membrane potential and decreasing ROS generation. Moreover, icariin also reduced iron overload-induced osteoclast differentiation and bone absorption.[84]

2.1.3 Flavone

Ovariectomy induced osteoporosis activity-guided isolation on the flowers of *Chrysanthemi indici* let the isolation of acacetin, apigenin, luteolin, and linarin (acacetin 7-*O*-rutinoside) (Fig. 1). All of four compounds promoted the differentiation and proliferation and the mineralization of MC3T3-E1 cells. They also upregulated the mRNA Runx2, OCN, OPN, and COL I. Among them, linarin, which was the only glycosylated flavonoid, exhibited the highest activity, whereas luteolin exhibited the lowest effect almost all tests in these four flavonoids.[85]

Linarin increased osteoblast proliferation and differentiation dose-dependently, also increased ALP activity and mineralization of extracellular matrix. Linarin also upregulated expressions of ALP, RUNX2, OCN, bone sialoprotein, and COL I in MC3T3-E1 cells. Pretreatment with noggin, BMP-2 antagonist, reduced effect of linarin. Linarin also increased the protein levels of BMP-2, RUNX2 and phosphorylation of SMAD1/5 in a dose-dependent manner, suggested that linarin promotes the osteoblast differentiation *via* the BMP-2/SMAD-dependent RUNX2 signaling pathway. Moreover, 50 and 150 mg/kg/day administration of linarin for 8 weeks prevented OVX-induced deterioration of trabecular microarchitecture *in vivo*.[86]

Luteolin increased OPG/RANKL, upregulated osteogenic markers, osterix, Col I, OCN, RUNX2. It inhibited osteoblastic apoptosis through increasing caspase-3, caspase-9 and Bax, decreasing Bcl-2. It also upregulated Lrp-5, β-catenin mRNA expressions, and p-ERK, p-GSK-3β protein expressions in both MC3T3-E1 cells and glucocorticoid-induced osteoporosis developed rats. ERK knockdown reduced the effect of luteolin. The treatment with a dose of 25–100 mg/kg significantly increased maximum load, maximum bending stress, coefficient of rigidity, and elasticity modules in a dose-dependent manner *in vivo*. These results indicated that luteolin increased bone formation through enhancing osteoblastic differentiation, mineralization associated with ERK/Lrp-5/GSK-3β pathway.[87] Luteolin 7-*O*-glycoside (5 mg/kg) administration for 4 weeks significantly inhibited the bone turnover markers plasma bone-specific alkaline phosphatase (BALP), plasma OCN, PINP, and C-telopeptide of type II collagen (CTX-II) levels compared with the control group in OVX rats.[88]

2.1.4 Flavonol

Kaempferol exhibited beneficial effects on osteoporosis with different mechanisms. It inhibited osteoclastic bone resorption by reducing TNF-α-induced secretion of key osteoclastogenic cytokines, monocyte chemoattractant protein-1 and IL-6, in MC3T3-E1 cells, and blocked TNF-α-induced translocation of NF-κB subunit p65 from the cytoplasm to the nucleus (Fig. 1). It also promoted differentiation and mineralization of osteoblasts. Kaempferol (5–20 μM) increased ALP activity, associated with mRNA expressions of Runx2, Osterix, Opn and bone sialoprotein. It promoted osteoblastic differentiation through the activation of the Wnt signaling pathway in SaOS-2 cells. Wnt signaling pathway inhibitor, FH 535, blocked kaempferol-induced ALP activity and ICI 182,780 (an inhibitor for estrogen receptors) inhibited kaempferol-induced activation of Wnt signaling pathway. These results may suggest that kaempferol stimulates estrogen signaling followed by Wnt signaling pathway activation to achieve its potential for anti-osteoporosis effects.[89,90] Astragalin is the 3-*O*-glucoside of kaempferol and a bioactive constituent from *Cuscuta chinensis* (Fig. 1). Astragalin (5–20 μM) upregulated mRNA and protein levels of osteoblastic marker genes ALP, Ocn, and Opn as kaempferol acts. Astragalin also enhanced the level of osteogenic differentiation related proteins, BMP-2 and p-Smad1/5/9. It also increased ALP activity. In turn, it increased the level of RUNX2 in MC3T3-E1 cells. It also upregulated MAPK pathway components, including Erk1/2, p-ERK, p38, p-p38, p-JNK. Astragalin (5–20 mg/kg) treatment for 4 weeks increased trabecular microarchitecture parameters. These results suggested that effect of astragalin on osteoporosis mediated by both BMP and MAPK pathways.[91] Similar to astragalin, 8 and 16 mg/kg kaempferitrin (7-*O*-glucopyranoside derivative of astragalin) also increased trabecular microarchitecture parameters *in vivo* (Fig. 1). In MC3T3-E1 cells, 0.001–1 μM kaempferitrin enhanced the activity of ALP, and upregulated p-ERK, like astragalin. However, it did not alter the levels of p-JNK and p-p38. It also diminished the activities of both cathepsin K and TRAP. Moreover, it downregulated the phosphorylation of IκB, indicating both MAPK and NF-κB pathway involved in the effect of kaempferitrin.[92]

Hyperoside (quercetin 3-*O*-β-galactopyranoside), quercitrin (quercetin 3-*O*-α-rhamnoside) and afzelin (kaempferol 3-*O*-α-rhamnoside) inhibited TRAP activity ($82.11 \pm 9.31\%$, $88.36 \pm 10.93\%$, $77.78 \pm 4.24\%$, respectively, at 10 μM) (Fig. 1). These results suggested that a OH group at C-4′ boost the TRAP inhibitory potential.[93]

2.1.5 Neoflavonoid

Oral administration of dalbergiphenol or dalsissooal 1 and 5 mg/kg/day for 6 weeks increased bone biomechanical strength and new bone formation rate in OVX mice, comparable with 5 μg/kg/day E2 treatment without inducing uterine hyperplastic effect (Fig. 1). Treatment restored bone volume/tissue volume, trabecular thickness, trabecular number at both doses. Treatment with dalbergiphenol or dalsissooal increased the expression of osteoblast-specific markers, including RUNX2, osterix, and Col I and effectively reduced expressions of mRNAs of osteoclast-specific marker TRAP and RANK expression in the femur of OVX mice compared to OVX control group. Dalsissooal also increased the activity of ALP. Moreover, treatment with dalbergiphenol at the dose of 5 mg/kg rescued osteoblast-specific markers better than E2, whereas effect of dalsissooal was comparable to E2. Interestingly, these compounds exhibited beneficial effects on osteoporosis without causing endometrial hyperplasia at 5 mg/kg/day.[94,95]

2.1.6 Isoflavonoids

Soy isoflavonoids (genistein, daidzein, glycitein and their glycosides as well as equol, the metabolite of daidzein in humans) mimic the actions of estrogen therefore, their effects on postmenopausal osteoporosis are well studied by *in vitro*, pre-clinical and clinical studies (Fig. 1). Isoflavonoids increased BMD through increasing bone formation markers, reducing bone resorption markers, rescuing deteriorated bone structure, quality and bone resistance to fracture. Isoflavonoids exhibited their effects by the genomic 'long-term' mechanism associated with the activation or inhibition of the estrogen receptor (ER), or through a variety of non-genomic 'short-term' mechanisms, including the inhibition of tyrosine kinase or topoisomerase II, peroxisome proliferator activated receptor-λ.

Daidzein prevents bone loss through increasing ALP, Col-I, osteocalcin, up-regulating expressions of RUNX2 and osterix.[96–98] Daidzein promoted proliferation and differentiation of human osteoblast-like MG-63 cells and prevented from cisplatin-induced cell apoptosis. The enhancement of the proliferation of MG-63 cells by daidzein are mediated primarily by ERα. Whereas prevention from cisplatin-induced cell apoptosis, effects on ALP activity, and increase in collagen content were mediated by both ERα and ERβ together. On the contrary, the effects of E2 on MG-63 cell proliferation and survival were mediated by both ER's, but the effects on osteoblastic differentiation were primarily mediated by ERα. Daidzein upregulated phosphorylation of Akt and ERK. These up-regulations are suppressed by the treatment with MEK1/2 specific inhibitor PD98059, or PI3K-specific inhibitor wortmannin. These results suggested that the effect of daidzein in MG-63 cells are associated with ER-dependent MEK/ERK and PI3K/Akt activation.[96] Subcutaneous injection 30 mg/kg daily of daidzein for 3 weeks stimulated C cells and increased trabecular bone mass. These results indicated that, in addition to direct action, daidzein may also affect bone structure indirectly, *via* enhancement of thyroid C cell activity in OVX rats.[99] Both daidzein and equol, metabolite of daidzein, increased osteoblast differentiation associated with up-regulation of proprotein convertase PACE4. While equol increased osteoblast differentiation more than daidzein, E2 was the strongest inducer. It attributed to the binding affinities of equol to ERα and ERβ which were higher than that of daidzein.[100]

Daidzin (5–20 μM) inhibited RANKL-induced osteoclast differentiation and the hydroxyapatite-resorbing activity of osteoclasts through inhibiting RANKL-induced NF-κB signaling pathway *in vitro* (Fig. 1). It also downregulated expressions of osteoclast marker genes, including nuclear factor of activated T cells cytoplasmic 1 (NFATc1), cellular oncogene (c-Fos), TRAP and cathepsin K. Moreover, 2.5 and 5 mg/kg daidzin administration for 7 days supressed LPS-induced calvarial osteolysis in mices.[101] Treatment with soy isoflavones (a mixture of genistein, daidzin and glycitein, accounting for 60%, 20% and 13%, respectively) for 10 days increased mineralized nodule formation. Treatment with this mixture increased the activity of ALP, and OPG levels ($P < 0.01$) and decreased RANKL levels in a dose-dependent manner in a 7-day culture. Besides, 0.1–0.5 M of this soy isoflavones mixture stimulated the proliferation of osteoblasts in culture, upregulated expression levels of the Wnt3a, β-catenin and Wnt7b proteins.[102] The same mixture of soy isoflavonoids also diminished osteoporosis in OVX rats at 56 mg/kg daily doses after 16 weeks treatment. It diminished bone mineral elements loss (calcium and potassium), improved biomechanics parameters in the similar manner with E2. It downregulated sost mRNA and sclfos (c-Fos), TRAP and CTSK, sclerosteosis protein expression in a dose-dependent manner, and upregulated β-catenin proteins expression in femur of OVX rats. These data suggest that isoflavonoids inhibited the canonical Wnt signal in OVX rats. Both studies suggested that Wnt3/β-catenin signaling pathway involving in beneficial effects of soy isoflavonoids in osteoporosis.[103]

2.2 Stilbenes

Osteogenic differentiation promoting activity-guided phytochemical studies on *Polygonum multiflorum* Thunb. let the isolation of 2,3,5,4′-tetrahydroxystilbene-2-*O*-β-D-glucoside (Fig. 2). This compound stimulated the ALP activity at concentrations of 1.56–25 μg/mL, but increased the content of OCN, 7 days after treatment with 6.25–25 μg/mL in rat mesenchymal stem cells. It also increased the mineralized matrix of the zebrafish head bone, dose-dependently and protected against osteopenia, induced

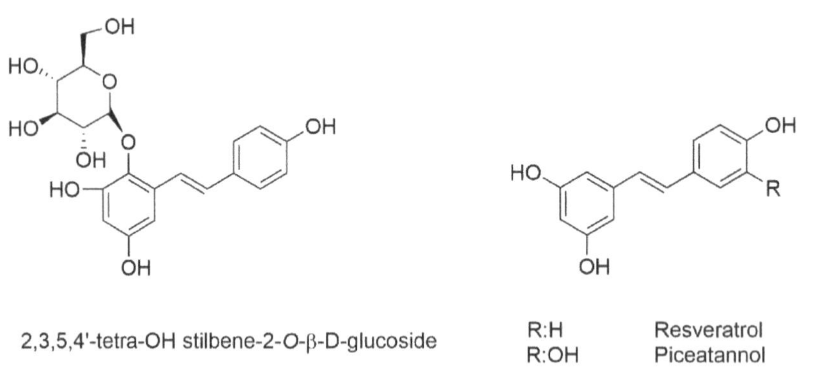

Fig. 2 Structures of selected stilbenes.

by dexamethasone *in vivo*.[104] 2,3,5,4'-Tetrahydroxystilbene-2-*O*-β-D-glucoside (50–400 μg/mL) inhibited nitric oxide (NO) and prostaglandin E2 (PGE2) production and iNOS and COX-2 expressions in LPS-stimulated RAW 264.7 cells in a dose-dependent manner. It also reduced PGE2 production, expressions of iNOS and matrix metalloproteinase-13 in interleukin-1-stimulated primary rat chondrocytes. Moreover, 10 mg/kg of 2,3,5,4′-tetrahydroxystilbene-2-*O*-β-D-glucoside decreased paw oedema in mono-iodoacetate-induced rat osteoarthritis model.[105] The same compound suppressed H_2O_2 induced cell death at 1–10 μM and improved antioxidant status in MC3T3-E1 cells. It also enhanced ALP activity, as well as the mineralization which was inhibited by H_2O_2. Moreover, this stilbene upregulated the expression of ALP, COL I, OCN compared with the group treated with H_2O_2. Further, it alleviated H_2O_2 induced IL-6 and RANKL levels in MC3T3-E1 cells.[106] 2,3,5,4′-Tetrahydroxystilbene-2-*O*-β-D-glucoside (1 and 10 μM) upregulated the expressions of TCF, AXIN2, ALP and OPG and downregulated expressions of FoxO3a, GADD45 in MC3T3-E1 cells. It also exhibited same effects on FoxO3a, GADD45, CAT, AXIN2 and TCF levels in the bone tissues of OVX mices. These results suggested that in addition to Wnt/FoxO3a signaling, the immune system and the chemokine signaling pathway may involve in the osteoprotective mechanism of 2,3,5,4′-tetrahydroxystilbene-2-*O*-β-D-glucoside.[107] Treatment with i.p. 1 and 3 μg/g 2,3,5,4′-tetrahydroxystilbene-2-*O*-β-D-glucoside for 6 weeks decreased serum Ca, inorganic phosphorus (IP), ALP, and total cholesterol compared to non-treated OVX mices. Moreover, this treatment induced bone growth as evidenced by the tibia lengths, as well as the tibia and femur weights, which were higher in the treatment group than in the non-treated OVX group. Even 1 μg/g 2,3,5,4′-tetrahydroxystilbene-2-*O*-β-D-glucoside alleviated serum TRAP activity in OVX mices. It also increased BALP at 3 μg/g compared to non-treated OVX mices. This compound also diminished the area of the TRAP-positive cells and increased the collagen area in the femur and tibia compared to non-treated OVX group. It also improved trabecular morphometric parameters in the distal femur of the treatment group.[108]

Resveratrol is the best-known stilbene and is also probably one of the most assessed natural products for osteoporosis (Fig. 2). Resveratrol inhibited the osteoclast activity and formation, stimulates osteogenic differentiation *in vitro* in different cell lines as well as *in vivo* and in OVX rats evidenced ameliorated decreased levels of OCN and RUNX2, OPG ALP activity and calcification ability increased RANKL, TRAP-b.[109–111] Interestingly, 40–100 μM resveratrol increased proliferation, inhibited apoptosis, stimulated pluripotency of BMS cells from patients with osteoporosis. It also increased the formation of mineralized nodules, upregulated mRNA expressions of ALP, bone morphogenetic protein 4 (BMP4), and osterix in turn, lead to the differentiation of BMS cells into osteoblasts.[112] The OVX rat model is the most used *in vitro* model to assess effect of resveratrol on osteoporosis. Notably, it did not exhibit hyperplastic effect on the uterus up to 80 mg/kg.[113] It was tested at the doses of 0.625–150 mg/kg in this model and reduced osteoporosis.[114,115] In addition to OVX model, the effect of resveratrol on osteoporosis was related to age, not to estrogen deficiency also tested. It seems that the effect of resveratrol on age-related osteoporosis changes according to the age of rats used in the test system, probably due to the bone loss that does not occur in the long bones of male rats until certain months (24–27 months). Resveratrol ameliorated age-related osteoporosis in rats older than 6 months.[113] Effect of resveratrol in other animal model also tested and 150 μM resveratrol also reduced dexamethasone-induced bone damage in zebra fish model evidenced by increased bone formation and mineralization.[116] Mechanistically, resveratrol exhibited these effects through SIRT1-NF-κB, AMPK/ROS as well as PI3K/AKT/FoxO1 pathways.[110,117–119] Resveratrol also suppressed miR-338-3p, followed by an increase in the expression of RUNX 2.[115] In addition to effect on burn turnover markers, resveratrol treatment at the dose of 625 μg/kg daily for 4 weeks decreased IL-23, IL-17A, IL-1β, and TNF-α cytokine expression levels; however, this decrease was less than that of 487.5 μg/kg daily treatment.[114]

Piceatannol inhibited osteoclast formation and bone resorption in a dose-dependent manner (Fig. 2). 40 μM piceatannol completely inhibited osteoclast formation and it decreased the TRAP activity in a dose-dependent manner. It also downregulated RANKL-induced increased mRNA expressions of NFATc1, DCSTAMP, CTSK, MMP-9 and TRAP. Further, it suppressed RANKL-induced phosphorylation of JNK, ERK, AKT, IkBa and p65 and promoted apoptosis of osteoclasts. These results suggested that MAPK, NF-κB and AKT signaling pathways involving in anti-osteoporosis effect of MAPK, NF-κB and AKT signaling pathways.[120]

2.3 Coumestans

Structure of coumestrol and psoralidin are very similar; the only difference between them is that psoralidin has a more isopentenyl group (Fig. 3). 10 mg/kg coumestrol or psoralidin suppressed OVX-induced osteoporosis

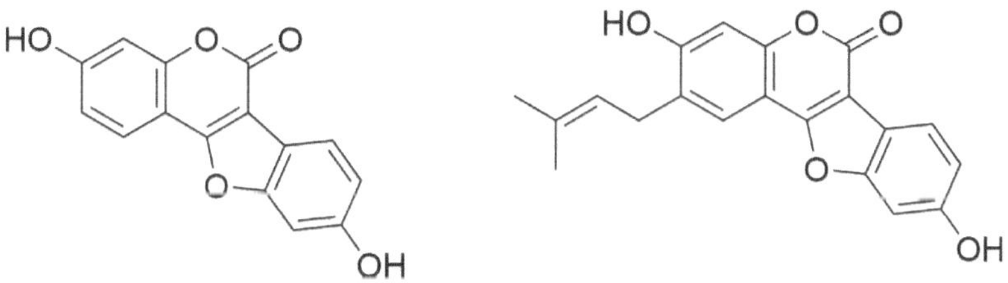

Coumestrol Psoralidin

Fig. 3 Structures of selected coumestans.

in rats. They improved total BMC, total BMD, mineral apposition rate, bone biomechanical properties, microarchitecture, and trabecular bone formation, and boosted osteogenic differentiation *in vivo* and the effect of psoralidin was higher than that of coumestrol. Both compounds also decreased the alleviated lipid peroxidation markers and psoralidin was a better suppressor than coumestrol. The body weight increase was lower than OVX groups. Coumestrol and psoralidin attenuated the increased BALP and PICP in the OVX group rats in the similar level with E2. Similarly, both compounds also suppressed increased TRACP-5b and C-terminal cross-linked telopeptides of collagen type I levels. These suppressions were lower than that of E2, and the levels in the psoralidin group were significantly lower than the coumestrol group. Moreover, they also suppressed adipogenic differentiation in BMS cells. Both compounds increased down-regulated phosphorylation of Akt, PI3K, GSK-3β. They also suppressed the increased β-catenin phosphorylation. They also reversed lowered OH-1 and phosphorylation of Nrf-2 and increased Keap-1 level. Psoralidin was more effective than coumestrol at all proteins tested. Furthermore, the osteo-protective effects of psoralidin and coumestrol were blocked by the treatment with ICI 182.780, an estrogen receptor antagonist. These results suggested that both compounds mechanistically possess protective effect on osteoporosis associated with PI3K/Akt pathway as well as ER pathway. Structurally, the prenyl group in psoralidin increased the activity.[121,122] Furthermore, psoralidin also inhibited the expression of some inflammatory cytokines, TNF-α and IL-6 which are associated with the inhibition of phosphorylation of some MAPK pathway components, including p38, ERK and JNK and supressed RANKL-induced NF-κB and c-Fos/NFATc1 activation. These results indicated that MAPK and NF-κB signaling pathways also involve bone-protective effect of psoralidin.[123]

2.4 Coumarins

Umbelliferone suppressed LPS-induced inflammatory bone loss *in vivo* evidenced by prevented trabecular bone matrix degradation and osteoclast formation in bone tissue by histological analysis (Fig. 4). It also altered RANKL-induced osteoclast differentiation. In molecular level, umbelliferone inhibited c-Fos and NFATc1 expressions also suppressed RANKL-induced Akt-c-Fos-NFATc1 signaling pathway and osteoclast-specific genes, such as TRAP, and OSCAR. However, it did not affect the expression of genes related to osteoblastic differentiation as RUNX2, ALP, Col IA1, and bone sialoprotein. These results suggested that umbelliferone possess bone-protective action through effecting osteoclast formation and function but not changing osteoblast function.[124]

Daphnetin which possess one more OH group at C-8 compared to umbelliferone, protected LPS-induced bone damage by inhibiting osteoclastogenesis (Fig. 4). It suppressed RANKL-induced bone resorption as well as expression of osteoclast-related proteins including TRAF6, c-Fos, NFATc1, and cathepsin K in bone marrow-derived macrophages cells in a dose-dependent manner. It reduced intracellular radical oxygen species generation and NADPH oxidase activity and decreased the expression of p-p65 and p-IκBα, p-Akt and p-GSK-3β in turn suppressed RANKL-induced NF-κB activation and Akt-GSK-3β pathway.[125]

Osthole (7-methoxy-8-isopentenoxy-coumarin) is one of the coumarins isolated from *Cnidium monnieri* (L.), which is used in Chinese medicine for the treatment of osteoporosis (Fig. 4). It promoted the viability of rat BMS cells and rat calvarial osteoblasts, and enhanced ALP activity in both cell lines, whereas 7-methoxy-coumarin did not. Both compounds upregulated mRNA expression of insulin-like growth factor-1, RUNX-2, osterix, OPG, and RANKL and this increase was higher in osthole treated cells comparing to cells treated with 7-methoxy-coumarin. While osthole also

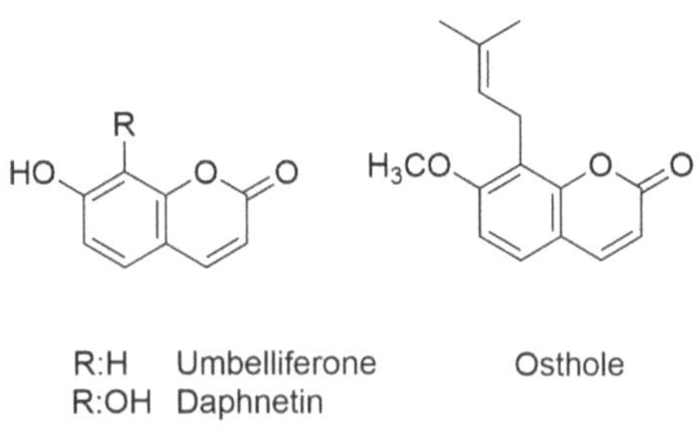

Fig. 4 Structures of selected coumarins.

boosted osteoblast mineralization and levels of Col I, BMP-2, OCN and OPN, 7-methoxy-coumarin did not. In addition, osthole increased the OPG/RANKL ratio in both cells more than that of 7-methoxycoumarin. These results indicated that though both compounds diminished the osteoclast formation and bone resorption, osthole was more potent. Hence, 8-isopentenoxy substitution, which means prenylation at C-8, improved the bone-protective effect.[126]

2.5 Lignans

Baicalensinosides A–C, 3,4-dibenzyltetrahydrofuran-type lignan glycosides (Fig. 5) which were isolated from *Scutellaria baicalensis* Georgi, increased the relative OPG level. Their level comparing to control were 1.83, 0.84, and 0.98 times higher. It seems that possessing two methoxy groups in both phenolic ring is important.[127] *Sambucus williamsii* Hance is used for the treatment of osteoporosis in Chinese medicine and some lignans, which were isolated from this plant, (7*R*,8*S*)-ficusal, (7*R*,8*S*)-ceplignan, (7*R*,8*S*)-dehydrodiconiferyl alcohol, (7*R*,8*S*)-dehydrodiconiferyl alcohol-γ′-methyl ether, and (7*S*,8*R*)-hierochin D, increased the proliferation of

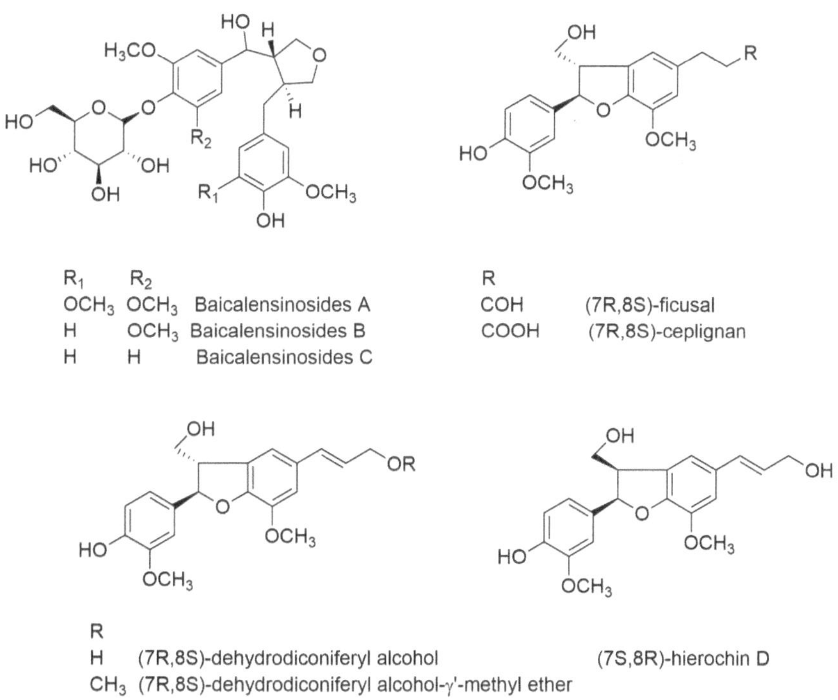

Fig. 5 Structures of selected lignans.

osteoblastic-like UMR 106 cells 31.3%, 24.4%, 16.6%, 25.1% and 16.0% at 10^{-10} M (Fig. 5). These results suggested that changing of aldehyde group with carboxylic acid diminished the effect but the absence of either aldehyde or carboxylic acid group at the same position lower the effect more.[128] Dehydrodiconiferyl alcohol stimulated BMP-2-induced differentiation of osteoblasts by upregulating BMP-2, ALP, osteocalcin and OPG in MC3T3-E1 cells. It increased phosphorylation of Smad1/5/9 and AMPK in turn, lead to BMP-2-induced RUNX2 production. Of note, it enhanced BMP-2-induced osteoblast differentiation through interacting with either ERα or ERβ receptors.[129] In addition to its osteoblastogenesis stimulating effect *in vitro*, dehydrodiconiferyl alcohol also inhibited RANKL-induced osteoclastogenesis *in vitro* in RAW264.7 and OVX rats *in vivo*. RANKL-induced NFATc1 and c-Fos expressions associated with inhibition of NF-κB and MAPK pathways and activation of AMPK. Dehydrodiconiferyl alcohol induced AMPK activation was related with ERα but not with ERβ.[130]

2.6 Diarylheptanoids

The bone-protective effect of curcumin has been well investigated and has been even proven by clinical studies. Curcumin (Fig. 6) ameliorated H_2O_2-induced apoptosis in osteoblasts (Saos-2 cells) through preserving mitochondrial functions and activation of Akt-GSK-3β signaling pathway.[131] Curcumin also protected H_2O_2-induced human mesenchymal stem cell death up to 20 μM. It stimulated cell death at 50 μM or higher concentrations. Curcumin reversed H_2O_2-induced alterations: enhanced ALP activity, calcium content, OPN, Col I levels, and decreased RANKL level and oxidation parameters up to 20 μM but it did not change OPG level. These effects were mediated by Wnt/β-catenin pathway.[132] Curcumin (9.5 μg/g) treatment for 8 weeks reduced osteoporosis in OVX rats shown by increased trabecular BMD, lower levels of RANKL-induced expressions of c-Fos and NFAT2, p-p38, p-ERK and p-Jun, IκBα p-IκBα. These results suggested that curcumin inhibited osteoclastogenesis *via* inhibition of MAPK and NF-κB pathways.[133,134] A dose of 40 mg/kg *in vivo* or 4 μM *in vitro* curcumin treatment also reduced microgravity induced bone loss possibly through suppressing oxidative stress and activating vitamin D receptor.[135]

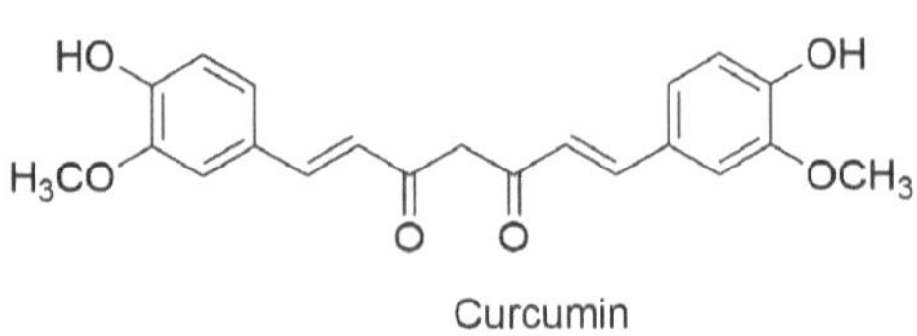

Fig. 6 Structure of curcumin.

2.7 Alkaloids

Neotuberostemonine (Fig. 7), which is an active alkaloid isolated from *Stemona tuberosa* Lour, inhibited RANKL—or cancer cell-mediated osteoclastogenesis *via* blockade of tumor necrosis factor receptor-associated factor (TRAF) 6 and NF-κB activation and impaired the formation of F-actin ring structure, an important feature of osteoclast differentiation and function in a dose-dependent manner.[136]The effect of lycorine which exists in the flowers and bulbs of Amaryllidaceae species, inhibited RANKL-induced osteoclastogenesis (Fig. 7). The TRAP-positive cell number decreased from 194.4/well (lycorine 0 μM) to 36.5/well (lycorine 0.4 μM), and the total osteoclast area from 43.3% (lycorine 0 μM) to 1.1% (lycorine 0.4 μM). This effect was associated with the inhibition of MAPK signaling. The MAPK agonist anisomycin could partially rescue the inhibitory effect of lycorine. Lycorine inhibited the osteoclast bone resorption *in vitro* (bone resorption was 30% less than the control group with 0.1 μM lycorine and completely abolished after treatment with 0.4 μM lycorine). Lycorine played a protective role in OVX-induced osteoporosis model and a titanium particle-induced osteolysis model. Lycorine effectively suppressed osteoclastogenesis and could be used to treat OVX-induced osteoporosis and wear particle-induced osteolysis.[137]

Avenanthramides are a group of alkaloid phenols, found in oats (Fig. 7). Three major synthetic avenanthramides (2c, 2f and 2p), which differ in the type of hydroxylcinnamic acid component (ferulic, caffeic or *p*-coumaric acid), to regulated osteoblastogenesis and osteoclastogenesis. Avenanthramide 2c and 2p (1, 5 and 10 μM and 1 and 5 μM, respectively) increased Col I expression. Lower doses of avenanthramides increased OPG in OB-6 osteoblastic cells, whereas 100 μM dose of 2f and all concentrations of 2c decreased RANKL gene expression in MLO-Y4 osteocytic cells. Avenanthramides did not change apoptosis of OB-6 osteoblastic cells or MLO-Y4 osteocytic cells; but they prevented apoptosis induced by the DNA topoisomerase inhibitor

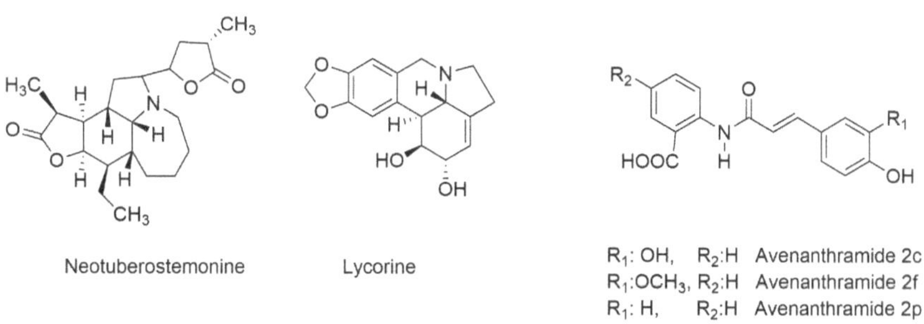

Fig. 7 Structures of selected alkaloids.

etoposide, the glucocorticoid dexamethasone, and hydrogen peroxide. Avenanthramides prevented apoptosis of wild type and Nrf2 Knockout osteoblasts, it was shown that avenanthramide-induced survival didn't require Nrf2 expression. While avenanthramides did not affect wild type osteoclasts, avenanthramide 2p reversed the low apoptosis of Nrf2 Knockout osteoclasts. Avenanthramides regulated the function of osteoblasts and osteocytes and prevented osteoblast/osteocyte apoptosis and increased osteoclast apoptosis; and all the regulatory actions were independent of Nrf2.[138]

2.8 Terpenes

Limonene (Fig. 8), which has a cyclic monoterpene structure and found in *Citrus* fruits, significantly increased ALP activity, calcium depositions and the expression of osteogenic specific genes such as myogenin, myogenic differentiation 1 (MyoD), ALP, RUNX2, OCN. Limonene stimulated osteoblast differentiation and glucose uptake through initiating p38 MAPK and Akt signaling pathways.[139]

Glycyrrhizic acid (Fig. 8) is a triterpenoid saponin present in *Glycyrrhiza glabra* L. and administration of glycyrrhizic acid to the OVX rats showed no

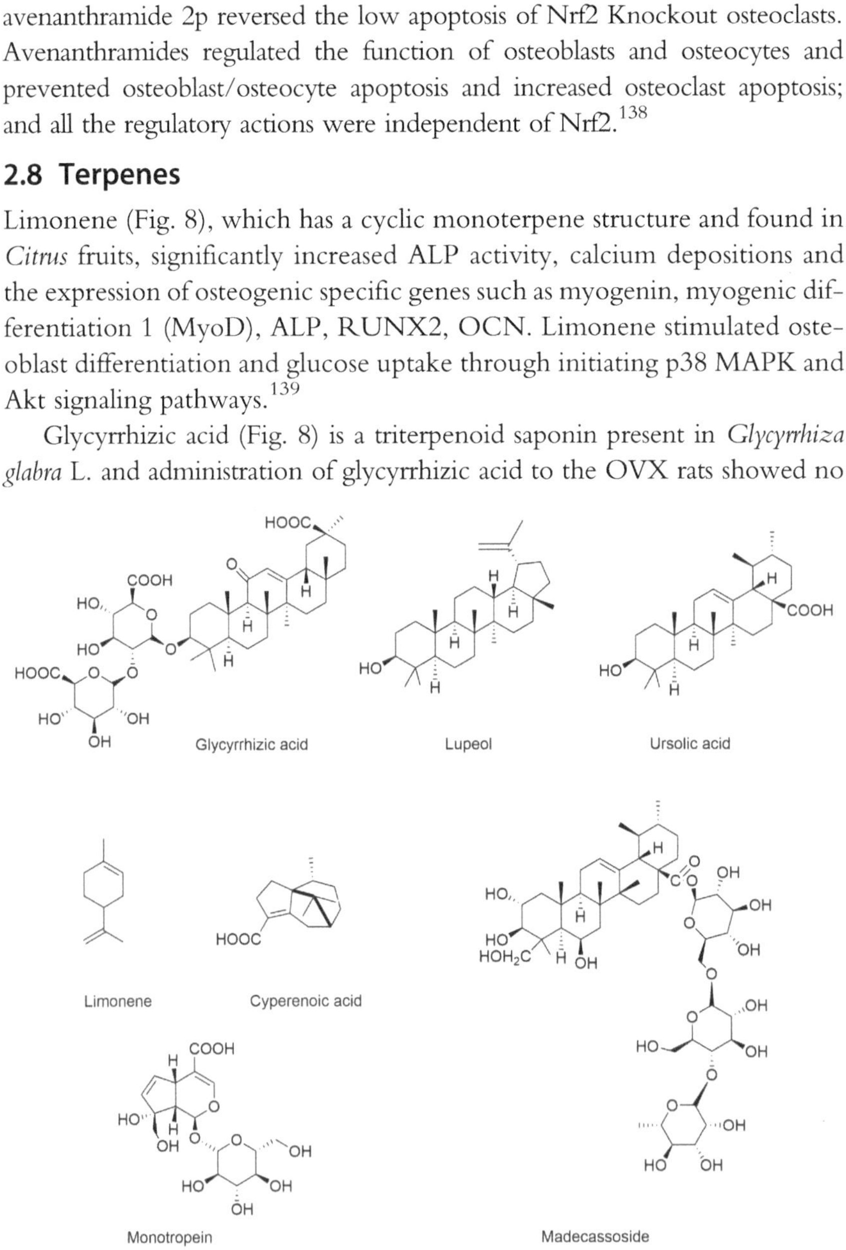

Fig. 8 Structures of selected terpenes.

statistically significant effect on body mass gain, uterus mass or the uterus mass/body mass ratio, tibia mass or the tibia mass/body mass ratio in comparison with the non-treated OVX rats. It decreased the width of the femoral epiphyseal cartilage (by 9.5%) and tibial endosteal 14.9% compared to the non-treated OVX rats.[140] Moreover, 18β-glycyrrhetinic acid restrained RANKL-induced osteoclastogenesis in RAW264.7 cells as well as bone marrow monocytes (BMMs), by inhibiting RANK expression in preosteoclasts and blocking the binding of RANK and TRAF6 (Fig. 8). RANK and TRAF6 lead to the inhibition of NF-κB and MAPK signaling pathways of which activation is vital for osteoclastogenesis.[141] Glycyrrhizin reduced the formation of ROS in osteoclasts by inducing AMPK phosphorylation and nuclear transfer of NRF2 and inhibited RANKL-induced osteoclastogenesis.[142]

Cyperenoic acid which is one of the terpenoids of the root of *Croton crassifolius*, suppressed RANKL-induced osteoclast differentiation (the IC_{50} value is 36.69 μM ± 1.02) by reducing the expression of two key transcription factors, NFATc1 and c-Fos (Fig. 8).[143] Lupeol is a lupane-type triterpene isolated from *Sorbus commixta* Hedl., inhibited osteoclast differentiation and bone resorption activity due to its effects on MAPKs and transcription factors (NF-κB, NFATc1, and c-Fos) downstream of the osteoclast differentiation factor receptor RANK (Fig. 8). *In vivo* efficacy of lupeol was also studied by using an animal model of hypercalcemic mediated bone loss. Lupeol attenuated osteoclast differentiation by increasing TRAP-positive osteoclasts. The lowering of serum RANKL levels by lupeol in hypercalcemic mice suggested that lupeol prevents bone loss by interfering with RANKL-induced signal transduction pathways.[144]

Ursolic acid, a pentacyclic triterpenoid that found in apple, rosemary, loquat and holey basil, inhibited osteoclast differentiation (with an IC_{50} value of 5.4 ± 0.96 μM) through targeting a nuclear export protein named exportin 5 (XPO5) (Fig. 8). Downregulation of XPO5 by ursolic acid downregulates the expression of mature microRNA let-7g-5p during osteoclast differentiation.[145] Another pentacyclic triterpenoid madecassoside, isolated from *Centella asiatica* (L.) Urban, inhibited the RANKL-induced formation and function of osteoclasts in a dose-dependent manner (at concentrations higher than 2.5 μM) by blocking Ca^{2+} oscillations and the NF-κB and MAPK pathways. Madecassoside suppressesed the activity of transcription factors NFATc1 and c-Fos, essential for osteoclast formation and also inhibited the expression of bone resorption-related genes and proteins (Acp5/TRAcP, CTSK, ATP6V0D2/V-ATPase-d2, and integrin β3).[146]

Monotropein, which is a bioactive iridoid glycoside from *Morinda officinalis* F.C.How, has been stated as a lead compound for anti-osteoporosis according to obtained results *in vivo* OVX-induced osteoporosis in mice and *in vitro* in MC3T3-E1 cells (Fig. 8). Monotropein administration at the doses of 40 or 80 mg/kg/day for 4 weeks increased BMC, BMD, bone volume fraction; improved bone microstructure and the parameters of biomechanical properties of femur, in addition, decreased the serum levels of IL-1, IL-6 and soluble receptor activator of NF-κB ligand (sRANKL) in OVX mice. It improved MC3T3-E1 cell proliferation at 10–100 μM. ALP activities were significantly increased after treatment with monotropein for 72 h.[147]

3. Mechanisms of action of anti-osteoporosis natural products

Osteoblasts form bone synthesize the organic matrix and regulate mineralization in bones. RUNX2 is a key transcription factor for osteoblastogenesis, regulates the expression of several osteoblast proteins including osterix, osteopontin, bone sialoprotein, type I collagen, osteocalcin. Bone morphogenic proteins (BMPs), especially BMP-2, play role in regulation of expression of RUNX2. BMP receptors are located on the surface of osteoblasts precursor cells. Smad proteins play an important role in mediating the effects of BMPs. Smad proteins transduce the signals from the cytoplasm to the nucleus and regulate the transcription of the target genes. BMPs bind to heterodimeric receptors and activate Smad proteins, which are transactive osteoblastogenic genes, either directly or through RUNX2. The low density lipoprotein (LDL)-related protein 5 receptor (LRP5) is also a cell surface receptor, important for bone formation. It activates the canonical Wnt signaling pathway.

In addition to Bmp/Smad pathway, Wnt/β-catenin pathway and Notch signaling also important regulator of osteoblast function and bone formation. β-Catenin, LRP-5, Runx2 and osterix involve to Wnt/β-catenin pathway and their upregulation enhance osteoblastogenesis. Although Notch signaling stimulates osteoblastogenesis, it doesn't necessarily enhance bone formation each time.[148] MAPK pathway is also involved in the induction of RUNX2 by BMP-2.[149] MAPKs are secondary messenger family, they transfer signals from the cell surface to nucleus. MAPKs including extracellular signal-related kinases (ERK), p38, c-Jun N-terminal kinases (JNK), contributes to cell proliferation, growth, differentiation, apoptosis and some other vital process.[150] During this events, osteoblastic phenotypic markers appear as accumulation of collagenous matrix, expression of ALP, osteocalcin and lead to mineralization of bone nodules.[148] RANKL is a member of TNF family and its activation is the main event for osteoclastogenesis. It activates intracellular signals of osteoclastogenesis, including MAPK, NF-κB. These two signaling pathways activate the nuclear factor of activated T cells cytoplasmic 1 (NFATc1), inducing the release of a number of osteoclast-specific genes, such as TRAP and cathepsin K. The OPG factor inhibits bone resorption and osteoclast differentiation. Hence, OPG/RANKL is an important indicator for bone mass and skeletal integrity.[151] Natural products improve osteoblastogenesis and inhibit osteoclastogenesis with these mechanisms.

4. Virtual screening of anti-osteoporosis natural products

Lepidium meyenii Walp. is used for the treatment of osteoporosis in folk medicine. Seventy-two compounds reported from this plant in earlier studies docked for the binding to ER. *N*-benzylpalmitamide had the lowest binding energy with high affinity to receptor. Further *in vitro* studies showed that *N*-benzyl-palmitamide upregulated osteogenesis-related genes, including BMP-2, Col I, and ALP and it increased proliferation, differentiation and mineralization of osteoblasts. The effects of *N*-benzyl-palmitamide on enhanced osteoblastogenesis mediated by ERs were evidenced by increased expressions of ERα and ERβ genes.[152]

17β-Hydroxysteroid dehydrogenase type 2 (17β-HSD2) coverts steroids hormones and plays role in regulation of cell and tissue specific steroid action. It is one of the target 17β-HSD2 for anti-osteoporotic treatment. To discover new hits from nature, a pharmacophore model based on 17β-HSD2 inhibitors was applied to a virtual screening of different natural products library and 36 compounds were selected for biological activity tests. Among them, nordihydroguaiaretic acid, (−)-dihydroguaiaretic acid, isoliquiritigenin, inhibited 17β-HSD2 with IC_{50} values of 0.38 ± 0.04, 0.94 ± 0.02 and 0.36 ± 0.08 μM with high selectivity.[153]

5. Anti-osteoporotic natural products in clinical trials (in the drug development chain)

Some examples of anti-osteoporotic natural products in clinical trials (in the drug development chain) are shown in Table 2.

Table 2 Anti-osteoporotic clinical studies on selected plant extracts.

Phyto constituents	Intervention	Number of patients	Duration	Result
Curcumin	Curcumin 110/mg/kg/day *versus* placebo	100	6 months	Significant increase was observed in BMD indicators of lumbar, femoral neck, and hip at curcumin group, significant difference between interventional and control groups for the mean BMD of femoral neck and hip at the end of study (0.718 ± 0.002 g/cm^2 *versus* 0.712 ± 0.003 g/cm^2 and 0.742 ± 0.031 g/cm^2 *versus* 0.692 ± 0.016 g/cm^2, respectively)[154]
Combined supplement (200 mg hop rho isoalpha acids, 100 mg berberine sulfate trihydrate, 500 IU vitamin D3, and 500 μg vitamin K 1)	Supplement twice daily *versus* placebo	33	14 weeks	31% mean reduction in serum osteocalcin and 13% increase in serum 25-hydroxyvitamin D[155]
Equol	Equol 2 mg/day *versus* equol 6 mg/day *versus* equol 10 mg/day *versus* placebo	93	12 months	Urinary deoxypyridinoline decreased significantly, with a −23.94% change in the group receiving equol supplement (10 mg/day) compared to a change of −2.87% in the placebo group ($P = 0.020$). BMD was maintained in the treated group, which decreased in the placebo group. Treatment with 10 mg/day of equol prevented the decrease in bone mineral density[156]

Red clover extract (isoflavone aglycones and probiotics)	60 mg isoflavone aglycones/d and probiotics *versus* placebo	78	12 months	Plasma concentrations of collagen type 1 cross-linked C-telopeptide was significantly decreased. The urinary 2-hydroxyestrone to 16a-hydroxyestrone ratio (*P*, 0.05), and equol-producer status (*P*, 0.05) was significantly increased. Bone turnover was improved[157]
Combined extract of Morus alba and Polygonum odoratum leaves	Combined extract at doses of 50 and 1500 mg/day *versus* placebo	45	8 weeks	Menopausal women in D2 group increased serum alkaline phosphatase, osteocalcin, and total phenolic compounds content but decreased CTX level[158]

6. Conclusions

Osteoporosis is a global burden and due to prolongation of human life span, it will be more common. According to current investigations, certain plants are very promising candidates for bone protection. There are many studies exploring the role of natural compounds in osteoporosis. It is difficult to make comments on structure activity relationships (SARs) due to limited studies, which mainly focus on structurally similar compounds; however, it can be concluded that prenylation generally increases bone remodeling effect. Still, *in silico*-supported SARs studies are needed more to achieve drugs candidates with optimized activity and safety profiles.

Some secondary metabolite groups including some flavonoids, stilbenes, coumestans and lignans exhibit promising bone-protective properties. Notably these group of compounds are also known as 'phytoestrogens.' Positive effect of these phytoestrogens may associated with reduced estrogen level which is a key factor leading to osteoporosis. Though most polyphenolic phytoestrogens have bioavailability problems, some of promising compounds exhibit their protective effect in osteoporosis at orally achievable concentrations, according to *in vitro* and *in vivo* studies. Furthermore, their effective potentials shown in clinical studies. However, more well-designed clinical trials with a higher number of patients enrolled are still needed.

References

1. Johnell, O.; Kanis, J. *Osteoporos. Int.* **2005**, *16*, 3–7.
2. Jolly, J. J.; Chin, K.; Alias, E.; Chua, K. H.; Soelaiman, I. N. *Int. J. Environ. Res. Public Health* **2018**, *15*, 963–979.
3. International Osteoporosis Foundation (IOF) *Facts and Statistics About Osteoporosis and Its Impact*. Available online https://www.iofbonehealth.org/facts-statistics#category-14; 2019. accessed on 26 October 2019.
4. Deal, C. *Nat. Rev. Rheumatol.* **2009**, *5*(1), 20–27.
5. Harvey, N. C.; Cooper, C. *Osteoporosis: A Lifecourse Epidemiology Approach to Skeletal Health*, 1st ed.; CRC Press, 2018.
6. Hough, F. S.; Brown, S. L.; Cassim, B.; Davey, M. R.; de Lange, W.; de Villiers, T. J.; Ellis, G. C.; Lipschitz, S.; Lukhele, M.; Pettifor, J. M. *S. Afr. Med. J.* **2014**, *104*, 279–282.
7. World Health Organization (WHO) Study Group *Assessment of Fracture Risk and Its Application to Screening for Postmenopausal Osteoporosis. Report No. 843*. World Health Organization: Geneva, Switzerland, 1994;1–134.
8. Osteoporosis prevention, diagnosis, and therapy. *NIH Consens. Statement* **2000**, *17*, 1–45.
9. Che, C.; Wong, M. S.; Lam, C. W. K. *Molecules* **2016**, *21*(3), 239.
10. Chen, X.; Wang, Z.; Duan, N.; Zhu, G.; Schwarz, E. M.; Xie, C. *Connect. Tissue Res.* **2018**, *59*(2), 99–107.

11. Swaminathan, R. *Clin. Chim. Acta* **2001**, *313*, 95–105.
12. Kuo, T.-R.; Chen, C.-H. *Biomark. Res.* **2017**, *5*, 18. https://doi.org/10.1186/s40364-017-0097-4.
13. Wei, J.; Karsenty, G. *Rev. Endocr. Metab. Disord.* **2015**, *16*(2), 93–98.
14. Denhardt, D. T. In: *Principles of Bone Biology*; Masaki, N., Bilezikian, J. P., Raisz, L. G., Martin, T. J., Eds.; Elsevier Academic Press, 2008, pp 351–366
15. Drake, M. T.; Clarke, B. L.; Oursler, M. J.; Khosla, S. *Endocr. Rev.* **2017**, *38*(4), 325–350.
16. Kostenuik, P. J.; Shalhoub, V. *Curr. Pharm. Des.* **2001**, 7(8), 613–635.
17. Kumar, S.; Maurya, R. In: *Natural Products and Drug Discovery*; Mandal, S. C., Mandal, V., Konishi, T., Eds.; Elsevier Ltd., 2018, pp 179–212
18. Su, Y.; Chen, Y.; Liu, Y.; Yang, Y.; Deng, Y.; Gong, Z.; Chen, J.; Wu, T.; Lin, S.; Cui, L. *J. Orthop. Translat.* **2016**, *4*, 75–91.
19. Li, S.; Chen, R.; Luo, K.; Guo, Y.; Xiao, M.; Du, G. *Exp. Ther. Med.* **2017**, *13*, 2893–2899.
20. Cho, H. J.; Kim, J. W.; Ju, S. Y.; Park, Y. K. *J. Nutr. Health Aging* **2016**, *20*(5), 534–539.
21. Moosa, S.; Kasonga, A. E.; Deepak, V.; Marais, S.; Magoshi, I. B.; Bester, M. J.; Kruger, M. C.; Coetzee, M. *Food Funct.* **2018**, *9*, 3301–3312.
22. Wong, S. K.; Chin, K.; Suhaimi, F. H.; Ahmad, F. *Bone* **2018**, *116*, 8–21.
23. Chauhan, S.; Sharma, A.; Upadhyay, N. K.; Singh, G.; Lal, U. R.; Goyal, R. *BMC Complement. Altern. Med.* **2018**, *18*, 233–245.
24. Yeom, M.; Kim, E.; Kim, J.; Jung, H.; Sohn, Y. *Front. Pharmacol.* **2018**, *9*, 1314. https://doi.org/10.3389/fphar.2018.01314.
25. Wu, X.; Xie, C.; Zhu, Q.; Wang, M.; Sun, B.; Huang, Y.; Shen, C.; An, M.; Zhao, Y.; Wang, X.; Sheng, J. *Food Nutr. Res.* **2018**, *62*, 1478. https://doi.org/10.29219/fnr.v62.1478.
26. Pal, S.; Kumar, P.; Ramakrishna, E.; Kumar, S.; Porwal, K.; Kumar, B.; Arya, K. R.; Maurya, R.; Chattopadhyay, N. *J. Ethnopharmacol.* **2019**, *235*, 8–18.
27. Hozayen, W. G.; El-Desouky, M. A.; Soliman, H. A.; Ahmed, R. R.; Khaliefa, A. K. *BMC Complement. Altern. Med.* **2016**, *16*, 165. https://doi.org/10.1186/s12906-016-1140-y.
28. Pujia, A.; Russo, C.; Maurotti, S.; Pujia, R.; Mollace, V.; Romeo, S.; Montalcini, T. *Nutrients* **2018**, *10*, 1305–1318.
29. Jeon, E. J.; Lee, D.; Kim, Y.; Ahn, J.; Kim, M. J.; Hwang, J.; Hur, J.; Kim, M.; Jang, Y.; Ha, T.; Seo, D.; Lee, J. S.; Sung, M. J.; Jung, C. H. *Mol. Nutr. Food Res.* **2016**, *60*, 2587–2601.
30. Dong, Y.; Feng, Z.; Chen, H.; Wang, F.; Lu, J. *Chinas Med.* **2018**, *13*, 34–54.
31. Lee, Y.; Kim, J.; Kim, K.; Cho, S.; Son, Y. *Evid. Based Complement. Alternat. Med.* **2018**, *2018*, 6302748. https://doi.org/10.1155/2018/6302748.
32. Lee, H.; Kim, M. H.; Choi, Y. Y.; Hong, J.; Yang, W. M. *Mol. Med. Rep.* **2018**, *17*, 3758–3762.
33. Ma, X.; Liu, J.; Yang, L.; Zhang, B.; Dong, Y.; Zhao, Q. *Life Sci.* **2018**, *209*, 140–148.
34. Karvande, A.; Khedgikar, V.; Kushwaaha, P.; Ahmad, N.; Kothari, P.; Verma, A.; Kumar, P.; Nagar, G. K.; Mishra, P. R.; Maurya, R.; Trivedi, R. *J. Pharm. Pharmacol.* **2017**, *69*(10), 1381–1397.
35. Suzuki, R.; Fukami, S.; Tomomura, M.; Tomomura, A.; Shirataki, Y. J. *Nat. Med.* **2019**, *73*, 331–337.
36. Son, Y.; Lee, E. M.; Lee, D. Y.; Lee, J. H.; Oh, S. *Phytomedicine* **2019**, *59*, 152910.
37. Han, N.; Xu, J.; Xu, F.; Liu, Z.; Yin, J. *J. Ethnopharmacol.* **2016**, *185*, 53–59.
38. Mou, M.; Li, C.; Huang, M.; Zhao, Z.; Ling, H.; Zhang, X.; Wang, F.; Yin, X.; Ma, Y. *Acta Pol. Pharm.* **2015**, *72*(6), 1073–1080.

39. Yang, X.; Chang, Z.; Ma, R.; Guo, H.; Zhao, Q.; Wang, X.; Kong, L.; Hao, D. *Int. J. Clin. Exp. Pathol.* **2017**, *10*(4), 4514–4521.
40. Xu, F.; Ding, Y.; Guo, Y.; Liu, B.; Kou, Z.; Xiao, W.; Zhu, J. *J. Ethnopharmacol.* **2016**, *177*, 148–160.
41. Ko, C. H.; Lau, K. M.; Chau, L.; Cheung, D. W. S.; Siu, W. S.; Wong, H. L.; Shum, W. T.; Gao, S.; Fung, K. P.; Leung, P. C.; Poon, S. K.; Lau, C. B. S. *J. Pharm. Pharmacol.* **2016**, *68*(6), 826–833.
42. Li, F.; Yang, X.; Bi, J.; Yang, Z.; Zhang, C. *Pharm. Biol.* **2016**, *54*(9), 1857–1864.
43. Idrus, R.; Sainik, N. Q.; Ansari, A. S.; Zulfarina, M. S.; Razali, R. A.; Nordin, A.; Saim, A. B.; Naina-Mohamed, I. *Sains. Malays.* **2018**, *47*(11), 2741–2755.
44. Tantowi, N. A. C. A.; Lau, S. F.; Mohamed, S. *Calcif. Tissue Int.* **2018**, *103*(4), 388–399.
45. Lucinda, L. M. F.; Aarestrup, B. J. V.; Reboredo, M. M.; Pains, T. D. A.; Chaves, R. Z.; Reis, J. E. P.; Louzada, M. J. Q.; Guerra, M. O. *An. Acad. Bras. Cienc.* **2017**, *89*(4), 2833–2841.
46. Desai, S.; Babaria, P.; Nakarani, M.; Shah, K.; Paranjape, A. *J. Ethnopharmacol.* **2017**, *199*, 1–8.
47. Keiler, A. M.; Helle, J.; Bader, M. I.; Ehrhardt, T.; Nestler, K.; Kretzschmar, G.; Bernhardt, R.; Vollmer, G.; Nikolic, D.; Bolton, J. L.; Pauli, G. F.; Chen, S.; Dietz, B. M.; Breemen, R. B.; Zierau, O. *Phytomedicine* **2017**, *34*, 50–58.
48. Prabha, P.; Rajkumar, J.; Karthikeyan, S. S. *Mater. Sci. Eng. C* **2018**, *90*, 657–663.
49. Satpathy, S.; Patra, A.; Hussain, M. D.; Ahirwar, B. *S. Afr. J. Bot.* **2018**, *117*, 247–255.
50. Nurdiana, N.; Mariati, N.; Noorhamdani, N.; Setiawan, B.; Budhiparama, N.; Noor, Z. *Clin. Nutr. Exp.* **2018**, *20*, 41–47.
51. Madzuki, I. N.; Lau, S. F.; Tantowi, N. A. C. A.; Ishak, N. I. M.; Mohamed, S. *Inflammopharmacology* **2018**, *26*(5), 1207–1217.
52. Jeong, D.; Kim, E.; Kim, J.; Lee, B.; Hong, S. Y.; Park, J. H.; Jung, H.; Sohn, Y. *Evid. Based Complement. Alternat. Med.* **2019**, *2019*, 3231784. https://doi.org/10.1155/2019/3231784.
53. Wattanathorn, J.; Thipkaew, C.; Thukham-mee, W.; Muchimapura, S.; Wannanon, P.; Tong-un, T. *Oxid. Med. Cell. Longev.* **2018**, *2018*, 1039364. https://doi.org/10.1155/2018/1039364.
54. Hussain, S.; Tamizhselvi, R.; George, L.; Manickam, V. *Int. J. Stem Cells.* **2016**, *9*(2), 221–229.
55. Xia, T.; Dong, X.; Lin, L.; Jiang, Y.; Ma, X.; Xin, H.; Zhang, Q.; Qin, L. *J. Pharm. Biomed. Anal.* **2019**, *166*, 336–346.
56. Liu, C.; Zhu, R.; Liu, H.; Li, L.; Chen, B.; Jia, Q.; Wang, L.; Ma, R.; Tian, S.; Wang, M.; Fu, M.; Niu, J.; Orekhov, A. N.; Gao, S.; Zhang, D.; Zhao, B. *Front. Pharmacol.* **2018**, *9*, 1239. https://doi.org/10.3389/fphar.2018.01239.
57. Lee, J. E.; Lee, H.; Kim, M. H.; Yang, W. M. *Biomed. Rep.* **2019**, *10*(1), 17–22.
58. Ahmad, N.; Chillara, R.; Kushwaha, P.; Khedgikar, V.; Karvande, A.; Choudhary, D.; Adhikary, S.; Maurya, R.; Trivedi, R. *Biomed. Pharmacother.* **2017**, *8*, 804–813.
59. Sharma, C.; Dixit, M.; Singh, R.; Agrawal, M.; Mansoori, M. N.; Kureel, J.; Singh, D.; Narender, T.; Arya, K. R. *J. Ethnopharmacol.* **2015**, *170*, 57–65.
60. Shen, G.; Ren, H.; Huang, J.; Zhang, Z.; Zhao, W.; Yu, X.; Shang, Q.; Qiu, T.; Zhang, Y.; Tang, J.; Liang, D.; Yang, Z.; Jiang, X. *Cell. Physiol. Biochem.* **2018**, *47*, 2307–2318.
61. Shang, Q.; Yu, X.; Ren, H.; Shen, G.; Zhao, W.; Zhang, Z.; Huang, J.; Yu, P.; Liang, D.; Tang, J.; Jiang, X. *Evid. Based Complement. Alternat. Med.* **2019**, *2019*, 6815620. https://doi.org/10.1155/2019/6815620.
62. Han, S.; Lee, K.; Kim, Y. *Int. J. Mol. Med.* **2018**, *42*(1), 331–345.

63. Zhou, M.; Li, J.; Wu, J.; Yang, Y.; Zeng, X.; Lv, X.; Cui, L.; Yao, W.; Liu, Y. *Exp. Ther. Med.* **2017**, *14*(3), 2445–2460.
64. Zhou, M.; Wu, J.; Yu, Y.; Yang, Y.; Li, J.; Cui, L.; Yao, W.; Liu, Y. *Mol. Med. Rep.* **2018**, *17*, 970–978.
65. Porwal, K.; Pal, S.; Dev, K.; China, S. P.; Kumar, Y.; Singh, C.; Barbhuyan, T.; Sinha, N.; Sanyal, S.; Trivedi, A. K.; Maurya, R.; Chattopadhyay, N. *J. Nutr. Biochem.* **2017**, *44*, 22–34.
66. Spilmont, M.; Leotoing, L.; Davicco, M.; Lebecque, P.; Miot-Noirault, E.; Pilet, P.; Rios, L.; Wittrant, Y.; Coxam, V. *Nutrients* **2015**, 7(11), 9265–9284.
67. Shaban, N. Z.; Talaat, I. M.; Elrashidy, F. H.; Hegazy, A. Y.; Sultan, A. S. *J. Nutr. Health Aging* **2017**, *21*(10), 1299–1306.
68. Lee, Y.; Jeong, M.; Kim, K.; Baek, S.; Hur, J.; Son, Y. *Biotechnol. Bioprocess Eng.* **2018**, *23*, 634–640.
69. Liu, C.; Ma, R.; Wang, L.; Zhu, R.; Liu, H.; Guo, Y.; Zhao, B.; Zhao, S.; Tang, J.; Li, Y.; Niu, J.; Fu, M.; Zhang, D.; Gao, S. *J. Ethnopharmacol.* **2017**, *198*, 351–362.
70. Xia, T.; Dong, X.; Jiang, Y.; Lin, L.; Dong, Z.; Shen, Y.; Xin, H.; Zhang, Q.; Qin, L. *Molecules* **2019**, *24*, 253. https://doi.org/10.3390/molecules24020253.
71. Kim, T.; Park, E. K.; Huh, M.; Kim, H. K.; Kim, S.; Lee, S. *Evid. Based Complement. Alternat. Med.* **2016**, *2016*, 3284704. https://doi.org/10.1155/2016/3284704.
72. Park, B.; Song, H. S.; Kwon, J. E.; Cho, S. M.; Jang, S.; Kim, M. Y.; Kang, S. C. *BMC Complement. Altern. Med.* **2017**, *17*, 545–560.
73. Kim, J.; Baek, J. M.; Ahn, S.; Cheon, Y.; Park, S.; Yang, M.; Choi, M. K.; Oh, J. *BMC Complement. Altern. Med.* **2016**, *16*, 301–320.
74. Shin, J.; Shin, E.; Jo, T. H.; Park, Y.; Hwang, I. *Food Sci. Biotechnol.* **2013**, *22*(2), 493–499.
75. Cheon, Y.; Baek, J. M.; Park, S.; Ahn, S.; Lee, M. S.; Oh, J.; Kim, J. *BMC Complement. Altern. Med.* **2015**, *15*, 280. https://doi.org/10.1186/s12906-015-0801-6.
76. Chiba, H.; Uehara, M.; Wu, J.; Wang, X.; Masuyama, R.; Suzuki, K.; Kanazawa, K.; Ishimi, Y. *J. Nutr.* **2003**, *133*, 1892–1897.
77. Gu, C.; Fu, L.; Yuan, X.; Liu, Z. *Molecules* **2017**, *22*(10), pii: E1735. https://doi.org/10.3390/molecules22101735.
78. Hong, G.; He, X.; Shen, Y.; Chen, X.; Yang, F.; Yang, P.; Pang, F.; Han, X.; He, W.; Wei, Q. *Stem Cell Res Ther* **2019**, *10*(1), 277.
79. Liu, Y. Q.; Han, X. F.; Liu, T.; Cheng, M. C.; Xiao, H. B. *Biotechnol. Lett.* **2015**, *37*(1), 219–226.
80. Song, L.; Zhao, J.; Zhang, X.; Li, H.; Zhou, Y. *Eur. J. Pharmacol.* **2013**, *714*(1–3), 15–22.
81. Zhai, Y. K.; Guo, X. Y.; Ge, B. F.; Zhen, P.; Ma, X. N.; Zhou, J.; Ma, H. P.; Xian, C. J.; Chen, K. M. *Bone* **2014**, *66*, 189–198.
82. Ye, Y.; Jing, X.; Li, N.; Wu, Y.; Li, B.; Xu, T. *Biomed. Pharmacother.* **2017**, *88*, 384–394.
83. Hsieh, T. P.; Sheu, S. Y.; Sun, J. S.; Chen, M. H. *Phytomedicine* **2011**, *18*(2–3), 176–185.
84. Jing, X.; Du, T.; Chen, K.; Guo, J.; Xiang, W.; Yao, X.; Sun, K.; Ye, Y.; Guo, F. *J. Cell. Physiol.* **2019**, *234*(7), 10123–10137.
85. Li, J.; Lin, X.; Zhang, Y.; Liu, W.; Mi, X.; Zhang, J.; Su, J. *Evid. Based Complement. Alternat. Med.* **2016**, *2016*, 2587201.
86. Li, J.; Hao, L.; Wu, J.; Zhang, J.; Su, J. *Int. J. Mol. Med.* **2016**, *37*(4), 901–910.
87. Jing, Z.; Wang, C.; Yang, Q.; Wei, X.; Jin, Y.; Meng, Q.; Liu, Q.; Liu, Z.; Ma, X.; Liu, K.; Sun, H.; Liu, M. *J. Cell. Physiol.* **2019**, *234*(4), 4472–4490.

88. Ammar, N. M.; El-Hawary, S. S.; Mohamed, D. A.; El-Halawany, A. M.; El-Anssary, A. A.; El-Kassem, L. T.; Hussein, R. A.; Jaleel, G. A.; El-Dosoky, A. H. *Phytother. Res.* **2016**, *30*, 768–773.
89. Miyake, M.; Arai, N.; Ushio, S.; Iwaki, K.; Ikeda, M.; Kurimoto, M. *Biosci. Biotechnol. Biochem.* **2003**, *67*, 1199–1205.
90. Sharma, A. R.; Nam, J.-S. *J. Nutr. Biochem.* **2019**, *74*, 108228.
91. Liu, L.; Wang, D.; Qin, Y.; Xu, M.; Zhou, L.; Xu, W.; Liu, X.; Ye, L.; Yue, S.; Zheng, Q.; Li, D. *Front. Endocrinol.* **2019**, *10*, 228.
92. Ma, X. Q.; Han, T.; Zhang, X.; Wu, J. Z.; Rahman, K.; Qin, L. P.; Zheng, C. J. *Phytomedicine* **2015**, *22*(13), 1159–1162.
93. Yang, S. Y.; Lee, S. H.; Tai, B. H.; Jang, H. D.; Kim, Y. H. *Molecules* **2018**, *23*(2), pii: E457. https://doi.org/10.3390/molecules23020457.
94. Gautam, J.; Kumar, P.; Kushwaha, P.; Khedgikar, V.; Choudhary, D.; Singh, D.; Maurya, R.; Trivedi, R. *Menopause* **2015**, *22*(11), 1246–1255.
95. Kushwaha, P.; Khedgikar, V.; Ahmad, N.; Karvande, A.; Gautam, J.; Kumar, P.; Maurya, R.; Trivedi, R. *Eur. J. Pharmacol.* **2016**, *788*, 65–74.
96. Jin, X.; Sun, J.; Yu, B.; Wang, Y.; Sun, W. J.; Yang, J.; Huang, S. H.; Xie, W. L. *Nutr. Res.* **2017**, *42*, 20–30.
97. Komrakova, M.; Sehmisch, S.; Tezval, M.; Schmelz, U.; Frauendorf, H.; Grueger, T.; Wessling, T.; Klein, C.; Birth, M.; Stuermer, K. M.; Stuermer, E. K. *J. Endocrinol.* **2011**, *211*, 157–168.
98. Sugimoto, E.; Yamaguchi, M. *Biochem. Pharmacol.* **2000**, *59*, 471–475.
99. Filipovic, B.; Sosić-Jurjević, B.; Ajdzanović, V.; Brkić, D.; Manojlović-Stojanoski, M.; Milosević, V.; Sekulić, M. *Osteoporos. Int.* **2010**, *21*(9), 1609–1616.
100. Kim, H.; Tabata, A.; Tomoyasu, T.; Ueno, T.; Uchiyama, S.; Yuasa, K.; Tsuji, A.; Nagamune, H. *J. Bone Miner. Metab.* **2015**, *33*(1), 30–39.
101. Wei, G.; Liang, T.; Wei, C.; Nong, X.; Lu, Q.; Zhao, J. *J. Cell. Biochem.* **2018**, *120*(4), 5304–5314.
102. Yu, F.; Liu, Z.; Tong, Z.; Zhao, Z.; Liang, H. *Gene* **2015**, *573*(2), 273–277.
103. Liang, H. D.; Yu, F.; Lv, P.; Zhao, Z. N.; Tong, Z. H. *Mol. Biol. Rep.* **2014**, *41*(7), 4447–4454.
104. Zheng, Y.; Li, J.; Wu, J.; Yu, Y.; Yao, W.; Zhou, M.; Tian, J.; Zhang, J.; Cui, L.; Zeng, X.; Liu, Y. *Exp. Ther. Med.* **2017**, *14*(4), 2845–2852.
105. Tsai, P. W.; Lee, Y. H.; Chen, L. G.; Lee, C. J.; Wang, C. C. *Molecules* **2018**, *23*(3). E571. https://doi.org/10.3390/molecules23030571.
106. Zhang, J. K.; Yang, L.; Meng, G. L.; Fan, J.; Chen, J. Z.; He, Q. Z.; Chen, S.; Fan, J. Z.; Luo, Z. J.; Liu, J. *Eur. J. Pharmacol.* **2012**, *689*(1–3), 31–37.
107. Zhang, J.; Li, S.; Wei, L.; Peng, Y.; Zheng, Z.; Xue, J.; Cao, Y.; Wang, B.; Du, J. *Int. J. Rheum. Dis.* **2018**, *21*(8), 1504–1513.
108. Kim, S. J.; Hwang, Y. H.; Mun, S. K.; Hong, S. G.; Kim, K. J.; Kang, K. Y.; Son, Y. J.; Yee, S. T. *Int. J. Mol. Sci.* **2018**, *19*(9). E2554https://doi.org/10.3390/ijms19092554.
109. Chen, X. H.; Shi, Z. G.; Lin, H. B.; Wu, F.; Zheng, F.; Wu, C. F.; Huang, M. W. *Eur. Rev. Med. Pharmacol. Sci.* **2019**, *23*(14), 6352–6359.
110. Feng, J.; Liu, S.; Ma, S.; Zhao, J.; Zhang, W.; Qi, W.; Cao, P.; Wang, Z.; Lei, W. *Acta Biochim. Biophys. Sin. Shanghai* **2014**, *46*(12), 1024–1033.
111. Mobasheri, A.; Shakibaei, M. *Ann. N. Y. Acad. Sci.* **2013**, *1290*, 59–66.
112. Li, J.; Xin, Z.; Cai, M. *J. Cell. Biochem.* **2019**, *120*(10), 16634–16642.
113. Tou, J. C. *Ann. N. Y. Acad. Sci.* **2015**, *1348*, 75–85.
114. Khera, A.; Kanta, P.; Kalra, J.; Dumir, D.; Thungapathra, M. *J. Women Aging* **2018**, *31*(6), 540–552.
115. Guo, D. W.; Han, Y. X.; Cong, L.; Liang, D.; Tu, G. J. *Mol. Med. Rep.* **2015**, *12*(2), 2098–2106.

116. Luo, Q.; Liu, S.; Xie, L.; Yu, Y.; Zhou, L.; Feng, Y.; Cai, D. *Front. Pharmacol.* **2019**, *10*, 736. https://doi.org/10.3389/fphar.2019.00736.
117. Feng, Y. L.; Jiang, X. T.; Ma, F. F.; Han, J.; Tang, X. L. *Int. J. Mol. Med.* **2018**, *41*(1), 202–212.
118. Wang, X.; Chen, L.; Peng, W. *Exp. Ther. Med.* **2017**, *14*(5), 5032–5038.
119. Zhou, T.; Yan, Y.; Zhao, C.; Xu, Y.; Wang, Q.; Xu, N. *Redox Rep.* **2019**, *24*(1), 62–69.
120. Yan, L.; Lu, L.; Hu, F.; Shetti, D.; Wei, K. *R. Soc. Open Sci.* **2019**, *6*, 190360. https://doi.org/10.1098/rsos.190360.
121. Zhai, Y.; Li, Y.; Wang, Y.; Cui, J.; Feng, K.; Kong, X. *Eur. J. Pharmacol.* **2017**, *801*, 62–71.
122. Zhai, Y.; Wang, Q.; Li, Y.; Cui, J.; Feng, K.; Kong, X.; Xian, C. J. *Biomed. Pharmacother.* **2018**, *102*, 1015–1024.
123. Kong, L.; Ma, R.; Yang, X.; Zhu, Z.; Guo, H.; He, B.; Wang, B.; Hao, D. *Int. Immunopharmacol.* **2017**, *51*, 31–39.
124. Kwak, S. C.; Baek, J. M.; Lee, C. H.; Yoon, K. H.; Lee, M. S.; Kim, J. Y. *Int. J. Biol. Sci.* **2019**, *15*(11), 2427–2437.
125. Liu, X.; Gao, X.; Liu, Y.; Liang, D.; Fu, T.; Song, Y.; Zhao, C.; Dong, B.; Han, W. *J. Cell. Biochem.* **2018**, https://doi.org/10.1002/jcb.27555.
126. Zhai, Y. K.; Pan, Y. L.; Niu, Y. B.; Li, C. R.; Wu, X. L.; Fan, W. T.; Lu, T. L.; Mei, Q. B.; Xian, C. J. *Int. J. Endocrinol.* **2014**, *2014*, 921954.
127. Long, H.; Zhang, H.; Deng, A.; Ma, L.; Wu, L.; Li, Z.; Zhang, Z.; Wang, W.; Jiang, J.; Qin, H. *Acta Pharm. Sin. B* **2016**, *6*(3), 229–233.
128. Xiao, H. H.; Dai, Y.; Wong, M. S.; Yao, X. S. *Fitoterapia* **2014**, *94*, 29–35.
129. Lee, W.; Ko, K. R.; Kim, H. K.; Lim, S.; Kim, S. *Biochem. Biophys. Res. Commun.* **2018**, *495*(3), 2242–2248.
130. Lee, W.; Ko, K. R.; Kim, H. K.; Lee, D. S.; Nam, I. J.; Lim, S.; Kim, S. *J. Nat. Prod.* **2018**, *81*(6), 1343–1356.
131. Dai, P.; Mao, Y.; Sun, X.; Li, X.; Muhammad, I.; Gu, W.; Zhang, D.; Zhou, Y.; Ni, Z.; Ma, J.; Huang, S. *Cell. Physiol. Biochem.* **2017**, *41*(2), 661–677.
132. Wang, N.; Wang, F.; Gao, Y.; Yin, P.; Pan, C.; Liu, W.; Zhou, Z.; Wang, J. *J. Pharmacol. Sci.* **2016**, *132*(3), 192–200.
133. Bharti, A. C.; Takada, Y.; Aggarwal, B. B. *J. Immunol.* **2004**, *172*(10), 5940–5947.
134. Kim, W. K.; Ke, K.; Sul, O. J.; Kim, H. J.; Kim, S. H.; Lee, M. H.; Kim, H. J.; Kim, S. Y.; Chung, H. T.; Choi, H. S. *J. Cell. Biochem.* **2011**, *112*(11), 3159–3166.
135. Xin, M.; Yang, Y.; Zhang, D.; Wang, J.; Chen, S.; Zhou, D. *Osteoporos. Int.* **2015**, *26*(11), 2665–2676.
136. Yun, J.; Lee, K. Y.; Park, B. *Biochimie* **2019**, *157*, 81–91.
137. Chen, S.; Jin, G.; Huang, K.; Ma, J.; Wang, Q.; Ma, Y.; Tang, X.; Zhou, Z.; Hu, Z.; Wang, J.; Qin, A.; Fan, S. *Sci. Rep.* **2015**, *5*, 12853. https://doi.org/10.1038/srep12853.
138. Pellegrini, G. G.; Morales, C. C.; Wallace, T. C.; Plotkin, L. I.; Bellido, T. *Nutrients* **2016**, *8*(7), 423. https://doi.org/10.3390/nu8070423.
139. Soundharrajan, I.; Kim, D. H.; Srisesharam, S.; Kuppusamy, P.; Sivanesan, R.; Choi, K. C. *Phytomedicine* **2018**, *45*, 41–48.
140. Klasik-Ciszewska, S.; Kaczmarczyk-Sedlak, I.; Wojnar, W. *Acta Pol. Pharm.* **2016**, *73*(2), 517–527.
141. Chen, X.; Zhi, X.; Yin, Z.; Li, X.; Qin, L.; Qiu, Z.; Su, J. *Front. Pharmacol.* **2018**, *9*, 647. https://doi.org/10.3389/fphar.2018.00647.
142. Li, Z.; Chen, C.; Zhu, X.; Li, Y.; Yu, R.; Xu, W. *Calcif. Tissue Int.* **2018**, *103*, 324–337.
143. Chawalitpong, S.; Chokchaisiri, R.; Suksamrarn, A.; Katayama, S.; Mitani, T.; Nakamura, S.; Athamneh, A. A.; Ritprajak, P.; Leelahavanichkul, A.; Aeimlapa, R.; Charoenphandhu, N.; Palaga, T. *Sci. Rep.* **2018**, *8*(1), 5625. https://doi.org/10.1038/s41598-018-23912-3.

144. Im, N. K.; Lee, D.; Lee, S.; Jeong, G. S. *J. Nat. Prod.* **2016**, *79*, 412–420.
145. Tan, H.; Zhao, C.; Zhu, Q.; Katakura, Y.; Tanaka, H.; Ohnuki, K.; Shimizu, K. *J. Agric. Food Chem.* **2019**, *67*(12), 3333–3340.
146. Wang, Q.; Yao, L.; Xu, K.; Jin, H.; Chen, K.; Wang, Z.; Liu, Q.; Cao, Z.; Keny, J.; Liu, Y.; Tickner, J.; Xu, H.; Xu, J. *J. Cell. Mol. Med.* **2019**, *23*(1), 380–394.
147. Zhang, Z.; Zhang, Q.; Yang, H.; Liu, W.; Zhang, N.; Qin, L.; Xin, H. *Fitoterapia* **2016**, *110*, 166–172.
148. Lin, G. L.; Hankenson, K. D. *J. Cell. Biochem.* **2011**, *112*(12), 3491–3501.
149. Lee, K.-S.; Hong, S.-H.; Bae, S.-C. *Oncogene* **2002**, *21*, 7156–7163.
150. Plotnikov, A.; Zehorai, E.; Procaccia, S.; Seger, R. *Biochim. Biophys. Acta* **2011**, *1813*(9), 1619–1633.
151. Soysa, N. S.; Alles, N.; Aoki, K.; Ohya, K. *J. Med. Dent. Sci.* **2012**, *59*(3), 65–74.
152. Liu, H.; Jin, W.; Fu, C.; Dai, P.; Yu, Y.; Huo, Q.; Yu, L. *Food Res. Int.* **2015**, *77*, 215–220.
153. Vuorinen, A.; Engeli, R. T.; Leugger, S.; Bachmann, F.; Akram, M.; Atanasov, A. G.; Waltenberger, B.; Temml, V.; Stuppner, H.; Krenn, L.; Ateba, S. B.; Njamen, D.; Davis, R. A.; Odermatt, A.; Schuster, D. *J. Nat. Prod.* **2017**, *80*, 965–974.
154. Hatefi, M.; Ahmadi, M. R. H.; Rahmani, A.; Dastjerdi, M. M.; Asadollahi, K. *World Neurosurg.* **2018**, *114*, 785–791.
155. Holick, M. F.; Lamb, J. J.; Lerman, R. H.; Konda, V. R.; Darland, G.; Minich, D.; Tripp, M. L. *J. Bone Miner. Metab.* **2010**, *28*, 342–350.
156. Tousen, Y.; Ezaki, J.; Fujii, Y.; Ueno, T.; Nishimuta, M.; Ishimi, Y. *Menopause* **2011**, *18*(5), 563–574.
157. Lambert, M. N. T.; Thybo, C. B.; Lykkeboe, S.; Rasmussen, L. M.; Frette, X.; Christensen, L. P.; Jeppesen, P. B. *Am. J. Clin. Nutr.* **2017**, *106*, 909–920.
158. Wattanathorn, J.; Somboonporn, W.; Sungkamanee, S.; Thukummee, W.; Muchimapura, S. *Oxid. Med. Cell. Longev.* **2018**, *2018*, 2091872.

CHAPTER TWELVE

Antidiabetic natural products

Pulok K. Mukherjee[a,b,*], Amit Kar[a], Subhadip Banerjee[a], C.K. Katiyar[c]
[a]School of Natural Product Studies, Department of Pharmaceutical Technology, Jadavpur University, Kolkata, India
[b]Institute of Bioresources and Sustainable Development, Imphal, India
[c]Health Care Division, Emami Limited, Kolkata, India
[*]Corresponding author: e-mail addresses: pulokm@gmail.com; director.ibsd@nic.in

Contents

1. Diabetes mellitus and its treatment

Diabetes mellitus (DM) is a chronic metabolic disorder, which has emerged as a global public health problem. The World Health Organization (WHO) reports a steady rise in incidence of DM patients up to 366 million by 2030.[1] Type-2 diabetes is a heterogeneous disorder, with a progressive decline in insulin action due to insulin resistance.[2] This insulin resistance is a situation, where, the insulin released from the pancreatic β-cells is unable to trigger glucose uptake in the muscle cells and adipose tissues. The reason behind this is the gradual down regulation of the insulin receptors in those cells, rather than insulin deficiency.[3] Therefore, β-cells work under stress to make up for the insulin resistance, thus giving rise to free radicals and oxidative stress. Due to insufficiency of the inherent antioxidant systems of the human body, the free radicals

Annual Reports in Medicinal Chemistry, Volume 55
ISSN 0065-7743
https://doi.org/10.1016/bs.armc.2020.06.002

(molecules possessing an unpaired electron) and the reactive oxygen species, i.e., the ROS gradually destroy the β-cells. This can result either from an overproduction of ROS or from the inactivation of the antioxidants (AO), thereby shifting the ROS/AO balance in favor of stress. Due to increasing oxidative stress on the pancreatic β-cells the free radicals are produced continuously in the blood hence the β-cells gradually get destroyed.[4] Various traditional medicinal plant species from Indian biosphere and their constituents have been shown to display potent hypoglycemic activity.[1] In Indian traditional medicine the history of the disease "*Madhumeha*" is well documented. The Sanskrit term "*Madhumeha*" is composed of *Madhu* + *Meha*. The word "*Madhu*" means sweet like honey and "*Meha*" refers to *atipravritti* (excessive urination).

In *Charaka Samhita* (1000 BC–4th century AD), the disease *Madhumeha* has been defined as the disease in which one passes urine as astringent, sweet and rough (Fig. 1). Disease concerning qualitative and quantitative change of urine is termed as *Prameha*. *Meha* denotes *mehana* that denotes flow or excretion of urine. When there is considerable qualitative and quantitative change of flow of urine, it is termed as *Prameha*. Eventually 20 types of *Prameha* changes to *Madhumeha* which is being attributed through metabolic changes.[2,3]

In DM, the postprandial glycemia hyperglycemia is one major problem. Novel therapeutic strategies aim to manage this situation with the help of

कषायमधुरं पाण्डु रूक्षं मेहति यो नरः।
वातकोपादसाध्यं तं प्रतीयान्मधुमेहिनम्॥४४॥

"Kasāyam madhuram pāndumrūksam mehati yonarah
Vātakopādasādhyam tam pratiyānmadhumehitam"

(*Caraka Samhita, Nidana Sthana*– Chapter 4/43)

Madhumeha **is the disease in which one passes urine as astringent, sweet and rough.**

Fig. 1 Excerpt from Charaka Samhita Regarding Diabetes mellitus.

drugs and herbal remedies which can interfere with the carbohydrate metabolizing machinery in our gut. These things have proved to be successful for preventing and treating chronic vascular complications.[4] Prior to drug therapy, doctors advise diet control and adequate physical exercise for management of diabetes. Exercise has a special advantage that during exercising, the glucose entry in the muscle cells is insulin sparing. So, besides drug therapy, this also serves as an important measure to control glycemia hyperglycemia. The American Diabetes Association, a specialist body of diabetologists, has suggested that there is no specific diet for controlling blood sugar level in a diabetes patient. The diet needs to be fixed as per the individual requirements and goals. In the Ayurvedic literature personalized food has been described which can also helpful in diabetic conditions.[2,5]

The α-glucosidase and α-amylase inhibitors have been proved to delay the glucose absorption through the intestinal epithelia. As a result, the undigested glucose gets transported to the distal jejunum, Ileum and colon region.[6] The result is the decrease in the post prandial glycemia hyperglycemia. As per the International Diabetes Federation (IDF) recommendations, α-glucosidase inhibitors (AGIs) are liable to be used as first-line, second-line or third-line treatment options. In clinical practice, AGIs may be combined with Metformin, Sulfonylurea, DPP4 inhibitors, thiazolidinediones and insulin therapy. Irrespective of the initial HbA1c level, AGIs may be used at all stages of diabetes, as recommended by the American Association of Clinical Endocrinology.[7] Though this class of molecules has been proved to be relatively safe, they are less potent than the standard therapeutic agents. Similar to the recommendations from Asian-Pacific panel, the National guidelines from China, Japan, Taiwan, Korea and Singapore—all support the use of AGIs.[8] Many clinical studies have been conducted on acarbose. The dose used in many of the studies was excessive. However, the dose of this molecule has been optimized and found to be 150–300 mg/day. Acarbose monotherapy has been compared with other oral antidiabetic agents in terms of blood sugar lowering capabilities.[9] Addition of α-glucosidase inhibitors to other oral hypoglycaemic agents triggers an improved glycaemic control, in comparison to the single drug therapeutic regimens. The addition of acarbose or miglitol to sulfonylurea-, metformin-, or insulin-treated patients with NIDDM results in improved glycemic control. Their addition to sulfonylureas reduces HbA1c, approximately 1.1%; to metformin, 0.8%; and to insulin, 0.5% here are two other factors regarding the antidiabetic activity of a medicinal agent.[10] They are postprandial GIP secretion and GLP-1 (glucagon like peptide) secretion, calculated 30–240 min postprandially. α-Glucosidase inhibitors reduce the GIP secretion, but increase GLP-1 secretion. The possible reason behind this effect may be the presence of carbohydrate in the Ileum during α-glucosidase inhibitor treatment.[11] The detailed GLP-1 level in the postprandial studies is diverse in nature. It is an important trigger in reducing the blood glucose level, through triggering insulin level from the pancreatic β-cells. Also, α-glucosidase inhibitors are able to bring down the serum triglyceride levels.[12] However, α-glucosidase inhibitors are not able to interfere with the primary cholesterol or triglyceride metabolism. They do not affect the body weight significantly. α-Glucosidase inhibitors are also having some additional metabolic benefit, which have been proved by some clinical studies.[13] Post prandial hyperglycemia is able to activate coagulation. This takes place through elevation in the plasma concentrations of prothrombin fragments 1 and 2, followed by the D-diamer of fibrinogen. Prevention of the α-glucosidase mediated carbohydrate metabolism is able to lower the postprandial glycemia hyperglycemia, and thus able to prevent the blood coagulation. Thus they are able to prevent the thromboembolic disorders.[10]

2. Carbohydrate metabolizing enzymes

The human diet includes plant starches, such as wheat, rice, flour, cereals, tubers, fruits and vegetables, as the chief source of glucose, the key metabolite needed for energy production. Starch, the most abundant storage polysaccharide in the plant body, is packaged as a semi-crystalline granule form. Starch is composed of mainly amylase and amylopectin. Amylose is composed of 500–600 glucose residues, arranged in the form of a long chain, joined by α-1,4 linked glucose units. Amylopectin it is a larger and more branched molecule. It has several α-1,4 linked glucose chains, each of which are made up of 12–70 glucose residues. Approximately 5% of the linkages are of α-1,6 types (Fig. 2). As the complex, large di- and oligo saccharides can't be transported across the intestinal epithelia, they need to be first converted in to simple glucose molecules. For this purpose, the intestinal enzymatic activities are crucial. The carbohydrates may be classified in two types, namely digestible and non-digestible carbohydrates.[14] Available carbohydrates include starches, simple mono-

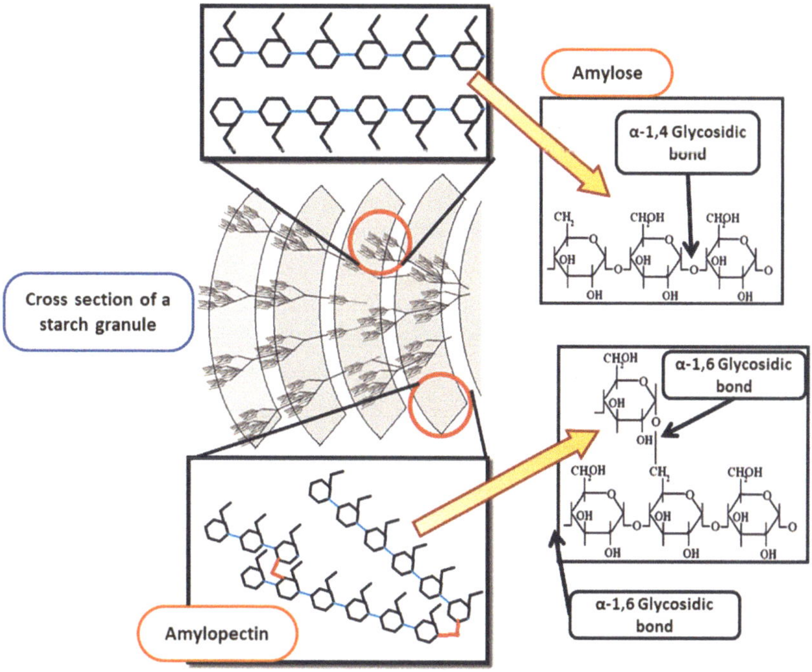

Fig. 2 Structure of a starch granule.

and di-saccharide sugars, such as glucose and sucrose, etc.[15] The disaccharides and larger molecules are converted into monosaccharide components and absorbed from upper intestinal tract. Non-available carbohydrates include fiber, polyols and insulins. These fibers can't be digested by endogenous intestinal enzymes. Instead, they get fermented by bacterial enzymes in the colon.[16]

Alpha-amylases and four α-glucosidases are involved in the complete digestion of starch in to glucose. Due to the action of different starch-digesting enzymes in the small intestine, namely sucrase-isomaltase (SI system) and maltase-glucoamylase (MGAM systems), the glucose gets extracted from the complex polysaccharides. It is then absorbed through the intestinal epithelium. These enzymes are the final catalysts required for the glucose release from these polysaccharides. MGAM and SI-both of the carbohydrate metabolizing systems in human small intestine are complex protienous structures, composed of N- and C-terminal domains. Their structure elucidation has revealed that these domains are complementary with the structures of the substrates, which are oligosaccharides in nature.[17] In order to prevent and alleviate the Type-2 diabetes through controlling of the postprandial blood glucose level, the α-glucosidase inhibitors have been proved to be quite useful.

The internal α-1,4 bonds of the amylose and amylopectin structures in the starch molecule are cleaved by the salivary and pancreatic α-amylases (EC 3.2.1.1.), which are endohydrolases in nature. After cleavage, they are converted into shorter linear and branched dextrin chains. This resultant mixture of dextrins is further converted into glucose through hydrolytic cleavage (Fig. 3) through the exohydrolytic cleavage activities of MGAM (maltase-glucoamylase; EC 3.2.1.20 and 3.2.1.3) and sucrase-isomaltase (SI; EC 3.2.148 and EC 3.2.10), before they are absorbed in to the bloodstream.[18] Sugar transport through the intestinal epithelia takes place through mainly two transporters, as sodium/glucose transporter SGLT1 and fructose transporter (GLUT 5).[4] The mechanism and site of the carbohydrate digestion have been schematically depicted in Fig. 3. Resistant starches and fibers are able to bypass small intestinal processing. They are further transported to colon and then fermented by bacteria.

2.1 α-Glucosidase in glucose metabolism

After the metabolism of the starch, the liberated digestion products are further metabolized by sucrase-isomaltase (SI) and maltase-glucoamylase (MGAM) in to glucose. As the α-amylases are able to cleave the α-1,4 bonds of the substrate, i.e., the complex polysaccharides, the α-glucosidases function as exohydrolases, i.e., cleave the limit dextrins from one end, one by one liberating the glucose residues from the substrate. Limit dextrins, also known as the α-limit dextrins, comprises maltose, maltotriose and various short, branched substrates.[19] As the α-amylase starch digestion in the small Intestine is not compartmentalized, the SI and MGAM, i.e., the brush border glucosidases are likely to be exposed to larger intermediates, in comparison to α-limit dextrins. It has been practically observed by the researchers that the substrate composition of the SI and MGAM varies throughout the length of the small intestine. The reason is that during the transit of the starch through the small intestine, it gradually is broken down in to limit dextrins.[16] So, as one proceeds through the intestine, the available substrate concentration gradually increases.

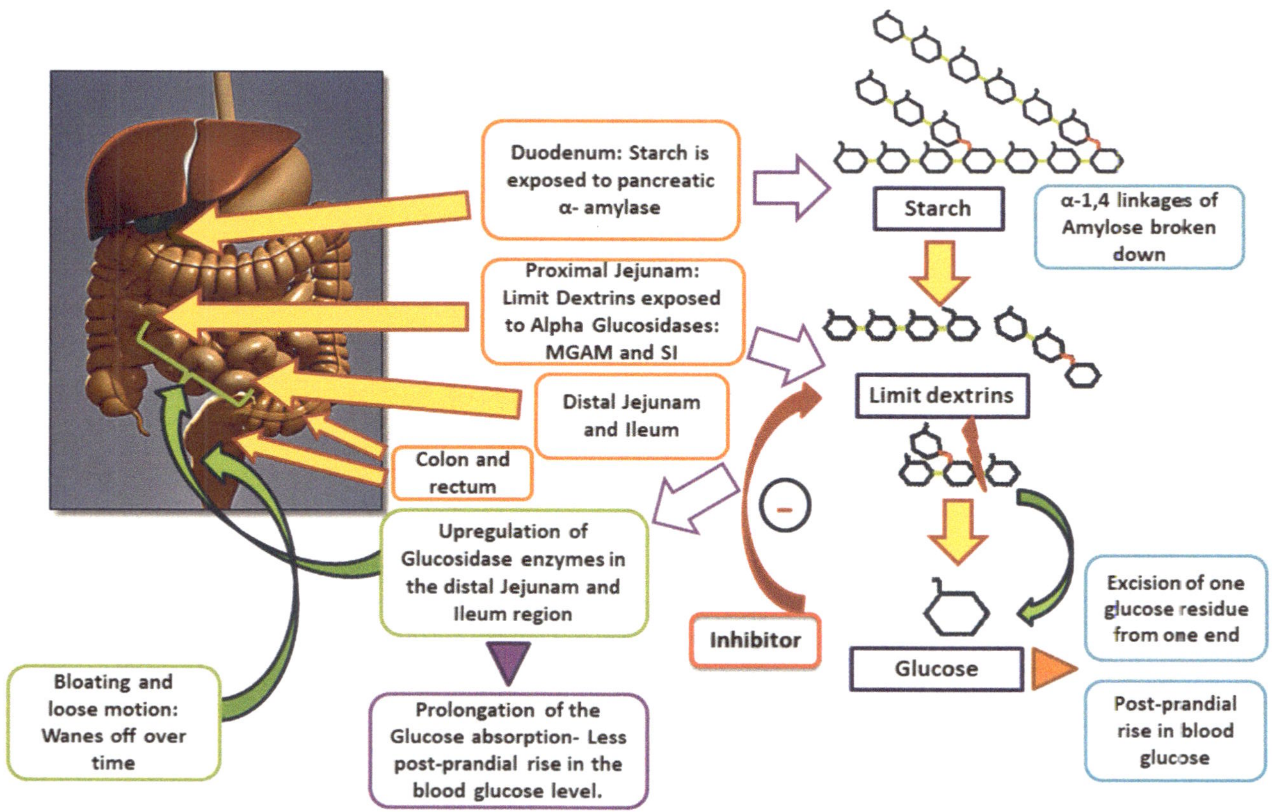

Fig. 3 Mechanism and sites of carbohydrate metabolism and absorption.

It has been shown that MGAM has very little activity for starch itself and but shows greatly enhanced hydrolytic activities with α-amylase pre-treated starch.

As the MGAM and SI are derived from the same superfamily of genes, they share some similar features. MGAM and SI are GH 31 super family of proteins, which are made up of N-terminal and C-terminal domains. They are stalked intrinsic membrane proteins in nature, who reside in the brush border of human small intestinal mucosa. The N-terminal region of the protein acts as an anchor, whereas the extracellular intestinal luminal side acts as the catalytic domain.[20] Conventionally, the denominations "sucrase" (α-1,2), "isomaltase" (α-1,6), "maltase" (α-1,4) and "glucoamylase" (α-1,4) were each associated with one of the GH31 catalytic domains.[21]

3. Importance of the inhibition of α-glucosidase enzyme

The α-glucosidase inhibitors are complex oligosaccharides in nature. They act as competitive inhibitors of the α-glucosidase enzyme, competing with the substrates. Thus they slow down the carbohydrate digestion, and thus modulate the postprandial glucose level spike. As the α-glucosidase inhibitors do not require the intervention of the insulin, the chances of hyperinsulinaemia is also less, as occurs in diabetic patients. They are also found to be not interfering with the weight of the patient, though, if used for years, they have been found to reduce weight in patients. A cochrane systematic review considering a meta-analysis of 41 studies, reports beneficial effects of AGIs on fasting glucose, post-load glucose, post-load insulin, and body mass index (BMI).[22] The highest dose suggested is 50 mg thrice daily. It has been shown in long-term studies that this glycemic control is multifaceted (fasting, postprandial glucose, and HbA1c control) and is sustained up to 5 years.[23] The short-term effect of these therapies is to decrease current blood glucose levels whereas the long-term effect is a small reduction in hemoglobin level.[24]

Modern system of medicine is very much evidence-based. If one has to bring a molecule in the market, that molecule needs to be supported by huge multicentric, randomized, controlled clinical trial data. The results obtained should be analyzed through a systematic review, if possible, with a meta-analysis. Regarding the use of α-glucosidase inhibitors, several guidelines have been laid down by different nations.[25]

4. The Ayurveda in the management of diabetes mellitus

The roots of the Ayurveda can be traced back to the beginning of human civilization. Belief-based Vedic Medicine transformed to the Ayurveda and was rationalized to experience based medicine. In the Ayurveda, several single herbs (51), bhasma metals and minerals (9), herbo-mineral formulations (30), herbal formulations (34), Ghritas (4), Taila (Oil) (5), Asava-aristas (6) and Guggulu preparations (5) have been mentioned for the management of diabetic conditions. According to the Indian mythology Lord *Gajanana* or *Ganesha* developed obesity due to his affinity toward sweets and sedentary lifestyle and eventually suffered from diabetes or *Madhumeha*. He was advised by his father Lord *Shiva* to take *Kapittha* (*Limonia acidissima*), Jambu (*Syzygium* sp.), *Shiva Gutika* (polyherbal preparation) to counter this dreadful disease.[2]

The Vedic literature is the oldest literature in the world. In *Atharva Veda* (2500 BC) the disease is mentioned by the term "*Aastravam*" which means *mutratisara*, i.e., excessive urination (polyurea). In *Charaka Samhita* (1000 BC–4th century AD) has mentioned the definition of *Madhumeha* as the disease in which one passes urine as astringent, sweet and rough Acharya *Sushruta* in *Sushruta Samhita* (1000 BC–5th century AD) has described *Madhumeha* by the term *Kshaudrameha* which means the urine in this condition resembles honey and acquires sweet taste. Acharya *Vagbhata* (7th century AD) has defined in a different way, which states that among 20 types of *Prameha* any type in which the urine become sweet like honey, the whole body turns sweet then it is to be named as *Madhumeha*.[2]

Table 1 enlists different herbs used for diabetes associated complications. The Ayurveda have empowered antidiabetic therapy further by many herbal formulations, *bhasmas* (calcined product of purified metals and minerals), herbo-mineral preparations, oils, *Asava-aristas* (self-generated alcoholic tincture), ghritas (milk fat based semisolid) and guggulu preparations (herbal preparation with *Commiphora* sp. as the base). There are several herbal preparations, herbo-mineral preparations have been practiced to manage *Madhumeha* or diabetes for ages listed in Table 2.

4.1 Ayurvedic prevention and management of diabetes

4.1.1 Early detection or pro-disease concept of diabetes in the Ayurveda

Madhumeha or diabetes is a complex metabolic disorder intricate with genetic variations. Early lipid material excretion (*Medadusti*) was detected

Table 1 Ayurvedic medicinal plants and formulations for diabetic complications.

Complications related to diabetes mellitus	Ayurvedic remedies
Neuropathy	Dashamula (10 herbs combinations like *Aegle marmelos*, etc.)
Cardio-vascular disorders	Kustham (*Saussurea lappa*)
Angiopathy	Manjistha (*Rubia cordifolia*) Kanchanar (*Toddalia aculeata*)
Retinopathy	Manjistha (*Rubia cordifolia*), Nagakesara (*Mesuafera*), Bala (*Sida cordifolia*) Saptamrita Lauha, Triphala Louha, Satavari (*Asparagus racemosus*)
Diabetic foot	Nirgundi (*Vitex negundo*), Manjistha (*Rubia cordifolia*), Ashwagandha (*Withania somnifera*) Daruharidra (*Berberis aristata*)
Nephropathy	Varun (*Crataeva nurvala*) Shigru (*Moringa oleifera*) Gokshura (*Tribulus terresteris*)

by mixing equal amount of Puskarmoola (*Inula racemosa*) and urine followed by boiling. If after boiling a red color is produced, presence of lipid was confirmed. This evidences along with the Ayurvedic traditional claims assert that urine associated lipid excretion parameters should be initiated as an biomarker to screen diabetic and pre-diabetic stages. The diagnostic parameters of diabetes mellitus or *Madhumeha* HbA_1c estimations must not be the only index. Other associated factors like *Medadhatudushti* (adipose tissue damage and dyslipidaemia) should be brought into consideration besides glycaemic index as we find molecular evidences that diabetes mellitus is associated with irregularities in signaling of lipid metabolism.[26] Selection of drugs in case of diabetes with specific *Prakriti* types (genomic traits) as enumerated in the Ayurveda may be logical. The principle of the Ayurvedic treatment depends on the body morphometric type symptom complex developed due to qualitative and quantitative variation of urine with involvement and excretion of *Medadhatu* (lipid) which is the pro-disease state of *Madhumeha* and termed as *Prameha* (before state of diabetes).[3,5] Extent of *Medadhatu* involvement could be early determined by the

Table 2 Formulation mentioned in Ayurveda for the management of diabetes mellitus and associated symptoms.

Sl. no.	Classical name	Major composition	Type of the formulations
1	Akulyadi Yoga	Herbal	Decoction
2	Akulyapaladi Yoga	Herbal	Decoction
8	BejakaVanaspati Kanda Sara Kwatha	Herbal	Decoction
9	Beejasaradi Kasaya	Herbal	Decoction
37	Mustadi Kwatha	Herbal	Decoction
61	Vyaghri Kwatha	Herbal	Decoction
7	Asanadi Yoga	Herbo-mineral	Decoction
33	Mahadadimbadya Ghrita	Herbal	Ghee preparation
16	Gokshuradi Guggulu	Herbal	Guggulu preparation
38	Navaka Guggulu	Herbal	Guggulu Preparation
3	Amritarnava Louha	Herbal	Iron preparation
46	Rohitakarista	Herbal	Liquid (Arista)
6	Arjunarishta	Herbo-mineral	Liquid (Arista)
27	Kumaryasava	Herbal	Liquid (Asava)
35	Madhvasava	Herbal	Liquid (Asava)
19	Haridradi Taila	Herbal	Oil preparation
41	Neem Taila	Herbal	Oil preparation
4	Aranyajiraka	Herbal	Powder
13	Chaturbeeja Churna	Herbal	Powder
21	Jambadi Yoga	Herbal	Powder
29	Madhumehari Churna	Herbal	Powder
39	Nagrodhadi Churna	Herbal	Powder
47	Salsaradi Yoga	Herbal	Powder
56	Triphala Churna	Herbal	Powder
57	Triphaladi Ghana Vati	Herbal	Powder
58	Trasyunadi Yoga	Herbal	Powder

Continued

Table 2 Formulation mentioned in Ayurveda for the management of diabetes mellitus and associated symptoms.—cont'd

Sl. no.	Classical name	Major composition	Type of the formulations
10	Bilwadi Yoga	Herbo-mineral	Powder
14	Chhinaruhadi Yoga	Herbo-mineral	Powder
18	Haridradi Amlaki	Herbo-mineral	Powder
24	Ketakadi Yoga	Herbo-mineral	Powder
25	Kiatadi Yoga	Herbo-mineral	Powder
26	Kshiradi Churna	Herbo-mineral	Powder
32	Madhumehari Kalpita Yoga	Herbo-mineral	Powder
34	Mamejjakadi Churna	Herbo-mineral	Powder
31	Madhumehari Vati	Herbal	Tablets
40	NeemGhanasatwaVati	Herbal	Tablets
42	NishaAmlaki Ghana	Herbal	Tablets
44	Phalatrikadi GhanasatwaVati	Herbal	Tablets
48	Salsaradi Ghana Vati	Herbal	Tablets
55	Sudamstadi Ghana Vati	Herbal	Tablets
45	Rajanitab	Herbo-mineral	Tablets
20	IndraVati	Herbal	Tablets (Rasa Yoga)
22	Kalyan Rasa	Herbal	Tablets (Rasa Yoga)
23	Kaspasthyadi Gutika	Herbal	Tablets (Rasa Yoga)
5	Arogyavardhini Vati	Herbo-mineral	Tablets (Rasa Yoga)
11	Bramhagajakeshari	Herbo-mineral	Tablets (Rasa Yoga)
12	Chandrakala Rasa	Herbo-mineral	Tablets (Rasa Yoga)
15	Devadarvadi Ghana Vati	Herbo-mineral	Tablets (Rasa Yoga)
17	Gulavela Sattavachi	Herbo-mineral	Tablets (Rasa Yoga)
28	Louhashiajit Vati	Herbo-mineral	Tablets (Rasa Yoga)
30	Madhumehari Rasayana	Herbo-mineral	Tablets (Rasa Yoga)
36	Mehamudgara Rasa	Herbo-mineral	Tablets (Rasa Yoga)
43	Pakshika Rasayana	Herbo-mineral	Tablets (Rasa Yoga)

Table 2 Formulation mentioned in Ayurveda for the management of diabetes mellitus and associated symptoms.—cont'd

Sl. no.	Classical name	Major composition	Type of the formulations
49	Salsaradi Shilajit	Herbo-mineral	Tablets (Rasa Yoga)
50	Samasevara Rasa	Herbo-mineral	Tablets (Rasa Yoga)
51	Shilajatu Swarna Makshika	Herbo-mineral	Tablets (Rasa Yoga)
52	Shilajatu Yasad Vati	Herbo-mineral	Tablets (Rasa Yoga)
53	Shilajityadi Vati	Herbo-mineral	Tablets (Rasa Yoga)
54	Shiva Gutika	Herbo-mineral	Tablets (Rasa Yoga)
59	Vangeswar Rasa	Herbo-mineral	Tablets (Rasa Yoga)
60	Vasanta Kusumakar Rasa	Herbo-mineral	Tablets (Rasa Yoga)
62	Yasada Amrita Yoga	Herbo-mineral	Tablets (Rasa Yoga)

Ayurvedic way of urine examination stated above. Calorie dense food, stressful lifestyle and sedentary habits are the reasons for the formation of *visamadharmimeda (*derogatory fat which can be correlated with present concept of high triglycerides or free fatty acids) which leads to development of *Prameha* proceeding to *Madhumeha*. So, early detection at prodisease state and intervention in high risk population can prevent diabetes development before its onset.

4.1.2 Preventive measures in the Ayurveda

Madhumeha being multi-factorial, therapeutic intervention using single target single drug are not tenable. Moreover, *Madhumeha* is defined as one of the *Vataja Prameha* (*Prameha* developed due to vitiation of *Vatadosa*) developed as a function of time over 20 stages (types of *Prameha*). In the Ayurveda, personalized diet and medicine is selected on the basis of *Trisutra* (three primary reasons) viz *Hetu* (causes), *Linga* (features) and *Ausadha* (therapeutics). The principles of treatment in a *Madhumeha* patient may be classified as (1) *Sthulapramehi* (obese) (2) *Krishapramehi* (lean). The treatments has been described as in *Sthula Prameha* (Obese)—in this type the treatment must be based on proper utilization of excess fat. He should be given

(i) *Shodhana* (purification process) (ii) *Apatarpana* (reduction of body wt. by way of diet control, drugs or exercise).

Krisha Prameha (Asthenic)—the treatment mainly based on the line of increasing stamina and vitality known as *Brimhana* using diet, and drugs. The patient should never be given excessive *Langhana* or *Apatarpana*, i.e., he should not be in fasting condition for prolonged period. It has been mentioned that as *Madhumeha* is developed due to disorder of *Agnashaya* (pancreas) till now no medicine is effective in total cure of this disease. Hence, this disease is called *Yapya Vyadhi*, i.e., may be under control but not curable. Thus preventive medicine should be stressed by detection at the *Prameha* stage only.[2]

4.1.3 Interventional herbal therapy

The Ayurveda states wide range of herbs, polyherbal formulations, herbo-mineral preparation and therapeutic measures (like panchakarma including meditation and Yogic exercises) to avert, alleviate and manage diabetes mellitus according to individual patients. Several Ayurvedic herbs like *Allium cepa, Allium sativum, Aloe vera, Cajanus cajan, Coccinia indica, Caesalpinia bonducella, Ficus benghalensis, Gymnema sylvestre, Momordica charantia, Ocimum sanctum, Pterocarpus marsupium, Swertia chirayita, Syzigium cumini, Tinospora cordifolia, Trigonella foenum-graecum, Withania somnifera, Murraya koenigii,* and *Brassica juncea,* have been widely reviewed in literature from time to time. These scholarly publications about these herbs explored their phytochemistry, antidiabetic effects and in some cases molecular mechanism of action have been reported.[2,3]

Plant-based folk wisdom does not show uniform hypoglycaemic activity in clinical settings. Individualized treatment is essential taking *Prakriti* (genomic trait of patient), *guna—karma* (quality and action relationship of herbal or herbo-mineral drug) and *dushya* (tissue level damage) relationships as guiding principles. In the treatment of diabetes mellitus or *Madhumeha,* use of *Rasayana* (immunomodulatory) drugs like Ashwagandha (*Withania somnifera*), Guduchi (*Tinospora cordifolia*), etc., are emphasized. We find most of the plants used for the treatment of Prameha/*Madhumeha*/diabetes mellitus are enlisted as *Rasayana.* Recent literature concerning Ayurgenomics supports this view of personalized medicine on the basis of Prakriti types. Combinations of drugs are logical. Here is a list of antidiabetic Ayurvedic herbs classified according to *Prakriti* types. (Table 3 compiled from *Charaka Samhita*).

Table 3 Antidiabetic Ayurvedic plants classified according to the *Prakriti* types.

Type	Scientific name	Rasa	Guna	Virya	Vipaka
Kapha	*Berberis aristata*	Katu-tikta	Ruksha	Ushna	Katu
	Embelia ribes	Tikta-kashya	Ruksha	Ushna	Katu
	Acacia catechu	Tikta-kashya	Laghu-Ruksha	Shita	Katu
	Anogeissus latifolia	Kashya	Laghu-Ruksha	Shita	Katu
Pitta	*Trichosanthes dioica*	Tikta	Laghu-Ruksha	Ushna	Katu
	Azadirachta indica	Tikta	Snigdha-Laghu	Ushna	Katu
	Phyllanthus emblica	Amla-madhur	Laghu-Ruksha	Shita Katu	Katu
	Tinospora cordifolia	Kashya-tikta	Guru-Snigdha	Ushna	Katu
Vata	*Cissampilos pariera*	Tikta	Laghu-Ruksha	Ushna	Katu
	Trichosanthes dioica	Tikta	Laghu-Ruksha	Ushna	Katu
	Terminilia arjuna	Kashya	Laghu-Ruksha	Ushna	Katu
	Azadirachta indica	Tikta	Snigdha-Laghu	Ushna	Katu

5. Natural α-glucosidase inhibitors in treatment of diabetes

Up to this point in time, more than 400 traditional herbal remedies have been used for diabetes management, of which only a limited number of plants have scientific validation. As per the WHO guidelines, the herbal drugs need further investigation, research and validation. Many Ayurvedic drugs, reported in the literature, have been used for the preparation of many modern drugs. Lack of proper standardization and animal study reports are the main obstacles in the way if proper validation and market approval, as far as the herbal medicines are concerned. Plants are known to produce a variety of secondary metabolites, mainly aimed toward providing protection from external insects and pathogens. Those metabolites are responsible for the α-glucosidase inhibitory property. Pancreatic and intestinal glucosidases are the key enzymes of dietary carbohydrate digestion and inhibitors of these enzymes may be effective in retarding glucose absorption to suppress PPHG. The liver glucosidase inhibitors, inhibits -1,6-glucosidase of glycogen digesting enzymes in the liver and decreases the glycogenolytic rate, which increases the accumulation of glycogen stores in the liver. Inhibition of these enzyme systems decreases the blood glucose levels in diabetic patient (as a short-term effect) and shows a small reduction in HbA_{1c} level.[25] Different important plants well-known for α-glucosidase inhibitory potential are as follows—*Momordica charantia* (Bitter melon, family-Cucurbitaceae),

Trigonella foenum-graecum (Fenugreek, family—Leguminosae), *Gymnema sylvestre* (Gudmar, family—Asclepidaceae), *Pterocarpus marsupium* (Vijayasar, family—Leguminosae), *Murraya koenigii* (Curry patta, family—Rutaceae), *Ocimum sanctum* (Tulsi, family—Lauraceae), *Tinospora cordifolia* (Seenthil, family—Menispermaceae), *Syzygium cumini* (*Eugenia jambolana*, family—Myrtaceae), *Zingiber officinale* (Ginger, family—Zingiberaceae), *Allium sativum* (Garlic, family—Liliaceae). The potential natural α-glucosidase inhibitors have been presented in Table 4.

α-Glucosidase inhibitors have been shown to reduce the postprandial hyperglycemia and also to reduce the difference between the maximum and minimum postprandial glucose levels. This reduction in the postprandial hyperglycemia results in a decrease in the HbA_{1c} level and is thus able to provide long-term benefit in patients with NIDDM. This control results in better glucose regulation in NIDDM patients.[24] The naturally obtained plant extracts and fractions, by virtue of containing a plethora of flavonoids and phenolics, which are important α-glucosidase inhibitors, serve as important and effective agents for the management of diabetes.[48] There are various phytochemicals with structural diversity which shows considerable inhibition property shown in Table 5. Phenolic compound like luteolin, chrysoeriol, rutin, hesperitin, quercetin, kaempferol, lutein, tannic acid, edgeworin, daphnoretin, edgeworthin, chlorogenic acid, resveratrol, oiceatannol, *trans*-pinostilbene and phenolic glycosides like edgeworoside A, edgeworoside C, 1,5 dicaffeoylquinic acid, quercitrin (quercetin3-*O*-rhamnoside), hyperin (quercetin3-*O*-rhamnoside).

Among phenylpropanoids, several honokiololigomers, were obtained from *Momordica charantia* which had good inhibitory effects on α-glucosidase with IC_{50} values ranging from 1.38 to 95.31 mol/L. Among all the tested compounds honokiol trimer (IC_{50} = 1.38 mol/L) exhibited the most significant inhibitory activity toward α-glucosidase in a concentration-dependent manner and was 128-fold more potent than that of honokiol (IC_{50} = 177.03 mol/L).[49] Curcuminoids curcumin, dimethoxycurcumin and bisdemethoxycurcumin isolated from *Curcuma longa* showed strong inhibitory activity on α-glucosidase (IC_{50} = 37.2, 42.7, and 23.0 mol/L, respectively).[50] Ellagitannins were identified from *Terminalia chebula* as chebulanin, chebulagic acid and chebulinic acid were shown to possess potent intestinal maltase inhibitory activity, with the IC_{50} values of 690, 97 and 36 μM.[51] Triterpene glycosides fraction from *Gymnema sylvestri* containing gymnemic acids I, IV, VII showed significant inhibition of

Table 4 Some α-glucosidase inhibitor plants.

Sl. no.	Name of the plant	Family	Parts used	IC_{50} value	References
1	*Passiflora subpeltata*	Passifloraceae	Fruit pulp	32.63 g/mL	27
2				0.93–1.06 μg/mL (*n*-butanol extract)	
3	*Myrcia rubella* Cambess.	Myrtaceae	Leaves	4.36 ± 0.08 μg/mL	28
4	*Myrcia vestita* DC.			13.60 ± 0.96 μg/mL	
5	*Myrcia virgata* Cambess.			6.29 ± 0.65 μg/mL	
6	*Myrcia guianensis*			7.80 ± 0.92 μg/mL	
7	*Myrcia variabilis*			3.17 ± 0.25 μg/mL	
8	*Myrcia torta*			5.31 ± 0.28 μg/mL	
9	*Xanthium strumarium* L.	Compositae	Fruits	15.25 g/mL	29
10	Black tea	Theaceae	Leaves	4.190 mL	30
11	Baked black tea	Theaceae	Leaves	6.56 mL	
12	*Phyllanthus nirruri* aqueous extract	Phyllanthaceae	Aerial part	3.7 ± 1.1 μg/mL	31
13	*Phyllanthus nirruri* methanolic extract	Phyllanthaceae	Aerial part	6.3 ± 4.8 μg/mL	
14	Raffia palm aqueous extract	Palmaceae	Leaves	99.96 ± 4.31 μg/mL	32
15	*Capparis spinosa* Salted buds	Capparaceae	Leaves and bud powder	21.09 ± 0.48 mmol Acarbose equivalent/g extract	33

Continued

Table 4 Some α-glucosidase inhibitor plants.—cont'd

Sl. no.	Name of the plant	Family	Parts used	IC_{50} value	References
16	Black turtle bean (*Phaseolus vulgaris*)	Fabaceae	Beans	$64.12 \pm 2.12\,\mu g/mL$	34
17	Soybean (*Glycine max*)	Fabaceae	Beans	$75.41 \pm 3.11\,\mu g/mL$	34
18	*Allophylus cominia*	Sapindaceae	Flavonoid rich extract	$86 \pm 0.3\%$ inhibition at $30\,\mu g/mL$	35
19	*Plectranthus esculenta*	Lamiaceae	Tubers	$128.89 \pm 4.85\,\mu g/mL$	36
20	*Empetrum nigrum* var. japonicum	Ericaceae	Aerial parts	$4.3 \pm 0.6\,\mu g/mL$	37
21	*Gossypium herbaceum*	Malvaceae	Leaf extract	$4.0\,g/mL$	38
22	*Poncirus trifoliata (L.) Raf.*	Rutaceae	Juice	$81.27 \pm 3.5\,\mu g/m$	39
23	*Swertia chirata.*	Gentianaceae	Leafy shoots	$4.72 \pm 0.02\,\mu g/mL$	40
24	*Swertia decussata.*	Gentianaceae	Leafy shoots	$4.62 \pm 0.00\,\mu g/mL$	
25	*Momordica charantia*	Cucurbitaceae	Leafy shoots	$66.64 \pm 2.94\%$ (at $1.25\,mg/mL$ concentration)	
26	*T. cucumerina*	Cucurbitaceae	Leafy shoots	$61.91 \pm 0.96\%$ (at $1.25\,mg/mL$ concentration)	
27	*L. siceraria*	Cucurbitaceae	Leafy shoots	$61.25 \pm 0.57\%$ (at $1.25\,mg/mL$ concentration)	
28	*Sechium edule*	Cucurbitaceae	Leafy shoots	$51.49 \pm 2.13\%$ (at $1.25\,mg/mL$ concentration)	

29	*Scabiosa arenaria Forssk.*	Dipsacaceae	Flowers and fruits	0.11–0.221 mg/mL	41
30	*Quercus glauca* Thunb	Fagaceae		44.7 μg/mL	
31	*Quercus gilva* Blume	Fagaceae		110.0 μg/mL	
32	*Mallotus japonicus*	Euphorbiaceae		8.4 μg/mL	
33	*Xylosoma congestum* Merr.	Salicaceae	Leaves	182.3 μg/mL	42
34	*Quercus phillyraeoides*	Fagaceae		9.8 μg/mL	
35	*Quercus dentate* Thunb	Fagaceae		42.2 μg/mL	
36	*Podocarpus macrophyllus* var maki	Podocarpaceae		45.2 μg/mL	
37	*Glycine max*	Fabaceae	Soyabeans	347.2 ± 8.2 μg/mL	43
38	*Echeveria subrigida*			25.21 ± 3.66 μg/mL	
39	*Echeveria craigiana*	Crassulaceae	Leaves	56.67 ± 4 μg/mL	44
40	*Echeveria kimnachii*			50.57 ± 1.48 μg/mL	
41	*Artocarpus heterophyllus*	Moraceae	Bark and fruit	76.90 g/mL	
42	*Artocarpus altilis*	Moraceae	Bark and fruit	129.85 g/mL	45
43	*Piper betel*	Piperaceae	Bark and fruit	96.56 μg/mL	
44	*Cinnamomum zeylanicum*	Lauraceae	Bark and fruit	140.01 μg/mL	
45	*Luffa acutangula*	Cucurbitaceae	Fruit	135 μg/mL	46
46	*Symplocos racemosa*	Symplocaceae.	Bark	8.16 ± 0.28 μg/mL	
47	*Plumbago zeylanica*	Plumbaginaceae.	Root	3.46 ± 0.53 μg/mL	47
48	*Terminalia arjuna*	Combretaceae.	Bark	0.69 ± 0.08 μg/mL	

Table 5 Some α-glucosidase inhibitor compounds obtained from natural source.

Sl no.	Name of compound	IC_{50}	References
1.	Arylated gymnemic acids	145–310 μg/mL	60
2.	Taraxiroside F	165.1 ± 0.6 μM	61
3.	Luteolin	39.8 ± 4.2 μM	
4.	1,2,5-tri-O-*p*-hydroxyphenylacetyl-L-chiro-inositol	61.2 ± 3.5 μM	
5.	Chrysoeriol	155.9 ± 3.2 μM	
6.	Rutin	26.31 ± 0.02 μg/mL	62
7.	Quercetin-3-O-(3″-O-galloyl)-β-galactopyranoside	0.97 ± 0.02 μM	63
8.	Quercetin3-O-(6″-O-galloyl)-β-glucopyranoside	1.35 ± 0.06 μM	
9.	Hesperitin	0.38 ± 0.05 mM	64
10.	Quercetin	66.8 μg/mL	65
11.		4.02 ± 1.09 μg/mL	66
12.	Kaempferol	109 μg/mL	65
13.	Luteolin	3.75 μg/mL	67
14.	Luteolin + lutein	6.47 μg/mL	67
15.	Methyl-3,4-di-O-caffeoylquinate	52.1 μM	68
16.	Methyl-4,5-di-O-caffeoylquinate	10.7 μM	68
17.	Methyl-3,5-di-O-caffeoylquinate	70.8 μM	68
18.	Rebaniolignan A	31.6 μM	69
19.	Rebaniolignan B	55.0 μM	
20.	Tannic acid	0.44 μg/mL	70
21.	Umbelliferone	107 μg/mL	
22.	Edgeworin	18.7 μg/mL	
23.	Daphnoretin	86.0 μg/mL	
24.	Edgeworthin	49.6 μg/mL	
25.	Edgeworoside A	780.0 μg/mL	
26.	Edgeworoside C	760.0 μg/mL	

Table 5 Some α-glucosidase inhibitor compounds obtained from natural source.—cont'd

Sl no.	Name of compound	IC_{50}	References
27.	1,5-Dicaffeoylquinic acid	0.66 mM	46
28.	Quercitrin (quercetin-3-O-rhamnoside)	0.4 mM	71
29.	Hyperin (quercetin-3-O-rhamnoside)	0.5 mM	71
30.	Chlorogenic acid	2.18–2.99 mM	55
31.	Resveratrol	0.091 mg/mL	
32.	Piceatannol	0.060 mg/mL	
33.	*trans*-Pinostilbene	0.028 mg/mL	72
34.	*trans*-Desoxy-rhapontigenin	0.042 mg/mL	
35.	3′-Hydroxy-pterostilbene	0.230 mg/mL	

α-gluosidase enzymes showing values of 3.16 ± 0.05 μg/mL.[52] The α-glucosidase inhibition potential of alkaloids mahanine was observed, obtained from the fruit pulp of *Murraya koenigii* with IC_{50} = ±0.4 mol/L, respectively, compared with acarbose (IC_{50} = 15.2 ± 0.6 mol/L).[53] Diasesartemin (IC_{50} = 60.6 ± 0.01 mol/L) were isolated from guggul, the oleogum resin of *Commiphora wightii* among which diasesartemin was found to be more potent than acarbose (IC_{50} = 92.94 ± 0.01 mol/L).

The Cucurbitaceae family of medicinal plants contains different gourds and pumpkins. These plants are generally consumed as vegetables. The fresh fruits, leaves, tendrils, seeds, fruit and seed coats of these plants are rich in phenolics, such as a. caffeic acid, b. chlorogenic acid, c. luteolin, d. apigenin, e. EGCG (epigallocatechin-3-gallate), f. β-sitosterol, g. gallic acid, h. cucurbitacins, i. quercetin, j. quercitrin, k. isoquercetin, l. ferulic acid, etc. Many of these compounds, such as chlorogenic acid, caffeic acid, luteolin, possess potent α-glucosidase inhibitory potential. Many of the Cucurbitaceae have additionally been shown to possess potent α-glucosidase inhibitory property.[54] The most well-known antidiabetic plants that have been shown to possess α-glucosidase inhibitory potential are—*Momordica charantia* (Karela, Bitter Melon), *Coccinia indica* (Tela-kucha). Antioxidant and α-glucosidase activities and total phenolic contents (TPC) in successive extracts of dried pulps from seven cucurbit fruit vegetables were determined

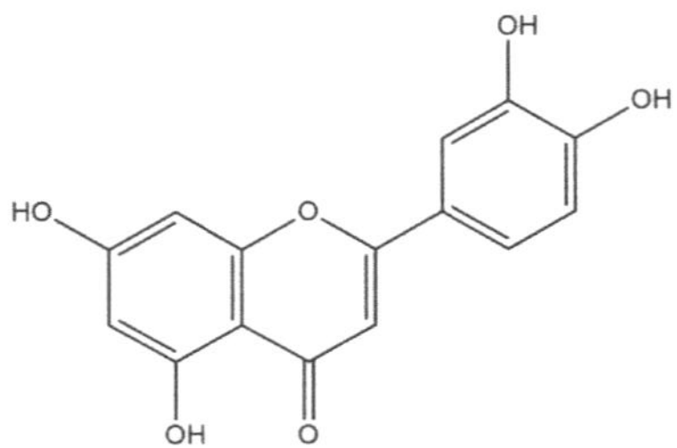	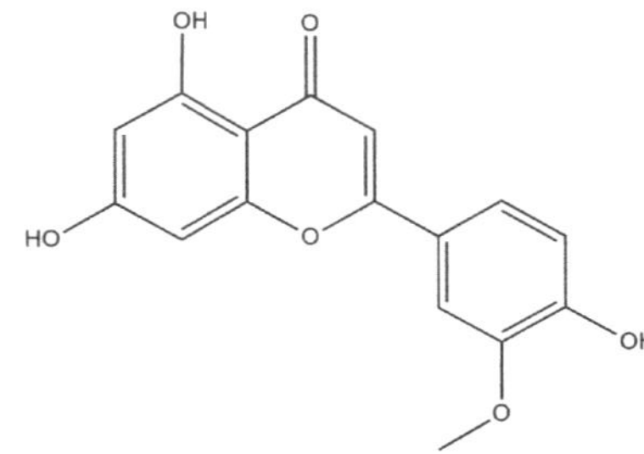
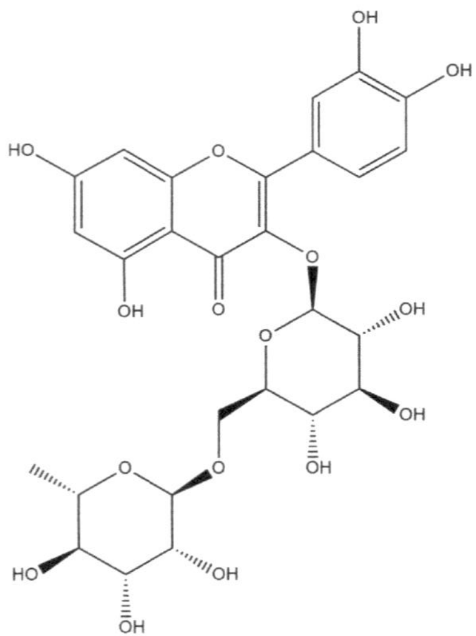	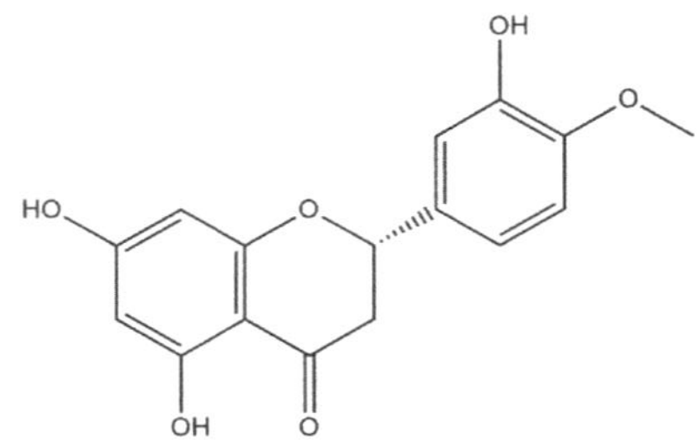

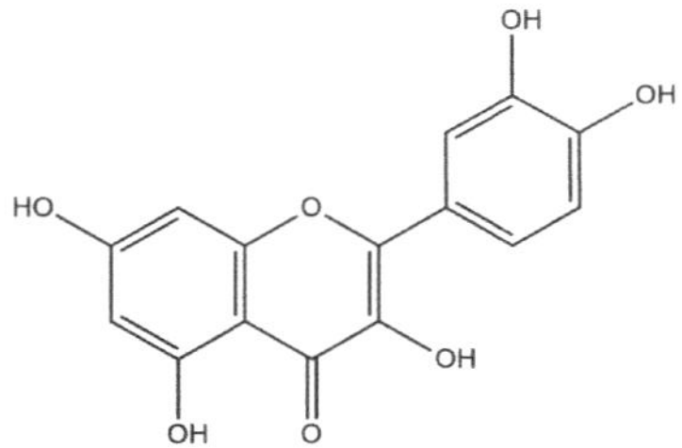

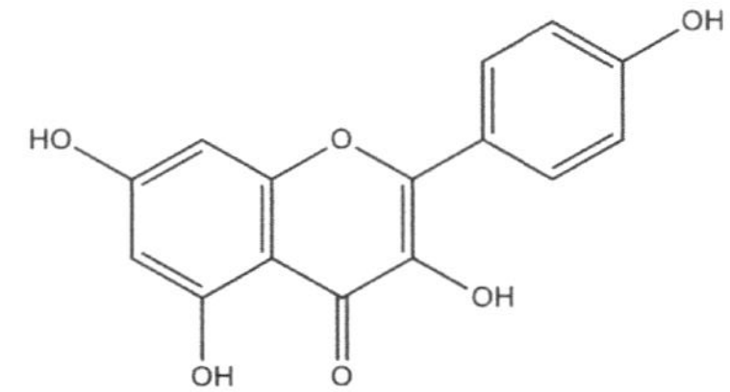

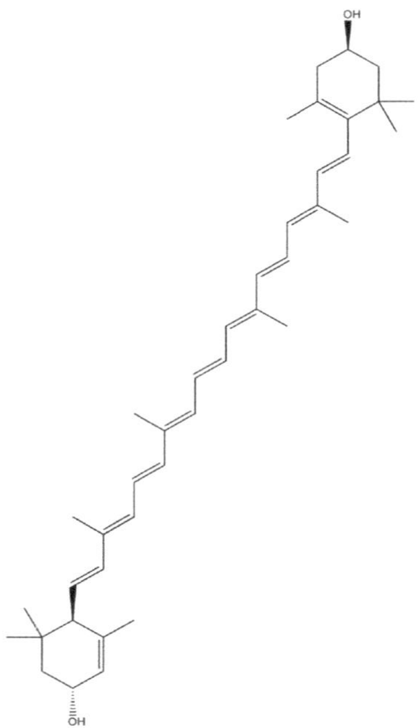

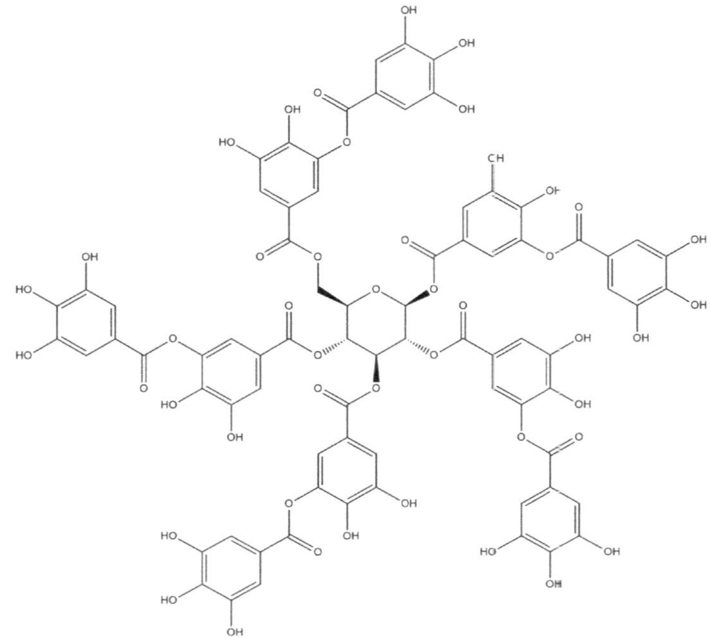

Continued

—cont'd

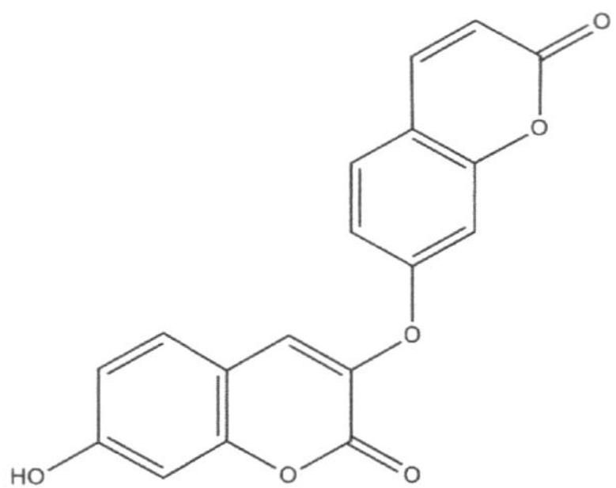	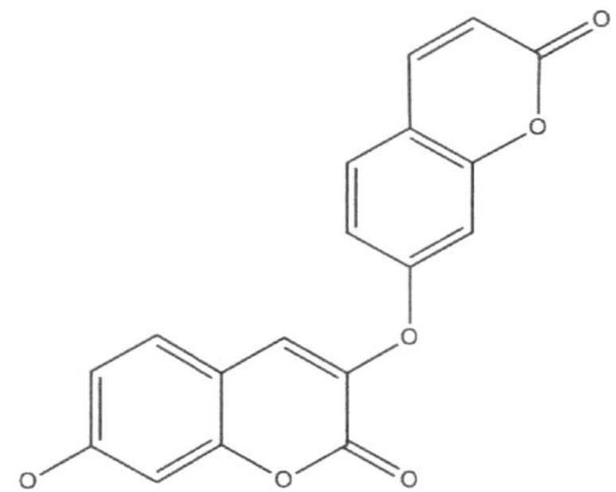
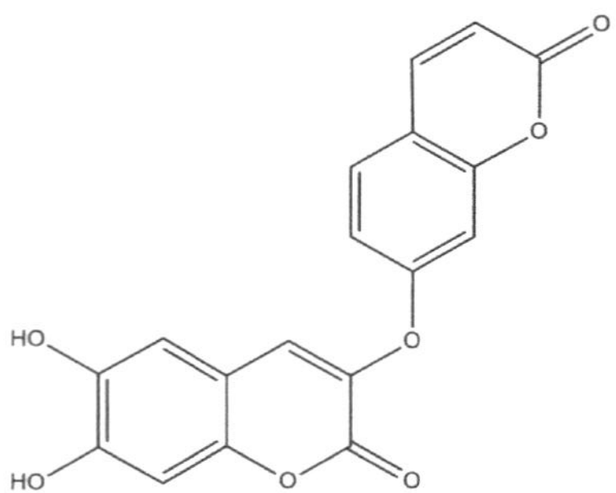	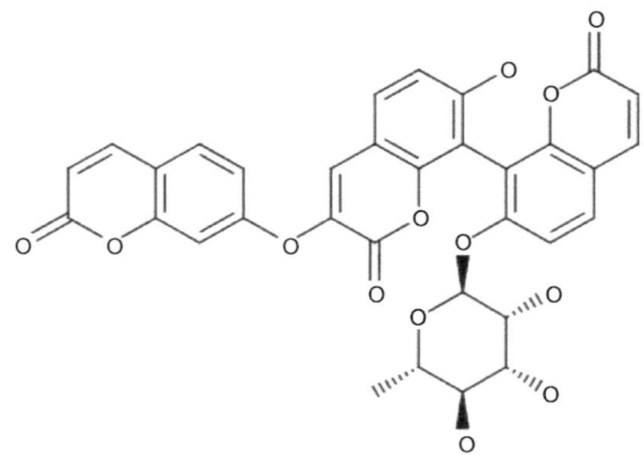

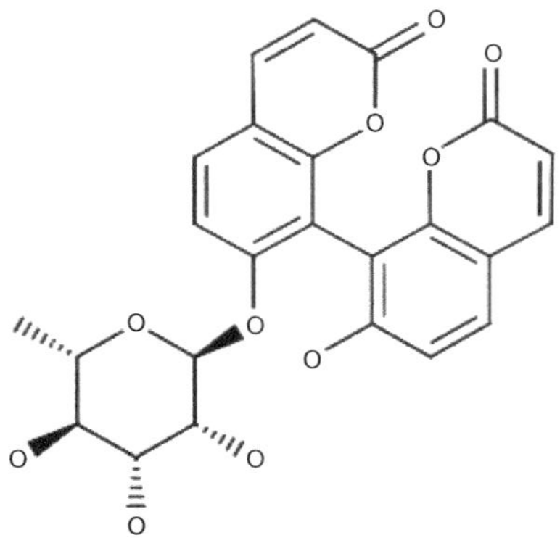

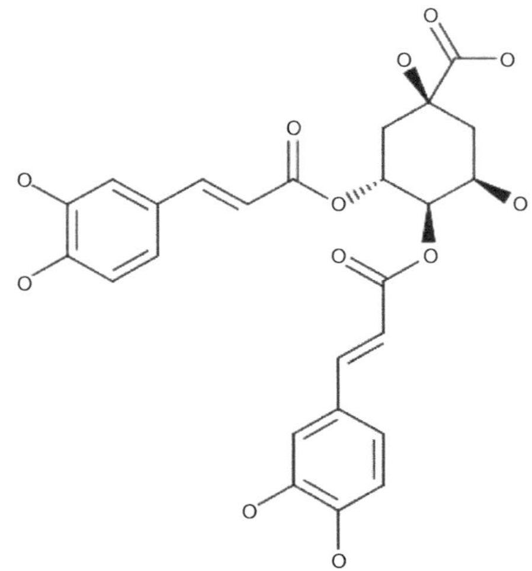

Continued

—cont'd

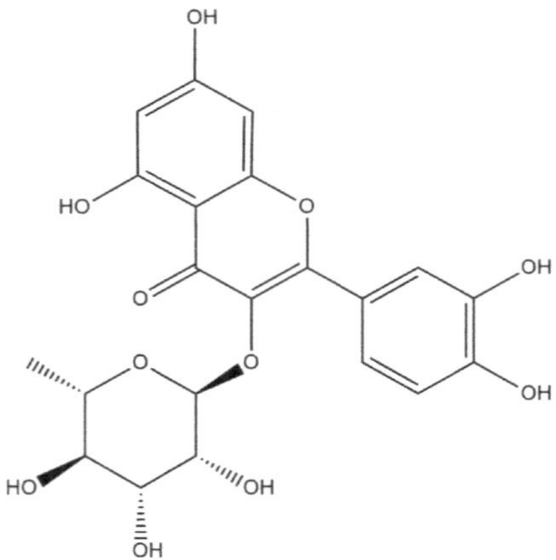

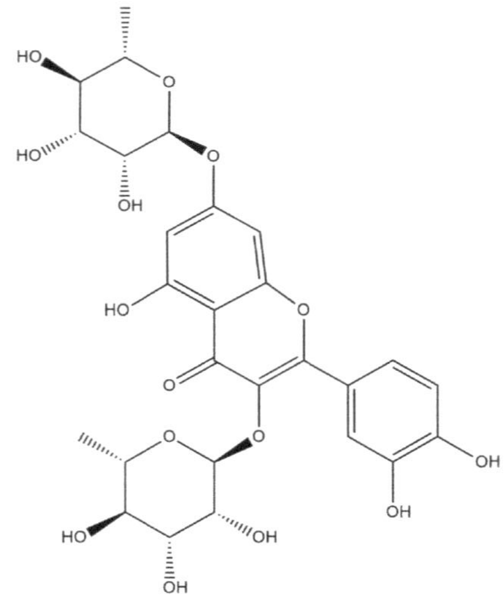

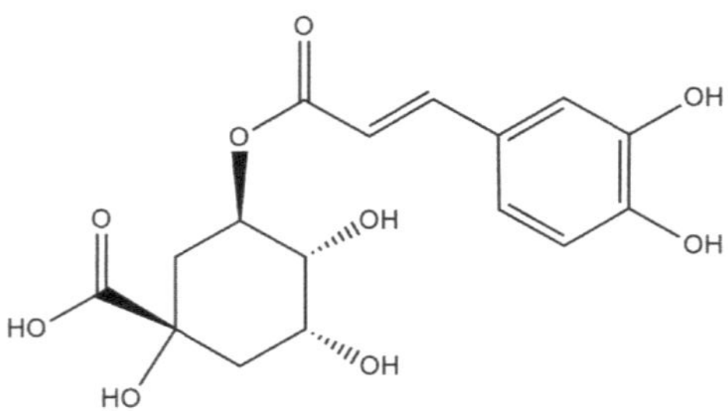

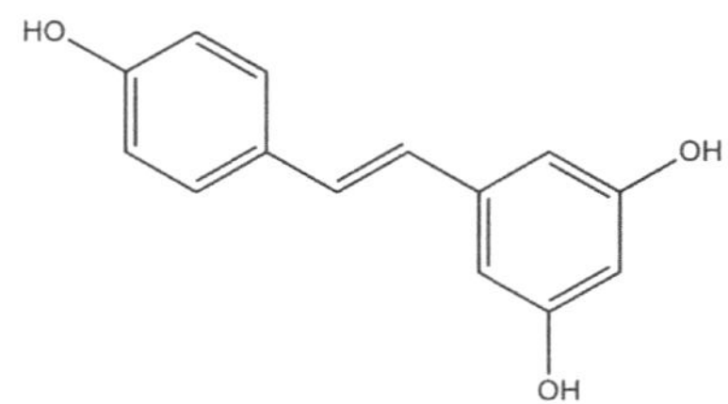

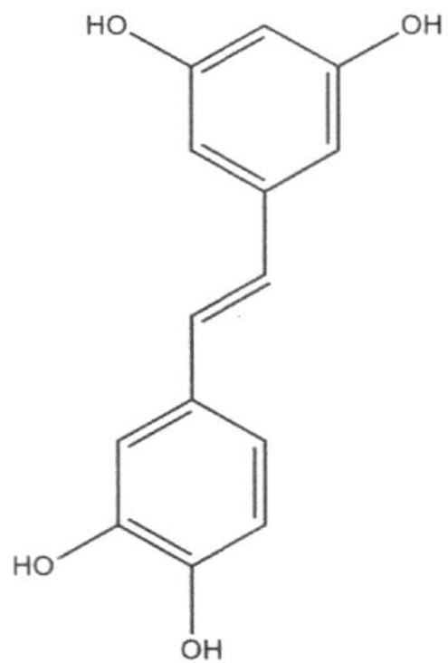	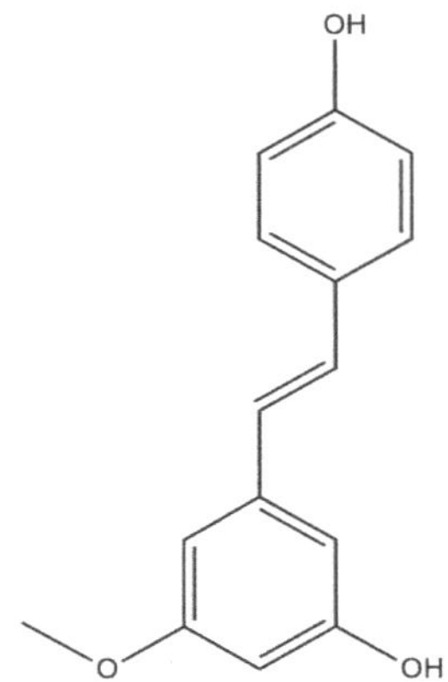
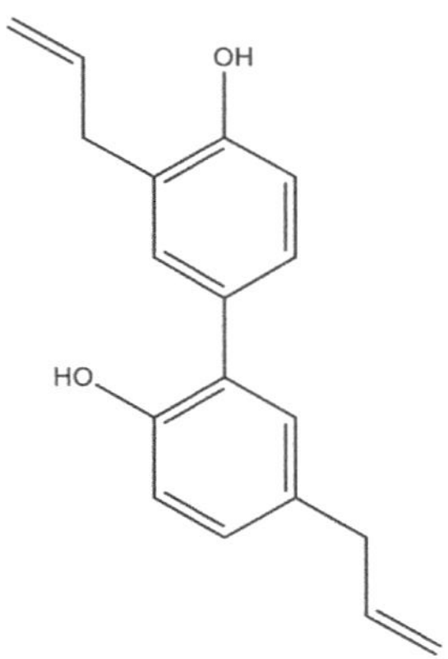	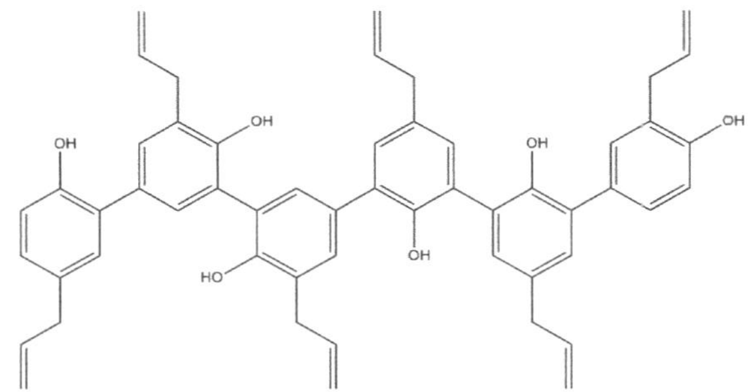

Continued

—cont'd

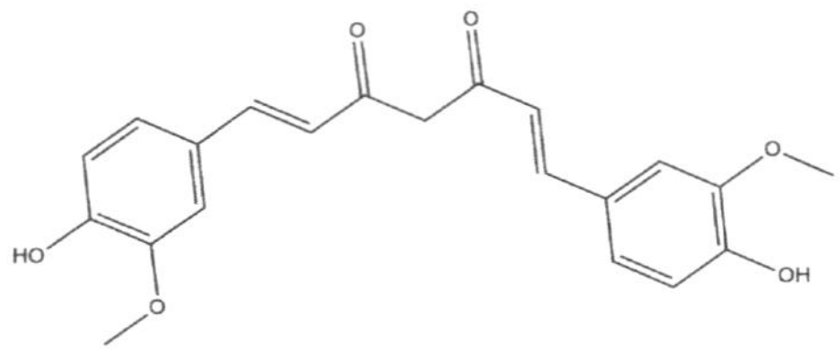	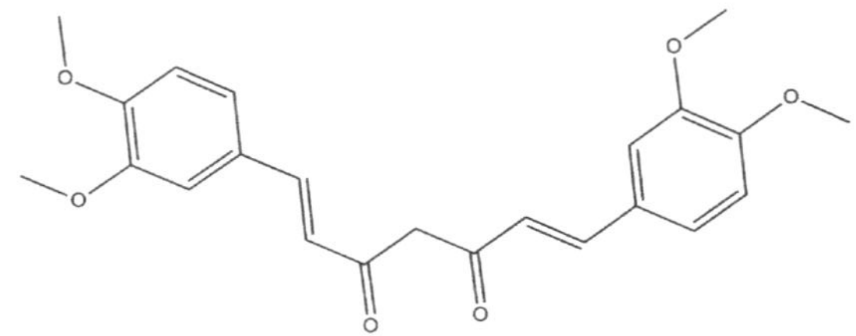
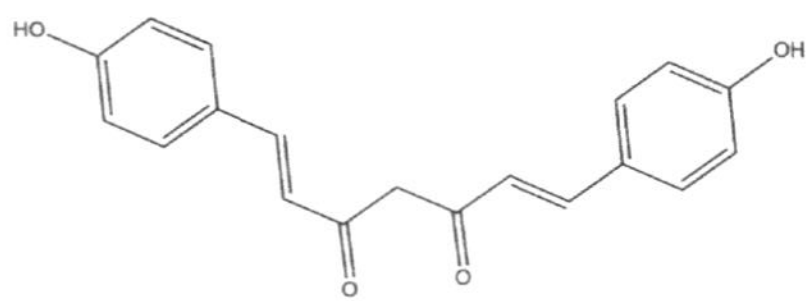	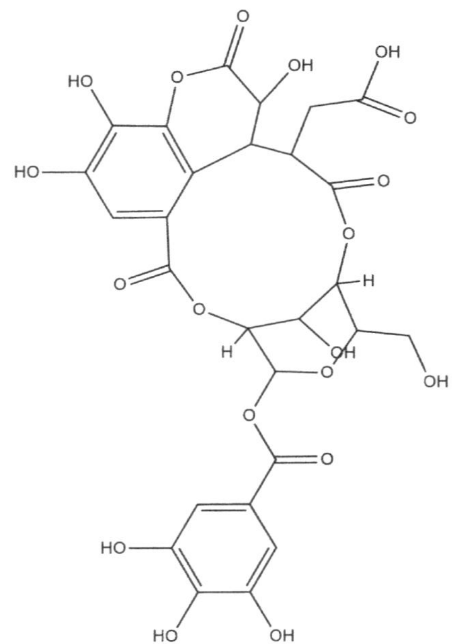

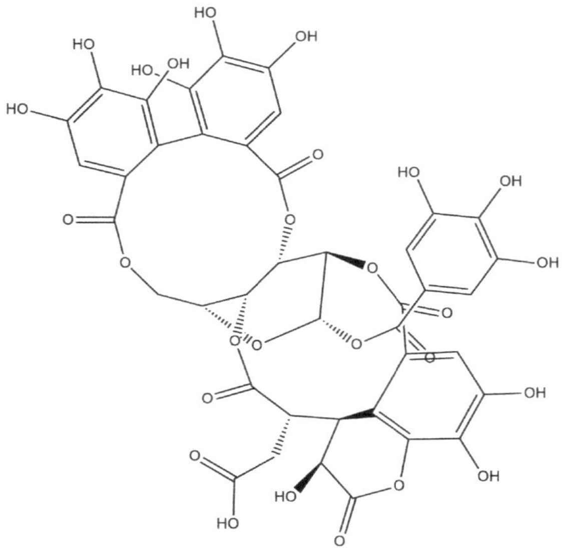

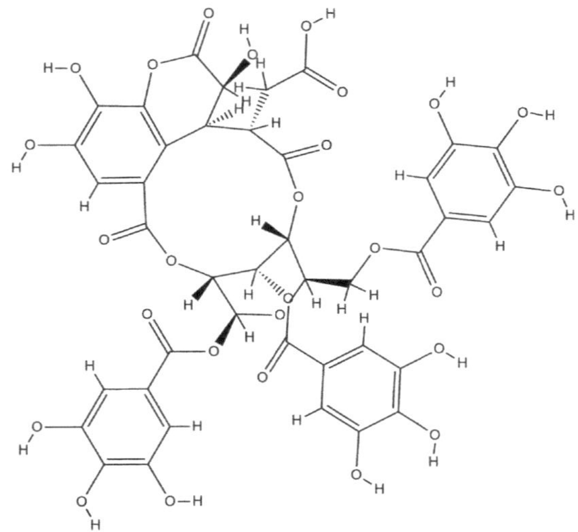

Continued

—cont'd

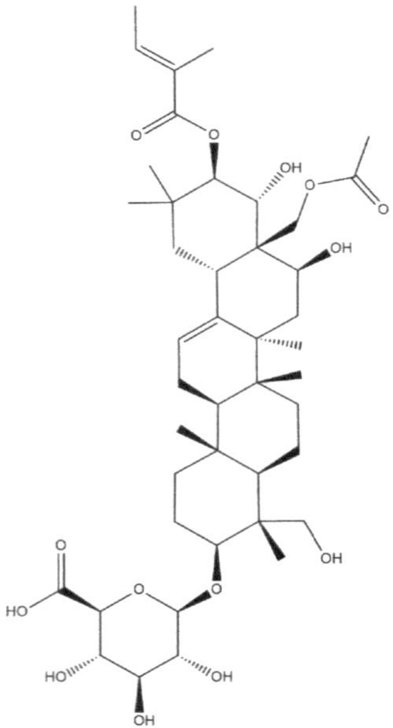

OH
O
O
OH
O
OH
OH
HO
O
O
HO
HO
OH
OH

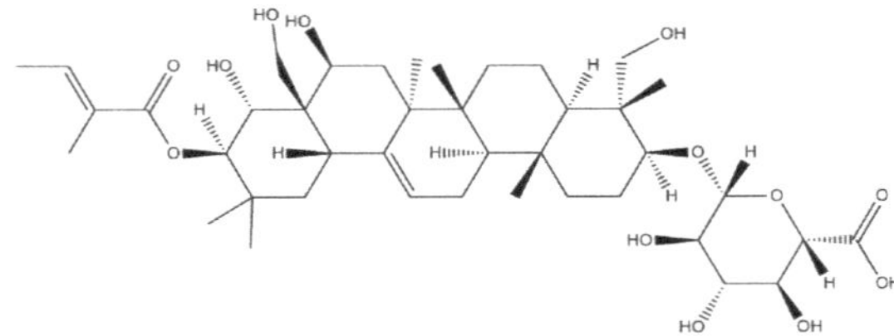

HO
HO
OH
HO
O
H
O
H
H
H
Hııı
O
H
H
O
O
HO
H
OH
HO
OH

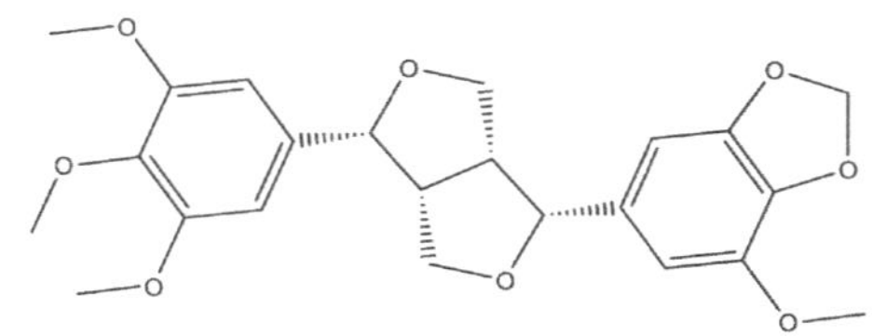

O
O
O
O
O
O
O
O

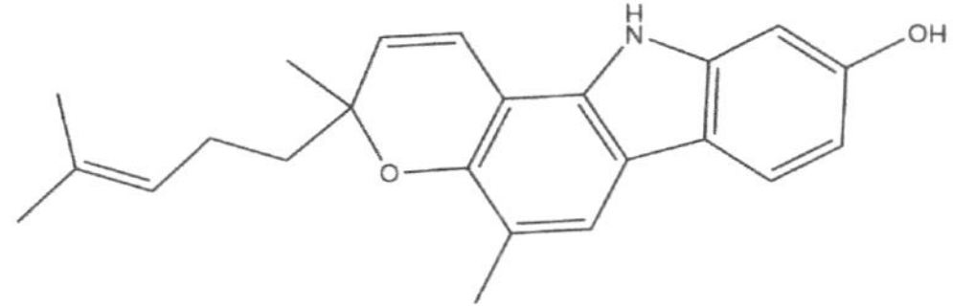

H
O
OH

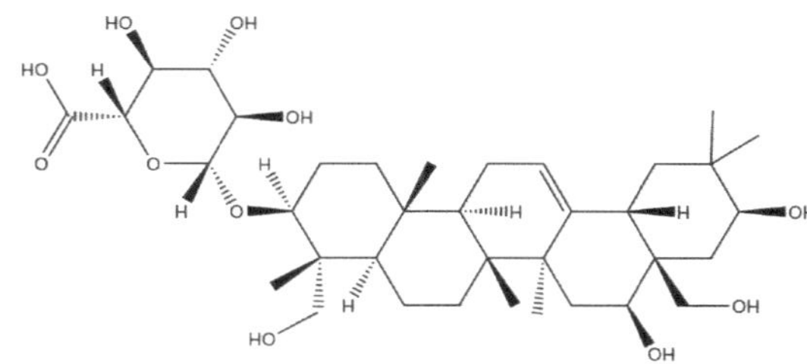

HO
OH
HO
H
OH
O
HO
O
OH
H
O
H
H
H
H
OH
HO
OH
OH

for the first time. The best TPC and metal chelating activity were obtained from the chloroform extracts of *Luffa acutangula* (28.04 ± 0.37 mg GAE/g extract) and *Benincasa hispida* (EC_{50} = 0.44 ± 0.03 mg/mL), respectively.[55] The ethyl acetate extract of *Sechium edule* showed the highest 1,1-diphenyl-2-picrylhydrazyl (DPPH) radical scavenging activity (951.73 ± 29.14 mm TE/g extract). The anti-α-glucosidase activities were observed in the methanol and ethyl acetate extracts of *Momordica charantia* (692.56 ± 43.38 mmAscAE/g extract; 66.64 ± 2.94%, respectively). In *Lagenaria siceraria,* Isoquercetin was found to be the main contributor to the activities. Gallic acid was identified as both the main and most active antioxidant constituent within the ethyl acetate extract of *S. edule.*[56]

6. Conclusions

Scientific validation and QC of TMs are critical and essential aspects to ensure therapeutic efficacy, safety and rationalization of their use in healthcare.[57–58] The markers used for the standardization, chemical and DNA fingerprinting, bioassays, metabolomics approach and the emerging field of phytomics provide mechanisms for assuring consistent quality and efficacy of herbal medicine.[59] In spite of the benefits of newer therapy, AGIs do not feature among the first-line or second-line drugs. Marketed α-glucosidase inhibitors such as Acarbose and Miglitol are reported to produce diarrhea, bloating and loose stool which are attributed to the undigested glucose in the colon. In randomized control trials, these undesirable side effects were proved to be the main causes behind the early discontinuation of this class of medicines. So, research on finding novel α-glucosidase inhibitors from plant sources is needed. The natural, plant based agents have been proven to be more safe and efficacious in comparison to the synthetic agents, with lesser side effects. Again, as the plant derived medicinal agents are rich in flavonoids and different phenolics, they are good antioxidant agents. For this reason, they also serve as free-radical scavenging agents in vivo. The main lacuna in the present treatments of diabetes mellitus has been the failure in managing exacerbations of the pathological changes. Traditional therapy mentioned in Ayurveda can provide ameliorative effects in diabetic patient in management and prevention in high risk populations. However in detailed research is required to further validate the multiple mechanisms of action and efficacy in clinical subjects.

Acknowledgments

Authors wish to acknowledge the financial support received from the Science and Engineering Research Board (SERB), Govt. of India, New Delhi, Vide File Ref. no. EMR/2016/007037. Thanks are also to Emami Ltd., Kolkata for their kind support.

References

1. Mukherjee, P. K.; Maiti, K.; Mukherjee, K.; Houghton, P. J. Leads From Indian Medicinal Plants With Hypoglycemic Potentials, *J. Ethnopharmacol.* **2006**, *106*(1), 1–28. https://doi.org/10.1016/j.jep.2006.03.021.
2. Banerjee, S.; Debnath, P.; Rao, P. N.; Tripathy, T. B.; Adhikari, A.; Debnath, P. K. Ayurveda in Changing Scenario of Diabetes Management for Developing Safe and Effective Treatment Choices for the Future, *J. Complement. Integr. Med.* **2015**, *12*(2), 101–110. https://doi.org/10.1515/jcim-2014-0012.
3. Banerjee, S.; Biswas, S.; Mukherjee, P. K.; Biswas, S.; Mukherjee, P. K. *Ayurveda: Evidence-Based Approach for Drug Development*, https://www.taylorfrancis.com/ (Accessed Feb 10, 2020) 2020. https://doi.org/10.1201/9781315187853-31.
4. Sim, L.; Willemsma, C.; Mohan, S.; Naim, H. Y.; Pinto, B. M.; Rose, D. R. Structural Basis for Substrate Selectivity in Human Maltase-Glucoamylase and Sucrase-Isomaltase N-terminal Domains, *J. Biol. Chem.* **2010**, *285*(23), 17763–17770. https://doi.org/10.1074/jbc.M109.078980.
5. Banerjee, S.; Debnath, P.; Debnath, P. K. Ayurnutrigenomics: Ayurveda-Inspired Personalized Nutrition From Inception to Evidence, *J. Tradit. Complement. Med.* **2015**, *5*(4), 228–233. https://doi.org/10.1016/j.jtcme.2014.12.009.
6. Chahal, S.; Vohra, K.; Syngle, A. Association of Sudomotor Function With Peripheral Artery Disease in Type 2 Diabetes, *Neurol. Sci.* **2017**, *38*(1), 151–156. https://doi.org/10.1007/s10072-016-2742-3.
7. Flint, H. J.; Bayer, E. A.; Rincon, M. T.; Lamed, R.; White, B. A. Polysaccharide Utilization by Gut Bacteria: Potential for New Insights From Genomic Analysis, *Nat. Rev. Microbiol.* **2008**, *6*(2), 121–131. https://doi.org/10.1038/nrmicro1817.
8. Henrissat, B. A Classification of Glycosyl Hydrolases Based on Amino Acid Sequence Similarities, *Biochem. J.* **1991**, *280*(Pt. 2), 309–316. https://doi.org/10.1042/bj2800309.
9. Bischoff, H. Pharmacology of alpha-glucosidase inhibition, *Eur. J. Clin. Invest.* **1994**, *24*(Suppl. 3), 3–10.
10. Lebovitz, H. E. Alpha-Glucosidase Inhibitors, *Endocrinol. Metab. Clin. North Am.* **1997**, *26*(3), 539–551. https://doi.org/10.1016/s0889-8529(05)70266-8.
11. Bayraktar, M.; Van Thiel, D. H.; Adalar, N. A Comparison of Acarbose Versus Metformin as an Adjuvant Therapy in Sulfonylurea-Treated NIDDM Patients, *Diabetes Care* **1996**, *19*(3), 252–254. https://doi.org/10.2337/diacare.19.3.252.
12. Pan, C.; Yang, W.; Barona, J. P.; Wang, Y.; Niggli, M.; Mohideen, P.; Wang, Y.; Foley, J. E. Comparison of Vildagliptin and Acarbose Monotherapy in Patients With Type 2 Diabetes: A 24-Week, Double-blind, Randomized Trial, *Diabet. Med.* **2008**, *25*(4), 435–441. https://doi.org/10.1111/j.1464-5491.2008.02391.x.
13. Ceriello, A.; Taboga, C.; Tonutti, L.; Giacomello, R.; Stel, L.; Motz, E.; Pirisi, M. Post-meal Coagulation Activation in Diabetes Mellitus: The Effect of Acarbose, *Diabetologia* **1996**, *39*(4), 469–473. https://doi.org/10.1007/BF00400679.
14. MacGregor, E. A.; Janecek, S.; Svensson, B. Relationship of Sequence and Structure to Specificity in the Alpha-Amylase Family of Enzymes, *Biochim. Biophys. Acta* **2001**, *1546*(1), 1–20. https://doi.org/10.1016/s0167-4838(00)00302-2.

15. Sajilata, M. G.; Singhal, R. S.; Kulkarni, P. R. Resistant Starch—A Review, *Compr. Rev. Food Sci. Food Saf.* **2006**, *5*(1), 1–17. https://doi.org/10.1111/j.1541-4337.2006.tb00076.x.
16. Ernst, H. A.; Lo Leggio, L.; Willemoës, M.; Leonard, G.; Blum, P.; Larsen, S. Structure of the Sulfolobus Solfataricus Alpha-Glucosidase: Implications for Domain Conservation and Substrate Recognition in GH31, *J. Mol. Biol.* **2006**, *358*(4), 1106–1124. https://doi.org/10.1016/j.jmb.2006.02.056.
17. Naumov, D. G. Structure and Evolution of Mammalian Maltase-Glucoamylase and Sucrase-Isomaltase Genes, *Mol. Biol. (Mosk)* **2007**, *41*(6), 1056–1068.
18. Nichols, B. L.; Avery, S.; Sen, P.; Swallow, D. M.; Hahn, D.; Sterchi, E. The Maltase-Glucoamylase Gene: Common Ancestry to Sucrase-Isomaltase With Complementary Starch Digestion Activities, *Proc. Natl. Acad. Sci. U. S. A.* **2003**, *100*(3), 1432–1437. https://doi.org/10.1073/pnas.0237170100.
19. Rossi, E. J.; Sim, L.; Kuntz, D. A.; Hahn, D.; Johnston, B. D.; Ghavami, A.; Szczepina, M. G.; Kumar, N. S.; Sterchi, E. E.; Nichols, B. L.; et al. Inhibition of Recombinant Human Maltase Glucoamylase by Salacinol and Derivatives, *FEBS J.* **2006**, *273*(12), 2673–2683. https://doi.org/10.1111/j.1742-4658.2006.05283.x.
20. Okuyama, M.; Okuno, A.; Shimizu, N.; Mori, H.; Kimura, A.; Chiba, S. Carboxyl Group of Residue Asp647 as Possible Proton Donor in Catalytic Reaction of α-Glucosidase From Schizosaccharomyces pombe, *Eur. J. Biochem.* **2001**, *268*(8), 2270–2280. https://doi.org/10.1046/j.1432-1327.2001.02104.x.
21. Lovering, A. L.; Lee, S. S.; Kim, Y.-W.; Withers, S. G.; Strynadka, N. C. J. Mechanistic and Structural Analysis of a Family 31 Alpha-Glycosidase and Its Glycosyl-Enzyme Intermediate, *J. Biol. Chem.* **2005**, *280*(3), 2105–2115. https://doi.org/10.1074/jbc.M410468200.
22. van de Laar, F. A. Alpha-Glucosidase Inhibitors in the Early Treatment of Type 2 Diabetes, *Vasc. Health Risk Manag.* **2008**, *4*(6), 1189–1195.
23. Kalra, S.; Chadha, M.; Sharma, S. K.; Unnikrishnan, A. G. Untapped Diamonds for Untamed Diabetes: The α-Glucosidase Inhibitors, *Indian J. Endocrinol. Metab.* **2014**, *18*(2), 138–141. https://doi.org/10.4103/2230-8210.129102.
24. Cai, S.; Tan, S.; Gluckman, P. D.; Godfrey, K. M.; Saw, S.-M.; Teoh, O. H.; Chong, Y.-S.; Meaney, M. J.; Kramer, M. S.; Gooley, J. J. Sleep Quality and Nocturnal Sleep Duration in Pregnancy and Risk of Gestational Diabetes Mellitus, *Sleep* **2017**, *40*(2). https://doi.org/10.1093/sleep/zsw058.
25. Chaudhury, A.; Duvoor, C.; Reddy Dendi, V. S.; Kraleti, S.; Chada, A.; Ravilla, R.; Marco, A.; Shekhawat, N. S.; Montales, M. T.; Kuriakose, K. Clinical Review of Antidiabetic Drugs: Implications for Type 2 Diabetes Mellitus Management, *Front. Endocrinol.* **2017**, *8*, 6. https://doi.org/10.3389/fendo.2017.00006.
26. Barrett, J. C.; Clayton, D. G.; Concannon, P.; Akolkar, B.; Cooper, J. D.; Erlich, H. A.; Julier, C.; Morahan, G.; Nerup, J.; Nierras, C.; et al. Genome-Wide Association Study and Meta-Analysis Find That Over 40 Loci Affect Risk of Type 1 Diabetes, *Nat. Genet.* **2009**, *41*(6), 703–707. https://doi.org/10.1038/ng.381.
27. Shanmugam, S.; Gomes, I. A.; Denadai, M.; dos Santos Lima, B.; de Souza Araújo, A. A.; Narain, N.; Neta, M. T. S. L.; Serafini, M. R.; Quintans-Júnior, L. J.; Thangaraj, P. UHPLC-QqQ-MS/MS Identification, Quantification of Polyphenols From Passiflora subpeltata Fruit Pulp and Determination of Nutritional, Antioxidant, α-Amylase and α-Glucosidase Key Enzymes Inhibition Properties, *Food Res. Int.* **2018**, *108*, 611–620.
28. de Cássia Lemos Lima, R.; Kato, L.; Kongstad, K. T.; Staerk, D. Brazilian Insulin Plant as a Bifunctional Food: Dual High-Resolution PTP1B and α-Glucosidase Inhibition Profiling Combined With HPLC-HRMS-SPE-NMR for Identification of Antidiabetic Compounds in Myrcia rubella Cambess, *J. Funct. Foods* **2018**, *45*, 444–451.

29. Ingawale, A. S.; Sadiq, M. B.; Nguyen, L. T.; Ngan, T. B. Optimization of Extraction Conditions and Assessment of Antioxidant, α-Glucosidase Inhibitory and Antimicrobial Activities of *Xanthium strumarium* L. fruits, Biocatal, *Agric. Biotechnol.* **2018**, *14*, 40–47.
30. Tong, D.-P.; Zhu, K.-X.; Guo, X.-N.; Peng, W.; Zhou, H.-M. The Enhanced Inhibition of Water Extract of Black Tea Under Baking Treatment on α-Amylase and α-Glucosidase, *Int. J. Biol. Macromol.* **2018**, *107*, 129–136. https://doi.org/10.1016/j.ijbiomac.2017.08.152.
31. Beidokhti, M. N.; Andersen, M. V.; Eid, H. M.; Villavicencio, M. L. S.; Stærk, D.; Haddad, P. S.; Jäger, A. K. Investigation of Antidiabetic Potential of *Phyllanthus niruri* L. Using Assays for α-Glucosidase, Muscle Glucose Transport, Liver Glucose Production, and Adipogenesis, *Biochem. Biophys. Res. Commun.* **2017**, *493*(1), 869–874 https://doi:10.1016/j.bbrc.2017.09.080.
32. Dada, F. A.; Oyeleye, S. I.; Ogunsuyi, O. B.; Olasehinde, T. A.; Adefegha, S. A.; Oboh, G.; Boligon, A. A. Phenolic Constituents and Modulatory Effects of Raffia palm Leaf (*Raphia hookeri*) Extract on Carbohydrate Hydrolyzing Enzymes Linked to Type-2 Diabetes, *J. Tradit. Complement. Med.* **2017**, 7(4), 494. https://doi.org/10.1016/j.jtcme.2017.01.003.
33. Mollica, A.; Zengin, G.; Locatelli, M.; Stefanucci, A.; Mocan, A.; Macedonio, G.; Carradori, S.; Onaolapo, O.; Onaolapo, A.; Adegoke, J.; Olaniyan, M.; Aktumsek, A.; Novellino, E. Anti-diabetic and Anti-hyperlipidemic Properties of Capparis spinosa L. In Vivo and In Vitro Evaluation of Its Nutraceutical Potential, *J. Funct. Foods* **2017**, *35*, 32–42.
34. Tan, Y.; Chang, S. K. C.; Zhang, Y. Comparison of α-Amylase, α-Glucosidase and Lipase Inhibitory Activity of the Phenolic Substances in Two Black Legumes of Different Genera, *Food Chem.* **2017**, *214*, 259–268.
35. Semaan, D. G.; Igoli, J. O.; Young, L.; Marrero, E.; Gray, A. I.; Rowan, E. G. In Vitro Anti-Diabetic Activity of Flavonoids and Pheophytins From Allophylus cominia Sw. on PTP1B, DPPIV, Alpha-Glucosidase and Alpha-Amylase Enzymes, *J. Ethnopharmacol.* **2017**, *203*, 39–46. https://doi.org/10.1016/j.jep.2017.03.023.
36. Eleazu, C. O.; Eleazu, K. C.; Iroaganachi, M. *In Vitro* Starch Digestibility, α-Amylase and α-Glucosidase Inhibitory Capacities of Raw and Processed Forms of Three Varieties of Livingstone Potato (*Plectranthus esculentus*), *Innov. Food Sci. Emerg. Technol.* **2016**, *37*, 37–43.
37. Hyun, T. K.; Kim, H.-C.; Ko, Y.-J.; Kim, J.-S. Antioxidant, α-Glucosidase Inhibitory and Anti-inflammatory Effects of Aerial Parts Extract From Korean Crowberry (Empetrum nigrum var. japonicum), *Saudi J. Biol. Sci.* **2016**, *23*(2), 181–188.
38. Olabiyi, A. A.; Smith, Y. R. A.; Babatola, L. J.; Akinyemi, A. J.; Oboh, G. Inhibitory Effect of Aqueous Extract of Different Parts of *Gossypium herbaceum* on Key Enzymes Linked With Type 2 Diabetes and Oxidative Stress in Rat Pancreas In Vitro, *Beni-Suef Univ. J. Basic Appl. Sci.* **2016**, *5*, 180–186.
39. Tundis, R.; Bonesi, M.; Sicari, V.; Pellicanò, T. M.; Tenuta, M. C.; Leporini, M.; Menichini, F.; Loizzo, M. R. *Poncirus trifoliata (L.)* Raf.: Chemical Composition, Antioxidant Properties and Hypoglycaemic Activity *via* the Inhibition of α-Amylase and α-Glucosidase Enzymes, *J. Funct. Foods* **2016**, *25*, 477–485. https://doi.org/10.1016/j.jff.2016.06.034.
40. Nag, G.; Das, S.; Das, S.; Mandal, S.; De, B. Antioxidant, Antiacetylcholinesterase and Antiglycosidase Properties of Three Species of Swertia, Their Xanthones and Amarogentin: A Comparative Study, *Phcog. J.* **2015**, 7, 117–123.
41. Hlila, M. B.; Mosbah, H.; Majouli, K.; Mssada, K.; Jannet, H. B.; Aouni, M.; Selmi, B. α-Glucosidase Inhibition by Tunisian Scabiosa arenaria Forssk. Extracts, *Int. J. Biol. Macromol.* **2015**, 77, 383–389.

42. Indrianingsih, A. W.; Tachibana, S.; Itoh, K. In Vitro Evaluation of Antioxidant and α-Glucosidase Inhibitory Assay of Several Tropical and Subtropical Plants, *Procedia Environ. Sci.* **2015**, *28*, 639–648.
43. Ademiluyi, A. O.; Oboh, G.; Boligon, A. A.; Athayde, M. L. Effect of Fermented Soybean Condiment Supplemented Diet on α-Amylase and α-Glucosidase Activities in Streptozotocin-Induced Diabetic Rats, *J. Funct. Foods* **2014**, *9*(1), 1–9.
44. López-Angulo, G.; Montes-Avila, J.; Díaz-Camacho, S. P.; Vega-Aviña, R.; Ahumada-Santos, Y. P.; Delgado-Vargas, F. Chemical Composition and Antioxidant, α-Glucosidase Inhibitory and Antibacterial Activities of Three Echeveria DC. Species From Mexico, *Arab. J. Chem.* **2019**, *12*, 1964–1973. https://doi.org/10.1016/j.arabjc.2014.11.050.
45. Nair, S. S.; Kavrekar, V.; Mishra, A. In Vitro Studies on Alpha Amylase and Alpha Glucosidase Inhibitory Activities of Selected Plant Extracts, *Eur. J. Exp. Biol.* **2013**, *3*(1), 128–131.
46. Pimple, B. P.; Kadam, P. V.; Patil, M. J. Antidiabetic and Antihyperlipidemic Activity of Luffa acutangula Fruit Extracts in Streptozotocin Induced NIDDM rats, *Asian J. Pharm. Clin. Res.* **2011**, *4*(Suppl. 2), 156–163.
47. Bachhawat, J. A.; Shihabudeen, M. S.; Thirumurugan, K. Screening of Fifteen Indian Ayurvedic Plants for Alpha-Glucosidase Inhibitory Activity and Enzyme Kinetics, *Int. J. Pharm. Pharm. Sci.* **2011**, *3*(4), 267–274.
48. Yin, Z.; Zhang, W.; Feng, F.; Zhang, Y.; Kang, W. α-Glucosidase Inhibitors Isolated From Medicinal Plants, *Food Sci. Human Wellness* **2014**, *3*(3), 136–174. https://doi.org/10.1016/j.fshw.2014.11.003.
49. Gao, H.; Huang, Y.-N.; Gao, B.; Li, P.; Inagaki, C.; Kawabata, J. Inhibitory Effect on α-Glucosidase by Adhatoda Vasica Nees, *Food Chem.* **2008**, *108*(3), 965–972. https://doi.org/10.1016/j.foodchem.2007.12.002.
50. Du, Z.; Liu, R.; Shao, W.; Mao, X.; Ma, L.; Gu, L.; Huang, Z.; Chan, A. S. C. α-Glucosidase Inhibition of Natural Curcuminoids and Curcumin Analogs, *Eur. J. Med. Chem.* **2006**, *41*(2), 213–218. https://doi.org/10.1016/j.ejmech.2005.10.012.
51. Gao, H.; Huang, Y.-N.; Xu, P.-Y.; Kawabata, J. Inhibitory Effect on α-Glucosidase by the Fruits of Terminalia chebula retz, *Food Chem.* **2007**, *105*(2), 628–634. https://doi.org/10.1016/j.foodchem.2007.04.023.
52. Shenoy, R. S.; Prashanth, K. V. H.; Manonmani, H. K. *In Vitro Antidiabetic Effects of Isolated Triterpene Glycoside Fraction From* Gymnema sylvestre, https://www.hindawi.com/journals/ecam/2018/7154702/ (Accessed Mar 8, 2020) 2020. https://doi.org/10.1155/2018/7154702.
53. Uvarani, C.; Jaivel, N.; Sankaran, M.; Chandraprakash, K.; Ata, A.; Mohan, P. S. Axially Chiral Biscarbazoles and Biological Evaluation of the Constituents from Murraya Koenigii, *Fitoterapia* **2014**, *94*, 10–20. https://doi.org/10.1016/j.fitote.2014.01.004.
54. Nhiem, N. X.; Kiem, P. V.; Minh, C. V.; Ban, N. K.; Cuong, N. X.; Tung, N. H.; Ha, L. M.; Ha, D. T.; Tai, B. H.; Quang, T. H.; et al. α-Glucosidase Inhibition Properties of Cucurbitane-Type Triterpene Glycosides From the Fruits of *Momordica charantia, Chem. Pharm. Bull. (Tokyo)* **2010**, *58*(5), 720–724. https://doi.org/10.1248/cpb.58.720.
55. Ishikawa, A.; Yamashita, H.; Hiemori, M.; Inagaki, E.; Kimoto, M.; Okamoto, M.; Tsuji, H.; Memon, A. N.; Mohammadio, A.; Natori, Y. Characterization of Inhibitors of Postprandial Hyperglycemia From the Leaves of Nerium indicum, *J. Nutr. Sci. Vitaminol. (Tokyo)* **2007**, *53*(2), 166–173. https://doi.org/10.3177/jnsv.53.166.
56. Sulaiman, S. F.; Ooi, K. L.; Supriatno Antioxidant and α-Glucosidase Inhibitory Activities of Cucurbit Fruit Vegetables and Identification of Active and Major Constituents From Phenolic-Rich Extracts of Lagenaria siceraria and Sechium edule, *J. Agric. Food Chem.* **2013**, *61*(42), 10080–10090. https://doi.org/10.1021/jf4031037.

57. Mukherjee, P. K. Qualitative Analysis for Evaluation of Herbal Drugs, In: *Quality Control and Evaluation of Herbal Drugs*; Mukherjee, P. K. Ed.; Elsevier, 2019; pp 79–149. Chapter 4. https://doi.org/10.1016/B978-0-12-813374-3.00004-1.
58. Mukherjee, P. K.; Harwansh, R. K.; Bahadur, S.; Banerjee, S.; Kar, A.; Chanda, J.; Biswas, S.; Ahmmed, S. M.; Katiyar, C. K. Development of Ayurveda—Tradition to Trend, *J. Ethnopharmacol.* **2017**, *197*, 10–24. https://doi.org/10.1016/j.jep.2016.09.024.
59. Gori, L.; Firenzuoli, F. Chapter 7—Pharmacovigilance: Tools in Establishing the Safety and Acceptability of the Natural Health Products—Clinical Evaluation, In: *Evidence-Based Validation of Herbal Medicine*; Mukherjee, P. K. Ed.; Elsevier: Boston, 2015, pp 165–174. https://doi.org/10.1016/B978-0-12-800874-4.00007-6.
60. Alkefai, N. H.; Ahamad, J.; Amin, S.; Sharma, M.; Mir, S. R. Arylated gymnemic acids from *Gymnema sylvestre* R.Br. as potential α-glucosidase inhibitors, *Phytochem. Lett.* **2018**, *25*, 196–202.
61. Choi, J.; Yoon, K. D.; Kim, J. Chemical constituents from *Taraxacum officinale* and their α-glucosidase inhibitory activities, Bioorg. Med. Chem. Lett. **2018**, *28*(3), 476–481.
62. Şöhretoğlu, D.; Sari, S.; Šoral, M.; Barut, B.; Özel, A.; Liptaj, T. Potential of *Potentilla inclinata* and its polyphenolic compounds in α-glucosidase inhibition: Kinetics and interaction mechanism merged with docking simulations, *Int. J. Biol. Macromol.* **2018**, *108*, 81–87.
63. Şöhretoğlu, D.; Sari, S.; Barut, B.; Özel, A. Discovery of potent α-glucosidase inhibitor flavonols: Insights into mechanism of action through inhibition kinetics and docking simulations, *Bioorg. Chem.* **2018**, *79*, 257–264. https://doi.org/10.1016/j.bioorg.2018.05.010.
64. Gong, Y.; Qin, X.-Y.; Zhai, Y.-Y.; Hao, H.; Lee, J.; Park, Y.-D. Inhibitory effect of hesperetin on α-glucosidase: Molecular dynamics simulation integrating inhibition kinetics, *Int. J. Biol. Macromol.* ***2017***, *101*, 32–39.
65. Chen, G.; Guo, M. l.; Guo, M. Q. Rapid screening for α- glucosidase inhibitors from *Gymnema sylvestre* by affinity ultrafiltration – HPLC-MS, *Front. Pharmacol.* **2017**, *8*, 228.
66. Muccilli, V.; Cardullo, N.; Spatafora, C.; Cunsolo, V.; Tringali, C. α-Glucosidase inhibition and antioxidant activity of an oenological commercial tannin. Extraction, fractionation and analysis by HPLC/ESI-MS/MS and (1)H NMR, *Food Chem.* **2017**, *215*, 50–60. https://doi.org/10.1016/j.foodchem.2016.07.136.
67. Collado-González, J.; Grosso, C.; Valentão, P.; Andrade, P. B.; Ferreres, F.; Durand, T.; Gil-Izquierdo, Á. Inhibition of α-glucosidase and α-amylase by Spanish extravirgin olive oils: The involvement of bioactive compounds other than oleuropein and hydroxytyrosol, *Food Chem.* **2017**, *235*, 298–307.
68. Van Nguyen Thien, T.; Huynh, V. H. T.; Vo, L. K. T.; Tran, N. T.; Luong, T. M.; Le, T. H.; Duc, T. P.; That, Q. T. Two new compounds and α-glucosidase inhibitors from the leaves of *Bidens pilosa* L, *Phytochem. Lett.* **2017**, *20*, 119–122.
69. Duong, T. T. T.; Do, T. N. V.; Nguyen, H. X.; Le, T. H.; Dang, P. H.; Nguyen, N. T.; Nguyen, T. N. T.; Nguyen, T. D.; Nguyen, M. T. T. α-Glucosidase inhibitors from the stem of *Mangifera reba, Tetrahedron Lett.* **2017**, *58*(23), 2280–2283.
70. Zhao, D.-G.; Zhou, A.-Y.; Du, Z.; Zhang, Y.; Zhang, K.; Ma, Y.-Y. Coumarins with α-glucosidase and α-amylase inhibitory activities from the flower of *Edgeworthia gardneri*, *Fitoterapia* **2015**, *107*, 122–127.
71. Zhou, H.; Xing, J.; Liu, S.; et al. Screening and determination for potential α-glucosidase inhibitors from leaves of *Acanthopanax senticosus* harms by using UF-LC/MS and ESI-MS (n), *Phytochem Anal.* **2012**, *23*(4), 315–323. https://doi.org/10.1002/pca.1360.
72. Zhang, A. J.; Rimando, A. M.; Mizuno, C. S.; Mathews, S. T. α-glucosidase inhibitory effect of resveratrol and piceatannol, *J. Nutr. Biochem.* **2017**, *47*, 86–93.

CHAPTER THIRTEEN

Anti-obesity natural products

Aditya Arya[a,b,*], Lutfun Nahar[b,c], Hidayat Ullah Khan[a], Satyajit D. Sarker[b]

[a]Department of Pharmacology and Therapeutics, School of Medicine, Faculty of Health and Medical Sciences, Taylor's University, Subang Jaya, Malaysia
[b]Centre for Natural Products Discovery (CNPD), School of Pharmacy and Biomolecular Sciences, Liverpool John Moores University, Liverpool, United Kingdom
[c]Laboratory of Growth Regulators, Institute of Experimental Botany ASCR & Palacký University, Olomouc, Czech Republic
*Corresponding author: e-mail address: aditya.arya@taylors.edu.my

Contents

1. Introduction

Obesity is an excessive body weight in the form of fat caused by an imbalance between energy intake and expenditure. Obesity is considered as a risk to global human health.[1–3] To evaluate either a person is obese or overweight, it is required to calculate the body mass index (BMI), which expresses body weight (in kilograms) as a function of body height (in meters^2). BMI is as a surrogate measure of body fatness, and is the most widely used and

Annual Reports in Medicinal Chemistry, Volume 55
ISSN 0065-7743
https://doi.org/10.1016/bs.armc.2020.02.006

accepted definition of obesity. Irrespective of gender, BMI in adults up to 25 kg/m^2 is considered normal, BMI ranging from 25.0 to 29.9 overweight, and obesity is described by a BMI of 30.0 or higher. The degree of obesity can be further subcategorized into class 1 (BMI of 30 to <35), class 2 (BMI of 35 to <40), and class 3 (BMI of >45). Assessing BMI in children requires some adjustments for both age and gender.[4]

Currently, it is reported that half of the adult population (13.1% of the global population) are overweight (BMI >30 kg/m^2).[5] Obesity is a complex disorder caused by the interaction of numerous genetic, dietary, lifestyle, and environmental factors.[6] Obesity and overweight are major contributors to the global burden of chronic diseases and complications, including cardiovascular diseases, diabetes, and cancer. In 2014, it was reported by World Health Organization (WHO) that more than 1.9 billion adults were overweight, and of those, over 600 million were obese. Once mainly a concern for higher income countries, overweight and obesity are also on the rise in low and middle income countries, especially in urban areas.[7] In addition to the social implications, excess body fat accumulation leads to several metabolic, mechanical and psychological complications such as insulin resistance and type 2 diabetes mellitus (T2DM), hypertension, hyperlipidemia, ischemic heart disease, obstructive sleep apnoea, non-alcoholic fatty liver disease, osteoarthritis, and several types of cancer.[8,9]

From an economic point of view, obesity and its related health consequences involve enormous costs currently and for future health care, such as physician visits, hospitalization, and other related expenses.[10] Furthermore, on the basis of medical expenditure panel surveys, the estimated medical cost associated with obesity in the United States alone was US$147 billion in 2008, which had increased from $78.5 billion in 1998.[11] Losing weight may help tackle obesity and its harmful effects on human heath, and prevent obesity-related diseases. Other than the over-the-counter (OTC) medicines, there are many different types of therapies for obesity treatments ranging from lifestyle modifications such as dietary control, increased physical activity, prescription weight-loss medications, and stomach surgeries for highly obese patients, either alone or in combination.[12,13] Although surgery and implantable devices could offer a solution for significantly obese patients, traditional drug based approaches are necessary to address the needs of a broader market.[14] According to the "Pharmacological Management of Obesity: An Endocrine Society Clinical Practice (ESCP) Guideline," recommended by the National Heart, Lung, and Blood Institute,[15] the most ideal treatment modality for weight loss should be appropriate dietary and lifestyle changes plus moderate-intensity exercise. However, many epidemic and clinical studies have shown that it is a great challenge to maintain long-term lifestyle modification.[16]

Centrally acting sympathomimetics, such as the amphetamine derivatives desoxyephedrine, diethylpropion and phentermine (Fig. 1), were among the earliest pharmacological agents used for weight loss.[17] They were popular in the 1950s and 1960s, but rising cardiovascular risks and drug abuse were the main reasons for their decline in use in early 1970s. Fenfluramine and dexfenfluramine, serotonin (5-HT)-releasing agents, replaced phentermine and diethylpropion in 1970s and 1980s, However, reports of cardiac valvulopathy, especially when combined with phentermine, led the manufacturers to withdraw fenfluramine and dexfenfluramine from the market.[18]

Until very recently, for the long-term clinical management of obesity and related metabolic syndromes, sibutramine (trade names Meridia® and Reductil®), rimonabant (Acomplia®) and orlistat (Xenical® and Alli®) were approved. Sibutramine, a dual monoamine (noradrenaline and serotonin)-reuptake inhibitor, was removed from the market in January 2010 due to severe cardiovascular adverse effects.[19]

Rimonabant, a cannabinoid CB1 receptor antagonist/inverse agonist, launched in the mid-1990s, but numerous reports directing its links to psychiatric complications (such as anxiety, depression and suicide) required this drug to withdraw from market in 2008.[20] Orlistat, a pancreatic lipases inhibitor, is the single weight loss agent approved by the Food and Drug Administration (FDA) for long-term (more than 12 weeks) use for the management of obesity. Compared with other agents, uncomfortable adverse effects associated with its use are limited.[21,22] Because of the limited success and noticeable adverse effects and toxicities of the marketed anti-obesity medications, alternative sources of weight loss drugs, such as natural products, are being investigated.[23] In fact, the use and popularity of natural remedies for weight loss have increased in recent years as there is an incorrect assumption that natural products are safe! However, some of these natural products have been shown to have significant weight-loss effects in various animal models as well as in human trials.

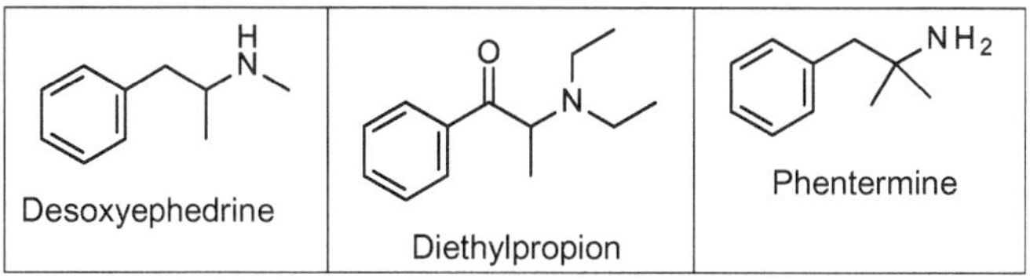

Fig. 1 Centrally acting sympathomimetics—the amphetamine derivatives.

2. Anti-obesity natural products

The growing threat of obesity to global health has encouraged scientists and researchers to put more efforts into finding new, effective and safe anti-obesity agents, especially from natural origins. Numerous potential materials from natural sources have been investigated along with their active ingredients. These natural products are mostly derived from plants, including fruits, vegetables, grains, and herbs. The biological benefits of these natural products are fundamentally contributed by the presence of an abundant amount of phytochemicals, fibers, and unsaturated fatty acids.[24] The anti-obesity products in the market can be classified into three categories: (1) food ingredients, (2) herbal ingredients, and (3) other functional supplements.

Developing functional products from what people usually consume in their daily life is probably the most popular segment in the functional supplement industry. Products made from fruits (citrus and berries), grains (soybean), vegetables, or beverage drinks (tea leaves) are relatively safer and more acceptable to consumers. Nowadays, traditional Chinese medical practitioners use herbal remedies that usually are combinations of different herbs, such as turmeric (*Curcuma longa*) and mulberry leaf (*Morus alba*), to treat obese patients. Recently, herbal therapies are not only popular in Asia, but also are becoming more and more common in the Western world. This is one of the reasons why herbal materials could be another major category of anti-obesity products.[25]

Some other materials, e.g., probiotic and calcium supplements, have also been proven to confer anti-obesity effects. Previous studies have explored the potential health benefits of fruits. The examples of potential health benefits include anticancer, anti-inflammatory, and anti-obesity effects.[26] Citrus fruits are one of the major categories used for the exploration and exploitation in new anti-obesity products. Phytochemicals including alkaloids, flavonoids and triterpenoids, are candidate ingredients that are found to be abundant in both the peel and pulp of citrus fruits. Cell and animal studies have shown the anti-obesity effects of citrus fruits extracts that help lower body weight gain and white adipose tissue weight. Leptin, which is a key hormone produced by adipocytes and functions in the regulation of food intake and energy expenditure, was found to be reduced by the intake of citrus fruits. This change in hormonal activity is desirable for the development of citrus-based anti-obesity product. In citrus fruits, methoxylated flavones and flavanone glycosides are the major bioactive flavonoid compounds capable of changing plasma leptin levels.

Anti-obesity products from green tea (*Camellia sinensis*) leaves are also popular in functional food market. The major bioactive constituents in green tea, accounting for up to 35% of the dry weight, are polyphenols. They may include flavonols, flavones, and flavan-3-ols (catechins). A number of clinical trials have revealed the beneficial effects of catechins (270–1200 mg/day), e.g., reduced body weight, lowered serum leptin levels, and reduced absorption of fatty acid. Another bioactive constituent in tea leaves is caffeine (Fig. 2), which influences somatic nervous system activity and acts synergistically with catechins to increase energy expenditure and fat oxidation. Apart from green tea, other herbal tea materials such as maté tea (*Ilex paraguariensis*), rooibos (*Aspalathus linearis*), and honey bush (*Cyclopia intermedia*) have also been studied for their roles in obesity prevention and lipid metabolism.[27] Owing to the wide array of anti-obesity natural products, and their structural complexities, it is difficult to elaborate on all of them in one study. Nonetheless, there are a considerable number of studies to determine anti-obesity properties of various natural products. It is, however, worth noting that the doses being studied in many of those experiments were usually much higher than the actual doses being used in commercial products.[28]

2.1 Plant derived anti-obesity natural products

Nowadays, changes in human lifestyle and high energy diet have increased the incidence of obesity and even have become a risk factor to the population of children.[29] Medicinal use of plants arises from ethnobotanical and ethnopharmacological approaches that test their therapeutic use in treating and preventing numerous diseases. These approaches include popular knowledge; thus anything "traditional" has become something of great importance to science. These approaches also contribute to the selection of species to be studied and the development of phytotherapeutic medicines based on ethnopharmacological investigation.[30,31] The potential of natural products for the treatment of obesity is still largely unexplored and can be an excellent alternative for the safe and effective development of anti-obesity drugs.[32] The following sections will provide some examples of plant-derived anti-obesity agents, either crude extracts or purified single compounds.

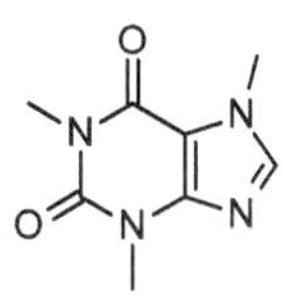

Fig. 2 Structure of caffeine.

2.1.1 Curcumin

Curcumin (Fig. 3), a diarylheptanoid, classified as phenolic compound, is a yellowish pigment from *Curcuma longa* rhizomes, commonly known as turmeric, and it can lower lipids and prevent obesity-associated complications by activating Wnt/β-catenin signaling and suppressing 3T3-L1 adipocyte differentiation.[33] Compared to the mice in the untreated high fat (HF) diet groups, those supplemented with 500 mg/kg dietary curcumin showed a significant

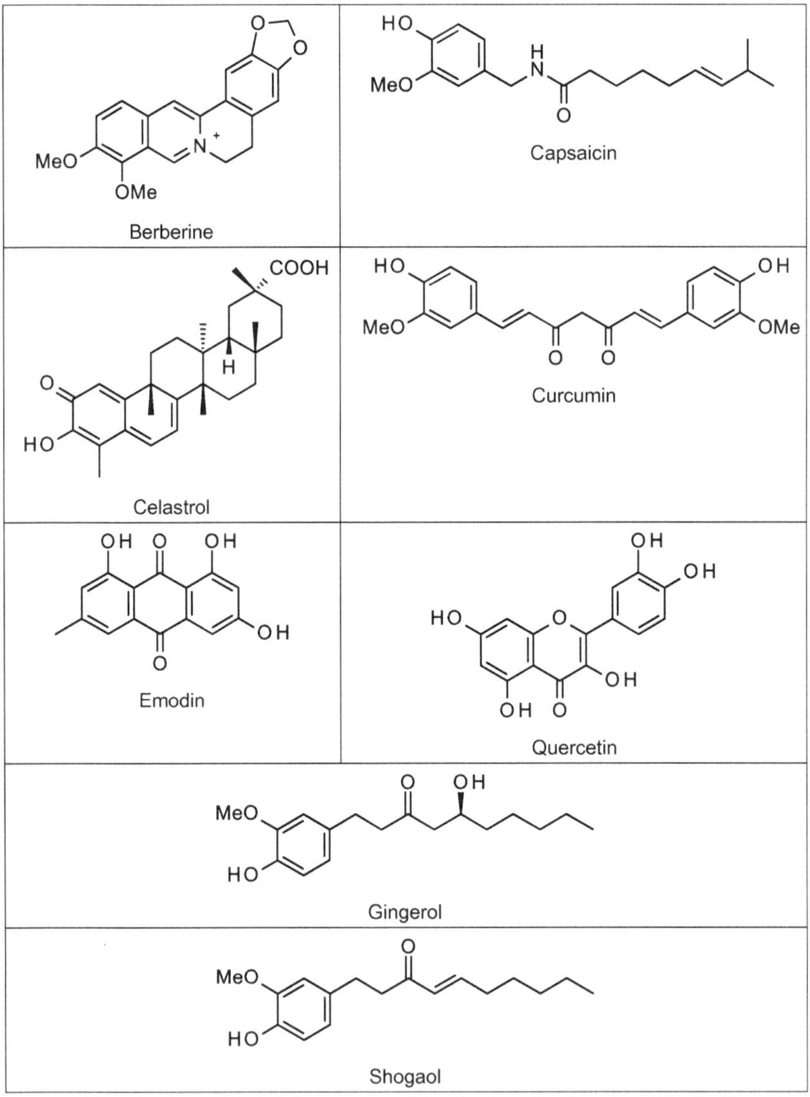

Fig. 3 Examples of some anti-obesity phytochemicals.

decrease in weight, total body fat, and serum cholesterol (all $P < 0.05$). *In vitro*, curcumin (5–25 μM) elevated 5′-AMP-activated protein kinase (AMPK) phosphorylation and carnitine palmitoyltransferase-1 expression ($P < 0.05$), but decreased glycerol-3-phosphate acyl transferase-1 expression. These actions reduced fatty acid esterification and enhanced fat oxidation ($P < 0.05$). In addition, curcumin significantly decreased the expression of PPARγ ($P < 0.05$) and CCAAT enhancer binding protein α (C/EBPα, $P < 0.05$), which are two key transcription factors in adipogenesis, and this effect modified lipid metabolism in adipocytes and inhibited white adipogenesis.[34]

2.1.2 Capsaicin

Capsaicin (Fig. 3) is the major ingredient in *Capsicum annuum*. This alkaloid increases the expression of browning-specific genes in subcutaneous white adipose tissue (WAT) and increases thermogenesis and mitochondrial biogenesis genes in brown adipose tissue (BAT).[35] Suppression of 3T3-L1 pre-adipocyte differentiation into adipocytes was observed in low-dose capsaicin (0.1–1 μM)-treated groups compared to the untreated control pre-adipocyte groups, suggesting an anti-adipogenic effect ($P < 0.05$) through the activation of the transient receptor potential vanilloid type-1 (TRPV-1) channel and induction of a brown-like phenotype.[36] Furthermore, a 0.01% capsaicin diet markedly upregulated UCP2 and UCP3 ($P < 0.01$ and < 0.05, compared to the control and HF diet groups, respectively) expression in mature adipocytes of visceral fat, promoting fat oxidation and energy expenditure.[37] Capsaicin combined with chitosan as a microsphere had additive obesity-reducing effects.[38]

2.1.3 Celastrol

Celastrol (Fig. 3), triterpene acid from the roots of *Tripterygium wilfordii*, is a potent anti-obesity agent as evident from animal studies. Celastrol (0.1 mg/kg) is a leptin sensitizer, which could enhance plasma leptin ($P < 0.001$), decrease appetite ($P < 0.001$), and promote a 45% weight loss ($P < 0.001$) in diet-induced obese mice by improving leptin sensitivity compared to the untreated vehicle groups.[39]

Moreover, thermogenesis (BAT level), white fat remodeling, and mitochondrial function in the fat ($P < 0.01$) and muscle ($P < 0.05$) were improved by celastrol (1 and 3 mg/kg) compared to the HF diet groups. This occurred by the activation of the HSF1-PGC1α transcriptional axis, which increased HSF1 ($P < 0.01$), a temperature sensor regulating energy metabolism, compared to that of the HSF1 knockout groups.[40]

2.1.4 Rhizoma coptidis *(Huang Lian)*

Rhizoma coptidis is derived from the roots of *Coptis chinensis*, *Coptis deltoidea* or *Coptis teeta.*[41] Among the alkaloids produced by these plants, berberine (Fig. 3) is the main active body weight reducing component of Rhizoma coptidis,[42] and it can lower lipids,[43] reduce lipid synthesis[44] and inhibit adipogenesis.[45] The anti-obesity mechanisms of Rhizoma coptidis involve decreasing degradation of dietary polysaccharides, lowering caloric intake, and systemically activating fasting-induced adipose factor (FIAF) protein and expression of genes related to mitochondrial energy metabolism.[46]

2.1.5 Panax ginseng *(Ren Shen)*

Ren Shen is derived from the dried roots and rhizomes of *Panax ginseng*.[47] Ginseng saponins (commonly known as ginsenosides) and polysaccharides are the central active components of this well-known medicinal plant.[48] Ginseng saponins can be subdivided based on structure into Rb1, Rb2, Rc, Rd, Re, and Rl.[49] This plant can reduce body weight, attenuate fat accumulation [50] and suppress lipid accumulation.[51] Oral administration of ginseng (0.5 g/kg diet) to high fat diet (HFD)-induced obese mice for 15 weeks significantly decreased body fat mass gain[52] and daily injections of 20 mg/kg Rb1 to diet-induced obese mice for 3 weeks, decreased significantly body weight, and lipid levels.[53]

2.1.6 Green tea (Lvcha)

Green tea (*Camellia sinensis*) contains high levels of tea polyphenols, catechins, amino acids and the xanthine alkaloid, caffeine. During the past decade, the health-promoting effects of green tea and its polyphenols have been intensively investigated.[54] Green tea is frequently used to ensure weight loss.[55] It induces weight loss in a variety of ways, such as activating the nuclear factor erythroid-2-related factor-2 (Nrf2) pathway,[56] prevention of gut dysbiosis, regulating metabolic balance in the body, inhibiting fat accumulation and cholesterol synthesis and reducing abdominal fat.[57] Oral administration of an HFD plus 0.25% (w/w) green tea extract for 12 weeks in diet-induced obesity (DIO) mice ameliorated obesity, hepatic steatosis, dyslipidemia and insulin resistance. Green tea extract contributed to the regulation of systemic metabolic homeostasis *via* transcriptional responses to lipid, glucose, and amino acid metabolism.[58] Marked decreased in body weight, cholesterol levels, and LDL levels were noted by supplementation with green tea extract powder and eriodictyol for 8 weeks.[59]

2.1.7 Ginger

Ginger (*Zingiber officinale*) has a dual anti-obesity effect. Gingerol and shogaol (Fig. 3) have been shown to increase the metabolic rate and thus to help to "burn off" excessive fat and to suppress the absorption of calorie-dense dietary fats from the intestines. The ginger extract inhibits the absorption of dietary fat by the intestine.[6]

2.1.8 *Stellaria media*

Stellaria media, commonly known as Chickweed, is a favorite salad herb widely distributed throughout the Himalayas upto an altitude of 4300 m.[60] It prevents high-fat-diet induced fat storage in adipose tissue by inhibiting the intestinal absorption of dietary fat and carbohydrates through inhibition of digestive enzymes.[61] The anti-obesity activity of this plant could be linked to some of its major phytochemicals of the classes of anthraquinones, flavonoids and triterpenoid saponins. Most important of them is emodin (Fig. 3), which has been shown in various animal studies to possess significant anti-obesity property.

2.2 Anti-obesity natural products from microbial sources

Microorganisms have great potential as natural sources of drugs for the treatment and prevention of infections and diseases like cancer, anemia, diarrhea, obesity, atopic dermatitis and Crohn's disease's. They are also potential sources of natural antioxidants, colors, immuno-suppressants, enzyme inhibitors, hypocholesterolemic agents, vitamins, enzymes, and obviously antibiotics.[62] The discovery of penicillin from *Penicillium notatum* by Alexander Fleming in 1928 marked a significant shift from plants to microorganisms as a source of bioactive natural products for new drug discovery.[63] Since then, microorganism-derived compounds have been utilized in medicine, agriculture and food industries.[64] Some examples of anti-obesity agents from microbial sources are discussed below.

2.2.1 Lipstatin

Microbial natural products also function as enzyme inhibitors. Lipstatin (Fig. 4) is a pancreatic lipase inhibitor produced by *Streptomyces toxytricini* that is used to combat obesity and diabetes by interfering with the gastrointestinal absorption of fat. It contains a beta-lactone structure that is likely responsible for irreversibly binding to the active site of lipase.[65]

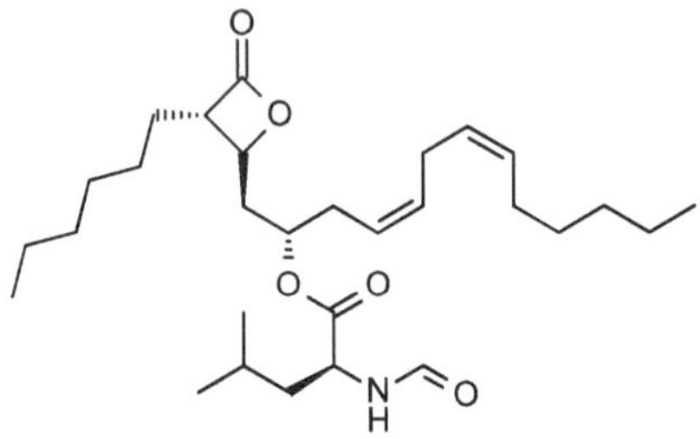

Fig. 4 Lipstatin, an anti-obesity agent from *Streptomyces toxytricini*.

2.2.2 Gut microbiota

The gut microbiota of an individual contains trillions of microorganisms that participate in various physiological functions, including vitamin production, maintenance of intestinal cells, development of the immune system and neutralization of pathogens, drugs and toxins.[66] The gut microbiota also has an important role in extracting energy from food and could be involved in the development of obesity.[67,68] In obese mice, the gut microbiota extracts more energy from food than in lean mice.[69] In humans with obesity, treatment with vancomycin for 1 week modulates the gut microbiota and reduces insulin sensitivity, compared with baseline levels.[70] Transfer of the gut microbiota from lean individuals to those with obesity improves insulin sensitivity in the recipients.[71] These results suggest that modulation of the gut microbiota could have beneficial effects on obesity.

2.2.3 Probiotic bacteria

The probiotic bacteria are used for the manufacture of a natural remedy, for controlling weight gain, preventing obesity, increasing satiety, prolonging satiation, reducing food intake, reducing fat deposition, improving energy metabolism, treating and enhancing insulin sensitivity and treating obesity. Recent studies suggest that manipulation of the composition of the microbial ecosystem in the gut might be a novel approach in the treatment of obesity. Such treatment might consist of altering the composition of the microbial communities of an obese individual by administration of beneficial microorganisms, commonly known as probiotics.[72] It has also been reported that the species from the genera *Lactobacillus* (*L. sporogenes* and *L. acidophilus* NCFB 1748) and *Bifidobacterium* may have a critical role in weight regulation as an anti-obesity effect in experimental models and humans, or as a growth-promoter effect in agriculture depending on the strains.[73]

2.3 Anti-obesity natural products from marine sponges

The world oceans cover about 70% of the earth surface, with more than 1 million multicellular organisms and 1 billion unicellular organisms, representing 34–36 Phyla, and represent an enormous resource for the discovery of potential therapeutic agents. During the last decades, numerous novel compounds have been found from marine organisms with interesting pharmaceutical activities.[74,75] Due to high taxonomic and ecosystem diversity, the marine world is believed to be a potential source to provide not only novel biologically active substances for the development of pharmaceuticals, but also essential compounds for human nutrition. Exploration of new and novel bioactive natural compounds from ocean looks to be an unlimited field.[76] Metabolites from sponges, algae and marine endophytic microbe extracts are reported to have antidiabetic, anti-obesity and antioxidants activities.[77] Research of bioactive compounds from marine organisms have developed rapidly.[78] Many marine natural products are efficient in preventing and treating obesity. With the development of marine technology and the growing interest in anti-obesity products, the current exploration of marine bioactive components will be tremendously expedited. This will continue to become a new trend and novel method for obesity treatment and prevention. Marine sponges have been considered as an excellent source of marine natural products since the 1950s, with about 4851 compounds described to date, contributing to nearly 30% of all marine natural products discovered so far.[79] Several sponges show anti-diabetes properties, e.g., inhibition of glycogen synthase kinase 3 beta (GSK-3B), a-glucosidase, PTP1B, dipeptidyl peptidase IV or protection of the beta pancreatic cells. Some of the marine natural products with anti-obesity potential are described below.

2.3.1 Palinurin

A sesquiterpene named palinurin (Fig. 5), found in the sponge *Ircinia dendroides*, and a phenylmethylene hydantoins, from the sponge *Hemimycale Arabica*, showed GSK-3B inhibitory activity, and assumed to have roles in preventing obesity. In 2007, a patent was published on GSK-3B inhibitors from the marine sponges *Ircinia dendroides, Ircinia variabilis* and *Ircinia oros* collected from the Mediterranean Sea.[80,81]

2.3.2 Callyspongynic acid

Callyspongynic acid (Fig. 5), isolated from sponge *Callyspongia truncate*, inhibited glucosidase,[82] and the galactosylceramide from the sponge

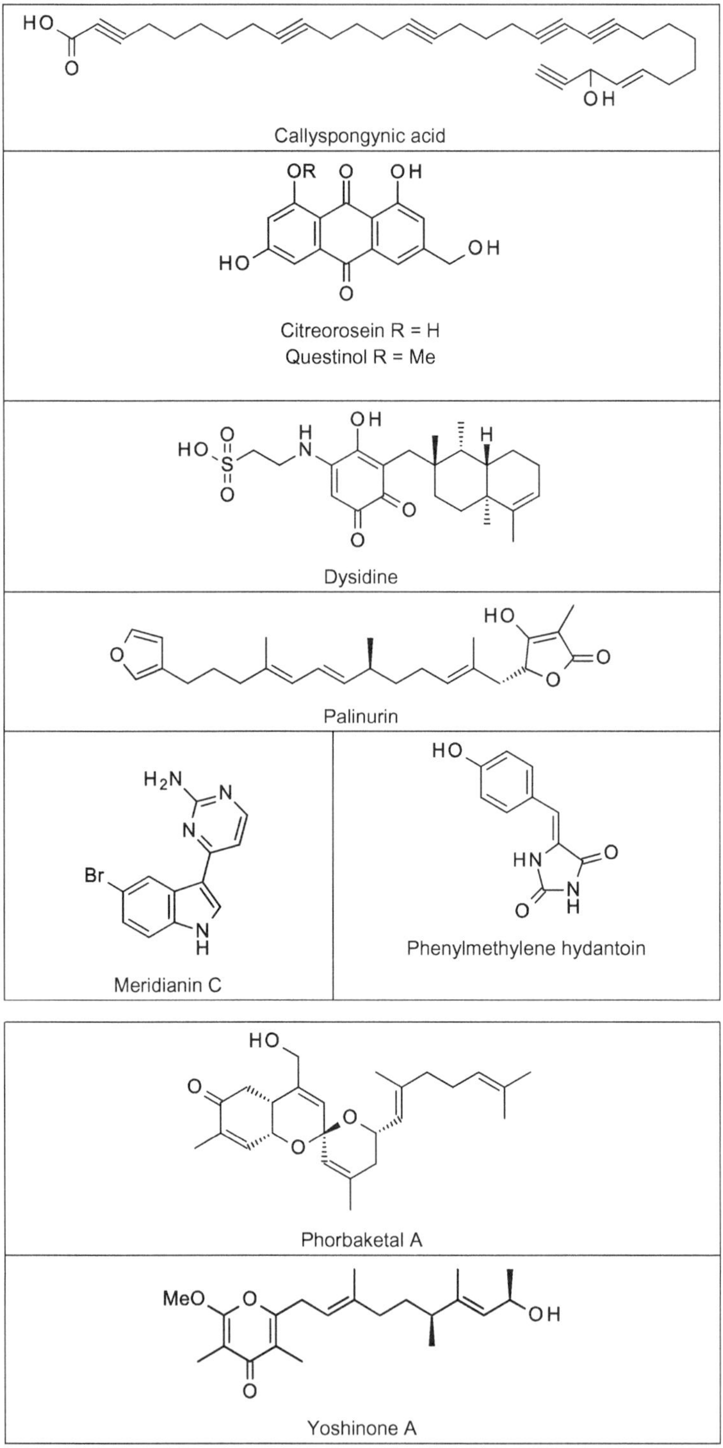

Fig. 5 Anti-obesity compounds from marine sources.

Agelas mauritianus[83] induced protection of pancreatic cells, whereas aqueous extract of the sponge *Xetospongia muta* inhibited dipeptidyl peptidase IV activity in *in vitro* models.[84]

2.3.3 Dysidine

Dysidine (Fig. 5) was found for the first time in a sponge at Lahdu (Santo) in Vanuatu in June 1996, identified as sponge *Dysidea* sp. (family Dysideidae, order Dictyoceratida)[85] Later, Noinart et al.[86] isolated dynosine from the Hainan sponge *Dysidea villosa* in the Chinese South Sea. The mechanism of action of dysidine was first studied by Zhang et al.[85] who found a strong PTP1B inhibition activity. Further cell based evaluation of this alkaloid indicated that (1) it could strongly promote membrane translocation of the glucose transporter 4 (GLUT4) in CHO-K1 (from *Cricetulus griseus* ovary) and 3T3-L1 (from *Musmusculus embryo*) cells, suggesting the involvement of GLUT4 in the promotion of glucose uptake; (2) dynosine activated the insulin receptor by modifying its phosphorylation (by PTP1B inhibition).[87]

2.3.4 Questinol and citreorosein

These anthraquinones were isolated from the marine sponge *Stylissa flabelliformis*, and showed strong anti-obesity activity. Questinol and citreorosein (Fig. 5) have similar structure, hydroxymethyl group on C-6 and a hydroxyl group on C-8. Replacing the hydroxyl group on C-1 by a methoxyl group, diminishes the activity whereas replacing the hydroxyl group on C-3 with a methoxyl group, completely subtracts the anti-obesity activity. The hydroxymethyl group on C-6 and the hydroxyl groups on C-3 and C-8 are necessary for the anti-obesity activity of these compounds. Thire anti-obesity activity were noted by using the zebrafish Nile red assay. Questinol and citreorosein reduced >60% and >90% of the stained lipids with the IC^{50} values of 0.95 and 0.17 μM, respectively.[88]

2.3.5 Brominated lipid

Brominated lipid, a new class of marine-derived lipid, is attracting attention for their health-promoting effect recently. In a report of brominated polyunsaturated lipids isolated form sponge *Xestospongia testudinaria*, methyl xestospongic ester was found to inhibit pancreatic lipase activity (IC_{50} = 3.1 μM), similar to that of positive control, orlistat (IC_{50} = 0.78 μM). Further investigation on lipase inhibition was evaluated with an acute oral lipid administration. The *in vivo* result showed a significant decrease of plasma TAG after lipid challenge, suggesting it could block dietary fat absorption and might be a potent candidate for obesity management.[89]

2.3.6 Phorbaketal A

Phorbaketal A (Fig. 5), isolated from the marine sponge *Phorbas* sp., showed a marked inhibition on adipogenesis during differentiation. Unlike other compounds targeting on the secondary stage of adipocyte differentiation, phorbaketal A suppressed adipogenic differentiation in mesenchymal stem cells, as indicated by less fat droplets and decreased adipogenic marker genes like PPARγ.[90]

2.3.7 Yoshinone A

Except macroalgae such as brown algae and green algae, microalgae also serve as a reservoir of many bioactive compounds. Some components from microalgae were reported to disturb the process of adipocyte differentiation. For example, yoshinone A (Fig. 5), a marine natural product of C-pyrone and a linear side chain, was isolated from the cyanobacterium *Leptolyngbya* sp. It was shown to suppress adipocyte differentiation without cytotoxicity, but its anti-adipogenesis mechanism is not well studied.[91]

2.3.8 Meridianin C

Meridianin C (Fig. 5), a marine indole alkaloid, was isolated from the macroalgae *Aplidium meridianum*. Meridianin C and its derivatives were investigated of their adipogenesis effects on 3T3-L1 cells. Although meridianin C inhibited lipid accumulation, it showed toxicity to 3T3-L1 adipocytes. However, its derivative strongly decreased TAG content in adipocytes without cytotoxicity. At the mechanistic level, its derivative suppressed adipogenesis by downregulation of C/EBPα and PPARγ, as well as the phosphorylated STAT-3 and STAT-5.[92]

3. Mechanisms of action of anti-obesity natural products

Weight loss can only occur if energy consumption is less than energy expenditure, thereby creating a negative energy balance. The obvious approach of life style management through exercise combined with dieting has a relatively low success rate and therefore alternative forms of obesity treatment, including surgery and medication, have become popular.[93] Natural anti-obesity preparations can stimulate weight loss *via* numerous mechanisms, which can be dived into following five categories.

3.1 Increase in energy expenditure

Excessive adiposity can primarily be attributed to an imbalance in energy homeostasis, in which the consequences of excessive food intake are not balanced by increased energy expenditure.[94] That is the reason why increasing energy expenditures to build a negative energy balance is crucial to weight management. Energy expenditure can be briefly classified into three categories: (1) physical activity, (2) obligatory energy expenditure, and (3) adaptive thermogenesis. Most anti-obesity products usually regulate body weight through an increase of obligatory energy expenditure. In the human body, the function of brown adipose tissue is to transfer energy from food into heat. Uncoupling protein 1 (also known as thermogenin) in brown adipose tissue plays a key role in thermogenic effect. Therefore, a material capable of upregulating uncoupling protein 1 gene expression could be a potential strategy for achieving an anti-obesity effect by increasing energy expenditure.[95]

3.2 Appetite suppressant effect

There has been a rise in satiety-enhancing products on sale in the food supplement marketplace. The biological mechanisms of appetite and satiety are regulated by a complex interaction of neurological and hormonal signals. Many studies have revealed that some food ingredients could provide satiety-enhancing effects and be beneficial for weight control.[94] The mechanism underlying the enhanced satiety includes an increase of noradrenaline level and subsequent activation of sympathetic nervous system activity, leading to an increase in satiety and energy expenditure, suppression of hunger, and also elevation in fat oxidation.[95] A line of evidence indicates that neural signal peptides like serotonin, histamine, dopamine, and their associated receptor activities are associated with satiety regulation. These neural signal peptides and their receptors could be potential target areas for the development of supplement products that treat obesity through energy intake reduction by increasing satiety.[96]

3.3 Inhibiting digestive enzyme activity

Most of the fat consumed in the Western diet comprises triglycerides (TGs) or esters of a single molecule of glycerol and three fatty acids, which are metabolized and absorbed directly from gastrointestinal tract.[97] Dietary fat that cannot be absorbed are broken down enzymatically by pancreatic lipase secreted from the pancreas to promote their absorption in human intestine.[98] Some natural products may inhibit pancreatic lipase activity, is therefore widely studied as one of the most important indicator for the determination of the anti-obesity potentials for weight loss treatment.[99] As a key pancreatic lipase inhibitor, tetrahydrolipstatin (orlistat), with a median inhibitory concentration (IC_{50}) of 0.7 μM, approved by the FDA for long-term clinical use for obesity treatment. Some substances have been confirmed to have capabilities to tetrahydrolipstatin (orlistat), a derivative of the naturally-occurring lipase inhibitor isolated from *Streptomyces toxytricini*.[100] It is a synthetic drug designed to act through a covalent bond to the active site serine of pancreatic lipase to block the absorption of dietary fat.[101,102] Due to numerous unacceptable side effects of orlistat's, discovering novel, effective pancreatic lipase inhibitors with less adverse effects from natural source are the needs of the day.

3.4 Regulatory effect on adipocyte differentiation

Adipocytes play a central role in the maintenance of lipid homeostasis and energy balance. They have a particular large capacity to store triglycerides as well as to release free fatty acids in response to changing energy demands. As adipocyte tissue growth is associated with both the hyperplasia and hypertrophy of adipocytes, this has led to the development of natural products in anti-obesity therapy that specifically target adipogenesis inhibition. Some research has also suggested that adipocyte differentiation could be inhibited by the blockade of several transcription factors.[103] A study on the inhibitory activity of *Sibiraea angustata* extracts on adipocyte differentiation has also revealed that the expression of both C/EBP-α and PPAR were significantly inhibited leading to a decrease in the lipid content of adipocyte cells.[104]

3.5 Regulatory effect on lipid metabolism

An increase in the rate of lipolysis stimulates triglyceride hydrolysis and hence diminishes fat storage and combats obesity. The key points to enhance lipolysis have been discussed in many studies. For instance, activation of β-adrenergic receptor initiatively triggers lipolysis in white adipocytes and non-shivering thermogenesis in brown fat.[105] An activation of adenosine monophosphate-activated protein kinase (AMPK) leading to the increase of fatty acid oxidation and glucose transport in skeletal muscle is another example.[106,107] Hence, transcription factors that can simulate lipolysis become an important feature in the development of anti-obesity products. It has been shown that flavonoids, specifically flavonols (e.g., quercetin), could activate lipoxygenase and attenuate adipogenesis through the upregulation of the AMPK pathway. Additionally, quercetin (Fig. 3) activates the apoptotic pathway in mature adipocytes through suppression of phosphorylation effect of signal-regulated kinases 1 and 2 (ERK1/2) and c-Jun N-terminal kinase (JNK), both belonging to subfamilies of mitogen-activated protein kinase (MAPK).[108]

4. Anti-obesity natural products in clinical trials (in the drug development chain)

So far, effective and safe pharmacological options for the prevention and treatment of obesity are elusive, and despite the growing amount of resources being applied to research in the development of anti-obesity drugs has been plagued by failures. One of the reasons for these failures might be that our understanding of the pathogenesis of obesity is inadequate. Alternatively, the molecular underpinning of biological processes that underlie the emergence of obesity may not lend themselves to selective manipulations that successfully reduce adipose tissue stores without impairing long-term health and survival.[109]

Indeed, many of the putative novel drugs for obesity have not achieved the level of clinical effectiveness that is required by regulatory authorities such as the US Food and Drug Administration (FDA), while the few that are sufficiently effective are associated with adverse side effects that preclude their long-term use in patients.[110] While there are several plant extracts currently undergoing pre-clinical animal studies and some human clinical trials, it appears that there are hardly any purified natural products currently considered for human clinical trials aiming at anti-obesity drug discovery and development. Some of the phytotherapeutic products that are in pre-clinical or clinical trials are summarized in Table 1.

5. Conclusions

Obesity is dangerous to human health and has severe consequences. Diet and exercise are the best for both prevention and treatment; unluckily, both require abundant discipline and are tough to maintain. Furthermore, the use of available drugs for treating obesity is inadequate by side effects, and the weight loss persists only as long as the drug is being taken, since as soon as the treatment is stopped, the weight is regained. Therefore, the development of alternative drugs from natural products is currently

Table 1 Phytotherapeutic products in pre-clinical and/or clinical trials for anti-obesity potential.

Phytotherapeutic products	Trial	Details
Seeds of *Carum carvi* L. (Apiaceae)	Human clinical trials	Reduces weight, body mass index, body fat percentage, and waist-to-hip ratio[109]
Cylaris formula containing *Cissus quadrangularis* L. (Vitaceae)	Human clinical trials	Anti-lipase and anorexiant properties that reduce the absorption of dietary fats and enhance satiation by increasing serum serotonin levels[110]
Fruits and leaves of *Citrus aurantium* L. (Rutaceae)	Pre-clinical trials	The active constituent *p*-synephrine increases metabolic rate, energy expenditure and increase in weight loss. The leaf extract downregulates the expression of C/EBPβ and subsequently inhibits the activation of PPARγ and C/EBPα. The anti-adipogenic activity of is mediated by the inhibition of Akt activation and GSK3β phosphorylation, which induces the downregulation of lipid accumulation and lipid metabolizing genes, ultimately inhibiting adipocyte differentiation[111,112]
Fruits and commercially available tablets of *Garcinia cowa* Roxb. ex Choisy (Clusiaceae)	Pre-clinical trials	Inhibits the enzyme ATP-dependent citrate lyase, which catalyzes the cleavage of citrate to oxaloacetate and acetyl-CoA. Serum apo A1 levels are increased by the plant and the serum total cholesterol levels[113,114]
Leaves of *Gynostemma pentaphyllum* (Thunb.) Makino (Cucurbitaceae)	Pre-clinical trials	Activation of AMPK by 5-aminoimidazole-4-carboxamide-1-β-D-ribofuranoside (AICAR) inhibits adipogenesis by downregulating PPARc and CEBP1a as well as lipogenic factors including fatty acid binding protein 4 and lipoprotein lipase. AMPK activation directly inactivates ACC through Ser79 phosphorylation, leading to decreased fat synthesis by reducing the production of malonyl-CoA from acetyl-CoA[115]

Table 1 Phytotherapeutic products in pre-clinical and/or clinical trials for anti-obesity potential.—cont'd

Phytotherapeutic products	Trial	Details
Oil of *Nigela sativa* L. (Ranunculaceae)	Human clinical trials	Reduces total cholesterol, low density lipoprotein (LDL) and fasting blood glucose, and is effective as an add-on therapy in patients with metabolic syndrome[116,117]
Roots of *Salacia reticulata* Wight (Celastraceae)	Pre-clinical trials	Weight and body-fat reduction and BMI reduction[117]
Seeds of *Trigonella foenumgraecum* L. (Leguminosae)	Human clinical trials	Daily fat consumption, expressed as the ratio fat reported energy intake/total energy expenditure (fat-REI/TEE), is decreased in overweight subjects administered the fenugreek seed extract. Significant decrease in the insulin/ glucose ratio in subjects treated with fenugreek seed extract[118]
Leaves and capsules of *Turnera diffusa* Willd. ex Schult. (Passifloraceae)	Human clinical trials	The herbal preparation capsules delays gastric emptying, reducing the time to perceived gastric fullness and induces weight loss[115]
Leaves of *Vernonia amygdalina* Delile (Compositae)	Human clinical trials	Lowers blood glucose level[116]
Fruits of *Ziziphus jujube* Mill. (Rhamnaceae)	Pre-clinical trials	Suppresses lipid accumulation and glycerol-3-phosphate dehydrogenase, and elicits the most inhibitory effect with attenuation of the expression of key adipogenic transcription factors, including PPAR-γ and CCAAT enhancer binding proteins (C/EBPs) at the protein level[118]

attracting considerable interests. As described in this chapter, various active ingredients from plants, marine sponges and microbial sources attract much attention for their health benefits, and contributed to promoting interest in the development of anti-obesity products. So far, a limited number of anti-obesity compounds were evaluated *in vitro* with cell lines only a few of them were tested in clinical trials. Therefore, further research is needed to investigate their efficacy and bioavailability in animal and human subjects.

Acknowledgment

L.N. gratefully acknowledges the financial support of the European Regional Development Fund—Project ENOCH (No. CZ.02.1.01/0.0/0.0/16_019/0000868).

References

1. Roh, C.; Jung, U. Screening of Crude Plant Extracts with Anti-Obesity Activity. *Int. J. Mol. Sci.* **2012**, *13*(2), 1710–1719.
2. Wakil, S. J.; Abu-Elheiga, L. A. Fatty Acid Metabolism: Target for Metabolic Syndrome. *J. Lipid Res.* **2009**, *50*(Supplement), S138–S143.
3. Wong, D.; Sullivan, K.; Heap, G. *The Pharmaceutical Market for Obesity Therapies*. Nature Publishing Group, 2012.
4. Schwartz, M. W.; Seeley, R. J.; Zeltser, L. M.; et al. Obesity Pathogenesis: An Endocrine Society Scientific Statement. *Endocr. Rev.* **2017**, *38*(4), 267–296.
5. Lee, P. C.; Dixon, J. Medical Devices for the Treatment of Obesity. *Nat. Rev. Gastroenterol. Hepatol.* **2017**, *14*(9), 553.
6. Mohamed, G. A.; Ibrahim, S. R.; Elkhayat, E. S.; El Dine, R. S. Natural Anti-Obesity Agents. *Bull. Facult. Pharm. Cairo Univ.* **2014**, *52*(2), 269–284.
7. Sun, N.-N.; Wu, T.-Y.; Chau, C.-F. Natural Dietary and Herbal Products in Anti-Obesity Treatment. *Molecules* **2016**, *21*(10), 1351.
8. Abente, E. J.; Subramanian, M.; Ramachandran, V.; Najafi-Shoushtari, S. H. MicroRNAs in Obesity-Associated Disorders. *Arch. Biochem. Biophys.* **2016**, *589*, 108–119.
9. Stevens, J.; Oakkar, E. E.; Cui, Z.; Cai, J.; Truesdale, K. P. US Adults Recommended for Weight Reduction by 1998 and 2013 Obesity Guidelines, NHANES 2007–2012. *Obesity* **2015**, *23*(3), 527–531.
10. Kazemipoor, M.; CWJWM, R.; Cordell, G. A.; Yaze, I. Potential of Traditional Medicinal Plants for Treating Obesity: A Review. *arXiv* **2012**, *39*, 1–6, preprint arXiv:12081923.
11. Finkelstein, E. A.; Trogdon, J. G.; Cohen, J. W.; Dietz, W. Annual Medical Spending Attributable to Obesity: Payer-and Service-Specific Estimates: Amid Calls for Health Reform, Real Cost Savings Are more Likely to Be Achieved Through Reducing Obesity and Related Risk Factors. *Health Aff.* **2009**, *28*(Suppl. 1), w822–w831.
12. Chandrasekaran, C.; Vijayalakshmi, M.; Prakash, K.; Bansal, V.; Meenakshi, J.; Amit, A. Herbal Approach for Obesity Management. *Am. J. Plant Sci.* **2012**, *3*(07), 1003.
13. Markt, P.; Herdlinger, S.; Schuster, D. Virtual Screening Against Obesity. *Curr. Med. Chem.* **2011**, *18*(14), 2158–2173.
14. Melnikova, I.; Wages, D. *Anti-Obesity Therapies*. Nature Publishing Group, 2006.
15. Apovian, C. M.; Aronne, L. J.; Bessesen, D. H.; et al. Pharmacological Management of Obesity: An Endocrine Society Clinical Practice Guideline. *J. Clin. Endocrinol. Metab.* **2015**, *100*(2), 342–362.
16. Thomas, J. G.; Bond, D. S.; Phelan, S.; Hill, J. O.; Wing, R. R. Weight-Loss Maintenance for 10 Years in the National Weight Control Registry. *Am. J. Prev. Med.* **2014**, *46*(1), 17–23.
17. Wilding, J. Clinical Investigations of Antiobesity Drugs. In: *Appetite and Body Weight*; Wilding, J. Ed.; Elsevier, 2007; pp337-III.
18. Rodgers, R. J.; Tschöp, M. H.; Wilding, J. P. Anti-Obesity Drugs: Past, Present and Future. *Dis. Model. Mech.* **2012**, *5*(5), 621–626.
19. James, W. P. T.; Caterson, I. D.; Coutinho, W.; et al. *N. Engl. J. Med.* **2010**, *363*(10), 905–917.

20. PJOwnatCafS *Obesity* **2008**, *3412*, 44–47.
21. Bray, G. A.; Greenway, F. L. *Pharmacol. Rev.* **2007**, *59*(2), 151–184.
22. Glandt, M.; Raz, I. *J. Obes.* **2011**, *2011*, 1–13.
23. Fu, C.; Jiang, Y.; Guo, J.; Su, Z. *J. Agric. Food Chem.* **2016**, *64*(51), 9571–9585.
24. Sasidharan, S.; Chen, Y.; Saravanan, D.; Sundram, K.; Latha, L. Y. *Afr. J. Tradit. Complement. Altern. Med.* **2011**, *8*(1), 1–10.
25. Park, J. P.; Kim, J. H.; Park, M. K.; Yun, J. W. *Biotechnol. Bioprocess Eng.* **2011**, *16*(6), 1065–1076.
26. Liu, R. H. *Adv. Nutr.* **2013**, *4*(3), 384S–392S.
27. Rains, T. M.; Agarwal, S.; Maki, K. C. *J. Nutr. Biochem.* **2011**, *22*(1), 1–7.
28. Bloomer, R. J.; Canale, R. E.; Blankenship, M. M.; Hammond, K. G.; Fisher-Wellman, K. H.; Schilling, B. K. *Lipids Health Dis.* **2009**, *8*(1), 1–9.
29. Klop, B.; Elte, J. W. F.; Cabezas, M. C. *Nutrients* **2013**, *5*(4), 1218–1240.
30. de Freitas Junior, L. M.; de Almeida, E. B., Jr. *Am. J. Transl. Res.* **2017**, *9*(5), 1–26.
31. de Albuquerque, U. P.; Monteiro, J. M.; Ramos, M. A.; de Amorim, E. L. C. *J. Ethnopharmacol.* **2007**, *110*(1), 76–91.
32. Bhutani, K. K.; Birari, R. B.; Kapat, K. *Nat. Prod. Commun.* **2007**, *2*(3), 331–338.
33. Ahn, J.; Lee, H.; Kim, S.; Ha, T. *Am. J. Physiol. Cell Physiol.* **2010**, *298*(6), C1510–C1516.
34. Ejaz, A.; Wu, D.; Kwan, P.; Meydani, M. *J. Nutr.* **2009**, *139*(5), 919–925.
35. Baboota, R. K.; Murtaza, N.; Jagtap, S.; et al. *J. Nutr. Biochem.* **2014**, *25*(9), 893–902.
36. Baboota, R. K.; Singh, D. P.; Sarma, S. M.; et al. Capsaicin Induces "Brite" Phenotype in Differentiating 3T3-L1 Preadipocytes. *PLoS One* **2014**, *9*(7), e103093. *PLoS One* **2014**, *9*(7), 1–12.
37. Ma, S. T.; Zhang, L. L.; Yang, D. C. *Prog. Nutr.* **2010**, *12*(1), 19–23.
38. Tan, S.; Gao, B.; Tao, Y.; Guo, J.; Su, Z.-Q. *J. Agric. Food Chem.* **2014**, *62*(8), 1866–1874.
39. Liu, J.; Lee, J.; Hernandez, M. A. S.; Mazitschek, R.; Ozcan, U. *Cell* **2015**, *161*(5), 999–1011.
40. Ma, X.; Xu, L.; Alberobello, A. T.; et al. *Cell Metab.* **2015**, *22*(4), 695–708.
41. Chinese Pharmacopoeia Commission; *Chinese Pharmacopoeia*. China Medical Science Press: Beijing, 2010; p2010–2228.
42. Ni, W. J.; Ding, H.-H.; Tang, L.-Q. *Eur. J. Pharmacol.* **2015**, *760*, 103–112.
43. Zou, Z. Y.; Hu, Y.-R.; Ma, H.; et al. *Fitoterapia* **2015**, *105*, 139–146.
44. Zhang, W. L.; Zhu, L.; Jiang, J. G. *Obes. Rev.* **2014**, *15*(12), 957–967.
45. Choi, J. S.; Kim, J.-H.; Ali, M. Y.; Min, B.-S.; Kim, G.-D.; Jung, H. A. *Fitoterapia* **2014**, *98*, 199–208.
46. Zhang, J.; Tang, H.; Deng, R.; et al. *PLoS One* **2015**, *10*(4), 1–16.
47. Li, Y.; Zhang, T. J.; Liu, S. X.; Wang, R. R.; Luo, J. *Chin. Tradit. Herb. Drugs* **2009**, *40*(1), 164–166.
48. Palaniyandi, S. A.; Suh, J.-W.; Yang, S. H. *Pharmacogn. Mag.* **2017**, *13*(Suppl. 1), 142–147.
49. Karu, N.; Reifen, R.; Kerem, Z. *J. Agric. Food Chem.* **2007**, *55*(8), 2824–2828.
50. Koh, E. J.; Kim, K.-J.; Choi, J.; Jeon, H. J.; Seo, M.-J.; Lee, B.-Y. *J. Ginseng Res.* **2017**, *41*(1), 23–30.
51. Cheon, J. M.; Kim, D.-I.; Kim, K.-S. *J. Ginseng Res.* **2015**, *39*(4), 331–337.
52. Li, X.; Luo, J.; Anandh Babu, P. V.; et al. *J. Med. Food* **2014**, *17*(12), 1287–1297.
53. Huang, J.; Wang, Y.; Xie, Z.; Zhou, Y.; Zhang, Y.; Wan, X. *Eur. J. Clin. Nutr.* **2014**, *68*(10), 1075–1087.
54. Türközü, D.; Tek, N. A. *CRC Crit. Rev. Food Sci. Nutr.* **2017**, *57*(2), 254–258.
55. Sampath, C.; Rashid, R. M.; Sang, S.; Ahmedna, M. *Biomed. Pharmacother.* **2017**, *87*(73), 73–81.

56. Cheng, M.; Zhang, X.; Miao, Y.; Cao, J.; Wu, Z.; Weng, P. *Food Res. Int.* **2017**, *92*, 9–16.
57. Choi, J. Y.; Kim, Y. J.; Ryu, R.; Cho, S.-J.; Kwon, E.-Y.; Choi, M.-S. *Nutrients* **2016**, *8*(10), 1–16.
58. Yamashita, M.; Kumazoe, M.; Nakamura, Y.; et al. *J. Nutr. Sci. Vitaminol.* **2016**, *62*(4), 249–256.
59. Ravindra, S. *Medicinal Plants of India: An Encyclopaedia*; Daya Publishing House, 2003.
60. Rani, N.; Verma, S.; Sharma, S. K. *BMC Complement. Altern. Med.* **2012**, *12*(1), 2–8.
61. Gupta, C.; Prakash, D.; Gupta, S. *J. Microbiol. Exp.* **2014**, *1*(1), 1–8.
62. Park, S. R.; Yoon, Y. J.; Pham, J. V.; et al. *Front. Microbiol.* **2019**, *10*, 1–27.
63. Sanches-TMS, S.; Ávalos, M.; Ruiz, B.; Rodríguez-Sanoja, R.; Jimenez-Estrada, M. *Microbial Nat. Product.* **2012**, *2012*, 65–108.
64. Lin, C. S.; Chang, C. J.; Lu, C. C.; et al. Impact of the gut microbiota prebiotics, and probiotics on human health and disease. *Biomed. J.* **2014**, *37*(5), 259–268.
65. Delzenne, N. M.; Neyrinck, A. M.; Bäckhed, F.; Cani, P. D. *Nat. Rev. Endocrinol.* **2011**, 7(11), 639–646.
66. Shen, J.; Obin, M. S.; Zhao, L. *Mol. Asp. Med.* **2013**, *34*(1), 39–58.
67. Turnbaugh, P. J.; Ley, R. E.; Mahowald, M. A.; Magrini, V.; Mardis, E. R.; Gordon, J. I. *Nature* **2006**, *444*(7122), 1027–1031.
68. Vrieze, A.; Out, C.; Fuentes, S.; et al. *J. Hepatol.* **2014**, *60*(4), 824–831.
69. Vrieze, A.; Van Nood, E.; Holleman, F.; et al. *Gastroenterology* **2012**, *143*(4), 913–916.
70. Mekkes, M.; Weenen, T. C.; Brummer, R. J.; Claassen, E. *Benefic. Microbes* **2013**, *5*(1), 19–28.
71. Kotzampassi, K.; Giamarellos-Bourboulis, E. J.; Stavrou, G. *ISRN Obes.* **2014**, 7(1), 1–8.
72. Vo, T. S.; Kim, S. K. *Mar. Drugs* **2010**, *8*(12), 2871–2892.
73. Ngo, D. H.; Vo, T. S.; Ngo, D. N.; Wijesekara, I.; Kim, S.-K. *Int. J. Biol. Macromol.* **2012**, *51*(4), 378–383.
74. Plaza, M.; Cifuentes, A.; Ibáñez, E. *Trends Food Sci. Technol.* **2008**, *3*(5), 544–563.
75. Debbab, A.; Aly, A. H.; Lin, W. H.; Proksch, P. *Microb. Biotechnol.* **2010**, *3*(5), 544–563.
76. Wang, J.; Zhang, L.; Teng, K.; Sun, S.; Sun, Z.; Zhong, J. *Appl. Environ. Microbiol.* **2014**, *80*(8), 2633–2643.
77. Esposito, G.; Teta, R.; Miceli, R.; et al. *Mar. Drugs* **2015**, *13*(1), 444–459.
78. Bidon Chanal, A.; Fuertes, A.; Alonso, D.; et al. *Eur. J. Med. Chem.* **2013**, *60*, 479–489.
79. Khanfar, M. A.; Asal, B. A.; Mudit, M.; Kaddoumi, A.; El Sayed, K. A. *Bioorg. Med. Chem.* **2009**, *17*(16), 6032–6039.
80. Nakao, Y.; Uehara, T.; Matunaga, S.; Fusetani, N.; van Soest, R. W. *J. Nat. Prod.* **2002**, *65*(6), 922–924.
81. Barde, S. R.; Sakhare, R. S.; Kanthale, S. B.; Chandak, P. G.; Jamkhande, P. G. *Asian Pac. J. Trop. Dis.* **2015**, S209–S213.
82. Pascual, I.; Lopéz, A.; Gómez, H.; et al. *Enzyme Microb. Technol.* **2007**, *40*(3), 414–419.
83. Giannini, C.; Debitus, C.; Lucas, R.; et al. *J. Nat. Prod.* **2001**, *64*(5), 612–615.
84. Li, Y.; Zhang, Y.; Shen, X.; Guo, Y.-W. *Bioorg. Med. Chem. Lett.* **2009**, *19*(2), 390–392.
85. Zhang, Y.; Li, Y.; Guo, Y.-W.; Jiang, H.-L.; Shen, X. *Acta Pharmacol. Sin.* **2009**, *30*(3), 333–345.
86. Noinart, J.; Buttachon, S.; Dethoup, T.; et al. *Mar. Drugs* **2017**, *15*(5), 1–12.
87. Hu, X.; Tao, N.; Wang, X.; Xiao, J.; Wang, M. *J. Funct. Foods* **2016**, *21*, 372–387.
88. Byun, M. R.; Lee, C. H.; Hwang, J.-H.; et al. *Eur. J. Pharmacol.* **2013**, *718*(1–3), 181–187.

89. Inuzuka, T.; Yamamoto, K.; Iwasaki, A.; et al. *Tetrahedron Lett.* **2014**, *55*(49), 6711–6714.
90. Park, Y.-K.; Lee, T.-Y.; Choi, J.-S.; et al. *Biochem. Biophys. Res. Commun.* **2014**, *452*(4), 1078–1083.
91. Vermaak, I.; Viljoen, A. M.; Hamman, J. H. *Nat. Prod. Rep.* **2011**, *28*(9), 1493–1533.
92. Aydin, S. *Peptides* **2014**, *56*, 94–110.
93. Kajimura, S.; Saito, M. *Annu. Rev. Physiol.* **2014**, *76*, 225–249.
94. Belza, A.; Frandsen, E.; Kondrup, J. Body Fat Loss Achieved by Stimulation of Thermogenesis by a Combination of Bioactive Food Ingredients: A Placebo-Controlled, Double-Blind 8-Week Intervention in Obese Subjects. *Int. J. Obes.* **2007**, *31*(1), 121; *International Journal of Obesity.* **2007**, i(1), 121–130.
95. Morton, G. J.; Meek, T. H.; Schwartz, M. W. *Nat. Rev. Neurosci.* **2014**, *15*(6), 367–378.
96. Buchholz, T.; Melzig, M. F. *Planta Med.* **2015**, *81*(10), 771–783.
97. Marrelli, M.; Loizzo, M. R.; Nicoletti, M.; Menichini, F.; Conforti, F. *Plant Foods Hum. Nutr.* **2013**, *68*(4), 340–346.
98. Zhu, T.; Wang, L.; Wang, W.; et al. *J. Gen. Appl. Microbiol.* **2014**, *60*(3), 106–111.
99. Mulzer, M.; Tiegs, B. J.; Wang, Y.; Coates, G. W.; O'Doherty, G. A. *J. Am. Chem. Soc.* **2014**, *136*(30), 10814–10820.
100. Tsujita, T.; Hiroe, T.; Takaku, T.; Aoyama, S.; Hiraki, J. *J. Lipid Res.* **2006**, *47*(8), 1852–1858.
101. Kang, H. J.; Hyun Ae, S.; Go, Y.; et al. *PLoS One* **2013**, *8*(4), 1–9.
102. Xiao, L.; Zhang, J.; Li, H.; et al. *Exp. Biol. Med.* **2010**, *235*(12), 1442–1449.
103. Bordicchia, M.; Pocognoli, A.; D'Anzeo, M.; et al. *J. Hypertens.* **2014**, *32*(2), 389–396.
104. Kola, B.; Grossman, A.; Korbonits, M. *Obes. Metab.* **2008**, *36*, 198–211.
105. O'Neill, H. M.; Holloway, G. P.; Steinberg, G. R. *Mol. Cell. Endocrinol.* **2013**, *366*(2), 135–151.
106. Ahn, J.; Lee, H.; Kim, S.; Park, J.; Ha, T. The anti-obesity effect of quercetin is mediated by the AMPK and MAPK signaling pathways. *Biochem. Biophys. Res. Commun.* **2008**, *373*(4), 545–549.
107. Ioannides-Demos, L. L.; Piccenna, L.; McNeil, J. J. *J. Obes.* **2010**, *2011*, 1–18.
108. Dietrich, M. O.; Horvath, T. L. *Nat. Rev. Drug Discov.* **2012**, *11*(9), 1–17.
109. Kim G-S; Park, H. J.; Woo, J.-H.; et al. Citrus Aurantium Flavonoids Inhibit Adipogenesis through the Akt Signaling Pathway in 3T3-L1 Cells. *BMC Complement. Altern. Med.* **2012**, *12*(1), 31; *BMC Complement. Altern. Med.* **2012**, *2*(1), 2–10.
110. Stohs, S. J.; Preuss, H. G.; Shara, M. *Int. J. Med. Sci.* **2012**, *9*(7), 527–538.
111. Altiner, A.; Ates, A.; Gursel, F. E.; Bilal, T. *J. Anim. Plant Sci.* **2012**, *22*, 872–877.
112. Márquez, F.; Babio, N.; Bulló, M.; Salas-Salvadó, J. *Crit. Rev. Food Sci. Nutr.* **2012**, *52*(7), 585–594.
113. Payab, M.; Hasani-Ranjbar, S.; Shahbal, N.; et al. Eothmioams Asramaoct PR. *Phytother. Res.* **2019**, *2019*, 1–20.
114. Razavi, B.; Hosseinzadeh, H. *J. Endocrinol. Investig.* **2014**, 7(11), 1031–1040.
115. Chevassus, H.; Gaillard, J.-B.; Farret, A.; et al. *Eur. J. Clin. Pharmacol.* **2010**, *66*(5), 449–455.
116. Szewczyk, K.; Zidorn, C. *J. Ethnopharmacol.* **2014**, *152*(3), 424–443.
117. Gao, Q.-H.; Wu, C.-S.; Wang, M. *J. Agric. Food Chem.* **2013**, *61*(14), 3351–3363.
118. Mostafa, U. E.-S.; Labban, L. *Adv. Med. Plant Res.* **2013**, *1*(3), 49–55.

CHAPTER FOURTEEN

Analgesic and antipyretic natural products

Jamil A. Shilpi*, Shaikh Jamal Uddin
Pharmacy Discipline, Life Science School, Khulna University, Khulna, Bangladesh
*Corresponding author: e-mail address: jamilshilpi@yahoo.com

Contents

1. Introduction

Almost every human being experiences pain or nociception in their lifetime, and in certain cases, it may require medical attention as well as long term therapeutic management. While pain could be a preventive manifestation to protect organs from injury or damage, maladaptive pain, in contrast, is the response to the injury or damage of the nervous system. Although treatment of the underlying cause would be the therapeutic goal, relief from pain is necessary; otherwise it can severely affect the quality of life. Physiologically, pain can be classified as nociceptive pain and neuropathic pain. Nociceptive pain, caused by external factors including

Annual Reports in Medicinal Chemistry, Volume 55
ISSN 0065-7743
https://doi.org/10.1016/bs.armc.2020.03.003

mechanical, chemical or thermal stimuli, resulting in sharp or dull, aching, nagging cramping, throbbing or pressure-like feeling and may include nausea and vomiting. Nociceptive pain is received by nociceptors and carried to the CNS through Aδ (myelinated) or C fibers (unmyelinated).[1] Nociceptive pain involves a number of mediators including prostaglandins, thromboxanes, cytokines, histamine, bradykinin and serotonin. Neuropathic or neurogenic pain is associated with central nerve tissues, and may occur due to nerve degeneration, diabetic neuropathy, postherpetic neuralgia, posttraumatic neuropathic pain, causalgia, reflex sympathetic dystrophy and osteoarthritic pain. Neuropathic pain is often dull in nature, persistent and may involve greater sensitivity toward pain stimuli.[2] Although pain has been classified in these two distinct categories, indeed there are some overlaps, such as neuronal inflammation leading to nerve damage. Nociceptive pain often responds to the commonly available nonsteroidal anti-inflammatory drugs (NSAIDs), but neuropathic pain is difficult to manage and opioid analgesics are the drug of choice. While aforementioned drugs are the mainstay for the treatment of pain, some other alternative methods are also practiced, which include physiotherapy, aromatherapy, acupuncture and electrical nerve stimulation. Pyresis is most often associated with infections by pathogens, but can also happen in some noninfectious cases. Algesia and pyresis are quite different in terms of their cause, but both shares some common pathways for their elicitation in the body and thus commonly available NSAIDs are effective both as analgesic and as antipyretic.

Use of natural products as analgesic or antipyretic is as old as that of using plants for the cure of diseases. Some of the early documentations record the use of various plants for the management of pain and pyresis as some of the most notable herbal practices. Isolation of natural products with analgesic and antipyretic activities are some of the first records in the field of medicinal chemistry. These include morphine and salicylic acid that ultimately created the two major classification of analgesic and antipyretic drugs. While a good number of natural products are currently in use, introduction of some new agents in recent years with novel mechanism of actions emphasized the importance of natural products not only for new drug discovery but also to dictate the basic research for the understanding of physiologic process involved in pain and pyresis.

2. Analgesic and antipyretic natural products

Morphine, hailed as God's own medicine by Sir William Osler, has been in use in the form of opium for over 3500 years. Isolation of morphine from opium by the German pharmacist Sertürner in 1805 was the first natural product obtained in pure form and this laid the foundation of modern-day pharmacology. Salicin from willow bark was isolated by a German pharmacologist Johann Andreas Büchner in 1828, which in the form of acetyl salicylic acid earned the title of first miracle drug, is still one of the most widely used drug. These two agents formed the basis of classifying analgesics into two main categories, the opioid analgesics and the NSAIDs. Later, some other classes were included among which cannabinoid receptor (CB) agonists, various ion channel blockers, interfering neurotransmitter functions are notable to mention. The Physicians' Desk Reference for Herbal Medicines includes more than hundred plants with the indication as analgesics. Some of most common natural products with extensive use in modern day medicine include cannabis, Jamaican dogwood, Jin Bu Huan (*Lycopodium serratum* or *Polygala chinensis*), poplar buds, rosemary leaves, veratrine seeds, cajuput oil.[3] Use of plant derived volatile oil as an analgesic was a common practice in various folklore medicine and they are still used in many cultures around the world. The analgesic activity of the essential oils is based on the analgesic effect of some of its components including menthol, eugenol, linalool, carvacrol, *p*-cymene, phellandrin, thymol and camphor.

2.1 Recently approved analgesic and antipyretic natural products as drug

Ziconotide (**1**) is the first marine derived natural product to be approved by FDA in 2005 (Prialt®) for its use as a nonopioid analgesic for neuropathic pain.[4] Ziconotide, also known as ω-conotoxin MVIIA (ω-MVIIA), is a component of the cone snail (*Conus magus*) venom consisting of 25 amino acids with 6 cysteine residues and three disulfide linkages. The discovery of ziconotide was through a basic research to find how the cone snails paralyze their fish of prey.[5] Among different peptides of the cone snail toxins, ziconotide was found to cause shaking in mice.[6] In fact it has a quite different effect in fish in which it causes delayed paralysis and interestingly the effect of ziconotide is also different in human due to its specific effect on N-type voltage gated calcium channels (VGCCs) $Ca_v2.2$. This ion channel is distributed in the PNS and CNS but of therapeutic importance is its presence in the laminae I and II of the spinal cord dorsal horn where ziconotide is administered to avoid side effects.[7] Ziconotide inhibits presynaptic VGCCs and prevents the passage of Ca^{2+}, resulting in the blockage of neurotransmitter release and thereby stops the synaptic transmission of pain sensation.[8] Ziconotide is believed to bind to the α_{1B} subunit of the $Ca_v2.2$ channel and causes an obstruction in the Ca^{2+} entry.[9] Currently ziconotide is available for intrathecal administration through a portable infusion pump for delivery direct into the dorsal horn of the spinal cord through a needle or catheter. Systemic administration of this compound is not indicated as it can cause orthostatic hypotension through the inhibition of norepinephrine release from the sympathetic nerves.[10]

Although, preparations containing capsaicin (**2**) was available in the market for a while, the FDA and the EU finally approved capsaicin 8% patch (Qutenza®) for neuropathic pain in 2009.[11,12] Previous preparations consisted of low concentrations of capsaicin (0.025–0.075%) in over the counter creams, lotions or patches for neuropathic pain conditions including diabetic neuropathy, postsurgical pain, osteoarthritis and rheumatoid arthritis. However, these preparations were of limited success due to the burning sensation, skin reaction in some patients and its requirement to apply several times a day for few weeks to attain therapeutic effect. This limitation was resolved by the use of high concentration of capsaisin. Capsaicin, a protoalkaloid is one of the active constituents of chilli pepper or paprika (*Capsicum*). Capsaicin is a highly selective and potent agonist of the transient receptor potential vanilloid-1 (TRPV-1) receptor. TRPV-1 is a non-selective ligand gated cation channel expressed in polymodal nociceptive nerve fibers, mainly the Aδ and C fibers. TRPV-1 receptors are activated by heat (>43 °C), proton/cation (pH < 6), various vanilloids, voltage, long chain unsaturated fatty acids and various inflammatory mediators resulting in the transient opening of ion channels and the influx of Na^+ and Ca^{2+}. Unlike the brief effects of other ligands, binding of capsaicin induces a persistent effect on TRPV-1 ion channel. Binding of capsaicin shifts the calcium: sodium threshold from 8:1 to 25:1 leading to an excessive influx of Ca^{2+}, causing the Ca^{2+} dependent release of protease enzyme which depolymerizes microtubules and breakdown of cytoskeletons.[13] In addition, direct inhibition of electron transport chain to cause mitochondrial dysfunction, edema by the simultaneous influx of chloride ion with cations ultimately results in the loss of cellular functions of the nociceptors.[14] Thus, treatment with capsaicin causes a temporary excitation followed by a long-lasting refractory state known as desensitization where the nerve fibers are no longer sensitive to its previous stimuli or other stimuli.[15,16]

Botulinum toxin type A (BoNT/A) is one of the seven neurotoxic proteins produced by *Clostridium botulinum* and related species. The FDA approved it for the treatment of chronic migraine and pediatric lower limb spasticity in 2010 and 2016, respectively. BoNT/A has the ability to inhibit the release of a number of neurotransmitters including acetylcholine, substance P, calcitonin gene-related peptide, vasoactive intestinal peptide and neuropeptide Y. BoNT/A has also been found to interfere with the functions of TRPV1, TRPA1 and Na^+ channel. The role of BoNT/A on the inflammatory mediators is still inconclusive. These findings helped in the understanding that the analgesic effect of BoNT/A is not by muscle relaxation but through centrally acting mechanisms. In fact, BoNT/A is effective in the treatment of various forms of pain including headaches, migraine, musculoskeletal pain and pains caused by hyperactive muscle.[17,18]

2.2 Analgesic and antipyretic natural products as drug candidates

The discovery of ziconotide inspired the search for more venoms both from marine or terrestrial organisms including cone snails, spiders, centipedes and scorpions for various pharmacological activities including analgesic activity. Some of the toxins that are under clinical trial for pain relief include ω-conotoxin CVID, χ-conotoxin MrIA, α-conotoxin VclA and contulakin-G.

Leconotide (**3**) (also known as CVID, AM336 and CNSB004), an ω-conotoxin peptide of *Conus catus* has close similarity to ziconotide both in terms of structure and mechanism of action.[19] Intrathecal administration of ziconotide requires expensive equipment and should be performed by only experts. The procedure carries the risk of infection at the site of surgery and might need removal of the pump in case of deep infection. The superiority of leconotide over ziconotide is that it is less toxic than ziconotide and can be administered intravenously.[20] Previously, intrathecal administration of leconotide in rats showed better ratio of efficacy over toxicity. Leconotide was also tested intrathecally to treat pain in cancer patients.[21]

MrIA (**4**) is a 13-amino acid containing χ-conotoxin peptide isolated from the marble cone, *Conus marmoreus*. Similar to other conotoxins of this class, MrIA has a hair-pin like structure formed by the two disulfide linkages between Cys1-Cys4 and Cys2-Cys3. MrIA reversibly and selectively inhibits norepinephrine uptake by norepinephrine transporter (NET) in a noncompetitive manner. Neuronal transmission of pain is thus inhibited by the depletion of norepinephrine in presynaptic neurones.[22] Xen2174 (**5**) is the synthetic analogue of MrIA differing only in the C-terminal where an amine group replaces the hydroxyl group. This change increases the affinity of the compound for NET receptor but does not hamper the stereochemistry of the compound.[23] Although in one of the recent clinical trial, intrathecal administration of Xen2174 did not show promising result, but as a representative of χ-conopeptides, MrIA might help in the development of novel analgesic.[24]

Contulakin G (**6**), a 16 amino acid containing glycopeptide of *C. geographus*, shows analgesic activity through the agonistic action at the neurotensin receptors (NRT). Contulakin G has a close structural similarity with that of endogenous neuropeptide neurotensin (13 amino acid residues) and binds to all of the subtypes of NRT receptor, i.e., NTSR1, NTSR2 and NTSR3. Posttranslational modification leads to a proglutamate residue at the *N*-terminus and a disaccharide (β-D-Gal-(1→3)-α-D-GalNAc-(1→) attached to Thr10 of contulakin G. Contulakin G earned orphan drug designation by FDA and entered clinical trial for intrathecal administration to treat chronic intractable pain.[25] More potent analgesic action of contulakin G than its affinity for NRT1 receptor as compared to neurotensin suggests the involvement of additional mechanisms for the analgesic action.[26]

The cone snail *Conus victoriae* derived 16-residue venom peptide Vc1.1 (**7**) is an α-conotoxin with amidated C-terminus, two disulfide linkages of Cys1-Cys3 and Cys2-Cys4 arrangement.[27] The Vc1.1 conotoxin exerts analgesic action through the antagonism of neuronal nicotinic acetylcholine receptors (nAChRs). The underlying mechanism of Vc1.1 mediated analgesia is through the activation of $GABA_B$ receptor and VGCCs ($Ca_v2.2$ and Ca_v 2.3).[28] This compound was in clinical trial for the management of neuropathic pain through subcutaneous administration, but later discontinued. A newly engineered cyclic Vc1.1 (**8**), with no change in three-dimensional structure or its biological activity, is now in preclinical trial that shows resistance toward hydrolysis and suitable for oral administration.[29]

Some other venom peptides that are in the preclinical trial are mambalgin-1 from black mamba snake venom (*Dendroaspis polylepis polylepis*),[30] APETx2 from sea anemone (*Anthopleura elegantissima*),[31] PcTX-1 (psalmotoxin 1) from South American tarantula (*Psalmopoeus cambridgei*)[32] all of which shows their analgesic activity solely or partly by inhibition of the acid sensing ion channels (ASICs).

The discovery of the involvement of TRPV-1 in pain and capsaicin (**2**) as the only known ligand paved the way to find its natural occurrence or natural analogues and to synthesize derivatives, which resulted in more than thousand patients.[16] Some notable natural capsaicinoids in different phases of clinical trials for the treatment of pain are *cis*-capsaicin, nonivamide (**9**) and capsiate (**10**). There are several topical and intranasal preparations containing *cis*-capsaicin (zucapsiacin) in Phase II and III clinical trials for the treatment of various pain conditions. In nature only the *trans*-isomer can be found and thus *cis*-capsaicin, also known as civamide need to be synthesized in the laboratory.[33] However, both isomers of capsaicin and their derivatives are no longer extracted from the plant and methods for their laboratory synthesis has been established. Nonivamide (**9**) is one of the alkamides present in the chilli pepper that differs from capsaicin in the side chain, which is devoid of the double bond making it more heat stable than capsaicin. Due to the presence of vanillyl moiety, it also acts as a TRPV-1 agonist. Topical preparation containing nicoboxil (2.5%) and nonivamide (0.4%) is in Phase III clinical trial for the treatment of nonspecific lower back pain.[34] Capsiate (**10**) is one of the analogues isolated from nonpungent cultivar of red pepper "CH-19 sweet" named as capsinoids. Capsinoids contains an ester linkage instead of amide linkage, which makes is less pungent, offering better patient acceptability than capsaicinoids.[35] Presence of ester linkage makes the capsinoids more prone to quick metabolism, but at the same time, they offer better absorption and transport to the site of action and more options of routes of administration. Capsiate (**10**) has a potency similar to that of capsaicin and thus holds a great promise for TRPV-1 based analgesic drug development.[36] Orally administered capsiate (**10**) preparation for the treatment of several conditions including analgesic and antiinflammatory action has completed its Phase I clinical trial.

Resiniferatoxin (RTX) (**11**), a cyclic diterpene, found in the resin of the Moroccan cactus *Euphorbia resinifera*, works the same way as that of capsaicin but 500–1000 times hotter than capsaicin on the Scoville scale. Although from a different source, and with a complex structure, same as the capsaicinoids or capsinoids, RTX possesses a vanilloid moiety that helps it interact with TRPV-1 receptors. Initially it was regarded as a nonpungent TRPV-1 agonist, but clinical data suggests its pungency slightly lower than that of capsaicin. RTX (**11**) rather kills the sensory neurones making them an important tool as molecular scalpel to destroy sensory neurones of a specific site to render analgesia. Subcutaneous administration of RTX (**11**) in experimental animals showed a complete loss of sensory neurones that started reappearing gradually reaching 30% of its original count in 4 weeks' time.[37] RTX (**11**) was in Phase II clinical trial for the treatment of neurogenic pain caused by interstitial cystitis (IC) but was the analysis of the report was declared ineffective, possibly due to the emerging success of botulinum toxin to treat it.[38] Currently RTX (**11**) is now in Phase I clinical trial for the treatment of pain associated with osteoarthritis and advanced cancer (Figs. 1 and 2).

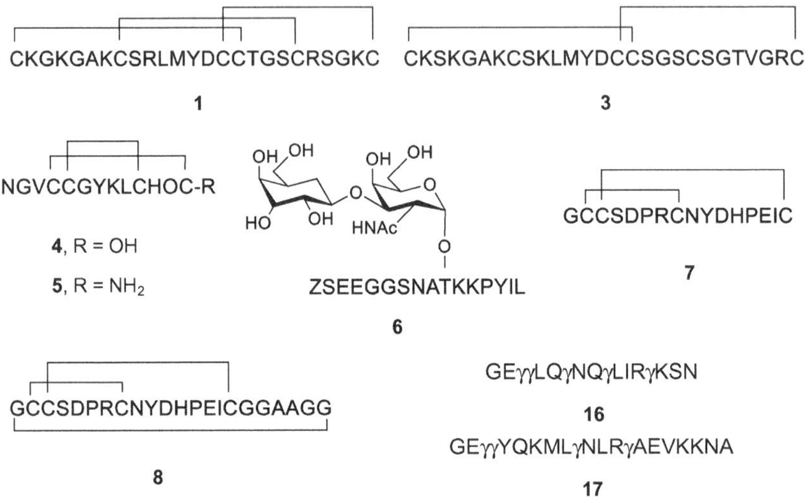

Fig. 1 Analgesic conotoxins (linked lines indicate disulphide linkages between cysteine residues; O in structures **4** and **5**, 4-*trans*-hydroxyproline; Z, pyroglutamate; Γ, γ-carboxyglutamic acid).

Tetrodotoxin (TTX) (**12**), a toxin usually found in the pufferfish is a potent inhibitor of the voltage gated sodium channels (VGSCs). Blocking of VGSCs results in the disruption of neuronal function and paralysis of its prey. The pufferfish poison tetrodotoxin is another natural product that is in Phase III clinical trial by WEX Pharmaceuticals Inc. (Tectin®) for its use as an analgesic in neuropathic pain.[4] Apart from pufferfish, TTX (**12**) has also been isolated from some other marine (red algae, copepods, gastropod, cephalopod and starfish) and terrestrial (frog) organisms, microbes harboring these organisms, but the original source of TTX (**12**) is still a controversy. Some research revealed that the tetrodotoxin can be produced by commensal microbes residing in a number of marine organisms or in sea sediments and thus gives a hope for mass production of this small but complex molecule rather in an easy way.[39] The underlying mechanism of TTX (**12**) makes it a good candidate as an analgesic drug. Being the most powerful neurotoxin known so far (1200 time more potent than cyanide), prevents the use of TTX (**12**) in human for therapeutic purposes, but offers a prototype for the development of analgesic drug.[40]

Last but not the least are the two important plants *Mitragyna speciosa* of the coffee family and *Salvia divinorum* of the mint family. *M. speciose*, native to South East Asia, popularly known by its Thai name "kratom" became available to the Western world with the recent means of globalization.

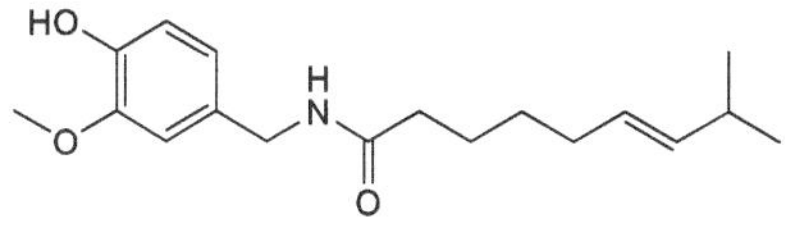

2

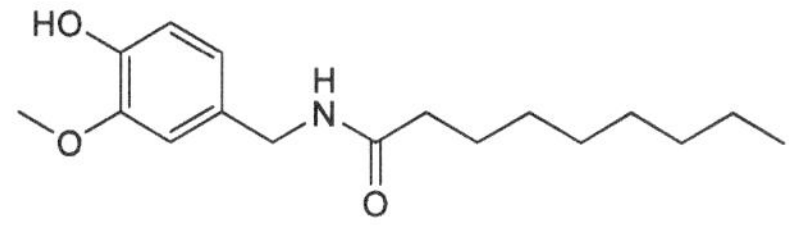

9

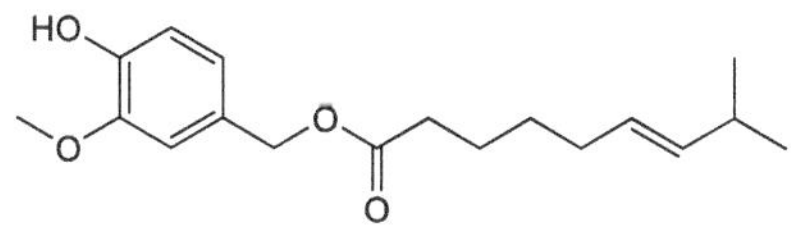

10

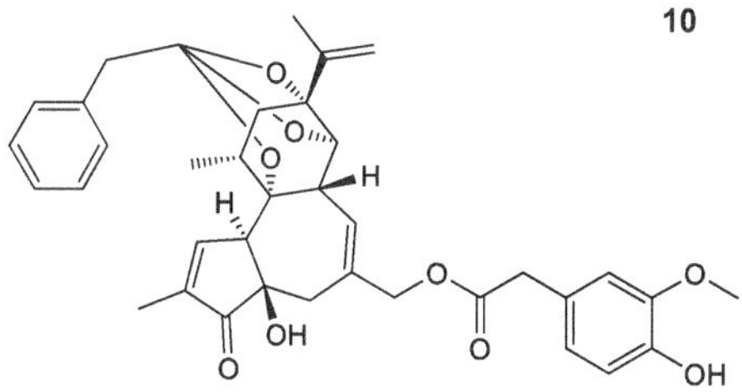

11

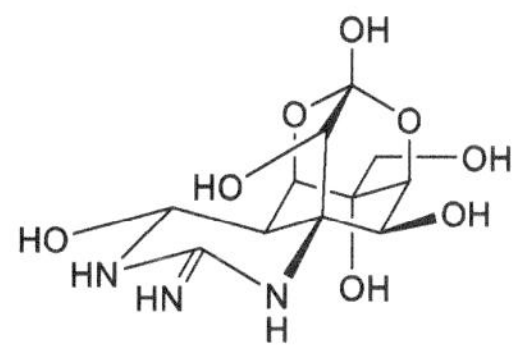

12

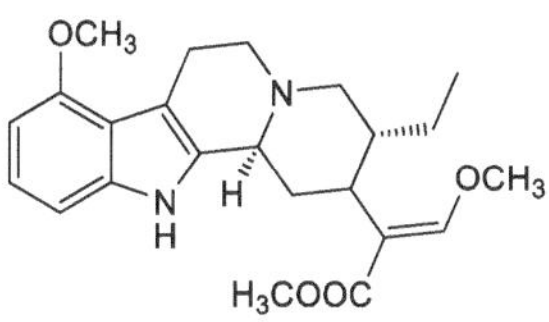

13

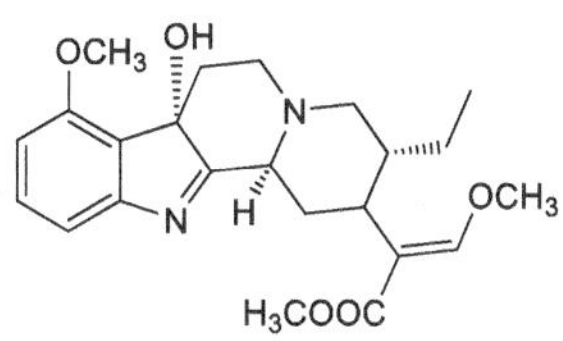

14

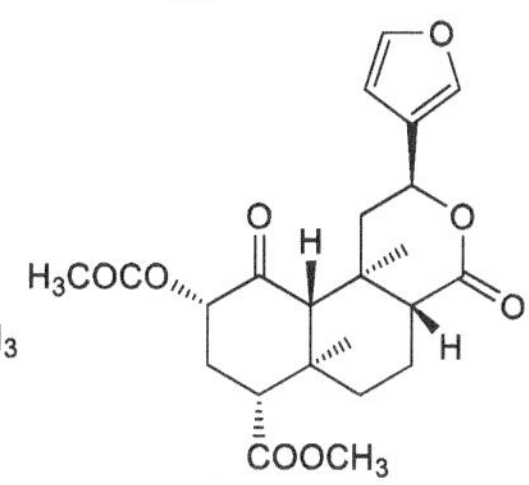

15

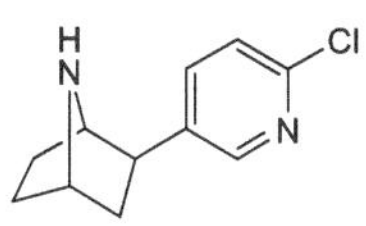

18

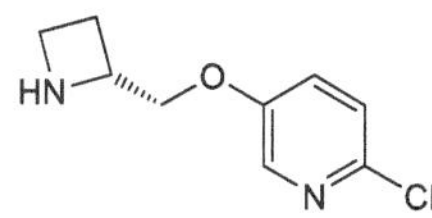

19

Fig. 2 Plant or animal derived analgesic compounds and analogues.

Two main indole alkaloids, namely mitragynine (**13**) and 7-hydroxymitragynine (**14**), of the leaves exert their analgesic action both centrally (mediated through μ-opioid receptor) and peripherally (mediated by α_2-adrenoreceptor and probably through serotonergic receptor). As pain often involves both central and peripheral mechanisms, the ability of kratom to work through both ways makes it an attractive alternative to present therapies. Moreover, it is a μ-opioid agonist at nanomolar affinity but with less respiratory, constipating and withdrawal symptoms than morphine. This is possibly because μ-opioid receptor binding of these kratom alkaloids do not recruit β-arrestin. Although contradictory opinions and some restrictions existed with kratom over the chance of abuse, the great potential of the therapeutic utility has turned FDA and other regulatory bodies for further research to fill the gap in the knowledge on kratom and to identify its suitability to use as a drug.[41–43] The Mexican sage bush *S. divinorum* has long been used by Mazatecs of Oaxaca for it hallucinogenic property. Salvinorin A (**15**), a neoclerodane diterpenoid is the active constituent responsible for the psychoactive, antinociceptive, antipruritic and antipruritic activity mediated through its potent and highly selective agonism of κ-opioid receptor.[44,45] High selectivity for κ-opioid receptor renders it nonaddicting opioid agonist making it an alluring molecule for researchers to design molecules with opioid receptor mediated therapeutic uses, i.e., analgesic, antipruritic, antidepressant actions. Although several hundred analogues has been synthesized so far, a well-defined structure activity relationship of salvinorin A (**15**) analogues is yet to be established. In addition to this, its short duration of action and related pharmacokinetic issues further hindering the development of salvinorin A (**15**) based new analgesic.[46] However, the current trend indicates more intensive research is going to continue with a view to exploit the pharmacological effects of salvinorin A (**15**).

2.3 Distribution of analgesic and antipyretic compounds in nature

Searching for analgesic and antipyretic agents in natural resources can be a tricky job as natural products with analgesic and antipyretic activity is widespread in nature. Moreover, such activity can involve synergism or a combined manifestation of several pharmacological effects. The interpretation of the experimental results becomes more complicated when the effect is associated with anxiolytic, sedative or other psychoactive effects. Chemical class that has the higher possibility to render compounds with analgesic and antipyretic effect includes essential oils, flavonoids, phenolics and sterols.

Plant essential oils have long been used in various traditional medicines as analgesics. Aromatherapy with essential oils has a major expected outcome of relief from chronic pain.[47] Essential oils are complex mixture of lipophilic and volatile compounds with characteristic aroma. Monoterpene is the major class of compounds of essential oils and the rest consists mainly of sesquiterpenes and phenylpropanoids. Some essential oils may also contain diterpenes. Because of the prominent analgesic activity, a good number of essential oil components were investigated and so far more than 50 compounds have been reported having analgesic activity (Table 1).[48,49] Some of the major essential oil components with potent analgesic activity are extensively used in alternative therapeutic practices and are included in some of reference books as supplementary medicine. These components are used alone or in combination with other topical analgesics in the form of balm, ointment or cream. Mechanism of action of these essential oil components are diverse but difficult to formulate them in suitable dosage form due to their volatile nature.

Flavonoids are another large class of natural products that attracted attention due to their wide biological activities including antioxidant property. A large number of flavonoids have been reported for their analgesic activity along with their mechanism of action (Table 2). Mechanisms identified so far indicate that the analgesic activity of flavonoids might involve more than one mechanism and inhibit both nociceptive and neuropathic pain. Analgesic activity of flavonoids is mainly due to their anti-inflammatory effect mediated through the inhibition of PEG_2 synthesis by COX-1 and COX-2 enzymes. As an inhibitor of NF-κβ, flavonoids can also reduce

Table 1 Essential oil constituents with analgesic activity.

Class	Compounds
Monoterpenes	(−)-α-Bisabolol, borneol, carvacrol, (−) and (+)-carvone, carvone epoxide, 1,8-cineol, citral (*cis*-isomer neral and *trans*-isomer geranial), (±)-citronellal, citronellol, citronellyl acetate, *p*-cymene, (−)-fenchone, hydroxydihydrocarvone, R-(+)-limonene, limonene oxide, (−)-linalool, linalyl acetate, myrcene, (−)-menthol, (+)-menthone, β-myrcene, myrtenol, nerol, α-phellandrene, (+)-pulegone, α-terpineol, (+)-3-thujone, (−) and (+)-isothujone, thymol, thymol acetate, thymoquinone, α and β-pinene, pulegone oxide, rotundifolone
Sesquiterpenes	Agarospirol, β-caryophyllene, caryophyllene oxide, costunolide, dehydrocostus lactone, farnesol, jinkoheremol, nerolidol (peruviol), α and β-santalol, zerumbone
Phenylpropanoids	*trans*-Anethole, cinnamaldehyde, eugenol, evodione, guaiacol, leptonol, ligustilide, methyleugenol, 1-nitro-2-phenylethane, vanillin

Table 2 Miscellaneous natural products with analgesic activity.

Class	Compounds
Flavonoids	Amentoflavone, apigenin, artemitin, baicalein, baicalin, fukugetin, ginkgetin, gossypin, hesperidin, hyperin, kaempferol, linarin, luteolin, mangiferin, myricetin, naringenin, pectolinarin, procumbentin, scandenone, quercertin, rutin, volkensiflavone
Alkaloids	Mitragynine (**13**),[41] nicotine[50]
Diterpenes	Hardwickiic acid,[51] salvinorin A (**15**)
Miscellaneous	Curcumin,[52] bromelain,[53] resveratrol,[54] ginsenosides Rc, Rd., Re and Rf[54]

the NF-κβ dependent COX-2 expression, shifting the selectivity of flavonoids more toward COX-2. However, since their activity is not exclusive to COX-2, they do not pose the threat of cardiovascular risk as that of the selective COX-2 inhibitors. Some of the flavonoids including baicalin have been found to inhibit the production of other inflammatory mediators including TNF-α, TNF-γ, IL-1β, IL-6, IL-8 and thus represent a good candidate for the treatment of rheumatoid arthritis.[55,56] Some of the extensively studied flavonoids for antinociceptive activity include quercetin, baicalin, hesperidin, and their glycosides. Investigation with quercetin revealed the involvement of D_2, α_2 and 5-HT_3 receptor.[57,58] In addition to the inhibition of inflammatory mediators including histamine, serotonin, cytokines, chemokines, cell derived factor (CDF), MIP-1β, MCP-2, C lymphotactin, flavonoids can reduce oxidative stress by inhibiting xanthine oxidase, NO and ROS and thereby contribute toward the antinociceptive activity.[59] Several bioflavonoids, e.g., cupressuflavone, ginkgetin, I3,II8-binaringenin, also reported to show analgesic activity in animal models and underlying mechanism involves similar to that of most other flavonoids, i.e., inhibition of PGE_2, TNF-a, IL-1b, and IL-6 production.[60,61]

Furthermore, many phytosterols (e.g., stigmasterol), triterpenoids (e.g., ursolic acid, betulinic acid, 24-hydroxytormentic acid, α and β amyrin acetate) phenolics (e.g., gallic acid derivatives) and phenyl propanoids can also show varying degrees of analgesic and antipyretic effect in experimental animals.[62] While phytosterols can show their analgesic effect through their anti-inflammatory effect, phenolics were suggested to exert their effect through the inhibition of oxygen species (e.g., NO), that have some role in inflammation or helps in neurotransmission.

However, therapeutic potential of aforementioned agents might not be promising due to their low potency but can help in the understanding of pain mechanisms and use them for structure-based drug design. However, identification of natural products with novel mechanisms of pain relief still continues and requires an in depth study of the ethnopharmacological uses to identify potential source of new analgesics. Same as that of morphine or salicin, discovery of ziconotide, capsaicin and many other analgesics lie in the wise judgment of their traditional use.

3. Mechanisms of action of analgesic and antipyretic natural products

3.1 Inhibition of inflammatory mediators

Injury or infection of tissue results in immune response characterized by accumulation of inflammatory mediators at the site. Most common inflammatory mediators include bradykinin, cytokines (TNF-α, IL-1, IL-6, IL-8), prostaglandins (PGE_2, PGI_2), growth factors (NGF), adenosine and endothelin, which either activate peripheral nociceptors or stimulate production of pain inducing agent (algogenic).[63,64] Apart from these, histamine, extracellular ATP, proteases, reactive oxygen species (ROS), leukotrienes, and lipoxins also have involvements in nociception. Commonly available NSAIDs including salicylates exert their antinociceptive activity through their inhibition of prostanoid synthesis by cyclooxygenase enzymes (COX-1 and or COX-2).

NSAIDs, including salicylates are also used as antipyretic agents to reduce fever caused by infection or as an inflammatory reaction. Cases like infection, tissue injury, graft rejection or malignancy triggers prostaglandin and cytokine release while cytokines can further induce synthesis of PGE_2 and this elevated level of PGE_2 increases cyclic AMP which in turn, causes hypothalamus to increase body temperature.[65] The underlying mechanism of most of the antipyretics lie in their ability to prevent prostanoid synthesis by inhibiting COX enzymes. Although recent studies suggests non-cyclooxygenase pathways involved in the action of antipyretic drugs which include inhibition of pyrogenic cytokine production, suppression of leucocyte migration and adhesion and enhanced production of endogenous antipyretics.[66] With the limited resource to reduce fever, use of medicinal plants by various cultures is a common practice irrespective of their active constituent is known or not.[3,67]

3.2 Inhibition of voltage gated calcium channels

Voltage gated calcium channels (VGCCs) (Ca_v) are transmembrane proteins that are activated by membrane depolarization and the activation allows the entry of Ca^{2+} in the cell important for the elicitation of secondary processes. There are six types of VGCCs of which, N-type VGCCs are located in the neurones and thus important in pain sensation.[68] Ziconotide was the first compound to be discovered to work through the inhibition of N-type VGCC. Binding of the opioid analgesics to opioid receptors results in the modulation of N and P/Q-type VGCCs leading to a disruption of Ca^{2+} influx in the cell.[69] In addition to neuronal cells, T-type VGCCs are also located in cardiac and smooth muscle tissues and distributed throughout the body. Distribution of P/Q-type is same as that of N-type VGCC. The actions of P/Q-type is not clearly understood while binding of ligands to R-type showed variable results.[70] Inhibition of T-type might work as an analgesic since mice lacking T-type $Ca_v3.2$ showed less response to pain stimulus[71] and showed a direct involvement in the processing of pain response especially in the dorsal horn.[72]

3.3 Inhibition of voltage gated sodium channels

Voltage gated sodium channels VGSCs (Na_v) are transmembrane proteins that are activated by membrane depolarization and results in transient influx of Na^+ before the pore is closed by an intracellular loop. VGSCs regulates the release of catecholamines and important for the modulation of a number of stimulus including touch, temperature and smell.[73,74] There are nine members of this family and names as $Na_v1.1$ to $Na_v1.9$. While $Na_v1.1$, $Na_v1.6$, $Na_v1.7$ and $Na_v1.9$ are associated with the modulation of different types of pain, $Na_v1.1$, $Na_v1.3$, $Na_v1.6$, $Na_v1.7$, $Na_v1.8$ and $Na_v1.9$ are involved in neuropathic pain. Thus, these ion channels are attractive therapeutic targets for both nociceptive and neuropathic pain. Among the toxins produced by the cone snails, μ-conotoxins are known for their pore blocking ability of these channels. TTX and some spider venom peptides are well known inhibitor of VGSCs and represents a valuable source for further drug development.[75]

3.4 Agonist and antagonist of transient receptor potential (TRP) channels

These are ion channels having nonselective permeability to cations, including Na^+, Ca^{2+} and Mg^{2+}. There are about 27 TRPs have been identified in human so far which shares close similarities, but vary greatly in function. Same as to their functional variability, members of this family also differ in their activator. For example, TRPV1, TRPV3, TRPV4 are activated by heat while TRPV4 is also activated by change in osmolarity, TRPM8 is activated by cold and TRPA1 is sensitive mainly to chemicals. Often these receptors can be activated by several physical and chemical factors and notable natural products that activate TRPs include capsaicin, piperine, gingerol, camphor, eugenol, menthol, eucalyptol and resiniferatoxin.[76] Discovery of the involvement of these receptors in pain coupled with a good number of natural products that interact with these channels inspired numerous studies for possible drug development. Unfortunately, capsaicin is the only one that got the approval of FDA. Several other natural products that act on these receptors are in use often as topical analgesic in alternative medicine or acting as the prototype for the synthesis of new drug molecules targeting mainly TRPV1, TRPA1 and TRPM8.[77]

3.5 Opioid receptor agonist

Opioid receptors (OR) are G protein coupled transmembrane proteins, which can be further classified into four subtypes, μ-OR, κ-OR, δ-OR and nociception peptide binding OR. Binding of endogenous peptides or exogenous agonists to these receptors results in the activation of heterotrimeric G-protein complex with then interacts with various ion channels including Ca^{2+} channel, Na^{+} channel and K^{+} channel. Activation of μ-OR, κ-OR and δ-OR can cause analgesia, but μ-OR agonism is linked to tolerance, dependence, constipation and respiratory depression, κ-OR agonism is responsible for hallucination, sedation, diuresis while δ-OR results in physical dependence and convulsion. Tolerance, respiratory depression and some other nonanalgesic effects of opioid agonists is due to the β-arrestin recruitment caused by alternative receptor conformation of OR.[78]

3.6 Cannabinoid receptor agonist

While CB_1 receptor is believed to be responsible for the marijuana-like effect, recent findings suggest that CB_2 receptor might contribute to anti-nociceptive effect. However, the role of CB_1 in the inhibition of pain cannot be ignored as it can contribute through the aversion of memories toward noxious stimulus,[79] inhibition of glutamate release.[80] Antinociceptive effect of CBs is through to their modulation of neurotransmitter release which include glycine, dopamine, endocannabinoids and endogenous opioids.[81]

Dibenzopyran derivatives from the plant *Cannabis sativa*, named as cannabinoids, later renamed as phytocannabinoids are the classical agonists of CB receptors, notable of which are Δ^9-tetrahydrocannabinol (Δ^9-THC) and cannabinol. Although a number of synthetic derivatives were developed, currently Δ^9-THC and its synthetic derivative nabilone was approved for the treatment of neuropathic pain, fibromyalgia, anti-emetic and some other conditions.

3.7 Miscellaneous targets for pain inhibition

3.7.1 Agonist of G-protein coupled receptors GPR18 and GPR55

Cannabinoids have been found to exert their effect, to some extent, though their interaction with some other G protein-coupled receptors namely, GPR18 and GPR55 and thus regarded as important targets for the development of new analgesic agents.[82]

3.7.2 Inhibition of serotonin–norepinephrine transporter

Serotonin and norepinephrine are two important neurotransmitters involved in the modulation of pain mainly in the descending pain pathways. Inhibition of serotonin and norepinephrine transporter can lead to the depletion of these neurotransmitters in the presynaptic neurones.[83] Some of the conotoxins, especially the χ-conopeptides show analgesic action through their highly specific inhibition of NET.[26]

3.7.3 NMDA receptor antagonists

NMDA is an ionotropic class of glutamate receptor, named according to its synthetic agonist *N*-methyl-D-aspartate. Activation of NMDA opens non-specific cationic channel that allows the entry of monovalent ions and Ca^{2+}. Although closed in normal physiologic condition, constant depolarization during pathological state induced Ca^{2+} entry causes phosphorylation of NMDA receptor resulting in the reduced blockage of the NMDA receptor by extracellular Mg^{2+}. Thus, NMDA has been linked to hyperalgesia and allodynia after injury.[84] Antagonism of NMDA can be achieved at different recognition sites of the receptor and classified as uncompetitive antagonists, glycine site antagonists and NR2B subunit binding antagonists.[85] Linalool, one of the component of plant essential oil, long known for its analgesic activity and reported for its analgesic activity in various pain models,[86] is a NMDA receptor antagonist.[87] Conantokins are 17–27 amino acid containing venom peptides produced by the fish hunting cone snail *Conus* that are known for their inhibition of NMDA receptors.[88] Conantokins G (**16**) and T (**17**), two most extensively studied conantokins of *C. geographus* show their analgesic activity through NMDA receptor. Conantokins G (**16**) and T (**17**) contain 17 and 21 amino acids respectively, and have a high proportion of the unusual amino acid γ-carboxyglutamate (5 and 4, respectively).[89] Conantokin G (**16**) binds to the NR2B subunit while conantokin T (**17**) binds to both NR2A and NR2B subunits of NMDA receptor. Intrathecal administration of these two peptides showed strong analgesic activity in several pain models in mice at a much lower concentration than the concentration required to impair motor function.[90]

3.7.4 Neurotensin receptor agonists

Binding of the tridecapeptide neurotensin (NT) to the neurotensin receptor (NTR) can modulate pain transmission in several regions of CNS, which include central amygdala, hypothalamic medial preoptic nucleus, certain thalamic nuclei, periaqueductal gray and rostroventral medulla.[91] So far, three types of NTR has been identified, two of which NTR1 and NTR2 are G-protein coupled receptors while NTR3 belongs to sortilin family with a transmembrane domain.[92] Investigation suggests that NTR2 is exclusively involved in the NT mediated antinociception and thus development of NTR2 agonist can be useful as nonopioid analgesic provided the compound has the ability to cross the blood brain barrier to reach these receptors in the CNS.[93]

3.7.5 Neuronal-type nicotinic acetylcholine receptor antagonist

Nicotinic acetylcholine receptors (nAChRs) are ligand gated cation channels located mainly at presynaptic sites where they modulate neurotransmitter release, cell excitability and neuronal integration. They are named according to the endogenous neurotransmitter (acetylcholine) or exogenous ligand (nicotine) binding to them.[94] Studies have shown that activation of nAChRs results in the release of multiple neurotransmitters in the spinal cord that activates descending inhibitory pathways of the brainstem. Involvement of these receptors in pain was first discovered when epibatidine (**18**), an alkaloid from Ecuadoran frog skin, showed its analgesic action partly through its action at nAChRs.[95] Based on the structures of nicotine or acetylcholine, several nAChRs antagonists were developed for the management of pain. ABT-594 (**19**), a 3-pyridyl ether, designed on the basis of the structure of nicotine is now in phase II clinical trial for the treatment of peripheral neuropathic pain.[95,96] Although a good number of nAChRs agonist and antagonist natural product has been reported so far, they lack their specificity and affects both muscle-type and neuronal-type nAChRs.[97] The α-conotoxin class of cone snail venom peptides show analgesic action through their highly specific antagonistic action against neuronal-type nAChRs and represents a potential candidate for new analgesic drug development.[27]

3.7.6 Antagonists of purinergic P2X receptors

The ionotropic P2X and metabotropic P2Y receptors are the subclasses of the ATP gated P2R family of ion channels, expressed along the central and peripheral nociceptive pathways of the spinal cord. Antagonists of P2X1, P2Y12, P2X7, and agonists of P2Y1 represents potential analgesic agent for neuropathic and inflammatory pain, respectively.[98] Although the role of inflammatory cytokines and PGE_2 on these receptors are not clear, but it is well documented that ATP released from sensory neurones or damaged cells mediate algesia through these receptors. Plant derived (emodin, amentoflavone, ligunstrazine and puerarin)[99] and animal derived (purotoxin-1 from Asian *Geolycosa* sp.,)[100] compounds showed their analgesic activity through P2X receptors indicating them a good target to develop new analgesics.

3.7.7 Antagonism of acid sensing ion channels

Acid sensing ion channels (ASICs) are cationic channels, activated by extracellular low pH (>6.9). ASICs are widely distributed in the peripheral and central nervous system and plays an important role in the modulation and transmission of pain. Signals.[101] Some of the currently available NSAIDs including aspirin and salicylic acid can reduce low pH induced pain as well as prevent the inflammation induced expression of ASICs in a COX-independent pathway indicating their analgesic activity is partly due to their interference with ASICs.[102] The subtype ASIC3 is important for inflammatory pain, visceral pain while ASIC1a participates in central pain processing and thus are important targets for both nociceptive and neuropathic pain.[101]

4. Screening of analgesic and antipyretic natural products

The need for better analgesics, coupled with the complexity in the underlying mechanisms led to the development of newer animal and human models for pain research in recent years but some of the old methods are still in use for their simplicity. Methods in pain research can be largely categorized in to two: somatic pain models and visceral pain models (Table 3).

Table 3 Commonly used animal and human models of pain.

Type of pain	Mode	Animal	Human
Somatic	Thermal	Hot plate test, tail-flick test,[103] UV radiation,[104] Hargreaves hindlimb-withdrawal test[105]	Cold, ice water, freeze lesion, contact heat, radiant heat,[106] burn injury[107]
	Chemical	Capsaicin, formalin,[103] carrageenan, Freund's complete adjuvant[106]	Capsaicin, mustard oil, bradykinin, serotonin, potassium chloride, glutamate, vit-C, acidic buffer, infusion of hypertonic saline (i.m.) buffer[108]
	Electrical	Electrical stimulation of tail, shuttle-box test, escape test[109]	Single stimuli, repeated stimuli[110]
	Mechanical	Pin prick, von Frey test using nylon rod or fiber[106]	Touch, pinprick, pressure, ischemic stimulation by pneumatic tourniquet, pressure algometry[110]
Visceral	Thermal	Thermal burns of skeletal muscle[111]	Thermal stimuli of the oesophagus, stomach and jejunum using recirculating hot water[112,113]
	Chemical	Intracolonic injections of mustard oil,[114] intraperitoneal injection of acetic acid[115]	Application of glycerol in large intestine, hypertonic saline in colon, capsaicin in ileum and urinary bladder[112]
	Electrical	Electrical stimulation of skeletal muscle[116]	Electrical stimulation of gut[116]
	Mechanical	Loose ligation of sciatic nerve, tight ligation of partial sciatic nerve, tight ligation of spinal nerves,[106] ballon distension in descending colon,[114] stone implantation in ureter	Hollow organ distension by ballon and barostat[116]

Unlike somatic pain that can be aroused simply by prick, press or application of chemicals, nature of stimuli for visceral pain is rather different which include abnormal distension or contraction of hollow organs of the body.[112,116,117] With the availability of crystal structures of the molecular targets coupled with the discovery of new natural products, receptor based and structure based drug design is increasingly becoming popular in recent years.

Induction of pyresis in test animal is based on the concept of introducing any foreign substance that can induce inflammatory response and raise the basal temperature of the body. Use of brewer's yeast (subcutaneous) is the most popular method due to its easy availability and storage. Some other agents used to induce pyresis include boiled cow milk, TAB (typhoid) vaccine and lipopolysaccharide (bacterial endotoxin).

5. Conclusions

Natural products have contributed enormously in the development of analgesic and antipyretic agents from the very beginning of therapeutic practice and they continue to bring new agents even today. Some serious side effects with the commonly available agents demand the search for suitable alternatives. Discovery of analgesic conotoxins was a breakthrough in discovering some important targets for analgesic drug development. Intensive study of their underlying mechanisms helped in the identification of important molecular targets which can help in receptor based drug design. However, their peptide nature remains one of major obstacle in their therapeutic application. The development of the antihypertensive drug captopril from peptide compound can be an inspiring story to follow for ligand based drug design from conotoxins. Capsaicin, mitragynine and salvinorin A offer interesting scaffolds to develop new analogues with better pharmacokinetic and pharmacodynamics properties.

References

1. Sweetman, S. C. *Martindale: The Complete Drug Reference*. Pharmaceutical Press: London, 2009;2.
2. Apkarian, A. V. In: *Raj's Practical Management of Pain*; Benzon, H., Rathmell, J. P., Wu, C. L., Turk, D. C., Argoff, C. E., Eds.; Elsevier: Philadelphia, 2008; pp 151–173.
3. Sifton, D. W. *Physicians' Desk Reference for Herbal Medicines*. Medical Economics Company: Montvale, NJ, 2001.
4. Newman, D. J.; Cragg, G. M. *Mar. Drugs* **2014**, *12*, 255–278.
5. Olivera, B. M.; McIntosh, J. M.; Clark, C.; Middlemas, D.; Gray, W. R.; Cruz, L. J. *Toxicon* **1985**, *23*, 277–282.

6. Olivera, B. M. In: *Drugs From the Sea*; Fusetani, N. Ed.; Karger: Basel, 2000; pp 74–85.
7. Olivera, B. M.; Gray, W. R.; Zeikus, R.; McIntosh, J. M.; Varga, J.; Rivier, J.; de Santos, V.; Cruz, L. J. *Science* **1985**, *230*, 1338–1343.
8. Wallace, M. S. *Expert Rev. Neurother.* **2006**, *6*, 1423–1428.
9. McGivern, J. G. *Neuropsychiatr. Dis. Treat.* **2007**, *3*, 69–85.
10. Klotz, U. *Int. J. Clin. Pharmacol. Ther.* **2006**, *44*, 478–483.
11. U.S. Food & Drug Administration, 2010 (accessed on 07 Mar 2020) https://www.accessdata.fda.gov/drugsatfda_docs/nda/2009/022395_qutenza_toc.cfm.
12. European Medicines Agency, 2009 (accessed on 07 Mar 2020) https://www.ema.europa.eu/en/medicines/human/EPAR/qutenza.
13. Chard, P. S.; Bleakman, D.; Savidge, J. R.; Miller, R. J. *Neuroscience* **1995**, *65*, 1099–1108.
14. Han, P.; McDonald, H. A.; Bianchi, B. R.; Kouhen, R. E.; Vos, M. H.; Jarvis, M. F.; Faltynek, C. R.; Moreland, R. B. *Biochem. Pharmacol.* **2007**, *73*, 1635–1645.
15. Baranidharan, G.; Das, S.; Bhaskar, A. *Ther. Adv. Neurol. Disord.* **2013**, *6*, 287–297.
16. Basith, S.; Cui, M.; Hong, S.; Choi, S. *Molecules* **2016**, *21*, 966.
17. Truong, D. D.; Jost, W. H. *Parkinsonism Relat. Disord.* **2006**, *12*, 331–355.
18. Matak, I.; Bolcskei, K.; Bach-Rojecky, L.; Helyes, Z. *Toxins* **2019**, *11*, 459.
19. Adams, D. J.; Smith, A. B.; Schroeder, C. I.; Yasuda, T.; Lewis, R. J. *J. Biol. Chem.* **2003**, *278*, 4057–4062.
20. Kolosov, A.; Aurini, L.; Williams, E. D.; Cooke, I.; Goodchild, C. S. *Pain Med.* **2011**, *12*, 923–941.
21. Sharp, D. *Lancet* **2005**, *366*, 439–440.
22. Sharpe, I. A.; Palant, E.; Schroeder, C. I.; Kaye, D. M.; Adams, D. J.; Alewood, P. F.; Lewis, R. J. *J. Biol. Chem.* **2003**, *278*, 40317–40323.
23. Sharpe, I. A.; Gehrmann, J.; Loughnan, M. L.; Thomas, L.; Adams, D. A.; Atkins, A.; Palant, E.; Craik, D. J.; Adams, D. J.; Alewood, P. F.; Lewis, R. J. *Nat. Neurosci.* **2001**, *4*, 902–907.
24. Okkerse, P.; Hay, J. L.; Sitsen, E.; Dahan, A.; Klaassen, E.; Houghton, W.; Groeneveld, G. J. *Br. J. Clin. Pharmacol.* **2017**, *83*, 751–763.
25. Lee, H. K.; Zhang, L.; Smith, M. D.; Walewska, A.; Vellore, N. A.; Baron, R.; McIntosh, J. M.; White, H. S.; Olivera, B. M.; Bulaj, G. *Front. Pharmacol.* **2015**, *6*, 11.
26. Lewis, R. J. *IUBMB Life* **2004**, *56*, 89–93.
27. Nevin, S. T.; Clark, R. J.; Klimis, H.; Christie, M. J.; Craik, D. J.; Adams, D. J. *Mol. Pharmacol.* **2007**, *72*, 1406–1410.
28. Castro, J.; Grundy, L.; Deiteren, A.; Harrington, A. M.; O'Donnell, T.; Maddern, J.; Moore, J.; Garcia-Caraballo, S.; Rychkov, G. Y.; Yu, R.; Kaas, Q.; Craik, D. J.; Adams, D. J.; Brierley, S. M. *Br. J. Pharmacol.* **2018**, *175*, 2384–2398.
29. Clark, R. J.; Jensen, J.; Nevin, S. T.; Callaghan, B. P.; Adams, D. J.; Craik, D. J. *Angew. Chem. Int. Ed. Engl.* **2010**, *49*, 6545–6548.
30. Diochot, S.; Baron, A.; Salinas, M.; Douguet, D.; Scarzello, S.; Dabert-Gay, A. S.; Debayle, D.; Friend, V.; Alloui, A.; Lazdunski, M.; Lingueglia, E. *Nature* **2012**, *490*, 552–555.
31. Jensen, J. E.; Mobli, M.; Brust, A.; Alewood, P. F.; King, G. F.; Rash, L. D. *Mar. Drugs* **2012**, *10*, 1511–1527.
32. Mazzuca, M.; Heurteaux, C.; Alloui, A.; Diochot, S.; Baron, A.; Voilley, N.; Blondeau, N.; Escoubas, P.; Gelot, A.; Cupo, A.; Zimmer, A.; Zimmer, A. M.; Eschalier, A.; Lazdunski, M. *Nat. Neurosci.* **2007**, *10*, 943–945.
33. Cordell, G. A.; Araujo, O. E. *Ann. Pharmacother.* **1993**, *27*, 330–336.
34. Gaubitz, M.; Schiffer, T.; Holm, C.; Richter, E.; Pisternick-Ruf, W.; Weiser, T. *Eur. J. Pain.* **2016**, *20*, 263–273.
35. Kobata, K.; Todo, T.; Yazawa, S.; Iwai, K.; Watanabe, T. *J. Agric. Food Chem.* **1998**, *46*, 1695–1697.

36. Iida, T.; Moriyama, T.; Kobata, K.; Morita, A.; Murayama, N.; Hashizume, S.; Fushiki, T.; Yazawa, S.; Watanabe, T.; Tominaga, M. *Neuropharmacology* **2003**, *44*, 958–967.
37. Szallasi, A.; Blumberg, P. M. *Neurosci. Lett.* **1992**, *140*, 51–54.
38. Bley, K. R. *Expert Opin. Invest. Drugs* **2004**, *13*, 1445–1456.
39. Chau, R.; Kalaitzis, J. A.; Neilan, B. A. *Aquat. Toxicol.* **2011**, *104*, 61–72.
40. Lago, J.; Rodriguez, L. P.; Blanco, L.; Vieites, J. M.; Cabado, A. G. *Mar. Drugs* **2015**, *13*, 6384–6406.
41. Raffa, R. B.; Pergolizzi, J. V.; Taylor, R.; Ossipov, M. H. *J. Clin. Pharm. Ther.* **2018**, *43*, 437–441.
42. Kruegel, A. C.; Grundmann, O. *Neuropharmacology* **2018**, *134*, 108–120.
43. Kruegel, A. C.; Uprety, R.; Grinnell, S. G.; Langreck, C.; Pekarskaya, E. A.; Le Rouzic, V.; Ansonoff, M.; Gassaway, M. M.; Pintar, J. E.; Pasternak, G. W.; Javitch, J. A.; Majumdar, S.; Sames, D. *ACS Cent. Sci.* **2019**, *5*, 992–1001.
44. Keasling, A. W.; Pandey, P.; Doerksen, R. J.; Pedrino, G. R.; Costa, E. A.; da Cunha, L. C.; Zjawiony, J. K.; Fajemiroye, J. O. *J. Psychopharmacol.* **2019**, *33*, 865–881.
45. Roth, B. L.; Baner, K.; Westkaemper, R.; Siebert, D.; Rice, K. C.; Steinberg, S.; Ernsberger, P.; Rothman, R. B. *Proc. Natl. Acad. Sci. U. S. A.* **2002**, *99*, 11934–11939.
46. Roach, J. J.; Shenvi, R. A. *Bioorg. Med. Chem. Lett.* **2018**, *28*, 1436–1445.
47. Buckle, J. *Altern. Ther. Health Med.* **1999**, *5*, 42–51.
48. Guimaraes, A. G.; Quintans, J. S.; Quintans, L. J., Jr. *Phytother. Res.* **2013**, *27*, 1–15.
49. de Cassia da Silveira, E. S. R.; Lima, T. C.; da Nobrega, F. R.; de Brito, A. E. M.; de Sousa, D. P. *Int. J. Mol. Sci.* **2017**, *18*, E2392.
50. de Moura, F. B.; Withey, S. L.; Bergman, J. *J. Pharmacol. Exp. Ther.* **2019**, *371*, 624–632.
51. Cai, S.; Bellampalli, S. S.; Yu, J.; Li, W.; Ji, Y.; Wijeratne, E. M. K.; Dorame, A.; Luo, S.; Shan, Z.; Khanna, M.; Moutal, A.; Streicher, J. M.; Gunatilaka, A. A. L.; Khanna, R. *ACS Chem. Neurosci.* **2019**, *10*, 1716–1728.
52. Sahebkar, A.; Henrotin, Y. *Pain Med.* **2016**, *17*, 1192–1202.
53. de Almeida, R. A. C.; de Sousa Lima, F. C. M.; do Vasconcelos, B. C. E. *Int. J. Oral Maxillofac. Surg.* **2019**, *48*, 651–658.
54. Calixto, J. B.; Scheidt, C.; Otuki, M.; Santos, A. R. *Expert Opin. Emerg. Drugs* **2001**, *6*, 261–279.
55. Verri, W. A., Jr.; Souto, F. O.; Vieira, S. M.; Almeida, S. C.; Fukada, S. Y.; Xu, D.; Alves-Filho, J. C.; Cunha, T. M.; Guerrero, A. T.; Mattos-Guimaraes, R. B.; Oliveira, F. R.; Teixeira, M. M.; Silva, J. S.; McInnes, I. B.; Ferreira, S. H.; Louzada-Junior, P.; Liew, F. Y.; Cunha, F. Q. *Ann. Rheum. Dis.* **2010**, *69*, 1697–1703.
56. Li, B. Q.; Fu, T.; Gong, W. H.; Dunlop, N.; Kung, H. F.; Yan, Y.; Kang, J.; Wang, J. M. *Immunopharmacology* **2000**, *49*, 295–306.
57. Naidu, P. S.; Singh, A.; Kulkarni, S. K. *Indian J. Exp. Biol.* **2003**, *41*, 1400–1404.
58. Lee, B. H.; Jung, S. M.; Lee, J. H.; Kim, J. H.; Yoon, I. S.; Lee, J. H.; Choi, S. H.; Lee, S. M.; Chang, C. G.; Kim, H. C. *Mol. Cells* **2005**, *20*, 69–73.
59. Verri, W. A.; Vicentini, F. T. M. C.; Baracat, M. M.; Georgetti, S. R.; Cardoso, R. D. R.; Cunha, T. M.; Ferreira, S. H.; Cunha, F. Q.; Fonseca, M. J. V.; Casagrande, R. In: *Studies in Natural Products Chemistry*; Rahman, A. Ed.; Vol. 36, Elsevier, 2012.
60. Al-Sayed, E.; Gad, H. A.; El-Shazly, M.; Abdel-Daim, M. M.; Nasser Singab, A. *Drug Dev. Res.* **2018**, *79*, 22–28.
61. Bittar, M.; De Souza, M. M.; Yunes, R. A.; Lento, R.; Monache, F. D.; Filho, V. C. *Planta Med.* **2000**, *66*, 84–86.
62. Dzubak, P.; Hajduch, M.; Vydra, D.; Hustova, A.; Kvasnica, M.; Biedermann, D.; Markova, L.; Urban, M.; Sarek, J. *Nat. Prod. Rep.* **2006**, *23*, 394–411.
63. Kidd, B. L.; Urban, L. A. *Br. J. Anaesth.* **2001**, *87*, 3–11.

64. Ricciotti, E.; FitzGerald, G. A. *Arter. Thromb. Vasc. Biol.* **2011**, *31*, 986–1000.
65. Brunton, L. L.; Lazo, J. S.; Parker, K. L. *Goodman and Gilman's the Pharmacological Basis of Therapeutics*, 11th ed.; McGraw-Hill, 2005.
66. Aronoff, D. M.; Neilson, E. G. *Am. J. Med.* **2001**, *111*, 304–315.
67. Sultana, S.; Asif, H. M.; Akhtar, N.; Ahmad, K. *Asian Pac. J. Trop. Dis.* **2015**, *5*, S202–S208.
68. Doan, L. *Tech. Reg. Anesth. Pain Manag.* **2010**, *14*, 42–47.
69. Bourinet, E.; Soong, T. W.; Stea, A.; Snutch, T. P. *Proc. Natl. Acad. Sci. U. S. A.* **1996**, *93*, 1486–1491.
70. Murakami, M.; Nakagawasai, O.; Suzuki, T.; Mobarakeh, I. I.; Sakurada, Y.; Murata, A.; Yamadera, F.; Miyoshi, I.; Yanai, K.; Tan-No, K.; Sasano, H.; Tadano, T.; Iijima, T. *Brain Res.* **2004**, *1024*, 122–129.
71. Choi, S.; Na, H. S.; Kim, J.; Lee, J.; Lee, S.; Kim, D.; Park, J.; Chen, C. C.; Campbell, K. P.; Shin, H. S. *Genes Brain Behav.* **2007**, *6*, 425–431.
72. Candelas, M.; Reynders, A.; Arango-Lievano, M.; Neumayer, C.; Fruquière, A.; Demes, E.; Hamid, J.; Lemmers, C.; Bernat, C.; Monteil, A.; Compan, V.; Laffray, S.; Inquimbert, P.; Le Feuvre, Y.; Zamponi, G. W.; Moqrich, A.; Bourinet, E.; Méry, P. F. *Sci. Rep.* **2019**, *9*, 3112.
73. Vranken, J. H. *Cent. Nerv. Syst. Agents Med. Chem.* **2012**, *12*, 304–314.
74. Yamamoto, R.; Yanagita, T.; Kobayashi, H.; Yokoo, H.; Wada, A. *J. Neurochem.* **1997**, *68*, 1655–1662.
75. Cardoso, F. C.; Lewis, R. J. *Br. J. Pharmacol.* **2018**, *175*, 2138–2157.
76. Nilius, B.; Szallasi, A. *Pharmacol. Rev.* **2014**, *66*, 676–814.
77. Moran, M. M.; Szallasi, A. *Br. J. Pharmacol.* **2018**, *175*, 2185–2203.
78. Che, T.; Majumdar, S.; Zaidi, S. A.; Ondachi, P.; McCorvy, J. D.; Wang, S.; Mosier, P. D.; Uprety, R.; Vardy, E.; Krumm, B. E.; Han, G. W.; Lee, M. Y.; Pardon, E.; Steyaert, J.; Huang, X. P.; Strachan, R. T.; Tribo, A. R.; Pasternak, G. W.; Carroll, F. I.; Stevens, R. C.; Cherezov, V.; Katritch, V.; Wacker, D.; Roth, B. L. *Cell* **2018**, *172*, 55–67.e15.
79. Marsicano, G.; Wotjak, C. T.; Azad, S. C.; Bisogno, T.; Rammes, G.; Cascio, M. G.; Hermann, H.; Tang, J.; Hofmann, C.; Zieglgansberger, W.; Di Marzo, V.; Lutz, B. *Nature* **2002**, *418*, 530–534.
80. Domenici, M. R.; Azad, S. C.; Marsicano, G.; Schierloh, A.; Wotjak, C. T.; Dodt, H. U.; Zieglgansberger, W.; Lutz, B.; Rammes, G. J. *Neuroscience* **2006**, *26*, 5794–5799.
81. Pertwee, R. G. *Handb. Exp. Pharmacol.* **2005**, *168*, 1–51.
82. Guerrero-Alba, R.; Barragan-Iglesias, P.; Gonzalez-Hernandez, A.; Valdez-Morales, E. E.; Granados-Soto, V.; Condes-Lara, M.; Rodriguez, M. G.; Marichal-Cancino, B. A. *Front. Pharmacol.* **2018**, *9*, 1496.
83. Marks, D. M.; Shah, M. J.; Patkar, A. A.; Masand, P. S.; Park, G. Y.; Pae, C. U. *Curr. Neuropharmacol.* **2009**, 7, 331–336.
84. Petrenko, A. B.; Yamakura, T.; Baba, H.; Shimoji, K. *Anesth. Analg.* **2003**, *97*, 1108–1116.
85. Parsons, C. G. *Eur. J. Pharmacol.* **2001**, *429*, 71–78.
86. Peana, A. T.; D'Aquila, P. S.; Chessa, M. L.; Moretti, M. D. L.; Serra, G.; Pippia, P. *Eur. J. Pharmacol.* **2003**, *460*, 37–41.
87. Silva Brum, L. F.; Emanuelli, T.; Souza, D. O.; Elisabetsky, E. *Neurochem. Res.* **2001**, *26*, 191–194.
88. Klein, R. C.; Prorok, M.; Galdzicki, Z.; Castellino, F. J. *J. Biol. Chem.* **2001**, *276*, 26860–26867.
89. Mena, E. E.; Gullak, M. F.; Pagnozzi, M. J.; Richter, K. E.; Rivier, J.; Cruz, L. J.; Olivera, B. M. *Neurosci. Lett.* **1990**, *118*, 241–244.

90. Malmberg, A. B.; Gilbert, H.; McCabe, R. T.; Basbaum, A. I. *Pain* **2003**, *101*, 109–116.
91. Kleczkowska, P.; Lipkowski, A. W. *Eur. J. Pharmacol.* **2013**, *716*, 54–60.
92. White, J. F.; Noinaj, N.; Shibata, Y.; Love, J.; Kloss, B.; Xu, F.; Gvozdenovic-Jeremic, J.; Shah, P.; Shiloach, J.; Tate, C. G.; Grisshammer, R. *Nature* **2012**, *490*, 508–513.
93. Dubuc, I.; Sarret, P.; Labbé-Jullié, C.; Botto, J. M.; Honoré, E.; Bourdel, E.; Martinez, J.; Costentin, J.; Vincent, J. P.; Kitabgi, P.; Mazella, J. *J. Neurosci.* **1999**, *19*, 503.
94. Gotti, C.; Zoli, M.; Clementi, F. *Trends Pharmacol. Sci.* **2006**, *27*, 482–491.
95. Decker, M. W.; Meyer, M. D.; Sullivan, J. P. *Expert Opin. Invest. Drugs* **2001**, *10*, 1819–1830.
96. Dutta, S.; Hosmane, B. S.; Awni, W. M. *AAPS J.* **2012**, *14*, 168–175.
97. Livett, B. G.; Sandall, D. W.; Keays, D.; Down, J.; Gayler, K. R.; Satkunanathan, N.; Khalil, Z. *Toxicon* **2006**, *48*, 810–829.
98. Ando, R. D.; Mehesz, B.; Gyires, K.; Illes, P.; Sperlagh, B. *Br. J. Pharmacol.* **2010**, *159*, 1106–1117.
99. Soares-Bezerra, R. J.; Calheiros, A. S.; Da Silva Ferreira, N. C.; Da Silva Frutuoso, V.; Alves, L. A. *Pharmaceuticals* **2013**, *6*, 650–658.
100. Savchenko, G. A.; Volkova, T. M.; Vasilevskii, A. A.; Korol'kova, Y. V.; Grishin, E. V.; Boichouk, Y. A.; Krishtal, O. A. *Neurophysiology* **2011**, *42*, 387–391.
101. Deval, E.; Gasull, X.; Noel, J.; Salinas, M.; Baron, A.; Diochot, S.; Lingueglia, E. *Pharmacol. Ther.* **2010**, *128*, 549–558.
102. Voilley, N.; de Weille, J.; Mamet, J.; Lazdunski, M. *J. Neurosci.* **2001**, *21*, 8026–8033.
103. Bannon, A. W.; Malmberg, A. B. *Curr. Protoc. Neurosci.* **2007**, *41, Chapter 8, 8.9.1–8.9.16.*
104. Sycha, T.; Gustorff, B.; Lehr, S.; Tanew, A.; Eichler, H. G.; Schmetterer, L. *Br. J. Clin. Pharmacol.* **2003**, *56*, 165–172.
105. Hargreaves, K.; Dubner, R.; Brown, F.; Flores, C.; Joris, J. *Pain* **1988**, *32*, 77–88.
106. Langford, D. J.; Mogil, J. S. In: *Anesthesia and Analgesia in Laboratory Animals*, 2nd ed.; Fish, R., Danneman, P., Brown, M., Karas, A., Eds.; Academic Press: Cambridge, Massachusetts, 2008; pp 549–560.
107. Ma, Z.; Li, Y.; Zhang, Y. P.; Shields, L. B. E.; Xie, Q.; Yan, G.; Liu, W.; Chen, G.; Zhang, Y.; Brommer, B.; Xu, X. M.; Lu, Y.; Chen, X.; Shields, C. B. *Funct. Neurol.* **2015**, *30*, 229–236.
108. Sluka, K. A.; Kalra, A.; Moore, S. A. *Muscle Nerve* **2001**, *24*, 37–46.
109. Mauderli, A. P.; Acosta-Rua, A.; Vierck, C. J. *J. Neurosci. Methods* **2000**, *97*, 19–29.
110. Handwerker, H. O.; Kobal, G. *Physiol. Rev.* **1993**, *73*, 639–671.
111. Perkins, M. N.; Campbell, E.; Dray, A. *Pain* **1993**, *53*, 191–197.
112. Staahl, C.; Drewes, A. M. *Basic Clin. Pharmacol. Toxicol.* **2004**, *95*, 97–111.
113. Drewes, A. M.; Gregersen, H.; Arendt-Nielsen, L. *Scand. J. Gastroenterol.* **2003**, *38*, 1115–1130.
114. Al-Chaer, E. D.; Kawasaki, M.; Pasricha, P. J. *Gastroenterology* **2000**, *119*, 1276–1285.
115. Martı´nez, V.; Coutinho, S. V.; Thakur, S.; Mogil, J. S.; Taché, Y.; Mayer, E. A. *Pain* **1999**, *81*, 179–186.
116. Ness, T. J.; Gebhart, G. F. *Pain* **1990**, *41*, 167–234.
117. Gebhart, G. F.; Basbaum, A. I.; Bird, S. J.; Flecknell, P.; Goodly, L. J.; Karas, A. Z.; Kelley, S. T.; Lacher, J.; Mason, G.; Sneddon, L. U.; Soriano, S. G.; Shafford, H. L. *Recognition and Alleviation of Pain in Laboratory Animals.* National Academies Press: USA, 2009.

CHAPTER FIFTEEN

Fertility regulating natural products

Sushmita Nath[a,*], Lutfun Nahar[a,b], Satyajit D. Sarker[a]
[a]Centre for Natural Products Discovery (CNPD), School of Pharmacy and Biomolecular Sciences, Liverpool John Moores University, Liverpool, United Kingdom
[b]Laboratory of Growth Regulators, Institute of Experimental Botany ASCR & Palacký University, Olomouc, Czech Republic
*Corresponding author: e-mail address: sushmitanath84@gmail.com

Contents

1. Introduction

Fertility is the natural capacity of a living being to establish the clinical pregnancy through an intricate biological process "reproduction" with the assistance of reproductive system to produce offspring, the real existence of a species. The system has unifying characteristics in males and females of a species including *Homo sapiens* (humans)—the highest ranked animal species. It controls the development of structural and functional differences between males and females, and influences their behavior.[1] Like the other

Annual Reports in Medicinal Chemistry, Volume 55
ISSN 0065-7743
https://doi.org/10.1016/bs.armc.2020.02.007

physiological systems in our body, reproductive system also needs efficacious way to care for and to maintain fecundity, which is generally obtained using fertility regulators. Usage of such regulators have risen at least threefold in recent years, where most of the commercially available regulators are synthetic chemicals, often have adverse effects and toxicity. This emphasizes the need for developing biologically active substances as potential fertility regulators from various natural sources. In this regard, the therapeutic potentials of the secondary metabolites or natural products, produced by living organisms in their natural environment, has been considered as one of the best choices. Over the last five decades, research on development of natural fertility regulation have come to the forefront of new therapeutics development in fertility regulation. It has been found that worldwide varied communities are chiefly practicing the plant-based traditional medicines for fertility regulation.[2] The purpose of this current chapter is to provide information about natural products involved in fertility regulation. While there are several reports, including many review articles, describing fertility regulatory properties of plants and plant extracts available in the literature, this chapter will only cover isolated natural products with fertility regulatory properties as evident from animal based *in vivo* studies, and wherever available from human clinical trials.

To understand potential involvement of natural products in the regulation of reproductive physiology and fertility, it is essential to have an idea about the human reproductive system. The following section will briefly discuss the reproductive system and its regulation, current fertility regulators, and inter-relationship between herbs and fecundity.

2. The human reproductive system and fertility

2.1 Glimpse of the male and female reproductive systems

The external male reproductive system comprises penis, scrotum and testicles. The internal male reproductive organs also known as accessory organs are (i) vas deferens—the transporter mature sperm to urethra for ejaculation, (ii) urethra—the urine carrier tube from the bladder to excrete and also ejaculate semen, (iii) seminal vesicles—produces fructose rich fluid to provide energy and motility to sperm, (iv) prostate gland—contributes additional fluid to the ejaculate and nourishes the sperm. These organs are specialized in (i) production, maintenance and transport of sperm and semen; (ii) sperm discharge; (iii) production and secretion of male sex hormones. The gonads, a pair of testes, made up of loops of convoluted seminiferous tubules governs the male reproductive system. The principal function of the male reproductive system is a production of functional spermatozoa, produced through spermatogenesis (process of development of haploid spermatozoa from germ cells in the seminiferous tubules of testis), to maintain male fertility.[3] Spermatozoa are super specialized and terminally differentiated motile cells transfer the paternal genetic material into the egg.[3,4]

Unlike male, the female reproductive system is much more complicated. The organs are situated both externally and internally. The external structure is genital, which enables sperm to enter the body and protects the internal organs from infectious organisms. The internal reproductive organs are (i) vagina the birth canal, (ii) uterus or womb the home to a developing fetus, (iii) ovaries the gonads, and (iv) fallopian tubes the traveling pathway for the ova from ovary to uterus. The principal functions of this system are to produce female gametes or ova for sperm fertilization and to provide appropriate conditions for embryo implantation, fetal growth and development, and birth.[4] Ovaries perform two main functions: production of oocytes (female gametes) and the secretion sex steroids or steroid hormones.[3,5] Verdict of life and death mechanism of a female gamete, through folliculogenesis (process of maturation of ovarian follicles) and ovulation (release of female gamete or ovum), is built upon the cross-dialog between ovarian cellular functions under the influence of steroids to establish fecundity.[5,6] Follicles, the principal functional unit of the ovary, mature by the process called folliculogenesis,[7] followed by the tissue remodeling event "ovulation" the oocyte escapes the confines of ovary to become available for fertilization.[8]

2.2 Endocrine regulation of the reproductive system

The reproductive system is principally governed by the endocrine hormones, signaling molecules that are transported through the blood stream to reach the target organs to regulate fertility. The level of each hormone associated with the reproductive system plays a crucial role to main fertility. The hypothalamic-pituitary-gonadal axis is in-charge of this endocrine regulation in both male and female reproductive systems by changing the pattern of hormone production and secretion by the respective gonads through negative feedback mechanism.[3,4] Females experience hormonal regulation in a cyclic manner, the menstrual cycle, once in a month. Every cycle prepares the woman's body for a potential pregnancy. In an average each cycle occurs in every 28 days in three Phases—(i) follicular Phase (egg development), (ii) ovulatory Phase (egg release) and (iii) luteal Phase (decrease of hormonal level in absence of egg implantation).

In the both male and female, gonadotropin-releasing hormone (GnRH) secreted in a pulsatile manner from the hypothalamus, the anterior pituitary gonadotropins, follicle-stimulating hormone (FSH) and luteinizing hormone (LH), and steroid hormones secreted by gonadal cells collectively control the individual reproductive functions.[3,9] GnRH stimulates and regulates LH/FSH release, sperm production, oocyte maturation and steroidogenesis.[10] LH stimulates the production of androgens in both male and female, and progestogen in female. It also acts as a key promoter of ovulation and luteinization in female and its receptors forms corpus luteum, cells responsible for progesterone (P4) production during early pregnancy, in response to mid-cycle surge of LH induces ovulation. FSH triggers Sertoli cells in the testis to activate sperm production in male and while, in female FSH is responsible for estrogen (E) production and stimulates folliculogenesis, the maturation of ovarian follicles. The FSH, LH, therefore, these gonadotropins are widely used to treat infertility, ovulation induction, embryo production and in assisted reproduction, but inactivation of LH receptors induces infertility.[11,12]

2.2.1 Androgen regulation in male and female

Androgens are steroid hormones synthesized both in testis and ovaries. Leydig cells present in the testis of male reproductive system under the influence of hypothalamo-pituitary-testicular axis, produces the major male androgen, testosterone (Fig. 1) approximately 6–7 mg/day. The regular concentration of testosterone gears up spermatogenesis, but its low-level leads to infertility. LH binds to LH receptors present in the Leydig cells and stimulates testosterone release until the release concentration reaches a critical threshold level in blood.

Once the threshold level is attained, testosterone itself binds to the androgen receptors of hypothalamus and pituitary to inhibit the synthesis and secretion of GnRH and LH *via* hormonal negative feedback mechanism.

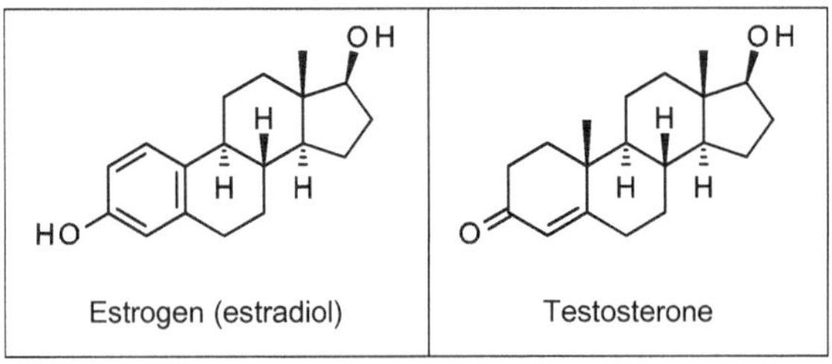

Fig. 1 Estrogen and testosterone, female and male sex hormone, respectively.

When there is decline in testosterone concentration in blood, it no longer interacts with the androgen receptors in a same intensity, as a result GnRH and LH start to release to stimulate testosterone production.[13]

Unlike male, the female reproductive endocrine system is more complex. In females, androgens are the precursor of the chief female reproductive hormone, estrogen (Fig. 1). Mainly the androstenedione (A4) and testosterone are produced by theca cells (a group of endocrine cells in ovary) in response to LH cholesterol gets converted to pregnenolone and act *via* cytoplasmic receptors localized in granulosa cells (somatic cells associated with the developing female gamete and responsible for the production of sex steroids in female) to regulate ovarian functions.[14] Fifty percent of testosterone produced in ovaries promote follicular proliferation, stimulate or suppress ovulation, enhance follicular atresia at different stages of folliculogenesis, and promote oocyte nuclear maturation.[5] Androgen receptors are essential for ovarian follicle development, ovulation, and female fertility.[15] *In vitro* studies suggest that androgens can directly promote ovarian folliculogenesis, gonadotropin reception, ovarian cell proliferation, follicular atresia, oocyte maturation, ovulation, corpus luteum development, and fecundity establishment.[15] Hyperproduction of androgens induces polycystic ovarian syndrome (PCOS), and antiandrogens can restore and ovulation and fecundity in PCOS patients. Hence, androgens can be practically used as fecundity regulator to treat reproductive disorders.[15]

2.2.2 Progestogen regulation in female

Progesterone (Fig. 2), the primary ovarian progestogen produced by granulosa and theca cells, is a major regulator of ovulation.[4] Progesterone induces ovulation by inducing pre-ovulatory LH surge, and promoting proteolytic enzyme production, the prerequisite for follicle rupture and protease release by granulosa cells.

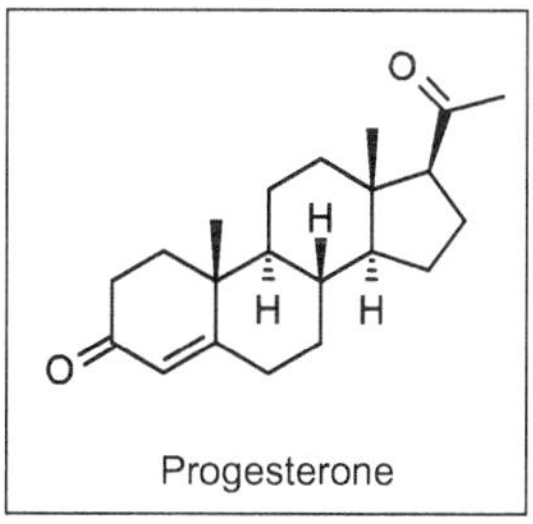

Fig. 2 Structures of progesterone, the major regulator of ovulation.

Progesterone inhibits granulosa cell mitosis and apoptosis, transfer primordial follicles to primary follicles,[5] promotes oocyte maturation and stimulates testosterone and estradiol (Fig. 1) production.[16] Progesterone receptors are involved in signal cascading that enhances survival and prevents apoptosis of garnulosa-lutein cells.[17] Apart from acting as a promoter of ovarian follicle luteinization and development progesterone is also responsible for gonadotropin surge, proliferation inhibition, stimulation of decidualization of endometrial bed and subsequent embryo implantation, and maintenance of pregnancy. Therefore, its chemical analog are used as principal ingredients in synthetic contraceptives to suppress mid-cycle LH surge, ovulation, and embryo implantation.[14,17]

2.2.3 Estrogen regulation in female

Estrogen, which refers to all of the chemically similar hormones in this group, plays an important role to influence fertility and infertility in females. It has both positive and negative feedback effects on the hypothalamic-pituitary axis. Due to its established effect on the uterus (secondary female sex-organ), hypothalamus, pituitary and local actions within the ovaries, made it an advantageous choice in the development of contraceptive pills.[18] Under the influence of FSH androgens produced by the theca cells are transported to granulosa cells aromatized to estrogen, which signals *via* two estrogen receptors (ER)—ERα and ERβ. The ERα inhibits ovulation *via* hypothalamo-pituitary axis, while ERβ stimulates follicular growth, oocyte release and induction of ovulation.[5] Naturally, there are three major types of estrogen occur in female: estrone, estradiol and estriol (Fig. 3). It stimulates granulosa cell proliferation, growth and differentiation of antral follicles and reduces follicular atresia. Whereas, ER if necessary, to maintain ovarian cycle, follicle selection, oocyte maturation, ovulation and fertility. Furthermore, estrogen is a potent regulator of GnRH, FSH, LH, and ovarian hormone release.

2.3 Prevalence and current available fertility regulators

Conventionally fertility regulation refers to intentionally adopted methods either by male or female to avoid pregnancy to maintain the number of offspring and timing of births. It has been observed that apart from birth control, some fertility regulating measures are used to manage reproductive health related problems, especially in females suffering from PCOS, irregular or absent menstrual cycle. Different fertility regulatory measures currently in use are[19]: combined oral contraceptive pills, progestin-only pills, emergency

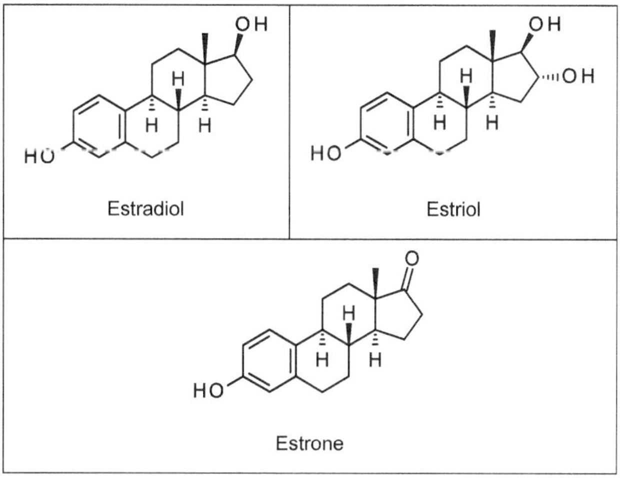

Fig. 3 Three different naturally produced estrogens.

contraceptive pills, progestin-only injectable, monthly-injectable, combined patch, combined vaginal ring, progesterone-releasing vaginal ring, implants, copper-bearing intrauterine device, levonorgestrel intrauterine device, female sterilization, vasectomy, male condoms, female condoms, spermicides and diaphragms, cervical caps, fertility awareness methods, withdrawal, and lactational amenorrhea method. It is to be noted that, though these measures are medically approved, effective and reversible, but none of them are free from side effects.[19]

Most of the pills and injectable are hormonal therapies that use progestin and estrogen either in combination or alone. Their mechanism of action is to prevent ovulation and thickening of cervical mucus. As side effects they cause irregular bleeding, lengthened postpartum amenorrhea, loss of bone density, upper respiratory injection, vaginitis, vaginal discharge, liver problem and in extreme deep vein thrombosis, stroke and heart attack.[19] Additionally, the insert methods also have same mode of actions, and they cause effects like enlarged ovarian follicles, ovarian cysts, infection at insertion site, perforation of the uterine wall, anemia, and pelvic inflammatory disease. Furthermore, frequent and long-term use of spermicides, nonoxynol-9, octoxynol-9, benzalkonium chloride, chlorhexidine, menfegol and sodium docusate, in any form are even not safe to use. These chemicals cause urinary tract infections and vaginal lesions.[19] The World Health Organisation (WHO) has mentioned that women using spermicides are more susceptible to become infected with HIV compared to non-spermicide using ones.[19]

2.4 Historical perspective of natural products in fertility regulation

Natural products are one of the major sources of modern drugs. The earliest records from Mesopotamia in 2600 BC depicted the use of natural products as medicine on clay tables in cuneiform. Since then its medicinal usage has been described throughout history in the form of traditional medication or folklore medicines.[20] Ancient systems of medicines have mentioned various plant-derived extracts or phytocomplexes to treat human ailments and enhancing human health and well-being.[21] The Egyptian Papyrus (1550 BC) and the Kahun Papyrus (1850 BC) have earliest documentation of fertility regulations by using honey, acacia leaves and lint as spermicide.[20] The Egyptian Papyrus (1550 BC) described the use of ground dates, acacia tree bark, and honey together as a pessary. In medieval Europe, women used lily roots and rue to prevent pregnancy. The oldest condom made up of animal gut dated back to 1640 has been recovered in the ruins of Dudley Castle in England. Some religious communities still use lemon juice as locally applied spermicide. Hippocrates has described the use of Queen Anne's lace also known as wild carrot's seed as contraceptive.[20] These seeds can block progesterone synthesis, disrupts implantation and can cause contraception within 8 h of exposure to sperm.

The Greeks and the Romans used pennyroyal plant to induce menstruation and abortion. Native Americans use blue cohosh as traditional birth control. Soranus, a gynecologist from 2nd century had described abortifacient activity of common rue. Evaluation of herbs for their effect on the reproductive system has been in progress worldwide. Scientists of the modern era are trying to formulate valid fertility regulators from the traditionally used medicinal plants considered under "village pharmacy." Traditional documentation represents the natural products' regulatory activity as inducer and suppressor. In this context, numerous experiments have already been done with traditionally documented plants that resulted in immense information on fecundity regulation. Pharmacological investigations have proven that an individual plant extract or combination of two or more plant's extracts can regulate fertility in various manner. Number of bioactive constituents have already been isolated and tested for various fertility regulatory activities and some are under clinical trials. Many commercialized fertility regulators from plant origin are available but they also contain synthetic compounds. The drugs containing levonorgestrel, desogestrel, norethindrone, norgestimate, drospirenone show severe side effects like liver dysfunction, genital bleeding, endometrial and breast cancer and decreased libido upon long-term exposure.

3. Fertility regulating natural products

Natural products are the basis of most of the medicines of this century. Natural products isolated from various natural sources, including terrestrial and marine animals, plants and microorganisms, have proven their therapeutic potentials against different human ailments. However, regarding fertility regulation, only natural products from plant origin have been taken into consideration and subjected to investigations in recent years.[22] These active phytoconstituents with fertility regulating potentials have been isolated from the traditionally used fertility regulatory medicinal plants. Examples of these fertility regulatory natural products isolated from medicinal plants (Fig. 4) may include achyranthine (**1**) from *Achyranthes aspera* L. (Amranthaceae),[23] anethole (**2**) from *Croton zehntneri* Pax & K. Hoffm. (Euphorbiaceae),[24] anonaine (**3**) from *Annona squamosa* L. (Annonaceae),[25] arecaidine (**4**) from *Areca catechu* L. (Arecaceae),[26] atropine (**5**) from *Annona squamosa* L. (Annonaceae) and *Atropa belladonna* L. (Solanaceae),[25] azadirachtin (**6**) from *Azadirachta indica* A. Juss. (Meliaceae),[27] β-carotene (**7**) from *Moringa oleifera* Lam. (Moringaceae),[28] β-sitosterol (**8**) from *Trichosanthes cucumerina* L. (Cucurbitaceae),[29] bisbenzylisoquinolines (**9**) from *Nelumbo nucifera* Gaertn. (Nelumbonaceae),[30] berberine (**10**) from *Tinospora cordifolia* (Thunb.) Miers. (Menispermaceae),[31] carvacrol (**11**) from *Ocimum sanctum* L. (Lamiaceae),[32] choline (**12**) from *Tinospora cordifolia* (Thunb.) Miers. (Menispermaceae),[31] crocin (**13**) and crocetin (**14**) from *Crocus sativus* L. (Iridaceae),[33] curcumin (**15**) from *Curcuma longa* L. (Zingiberaceae),[34] cucurbitacin B (**16**) from *Trichosanthes cucumerina* L. (Cucurbitaceae),[29] dimethyl crocetin (**17**) from *Crocus sativus* L.,[33] diosgenin (**18**) from *Balanites roxburghii* Planch. (Zygophyllaceae),[35] embelin (**19**) from *Embelia ribes* Burm. f. (Primulaceae),[22] epicatechin (**20**) from *Amburana cearensis* (Allemao) A. C. Sm. (Fabaceae),[36] ferulic acid (**21**) and kaempferol (**22**) from *Allium cepa* L. (Amaryllidaceae),[37] gitogenin (**23**) from *Yucca schidigera* Roezi ex Ortgies (Asparagaceae),[38] gossypol (**24**) from *Gossypium* species L. (Malvaceae),[39] lepidine (**25**) from *Lepidium sativum* L. (Brassicaceae),[40] markogenin (**26**) from *Yucca schidigera* Roezi ex Ortgies,[38] naringin (**27**) from *Citrus x paradise* Macfad. (Rutaceae),[41] noscapine (**28**) from *Papaver somniferum* L. (Papaveraceae),[42] oleanolic acid (**29**) from *Carica papaya* L. (Caricaceae) and *Lantana camara* L. (Verbenaceae),[43,44] palmatine (**30**) from *Tinospora cordifolia* (Thunb.) Miers,[31] pilocarpine (**31**) from *Areca catechu* L.,[26] plumbagin (**32**) from *Plumbago rosea* L. (Plumbaginaceae),[45] protocatechuic acid (**33**) from

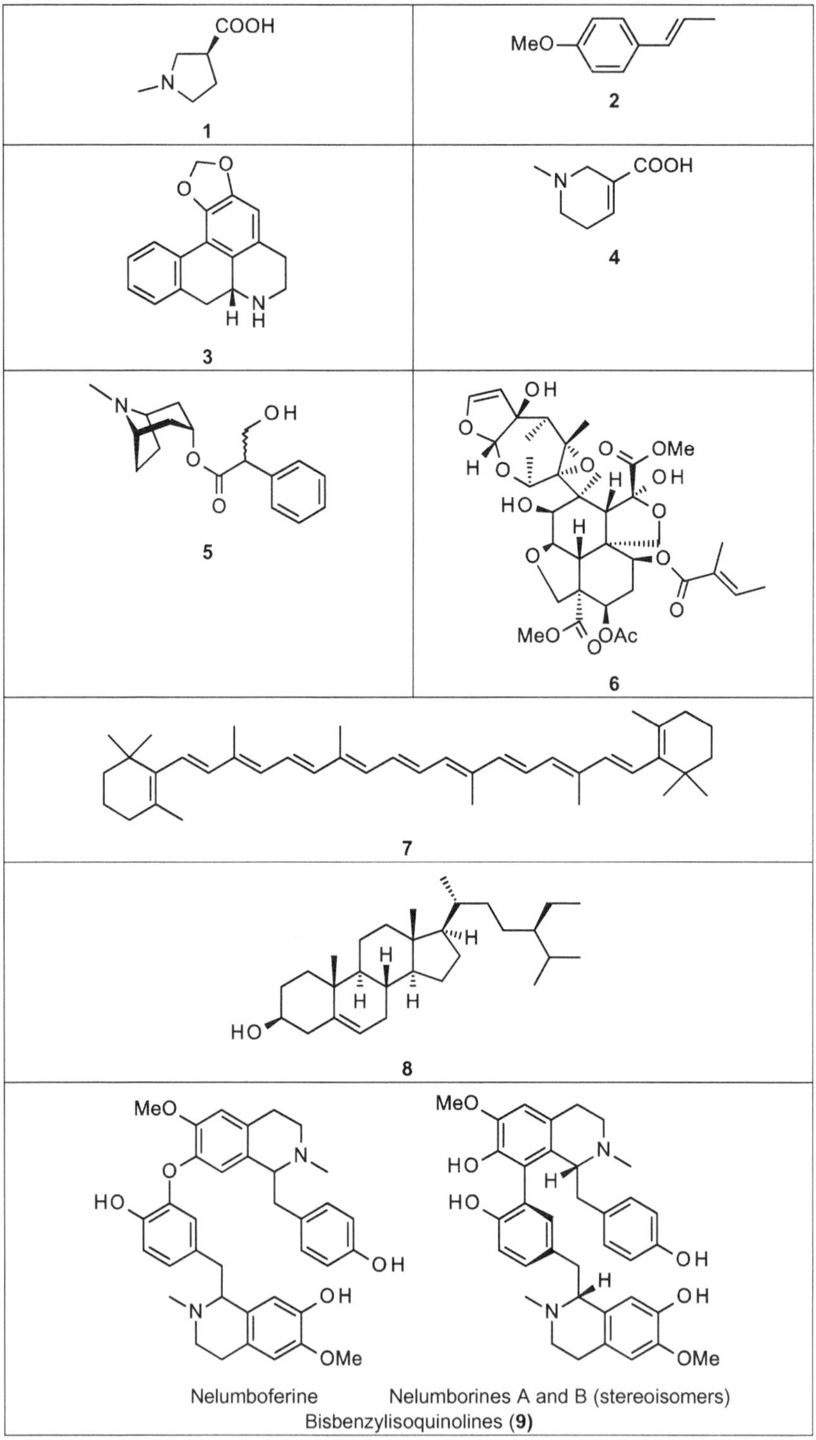

Fig. 4 Fertility regulating phytochemicals.

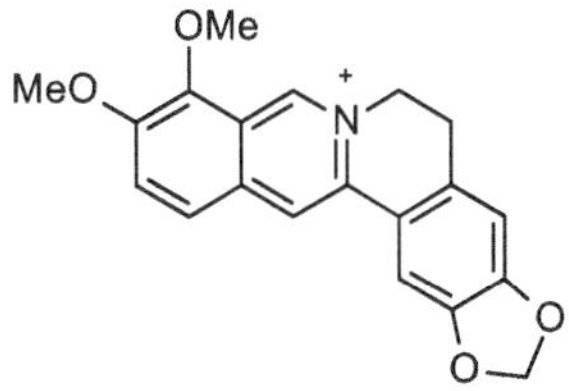

10

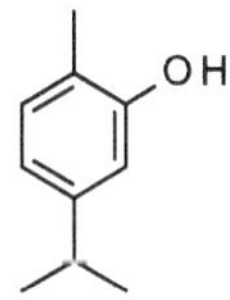

11

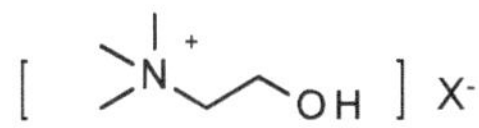

12

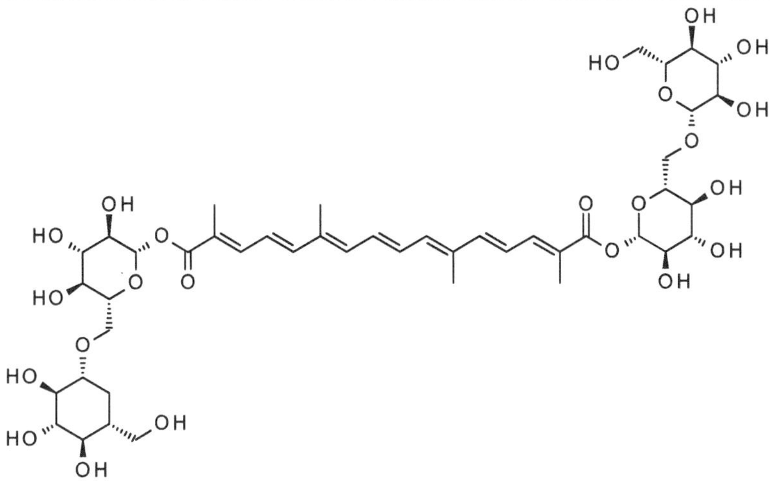

13

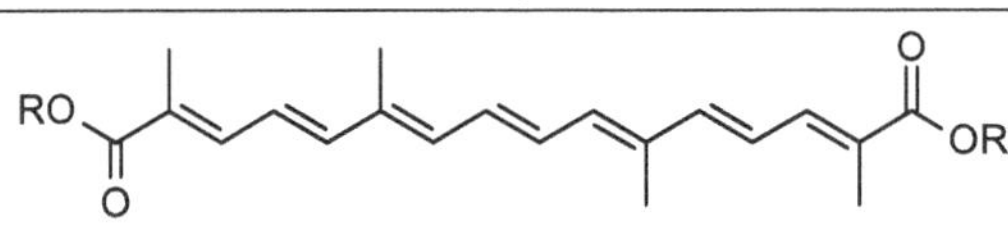

14 R = H
17 R = Me

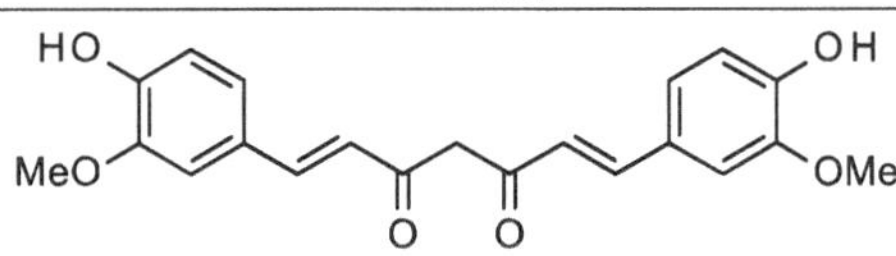

15

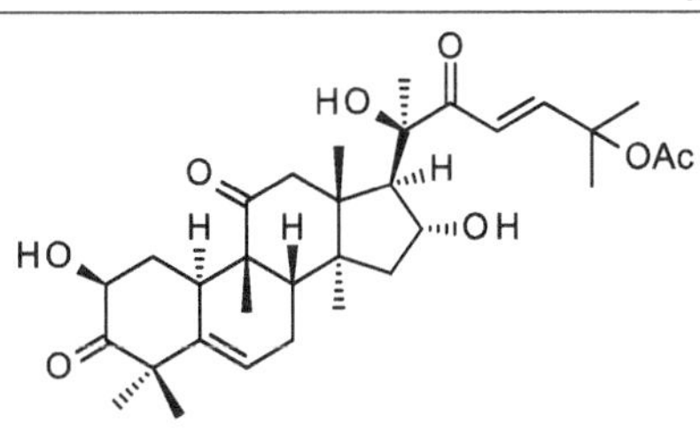

16

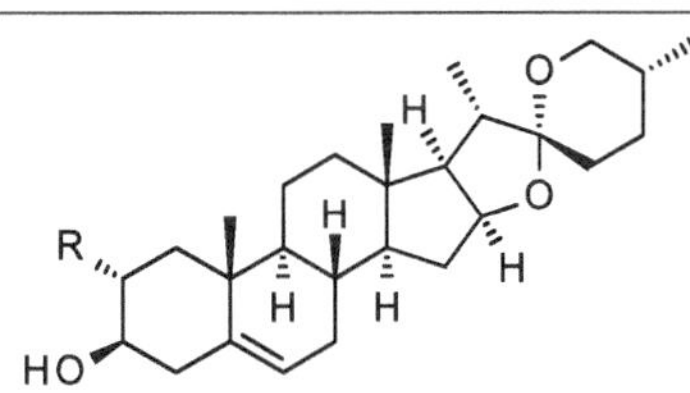

18 R = H
23 R = OH

Fig. 4—Cont'd

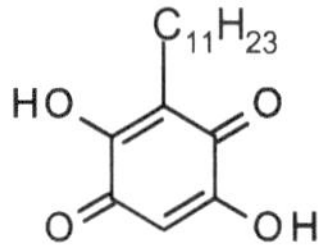

19

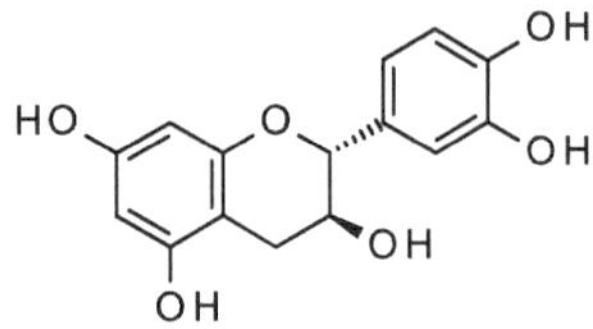

20

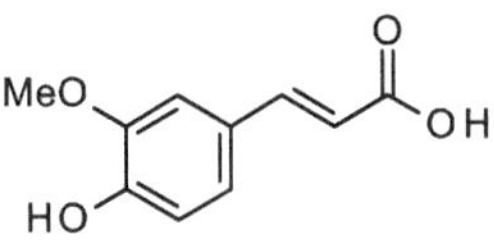

21

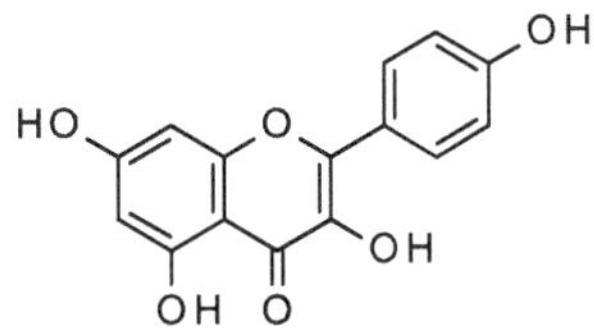

22

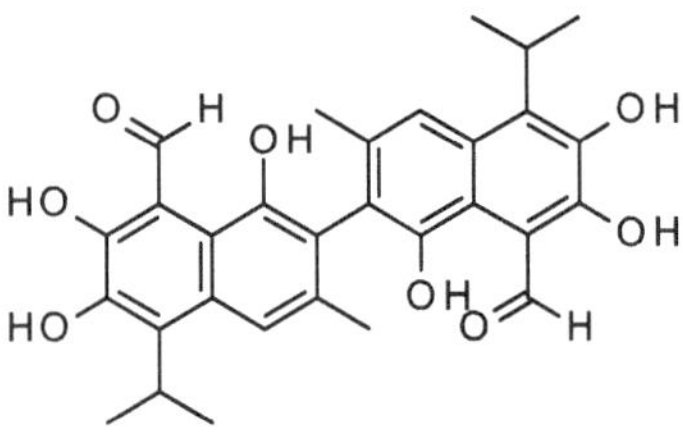

24

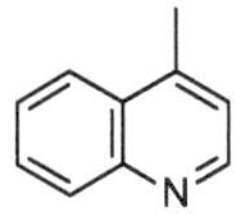

25

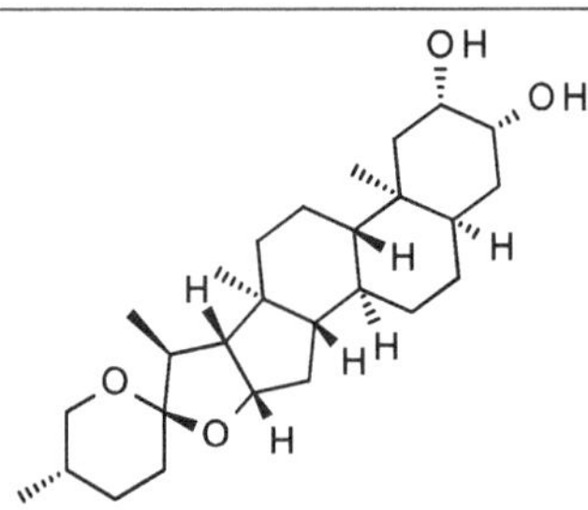

26

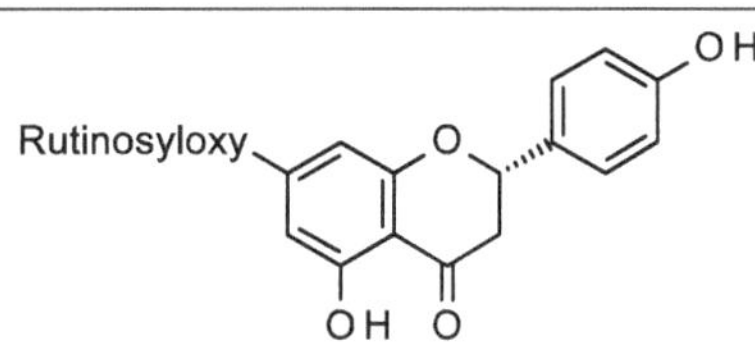

27

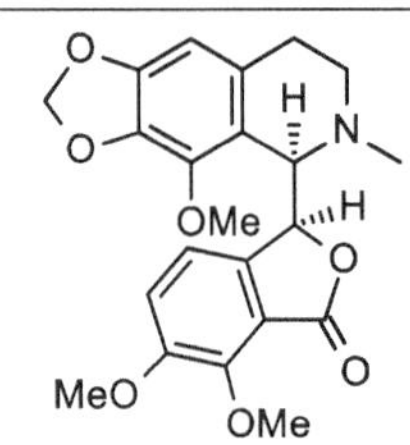

28

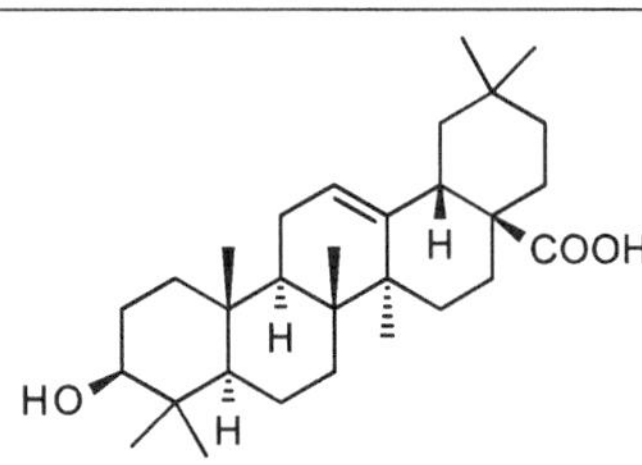

29

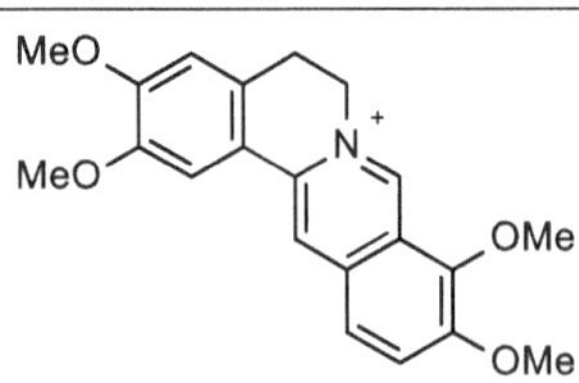

30

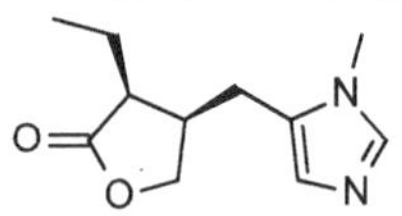

31

Fig. 4—Cont'd

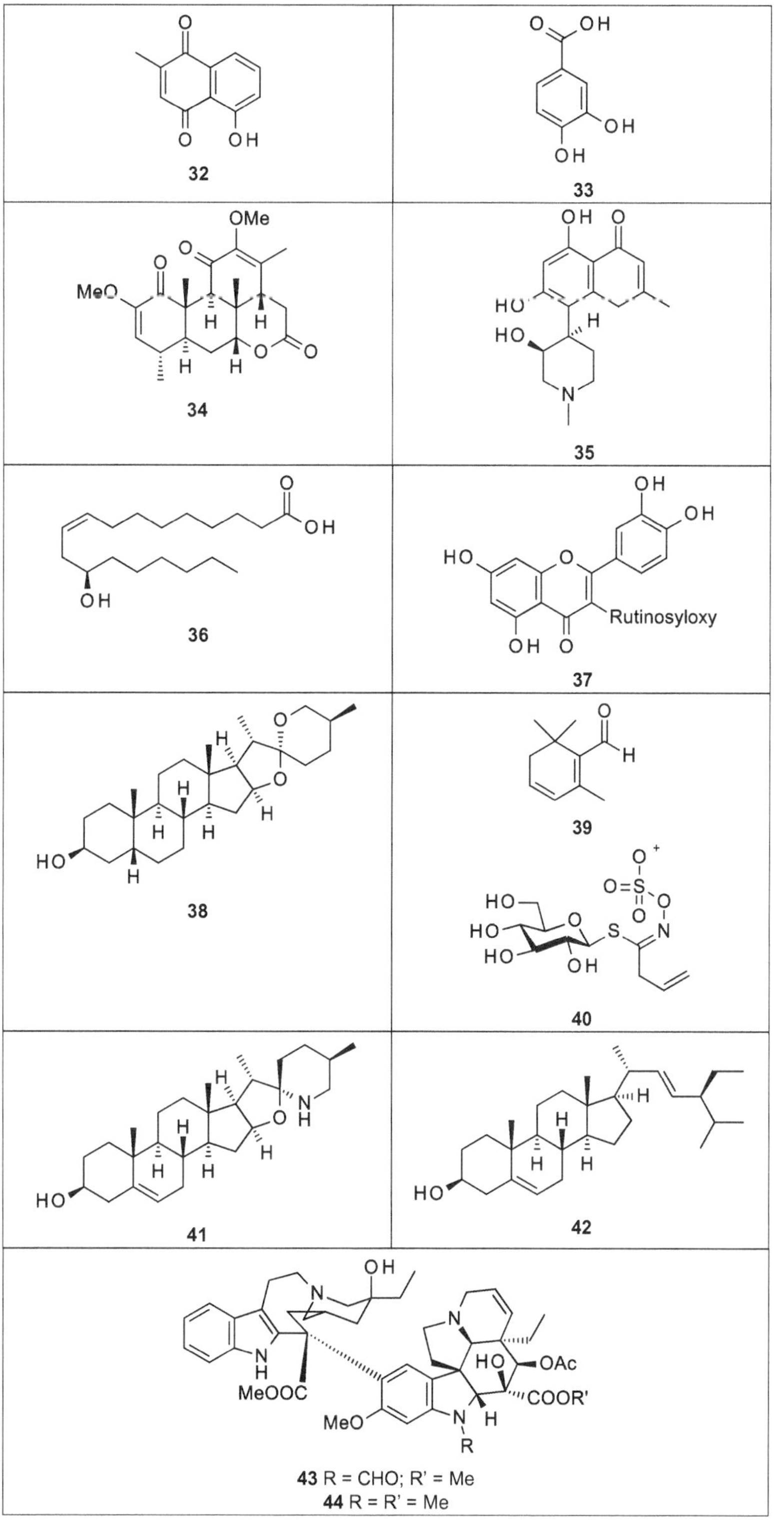

Fig. 4—Cont'd

Amburana cearensis (Allemao) A. C. Sm. (Fabaceae),[36] quassinoids (quassin, **34**) from *Ailanthus excelsa* Roxb. (Simaroubaceae),[46] rohitukine (**35**) from *Dysoxylum binectariferum* Hook.f. (Meliaceae),[47] ricinoleic acid (**36**) from *Ricinus communis* L. (Euphorbiaceae),[48] rutin (**37**) from *Ruta graveolens* L. (Rutaceae),[41] sarsapogenin (**38**) from *Yucca schidigera* Roezi ex Ortgies,[38] safranal (**39**) from *Crocus sativus* L.,[33] sinigrin (**40**) from *Carica papaya* L.,[43] solasodine (**41**) from *Solanum xanthocarpum* Schrad. and Wendl. (Solanaceae),[22] stigmasterol (**42**) from *Trichosanthes cucumerina* L.,[29] and vincristine (**43**) and vinblastine (**44**) from *Catharanthus roseus* (L.) G. Don. (Apocynaceae).[49]

These phytochemicals have been shown to possess distinct effects on the ovary, uterus, hormonal level, implantation pattern, sperm production and sperm morphology. Some of them create a protective layer around the egg to prevent fertilization.[25] The components that affects fertility are broadly classified as abortifacient (causes early abortion), anti-estrogenic (estrogen antagonists), antifertility (prevents fertilization), anti-implantation (blocks implantation), anti-ovulatory (inhibit ovulation), antispermatogenic (reduces sperm production), antisteroidogenic (reduces steroidogenesis) and spermicidal (destroys sperm). Moreover, some of these phytochemicals can also modulate the fertility cycle, ovarian cells, steroidogenesis and sperm functions without blocking or preventing the process of fertilization. Furthermore, a few compounds help in assisted reproductive technologies. The fertility modulating activities of these active ingredients are summarized in Table 1.

Table 1 Plant-derived fertility regulating natural products and their activities.

Phytochemicals	Activities
Achyranthine (**1**)	Anti-implantation, anti-ovulatory and spermicidal[23,25]
Anethole (**2**)	Improves oocyte maturation rate of pre-antral follicles[24]
Anonaine (**3**)	Abortifacient[25]
Arecaidine (**4**)	Anti-ovulatory and spermicidal[25,26]
Atropine (**5**)	Abortifacient[25]
Azadirachtin (**6**)	Antifertility and spermicidal[27]
β-Carotene (**7**)	Improves oocyte maturation rate[28]
β-Sitosterol (**8**)	Anti-ovulatory[29]
Bisbenzylisoquinolines (**9**)	Anti-estrogenic[30]

Table 1 Plant-derived fertility regulating natural products and their activities.—cont'd

Phytochemicals	Activities
Berberine (**10**)	Antifertility and spermicidal[25,31]
Carvacrol (**11**)	Antifertility[32]
Choline (**12**)	Antifertility and spermicidal[31]
Crocin (**13**)	Improves *in vitro* early embryo development[33]
Crocetin (**14**)	Improves *in vitro* early embryo development[33]
Curcumin (**15**)	Antifertility and anti-ovulatory[34]
Cucurbitacin B (**16**)	Anti-ovulatory[29]
Dimethylcrocetin (**17**)	Improves *in vitro* early embryo development[33]
Diosgenin (**18**)	Abortifacient, anti-implantation and spermidice[25,35]
Embelin (**19**)	Spermicidal[22]
Epicatechin (**20**)	Maintain follicular survival and promotes development of secondary follicles[36]
Ferulic acid (**21**)	Anti-implantation[37]
Kaempferol (**22**)	Anti-implantation[37]
Gitogenin (**23**)	Regulates ovarian cell functions and steroidogenesis[38]
Gossypol (**24**)	Antifertility and spermicidal[39]
Lepidine (**25**)	Anti-ovulatory[40]
Markogenin (**26**)	Regulates ovarian cell functions and steroidogenesis[38]
Naringin (**27**)	Improves sperm motility affected by metal exposure[41]
Noscapine (**28**)	Anti-ovulatory[42]
Oleanolic acid (**29**)	Antifertility and antispermatogenic[43,44]
Palmatine (**30**)	Antifertility and spermicidal[31]
Pilocarpine (**31**)	Anti-ovulatory[26]
Plumbagin (**32**)	Anti-estrogenic and anti-ovulatory[45]
Protocatechuic acid (**33**)	Maintain follicular survival and promotes development of secondary follicles[36]
Quassinoids (quassin, **34**)	Anti-implantation[46]
Rohitukine (**35**)	Antifertility and anti-implantation[47]
Ricinoleic acid (**36**)	Antifertility[48]

Continued

Table 1 Plant-derived fertility regulating natural products and their activities.—cont'd

Phytochemicals	Activities
Rutin (**37**)	Improves sperm motility affected by metal exposure[41]
Sarsapogenin (**38**)	Regulates ovarian cell functions and steroidogenesis[38]
Safranal (**39**)	Improves *in vitro* early embryo development[33]
Sinigrin (**40**)	Antispermatogenic[43]
Solasodine (**41**)	Affects sperm function and sperm morphology[22]
Stigmasterol (**42**)	Anti-ovulatory[29]
Vincristine (**43**)	Degenerates A4 spermatogonia and changes sperm morphology[49]
Vinblastine (**44**)	Degenerates A4 spermatogonia and changes sperm morphology[49]

4. Mechanisms of action of fertility regulating natural product

The plant-derived natural products are reported to enhance as well as hinder the reproductive process due to their specific mechanisms of actions. The secondary metabolites present in plants with fertility regulatory property generally belongs to the classes—alkaloids, coumarins, flavonoids, glycosides, isoflavonoids, phenolics and terpenoids. Each class of compounds exhibit different mechanisms of action. For example, alkaloids inhibit the secretion of GnRH to suppress gonadotropin production and temporarily inhibit folliculogenesis and ovulation. The gonadotropins act on the receptors located on the granulosa cells to stimulate the growth and maturation of ovarian follicles. Alkaloids exert their effect on the hypothalamus to reduce the secretion of LH/FSH. This reduced level adversely affects the folliculogenesis. Likewise, inhibition of LH release provokes ovulation disruption by decreasing the number of mature follicles. The alkaloids, *e.g.*, anonaine (**3**), which show abortifacient activity interfere with the mitotic division of the developing fetus, disrupt the endometrial uterine lining and alters the hormonal balance. Whereas berberine (**10**) causes prolonged uterine contraction.[42,50] In males, alkaloids cause sperm membrane disruption, distorted sperm morphology, reduced sperm motility and viability, example arecaidine (**4**).[51]

Phenolic compounds like flavonoids and lignans show estrogenic activities. They bind with cytochrome P450 aromatase and 17β-hydroxysteroid dehydrogenase (key enzyme pf steroidogenesis) to mimic the biological actives of the endogenous hormones. These compounds *via* the hypothalamus-pituitary complex stimulates FSH to induce ovarian steroidogenesis.[52] Likewise, phytoestrogens induce decrease of P4 release in granulosa cells by inhibiting the 3β-hydroxysteroid enzyme.[53] Additionally, flavonoids increase sperm motility (in the cases of occupational hazards) by reducing malondialdehyde (MDA) production.[41]

Glycosides reduces spermatogenesis and inhibits male steroid production.[43] Glycosides, saponins, tannins, terpenoids and triterpenoids directly acts on ovarian morphology, decreases folliculogenesis, increases atretic follicles and causes irregular cycles.[25,40] The sesquiterpene and hydrocarbons disrupt the sperm membrane potential to decrease sperm motility.[32] Alkaloids, flavonoids, glycosides and coumarin in combination decreases P4 and induces E2 synthesis.[54] Studies show that flavonoids can be used as candidates for *in vitro* fertilization techniques, where they stimulate the early embryo development and maturation.[33] Furthermore, epicatechin (**20**) and other phenolics have direct regulatory effect on ovarian cell functions—proliferation and apoptosis, and their secretory activities.[38,55] Famous natural products like azadirachtin (**6**), gossypol (**24**), vincristine (**43**) and vinblastine (**44**) affect the seminiferous tubules to degenerate germ cells and blocks the stage XII of late spermatids and sometimes interferes in steroid aromatization.[22,49] Vincristine (**43**) and vinblastine (**44**) are antimicrotubule agents, cause cell arrest in the metaphase stage to block mitosis. These alkaloids through mitotic/meiotic arrest, act at all the stages in seminiferous epithelium, decline primary spermatocytes production and elongate the spermatids.[49]

5. Virtual screening of fertility regulating natural products

Natural products are the biggest repository of large-scale structurally diverse compounds. Efficient and effective access to this diverse chemical space can assist in new drug development with predicted targets and mechanisms of actions. Virtual screening came out as a great promise in this regard to identify lead compounds from natural products with relevant bioactivity retrieved from the literature. These screening methods are the *in silico* approaches to rationalize the bioactive component to putative protein-ligand binding characterization.

Ligand rohitukine (**35**) was evaluated by docking against reference protein of ERα which appears to fit ideally into the hydrophobic pocket of ER. Absence of basic sidechain toward the anti-estrogen binding site on the receptors predicts estrogen antagonistic activity of the ligand.[47] Experimental evidence suggests structural modification of the ligand would be promising to enhance anti-implantation/estrogen antagonistic activity.[47] Another *in silico* study conducted with ricinoleic acid (**36**) as ligand against human acrosin (prepared by homology modeling) have shown the acrosin blocking activity of the ligand. Results suggests the probable development of a new spermicidal drug.[56]

6. Fertility regulatory natural products in clinical trials (in the drug development chain)

The pharmaceutical industries have witnessed massive changes in the past decade, in relation to new approaches to new drug discovery and development. Identification of novel lead compounds predominantly relies upon high throughput screening, where natural products screening can play an important role. Plant-derived clinical candidates for fertility regulatory drug development that are in the development chain are discussed in this section.

Consap: The Central Drug Research Institute, Lucknow, India has developed a contraceptive cream "consap" using saponins obtained from *Sapindus mukorossi*. Consap alters the glycoprotein associated with the lipid bilayer of spermatozoa plasma membrane. It can change the sperm membrane during sperm maturation without affecting the testicular and epididymal morphology. It has proven its effective fertility regulatory activity through Phase I, II and III clinical trials. Consap has been licensed to the Hindustan Latex Limited, Mumbai, India to market as non-hormonal herbal spermicidal contraceptive.[57]

Pippaliyadi vati: Clinical trials are going on with "pippaliyadi vati." This is an Ayurvedic contraceptive combination of powdered fruits berries of *Embelia ribes*, *Piper longum* and borax. The Postgraduate Institute of Medical Education and Research, Chandigarh, India; the JIPMER, Pondicherry, India, and the KEM Hospital, Mumbai, India have completed the Phase I clinical trial for pippaliyadi vati. The trials have been conducted in the Central Council for Research in Ayurveda and Sidha, Kolkata and Hyderabad, India. Its acute, subacute toxicology and teratological examinations have been carried out at the National Institute of Immunology, New Delhi, India.[58] Reports have deduced that *in utero* exposure to pippaliyadi vati do not manifest in any form of adverse effects on postnatal development and reproductive performance of the F_1 progeny.[59]

Praneem: "Praneem," emerged as an intravaginal polyherbal (PHH) cream and pessary at the National Institute of Immunology, New Delhi, India, is a combination of seed extract of *Azadirachta indica*, fruit extract of *Sapindus mukorossi* and *Mentha citrate* oil. The formulation has been found to be safe in subacute toxicity studies and its intravaginal application in rabbits and monkeys exhibited effective contraceptive efficacy. Praneem has undergone Phase II clinical trials and has been accepted by the participants in Phase II safety and acceptability study.[60] Clinical trials have been conducted in Postgraduate Institute of Medical Education and Research, Chandigarh, India; Safdarjung Hospital, New Delhi, India and Kamla Nehru Memorial Hospital, Allahabad, India. Intravaginal use during clinical investigations was found to be non-toxic. As a contraceptive and to manage abnormal vaginal discharge syndrome a microbicide praneem polyherbal tablet (PPHT) has successfully passed through Phase III clinical trials and has been licensed to M/s Panacea Biotec, India.[60]

7. Conclusions

Several natural products have been isolated from a wide range of plants and evaluated for their fertility regulatory activity. But most of them are yet to get clinical clearance to become a marketed drug. Moreover, reports on virtual screening for herbal fertility regulators are just a few. Traditional documentations often refer to many plants with fertility regulatory activity, but not much systematic studies aiming at isolation of active phytochemicals from these plants and working out their mechanism of action have been performed.

Acknowledgments

L.N. gratefully acknowledges the financial support of the European Regional Development Fund—Project ENOCH (No. CZ.02.1.01/0.0/0.0/16_019/0000868), and S.N. thanks the Commonwealth Commission for the award of the Rutherford Post-doctoral Fellowship.

References

1. Seeley, R.; VanPutte, C.; Regan, J.; Russo, A. *Seeley's Anatomy & Physiology*. McGraw-Hill: New York, 2011.
2. Lee, F.-K.; Tsui, K.-H.; Wang, P.-H. *Plant Products Impairing Fertility of Animals*. LWW, 2015.
3. Knobil, E. *Knobil and Neill's Physiology of Reproduction*. Vol. 2. Gulf Professional Publishing, 2006.
4. Molina, P. E.; Education, M.-H. Endocrine physiology. In *Lange Medical Books*, McGraw-Hill, 2006.

5. Sirotkin, A. V. *J. Cell. Physiol.* **2011**, *226*(9), 2222–2225.
6. Conneely, O. M.; Mulac-Jericevic, B.; Demayo, F.; Lydon, J. P.; O Malley, B. W. *Recent Prog. Horm. Res.* **2002**, *57*, 339–356.
7. Tingen, C.; Kim, A.; Woodruff, T. K. *Mol. Hum. Reprod.* **2009**, *15*(12), 795–803.
8. Aerts, J.; Bols, P. *Reprod. Domest. Anim.* **2010**, *45*(1), 171–179.
9. Gurung, P.; Jialal, I. Physiology, male reproductive system. In *StatPearls*, StatPearls Publishing, 2019; [Internet], PMID: 30860700.
10. Leung, P.; Cheng, C. *The Ovary*. Elsevier Academic Press: Amsterdam, 2004; pp. 289–304.
11. Jiang, J.-Y.; Cheung, C.; Wang, Y.; Tsang, B. K. *Front. Biosci.* **2003**, *8*(1), 222–237.
12. Mihm, M.; Evans, A. *Reprod. Domest. Anim.* **2008**, *43*, 48–56.
13. Molnar, C.; Gair, J. *Concepts of Biology*, 1st Canadian ed.; BC Campus, 2015.
14. Drummond, A. E. *Reprod. Biol. Endocrinol.* **2006**, *4*(1), 16. https://doi.org/10.1186/1477-7827-4-16.
15. Walters, K.; Allan, C.; Handelsman, D. *Biol. Reprod.* **2008**, *78*(3), 380–389.
16. Deng, J.; Carbajal, L.; Evaul, K.; Rasar, M.; Jamnongjit, M.; Hammes, S. R. *Steroids* **2009**, *74*(7), 595–601.
17. Peluso, J. J. Non-genomic actions of progesterone in the normal and neoplastic mammalian ovary. *Semin. Reprod. Med.* **2007**, *25*, 198–207; Thieme Medical Publishers, Inc., 333 Seventh Avenue, New York.
18. Findlay, J.; Liew, S. H.; Simpson, E. R.; Korach, K. S. Estrogen signaling in the regulation of female reproductive functions. In: *Fertility Control*; Habenicht, U.-F., John Aitken, R., Eds.; Springer, 2010; pp 29–35.
19. World Health Organization *Family Planning: A Global Handbook for Providers*. World Health Organization, 2018.
20. Dias, D. A.; Urban, S.; Roessner, U. *Metabolites* **2012**, *2*(2), 303–336.
21. Castellanos, J. R. G.; Prieto, J. M.; Heinrich, M. *J. Ethnopharmacol.* **2009**, *121*(1), 1–13.
22. Kamal, R.; Gupta, R.; Lohiya, N. *Phytother. Res.* **2003**, *17*(6), 579–590.
23. Varuna, K. M.; Khan, M. U.; Sharma, P. K. *J. Pharm. Res.* **2010**, *3*(4), 714–717.
24. Sá, N.; Araújo, V.; Correia, H.; Ferreira, A.; Guerreiro, D.; Sampaio, A.; Escobar, E.; Santos, F.; Moura, A.; Lôbo, C. *Theriogenology* **2017**, *89*, 226–234.
25. Bala, K.; Arya, M.; Katare, D. P. *World J. Pharm. Pharm. Sci.* **2014**, *3*, 305–1326.
26. Shrestha, J.; Shanbhag, T.; Shenoy, S.; Amuthan, A.; Prabhu, K.; Sharma, S.; Banerjee, S.; Kafle, S. *Indian J. Pharmacol.* **2010**, *42*(5), 306–311.
27. Aladakatti, R. H.; Sukesh, B.; Jadaramkunti, U. C.; Hiremath, M. *Recent Res. Sci. Technol.* **2011**, *3*(2), 34–46.
28. Barakat, I. A.; Khalil, W. K.; Al-Himaidi, A. R. *Small Rumin. Res.* **2015**, *130*, 67–75.
29. Sandhya, S.; Vinod, K. R.; Chandra, J. S.; Aradhana, R.; Nath, V. S. *Int. J. Pharm. Sci. Rev. Res.* **2010**, *1*(2), 56–60.
30. Bhardwaj, A.; Modi, K. P. *Int. J. Phram. Sci. Res.* **2016**, 7(1), 42–54.
31. Spandana, U.; Ali, S. L.; Nirmala, T.; Santhi, M.; Babu, S. *Int. J. Curr. Pharm. Rev. Res.* **2013**, *4*(2), 61–68.
32. Pandey, G.; Madhuri, S. *Int. J. Pharm. Sci. Rev. Res.* **2010**, *5*(1), 61–66.
33. Maleki, E. M.; Eimani, H.; Bigdeli, M. R.; Ebrahimi, B.; Shahverdi, A. H.; Narenji, A. G.; Abedi, R. *Taiwan. J. Obstet. Gynecol.* **2014**, *53*(1), 21–25.
34. Mohebbati, R.; Anaeigoudari, A.; Khazdair, M. *Endocr. Regul.* **2017**, *51*(4), 220–228.
35. Hostettmann, K.; Marston, A. *Saponins*. Cambridge University Press: United Kingdom, 2006.
36. Barberino, R.; Barros, V.; Menezes, V.; Santos, L.; Araújo, V.; Queiroz, M.; Almeida, J.; Palheta, R.; Matos, M. *Zygote* **2016**, *24*(2), 277–285.
37. Thakare, V. N.; Kothavade, P. S.; Dhote, V. V.; Deshpande, A. D. *Int. J. Pharm. Technol. Res.* **2009**, *1*(1), 73–78.

38. Štochmaľová, A.; Kadasi, A.; Alexa, R.; Grossman, R.; Sirotkin, A. *Potravinarstvo Slovak J. Food Sci.* **2014**, *8*(1), 87–91.
39. Coutinho, E. M. *Contraception* **2002**, *65*(4), 259–263.
40. Pande, D.; Malik, S.; Bora, M.; Srivastava, P. *In Vitro Cell Dev. Biol. Plant* **2002**, *38*(5), 451–455.
41. Jamalan, M.; Ghaffari, M. A.; Hoseinzadeh, P.; Hashemitabar, M.; Zeinali, M. *Int. J. Fertil. Steril.* **2016**, *10*(2), 215.
42. Kumar, C. P.; Sachin, J. *Int. J. Pharm. Sci. Rev. Res.* **2013**, *23*(2), 98–107.
43. Lakshman, J.; Changamma, C. *Int. J. Integr. Sci. Innov. Technol.* **2013**, *4*(2), 6–10.
44. Mello, F. B.; Jacobus, D.; Carvalho, K.; Mello, J. R. *Toxicon* **2005**, *45*(4), 459–466.
45. Sheeja, E.; Joshi, S.; Jain, D. *Indian J. Pharmacol.* **2009**, *41*(6), 273–277.
46. Tamboli, S.; Konadawar, M. *Int. J. Pharm. Pharm. Sci.* **2013**, *5*(4), 128–129.
47. Keshri, G.; Oberoi, R. M.; Lakshmi, V.; Pandey, K.; Singh, M. M. *Contraception* **2007**, *76*(5), 400–407.
48. Han, F.; Zhang, X.-X.; Pi, N.-N.; Wang, J.; Ding, X.; Su, D.-H.; Gao, P. *Sichuan J. Zool.* **2008**, *27*(3), 390–395.
49. D'cruz, S. C.; Vaithinathan, S.; Jubendradass, R.; Mathur, P. P. *Asian J. Androl.* **2010**, *12*(4), 468.
50. Soni, P.; Siddiqui, A. A.; Dwivedi, J.; Soni, V. *J. Appl. Pharm. Sci.* **2013**, *3*(8), 1–12.
51. Sharma, N.; Jocob, D. *J. Ethnopharmacol.* **2001**, *75*(1), 5–12.
52. Mbemya, G. T.; Vieira, L. A.; Canafistula, F. G.; Pessoa, O. D. L.; Rodrigues, A. P. R. *Reprod. Clim.* **2017**, *32*(2), 109–119.
53. Santini, S. E.; Basini, G.; Bussolati, S.; Grasselli, F. *Biomed. Res. Int.* **2009**, *2009*, 41989155. https://doi.org/10.1155/2009/419891.
54. Telefo, P.; Moundipa, P.; Tchouanguep, F. *J. Ethnopharmacol.* **2004**, *91*(2–3), 225–230.
55. Kádasi, A.; Kolesárová, A.; Maruniaková, N.; Grossmann, R.; Stochmalová, A.; Alexa, R.; Sirotkin, A. V. *J. Microbiol. Biotechnol. Food Sci.* **2014**, *3*, 107–111.
56. Nath, S.; Choudhury, M. D.; Roychoudhury, S.; Talukdar, A. D.; Sirotkin, A. V.; Baková, Z.; Kádasi, A.; Maruniaková, N.; Kolesárová, A. *J. Microbiol. Biotechnol. Food Sci.* **2011**, *1*(2), 236–246.
57. Bhosale, V. V.; Asthana, O.; Gaur, S. *Curr. Sci.* **2013**, *104*(12), 1701–1703.
58. Chaudhury, R. R. Traditional methods of fertility regulation. In: *Current Status in Fertility Regulation: Indigenous and Modern Approaches*; Chaudhury, S. R., Gupta, C. M., Kamboj, V. P., Eds.; Central Drug Research Institute: Lucknow, India, 2001; pp 11–21.
59. Balasinor, N.; Bhan, A.; Paradkar, N. S.; Shaikh, A.; Nandedkar, T. D.; Bhutani, K.; Roy-Chaudhury, M. *J. Ethnopharmacol.* **2007**, *109*(3), 406–411.
60. Joglekar, N.; Joshi, S.; Deshpande, S.; Parkhe, A.; Katti, U.; Mehendale, S. *Trans. R. Soc. Trop. Med. Hyg.* **2010**, *104*(6), 412–415.

Further reading

61. Hutt, K. J.; Albertini, D. F. *Reprod. Biomed. Online* **2007**, *14*(6), 758–764.

CHAPTER SIXTEEN

Antipsychotic natural products

Krystyna Skalicka-Woźniak[a,b,*], Jürg Gertsch[b]
[a]Independent Laboratory of Natural Products Chemistry, Medical University of Lublin, Lublin, Poland
[b]Institute of Biochemistry and Molecular Medicine, University of Bern, Bern, Switzerland
[*]Corresponding author: e-mail address: kskalicka@pharmacognosy.org

Contents

1. Introduction

"The road to creativity passes so close to the madhouse and often detours or ends there."

Ernest Becker, The Denial of Death

Neuropsychiatric diseases (NDs) occur in all societies with an equal prevalence across countries and ethnicities, with few debatable exceptions.[1] NDs that embrace psychosis as a symptom comprise schizophrenia, schizophreniform disorder, affective (mood) disorders, including major depression, severe depression or mania in bipolar disorder (manic depression) and schizoaffective disorders.[2] Despite a lack of insights from well-designed empirical studies, it is generally accepted that creative minds more likely suffer from psychotic NDs, or that the predisposition of this mental disease positively correlates with creativity, e.g., in art.[3,4] Thus, psychosis potentially occurs as collateral evolutionary effect of recombination if selection favors creativity. Different genetic predispositions are associated with psychotic disorders,[5] but also lifestyle factors play a role, in particular related to substance-induced first episodes of psychoses associated to recreational psychoactive drugs.[6] It is debatable whether diet plays an underlying role.

Annual Reports in Medicinal Chemistry, Volume 55
ISSN 0065-7743
https://doi.org/10.1016/bs.armc.2020.03.004

A recent review by Aucoin et al.[7] discusses the observational evidence regarding the connection between the occurrence of psychotic episodes in NDs and poorer quality dietary patterns, e.g., a lower intake or fibers, ω-3 fatty acids, vegetables, fruits, vitamins and minerals, as well as an altered microbiome composition. Moreover, ketogenic diet (low carbohydrate intake) was mentioned to reduce psychotic episodes in mood disorders.[7] As reviewed by Beer (1996),[8] the medical concept of psychosis can be traced back to the mid-19th century, thus prior to the description of schizophrenia as a mental disease and the worldwide onset of illicit drug abuse after the world wars.

The Swiss psychiatrist Eugen Bleuler coined the term "schizophrenia" in 1911 describing the symptoms as positive and negative. Positive symptoms included episodes of psychosis. In the Diagnostic and Statistical Manual of Mental Disorders DSM-III, the types of schizophrenia are delineated as disorganized, catatonic, paranoid, residual, and undifferentiated. As pointed out in a recent review by Loch,[9] psychotic symptoms are clearly not exclusive to schizophrenia, but are also present in most nonpsychotic psychiatric diseases; being a marker of severity. He argues that hallucinations and delusions appear to have distinct, independent biological underpinnings, in the general population, in psychotic, and in nonpsychotic disorders as well. In all contexts, psychosis results in altered perception of reality. Symptoms include delusions and hallucinations, incoherent speech and behavior that is inappropriate for the situation. Moreover, insomnia, social withdrawal and lack of motivation lead to difficulties in carrying out daily activities. While in Western societies psychotic behavior is clearly demarcated as illness, in some traditional societies schizophrenia and substance-induced hallucinations are regarded as supernatural (divine) phenomena. The manifestations of psychosis, and the ancient use of psychedelic drugs in these societies, is closely associated with shamanism and animism.[10,11] Likewise, animistic thinking is more pronounced in psychotic patients.[12]

From an ethnopharmacological point of view, besides alcohol, plant and fungal psychoactive drugs like opium (*Papaver somniferum* L.), fly agaric (*Amanita muscaria* (L.) Lam), ayahusaca (herbal drug mixture containing both dimethyltryptamine and monoamine oxidase (MAO) inhibitors), peyotl (*Lophophora williamsii* Lem. ex Salm-Dyck), or teonanácatl (*Psilocybe* spp.) have influenced humans over millennia in their intellectual and cultural evolution.[13–15] Archeology has uncovered hallucinogens in ancient tombs in the big civilizations in Latin America,[16] described petroglyphs portraying rituals with possibly psychoactive fungi,[17] or unearthed psychoactive

Cannabis sativa L. in the desert of Gobi.[18] There is nothing more intriguing to humans than psychoactive drugs because they act on the mind, i.e., the most recently evolved higher cognitive brain areas affecting cortico-subcortical processing related to neural processes of self-reflection. These neurobiological correlates show similarities to the processes observed in schizophrenia.[19] As will be discussed, certain botanical drugs that exhibit hypotensive or cholinergic effects, such as Rauwolfiae radix (*Rauwolfia* spp.) or *Physostigma venenosum* Balf., also exert clinically significant antipsychotic effects.[20–22] The quest for psychoactive NPs started with the seminal isolation of the highly effective analgesic morphine (**1**, Table 1) from opium by Sertürner at the beginning of the 19th century. While the habitual use of morphine does not cause psychosis, forms of psychosis due to idiosyncrasy, chronic use, over dosage or withdrawal or are documented.[23] Opium preparations have been known for millennia to exert strong analgesic, sedative, but also psychedelic effects, which may be either weakly pro- or antipsychotic, depending on dose and drug preparation.[24,25] In opium, alkaloids such as the antipsychotic nonopioid phosphodiesterase inhibitor papaverine (**2**) can further modify the effects of morphine, thus altering the overall pharmacological properties of opium.[26] Today, certain opioids are believed to be the only sedative drugs that possess antipsychotic properties.[27]

In agreement, individuals suffering from mood and anxiety disorders like bipolar, panic and major depressive disorders more likely to abuse opioids as a form of self-medication, a factor potentially also related to the current opioid crisis in the US.[28,29] In the 19th century, pharmacology emerged as a scientific discipline which exposed how dissimilar the effects of pure NPs compared to botanical drugs can be, exemplified by the distinctly different effects of cocaine salts or free base (**3**) and leaves of *Erythroxylum coca* Lam. (cocae folium). Cocaine and amphetamine (**4**) abuse is among the most frequent causes for psychosis,[30] whereas coca leaf consumption does not show such effects. The generation of centrally active synthetic molecules like LSD (**5**), or the benzodiazepine chlordiazepoxide (Librium) (**6**), completely changed our understanding of molecular neuropharmacology,[31,32] and opened new rational avenues to explore psychosis. With the seminal discoveries of neurotransmitter receptors mediating psychopharmacological effects in the 1970s,[33–35] novel and more targeted synthetic psychoactive agents could be developed.[36] As sedative-hypnotic or anxiolytic agents were not suitable or ineffective to treat the positive symptoms in schizophrenia patients, the need for novel medicines became a topic of intense research. Nevertheless, the first synthetic antipsychotic (i.e., neuroleptic) drugs

Table 1 Structures of compounds.

No	Name of the compounds	Structures
1	Morphine	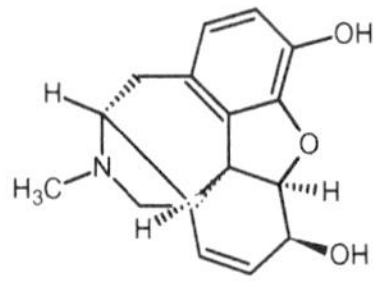
2	Papaverine	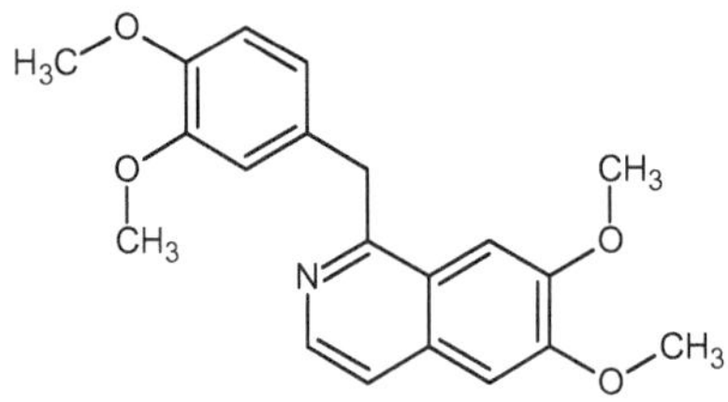
3	Cocaine	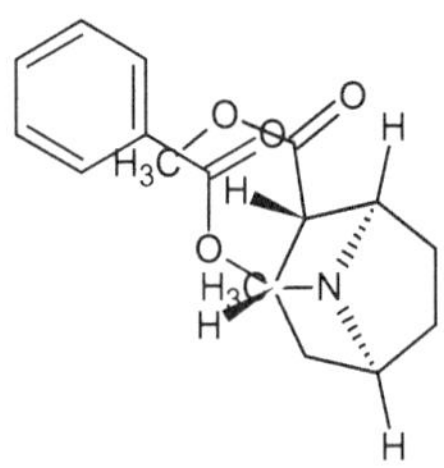
4	Amphetamine	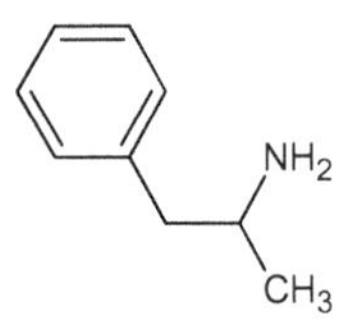
5	LSD	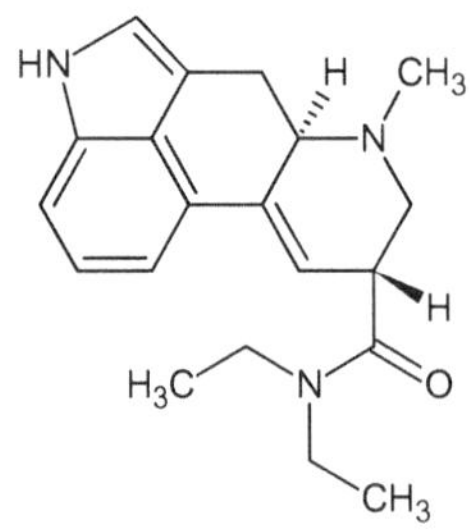

Table 1 Structures of compounds.—cont'd

No	Name of the compounds	Structures
6	Chlordiazepoxide	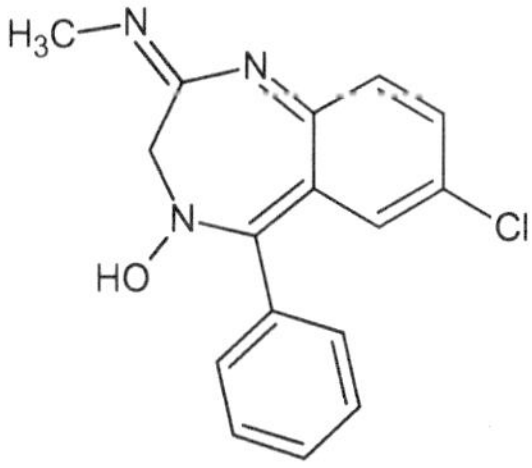
7	Chlorpromazine	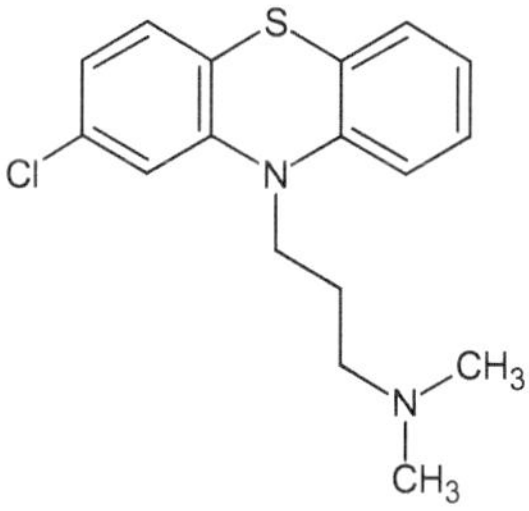
8	Haloperidol	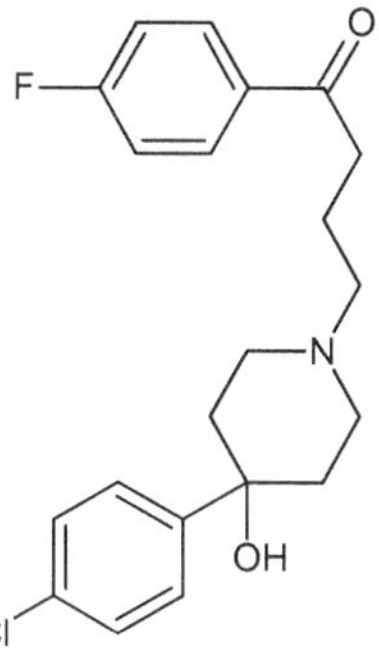
9	AMB-FUBINACA	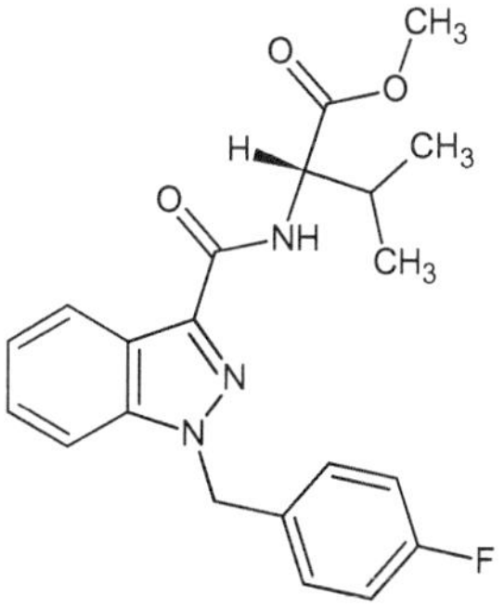

Continued

Table 1 Structures of compounds.—cont'd

No	Name of the compounds	Structures
10	Grayanotoxin	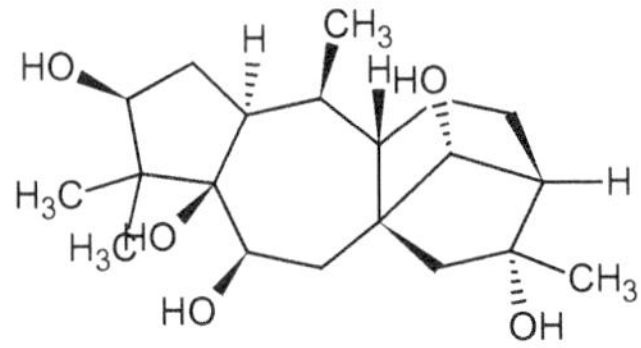
11	Ergine	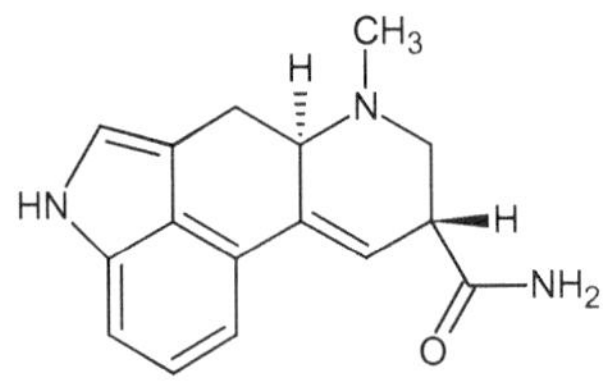
12	Mescaline	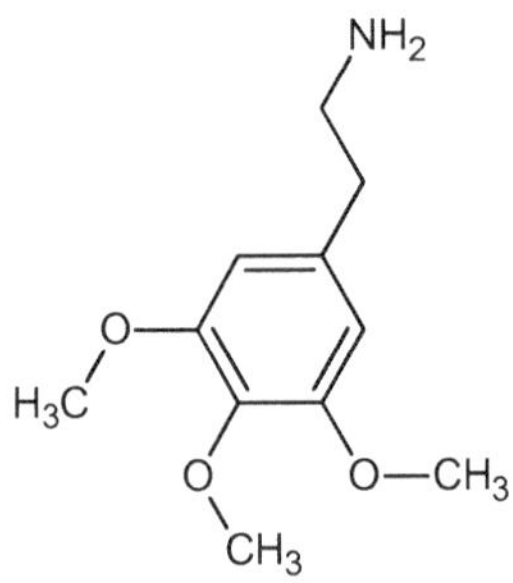
13	Dimethyltryptamine	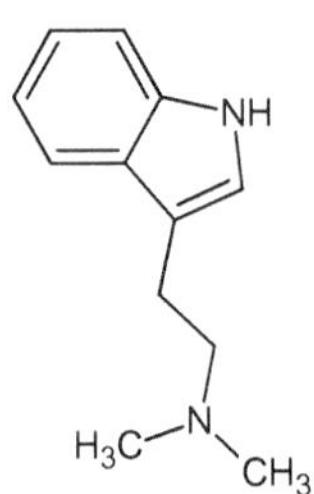
14	Psilocybin	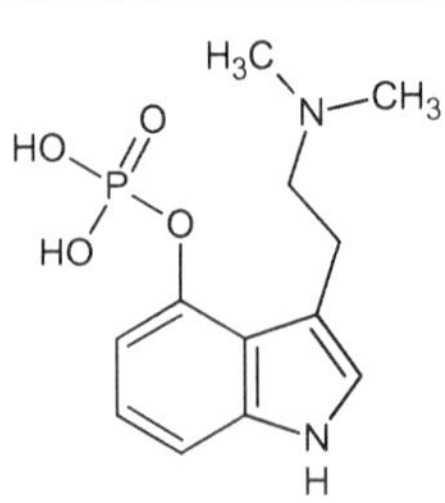

Table 1 Structures of compounds.—cont'd

No	Name of the compounds	Structures
15	Ibotenic acid	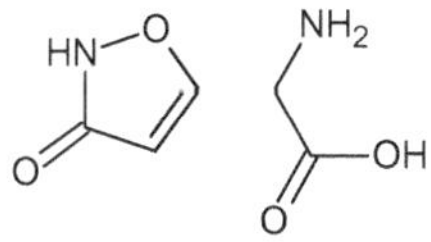
16	Muscimol	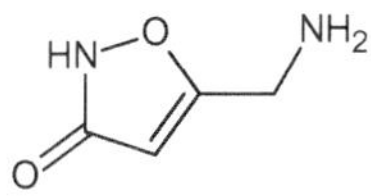
17	S-Hyoscyamine	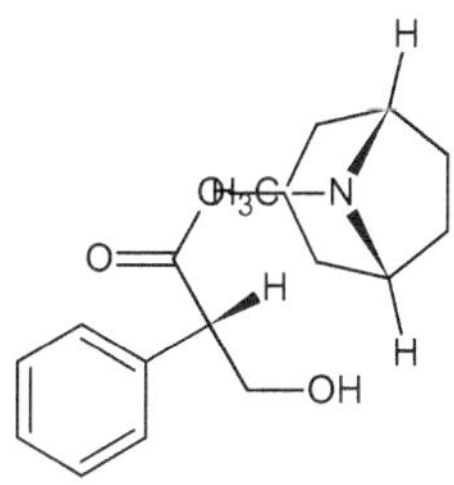
18	Scopolamine	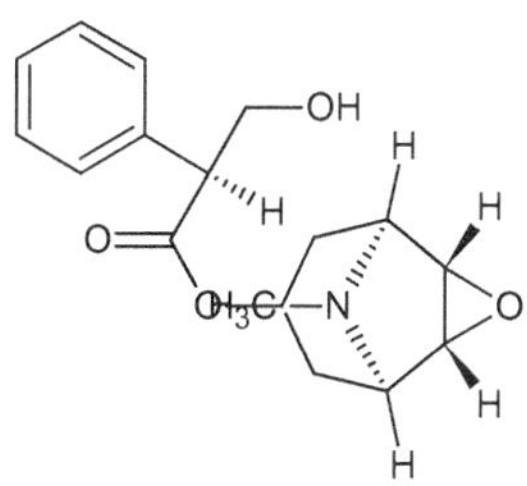
19	Ibogaine	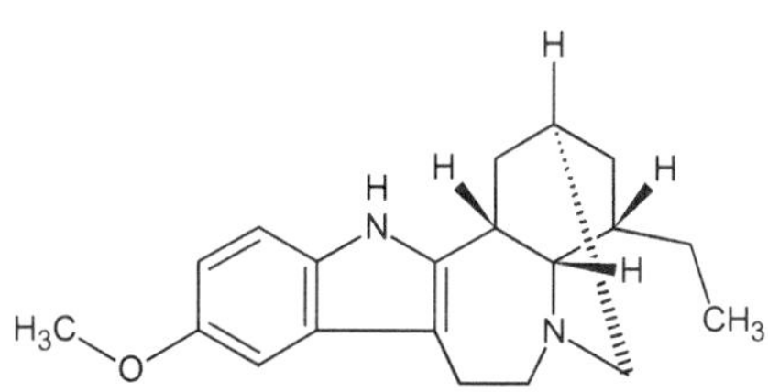
20	Salvianorine A	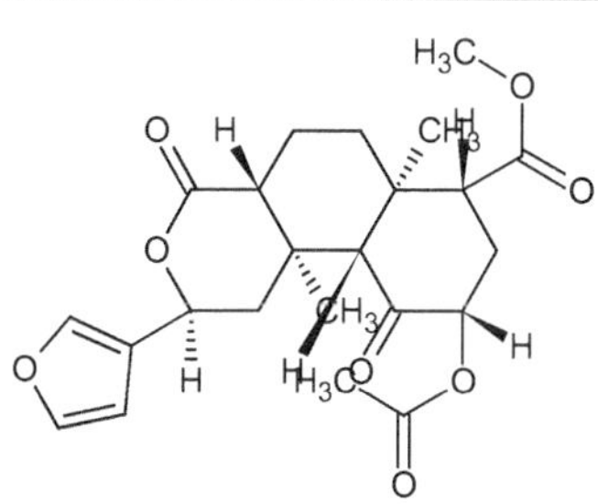

Continued

Table 1 Structures of compounds.—cont'd

No	Name of the compounds	Structures
21	Δ^9-tetrahydrocannabinol (THC)	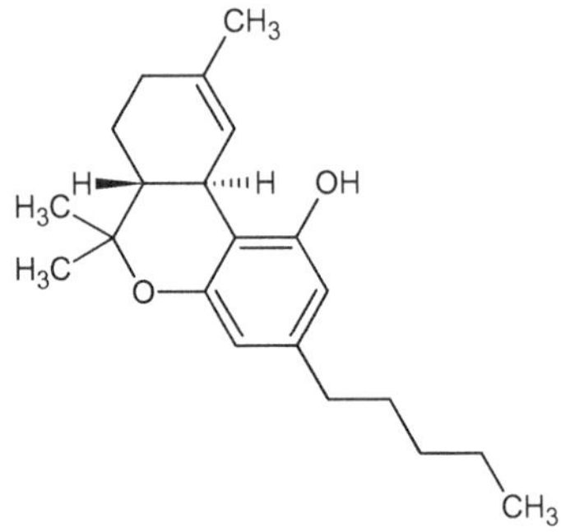
22	Cannabidiol (CBD)	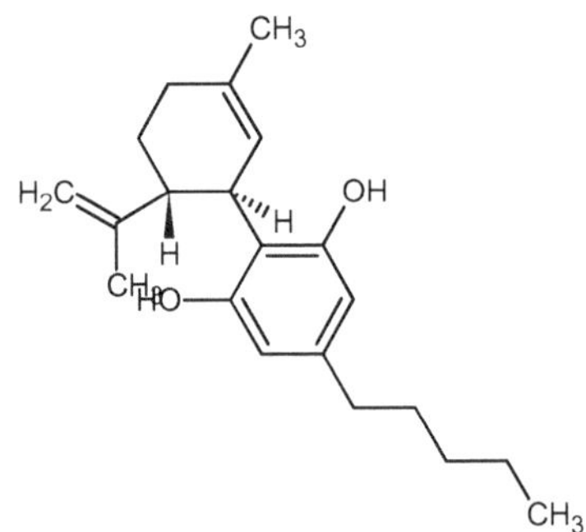
23	β-Caryophyllene (BCP)	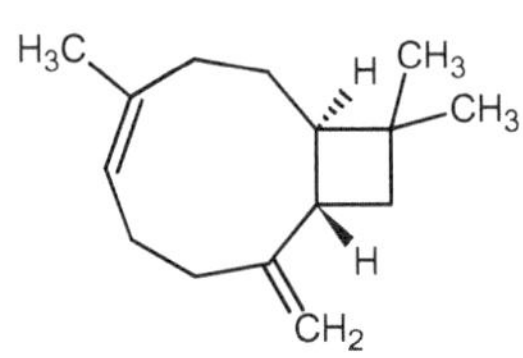
24	Ephedrine	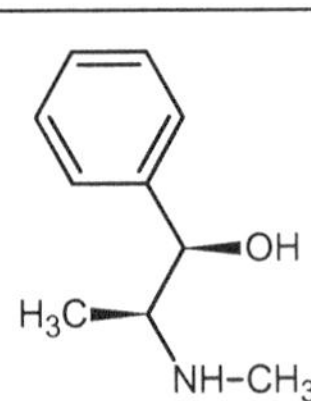
25	Arecaidine	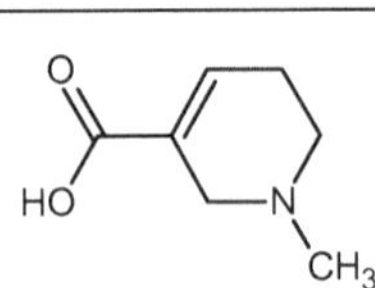

Table 1 Structures of compounds.—cont'd

No	Name of the compounds	Structures
26	Arecoline	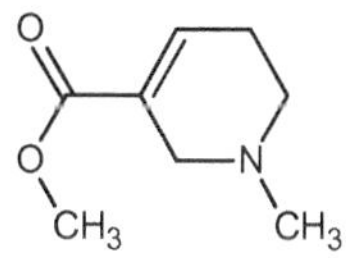
27	Mitragynine	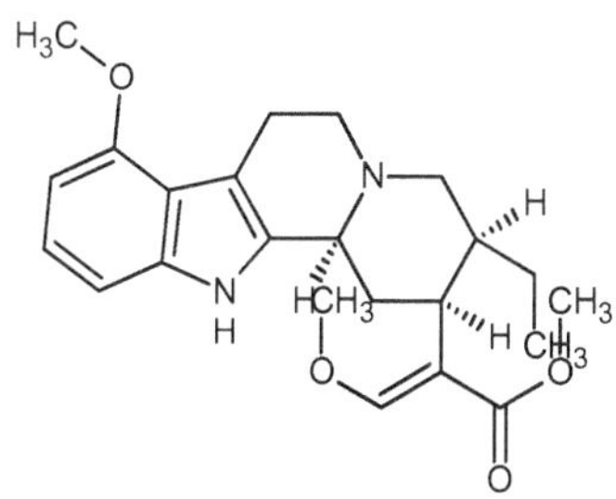
28	Yohimbine	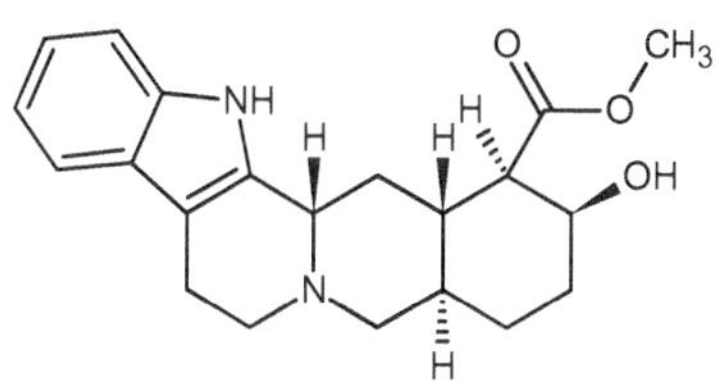
29	Reserpine	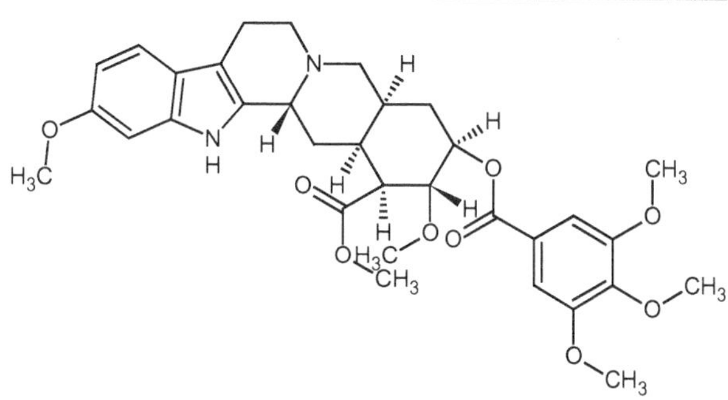
30	Physostigmine	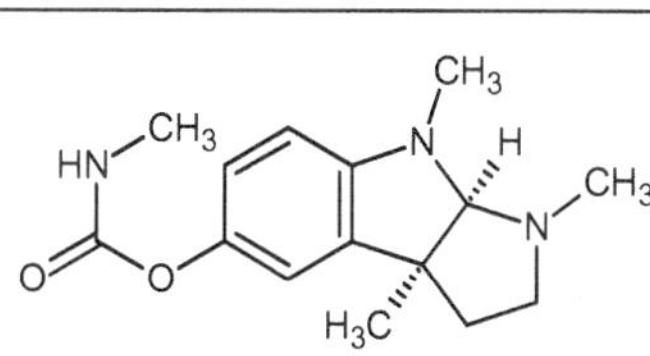

Continued

Table 1 Structures of compounds.—cont'd

No	Name of the compounds	Structures
31	Galantamine	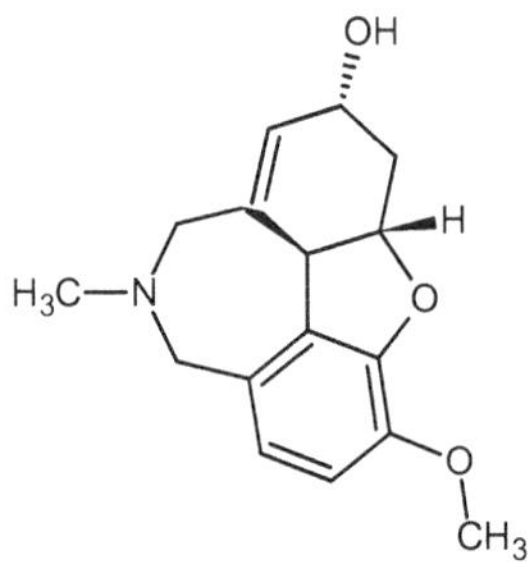
32	Lobeline	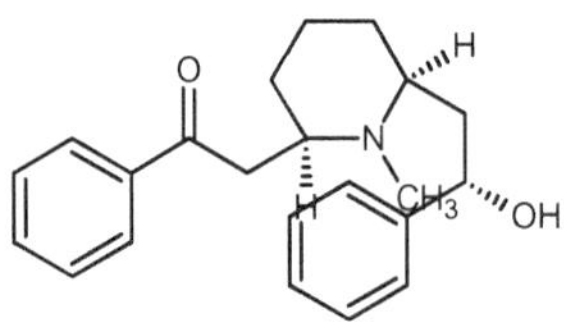
33	Anabasine	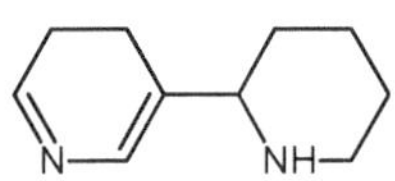
34	(−)-Stepholidine	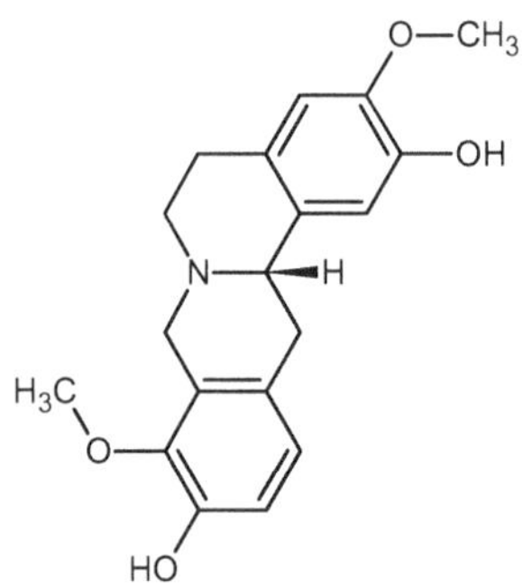
35	Scopoletin	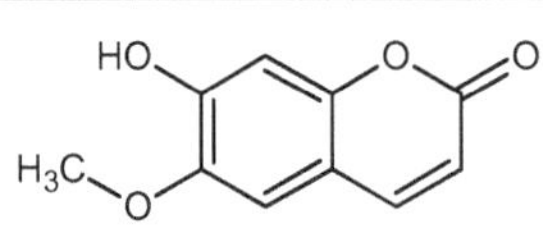

Table 1 Structures of compounds.—cont'd

No	Name of the compounds	Structures
36	Rutin	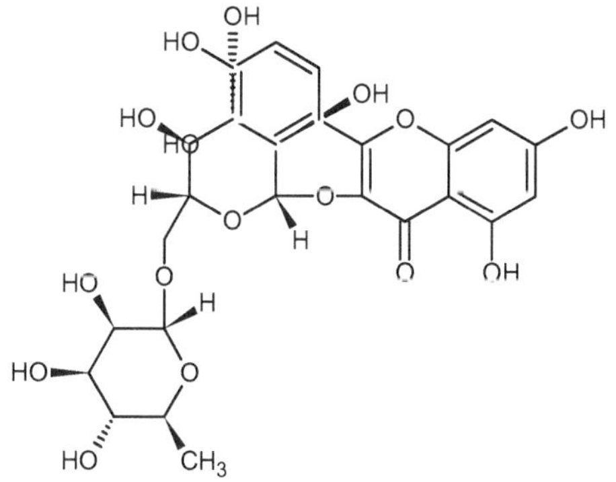
37	Apomorphine	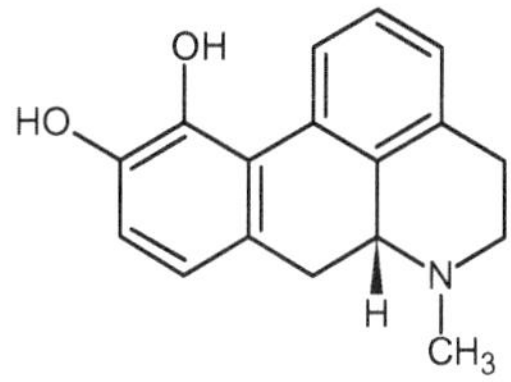
38	Emodin	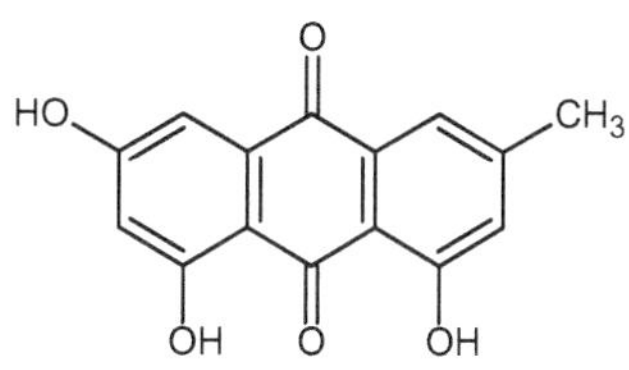
39	Curcumin	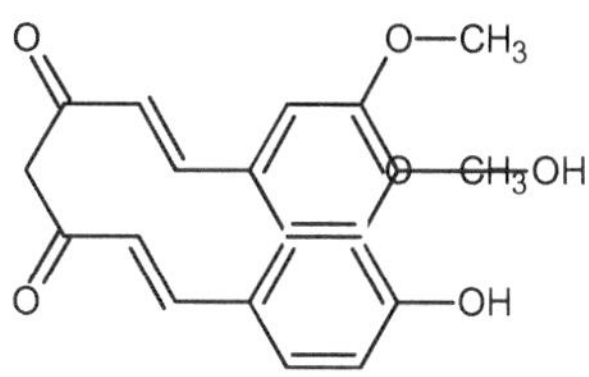
40	Salidroside (rhodioloside)	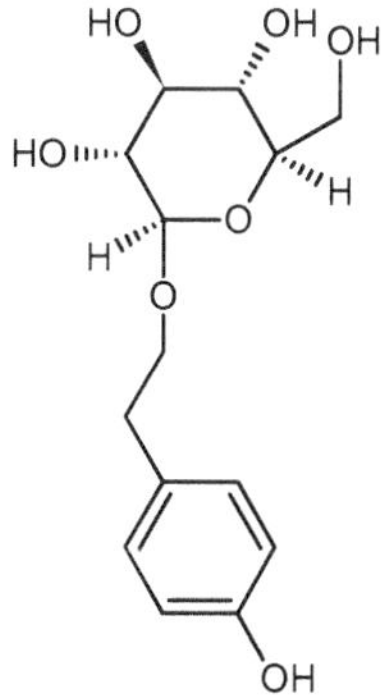

Continued

Table 1 Structures of compounds.—cont'd

No	Name of the compounds	Structures
41	Withaferin A	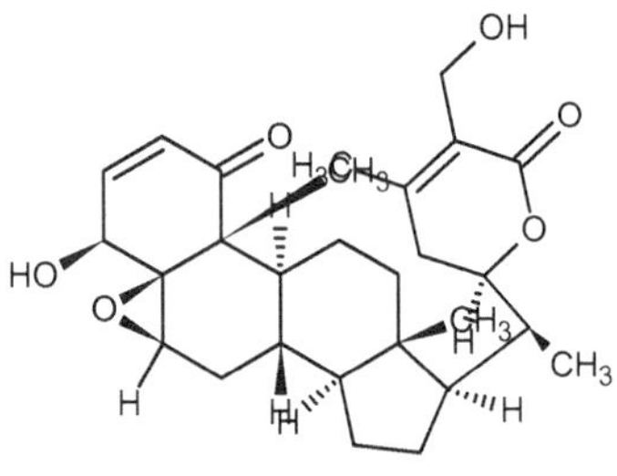
42	Hypericin	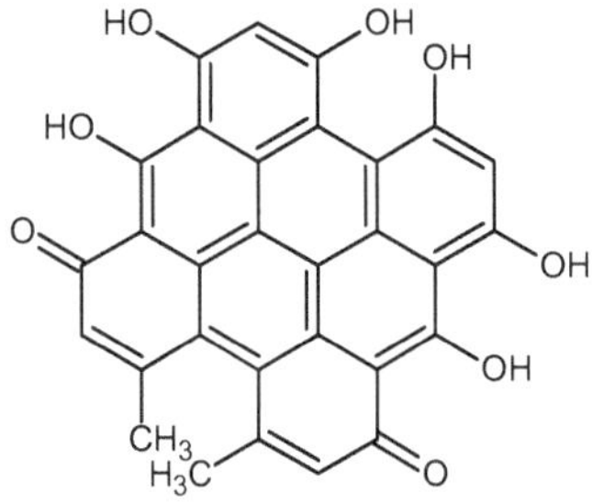
43	Stigmasterol	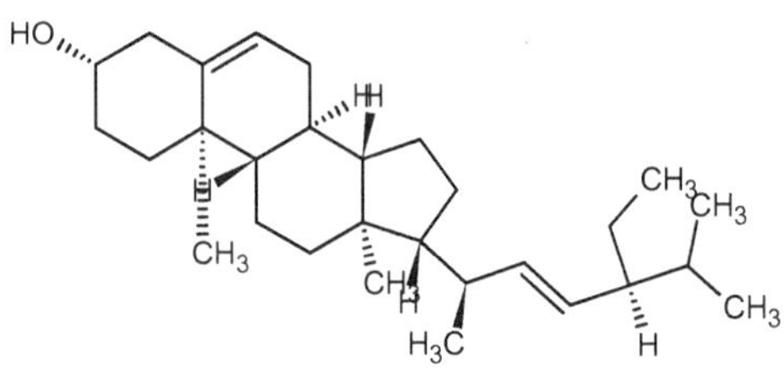
44	Guineensine	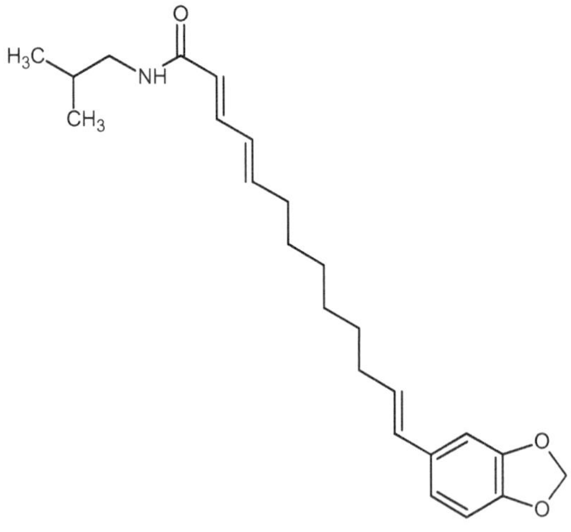

chlorpromazine (**7**) and haloperidol (**8**), potent antagonists of dopamine receptors, were already introduced in the 1950s. In subsequent decades, both medicinal chemistry focusing on potency and curiosity-driven projects led to numerous new psychoactive substances, most of which were never tested in clinical trials. The latter is impressively illustrated by the numerous psychoactive phenylethlyamines and tryptamines synthesized by Alexander Shulgin, who also described their effects based on self-experimentation,[37] or the highly potent synthetic cannabinoids like AMB-FUBINACA (**9**) that act as full agonists on cannabinoid CB1 receptors, and may cause dramatic motor suppression and psychosis.[38] Most NPs (and synthetic compounds) that induce hallucinations, but also psychostimulants like cocaine, amphetamines and MDMA, among many others, can trigger psychosis, especially in individuals with a predisposition for schizophrenia. Several of the antipsychotic drugs have paradoxical case reports, thus demonstrating the importance of inter-individual differences in neurobiochemical homeostasis, bell-shaped dose-response effects of drugs, and the inherent danger of oversimplifying complex psychopharmacological principles. The impact of plant secondary metabolites on psychosis seems to be a largely underestimated area of research, especially in the context of diet, recreation, herbal medicine and drug discovery.

2. Drunkenness, pontic honey, dancing plagues and natural product-induced psychosis

"One sip of this will bathe the drooping spirits in delight, beyond the bliss of dreams."

—John Milton, Comus

Mammalian brains reflect and are hyperexplorative, sensing and seeking small molecules with perceivable pharmacological effects. The human mind evolved under molecular selection pressures like nutrients and dietary metabolites. It is thus not surprising that *Homo sapiens* employ psychosis modulating substances in the context of uncovering and modifying the *conditio humana*. The association of acute neurotoxicity by certain natural products (NPs) is extensively illustrated in the literature describing voluntary or involuntary substance-mediated psychosis upon intoxication with a wide range of different agents.[39–41] Because many recreational xenobiotics can trigger psychosis there could be evolutionary implications for their use. Recent findings do not support previous hypotheses stating that schizophrenia risk genes have an evolutionary advantage,[42] and substance-induces psychosis may be hazardous, thus indicating high risk with uncertain benefit. The use of excessive alcohol is probably the oldest proven cause of substance-induced psychosis, dating back at least 9000 years, to when humans started to systematically generate alcoholic beverages from grains, honey and fruits.[43,44] Approximately three percent of people who are suffering from alcoholism experience psychosis during acute intoxication or withdrawal.[45] Although daily tobacco use is associated with increased risk of psychosis and an earlier age at onset of psychotic illness,[46] it does not appear that the cholinergic drug nicotine, a uniquely addictive psychoactive substance, can generate psychosis like other psychostimulants. The most widespread purinergic psychostimulant caffeine (discovered independently on three continents in *Coffea arabica* L., *Camellia sinensis* (L.) Kuntze, *Cola accuminata* (P.Beauv.) Schott & Endl., *Theobroma cacao* L., *Ilex paraguariensis* A.St.-Hil., *and Paullinia cupana* Kunth) may affect psychotic behavior in schizophrenia patients,[47] also in combination with smoking.[48] High doses of caffeine are particularly problematic for individuals with schizophrenia as this purine alkaloid alters sleep patterns and dopaminergic activity at postsynaptic neurons through its actions at adenosine A2A receptors, potentially exacerbating positive symptoms like delusions and hallucinations. In reported cases, chronic excessive intake of caffeine ($\geq$200 mg daily) combined with stress caused psychosis in healthy individuals.[49] Although virtually all illicit recreational drugs potentially trigger psychosis, acutely propsychotic food intoxications are rare, for example with the toxic "Pontic honey" first described by Pliny (23–79 CE) in the region of the Pontus Empire at the Black Sea.[50] To date, intoxications from such honeys are reported from Turkey and the Himalayans where bees feeding on *Rhododendron* species and plants from the Ericaceae transmit grayanotoxins (**10**) through their pollen to honey.[50] Grayanotoxins act on the autonomous nervous system or CNS by blocking voltage-gated sodium (NaV) channels, leading to severe overall intoxications.[51] Pliny also described another type of toxic honey, which he called "meli maenomenon," associated with *Nerium* spp. (oleander) flowers (i.e., pollen containing cardiac glycosides (cardenolides)).[52] In agreement, certain cardiac glycoside intoxications can lead to blockage of N^{+}/K^{+} ATPase not just in the heart but acutely also in the CNS, thus modulating the resting membrane potential required for the proper functioning of monoamine neurotransmitter transporters at chemical synapses. Cardenolide intoxications include symptoms like confusion, delirium, seizures, psychosis, but also abnormal dreams and color vision disturbance (xanthopsia).[53] However, the most common form of food-related psychosis in the past was related to ergot indole alkaloid poisoning (ergotism) through ingestion of the ergot alkaloids produced by the fungus *Claviceps purpurea* (Fr.) Tul in rye and contaminated flour. For centuries, epidemics of ergotism were common in Europe.[54] While the gangrenous ergotism resulting from peripheral 5HT1D serotonin receptor activation and associated vasoconstriction was likely mediated by ergotamine or ergocristine, the convulsive form of ergotism with seizures paresthesia, mania or psychosis rather resulted from the psychoactive d-lysergic acid amide called ergine (**11**) that potently activates central 5HT1A receptors. Some historians describe ergot poisoning as a possible cause in the dancing plagues reported in the late medieval time and Renaissance, and according to Sacks, even in the Salem Witch hysteria.[55] Ergotism is well documented and illustrated also in art (Fig. 1). However, to which degree ergotism could facilitate such a mass phenomenon remains unclear. The dancing mania or tarantism described in Southern Italy could also be associated to intoxication with spider toxins, maybe in combination with nutritional deficits.[56] The fact that the "dancing mania" only occurred during the summer months is clearly arguing against ergotism. Moreover, European spider toxin poisoning is not known to cause psychosis, despite the fact that α-latrotoxin triggers massive release of neurotransmitters in vitro via a complex mechanism involving activation of receptor(s) and pore formation.[57] The phenomenon of "dancing mania" is most consistent with mass psychogenic illness as discussed in the excellent review by Bartholomew (2000).[58] Indeed, strong religious emotions (religious delusions), but also pathological love emotions (erotomania or De Clérambault's syndrome), are closely

Fig. 1 Erogtism induced mania and psychosis. (A) Painting by Matthias Grünewald "The Temptation of St. Anthony" illustrating psychosis. (B) Peter Bruegel the Elder painted "The Beggars" in 1568. The figures in this detail may depict survivors of St. Anthony's fire. Oil on wood, Louvre Museum, Paris.

associated with dopaminergic and serotonergic signaling in the brain, potentially resembling the manic phase in bipolar depression.[59,60]

Early research on vasoactive ergot alkaloids at Sandoz (now Novartis) let to the synthesis of LSD by Albert Hofmann in 1938 and the subsequent discovery of its highly potent psychomimetic properties in 1943.[61] Hofmann discovered that also the seeds of Hawaiian baby woodrose (*Argyreia nervosa* (Burm. f.) Bojer.), a plant of Indian origin, and the morning glory (*Turbina corymbosa* (L.) Raf., syn. *Rivea corymbosa*) contained the alkaloid ergine.[62] Between 1947 and 1965 LSD was used clinically under the name Delysid for various psychiatric conditions.[63] Reports were highly variable but promising, especially for psychotherapy and minor personality disorders.[64] If convulsive ergotism was the result of brute food intoxication, LSD was the most potent hallucinogen ever described, and presented a new form of controllable substance-mediated experimental psychosis in which a healthy person knows that the hallucinations are not real. The molecular principles in all major hallucinogenic botanical drugs known to date were discovered, including mescaline (**12**) from psychoactive cacti (*L. williamsii* and *Trichocereus* spp), dimethyltryptamine (DMT, **13**) from hallucinogenic snuffs (e.g. *Anadenanthera peregrina* (L.) Speg.), or orally bioavailable preparations like ayahuasca (mixture of e.g. *Banisteriopsis caapi* (Spruce ex Griseb.) Morton and *Psychotria* spp.).

The isolation of psilocybin (**14**) from the Mexican psychoactive mushrooms (*Psilocybe* spp) was another milestone as it uncovered the pharmacological convergence between LSD, DMT and psilocin.[65] Mushroom poisoning can cause psychosis by distinctly different mechanisms as shown by the pharmacology of the prodrug ibotenic acid (**15**), the active principle in the fly agaric (*A. muscaria*) that is decarboxylated to muscimol (**16**). Fly agaric frequently triggers liliputhian hallucinations (environment turns small and dwarfs appear).[66] Fly agaric ingestion was a common hallucinogen among shamans in Eastern Siberia, but it can induce persistent psychosis in recreational us.[67] Fly agaric is special as it is the only hallucinogen to target $GABA_A$ receptors.[68] The tropane alkaloids from the Solanaceae like *Datura stramonium* L., *Atropa belladonna* L., *Mandragora officinarum* L., and *Hyoscyamus niger* L. have been known for millennia to induce intoxications with psychosis-like hallucinations and delusions, and retrograde amnesia.[69] Unlike "real serotomimetic hallucinogens" such as DMT, LSD or psilocybin, the Solanaceae tropane alkaloids strongly effect cognitive control, consciousness and short term memory.[70] S-Hyoscyamine (**17**) or its racemic mixture atropine, in combination with other Solanaceae alkaloids can trigger severe substance-induced psychosis, characterized by paranoia and strong delusions.[70,71] The German artist Hieronymus Bosch (1450–1516)

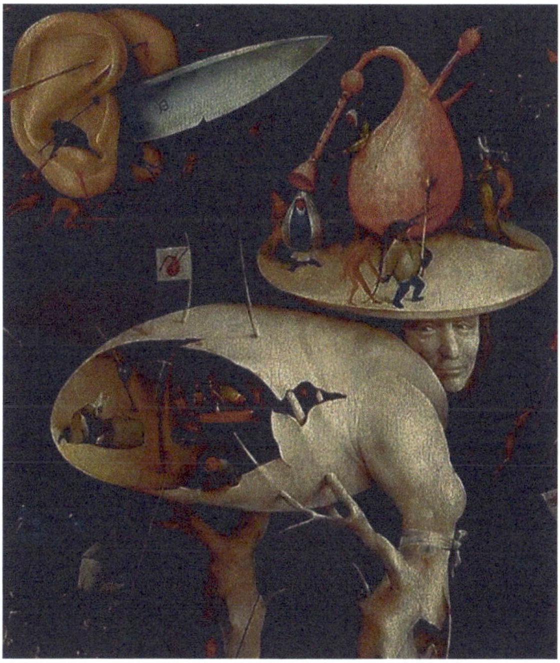

Fig. 2 The Renaissance painting "der Garten der Lüste [the garden of earthly delights]" by Hieronymus Bosch (1450–1516) points toward an experience of Solanaceae tropane alkaloid-induced psychosis. The circle shows the fruit of *Datura stramonium* L. which has been known for centuries to induce hallucinations and psychosis-like mental states.

impressively illustrated psychosis that was possibly caused by *D. stramonium* L. intoxication (Fig. 2). Levsin® (S-hyoscyamine sulfate) is used as adjunctive therapy in the treatment of peptic ulcer or to control gastric secretion, visceral spasm, spastic bladder, and rhinitis, among others. Levisin® associated psychosis is described as a potential adverse effect. The related alkaloid scopolamine (**18**), which is used for travel sickness at low doses, has strong propsychotic effects at high doses.[72] In Colombia, herbal preparations containing scopolamine, referred to as burundanga, are abused in crime with thousands of reported cases annually.[73] The most intriguing phenomenon seen in burundanga intoxication is not related to delusions or hallucinations, but the submissive behavior of the victim, caused by a reduction in declarative memory.[74] In some individuals, transdermal scopolamine psychosis has been reported.[75] The CNS signs and symptoms are diverse and include confusion, disorientation, hallucinations, dysarthria, ataxia, euphoria, fatigue, insomnia, agitation and retrograde amnesia. All botanical drugs containing antimuscarinic tropane alkaloids can cause anticholinergic psychosis, including *Cestrum nocturnum* L., *Brugmansia* spp., and *Solanum* spp.

Anticholinergic psychosis and delusions can be very strong without conscious control during the intoxication, thus frequently leading to accidents.[71]

Other propsychotic mechanisms of psychoactive drugs include NMDA receptor antagonism by ibogaine (**19**) from *Tabernanthe iboga* Baill.[76] Ibogaine induces oneirophrenia, a hallucinatory, dream-like state.[76] The Mexican sage species *Salvia divinorum* Epling & Játiva was traditionally used as ritual hallucinogen by indigenous ethnic groups in Oaxaca, possibly already in pre-Columbian times.[77] The leaves, which are either smoked or ingested at high doses, contain the diterpene salvinorin A (**20**), a potent kappa opioid agonist. It was postulated that consumption of *S. divinorum* could be linked to the development of psychotic disorders.[78] In recent years, attention has turned to the impact of various cannabinoids on risk of poor mental health. In particular, there is some suggestion from short-term experimental intoxication studies that ratios of Δ^9-tetrahydrocannabinol (THC) (**21**) to cannabidiol (CBD) (**22**) may have an impact on risk of psychotic-like experiences,[79] with some emerging evidence even suggesting that CBD might be antipsychotic (vide infra).[80] Psychoactive *C. sativa* (marihuana, hashish) with high amounts of THC is strongly correlated with psychosis in schizophrenia patients.[81] Evidence for an association between cannabis and psychosis has been documented in literature in many forms including experimental studies, epidemiological data, and numerous case studies.[82] THC produces a broad range of transient symptoms, behaviors, and cognitive deficits in healthy individuals that resemble aspects of endogenous psychosis. However, evidence for cannabis acting as a causal factor for schizophrenia has so far not been established.[83] Depending on the dose, THC may also exert antipsychotic (low THC dose is anxiolytic) effects, and potentially antipsychotic constituents in cannabis such as CBD or β-caryophyllene (BCP) (**23**) may further modulate the effect of cannabis herbal drugs.[84,85] In addition to hallucinogens, the majority of psychostimulants can potentially induce psychosis, including cocaine, amphetamines, or cathinones from *Catha edulis* (Vahl) Endl.[86,87] which are regulated substances in most countries. Yet, other psychostimulants remain legal. For instance, the monoaminergic ephedrine (**24**) from *Ephedra* spp. or the muscarinic cholinergic alkaloids arecaidine (**25**) and arecoline (**26**) present in *Arecea catechu* L., a botanical stimulant used by millions in Southeast Asia.[87,88] The excessive use of *Ephedra* spp. (called Ma Huang in China) is clearly associated with an increased risk for psychosis.[89] Likewise, there are numerous reported cases on betel induced psychosis.[90]

Paradoxically, in one study male high consumers of betel had significantly lower positive symptoms than low consumers or nonbetel users.[88] Kratom (*Mitragyna speciose* (Korth.) Havil.) is a tropical tree and its leaves are widely used in Malaysia and Thailand.[91] Kratom acts as a weak stimulant but when taken in higher doses acts as an analgesic, causing euphoria and sedation. The major indole alkaloid mitragynine (**27**) and its metabolite 7-hydroxymitragynin activate μ-opioid receptors.[92]

Kratom associated psychosis is well documented.[93] Despite its widespread use and statements by regulatory agencies on the potential for kratom-associated psychosis, there is little information regarding the prevalence of psychotic symptoms among kratom users. Overall, numerous centrally active herbal drugs have been reported to potentially trigger psychosis,[94] including ginseng (*Panax ginseng* C.A.Mey).[95] Although ginseng will be discussed as potentially antipsychotic (vide infra), Vázquez and Agüera-Ortiz reported the case of a 56-year-old woman with previous major depressive disorder who presented a manic episode during ginseng intake.[96] Similar cases were described in independent reports.[97] The indole alkaloid yohimbine (**28**) present in *Pausinystalia* and *Rauwolfia* species, exerts both cholinergic and adrenergic-type autonomic effects, potentially triggering an anxiety state in humans.[98] However, the degree of propsychotic response to yohimbine correlates highly with the basic level of emotional reactivity. It is a centrally active adrenoceptor α2 subtype inhibitor and used as aphrodisiac in men to increase sexual incentive in the context of erectile dysfunction, showing fear extinction in rodents.[99]

3. The discovery of antipsychotic natural products—From toxic drugs to healthy diet

"The heart, like the stomach, wants a varied diet."

Gustave Flaubert

While the vast majority of recreational drugs are more likely propsychotic (vide supra) antipsychotic botanical drugs are rare. The Indian snakeroot (*Rauwolfia serpentina* L.) is an exception. It has been used traditionally in India, China and Southeast Asia as a relatively safe botanical drug to reduce blood pressure and as tranquilizer, but also to reduce anxiety (states of fear, tension) and to treat psychotic conditions.[100] *Rauwolfiae radix* was used throughout India in the 1940s and worldwide in the 1950s.[101] Its major bioactive constituents are indole alkaloids, mainly reserpine (**29**), but also lower amounts of ajmaline, ajmalicine (raubasine) and yohimbine.[102] The isolation of reserpine led to extensive pharmacological investigations and clinical trials in the 1950s. Sandoz was marketing reserpine (4 mg) for hypertension treatment by 1954 under the name Serpasil.[103] Serpasil was tested also on patients suffering from psychosis and schizophrenia in comparison to the hitherto newly launched chlorpromazine.[104] The drugs differed in their side-effects, but they were useful to the psychiatrist to control psychosis and mania regardless of the underlying etiology. The initial enthusiasm regarding Serpasil as a treatment option for neuropsychiatric patients in combination with, at that time common, electroconvulsive therapy, was shadowed when side-effects were reported.[105] Long-term treatment with high doses of reserpine induced depression and reversible Parkinson's disease-like symptoms (possibly extrapyramidal side-effects).[106] Interestingly, local *Rauwolfia* species have been used in Africa to treat neuropsychiatric conditions including psychotic disorders with no reports about adverse effects.[107] Reserpine is a potent inhibitor of vesicular monoamine uptake and blocks the vesicular monoamie transporter 2 (VMAT2, SLC18A2).[108] Inhibition of VMAT2 reduces monoamine (mainly dopamine and noradrenaline) stimulation of sympathetic efferents, leading to sympatholytic actions and a reduction of blood pressure. The effectiveness of reserpine in psychosis is in agreement with the formulated dopamine hypothesis, which assumes hyperactive dopaminergic signaling in the mesolimbic system.[109] The typical antipsychotics (e.g., chlorpromazine and haloperidol) inhibit dopaminergic signaling via inhibition of D2 receptors, the atypical antipsychotics (e.g., risperidone and asenapine) act also on other monoamine signaling pathways, such as serotonin receptors, with the aim to reduce the extrapyramidal side-effects.

The cholinergic hypothesis of schizophrenia is substantiated by the psychosis triggered by antimuscarinic drugs and the opposite antipsychotic actions of acetylcholine (ACh) esterase inhibitors.[110] The discovery of such inhibitors has its origin in the ethnopharmacological use of the seeds of *P. venenosum* (Calabar bean), native of tropical Africa. The Calabar bean was used by different ethnic groups in ordeal rituals where they were administered to persons accused of witchcraft or other crimes.[111] In cases where deadly intoxication occurred, the result was considered to indicate guilt and, at the same time, to be a rightful punishment. The toxicity of the seeds stems from the indole alkaloid content related to calabrine and physostigmine (**30**). Currently, the reversible ACh esterase inhibitor physostigmine is medically used for the management of severe anticholinergic syndrome

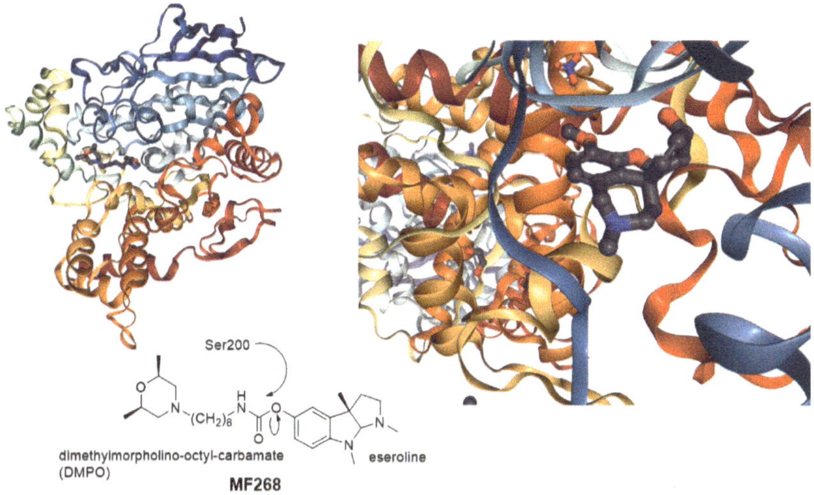

Fig. 3 ACh esterase is one of the fastest enzymes known, even though the active site is buried inside the protein at the end of a 20-A deep narrow gorge. In the X-ray structure, the dimethylmorpholinooctylcarbamic moiety of the physostigmine derived pseudo-irreversible inhibitor MF268 is covalently bound to the catalytic serine, which is located at the bottom of a long and narrow gorge. Surprisingly, the relatively bulky eseroline is not found in the crystal structure, thus implying the existence of an alternative route for its clearance. The images were generated in ChimeraX based on the PDB file 1OCE.

typically induced by antimuscarinic agents (e.g., benztropine, Solanaceae tropane alkaloids or diphenhydramine).[112] Physostigmine also inspired mechanistic drug discovery. For instance, its analogue 8-(cis-2,6-dimethylmorpholino)octylcarbamoyleseroline (MF268) helped to understand how ACh esterase works. The crystal structure of *Torpedo californica* (Tc) acetylcholinesterase (AChE) carbamoylated by the physostigmine analogue MF268 was reported at 2.7 A resolution[113] (Fig. 3). Physostigmine can also be used diagnostically to differentiate functional psychosis from anticholinergic delirium.[114] In the 1970s, different studies were performed to test the hypothesis that physostigmine and the synthetic ACh esterase inhibitor neostigmine could have beneficial effects in the cognitive impairment in schizophrenia patients. This was supported by the positive association between physostigmine and lithium treatment in schizophrenia patients that did not respond to neuroleptic treatment.[115] In the early 1970s, Janowsky and colleagues postulated that physostigmine might suppress delusions and turn off mania and psychosis.[116] They found that physostigmine attenuated psychotic behaviors in several patients fast and efficiently, and proposed the adrenergic-cholinergic hypothesis of mania and depression in

1972.[117] However, in controlled studies with schizophrenia patients, physostigmine showed significant depressiogenic effects.[118] The reversible ACh esterase inhibitor galantamine (**31**), which was first isolated from snowdrop (*Galanthus nivalis* L.) and is clinically used to improve symptoms in Alzheimers disease,[119] showed some promise as antipsychotic agent in monkeys.[120]

A therapeutic role of a galantamine-memantine combination as novel treatment option in schizophrenia patients to reduce the negative symptoms was proposed by Koola.[121] Adjunctive treatment with galantamine improved memory and attention in patients with schizophrenia, who were stabilized on risperidone.[122] Yet, the exact role of increasing CNS levels of ACh in psychosis is convoluted by the complex ACh receptor pharmacology. More selective ACh receptor interventions were described. For instance, lobeline (**32**) from the plant *Lobelia inflata* L. antagonizes α3β2, α4β2 and α7β2 nicotinic acetylcholine (ACh) receptor subtypes.[123]Lobeline was cocrystalized with α7 subunits by different groups and the receptor bound conformation could be established[124] (Fig. 4). *Lepidostoma fimbriatum* has been used as emetic and as a respiratory stimulant in the early 1800s.[125] Lobeline is discussed as an aid in smoking cessation, but evidence from such trials suggests that it has no benefit.[126] Lobeline was also tested as a possible treatment of drug abuse (amphetamine) and in some neurological disorders.[123] Since lobeline also moderately inhibits VMAT2, but does not trigger depression, there may be a potential of this natural product as antipsychotic agent. In humans, lobeline

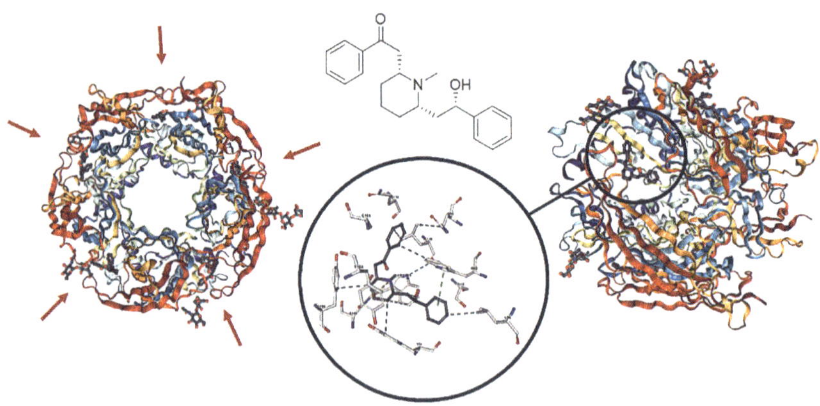

Fig. 4 Visualization of conformation of lobeline bound to alpha7-AChBP and fragment 3. The images showing the pentameric structure of the protein with five lobeline molecules in complex with the protein and the 3D conformation were generated in ChimeraX based on the PDB file 5AFL.

(15–30 mg) could modestly improve working memory in adults with ADHD, but no significant improvement in attention was observed.[127] Interestingly, the selective α7 ACh receptor agonist anabasine (**33**) from tobacco (*Nicotinia tabacum* L.) showed antipsychotic effects in a mouse model of MK-801-elicited popping behavior.[128] The presence of α7 ACh receptor agonist in tobacco is noteworthy as it could link the excessive smoking behavior with a potential self-medication (vide supra). Over decades, efforts have been undertaken to uncover novel NPs acting selectively at dopamine receptors or as new cholinergics, e.g., in the CDRI Program thousands of plant extracts were screened on CNS targets.[129] Likewise, botanical drugs from traditional Chinese medicine (TCM) have been screened. (−)-Stepholidine (**34**), an active ingredient of the TCM plant *Stephania* spp., was the first compound with a reported dual function as a dopamine receptor D1 agonist and D2 antagonist, showing promising antipsychotic activity in animal studies.[130] In another study, (−)-stepholidine was shown to be a moderately potent pan-dopamine receptor antagonist.[131] The antipsychotic-like activity of the relatively widespread NPs scopoletin (**35**) and rutin (**36**), the prodrug of quercetin, were assessed using apomorphine (**37**)-induced climbing behavior and methamphetamine induced stereotypy in mice.[132] Though disputed, both applied models are widely used to screen for potential antipsychotic-like activity in terms of their effects on positive symptoms of schizophrenia. Apomorphine is a D1 and D2 agonist, and methamphetamine induces positive symptoms of psychosis by reversing the action of the dopamine transporter DAT.[133] Noteworthy, apomorphine does not induce psychosis in healthy humans which may be a limitation of the model. Both scopoletin and rutin reduced apomorphine/methamphetamine induced climbing and stereotyped behaviors, but only at the lowest dose 0.1 mg/kg.[132] In both cases, higher doses did not exhibit antipsychotic activity, somewhat in agreement with the previously published finding that scopoletin at a high dose (80 mg/kg, i.p.) increased dopamine by inhibiting both A and B MAO isoforms, thus rather stabilizing propsychotic effects.[134]

A potential antipsychotic activity was also reported for emodin (**38**), an anthraquinone from *Aloe vera* (L.) Burm. f.[135] Experiments were performed in an animal model for schizophrenia-like behavior in which epidermal growth factor (EGF) was administered to neonatal rats. Subchronic oral administration of emodin (50 mg/kg) suppressed acoustic startle responses and abolished prepulse inhibition (PPI) deficits in this model.[135] In an independent study an *A. vera* extract showed antipsychotic and anxiolytic effects in rats.[136] The nor-seco-triterpenoid galphimine B from *Galphimia glauca*

Bartl., a plant used in Mexican traditional medicine for treating mental diseases, potentiated the cataleptic effect induced with haloperidol and exhibited a protective effect on some of the schizophrenia-like symptoms generated by ketamine and those induced by apomorphine in mice.[137,138] Despite the poor bioavailability of CBD (**22**), there is increasing evidence that this cannabinoid could have certain antipsychotic effects.[80] The effects of CBD on psychiatric symptoms, cognition, and marihuana consumption in marihuana users with recent-onset psychosis is currently being investigated.[139] Much of our knowledge of the endocannabinoid system (ECS) in schizophrenia comes from behavioral measures in rodents, including prepulse inhibition of the acoustic startle and open-field locomotion. However, the molecular target of CBD seems unrelated to the ECS and remains elusive.

Most schizophrenia patients do not achieve a satisfactory treatment response with their initial antipsychotic drug treatment. Sometimes botanical drugs are added to the neuroleptic treatment. As recently reviewed by Hoenders et al.,[140] adding TCM or Ayurvedic herbs to antipsychotics, the psychopathology may improve, but no conclusive evidence for this assumption can be provided at this point. The trend that herbal drugs improve the negative symptoms in schizophrenia, and in few cases also reduce psychosis, may be due to the fact that plant NPs generally exert anti-inflammatory effects,[141] and that inflammation (including high histamine and dysfunctional hypothalamus-pituitary-adrenal axis) is a risk factor in most neuropsychiatric disorders including psychotic diseases.[142]

Recently, Yadav et al.[143] reviewed 48 botanical drugs that possess antipsychotic activity. However, so far, none of these drugs have been studied further and the evidence remains anecdotal. Nevertheless, in rodent studies the aqueous pyrone-free extract and the lipid-soluble extract (kava resin) of kava-kava (*Piper methysticum* G. Forst. roots), a South Pacific recreational botanical drug with proven anxiolytic activity in men, showed antipsychotic like effects in rats. Both kava-kava extracts (30–150 mg/kg) reduced amphetamine-induced hypermotility in rats, while higher doses caused ataxia and sedation, with overall comparable effects to standard antipsychotic drugs chlorpromazine and haloperidol.[144] The central effects or kava-kava are attributed to the modulation of $GABA_A$ receptor activity, MAO B inhibition, and noradrenaline and dopamine reuptake inhibition.[145] Unfortunately, kava-kava extracts have been clinically tested primarily for their anxiolytic effects in treatment of nonpsychotic anxiety disorders and further research was hampered by the potential liver toxicity problems.

It thus remains unclear whether the centrally active kava-kava drugs have either anti- or propsychotic effects (vide infra).

Curcumin (**39**), the major constituent in *Curcuma longa* L. (Zingiberaceae) rhizome, is a pan-assay interference compound in vitro that is poorly bioavailable in vivo.[146] Curcuma is traditionally used in Asian countries as spice, but also because of its medicinal properties.[147] From in vitro and preclinical studies a huge variety of pharmacological effects have been reported, including anti-inflammatory, antioxidant, and neuroprotective.[147] In support of the proposed role of curcumin as a pan-histone deacetylase (HDAC) inhibitor also in the brain, regulating the expression of genes involved in inflammation and glutamatergic signaling related to schizophrenia,[148] its efficacy as an add-on agent to regular antipsychotic medications in patients with chronic schizophrenia was examined. A small 24-week double-blind, randomized and placebo-controlled study was designed with 38 hospitalized patients suffering from chronic schizophrenia, receiving their regular antipsychotic medications combined with either 3 g daily curcumin or placebo.[149] A beneficial effect of curcumin on negative symptoms in patients with schizophrenia was noticed, what is in agreement with previously small trials.[150] Although these data are encouraging, larger GCP trials would be necessary to generate conclusive clinical data.

Rhodiolae radix and rhizome of *Rhodiola rosea* L. (Crassulaceae) has reported adaptogenic effects on stress symptoms. Dry extracts have been used traditionally for relief of mental and physical symptoms of stress, such as fatigue, weakness, exhaustion, irritability and weak anxiety.[151,152] *R. rosea* extract was first recommended in 1969 by the Pharmacological Committee of the Ministry of Health of the USSR for use as a stimulant against fatigue. In psychiatric practice, it is applied in cases of borderline nervous-mental diseases, neuroses and neurotic disorders. In mice, the extract containing salidroside (rhodioloside) (**40**) showed antidepressant-like effects as measured in the Porsolt behavioral despair assay, while a fixed combination of salidroside, rosavin, rosarin and rosin was more active than any of the individual components alone.[153] Unfortunately, there is not sufficient evidence for the clinical efficacy of herbal preparations of *R. rosea* for the treatment of symptoms of fatigue or mental weakness (EMA 2012),[152] but the phase III clinical trial designed by Olsson et al. should be mentioned.[154] Repeated administration of the extract (576 mg extract/day) has an antifatigue effect and increases mental performance, particularly the ability to concentrate superior to placebo. *R. rosea* root extracts may also have antipsychotic effects. In a preclinical study, prepulse inhibition deficits induced by psychostimulants in rodents, as an established operational measure of sensorimotor gating was applied (impaired in schizophrenia and other psychotic disorders). While prepulse inhibition deficit was induced wither by apomorphine or the NMDA receptor antagonist dizocilpine, pretreatment with the *R. rosea* extract significantly restored both.[155] Additional studies are required to uncover the mechanisms underlying these effects.

Bacopa monnieri (L.) Wettst. (Brahmi) leaf is an Ayurvedic botanical drug widely used as memory enhancer and neuropharmacological disorders like insomnia, insanity, depression, anxiety, psychosis, epilepsy and stress.[156] An add-on effect of *B. monnieri* has been found in management of schizophrenia.[156] The pharmacological properties of *B. monnieri* were attributed mainly the characteristic saponins called bacosides. Bioactive bacosides A and B, bacopasides I and II and bacopasaponin C exhibited memory enhancing and antidepressant activity when tested on experimental animals.[157,158] In a rat model of schizophrenia *B. monnieri* extract improved novel object recognition, which let the authors to conclude that Brahmi could restore and prevent negative symptoms in schizophrenia.[159]

Withania somnifera (L.) Dunal (Ashwagandha) root is a key Ayurvedic botanical drug used widely in India. *W. somnifera* root extracts have many reported pharmacological effects, such as anxiolytic, antidepressant, cardioprotective and neuroprotective, generally attributed to alkaloids, steroidal lactones (withanolides and withaferins) and saponins.[160] Sitoindosides VII–X and withaferin A (**41**) showed significant stress reducing activity against acute models of experimental stress.[161] In a randomized, double-blind, placebo-controlled study, adjunctive use of a standardized extract of *W. somnifera* showed positive effects in the treatment of symptom exacerbation in schizophrenia.[162] Further studies would be needed to substantiate these findings.

Hypericin (**42**) and pseudohypericin from *Hypericum perforatum* L. (hyperici herba), a botanical drug widely used for its antidepressant effects, potently binds to D3/4 receptor subtypes.[163] Although the antagonism was not determined, the affinity of hypericin to the rD3 receptor subtype was much higher than that of the atypical antipsychotic clozapine.[163] Thus, if high enough brain exposure can be achieved, napthoquinones may potentially act in concert with antipsychotics. The *H. perforatum* extract LI160 could reduce the auditory evoked potentials induced by S-ketamine in humans.[164]

Ginkgo biloba L. has been used extensively for the treatment of CNS diseases including neurosensory disorders, cerebrovascular insufficiency, peripheral vascular disturbances, and degenerative dementia.[165] The widely studied *G. biloba* standardized extract EGb 761 has also been suggested for the treatment of psychiatric disorders such as anxiety and depression, but also treat patients with schizophrenia, usually as an assistant therapy. The flavonoids and terpenoids are proposed as the biologically active compounds of EGb 761 with pharmacological activity and different properties.[165] Adjuvant prescription of *G. biloba* drugs to antipsychotics in chronic schizophrenia treatment was evaluated in a meta-analysis.[166] Eight randomized, double-blind, placebo-controlled trials involving 1033 patients were assessed (with 571 cases in EGb group and 462 in placebo). The result indicated that ginkgo extract exhibited a significant difference in ameliorating total and negative symptoms of chronic schizophrenia as an adjuvant therapy to antipsychotics. It is important to notice that all eight trials were carried out in China and further transcultural research is needed.[166] It was shown previously that a combination of Ginkgo extract and haloperidol (360 mg/day plus 0.25 mg/kg/day) could enhance the effectiveness of antipsychotic drugs and reduce their extrapyramidal side effects.[167] However, as reported above, ginseng can also trigger psychosis, reflecting the complexity of the neurobiological treatment context and overall clinical challenge for neuropharmacological interventions.

Given the limited success in identifying novel antipsychotic NPs, one of the most encouraging fields is the impact of differential diet on psychotic disorders. Dietary NPs are likely to have chronic effects on the brain by weakly but differentially modulating biochemical systems as reviewed previously.[168,169] Much research has been dedicated toward preventing the transition from ultra-high risk status to psychotic illness.[170] From these trials, polyunsaturated long-chain ω-3 fatty acids (PUFAs) emerged as apparently effective first-line nutraceutical treatment.[171] PUFAs were found to be effective not only during the period of administration,[172] but also up to several years later.[173] Since conventional antipsychotics carry undesirable side-effects and an increased risk of adverse events these findings are intriguing. In a newly published study, the authors attempted to replicate the first findings in a slightly modified trial.[174] However, ω-3 PUFAs were not found to be more effective than placebo in the prevention of psychosis at the 6 or 12-month follow-ups, thus essentially failing to replicate the previous trial.[171] In conclusion, ω-3 PUFAs are not effective for preventing the transition to psychosis, at least in ultra-high risk individuals.

Recently, attention was put on the role of stigmasterol (**43**), a naturally occurring phytosteroid found abundantly in vegetable diet.[175] Stigmasterol is bioavailable and easily crosses the blood brain barrier. In different models, this phytosteroid exhibits neuroprotective effects, enhances memory, and weakly inhibits ACh esterase.[175] In ketamine-induced psychosis in mice, stigmasterol (5, 10 and 20 mg/kg, *p.o.*) significantly attenuated the ketamine associated behavior.[176] Moreover, stigmasterol effectively increased GABA and diminished dopamine levels in the brain.[176] Similar preclinical experiments were performed by the same group on the seed extract of a commonly used vegetable—spinach (*Spinacia oleracea* L.). The tested extract (50–200 mg/kg, *p.o.*) containing polyphenols decreased ketamine-induced locomotor counts and stereotypic behaviors in mice. In the same model, the extract also improved memory, learning and cognitive function.[177]

The widespread dietary sesquiterpene BCP (**23**), which is present in many spices such as black pepper (*P. nigrum* L.), acts as a cannabinoid CB2 receptor agonist in vivo.[178] BCP was shown to exert general neuroprotective effects and reduce depressive-like behaviors in mice.[179] Based on patent literature, CB2 receptor agonists may be a potential antischizophrenic agents,[84] in agreement with the increased risk of schizophrenia for people with low CB2 receptor function.[180] The use of *Piper retrofractum* Vahl. leaves as antipsychotic drug was reported in Bangladesh.[181] Among other phytochemicals, *Piper* species including black pepper contain various amounts of guineensine and related *N*-isobutylamides which are effective in inhibiting endocannabinoid reuptake,[182] which is a potential antipsychotic mechanism.[183] Guineensine (**44**) is an effective anti-inflammatory NP which shows strong catalepsy in mice, possibly via dopamine transporter (DAT).[184] In addition to the possible relevance of NPs in vegetable food, recent clinical data further suggest that ketogenic diet could present a therapeutic approach in schizophrenia through restoring brain energy metabolism.[185] Overall, more and better GCP clinical trials are needed to further show the efficacy of dietary NPs and ketogenic diet as a cotreatment to manage both the clinical symptoms and metabolic abnormalities inherent to the disease and presented by antipsychotic treatment.

4. Conclusion

With the dramatic rise of drug abuse in the late 1950s, most recreational drugs were found to mediate stimulant psychosis. On the other hand, reserpine and physostigmine showed pronounced antipsychotic effects in patients. The pharmacological mechanisms of action of these NPs have inspired hypotheses regarding the pathophysiology of the positive symptoms in psychotic disorders, especially schizophrenia. Understanding the interactions between the different neurotransmitter systems and their relationship with symptoms will be an important step towards building a coherent hypothesis for the pathogenesis of schizophrenia. Although some NPs were reported to show clear anti-psychotic effects in animal models, the translational relevance of these findings remains largely unclear. In developing countries, patients with psychotic disorders more often use herbal medicines to manage psychosis, although the underlying mechanisms of action of these herbal medicines are poorly understood and no clinical data exist to substantiate the ethnopharmacology at this point. Retrospective treatment outcome, as proposed by Graz and colleagues may be a way forward.[186] Moreover, epidemiological studies focusing on the prevalence of schizophrenia and psychotic disorders could unveil a potential relationship with dietary and/or drug use habits. For instance, it is noteworthy that in Oceania and Indonesia the occurrence of schizophrenia is generally higher than in other Asian countries. The DALY rate of schizophrenia in Indonesia nearly doubles that of Australia.[187] Since betel consumption, which is frequent in these countries, may or may not be associated with schizophrenia,[188] it would be interesting to assess also the role of kava-kava usage in in these countries. The integration of observations from epidemiological studies, neurobiochemical mechanisms with NPs, drug discovery screening campaigns with NPs, and a better translation of animal models to patient treatment may facilitate the discovery of novel antipsychotic NPs. The prospect that food impacts neurobiological mechanisms related to psychotic behavior further provides a new frontier for research with dietary NPs largely underestimated in drug development.

References

1. Schwartz, R. C.; Blankenship, D. M. *World J. Psychiatry* **2014**, *4*, 133–140.
2. Anderson, K. K. *Lancet Public Health* **2019**, *4*, e212–e213.
3. Reddy, I. R.; Ukrani, J.; Indla, V.; Ukrani, V. *Indian J. Psychiatry* **2018**, *60*, 168–174.
4. Fink, A.; Benedek, M.; Unterrainer, H. F.; Papousek, I.; Weiss, E. M. *Front. Psychol.* **2014**, *5*, 1–4.
5. Bearden, C. E.; Forsyth, J. K. *F1000Res.* **2018**, *3*, https://doi.org/10.12688/f1000research.16574.1. (eCollection 2018).
6. Wilson, L.; Szigeti, A.; Kearney, A.; Clarke, M. *Schizophr. Res.* **2018**, *197*, 78–86.
7. Aucoin, M.; LaChance, L.; Cooley, K.; Kidd, S. *Neuropsychobiology* **2018**, *25*, 1–23.
8. Beer, M. D. *Compr. Psychiatry* **1996**, *37*, 273–291.
9. Loch, A. A. *Front. Psych.* **2019**, *10*, 328.
10. Polimeni, J.; Reiss, J. P. *Med. Hypotheses* **2002**, *58*, 244–248.
11. Singh, M. *Behav. Brain Sci.* **2017**, *6*, 1–83.
12. Atlas, J. A.; Miller, A. L.; Arsenio, W. F. *Psychol. Rep.* **1993**, *73*, 611–614.
13. Wasson, R. G.; Hofmann, A.; Ruck, C. A. P. *The Road to Eleusis. Unveiling the Secret of the Mysteries.* Harcourt Brace Jovanovich: New York/London, **1978**.
14. Perreault, C. *PLoS One* **2012**, *7, e45150.*

15. McKenna, T. *Food of the Gods. The Search for the Original Tree of Knowledge A Radical History of Plants, Drugs, and Human Evolution.* Bantam: Spokane, WA, USA, **1992**.
16. Carod-Artal, F. J. *Neurologia* **2015**, *30*, 42–49.
17. Lewis-Williams, D. *Camb. Archaeol. J.* **2004**, *14*, 107–111.
18. Russo, E. B.; Jiang, H. E.; Li, X.; Sutton, A.; Carboni, A.; del Bianco, F.; Mandolino, G.; Potter, D. J.; Zhao, Y. X.; Bera, S.; Zhang, Y. B.; Lü, E. G.; Ferguson, D. K.; Hueber, F.; Zhao, L. C.; Liu, C. J.; Wang, Y. F.; Li, C. S. *J. Exp. Bot.* **2008**, *59*, 4171–4182.
19. van der Meer, L.; Costafreda, S.; Aleman, A.; David, A. S. *Neurosci. Biobehav. Rev.* **2010**, *34*, 935–946.
20. Kline, N. S. *Ann. N. Y. Acad. Sci.* **1954**, *30*, 107–132.
21. Burton-Bradley, B. G. *Med. J. Aust.* **1974**, *17*, 17–19.
22. Iwu, M. M. *Ciba Found. Symp.* **1994**, *185*, 116–126.
23. Pfeffer, A. A. *Arch. Neurol. Psychiatry* **1947**, *58*, 221–226.
24. Allan, J. *Edinb. Med. Surg. J.* **1823**, *19*, 538–545.
25. Weber, M. M.; Emrich, H. M. *Int. Clin. Psychopharmacol.* **1988**, *3*, 255–266.
26. Siuciak, J. A.; Chapin, D. S.; Harms, J. F.; Lebel, L. A.; McCarthy, S. A.; Chambers, L.; Shrikhande, A.; Wong, S.; Menniti, F. S.; Schmidt, C. J. *Neuropharmacology* **2006**, *51*, 386–396.
27. Maremmani, A. G.; Rovai, L.; Rugani, F.; Bacciardi, S.; Dell'Osso, L.; Maremmani, I. *Eur. Rev. Med. Pharmacol. Sci.* **2014**, *18*, 287–302.
28. Gold, M. S.; Pottash, A. C.; Sweeney, D.; Martin, D.; Extein, I. *Ann. N. Y. Acad. Sci.* **1982**, *398*, 140–150.
29. Martins, S. S.; Fenton, M. C.; Keyes, K. M.; Blanco, C.; Zhu, H.; Storr, C. L. *Psychol. Med.* **2012**, *42*, 1261–1272.
30. Fiorentini, A.; Volonteri, L. S.; Dragogna, F.; Rovera, C.; Maffini, M.; Mauri, M. C.; Altamura, C. A. *Curr. Drug Abuse Rev.* **2011**, *4*, 228–240.
31. Nichols, D. E. *ACS Chem. Neurosci.* **2018**, *17*, 2331–2343.
32. Zbinden, G.; Randall, L. O. *Adv. Pharmacol.* **1967**, *5*, 213–291.
33. McIlwain, H. *Biochem. Soc. Trans.* **1975**, *3*, 579–590.
34. Chau-Pham, T. T. *Drug Metab. Rev.* **1978**, 7, 255–294.
35. Lowney, L. I.; Schulz, K.; Lowery, P. J.; Goldstein, A. *Science* **1974**, *183*, 749–753.
36. Marley, E. *Psychol. Med.* **1972**, *2*, 208.
37. Jacob, P., 3rd; Shulgin, A. T. *NIDA Res. Monogr.* **1994**, *146*, 74–91.
38. Gerona, R.; Adams, A.; Banister, S. D. *N. Engl. J. Med.* **2017**, *20*, 1597–1598.
39. Lanska, D. L. *Front. Neurol. Neurosci.* **2018**, *42*, 132–141.
40. Müller, J. *Gesnerus* **1998**, *55*, 205–220.
41. Becker, H. S. *J. Health Soc. Behav.* **1967**, *8*, 163–176.
42. Scheepers, F. E. *Front. Psych.* **2018**, *9*, 237.
43. Wang, J.; Liu, L.; Ball, T.; Yu, L.; Li, Y.; Xing, F. *Proc. Natl. Acad. Sci. U. S. A.* **2016**, *113*, 6444–6448.
44. Aouizerat, T.; Gutman, I.; Paz, Y.; Maeir, A. M.; Gadot, Y.; Gelman, D.; Szitenberg, A.; Drori, E.; Pinkus, A.; Schoemann, M.; Kaplan, R.; Ben-Gedalya, T.; Coppenhagen-Glazer, S.; Reich, E.; Saragovi, A.; Lipschits, O.; Klutstein, M.; Hazan, R. *mBio* **2019**, *10, e00388-19.*
45. Cohen, S. I.; Johnson, K. *BMJ* **1988**, *297*, 1270–1271.
46. Gurillo, P.; Jauhar, S.; Murray, R. M.; MacCabe, J. H. *Lancet Psychiatry* **2015**, *2*, 718–725.
47. Hughes, J. R.; McHugh, P.; Holtzman, S. *Psychiatr. Serv.* **1998**, *49*, 1415–1417.
48. Adolfo, A. B.; AhnAllen, C. G.; Tidey, J. W. *Schizophr. Res.* **2009**, *107*, 192–197.
49. Temple, J. L.; Bernard, C.; Lipshultz, S. E.; Czachor, J. D.; Westphal, J. A.; Mestre, M. A. *Front. Psych.* **2017**, *8*, 80.

50. Jansen, S. A.; Kleerekooper, I.; Hofman, Z. L.; Kappen, I. F.; Stary-Weinzinger, A.; van der Heyden, M. A. Grayanotoxin Poisoning: 'Mad Honey Disease' and Beyond. *Cardiovasc. Toxicol.* **2012**, *12*, 208–215.
51. Wang, S. Y.; Wang, G. K. *Cell. Signal.* **2003**, *15*, 151–159.
52. Forbes, R. J. *Studies in Ancient Technology*. Band 6, EJ Brill: Leiden, **1966**.
53. Mahdyoon, H.; Battilana, G.; Rosman, H.; Goldstein, S.; Gheorghiade, M. *Am. Heart J.* **1990**, *120*, 1189–1194.
54. Eadie, M. J. *Lancet Neurol.* **2003**, *2*, 429–434.
55. Sacks, O. *Hallucinations*. Knopf/Picador, A Division of Random House, 2012.
56. Richardson-Boedler, C. *Br. Homeopath. J.* **2001**, *90*, 44–49.
57. Martinez, A. F.; Muenke, M.; Arcos-Burgos, M. *Am. J. Med. Genet. B Neuropsychiatr. Genet.* **2011**, *156*, 1–10.
58. Bartholomew, R. E. *Psychol. Med.* **1994**, *24*, 281–306.
59. Iyassu, R.; Jolley, S.; Bebbington, P.; Dunn, G.; Emsley, R.; Freeman, D.; Fowler, D.; Hardy, A.; Waller, H.; Kuipers, E.; Garety, P. *Soc. Psychiatry Psychiatr. Epidemiol.* **2014**, *49*, 1051–1061.
60. Guirguis, W. R. *Br. J. Psychiatry* **1981**, *138*, 139–140.
61. Sticher, O.; Hamburger, M. *Planta Med.* **2008**, *74*, 791–793.
62. Hofmann, A. *MAPS.org*, 4th ed.; **2009**.
63. Dobkin, A. B.; Pearson, J. L.; Barr, J. S. *Can. Anaesth. Soc. J.* **1960**, *7*, 16–20.
64. Vollenweider, F. X.; Kometer, M. *Nat. Rev. Neurosci.* **2010**, *11*, 642–651.
65. Hofmann, A.; Heim, R.; Brack, A.; Kobel, H. *Experientia* **1958**, *14*, 107–109.
66. Satora, L.; Pach, D.; Butryn, B.; Hydzik, P.; Balicka-Slusarczyk, B. *Toxicon* **2005**, *45*, 941–943.
67. Brvar, M.; Mozina, M.; Bunc, M. *Wien. Klin. Wochenschr.* **2006**, *118*, 294–297.
68. Olsen, R. W. *Neuropharmacology* **2018**, *136*, 10–22.
69. Lakstygal, A. M.; Kolesnikova, T. O.; Khatsko, S. L.; Zabegalov, K. N.; Volgin, A. D.; Demin, K. A.; Shevyrin, V. A.; Wappler-Guzzetta, E. A.; Kalueff, A. V. *ACS Chem. Neurosci.* **2019**, *10*, 2144–2159.
70. Glatstein, M.; Alabdulrazzaq, F.; Scolnik, D. *Am. J. Ther.* **2016**, *23*, e74–e77.
71. Marneros, A.; Gutmann, P.; Uhlmann, F. *Eur. Arch. Psychiatry Clin. Neurosci.* **2006**, *256*, 458–459.
72. Sandia, S. I.; Ramírez, V. J.; Piñero, A. J.; Baptista, T. T. *Eur. Neuropsychopharmacol.* **2017**, *27*, 833–834.
73. Sáiz, J.; Mai, T. D.; López, M. L.; Bartolomé, C.; Hauser, P. C.; García-Ruiz, C. *Sci. Justice* **2013**, *53*, 409–414.
74. Reichert, S.; Lin, C.; Ong, W.; Him, C. C.; Hameed, S. *Can. Fam. Physician* **2017**, *63*, 369–370.
75. MacEwan, G. W.; Remick, R. A.; Noone, J. A. *CMAJ* **1985**, *133*, 431–432.
76. Wasko, M. J.; Witt-Enderby, P. A.; Surratt, C. K. *ACS Chem. Neurosci.* **2018**, *9*, 2475–2483.
77. Casselman, I.; Nock, C. J.; Wohlmuth, H.; Weatherby, R. P.; Heinrich, M. *J. Ethnopharmacol.* **2014**, *151*, 768–783.
78. El-Khoury, J.; Sahakian, N. *J. Psychoactive Drugs* **2015**, *47*, 286–292.
79. Gage, S. H. *Lancet Psychiatry* **2019**, *6*, 364–365.
80. Davies, C.; Bhattacharyya, S. *Ther. Adv. Psychopharmacol.* **2019**, *9*, 1–16.
81. Black, N.; Stockings, E.; Campbell, G.; Tran, L. T.; Zagic, D.; Hall, W. D.; Farrell, M.; Degenhardt, L. *Lancet Psychiatry* **2019**, *6*, 995–1010.
82. Hasan, A.; von Keller, R.; Friemel, C. M.; Hall, W.; Schneider, M.; Koethe, D.; Leweke, F. M.; Strube, W.; Hoch, E. *Eur. Arch. Psychiatry Clin. Neurosci.* **2019**, https://doi.org/10.1007/s00406-019-01068-z.
83. Hamilton, I.; Monaghan, M. *Curr. Psychiatry Rep.* **2019**, *21*, 48.

84. Anavi-Gofffer, S.; Gertsch, J. Treatment of Schizophrenia Using Beta-Caryophyllene and CB2 Receptor Agonists. *United States Patent Application 20170151182 Kind Code: A1* **2016**.
85. Schubart, C. D.; Sommer, I. E.; van Gastel, W. A.; Goetgebuer, R. L.; Kahn, R. S.; Boks, M. P. *Schizophr. Res.* **2011**, *130*, 216–221.
86. Lappin, J. M.; Sara, G. E. *Addiction* **2019**, *114*, 2065–2077.
87. Patel, N. B. *Int. Rev. Neurobiol.* **2015**, *120*, 235–255.
88. Coppola, M.; Mondola, R. *Nord. J. Psychiatry* **2012**, *66*, 73–78.
89. Maglione, M.; Miotto, K.; Iguchi, M.; Jungvig, L.; Morton, S. C.; Shekelle, P. G. *Am. J. Psychiatry* **2005**, *162*, 189–191.
90. Peng, W.; Liu, Y. J.; Wu, N.; Sun, T.; He, X. Y.; Gao, Y. X.; Wu, C. J. *J. Ethnopharmacol.* **2015**, *164*, 340–356.
91. Warner, M. L.; Kaufman, N. C.; Grundmann, O. *Int. J. Leg. Med.* **2016**, *130*, 127–138.
92. Raffa, R. B.; Pergolizzi, J. V.; Taylor, R.; Ossipov, M. H. *J. Clin. Pharm. Ther.* **2018**, *43*, 437–441.
93. Fluyau, D.; Revadigar, N. *Front. Psych.* **2017**, *8*, 62. https://doi.org/10.3389/fpsyt.2017.00062.
94. Wong, M. K.; Darvishzadeh, A.; Maler, N. A.; Bota, R. G. *Prim. Care Companion CNS Disord.* **2016**, *18,* https://doi.org/10.4088/PCC.15br018569.
95. Norelli, L. J.; Xu, C. *J. Diet. Suppl.* **2015**, *12*, 119–125.
96. Vázquez, I.; Agüera-Ortiz, L. F. *Acta Psychiatr. Scand.* **2002**, *105*, 76–77.
97. Joshi, K. G.; Faubion, M. D. *Psychiatry* **2005**, *2*, 56–61.
98. Holmberg, G.; Gershon, S. *Psychopharmacologia* **1961**, *2*, 93–106.
99. Holmes, A.; Quirk, G. J. *Trends Pharmacol. Sci.* **2010**, *31*, 2–7.
100. Stitzel, R. E. *Pharmacol. Rev.* **1976**, 28, 179–208.
101. Lobay, D. *Integr. Med.* **2015**, *14*, 40–46.
102. Holt, W. L.; Costello, C. H. *J. Am. Pharm. Assoc.* **1954**, *43*, 144–148.
103. Wilkins, R. W.; Judson, W. E.; Stone, R. W.; Hollander, W.; Huckabee, W. E.; Friedman, I. H. *N. Engl. J. Med.* **1954**, *18*, 477–478.
104. Cowden, R. C.; Zax, M.; Sproles, J. A. *A.M.A. Arch. Neurol. Psychiatry* **1955**, *74*, 518–522.
105. Foster, M. W.; Gayle, R. F. *JAMA* **1955**, *17*, 1520–1522.
106. Rashkis, H. A.; Smarr, E. R. *Am. J. Psychiatry* **1957**, *113*, 1116.
107. Ayokun-nun, A. A.; Alimi, O. A.; Adekilekun, S. *Trans. R. Soc. S. Afr.* **2017**, *73*, 33–41.
108. Erickson, J. D.; Schafer, M. K.; Bonner, T. I.; Eiden, L. E.; Weihe, E. *Proc. Natl. Acad. Sci. U. S. A.* **1996**, *14*, 5166–5171.
109. Csernansky, J. G.; Murphy, G. M.; Faustman, W. O. *Biol. Psychiatry* **1991**, *15*, 383–400.
110. Davis, K. L.; Berger, P. A. *Biol. Psychiatry* **1978**, *13*, 23–49.
111. Proudfoot, A. *Toxicol. Rev.* **2006**, *25*, 99–138.
112. Boley, S. P.; Stellpflug, S. J. *Ann. Pharmacother.* **2019**, *53*, 1026–1032.
113. Bartolucci, C.; Perola, E.; Cellai, L.; Brufani, M.; Lamba, D. *Biochemistry* **1999**, *38*, 5714–5719.
114. Lipowski, Z. J. *JAMA* **1987**, *258*, 1789–1792.
115. Edelstein, P.; Schultz, J. R.; Hirschowitz, J.; Kanter, D. R.; Garver, D. L. *Am. J. Psychiatry* **1981**, *138*, 1078–1081.
116. Leong, S. S.; Brown, W. A. *J. Neural Transm.* **1987**, *70*, 295–312.
117. Janowsky, D. S.; el-Yousef, M. K.; Davis, J. M.; Sekerke, H. J. *Lancet* 1972, 23, 632–635.
118. Modestin, J.; Hunger, J.; Schwartz, R. B. *Arch. Psychiatr. Nervenkr.* **1973**, *218*, 67–77.
119. Mucke, H. A. *Future Sci. OA* **2015**, *3*, FSO73.

120. Andersen, M. B.; Werge, T.; Fink-Jensen, A. *Pharmacol. Exp. Ther.* **2007**, *321*, 1179–1182.
121. Koola, M. M. *Mol. Neuropsychiatry* **2018**, *4*, 134–148.
122. Schubert, M. H.; Young, K. A.; Hicks, P. B. *Biol. Psychiatry* **2006**, *15*, 530–533.
123. Dwoskin, L. P.; Crooks, P. A. *Biochem. Pharmacol.* **2002**, *15*, 89–98.
124. Spurny, R.; Debaveye, S.; Farinha, A.; Veys, K.; Vos, A. M.; Gossas, T.; Atack, J.; Bertrand, S.; Bertrand, D.; Danielson, U. H.; Tresadern, G.; Ulens, C. *Proc. Natl. Acad. Sci. U. S. A.* **2015**, *12*, E2543–E2552.
125. Bradley, P. *British Herbal Compendium*. Vol. I. British Herbal Medicine Association: Dorset (Great Britain), **1992**; Vol. I149–150.
126. Stead, L. F.; Hughes, J. R. *Cochrane Database Syst. Rev.* **2012**, 15(2), CD000124.
127. Martin, C. A.; Nuzzo, P. A.; Ranseen, J. D.; Kleven, M. S.; Guenthner, G.; Williams, Y.; Walsh, S. L.; Dwoskin, L. P. *J. Atten. Disord.* **2018**, *22*, 1361–1366.
128. Mastropaolo, J.; Rosse, R. B.; Deutsch, S. I. *Behav. Brain Res.* **2004**, *31*, 419–422.
129. Kanba, S.; Richelson, E. *Herbal Medicines for Neuropsychiatric Diseases: Current Developments and Research*. Psychology Press, **1999**; 75–76. 144 Seiten.
130. Natesan, S.; Reckless, G. E.; Barlow, K. B.; Odontiadis, J.; Nobrega, J. N.; Baker, G. B.; George, S. R.; Mamo, D.; Kapur, S. *Psychopharmacology* **2008**, *199*, 275–289.
131. Meade, J. A.; Free, R. B.; Miller, N. R.; Chun, L. S.; Doyle, T. B.; Moritz, A. E.; Conroy, J. L.; Watts, V. J.; Sibley, D. R. *Psychopharmacology* **2015**, *232*, 917–930.
132. Pandy, V.; Vijeepallam, K. *Exp. Anim.* **2017**, *66*, 417–423.
133. Rothman, R. B.; Baumann, M. H.; Dersch, C. M.; Romero, D. V.; Rice, K. C.; Carroll, F. I.; Partilla, J. S. *Synapse* **2001**, *39*, 32–41.
134. Basu, M.; Mayana, K.; Xavier, S.; Balachandran, S.; Mishra, N. *Neurochem. Int.* **2016**, *93*, 113–117.
135. Mizuno, M.; Kawamura, H.; Takei, N.; Nawa, H. *J. Neural. Transm.* **2008**, *115*, 521–530.
136. Donthula, K. K.; Dholi, S. K. *World J. Pharm. Pharm. Sci.* **2018**, 7, 889.
137. Santillán-Urquiza, M. A.; Herrera-Ruiz, M.; Zamilpa, A.; Jiménez-Ferrera, E.; Román-Ramos, R.; Tortoriello, J. *Biomed. Pharmacother.* **2018**, *103*, 879–888.
138. Santillán-Urquiza, M. A.; Herrera-Ruiz, M.; Zamilpa, A.; Jiménez-Ferrera, E.; Román-Ramos, R.; Alegr-a-Herrera, E. Y.; Tortoriello, J. *Evid. Based Complement. Alternat. Med.* **2019**, 8404258.
139. https://clinicaltrials.gov/ct2/show/NCT03883360. (accessed 21.12.2019).
140. Hoenders, H. J. R.; Bartels-Velthuis, A. A.; Vollbehr, N. K.; Bruggeman, R.; Knegtering, H.; de Jong, J. T. V. M. *J. Nerv. Ment. Dis.* **2018**, *206*, 81–101.
141. Gertsch, J.; Viveros-Paredes, J. M.; Taylor, P. J. *J. Ethnopharmacol.* **2011**, *136*, 385–391.
142. Aricioglu, F.; Ozkartal, C. S.; Unal, G.; Dursun, S.; Cetin, M.; Muller, N. *Bull. Clin. Psychopharmacol.* **2016**, *26*, 329–444.
143. Yadav, M.; Parle, M.; Kadian, M.; Sharma, K. *Asian J. Pharm. Clin. Res.* **2015**, *8*, 24–28.
144. Duffield, P. H.; Jamieson, D. D.; Duffield, A. M. *Arch. Int. Pharmacodyn. Ther.* **1989**, *301*, 81–90.
145. Singh, Y. N.; Singh, N. N. *CNS Drugs* **2002**, *16*, 731–743.
146. Shen, L.; Liu, C. C.; An, C. Y.; Ji, H. F. *Sci. Rep.* **2016**, *6*, 20872.
147. Kunnumakkara, A. B.; Bordoloi, D.; Padmavathi, G.; Monisha, J.; Roy, N. K.; Prasad, S.; Aggarwal, B. B. *Br. J. Pharmacol.* **2017**, *174*, 1325–1348.
148. Lee, S. J.; Krauthauser, C.; Maduskuie, V.; Fawcett, P. T.; Olson, J. M.; Rajasekaran, S. A. *BMC Cancer* **2011**, *18*, 144.
149. Miodownik, C.; Lerner, V.; Kudkaeva, N.; Lerner, P. P.; Pashinian, A.; Bersudsky, Y.; Eliyahu, R.; Kreinin, A.; Bergman, J. *Clin. Neuropharmacol.* **2019**, *42*, 117–122.

150. Wynn, J. K.; Green, M. F.; Hellemann, G.; Karunaratne, K.; Davis, M. C.; Marder, S. R. *Schizophr. Res.* **2018**, *195*, 572–573.
151. European Medicines Agency *Assessment Report on* Rhodiola rosea *L., rhizoma et radix.* EMA/HMPC/232100/2011. http://www.ema.europa.eu/docs/en_GB/document_library/Herbal__HMPC_assessment_report/2012/05/WC500127861.pdf; 2012.
152. Panossian, A.; Wikman, G.; Sarris, J. *Phytomedicine* **2010**, *17*, 481–493.
153. Panossian, A.; Nikoyan, N.; Ohanyan, N.; Hovhannisyan, A.; Abrahamyan, H.; Gabrielyan, E.; Wikman, G. *Phytomedicine* **2008**, *15*, 84–91.
154. Olsson, E. M.; von Schéele, B.; Panossian, A. G. *Planta Med.* **2009**, *75*, 105–112.
155. Coors, A.; Brosch, M.; Kahl, E.; Khalil, R.; Michels, B.; Laub, A.; Franke, K.; Gerber, B.; Fendt, M. *J. Ethnopharmacol.* **2019**, *235*, 320–328.
156. Sarkar, S.; Mishra, B. R.; Praharaj, S. K.; Nizamie, S. H. *J. Ayurveda Integr. Med.* **2012**, *3*, 223–225.
157. Al Snafi, A. E. *Int. J. Pharm. Sci. Res.* **2013**, *4*, 154–159.
158. Kishore, K.; Singh, M. *Indian J. Exp. Biol.* **2005**, *43*, 640–645.
159. Piyabhan, P.; Wannasiri, S.; Naowaboot, J. *Clin. Exp. Pharmacol. Physiol.* **2016**, *43*, 1234–1242.
160. Ng, Q. X.; Loke, W.; Foo, N. X.; Tan, W. J.; Chan, H. W.; Lim, D. Y.; Yeo, W. S. *Phytother. Res.* **2020**, *34*, 583–590. https://doi.org/10.1002/ptr.6552.
161. Brahmacha, G. *Neuroprotective Natural Products: Clinical Aspects and Mode of Action.* Wiley-VCH Verlag GmbH & Co. KGaA: Weinheim, Germany, 2017;314–318.
162. Chengappa, K. N. R.; Brar, J. S.; Gannon, J. M.; Schlicht, P. J. *J. Clin. Psychiatry* **2018**, 79, 17m11826.
163. Butterweck, V.; Nahrstedt, A.; Evans, J.; Hufeisen, S.; Rauser, L.; Savage, J.; Popadak, B.; Ernsberger, P.; Roth, B. L. *Psychopharmacology* **2002**, *162*, 193–202.
164. Murck, H.; Spitznagel, H.; Ploch, M.; Seibel, K.; Schaffler, K. *Biol. Psychiatry* **2006**, *59*, 440–445.
165. Montes, P.; Ruiz-Sánchez, E.; Rojas, C.; Rojas, P. *CNS Neurol. Disord. Drug Targets* **2015**, *14*, 132–149.
166. Chen, X.; Hong, Y.; Zheng, P. *Psychiatry Res.* **2015**, *228*, 121–127.
167. Zhang, X. Y.; Zhou, D. F.; Zhang, P. Y.; Wu, G. Y.; Su, J. M.; Cao, L. Y. *J. Clin. Psychiatry* **2001**, *62*, 878–883.
168. Gertsch, J. *Planta Med.* **2016**, *82*, 920–929.
169. Gertsch, J. *J. Pharmacol.* **2017**, *174*, 1464–1483.
170. Yung, A. R.; Phillips, L. J.; Yuen, H. P.; McGorry, P. D. *Schizophr. Res.* **2004**, *1*, 131–142.
171. Joy, C. B.; Mumby-Croft, R.; Joy, L. A. *Cochrane Database Syst. Rev.* **2000**, *2*, CD001257.
172. Amminger, G. P.; Schäfer, M. R.; Papageorgiou, K.; Klier, C. M.; Cotton, S. M.; Harrigan, S. M.; Mackinnon, A.; McGorry, P. D.; Berger, G. E. *Arch. Gen. Psychiatry* **2010**, *67*, 146–154.
173. Amminger, G. P.; Mechelli, A.; Rice, S.; Kim, S. W.; Klier, C. M.; McNamara, R. K.; Berk, M.; McGorry, P. D.; Schäfer, M. R. *Transl. Psychiatry* **2015**, *13*, e495.
174. McGorry, P. D.; Nelson, B.; Markulev, C.; Yuen, H. P.; Schäfer, M. R.; Mossaheb, N.; Schlögelhofer, M.; Smesny, S.; Hickie, I. B.; Berger, G. E.; Chen, E. Y.; de Haan, L.; Nieman, D. H.; Nordentoft, M.; Riecher-Rössler, A.; Verma, S.; Thompson, A.; Yung, A. R.; Amminger, G. P. *JAMA Psychiat.* **2017**, *74*, 19–27.
175. Kaur, N.; Chaudhary, J.; Jain, A.; Kishore, L. *Int. J. Pharm. Sci. Res.* **2011**, *2*, 2259–2265.
176. Yadav, M.; Parle, M.; Jindal, D. K.; Dhingra, S. *Pharmacol. Rep.* **2018**, *70*, 591–599.

177. Yadav, M.; Parle, M.; Sharma, N.; Jindal, D. K.; Bhidhasra, A.; Dhingra, M. S.; Kumar, A.; Dhingra, S. *Biomed. Pharmacother.* **2018**, *105*, 1015–1025.
178. Gertsch, J.; Leonti, M.; Raduner, S.; Racz, I.; Chen, J. Z.; Xie, X. Q.; Altmann, K. H.; Karsak, M.; Zimmer, A. *Proc. Natl. Acad. Sci. U. S. A.* **2008**, *105*, 9099–9104.
179. Machado, K. D. C.; Islam, M. T.; Ali, E. S.; Rouf, R.; Uddin, S. J.; Dev, S.; Shilpi, J. A.; Shill, M. C.; Reza, H. M.; Das, A. K.; Shaw, S.; Mubarak, M. S.; Mishra, S. K.; Melo-Cavalcante, A. A. C. *Phytother. Res.* **2018**, *32*, 2376–2388.
180. Ishiguro, H.; Horiuchi, Y.; Ishikawa, M.; Koga, M.; Imai, K.; Suzuki, Y.; Morikawa, M.; Inada, T.; Watanabe, Y.; Takahashi, M.; Someya, T.; Ujike, H.; Iwata, N.; Ozaki, N.; Onaivi, E. S.; Kunugi, H.; Sasaki, T.; Itokawa, M.; Arai, M.; Niizato, K.; Iritani, S.; Naka, I.; Ohashi, J.; Kakita, A.; Takahashi, H.; Nawa, H.; Arinami, T. *Biol. Psychiatry* **2010**, *15*, 974–982.
181. Ahmed, M. N.; Kabidul Azam, M. N. *Schizophr. Res. Treatment* **2014**, *2014*, 679810.
182. Nicolussi, S.; Viveros-Paredes, J. M.; Gachet, M. S.; Rau, M.; Flores-Soto, M. E.; Blunder, M.; Gertsch, J. *Pharmacol. Res.* **2014**, *80*, 52–65.
183. Parolaro, D.; Realini, N.; Vigano, D.; Guidali, C.; Rubino, T. *Exp. Neurol.* **2010**, *224*, 3–14.
184. Reynoso-Moreno, I.; Najar-Guerrero, I.; Escareño, N.; Flores-Soto, M. E.; Gertsch, J.; Viveros-Paredes, J. M. *J. Agric. Food Chem.* **2017**, *1*, 9435–9442.
185. Palmer, C. M.; Gilbert-Jaramillo, J.; Westman, E. C. *Schizophr. Res.* **2019**, *208*, 439–440.
186. Graz, B.; Elisabetsky, E.; Falquet, J. *J. Ethnopharmacol.* **2007**, *25*, 382–386.
187. Ayuso-Mateos, J. L. **2013**. Global Burden of Schizophrenia in the year 2000. World Health Organization. 27.02.2013.
188. Udupi, S. A.; Kumar, C. N.; Suresh, K. K.; Arunachala, U.; Thirthalli, J. *Asian J. Psychiatr.* **2018**, *38*, 60–64.